Discovering the Cosmos

R. C. BLESS

UNIVERSITY OF WISCONSIN–MADISON

University Science Books
Sausalito, California

University Science Books
55D Gate Five Road
Sausalito, CA 94965
Fax: (415) 332-5393

Production Manager: *Susanna Tadlock*
Manuscript Editors: *Aidan Kelly and Hope Steele*
Designer: *Robert Ishi*
Illustrators: *John and Judy Waller*
Compositor: *Wilsted & Taylor*
Printer & Binder: *Edwards Brothers*

This book is printed on acid-free paper.

Library of Congress Cataloging-in-Publication Data

Bless, R. C., 1927–
    Introductory astronomy / R. C. Bless.
        p.   cm.
    Includes index.
    ISBN 0-935702-67-9
    1. Astronomy.   I. Title.
QB43.2.B585   1995
520—dc20   94-37452
                                                              CIP

Printed in the United States of America
10 9 8 7 6 5 4 3 2

Some mathematical symbols		Greek alphabet	
∝	Proportional to	α	Alpha
≈	Approximately equals	β	Beta
		γ	Gamma
>	Greater than	δ	Delta
<	Less than	ε	Epsilon
		ζ	Zeta
±	Plus or minus	η	Eta
		θ	Theta
		ι	Iota
		κ	Kappa
		λ	Lambda
		μ	Mu
		ν	Nu
		ξ	Xi
		o	Omicron
		π	Pi
		ρ	Rho
		σ	Sigma
		τ	Tau
		υ	Upsilon
		φ	Phi
		χ	Chi
		ψ	Psi
		ω	Omega

Discovering the Cosmos

*In Nature's infinite book of secrecy
A little can I read.*

—Shakespeare, *Antony and Cleopatra*

Contents in Brief

Contents

Comments to the Instructor

Over a three-decade period at the University of Wisconsin–Madison, I have taught the elementary one-semester survey course in astronomy for nonscience students many times. My primary aim has been to describe the developing views of the astronomical universe, and to communicate to students a sense of the incredible growth our science has undergone over the last half century. An important secondary goal has been to give students some understanding of the methods, goals, and limitations of science.

For several reasons, the study of astronomy is a wonderful way of introducing science to students. To begin with, it is the oldest of the sciences and well exemplifies the way science became defined and focused, how it developed and how it is done. Furthermore, even in an introductory course, the student can be taken right up against the current limits of astronomy and understand, qualitatively at least, what those limits are and how they might be overcome. Also, many of the objects of astronomy—planets, the Sun and other stars, comets, etc.—are familiar to everyone in a way that an electron or a molecule is not. It is easier to connect with the objects we study than is often the case in other physical sciences. Finally, ever since the Second World War, astronomy has been bubbling with new techniques, new discoveries, and new concepts that have fundamentally altered our views of much of the universe. The remarkable qualities of these concepts appeals to most people.

Given the increasingly important roles that science and technology play in our lives, it is important that as many people as possible have some understanding of the techniques, the ultimate aims and limitations of science. To these ends, this book gives considerably more emphasis than do most other texts to the early development of astronomy. Though the growth of astronomy through Newton is given the most space, the history does not end there. For example, the detour-filled road to our understanding of what spiral nebulae are is sketched, as well as attitudes—ancient and modern—toward life elsewhere. In addition, several examples of the crucial role played by new technology are described.

Another aspect stressed in this book is *how* we know what we know, not just *what* we know. The application of physics to astronomy has transformed our field and will continue to do so. The connection between cosmology and particle physics is just one of the most recent examples. Within appropriate limits, one of which is little use of mathematics, this book provides relevant physical explanations, sometimes in greater detail than in other texts. I know that some instructors will feel that there is too much detail and that this makes it difficult to cover all topics desired. This may be so, but if in this way students can acquire some understanding of the sort of proof required by science (and can compare it with that of pseudo-science), then it seems to me that teaching the *how* of what we know is at least as important as teaching *what* we now know. After all, this book is not intended for students who plan to become professional astronomers; so must all of astronomy be covered? When time limits require cutting, I strongly recommend that an instructor omit a topic altogether rather than selectively omitting the *how* sections from several topics.

These features—larger servings of history and emphasis on nonmathematical physical explanations—differentiate this book from most other elementary texts. To allow the instructor greater flexibility, somewhat more material has been included than can be covered comfortably in one semester. To keep the book within reasonable limits, however, several topics have been omitted; for example, descriptions of most variable stars, of the spiral structure of the Milky Way as suggested by 21-cm observations, and of superluminal expansion have not been included. Stellar magnitudes are not used, nor are astronomical coordinate systems described in any detail. I use Angstroms for wavelengths rather than nanometers because I think that remembering a pioneer is more appropriate than adherence to an arbitrary convention. Physical units are rarely used in the book, and on the few occasions that they are, they are those of the cgs system. Some of my younger colleagues may be impatient with this, but most quantities in the text or in the problems are given in relative terms—solar masses or solar luminosities, for example—since dynes (or even Newtons!) are needless complications and have little significance for nonscientists.

An important point to note is that this book is not as long as it may seem. Had its format been similar to that of most other astronomy texts, namely two columns per page, the number of pages would have been significantly diminished. In addition, about one-seventh of the text is historical material that allows for easy reading. We all have our preferences, however, and some instructors might wish to shape a course in a particular manner, for example with a greater or lesser emphasis on planetary astronomy. If less emphasis was felt to be appropriate, the first part of Chapter 23 could be dropped and a few examples from Chapter 24 could be used to illustrate the principles given in the second part of Chapter 23. (Note that the solar system is not described planet-by-planet, but rather in terms of processes and concepts applied to groups of objects.) Also, some instructors might feel that the subject matter of Chapter 25 does not warrant as full a treatment as is given. On the other hand, if solar system astronomy is particularly appealing, all of Chapters 23–25 could be included and material in other chapters omitted, for example, some of the astrophysics in Chapters 11 and 14.

An obvious feature of this book is the near-absence of color photographs. There are two reasons for this. One is to keep the cost of the book as low as possible. The second reason is a bit more complicated. Consider false-color images first. It seems to me that they can easily mislead the student. Some are rather complicated and difficult to interpret, as when colors are used to represent velocities in a complex gas cloud. Representing images taken in wavelengths outside the visible band is another problem; for example, the Magellan radar images of Venus are often shown in orange. Radio images, color-coded for intensity, can too easily suggest to the student that these objects are somehow visible to the naked eye. Black-and-white representations of radio data can also be misinterpreted, but perhaps less easily so.

Color images taken in visible light are attractive, often even spectacular, but what do they mean? It is often not apparent what the intent was; witness, for example, the rather different color images of commonly illustrated objects like the Orion nebula or M31. Differences in the color systems used make comparisons of such images from a variety of sources misleading at best. Also, there is a disconnection between such images and what a student would see in a telescope, where the Orion nebula appears greenish and Jupiter appears in subdued browns and oranges. There is, of course, a disconnection between what the naked eye sees and a long exposure image, but that is not as fundamental as adding color, which seems to me to compound the question of what is real. Of course color is sometimes used to indicate the distribution of various ions in a nebula, for example. This is useful information for an astronomer, but not for a nonastronomy student. Color images can also be useful to illustrate a point where the effect is not subtle, for example, the striking difference in color between an emission and reflection nebula. Apart from such examples, however, few color illustrations are used in this book.

The reader is reminded that the phenomena and objects described in the text really do exist in the sky. This is done by a variety of means—including the simple naked-eye observing projects at the end of Chapter 3, dates of various celestial events, and star charts showing the locations of objects mentioned. I hope that you will urge your students to take advantage of whatever opportunities they may have to become familiar with the sky.

I am grateful to many colleagues for comments and suggestions that have helped to improve this text. These include Robert Allen, University of Wisconsin–La Crosse; Lawrence Anderson, University of Toledo; Martha Hanner, University of Hawaii; Kenneth Janes, Boston University; Charles Lada, Smithsonian Astrophysical Observatory; Jim Lattis, University of Wisconsin–Madison; Gordon MacAlpine, University of Michigan; John Mathis, University of Wisconsin–Madison; Joseph Miller and Donald Osterbrock, Lick Observatory; Anthony Nicastro, West Chester University; Joseph Patterson, Columbia University; Richard Pogge, The Ohio State University; John R. Thorstensen, Dartmouth University; and Alma Zook, Pomona College. My ability to make mistakes exceeds theirs to catch them, however, and I must take responsibility for errors that may still remain in the text.

I must also express appreciation to my editor, Jane Ellis, especially for her gentle and tactful reminders that a book really should be finished, and to Judy and John Waller for converting my rough sketches to intelligible diagrams.

I hope you find this text useful. I would greatly appreciate receiving corrections, criticisms, and suggestions for improvement.

Bob Bless, Madison

The Earth as seen from lunar orbit.

"We travel together, passengers in a little space-ship, dependent on its vulnerable supplies of air and soil."

—Adlai E. Stevenson, 1965; U.S. politician

CHAPTER 1

Introduction

This book has two objectives: first, to describe the leading ideas and concepts of modern astronomy; second, to indicate how astronomy in particular and science in general have developed, and the methods, goals, and limitations shared by both.

The text is organized around three broad astronomical topics presented in four parts, in the following order.

Part I: Cosmology: The Beginnings Through Newton
Part II: The Life and Death of Stars
Part III: Cosmology: Herschel to the Present
Part IV: Worlds Beyond the Earth

The first general topic (given in Parts I and III) is the origin and evolution of the physical universe, which includes all matter and energy that we can now detect and, in fact, all that might ever be detected by anyone; today this study is commonly called cosmology.[1] The second topic deals with the birth, life, and death of stars or, more succinctly, with stellar evolution. Finally, the last general topic considers the physical characteristics of the planets of our solar system, and the possibility that other planetary systems exist and might even be abodes of life.

I hope you will read this introduction carefully, not only now, but occasionally while studying the rest of the book. It will remind you of its overall organization and intent.

The Astronomy You Will Learn
Cosmology: The Beginnings Through Newton (Part I)

How did the universe of stars and galaxies we see around us come to be? What is our place among them? What processes were important in shaping the universe? What will the universe be like billions of years from now? Questions of this sort—some of the grandest we can pose—have been asked in one way or another since we became conscious of ourselves in our environment. The answers given throughout history not only show how our astronomical ideas have developed and how modern science grew out of them, but also tell us a good deal about ourselves—our fears, our needs, and our attitudes toward nature and our place in it. For these reasons, this topic will be presented historically, with Part I dealing with developments from earliest times up

---

[1] The original meaning of the word had to do with the study of, or the attempt to find, order in the universe. This older meaning is particularly appropriate for Part I of this book.

through Newton. During this period the cosmological focus was almost entirely on our own planetary system, since it, along with the apparently unchanging (and hence rather uninteresting) stars, was thought to make up the universe.

We will begin with the earliest attempts, based on naked-eye observations, to find a measure of order (and hence security) in a world that most people thought was ruled by the whims of gods. Regularities in the sky, such as the daily appearance of the Sun, the monthly cycle of the Moon, and the reappearance of the stars year after year, perhaps suggested that a corresponding order might exist on the Earth. Furthermore, a few people felt that this order might yield some of its secrets to human thought and reasoning. This remarkable idea, that the world is *knowable*, that it could be understood, was the crucial first step toward the attitudes and methods we call science.

Early views of the Sun, Moon, and planets eventually evolved into the so-called Ptolemaic model of the astronomical universe (an elaborate Earth-centered system in which the movements of stars and planets were represented by combinations of circular motions). This was the standard model for the first 1,500 years of the Christian era. It formed the astronomical backdrop against which Western thought and attitudes developed. Beginning in the twelfth century and especially with the coming of the Renaissance, however, many old notions were questioned, and were often replaced by new knowledge, new ideas, new ways of looking at the world. The astronomical work of Copernicus, Tycho Brahe, Kepler, and Galileo in the sixteenth and seventeenth centuries was part of this general movement. Their work resulted in a drastically different picture of the physical world, one centered on the Sun rather than on the Earth. This new model and the circumstances surrounding its development were major factors in the transformation of Western medieval attitudes to those of the modern world. Perhaps the most dramatic of these changes in attitude is in our view of our relation to the universe. Ancient and medieval people thought themselves to be literally at the center of the universe, the purpose, in fact, for which it existed. By contrast, contemporary cosmological views relegate us to being little more than a recent surface phenomenon on a tiny speck orbiting an ordinary star. What significance we may have is not mirrored by any special relation we bear to the physical universe.

The work of Copernicus and his followers, though based on observations and measurements, lacked unifying concepts or theories about what physical causes might underlie this new model of the planetary system. These were provided by Newton with his laws of motion and of universal gravitation. Their triumphant application in quantitative detail to all aspects of motion in the solar system solved what had been the central problem of astronomy for 2,000 years. The prestige of the Newtonian system quickly led to a new worldview in which the cosmos was likened to a huge machine or clock, the operation of which was described by Newton's laws. His work can be taken to mark the beginning of modern science, and is still often offered as the best example of the methods of science.

Before describing more recent ideas and discoveries, however, we will interrupt the cosmological story in order to describe our current understanding of the nature of stars and of interstellar matter. Not only is this necessary to comprehend modern developments in cosmology, but we have learned a lot about stellar evolution; it is a fascinating story in its own right.

The Life and Death of Stars (Part II)

Our understanding of this topic, essentially the life histories of stars, is of rather recent origin, and will not be presented chronologically. Since very nearly all the observational data we have on stars comes to us in the form of electromagnetic radiation (electrically charged particles from the Sun, cosmic rays—which are high energy particles from the Sun and the Galaxy—and neutrinos are the only exceptions), we must first become familiar with some of the properties of light and understand how we can extract information from this radiant energy. We will also have to learn about the most important instrument of the astronomer, the telescope, as well as become acquainted with the instruments used to analyze the radiation it collects. We will take a look at some of the methods by which the physical properties of stars are measured. These properties include mass, diameter, temperature, and chemical composition. With these theoretical and observational tools in hand, we will be able to see how to construct an extremely useful and concise way of displaying many of the properties of stars: the Hertzsprung-Russell diagram.

After these preparatory steps we will try to answer such questions as: How and from what are stars formed? Are they being formed now? What are the sources of their enormous energy outputs? How long can they live, and what happens when they start to run out of energy? By what processes do they die, and what are the properties of stellar corpses? What will be the fate of our Sun and our planet? We will encounter many exotic creatures in the stellar zoo—red giants, white dwarfs, and black holes, neutron stars, pulsars, and supernovae—all of which show characteristics far removed from our direct experience of the physical world.

Cosmology: Herschel to the Present (Part III)

Next, we return to cosmology, beginning with the attempts by a few astronomers (most notably William Herschel in late eighteenth-century England) to explore the universe beyond the solar system. Herschel painstakingly used his excellent telescopes to take inventories of stars and gas clouds in the first real attempt to understand their nature and discover their distribution in space. He also attempted to deduce where the Sun was located with respect to the Milky Way as a whole. By the mid-1800s fuzzy objects showing spiral structure caught astronomers' attention. Were they gas clouds within the Milky Way, or were they separate systems of stars (which we now call galaxies) far beyond it? Was the Milky Way the only grouping of stars in the universe, or was it just one of many? Telescopic observations ultimately led to the recognition that, contrary to what had been thought, the Sun was not located at the center of our Milky Way Galaxy, but was far out toward its edge.

Soon after that, it was shown that the Milky Way was not the only system of stars and gas in the universe, but that the spirals were indeed huge separate systems, galaxies like our own Milky Way. The application of spectroscopy[2] to astronomy allowed the line-of-sight velocities (toward or away from us) of these galaxies to be measured. This in turn led to the recognition by about 1930 that the universe, once taken to be the very model of stability, is apparently flying apart, extending itself to inconceivably great distances. This first cosmological clue has recently been supplemented by two more: measurement of the relative amounts of the lightest chemical elements like helium and lithium; and the discovery of the so-called cosmic background radiation, a faint whisper of energy from the beginning of the universe. As we shall see, all three clues support the so-called Big Bang model of the universe, that is, the idea that roughly 15 billion years ago, the universe "exploded" from an incredibly hot and dense state, and evolved to what we now see around us.

We will look at recently recognized phenomena collected under the term active galactic nuclei, with an eye toward what they might tell us about cosmological prob-

---

[2] Spectroscopy is the technique whereby light is decomposed into its constituent colors or wavelengths.

lems. Several kinds of galaxies show a certain kind of activity, which is understood to be the result of extremely energetic events taking place in their centers. They may give us insights into how galaxies change with time, and possibly enable us to probe the distant early universe.

Astronomers have attempted to put all of this together into a model of the universe. Since such models are constructed in the context of Einstein's theory of gravity, we will devote a little time to some aspects of his theory of general relativity before we present current views on the structure and history of the universe. In a remarkable development, recent theories link the large-scale characteristics of our universe with events that occurred immediately after its beginning, when the universe was inconceivably young and small. That is, there seems to be an intimate connection between the sub-nuclear microworld and the macroworld of galaxies, between sub-nuclear particles and the large-scale structure of the universe!

Finally, we will see that some answers can be given to questions that might at first glance appear to be unrelated to this topic, but in fact are intimately connected with it. For example, why is there so much hydrogen and helium in the universe, but so little uranium and lead? Where do the chemical elements come from, anyway? Does the composition of matter itself evolve (that is, change) with time? And what does all this have to do with us? A lot, as it turns out!

Not only can we say something about the changing *composition* of matter in the universe, we can also discuss the changing *form* of that matter, from intergalactic gas to galaxies, from interstellar matter to stars. We are pretty sure that we know what the night sky will look like next Tuesday, but how about a billion years from now? Will our Galaxy still be shining five billion years from now? That we are now able to describe, at least in outline form, how matter itself has evolved is one of the dramatic developments of modern astronomy.

Worlds Beyond the Earth (Part IV)

In the last part of this book, we will discuss the physical characteristics of the planets with an eye on their ability to nurture life. The advent of the space age has had an enormous impact on solar-system astronomy. Before 1960, relatively few astronomers were interested in studying the planets and satellites of our system. Now that we can send instrumented probes on long journeys to fly by distant planets or even to land on their surfaces, solar-system astronomy has been booming. Astronomers can now answer questions that couldn't even have been asked only a decade or two ago. This has emboldened us to consider more seriously than ever before the possibility that there may be other life-bearing worlds. Though we have no answers as yet to this enormous question, we can speculate about the likelihood of life elsewhere from a much broader awareness of the relevant problems and possibilities. In this part of the text we will sketch some of these new ideas.

Though the technical problems that space vehicles must overcome are formidable, the basic principles of interplanetary flight are simple, and we will first describe how a trip to Mars might be made. You will also learn a few of the techniques of planetary exploration by space vehicles. An overview of the solar system, and of the major processes that shape planetary surface and interiors, will prepare you for a comparative look at the planets with all their fascinating variety. You will find that though the planets and moons of our solar system share many properties, the ability to support life is not one of them.

Our look at the planets will begin with that environment most familiar to us—the Earth—as the best-understood example of the planets of the inner solar system, and will use it as a benchmark when describing the other Earth-like, or terrestrial, planets.

We will consider why Venus, in many ways so similar to Earth, has such an incredibly hostile environment. Though Mars appears to be biologically dead now, were conditions there more favorable for the development of life many years ago? What do meteorites tell us about the early history of our planetary family? How common are biologically interesting molecules in the solar system?

We will follow this with a description of the physical conditions on the larger, more distant planets, the giants of the solar system. We will try to understand how the striking differences between the giant and terrestrial planets might have arisen. Recent results, especially from the Voyager space probes, have shown how varied the conditions can be among the several satellites of even one planet, Jupiter. The variety of physical conditions among all the planetary objects of our solar system is far broader (and more bewildering) than had been realized only a decade or so ago. This diversity has implications that must be considered when we attempt to tackle the extremely difficult question of whether other planets exist, and if so, whether any of these might have environments suitable for the development of life.

The last chapter of the text begins with a brief account of some of the thoughts that people have had over the centuries about life elsewhere. This will be followed by a quick overview of the most general biological requirements for life and of the physical conditions under which it might develop. Using insights gained earlier from our study of stellar evolution, we will consider the astronomical circumstances favorable for this development, and attempt to assess the likelihood of such circumstances existing outside the solar system. Not only should this bring home to you the relevant factors involved, but also the huge gaps—chasms, really—remaining in our knowledge. We will conclude with brief accounts of how life—if indeed it exists elsewhere—might be detected, the searches so far mounted, and what might be the consequences of success.

Astronomy as a Science

Throughout this book we will emphasize not only *what* we think we know, but also *why* we think it is so. Even in an elementary textbook like this one, we get to the very edge of our understanding of the universe. Consequently, you may be surprised occasionally by the weakness of our evidence and the corresponding tentativeness of our conclusions; in one or two instances we may be able to do little more than speculate. The point is, of course, that our knowledge is far from complete, that some of what we claim to know is probably wrong, that today's facts may be tomorrow's follies.

On a more general note, science is no more a collection of facts than is literature a collection of words. Rather, physical science is a continuing attempt to organize the experimentally verified data of the physical world within a framework built from as few basic assumptions—physical laws—as possible. Note that there are two elements here: experimental (or observational) data—the "facts" of nature—and a set of ideas and concepts that we call theories. It is the interplay of data and theory that is the unique characteristic of modern science. One without the other is ultimately sterile.

Because of the long history of astronomy, its effect on the way we look at the physical world, and its continuing vitality, the development of astronomy affords a good view of the origins and early development of science and of scientific attitudes. Furthermore, astronomy provides many examples of the broader relationships of science to other aspects of our experience. Whether we like it or not, science plays a large role in our lives. That this role is often poorly understood is lamentable, since science and technology, along with nationalism, are probably the greatest influences on our lives today. Only a few decades ago we had the overly optimistic view that science would be our infallible guide to an ever more-perfect society. Nowadays, however, some people have the unduly suspicious and fearful attitude that science is to blame for nearly all our problems. Neither view is correct or fruitful; each at least in part results from misunderstanding the aims, consequences, and limitations of science. Thus, in studying this book I hope that you will get some idea of what science is about, what it is and what it isn't, what it can do well and what it can't do at all; also, that you understand how science is done, what its methods, techniques, and values are; and that you consider whether the nature of the subject matter that is treated "scientifically" is as important to the success of the scientific endeavor as the so-called scientific method itself. In other words, how are such disciplines as political science, social science, or educational science "scientific" in the same ways as are the physical sciences? How do they differ? How profound are the similarities in the methods employed and in the character of the results obtained?

You should also gain an appreciation of how science affects us, and not just in the obvious ways of fluorides in toothpaste or x-rays of broken bones. Science influences us in the more subtle and basic ways of changing our view of the world and of our relationship to it. Ultimately, it is one of the forces that shapes our view of ourselves. In many ways we differ profoundly from our ancestors of only a few hundred years ago. We have already mentioned that what we take to be our relationship to the physical world and our place in nature is markedly different from the attitudes held by Europeans in the Middle Ages. You will see that astronomy played a significant role in this transformation.

Finally, I hope that in reading this book you will discover that astronomy is enjoyably mind expanding and even exciting. This is not, however, "astronomy without tears." Some sections are difficult, and require careful reading and a little thought. But when you are finished, I hope that you will have found the experience worth the effort and, as at the end of a long journey to unfamiliar places, that you will be a somewhat different person for it.

Suggestions for Further Reading

At the end of every chapter there will be a brief listing of books and magazine articles where further information can be found on the topics discussed in that chapter.

Probably the two best-known monthly astronomy magazines available to amateurs at newsstands, by subscription, or in libraries are: *Astronomy*, Kalmbach Publishing Co., P.O. Box 1612, Waukesha, Wisconsin 53187 and *Sky & Telescope*, Sky Publishing Co., P.O. Box 9111, Belmont, Massachusetts 02178. Both magazines contain beautiful astronomical photographs and articles on all aspects of astronomy as well as advertisements for all sorts of astronomical equipment and publications for the amateur. *Sky & Telescope* is somewhat more advanced than *Astronomy*.

A bi-monthly non-technical magazine available by subscription or from libraries is *Mercury*, published by the Astronomical Society of the Pacific, 390 Ashton Ave., San Francisco, California 94112.

Scientific American, which is also available at newsstands and libraries, often has articles on astronomy.

Journal for the History of Astronomy. Science History Publications, Cambridge, England. Publishes articles on all aspects of astronomical history, including archaeoastronomy.

Some general reference books that are often helpful.

DeVorkin, D., *The History of Modern Astronomy and Astrophysics: A Selected, Annotated Bibliography*. New York: Garland Publishing, Inc., 1982. A bibliography of books and major articles on the history of astronomy from the time of the telescope through 1980.

Hetherington, N., *Encyclopedia of Cosmology: Historical, Philosophical and Scientific Foundations of Modern Cosmology*. New York: Garland Publishing, Inc., 1993. The subtitle describes this volume well. Many articles have good lists of additional readings.

Maran, S., *The Astronomy and Astrophysics Encyclopedia*. New York: Van Nostrand Reinhold, 1992. An up-to-date and extensive collection of articles, many of which are rather technical in nature, although others are not. It can be a useful reference for beginning astronomy students.

Moore, P. (ed.), *The International Encyclopedia of Astronomy*. New York: Orion Books, 1987. Less technical than the volume above, but more comprehensive than the Oxford volume below.

Roy, A. (ed.), *The Oxford Illustrated Encyclopedia of Astronomy*. Oxford: Oxford University Press, 1992. A concise, elementary, and well-illustrated volume, this reference can be a good place to begin looking for supplemental material.

Two collections of short articles on a wonderfully wide variety of obscure, bizarre, as well as standard astronomical topics are given in the following:

Ashbrook, J., *The Astronomical Scrapbook: Skywatchers, Pioneers and Seekers in Astronomy*. Cambridge: Cambridge University Press, 1984.

Moore, P., *Fireside Astronomy*. New York: John Wiley and Sons, Inc., 1991.

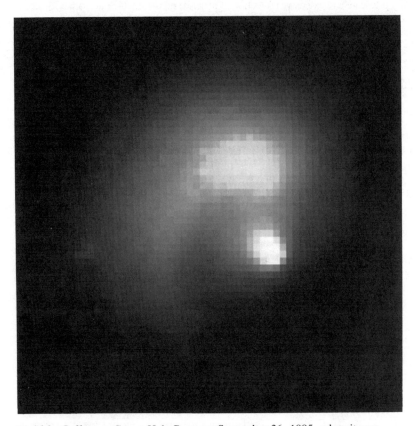

Hubble Gallery. Comet Hale-Bopp on September 26, 1995, when it was about one billion kilometers from Earth. Note the bright chunk of material apparently ejected from the nucleus. It was too far from the Sun to have developed a tail.

PART I

Cosmology: The Beginnings Through Newton

Halley's comet (lower left) in November, 1985. The exposure is too short to show its tail. The Pleiades are in the upper right.

"The light which comes to us from the starry skies is very weak. But what would human thought have been like if we could not perceive the stars, as would have been the case if the Earth had been, like our sister Venus, enveloped in a cloak of cloud?"

—Paul Couderc, French astronomer

CHAPTER 2

The Naked-Eye Sky

The astronomical universe is composed of individual objects—like planets and stars—and systems or collections of objects—the solar system and star clusters, for example. These generally differ enormously in their sizes, masses, distances, energy outputs, etc., from those of our everyday experience. In addition, there is a considerable range in the "size" of these quantities from one kind of object to another. For example, astronomical distances in absolute terms are far larger than distances we ordinarily deal with, even within our relatively tiny solar system. Earth is about 93,000,000 miles from the Sun, and Jupiter is roughly 500,000,000 miles from the Sun. In terms of our ordinary experience, such distances will not really mean anything to you (nor to astronomers, for that matter). You will soon find, however, that *relative* distances—Jupiter is about five times farther from the Sun than is the Earth—will become comprehensible and significant. One of the important (and fascinating) things for you to learn from your study of astronomy is an appreciation of the range of these physical quantities, that is, about how many times more massive or larger one kind of object is than another. You will be given help in doing this whenever we first encounter new kinds of objects.

Since this part of the book is concerned with the solar system, it is appropriate for you to acquire a good idea of its content and scale.

The Size of the Solar System

The **solar system** is the collection of objects—most prominently the planets and their satellites, but also asteroids (small planets), meteoroids (boulders to grains in size), and comets—that are in orbit around a very ordinary star, the Sun. An easy visualized model of this system can be made by imagining the Sun, a star 864,000 miles in diameter (= 1,380,000 km),[1] shrunk to a diameter of one inch, or to a bit smaller than the size of a ping-pong ball. With all other sizes and distances reduced by the same factor, the Earth would be a speck 0.01 inches in diameter and about nine feet away from the ping-pong-ball Sun. Our Moon would be only about 0.0025 inches in diameter (the thickness of a human hair) and only a little over one-quarter of an inch from

---

[1] By the end of this chapter we will be using scientific notation to express the very large (and very small) numbers that are encountered in astronomy. You will find this reviewed in Appendix A, "A Little Arithmetic."

Table 2.1.   Model of solar system

Object	Diameter (inches)	Distance
Sun	1.0	—
Earth	0.01	9.0 feet from Sun
Moon	0.0025	0.27 inches from Earth
Jupiter	0.1	46 feet from Sun
Pluto	0.00083	355 feet from Sun
Nearest star	1	360 miles

the Earth. Table 2.1 gives these and a few other dimensions using this model (see Figure 2.1). It also shows several interesting relations that you should remember. The Sun is by far the largest object in the solar system (and is also the most massive, constituting about 99.9 percent of the total). It is one hundred times Earth's diameter and ten times Jupiter's (the largest planet in our system). Hence Jupiter, about the size of a small pea on this scale, is ten times larger than Earth, which would be about the size of a period on this page.

With Pluto as the outermost planet, the whole range of planetary distances on this scale could be laid out on a football field, with the Sun being a ping-pong ball at one goal line and Pluto being a mere speck just beyond the other end zone. The most distant objects belonging to our solar system (that is, whose motions are governed primarily by the gravitational force of the Sun) are thought to be comets in a "reservoir" about 15 or 20 miles away on this scale. These objects, however, contribute a totally negligible amount of mass to the rest of the solar system; so for most purposes it is not unreasonable to take the radius of our system on this scale to be encompassed by a football field. Big as the solar system is, we can still get some feel for its size with a model such as this.

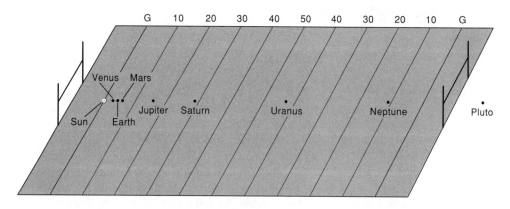

Figure 2.1.   If the Sun were about the size of a ping-pong ball, the solar distances of the planets on a football field would be as shown.

The Nearest Star

Now, still assuming our ping-pong-ball Sun, how far would we have to travel to reach the nearest star, α (alpha) Centauri?[2] Three hundred and sixty miles or about the distance between Chicago and Minneapolis! (See Figure 2.2.) Thus, although exploration of the solar system by space vehicles is well under way, a journey to even the nearest star involves traveling distances thousands of times greater than that required by Voyager's trip to Neptune. Tickets for a trip to α Centauri won't be on sale in your lifetime.

Distances within the solar system are not generally given in miles or kilometers, much less by models that make the Sun a ping-pong ball. Rather, they are given in terms of the **astronomical unit (AU)**, which is defined as the *average* distance of the Earth from the Sun. (Since the Earth's orbit is not quite circular but slightly elliptical, the Earth–Sun distance varies slightly; hence the definition of the AU in terms of the average distance.) One AU = 92,955,800 miles, which we will round off to

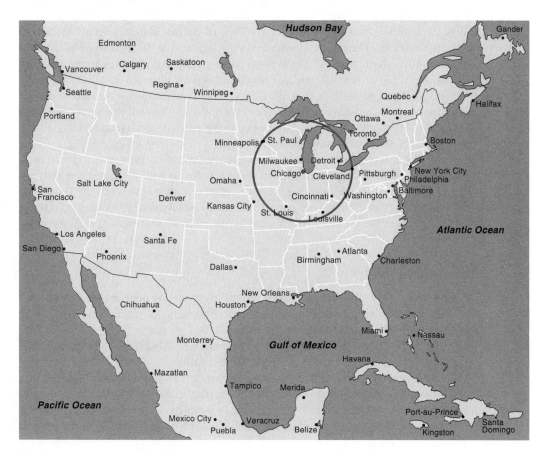

Figure 2.2. With the Sun still about the size of a ping-pong ball and located in Chicago, the nearest star would be in Minneapolis (or Cleveland or just beyond St. Louis). Stars fill only a tiny fraction of the volume of space!

---

[2] Bright stars in a constellation are generally denoted by letters of the Greek alphabet, α being the brightest star, β (beta) the next brightest and so on, followed by the Latin possessive form of the constellation name. Alpha Centauri is the brightest star in the constellation of Centaurus. Stars in a constellation are also numbered, and if one is too faint to be denoted by a Greek letter, it may be referred to by its number, for example, 61 Cygni. We will encounter many other cataloging schemes as well.

93,000,000 miles = 149,000,000 kilometers. On this scale Jupiter is 5.2 AU from the Sun, Pluto is about 40 AU from it, and the nearest star is about 270,000 AU away. Thus our solar system is a tiny island in a vast sea.

Motions of the Sun and Stars

Ever since our earliest days in school we have all been taught that the Sun is near the center of our solar system and that the planets, including the Earth, revolve around it in nearly circular orbits. We have also been told that the Earth is very nearly spherical and rotates on its axis once every 24 hours. All this is true, of course, but suppose we didn't know any of that; suppose, in fact, that we knew nothing about the motions of the Sun and the Moon, the planets and the stars. In addition, suppose we had no telescopes, clocks, or any other instruments except simple ones that we could make ourselves. What could we learn about the sky through naked-eye observations made from what appears to be an unmoving Earth? What would we see, and how would we be most likely to interpret our observations? This, of course, is the situation our ancient ancestors were in. The primary purpose of this chapter is to describe some of the simple naked-eye observations of the sky that can be made and how they might be interpreted. Also, I hope you will become convinced that the idea of a stationary, centrally located Earth is a perfectly reasonable assumption, even though we now know it to be incorrect. In addition, from some of these observations you will get hints about how we learned to tell time, find directions, identify the seasons, etc. These techniques will be more fully developed in the next chapter.

Our first reaction to the night sky would be like that of our ancestors—one of pleasure and awe at its beauty and mystery, coupled, perhaps, with bewilderment at its apparent complexity. However, if we carefully observed the positions of the Moon and the planets night after night, our persistence would be rewarded, and we would begin to see the order, the regularities exhibited by the motions of these celestial objects.

Some Useful Terms

After a time, we would probably find it useful, in organizing our observations and describing them to others, to define a few terms. For example, when we are in a large open field or on a calm sea, the part of the Earth that we see would appear to be flat. The intersection of the flat Earth and the sky would be our **horizon** (see Figure 2.3A). "Up" or more exactly, "directly overhead" (the **zenith**) would be above our heads, in a direction perpendicular to the flat surface on which we were standing. "Down" (the **nadir**) would, of course, be in the opposite direction. This up-down line is essentially just the direction of gravity, and we could easily establish it by noting the direction in which a weighted string hangs (see Figure 2.3B). The horizon would everywhere be just 90° from the zenith.[3] The sky itself would appear to be a huge inverted bowl, a hemisphere; in fact, we still call the sky the **celestial sphere**.

Motion of the Sun

Moving across this apparent sphere are many objects, some mere points of light, others much larger and brighter. The most obvious of these, of course, is the Sun. It rises

---

[3] If you are unfamiliar with angular measure—degrees, minutes, seconds—you should read "Angular Measure" in Appendix A.

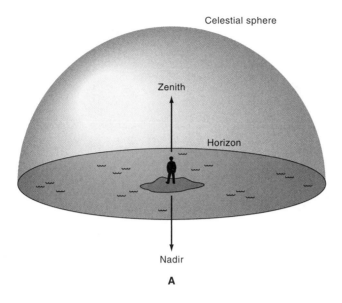

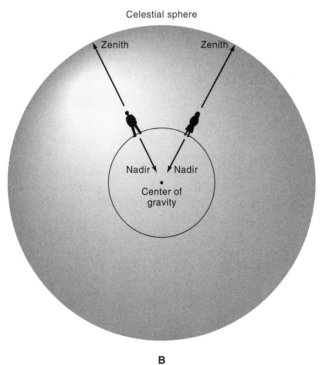

Figure 2.3. (A) Your horizon is the imaginary intersection of the celestial sphere and a flat surface everywhere perpendicular to your zenith. (B) Since your zenith and nadir are defined by the direction of gravity where you are standing, strictly speaking someone next to you would have a different zenith and nadir.

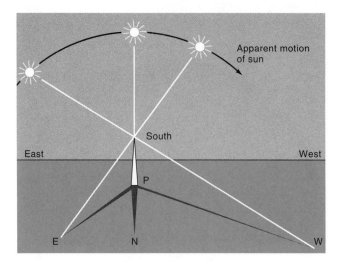

Figure 2.4. The length and direction of the shadow cast by a vertical stick change throughout the day. In the morning the shadow is long and points toward the west (PW), whereas at noon the shadow is shortest and points due north (PN). As the day goes on the shadow becomes longer, and points more and more to the east (PE).

from a generally easterly direction, moves across the sky, reaching its greatest angular altitude above the horizon at what we would call **local noon**, and then sets in the west. At local noon it is directly to the south for an observer north of 23.5° latitude. (See Question 16 for more on this and related points.) We could establish the approximate time of local noon by watching the shadow cast by a rod stuck perpendicularly into the ground. At local noon, when the Sun is highest in the sky, the shadow will be shortest (Figure 2.4). Furthermore, at this moment the shadow is pointing directly to the north. This north-south line, extended to the sky and passing through our zenith, cuts the sky into two equal parts. The dividing line is called the **meridian**. Morning and afternoon hours (determined by the position of the Sun) are denoted "AM" or "PM," respectively, from the Latin *ante* (before) and *post* (after) *meridiem* (noon).

Motions of the Stars

When we looked at the sky at night we would see that stars, except those near the north celestial pole, follow the same pattern as the Sun (Figure 2.5). They rise in the east, climb until they reach maximum angular altitudes above our horizon (when they would be crossing our meridian), and finally set in the west. If we watched toward the north we would notice that the stars describe circular arcs around a point in the sky (called the **north celestial pole** or **NCP**) now about one degree from the star Polaris (Figure 2.6). If we watched all night we would observe that some of the stars never disappear below our horizon, but make complete circles around the NCP. These are called **circumpolar** stars. After a little thought we would conclude that the Sun and the stars move in circular paths all centered on the NCP and that as we look southward an increasing part of this circular motion takes place below our horizon. These motions would all be parallel to the **celestial equator**, that is, parallel to the circle on the celestial sphere that is everywhere 90° from the NCP. We know (but our ancient ancestors didn't) that the apparent daily motions of the Sun and stars across the sky are simply

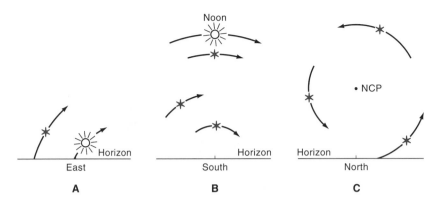

Figure 2.5. Like the Sun, stars rise in the east (A); looking toward the south you would see them reach their maximum angular altitude above the horizon (B) when they cross the meridian (the north-south line passing through your zenith, cutting the sky into two equal parts); later they set in the west. If you watched toward the north for several hours you would see not only stars rising in the east and setting in the west, but some that do not rise and set at all, but follow circular paths around a common central point, the north celestial pole (C). These are called circumpolar stars.

Figure 2.6. The trails of circumpolar stars seen over Kitt Peak National Observatory in Arizona. The tall building contains a 4-meter telescope, the largest on the mountain. The lights of Phoenix, more than 100 miles away, are visible in the background.

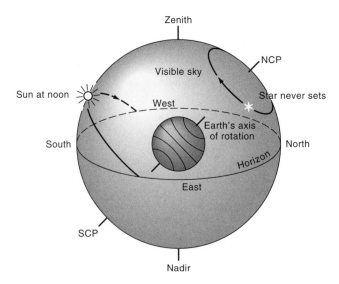

Figure 2.7. The celestial sphere, combining several features we have discussed separately: the zenith, nadir, and horizon; the geographic and celestial poles; circumpolar stars; and the Sun's path over the course of a day. Note that the daily motions of the Sun and stars are produced by the Earth's rotation, and so are all parallel to each other and to the celestial equator.

consequences of the Earth's rotation on its axis. (Even though some of the ancients were aware that these phenomena could be explained by a moving and rotating Earth, taking the Earth to be stationary and the sky to be in motion seemed much more reasonable, more commonsensical to them.) Hence the north celestial pole is just the projection on the sky of the Earth's geographic pole, and the celestial equator is the Earth's geographic equator extended to the sky (see Figure 2.7).

The Sphericity of the Earth

This last comment implies that the Earth is spherical. If we lived on the sea coast, the disappearance of ships over the horizon—hull first, then sails, then tops of the masts—might suggest that the Earth is a sphere (Figure 2.8A). (How would a ship disappear if the Earth were flat?) Perhaps a more convincing case could be made if we traveled north or south and measured the angular height of Polaris above our horizon. What we would find is shown in Figure 2.8B. As we traveled north, we would discover that Polaris was higher above the horizon than it was when we started out. A spherical Earth provides the simplest explanation of this effect (though it might present a problem in understanding why people on the "other side" of the Earth didn't fall off). Also, during a lunar eclipse (described later in this chapter) the shadow cast by the Earth on the Moon is always circular. Only a spherical object can always cast a circular shadow regardless of its orientation. All of these are simple phenomena; their interpretation requires a little thought, however.

Let's travel in our imagination to the north geographic pole of the Earth. The NCP would then be directly overhead at our zenith. Therefore, the daily paths of the stars would be parallel to the horizon (which would also represent the direction of the celestial equator) and none would rise or set. If we traveled 90° south to the equator, our zenith and horizon would, of course, also move 90° (see Figure 2.9). The NCP would

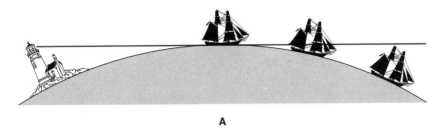

A

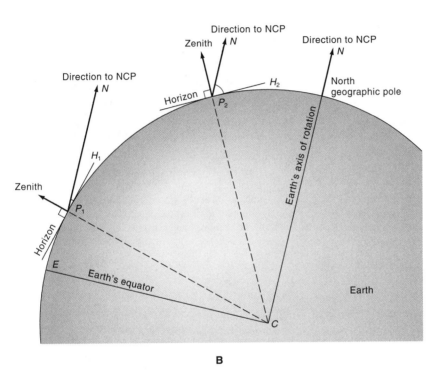

B

Figure 2.8. (A) The way a ship disappears over the horizon indicates that the Earth must be spherical. (B) The change in the angular altitude above your horizon of the direction to the celestial pole (angle H_1P_1N or H_2P_2N) as you move north or south shows that we live on a sphere. Note that the angle ECP_1 is equal to the angle H_1P_1N. What does this say about the relation between your geographic latitude and the altitude of the pole?

be the north point on our horizon and the celestial equator would cross the zenith. Hence, the stars would rise and set along paths that are perpendicular to the horizon.

It is crucial to realize that nothing in our immediate experience requires or even strongly suggests that the Earth moves. Quite the contrary: plain "common sense" tells us that the Earth is fixed. Thus the simplest explanation of these observations of the Sun and stars is that we are standing on a central, unmoving Earth. The Sun and the stars are fixed to the celestial sphere, which rotates from east to west once about every 24 hours about an axis running through the north and south celestial poles. A much less obvious interpretation (but the correct one) is that it is the Earth, not the sky, which rotates once every 24 hours from west to east. The Earth's axis of rotation, extended to the sky, "intersects" it at the celestial poles.

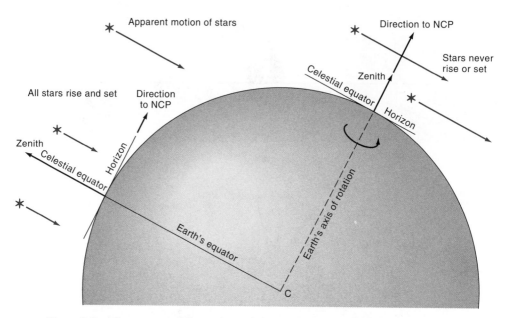

Figure 2.9. The apparent daily motions of stars as seen at the Earth's equator and at the North Pole. Stars neither rise nor set (that is, all stars are circumpolar) at either geographic pole because the polar zenith points to the celestial pole, and so the horizon, 90° from the zenith, is parallel to the celestial equator. At locations closer and closer to the geographic equator, fewer and fewer stars are circumpolar until, at the equator itself, where the celestial equator is perpendicular to the horizon, all stars rise and set; none are circumpolar.

Relative Motion of Sun and Earth

That neither of these explanations can be the whole story, however, would become apparent from a simple observation we could make over the course of a year: different stars are out at different seasons of the year. For example, the Pleiades are visible in the evening in fall and winter, but not in the spring or summer; Orion is a prominent constellation in the winter sky but is not visible in the summer. This phenomenon can be explained by invoking *relative* motion between the Sun and the Earth. As before, if we had not been told the contrary, it would seem far more reasonable to assume that it was the Sun that moved and not the Earth. A sketch like Figure 2.10 would give a perfectly plausible interpretation of these observations. When the Sun is at *A*, stars in

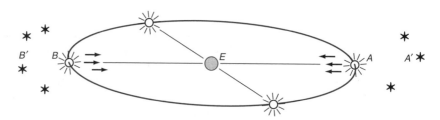

Figure 2.10. Different stars are seen in the night sky at different times of the year. The sketch shows how this phenomenon can be explained by the Sun moving around a fixed Earth. When the sun is at *A*, only stars in the direction of *B'* will be visible at night; six months later the opposite is the case. That is, an Earth moving about a central, fixed Sun is not required to explain this observational fact.

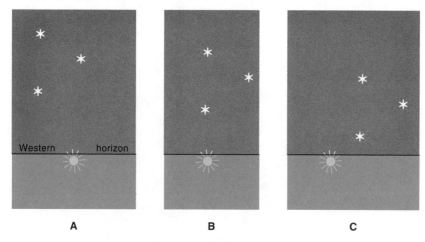

Figure 2.11. Suppose some bright stars could be seen in the western sky soon after sunset (A); just after sunset a few days later, they would be located as shown in (B), and some days later yet their sunset position with respect to the Sun would be as in (C). Thus there is relative motion between the Sun and the stars, and it is not unreasonable (especially if it seems hard to imagine that the Earth could be moving) to assume that the Sun moves eastward with respect to the stars. In fact, of course, this motion is really a "reflection" of the Earth's motion around the Sun.

the direction EA' are not visible, because they are in the daytime sky. Stars in the opposite direction, EB', can be seen at night, because now the observer is looking out away from the Sun into the night sky. Six months later, when the Sun is at B, just the opposite is true, and stars toward A' are visible. The correct, but less obvious, interpretation is given by interchanging the Earth and the Sun in the sketch. In either case, there is a yearly cycle of relative motion between the stars and the Sun (or the Earth, depending on your point of view).

An equivalent but more subtle kind of observation involves noting the angular altitude above the horizon of a star that appears in the west just after sunset, as shown in Figure 2.11.[4] After a few days, we would find that the pattern of stars has slowly shifted in position relative to the Sun. Stars set about four minutes earlier each day, that is, sooner after sunset. Or, looking at it from another point of view, the Sun appears to move *eastward* each day with respect to the stars. (Make sure you understand why these two statements are equivalent.) As time goes by, different stars appear just after the Sun sets. After one year, the original pattern of stars reappears around the sunset point, and the cycle repeats itself. Since there are 360° in a circle and about 365 days in a year, the Sun appears to move about one degree each day with respect to the stars, or about twice its angular diameter every day.

The apparent path of the Sun against the stars is called the **ecliptic**. If we carefully plotted this path, we would find that it was inclined at an angle of about 23.5° to the celestial equator. In other words, the north celestial pole is tipped 23.5° away from the north pole of the ecliptic, as shown in Figure 2.12. The ecliptic is really the Earth's

---

[4] Since at sunset the eastern sky is much darker than that in the west, it would be easier to observe the locations of stars near the eastern horizon. Better yet, wait until midnight, say, and observe the positions of stars with respect to the meridian. Noting the altitudes of stars that come out just after sunset is perhaps easier to visualize, however.

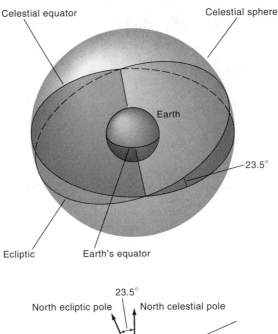

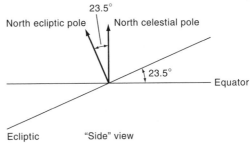

Figure 2.12. This diagram shows the relation between the planes containing the celestial equator and the ecliptic (the apparent path of the Sun against the stars). They make an angle of 23.5° to each other.

orbital plane around the Sun, and not the other way around; the apparent motion of the Sun is just the reflection of the Earth's own motion. None of the observations we have made so far, however, rules out the possibility that it is the Sun that is in motion and not the Earth. I think you would agree that our first impression would be that the solid Earth is stationary, just as our ancestors believed until only about 400 years ago.

The Seasons

The fact that we have seasons also requires that the Earth and the Sun mutually revolve. The axis of rotation of the Earth, like that of a spinning top, keeps pointing in approximately the same direction in space (toward the NCP) as the Earth travels around the Sun (see Figure 2.13). In a northern-hemisphere winter, the Sun's rays strike the Earth's surface (for example, at the point marked *M*) more obliquely (at more of a grazing angle) than they do in summer. In summer the northern hemisphere is tipped toward the Sun, and so the Sun appears higher in the sky, the days are longer, and the weather is warmer.

Near the first day of winter the northern-hemisphere day is the shortest of the year and the southern-hemisphere day is the longest. Six months later the opposite is true.

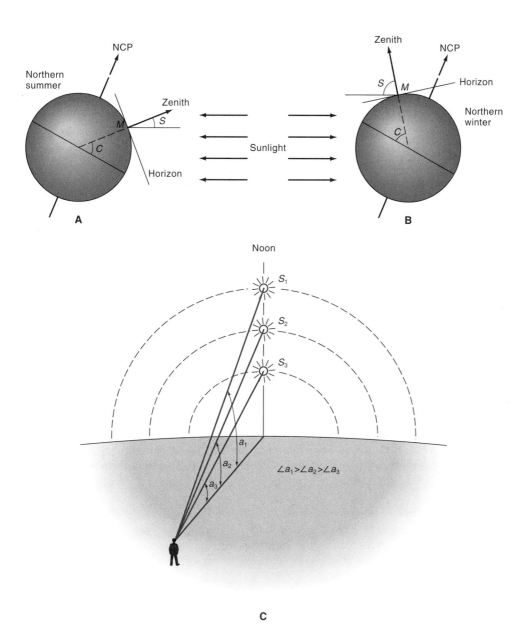

Figure 2.13. As the Earth travels around the Sun, its axis of rotation keeps pointing in the same direction in space. Northern summers occur when the northern hemisphere is tipped toward the Sun (A); six months later (B) the northern hemisphere is tipped away from the Sun, and it's winter. The southern hemisphere experiences the opposite seasons. Note that the direction of the Sun is near the zenith at noon in the northern summer, but is low in the sky in winter. This is sketched in (C), where S_1 shows the altitude a_1 above the horizon of the summer Sun at noon and S_3 the altitude a_3 of the Sun in winter. The point M is a mid-northern latitude location.

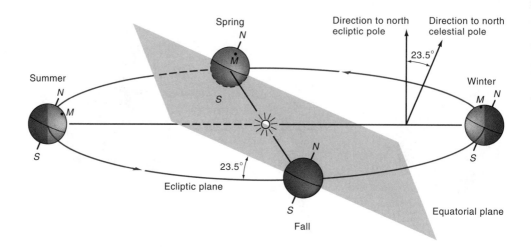

Figure 2.14. All four seasons (for the northern hemisphere) are shown. The Earth travels around the Sun in the ecliptic plane, which is tipped by 23.5° to the equatorial plane. Notice that more of the northern hemisphere is illuminated by sunlight in the summer than in the winter—the summer days are longer than those in winter. At the spring and fall equinoxes, both hemispheres receive sunshine for equal times, and the lengths of day and night are equal. The point *M* is a mid-northern latitude location.

Hence there must be a time about halfway between when the length of the day is the same in both hemispheres. This occurs on the first day of spring and again six months later on the first day of autumn, that is, at the **vernal** and **autumnal equinoxes**. (*Equinox* means "equal night.") At those times the Earth's axis of rotation is tipped neither toward nor away from the Sun, and the Earth is at one of the two points in its orbit where the line of intersection of the equatorial and ecliptic planes passes through the Sun (see Figure 2.14). The direction of the Sun (against the background stars) as seen from the Earth on the first day of spring has been made a reference direction against which to measure other directions in the sky. However, this direction has no intrinsic significance; it's just a convenient reference point.

Now, suppose the Earth's axis of rotation were not tipped, but instead were perpendicular to the plane of the ecliptic (which would put the equatorial plane in the ecliptic). As the Earth revolved around the Sun, the orientation of a given point on the Earth with respect to the Sun would remain the same; the altitude of the Sun above the horizon at noon would always be the same; the length of the day would not change during the year; and there would be no seasons: all in all, a pretty boring state of affairs! Suppose you lived in Chicago: to experience hot weather at any time of the year you would go south, and to go skiing you would go north. The contrast of our seasons depends on the completely accidental inclination of Earth's equator with respect to its orbital plane. Had the angle between the two planes been larger, the seasonal contrast would have been greater, and vice versa. Thus, for us to have seasons the celestial equator and the ecliptic plane must not coincide, and there must be *relative* motion of the Sun and the Earth. None of the observations made so far, however, *requires* that it is the Earth that is moving.

Interpretation of These Observations

So how might we account for the motions we have so far discovered, making the perfectly reasonable—but incorrect—assumption that motion takes place around a fixed, central Earth? Just one sphere won't do, because the Sun and the stars do not move exactly together. (Remember that with respect to the stars, the Sun appears to move eastward about one degree each day.) Two spheres are necessary, both centered on the Earth. One takes the stars completely around the sky from east to west every 24 hours. The second one, carrying the Sun, also moves east to west with the first. At the same time, however, it rotates about its own axis (inclined about 23.5° with respect to the axis of the stellar sphere), slipping very slightly to the east, about one degree each day, with respect to the sphere of the stars. That is, when the stellar sphere has completed one daily rotation of 360°, the solar sphere will have turned only 359°. In this way our ancestors accounted for the daily motions of the stars and for the daily and yearly movement of the Sun. You should understand, however, that these motions attributed to the Sun are actually a consequence of two motions of the Earth: its yearly trip around the Sun along the ecliptic, and its daily 24-hour rotation period. This last is about an axis (running through the Earth's geographic poles) that is tipped 23.5° with respect to the perpendicular to the ecliptic plane.

Motion and Appearance of the Moon
The Moon's Motion

Next, let us consider the motion of the Moon around the Earth. We would quickly discover that the Moon rises in the east and sets in the west, just like the Sun and the stars. Superposed on this motion is an eastward drift of the Moon with respect to the background stars. Thus, the Moon's motion around the Earth is analogous to that of the Sun's, but its eastward movement is much more rapid than the corresponding motion of the Sun. The Moon takes only about 27 days to make a complete trip around the sky with respect to the stars, rather than 365 days. This average daily eastward motion of about 13° (or about one lunar diameter per hour, since the angular diameter of the Moon is about 0.5°) is easily detected in the course of just one night. Furthermore, if we tracked the lunar motion against the stars, we would find that its path was inclined at an angle of about 5° with respect to the ecliptic. This small angle of inclination has an important consequence for eclipses, as we shall soon see.

Phases of the Moon

We know that at times the Moon appears as a bright circular disk ("full Moon"), or perhaps as a half-illuminated disk ("quarter Moon"); at other times it is a thin arc ("crescent Moon"). We call these different appearances the **phases** of the Moon. These regularly changing aspects are the most striking feature of the Moon. Understanding how they come about must have been very difficult when people first wondered about them. A first step toward this understanding is taken when we realize that the Moon is not self-luminous, but must shine by reflected light. (What simple observations would suggest this?) Then a sketch like that in Figure 2.15 enables us to see how lunar phases occur. The Sun is assumed to be so far away that its rays are essentially parallel by the time they reach us. (The Earth and Moon are not drawn to scale.) The hemisphere of the Moon facing the Sun is always illuminated, of course, but we see varying fractions

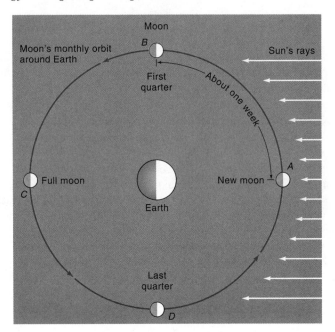

Figure 2.15.   This is a view looking down on the Moon's orbital plane from the north. The Sun is at a great distance to the right and illuminates the hemisphere of the Moon facing it. The unilluminated hemisphere is shaded. We on Earth see the hemisphere of the Moon facing us. At new Moon the hemisphere we see is unilluminated, but about a week later at first quarter, half of the face we see is illuminated.

of it depending on the relative orientation of the Earth, Moon, and Sun. For example, when the direction to the Sun and Moon coincide (as at *A*), we see none of the Moon's illuminated half, and call the Moon new. When the Sun and the Moon are in opposite directions (*C*), we see all of the Moon's illuminated hemisphere, and say that it is full. When the Moon has traveled halfway between these two orientations (*B*), only half the face that we see is illuminated, giving us a quarter Moon. Figure 2.16 shows various phases of the Moon.

Figure 2.15 also shows the position of the Moon at various times during the month. At *A*, when the Moon is new, placing it in the same direction as the Sun (making the former difficult to see), the Moon will set with the Sun; the full Moon (at *C*), 180° from the direction to the Sun, rises at sunset and sets at sunrise. At *B* the direction between the Sun and the Moon as seen from the Earth is 90°. Thus, at noon the first-quarter Moon is rising, and at sunset it is on the meridian. When does the third-quarter Moon rise?

Lunar Eclipses

When the Moon is full, it would appear from Figure 2.15 that it must fall within the shadow cast by the Earth and consequently be **eclipsed**. (Remember that the Earth's diameter is four times that of the Moon.) So why isn't there a lunar eclipse every month? There would be if the orbit of the Moon were in the ecliptic, but as we saw earlier, it is inclined to it at an angle of about 5°. Consequently, the full Moon is gen-

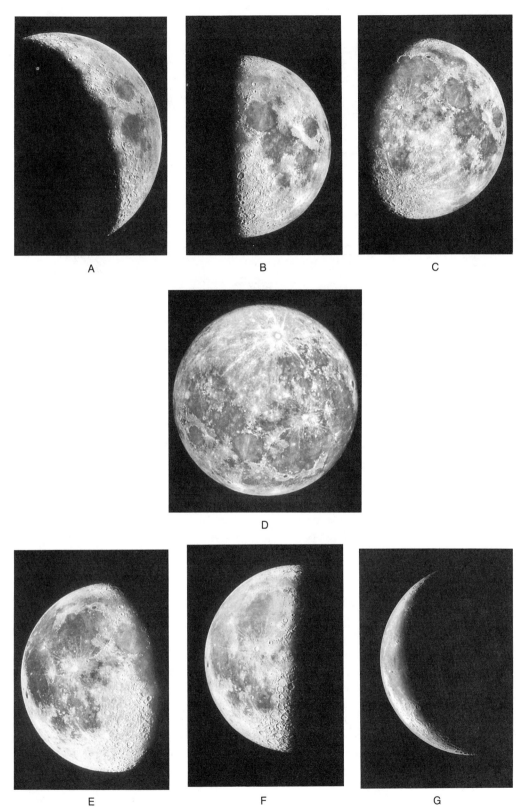

Figure 2.16. The Moon at several phases. Starting from new, the age of the Moon is (A) 4 days, waxing (increasing) crescent; (B) 7 days, first quarter (B on Figure 2.15); (C) 10 days, waxing gibbous; (D) 14 days, full (C on Figure 2.15); (E) 20 days, waning (decreasing) gibbous; (F) 22 days, last quarter; (G) 26 days, waning crescent.

27

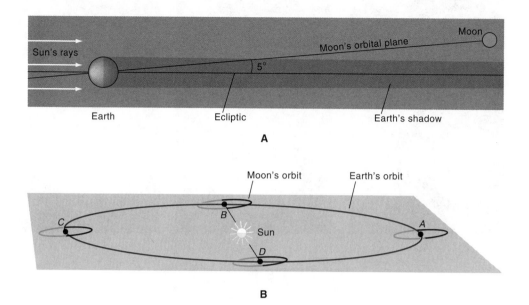

Figure 2.17. This diagram shows why there is not an eclipse of the Sun at every new Moon, and why the Moon is not eclipsed every time it is full. The Moon's orbital plane is inclined 5° to the ecliptic; so generally the Moon is not in the Earth's shadow when it is full, and therefore is not eclipsed (A). Similarly when the Moon is new it will fall north or south of the Earth-Sun line and so will not eclipse the Sun. Only when the Moon is in the plane of the ecliptic when it is new or full will there be a solar or lunar eclipse, respectively (B).

erally not in the Earth's shadow, but is north or south of it; so no eclipse occurs (see Figure 2.17A). Now, the Moon must cross the ecliptic plane twice each month, and if this crossing occurs at the time of a full Moon, a total lunar eclipse takes place (Figure 2.17B). Such an eclipse can be seen by anyone on the Earth who can see the Moon at that time, so most people have seen lunar eclipses (see Figure 2.18).

Solar Eclipses

Solar eclipses are a different matter, however. These magnificent spectacles occur only because of a most remarkable coincidence, namely, that the average *angular* diameters of both the Sun and the Moon just happen to be about 31 minutes of arc, or slightly more than 0.5°. In other words, the huge Sun is so far away from the Earth that it appears to be no larger in the sky than the small, relatively nearby Moon. Consequently, when the new Moon is in the ecliptic plane, it may just block out the Sun as seen from the Earth and cause a total solar eclipse to occur (Figure 2.19). Such an eclipse does not take place at every new Moon for the same reason that a lunar eclipse does not happen at every full Moon: the Moon's orbital plane is inclined to that of the ecliptic. (Incidentally, the origin of the name "ecliptic" should be clear to you now.)

There is an additional wrinkle, however. Some solar eclipses are not total but instead are **annular** (see Figure 2.20). Though as seen from the Earth the center of the Moon lines up with the center of the Sun, not all of the Sun is covered; instead, a thin ring—an annulus—of sunlight is seen. Careful measurements of the angular diameter of the Moon during total and annular eclipses show that it can be slightly larger or slightly smaller than average. We can understand this easily by supposing that the

A

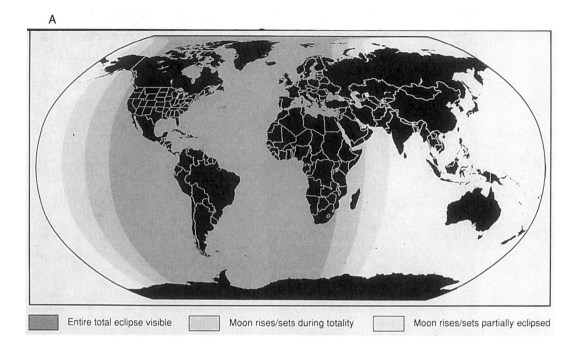

Entire total eclipse visible　　Moon rises/sets during totality　　Moon rises/sets partially eclipsed

B

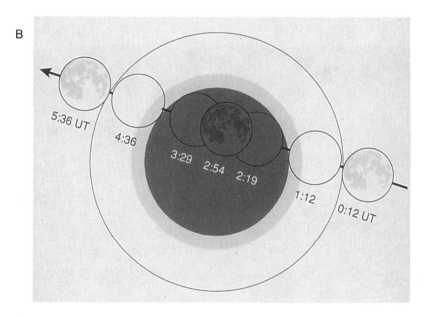

Figure 2.18. The entire total lunar eclipse (A) of September 27, 1996, will be visible from all of Central and South America, and from most of North America, Europe, and Africa. The Moon's position (B) with respect to the Earth's shadow during the September 27, 1996, lunar eclipse. Outside the outer circle no sunlight is blocked by the Earth; within the outer circle (the penumbra) an increasing fraction of light is blocked by the Earth until, within the dark circle (the umbra), none of the Sun is visible. A total eclipse occurs when all of the Moon passes within the Earth's shadow, the umbra; if only part of it does, a partial eclipse takes place. Although the Moon's brightness decreases somewhat when it enters the penumbra, the change is hard to detect. (UT is Universal Time, kept in the time zone centered on Greenwich, England. To convert UT to CST, for example, subtract 6 hours.)

Figure 2.19. The faint glow of the corona is visible only during a total eclipse when the light from the Sun is blocked by the Moon as it was in the eclipse of March 1970.

Figure 2.20. The annular eclipse of January 4, 1992, photographed from San Diego. The Moon was far enough from the Earth that its angular diameter was slightly smaller than the Sun's, making a total eclipse impossible.

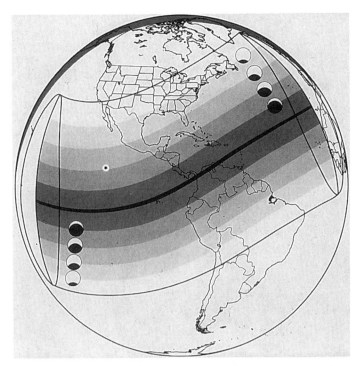

Figure 2.21. The path of the total solar eclipse of February 26, 1998, will begin at sunrise in the Pacific and, moving about 1600 kilometers/hour, will be seen around noon in northern South America. The eclipse path is at most 151 kilometers wide, giving a totality maximum of about 4.1 minutes. Half of the Sun will be eclipsed as seen from Miami and one-quarter eclipsed from New Orleans and Atlanta.

Moon's orbit around the Earth is not circular, but elliptical; then the Earth–Moon distance would change, and with it the apparent diameter of the Moon. This is in fact what is happening; the maximum distance of the Moon from the Earth is about 12 percent greater than its minimum distance. Consequently, even if the new Moon were in the ecliptic, a total eclipse might not occur. This is so because if the long axis of the Moon's elliptical orbit is in the direction to the Sun, that is, in the ecliptic plane, the Moon can be at its maximum distance from the Earth, so that its angular diameter is slightly smaller than the Sun's, resulting in an annular eclipse. Furthermore, because, under the best of circumstances, the angular diameter of the Moon is just barely large enough to cover the Sun, a total eclipse can be seen from only a narrow band on the Earth, typically about 150 kilometers wide (but thousands of kilometers long). The maximum width of the Moon's shadow within which the Sun appears to be totally eclipsed is about 267 kilometers, producing almost seven and one-half minutes of totality. On either side of the band the eclipse will be **partial**, not total; that is, only a part of the Sun will be eclipsed (see Figure 2.21). A good way of viewing the Sun in eclipse or out is shown in Figure 2.22.

Even though a total solar eclipse occurs twice every three years on the average, one is visible from a given place on Earth only once in about 360 years. That is why most people have not seen a total solar eclipse. The last total solar eclipse visible in the continental U.S. was in 1979; none will occur in the continental U.S. for the rest of this century, but one was visible from Hawaii and Baja California, Mexico, on July 11, 1991. Totality lasted an unusually long 7.1 minutes. Paths across the Earth's surface of several of the next solar eclipses are given in Figure 2.23.

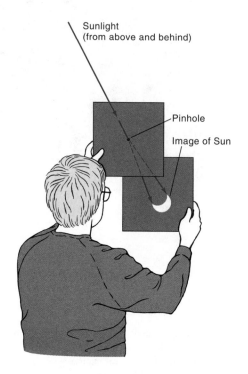

Figure 2.22. A safe way to observe the partial phases of a solar eclipse is to pass sunlight through a small pinhole in a card and project the image on a second card.

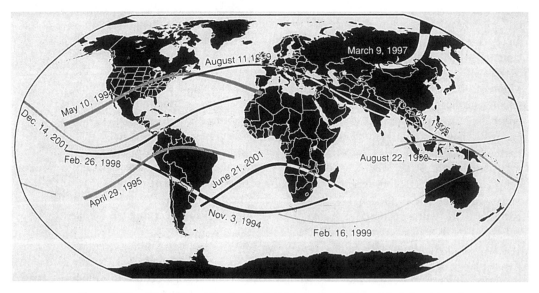

Figure 2.23. The paths on the Earth of all total solar eclipses (black bands) and annular solar eclipses (gray bands) through 2001.

Motions of the Planets

The Zodiac

The motions so far considered are fairly easy to understand, and can be conveniently modeled assuming a fixed, central Earth. The motions of the planets, however, are much more puzzling. (The planets visible to the naked eye are Mercury, Venus, Mars, Jupiter, and Saturn; Uranus, Neptune, and Pluto were discovered with the telescope.) If we plot the paths of the naked-eye planets against the stars, we find that these paths are not randomly distributed across the sky. Instead they all fall within a band on the sky only about 16° wide, centered on the ecliptic (see Figure 2.24). This band is called

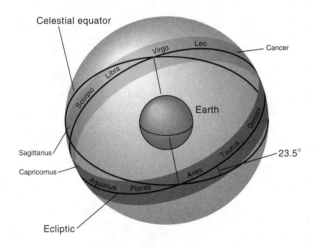

Figure 2.24. The zodiac is a band in the sky about 16° wide, centered on the ecliptic. The naked-eye planets and the Moon all travel within it. About 2,500 years ago the zodiac was divided into twelve segments, each named for the constellation that happened to fall within it at that time.

the **zodiac**, which comes from a Greek word meaning "carved figures." Apparently it was first recorded by the Babylonians. About 2,500 years ago, they showed the zodiac divided into twelve segments, each one extending 30° along the band. Each section is denoted by a **sign** named for the constellation that happened to be located in that particular division at that time. In present usage **constellation** simply refers to one of the 88 areas into which the whole sky is divided; each is named after the figure or object associated with the brightest stars located within it. When we say, for example, that Jupiter is in Taurus (the bull), we mean that Jupiter is in the area of the sky defined by the constellation Taurus (Figure 2.25), just as we would say that Oshkosh is in Wisconsin. Most of the bright stars delineating the "object" of a given constellation have no physical relation with each other whatsoever; they just happen to be in the same general direction. See Appendix D for more about constellations.

Retrograde Motion Like the other celestial objects, the planets appear to move from east to west every day. Just as with the Sun and the Moon, superposed on the planets' daily motions is a slow eastward drift against the background stars. In this respect the five naked-eye planets, the Sun, and the Moon behave similarly; they all

34

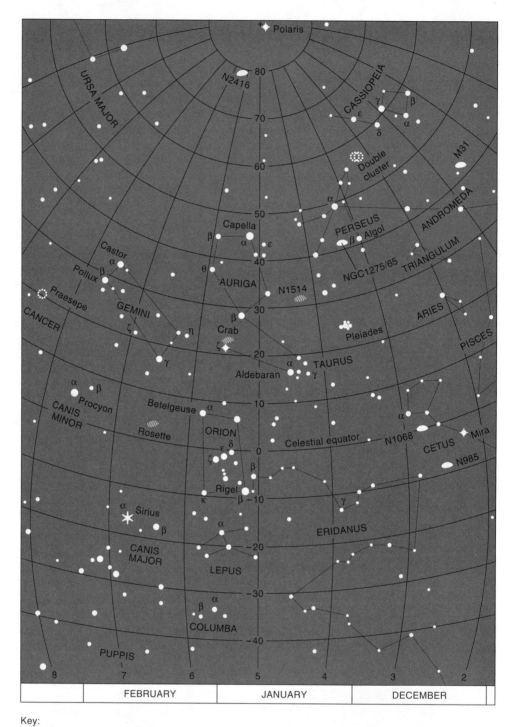

| FEBRUARY | JANUARY | DECEMBER |

Key:

Variable Faint ⟶ Bright   Open   Double   Globular   Galaxy   Nebula
                                     cluster   cluster   cluster

Stars

Figure 2.25.   This shows the brighter stars visible in the (northern) winter sky. Hold the chart in front of you while you are facing south. At around 9 P.M. at the end of January and the beginning of February, Orion would be found on your meridian (the north-south line through your zenith). Around mid-February, the brightest star in the sky, Sirius, would be on your meridian at 9 P.M., and Orion would be to the west. Note that the chart extends to the pole star, Polaris, so many stars would be behind you. Similarly, some stars would be below your horizon.

How are positions of stars specified in the sky? Just imagine the latitude and longitude grid on the Earth extended to the sky. On the chart above, the vertical row of numbers starting with "80" just below Polaris and extending to the south corresponds to latitude on the Earth—90° at the celestial pole, 0° at the celestial equator, etc. Celestial latitude is called **declination**. Similarly, the numbers extending across the bottom of the chart correspond to longitude on the Earth. On the sky these are called **right ascension**. Just as longitude and latitude specify a particular point on the Earth, right ascension and declination specify a particular point in the sky.

move with respect to the stars. Because of this characteristic, the Greeks referred to them all as *planetoi*, "wanderers," from which our word planet comes. Unlike that of the Sun and the Moon, however, the planets' eastward motion within the zodiac occasionally slows, stops, and reverses itself. For a time they appear to move slowly *westward*—backward—against the stars before reversing direction again and continuing their slow travel to the east, as seen in Figure 2.26. The slow eastward motion is said to be **direct**; the strange and unexpected westward motion is called **retrograde**. (Remember, these slow eastward and westward motions are superposed on the planets' daily travels east-to-west.) These irregular, looping motions are in striking and even bizarre contrast to the constant, regular motions of the stars.

Careful observations reveal other strange features. Planets don't shine with a constant brightness; Mars, Jupiter, and Saturn are brightest when in retrograde motion! Furthermore, they can be seen at any angular distance from the Sun. That is, they could be in the direction toward the Sun, in the opposite direction, or anywhere in between.

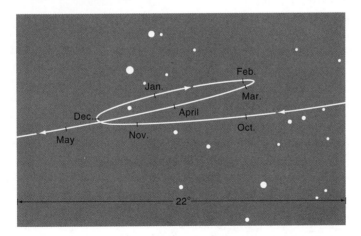

Figure 2.26.   The retrograde motion of Mars in the fall and winter of 1992/93, when it was in the constellation of Gemini. Mars began its retrograde motion (toward the west) in early December and resumed its direct (eastward) motion in mid February. The retrograde loop is about 10° across.

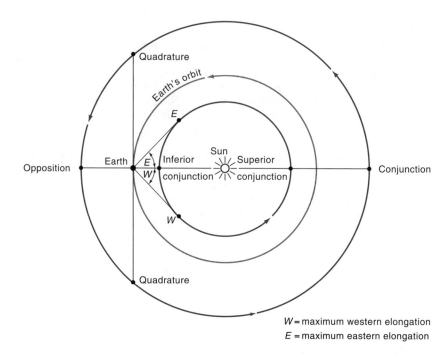

W = maximum western elongation
E = maximum eastern elongation

Figure 2.27. The various arrangements (configurations) of the Earth, Sun, and planet, which are given special names. For example, Mars is closest to the Earth when it is in opposition, but Venus or Mercury are closest when they are at inferior conjunction.

In striking contrast, Mercury and Venus are never found more than about 28° and 47°, respectively, from the Sun. (We call Mercury and Venus' greatest angular distances from the Sun as seen from the Earth their **maximum elongations**. See Figure 2.27.) These two planets, therefore, can never be found on the meridian at midnight or at sunset, for example, but appear only in the early evening or pre-dawn sky; consequently, they are sometimes called "evening stars" or "morning stars." Finally, since the planets appear to change the direction of their slow motion across the sky (from direct to retrograde and back to direct again), their motions can't be uniform. That is, each cannot move a fixed number of degrees per day with respect to the stars. Careful naked-eye observations show that the eastward motions of the Sun and Moon are also nonuniform (but only slightly so). The situation becomes more and more puzzling.

We have described most of the naked-eye motions of this chapter in terms of a fixed Earth because that is what our ancestors did, and that is just what we would have done had we been in their shoes. To understand how early astronomy developed we must constantly remind ourselves just how absurd the notion of an Earth rotating on its axis and revolving around the Sun (as it is in fact doing) appeared to our forebears. ("How could objects stay fixed to the Earth? Why wouldn't everything fly off? Where is any evidence, any feeling of this supposed motion? Are you crazy or what?")

As I think you can easily imagine, retrograde motion in particular, when assumed to take place about a central, fixed Earth, requires a much more elaborate system of spheres for its representation than we have so far considered. Obviously, a single, uniformly moving sphere for each planet won't reproduce retrograde motion. For more than 2,000 years the central problem of astronomy was how to account for these strange planetary motions observed from a stationary, central Earth. Their modern interpretation in terms of an Earth moving about the Sun was finally given in the sixteenth and early seventeenth centuries by Copernicus and Kepler.

With a little care and persistence, you can make most of the observations described in this chapter yourself; I urge you to do so. It is remarkable how much can be learned just by looking at the sky and making a few simple measurements. (Actually you must learn to *see* the sky and not just look at it, to note the relative positions and brightnesses of objects from night to night, for example.) These observations provide the raw materials from which models of the naked-eye universe were fashioned, models that left their imprint on our way of looking at the world for 2,000 years.

Terms to Know

Horizon; zenith; nadir; celestial sphere; celestial poles and equator; circumpolar; meridian; local noon; direct and retrograde motions; ecliptic; zodiac; constellation; equinox; lunar phases; partial, total, and annular eclipses; solar system; astronomical unit; angular separation; angular motion; maximum elongation; inferior and superior conjunction; opposition.

Ideas to Understand

Explanation of day and night, the motions of the Moon and stars, and the seasons in terms of both a Sun-centered and an Earth-centered solar system; the causes of the different types of solar and lunar eclipses, and the phases of the Moon.

Questions

1. (a) Suppose you are 5 feet 8 inches tall; what is your height in centimeters? in meters?

(b) What is the mass in grams and in kilograms of a 10-pound ball?

(c) Express the following in scientific notation: the Astronomical Unit in both miles (about 93,000,000) and kilometers (149,000,000); the number of minutes of arc in 90° and the number of seconds of arc in one radian (= 57.3°); the velocity of light, c = 300,000 km/sec.

2. (a) The Moon orbits the Earth in a nearly circular orbit at a distance of about 240,000 miles. How many kilometers does the Moon travel in one trip around the Earth?

(b) If the Moon takes about 28 days for one round trip, what is its average velocity in km/hr?

3. (a) How many times larger in diameter is the Sun compared to the height of a six-foot-tall person?

(b) How many times larger is the height of the six-foot-tall person than the diameter of an atom (which is about 10^{-7} cm across)?

(c) The diameter of the planetary portion of our solar system is about 10 light hours (that is, it takes a light wave 10 hours moving at 300,000 km/hr to travel that distance); the diameter of our Galaxy is about 100,000 light years. How many times larger is the Galaxy's diameter than the diameter of our planetary system? Compare and comment on the ratios found in (a), (b), and (c).

4. (a) Do we ever see Venus and the full Moon near each other in the sky (that is, in about the same direction)? Explain your answer with a sketch of the orbits of Earth, Moon, and Venus.

(b) Do we ever see the new Moon on the meridian at sunset? the full Moon rising at midnight? Explain with a sketch of the Moon's orbit around the Earth.

(c) What simple observations can you make to convince yourself that the Moon is not self-luminous—that is, that it shines by reflected sunlight, not by its own light—and that the Sun *is* self-luminous?

5. Make a sketch of the orbits of Mercury, Venus, Earth, and Mars around the Sun, as seen from the north ("above" the plane of the ecliptic). Explain your answers with the sketch.

(a) As seen from the Earth, where in its orbit would Mercury appear largest, at new or at full phase? Where would it appear smallest?

(b) Could Venus be seen on your meridian at midnight? Could Mars?

(c) Why is Mercury always seen either soon after sunset (as the "evening star") or before sunrise (the "morning star"), but never at midnight?

6. You are at the seashore watching a ship disappear over the horizon. How does it disappear? How would it disappear if the Earth were flat?

7. Suppose that the Earth's axis of rotation were tipped over, so that it was in the ecliptic (very nearly like Uranus). What would the seasons be like in both hemispheres over the course of a year?

8. Refer to Figure 2.18B in the text. Give times in UT and for your time zone.

(a) How long will it take for the Moon to travel through the penumbra? Will its appearance change much during this time?

(b) At what time will the Moon begin to be eclipsed? At what time will the eclipse be total?

(c) How long will totality last? About how much longer would it have lasted if the eclipse had been central (that is, if the Moon had traveled through the center of the Earth's shadow)?

9. (a) Suppose the Moon's diameter were twice as large as it actually is. Would total solar eclipses be more or less common? Why? Would total lunar eclipses be more or less common? Why?

(b) Suppose the orbit of the Moon were in the plane of the ecliptic. Would there be more, fewer, or the same number of solar eclipses? Why?

(c) Could an eclipse of the Moon occur the day following an eclipse of the Sun? Explain your answer.

10. Show that the Moon's eastward motion against the stars, about 13°/day, amounts to about one lunar diameter per hour.

11. (a) What is the phase of the Moon just before it is totally eclipsed?

(b) If you were on the Moon looking at the Earth during a total lunar eclipse, what would be the phase of the Earth? Would you see the Sun? Explain your answers with a sketch.

12. (a) Since the Moon rotates on its axis in the same time that it revolves around the Earth, the same hemisphere is always facing us. Does this mean that an astronaut on the Moon would also see only one hemisphere of the Earth? Explain.

(b) Suppose the Moon were self-luminous. What phase(s) would we see? Explain your answer.

13. (a) The length of the shadow at noon cast by a flagpole depends on the time of the year. Explain why.

(b) Away from city lights it is easy to see the shadow cast by the flagpole illuminated by the full Moon. At about what time of night is the shadow shortest? At what time of year is the shadow shorter than at any other time?

(c) Suppose it were possible to measure the length of the shadow cast by the pole illuminated by a bright star when it is highest in the sky. Would the length of the shadow change during the year?

14. From where on Earth can you see all the stars during the course of a year? Where would you be limited to seeing only half the stars? Explain your answers.

15. (a) Suppose you point a 35-mm camera at the north celestial pole at night, open its shutter, and expose the film for six hours. Sketch what you would expect to see on the developed photograph.

(b) Estimate how long the exposure was in Figure 2.6.

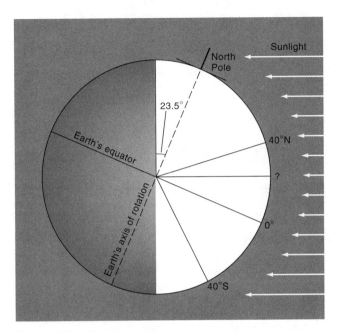

Figure 2.28. Use this diagram in answering Question 16.

16. Figure 2.28 shows the Earth at about June 25th, as seen from a point in the ecliptic. Refer to the figure to answer the following.

(a) What season is it in the southern hemisphere? What does the shaded area of the Earth signify?

(b) What is the angular distance of the noon-day Sun from the zenith at 40° latitude north and south of the equator?

(c) At what latitude in the northern hemisphere is the Sun directly overhead (at the zenith) at noon? (This latitude is called the Tropic of Cancer.)

(d) At the equator what is the angular distance of the noon-day Sun from the zenith? Is it north or south of the zenith?

(e) At what northern latitude would the Sun be just above the horizon for 24 hours? (This latitude is called the Arctic Circle.)

(f) What is the angular altitude of the Sun at noon at the North Pole?

Suggestions for Further Reading

Below are listed just a few of the many books useful to the beginning observer.

Berry, R., *Discover the Stars*. New York: Crown Publishing Group, 1987.

Editors of *Astronomy*, "How to Buy Your First Telescope," *Astronomy*, **19**, special section, November 1991. Answers many of the questions asked by the beginner about telescopes.

Eicher, D., *Beginner's Guide to Amateur Astronomy*. Waukesha, WI: Kalmbach Publishing Co., 1993. Covers topics of interest for beginning amateurs.

Forey, P. and Fitzsimons, C., *An Instant Guide to Stars and Planets*. New York: Crescent Books, 1988. A pocket-size guide.

Matloff, G., *The Urban Astronomer: A Practical Guide for Observers in Cities and Suburbs*. New York: John Wiley and Sons, Inc., 1991.

Moore, P., *Exploring the Night Sky with Binoculars*. Cambridge: Cambridge University Press, 1986. Don't overlook binoculars to explore the sky. They give a much wider view of the sky than do telescopes, although faint objects are more difficult to see.

Norton, A., *Norton's Star Atlas and Telescopic Handbook*. Cambridge, MA: Sky Publishing Corp., often reprinted. One of the standard atlases with much additional information for the amateur.

Reddy, F. and Walz-Chojnacki, G., *Celestial Delights*. Berkeley, CA: Celestial Arts Publishing, 1992. A useful, well-illustrated guide to naked-eye astronomy.

Schaaf, F., *Wonders of the Sky*. Mineola, NY: Dover Publications, Inc., 1983. A helpful guide for the naked-eye observer.

Tirion, W., *Star Atlas 2000.0*. Cambridge, MA: Sky Publishing Corp., 1981. A good star atlas available in several formats.

Whitney, C., *Whitney's Star Finder*. New York: Random House, 1990 (new editions appear regularly).

The following describe various aspects of observing the Sun and eclipses.

Dyer, A., "How to Photograph the Eclipse," *Astronomy*, **19**, p. 68, April 1991.

Hill, R., "Equipped for Safe Solar Viewing," *Astronomy*, **17**, p. 66, February 1989.

Littmann, M. and Wilcox, K., *Totality: Eclipses of the Sun*. Honolulu: University of Hawaii Press, 1991.

Stephenson, F., "Historical Eclipses," *Scientific American*, **247**, p. 170, October 1982.

Taylor, P., "Watching the Sun," *Sky & Telescope*, **77**, p. 220, February 1989.

Hubble Gallery. This HST image of Saturn, made on December 1, 1994, shows a large storm (the light area at the center of the planet's disk, just above the rings), which was discovered three months earlier. The storm clouds are ammonia ice crystals that are formed when warmer gases well up from below. Easily visible with the naked eye, Saturn is a spectacular sight in even a small telescope.

> "Being before the time, the astronomers are to be killed without respite; and being behind the time, they are to be slain without reprieve."
>
> —Shu Ching, before 250 B.C.E., on the fate of Chinese astronomers if their eclipse predictions were wrong

CHAPTER 3

The Origins of Astronomy

We have described the simpler kinds of observations that can be made with the naked eye. These reveal both regularities and apparent irregularities in the motions of celestial objects. In particular, we found that some of the brightest objects in the sky—the planets—move in very strange ways indeed. With this background we'll now look at some of the reasons why such observations were made in the first place.

There is good evidence that people were studying the skies carefully a very long time ago, at least by 4000 B.C.E. ("Before the Common Era") in the Near East and nearly that long ago in Western Europe. In many parts of the world—Western Europe, Egypt, and Central America, for example—there exist stone structures that testify eloquently to the importance and fascination celestial objects held for our ancestors. Understand that these observations were not easy. It must have taken centuries simply to establish that there were five objects that were star-like, yet moved independently of the stars, always within a narrow band extending across the sky. Why did our ancestors go to great trouble to observe and to note the motions of the Sun, Moon, planets, and stars?

There were probably several reasons. The simplest of these, which we shall describe first, has to do with practical matters like finding directions and keeping time—daily, monthly, and yearly. Much more profound and complicated motivations were connected with the religious ideas and practices that originally were probably the basis for astrology. Finally, running through all of these must surely have been feelings of wonder and pleasure at the mystery and magnificence of the night sky.

Direction Finding

One of the most obvious uses to which astronomical observations can be put is in establishing direction. Until the discovery of the magnetic compass, celestial objects provided the only means of direction finding, apart from local landmarks. In fact, the origin of the word "orientation" is related to alignment with the rising Sun (the "Orient" still refers to the eastern part of Asia). As we saw in Chapter 2, a fairly accurate north–south line can be laid out by noting the direction of the shortest shadow cast by a stick set in the ground (that is, its shadow at noon). At night we can find the general direction of north by noting the point in the sky about which the stars appear to revolve. For many purposes, Polaris now serves as an adequate marker, since it is only about 50 minutes of arc from the NCP (see Figure 3.1).

44

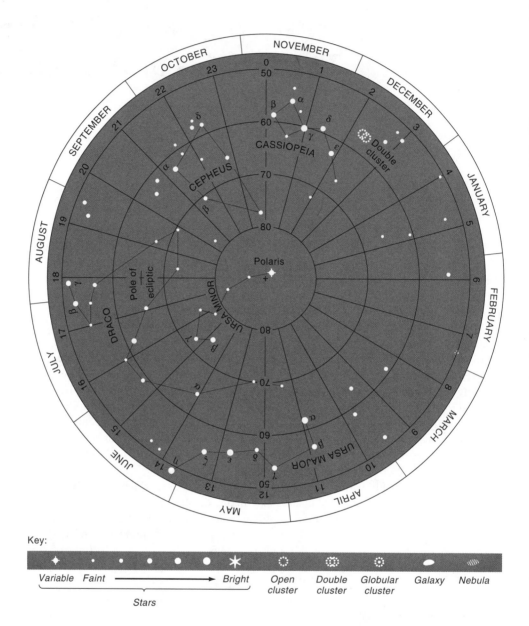

Figure 3.1. The northern cap of the sky centered on Polaris and extending 40° around it. To use this chart face north. The angular altitude of Polaris above your horizon should equal your geographic latitude. Turn the map so that the date is directly "up," and the vertical line will be your meridian at about 9 P.M. For example, in late October α and β Ursae Majoris (the two stars of the Big Dipper that point to Polaris) will be on your meridian at about 9 P.M.

Figure 3.2. The pyramid of Cheops (or Khufu) is the largest of those at Gizeh (on the opposite bank of the Nile from Cairo). The Great Sphinx is in the foreground.

The **cardinal points**—north, south, east, and west—were considered to be very significant in many ancient cultures. Sometimes great pains were taken to align major buildings and monuments along these directions. For example, the great pyramid of Cheops in Egypt (about 3000 B.C.E.) is aligned quite accurately along the cardinal directions (Figure 3.2). The largest misalignment of one side is only 5 arc minutes, and two sides are aligned to within 2.5 arc minutes. (To acquire a feel for angular sizes, see Question 5 at the end of this chapter.)

Megalithic Structures

Some of the tens of thousands of **megalithic** (large stone) structures in Western Europe (see Figure 3.3), including the best-known of these—Stonehenge in southern England—give evidence of alignments toward the directions of various celestial events, such as sunrise on the first day of summer (see Figure 3.4). On the first day of spring,

Figure 3.3. The Stonehenge that captures our attention today—the stones making up the horseshoe and its enclosing circle—are only the most recently constructed portion of this monument. The central stone circle originally consisted of 30 standing stones topped by connecting lintels. The interior horseshoe consisted of five groups of three stones each—two uprights and a connecting lintel. The most massive stone (in the horseshoe) weighs about 50 tons and is the largest prehistoric hand-worked stone in Britain. It is interesting to speculate what structures of our civilization will be recognizable 4,000 years from now.

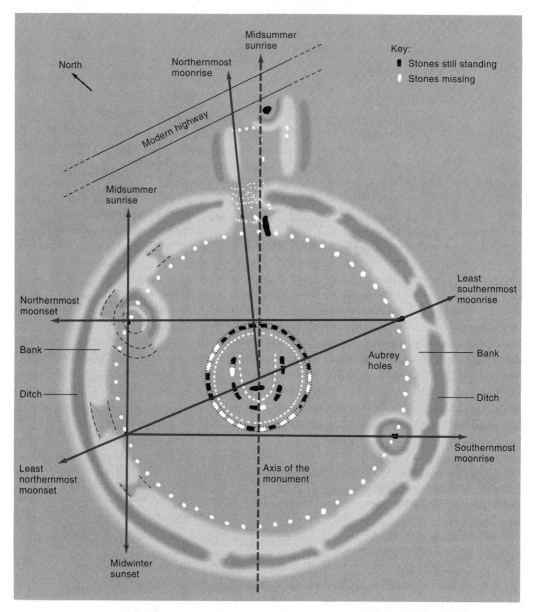

Figure 3.4. Surrounding the stones making up the horseshoe and the enclosing circle are a circle of holes (the so-called Aubrey Holes) and a mound and ditch, as well as many other smaller constructions. The bank is about 100 meters in diameter. Stones still standing are shown in black. The surrounding mound is cut in the direction of the sunrise at the summer equinox, which is the axis of symmetry of the monument. The least controversial alignments are shown.

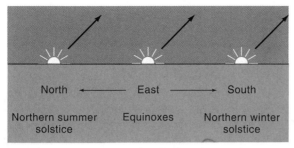

Figure 3.5. The point on the horizon at which the Sun rises (and sets) moves toward the north or south with the seasons. In the northern hemisphere the rise and set points are most northerly and the days are longest at the summer solstice; they are due east and west at the equinoxes, and reach south-most points at the winter solstice, when the days are shortest.

the Sun's rising and setting points on the horizon are due east and due west, respectively. As summer approaches in the northern hemisphere, the rising and setting points move toward the north (see Figure 3.5) until they reach a maximum northerly position at the summer solstice, the longest day of the year. At this time the daily change in the rising and setting points is very small, as reflected in the meaning of the word **solstice** ("Sun stands still"). A similar effect occurs at the winter solstice, when the rising and setting points reach their maximum southerly positions. These and other directions are indicated by megaliths such as those at Stonehenge.

Some workers in the new field of archaeoastronomy[1] interpret alignments found in some of the British megalithic structures as indicating a nearly incredible degree of measurement accuracy and technical sophistication. They infer directional precisions of a few arc minutes and the ability to predict eclipses, for example, in the astronomical "records" of these people who lived more than 4,000 years ago and about whom little is known. Unfortunately, when there are many possible alignments at a given site, it may not be obvious which were intended to mark an event and which are just accidental. Even if some of the interpretations prove to be incorrect, however, there seems to be little doubt that these early people valued their astronomical observations. The exact significance of their measurements or the role these played in the culture of these people is not yet clear. Nevertheless, they observed the sky carefully and systematically for long periods of time. Stonehenge, for example, was built over a period of approximately 1,000 years, beginning about 2800 B.C.E. Apparently the people who built Stonehenge were members of a relatively stable and organized society. They must have had an effective method for recording their astronomical observations, though there is no evidence for a written language. Though we have little idea what their motivations were, they must have expended a significant fraction of what we would now call their gross national product to record the results of some of these observations in impressive stone structures.

Astronomical Structures in Asia and the Americas

Astronomical observations have been made in Asia for thousands of years. For example, the earliest recorded observations of solar eclipses by Korean astronomers go back to 53 B.C.E. On royal order, an observing tower was built in 647 near the Korean

---

[1] Archaeoastronomy is the study of astronomy in ancient societies.

Figure 3.6. Sightlines from Pyramid VII at Uaxactun to structures I, II, and III define the direction of sunrise at the summer solstice, the equinoxes, and the winter solstice, respectively.

town of Kyangju, about 200 miles southeast of Seoul. This 30-foot tower, topped with a 10-foot-diameter observing platform, is possibly the oldest existing observatory in the world.

The Mayas regularly observed celestial phenomena and built large-scale structures to mark various seasonal events. Venus, the brightest object in the sky after the Sun and the Moon, was particularly important in many aspects of Mayan life. Their long-term observations of Venus did not lead to any theory of orbital motion of the planet, but rather to various numerical rules for describing its motion. In this they were quite successful, since they could predict Venus' position in the sky relative to its true position to within two hours after 480 years!

At Uaxactun (in present-day Guatemala) they built three temples in a north–south line that, when properly sighted from the steps of a fourth temple to the west, indicated the direction of the rising Sun at the summer and winter solstices and at the equinoxes, as shown in Figure 3.6.

Closer to home, several Native American structures in the West and Southwest show possible astronomical alignments. Among the more convincing of these are the so-called medicine (magic) wheels, for example, the one in the Bighorn Mountains of Wyoming (Figure 3.7A). The 90-foot-diameter wheel and its 28 spokes are outlined by stones, and its hub is marked by a cairn about two feet high. Five other cairns are distributed around the circumference of the wheel, while a sixth, to the southwest, is well outside it. Possibly the cairns originally supported vertical sighting posts. The sightline from the offset cairn to the one at the center is within 0.2° of the direction of the sunrise at the summer solstice. Another cairn along with the central one defines the direction of sunset on the same day (Figure 3.7B). That the summer solstice should

A

B

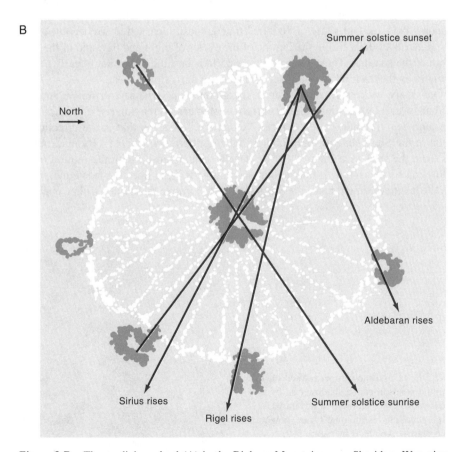

Figure 3.7. The medicine wheel (A) in the Bighorn Mountains near Sheridan, Wyoming. The wheel, spokes, and the central and surrounding cairns are all made of stones, though perhaps the latter supported posts when built. Schematic diagram (B) of the medicine wheel showing the alignments defined by the stone cairns.

be so marked is not too surprising, since Native Americans in the area held major ceremonies at this time of the year. Other alignments established by pairs of cairns apparently refer to the rising of Aldebaran, the brightest star visible just at dawn at the solstice. As we shall see in Chapter 5, the phenomenon of precession changes the positions of the stars with respect to the Sun. It is interesting, therefore, that Aldebaran indeed rose with the Sun—a so-called **heliacal** rising (from "Helios" or Sun—around 1500, when the medicine wheel is thought to have been built. Two other alignments point to the heliacal risings of Rigel, 28 days after the solstice, and of Sirius, another 28 days later. Whether these numbers and the 28 spokes of the medicine wheel are related is unknown.

Timekeeping

Timekeeping was another impetus to astronomy. In contrast with many other animals, we have only a rudimentary internal sense of time. Experiments in which people have lived for a long time isolated from all external influences (in caves, for example) show that we adapt our lives to a cycle of anywhere from about 16 to 25 hours per "day." We are all familiar with our own "personal" time, which runs very irregularly, sometimes barely dragging along (as in a dull class), at other times rushing by (as in your astronomy course). Our sense of "public" time is mostly learned as we become aware of the patterns of daily life in the family, of the cycles of night and day, and of the slow rhythm of the seasons. These recurring time periods became important to early people especially as they settled in one place and took up agriculture.

Our trusty stick stuck straight up in the ground can serve as a primitive sundial. The shadow cast by the stick will travel from west to east as the Sun rises, reaches local noon, and sets. (Hands on a clock run "clockwise" because that is the direction of motion of the Sun's shadow on a sundial in the northern hemisphere.) By marking off lines from the stick (see Figure 3.8) we have a scale against which the motion of the shadow can be measured. If the angular separation of these lines is made equal, however, the length of time it takes the Sun's shadow to traverse these lines will be unequal.

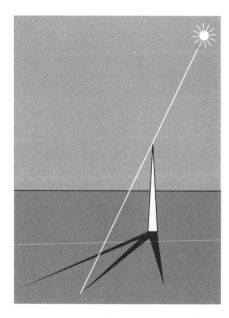

Figure 3.8. Day time can be marked off by noting the position of the shadow cast by a vertical stick. The shadow will not move through equal angles in equal times, however, for the reason explained in Figure 3.9.

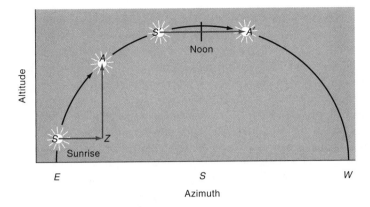

Figure 3.9. The Sun moves at a constant rate in its path (*SS'*) across the sky. It does not, however, move at a constant rate along the horizon (azimuth) or in altitude. The equally long arrows *SA* and *S'A'* represent the motion of the Sun during one hour (say) near sunrise and one near noon, respectively. Near sunrise much of the Sun's motion is in altitude (*ZA*), and only a relatively small part is in azimuth (*SZ*), whereas at noon essentially all its motion is in azimuth. Hence the shadow of the stick in Figure 3.8 will move through a smaller angle in one hour at sunrise than it will during an hour at noon.

Figure 3.9 shows why this is so. As the Sun rises, it moves not only in **azimuth** (that is, along the horizon), but also in **altitude** (perpendicular to the horizon). Near noon the Sun's motion is nearly entirely in azimuth and does not change much in altitude. Since it is the Sun's *total* apparent angular motion that is (nearly) uniform from one day to the next, the Sun will not sweep out equal angles *along the horizon* in equal units of time. Thus, to make a simple sundial that keeps uniform time, we must have unequal angles between the marks.

Until fairly recently, however, people were satisfied with marking time quite unevenly. The Babylonians and later the Romans divided the daytime from sunrise to sunset into 12 hours and the night from sunset to sunrise into 12 hours. (Our practice of dividing the 24-hour day into two sets of 12 hours is a relic of this old tradition.) Therefore, the hours during summer days were longer than those at night; in the winter the opposite was true. Considerable pains and ingenuity were taken to make the time marked by water clocks run in the same irregular fashion! It was only in the fourteenth century that uniform timekeeping, provided by the newly invented mechanical clock, was generally established. The invention of the mechanical clock, along with its many offshoots, like the wrist watch and alarm clock, has freed us from the vagaries of nonuniform time, but it has also made us slaves to time.

Calendars

The Length of the Year

As people became more dependent on agriculture, it was important to know such things as when to plant and when the rainy season or the first frost could be expected. Also, as societies became more structured and complex and as kingdoms grew in size, a good yearly calendar became necessary in order to set the dates of civic events such as festivals or tax-paying times. What simple methods can be used to measure the length of the year? Once again, with our stick in the ground we can count the number of days it takes for its shadow to be shortest (which would be at noon on the first day of summer) to the next time that same length was observed. Or we could find when the direction of the shadow was most extreme, either southernmost or northernmost at sunrise at the summer or winter solstice, respectively, and wait until that extreme position of shadow was reached again. Unfortunately, none of the methods making such simple use of a stick is very accurate. A somewhat better method is to note a particular star that rises just as the Sun does. The Sun, moving with respect to the stars, will again rise with that star one year later. The Egyptians used this method and attached great importance to the heliacal rising of the bright star Sirius, because at about this time (early July) the Nile was in flood. Since the Egyptian economy was based on the agricultural consequences of this annual flooding, it was a particularly significant event for them. In fact, they took it as the beginning of their year.

The Lunar Calendar

Now, how can we mark the time intervals within a year? The most obvious marker and the first one used was the Moon, from which our word "month" comes. In the Middle East the first appearance of the crescent Moon immediately after sunset (that is, the first visibility of the Moon after it is new) marked the beginning of the month. This is still the basis for the Jewish and Islamic religious calendars. There are obvious practical difficulties, because the Moon, when new, will be very close to the Sun and difficult to see before it sets.

A more fundamental problem with using the Moon for marking the months, however, is that the lunar month varies in length between 29 and 30 days, with an average of about 29.5 days. The variability in the length of the lunar month is due not to any randomness in the Moon's motion, but to the definition of the month. A month defined by lunar phases (for example, from one new Moon to the next) is the interval from a particular orientation of the Sun, the Moon, and the Earth to the next time this same arrangement occurs. The period of time is, therefore, affected by the Earth's motion as well as by the Moon's. An interval defined like this in terms of a particular configuration of Earth, Sun, and third object (a planet or the Moon) is called the **synodic period** of that object. The synodic period of the Moon (one full Moon to the next, for example) is longer than the 27 days it takes for the Moon to make one trip around the Earth with respect to the stars. This time interval is called the Moon's **sidereal period**, that is, the Moon's period as measured against the fixed background of the stars. The sidereal period depends on only the Moon's motion.

The obvious difficulty of keeping a lunar calendar in step with the seasonal year arises because 365 days cannot be evenly divided by 29.5. Twelve lunar months are about 354 days, or 11 days shorter than the seasonal year, but 13 months would be about 19 days longer. Hence a lunar-based calendar will slowly get out of step with the seasons, so that the hottest time of the year would eventually occur during months previously associated with winter, and vice versa.

Many people—the Jews, for example—got around this difficulty by inserting an extra, thirteenth month every few years, so that the seasonal and calendrical months stayed fairly close together.[2] The Babylonians discovered the so-called Metonic cycle

---

[2] This special thirteenth month may be the origin of 13 being an "unlucky" number. Whatever its origin, its "threat" is such that many people still can't stare 13 in the face; just notice if there is a thirteenth floor the next time you are in a tall building. The association of Friday with the thirteenth as particularly ominous may stem from the tradition that Adam and Eve were ejected from Eden on a Friday, or that the crucifixion of Jesus took place on a Friday; but no one really knows.

(after Meton, a fifth-century B.C.E. Athenian astronomer, who may have independently discovered the cycle) in which 235 lunar months very nearly equal 19 solar years. By giving 13 months to 7 years and 12 months to the remaining 12 years of each cycle, they could keep the lunar and solar calendars in good accord. In fact, the Babylonians managed their calendar so well that for the last seven centuries before the Christian era, we can date their history with an uncertainty of only about one day. Such a precision is impossible with the various calendars of the Greek city-states, which were also based on 12 lunar months per year, but in which the necessary additional months were not as well scheduled. To complicate things further, the names of the months, as well as their order, differed from one city to the next, as did the time of the new year. (In writing his history of the Peloponnesian War between Athens and Sparta in the fifth century B.C.E., Thucydides had to devise a scheme of his own to keep track of the dates of events in that long conflict.)

The Egyptians took a more radical approach: they gave up entirely on a lunar-based calendar and instead adopted one based on the Sun. As early as the fifth millennium B.C.E. they divided the year into 12 months of 30 days each, making a total of 360 days. At the end of every year they added 5 days, making the total a fair approximation to the **tropical** (seasonal) year of 365 days, 5 hours, 48 minutes, 46 seconds. This is the time required for the Sun to travel from one vernal equinox (the first day of spring) to the next. By the third century B.C.E. they had measured the length of the year to be 365.25 days, quite close to the correct value.

The Julian Calendar

For a long time the Romans did not manage their calendar well at all. Given their reputation as practical, well-organized people, it is surprising that until the time of Julius Caesar their calendar was the mess it was. It had apparently once consisted of 10 lunar months, with March being the first month. In fact, the names September, October, November, and December are based on their being the seventh, eighth, ninth, and tenth months, respectively. Extra months were inserted now and then in a failing attempt to keep the seasonal and calendrical years approximately in phase. Caesar asked the Alexandrian astronomer Sosigenes to make a solar-based calendar for Rome. The resulting **Julian calendar**, instituted in 45 B.C.E. (the year before Caesar's assassination), was uncoupled from the Moon. The new calendar divided the year into 12 almost equal months totaling 365 days. Caesar apparently felt it was appropriate for him to be commemorated in the calendar and changed the name of the fifth month, Quintilis, to Julius. Later Augustus, following his great-uncle's example, renamed the next month, Sextilis, for himself. Not surprisingly, other Emperors tried to memorialize themselves in this way, but none of their names stuck. Now, since it was known that the average length of the year is about 365.25 days, one extra day was added to February every fourth, or leap, year. Finally, in order to get the new calendar in step with various festivals that had seasonal associations, Caesar had to make the first year of the new system 445 days long; the Romans apparently referred to it as the "year of total confusion."

The Gregorian Calendar

Though the Julian year was a huge improvement, it too got out of step with the seasons, because it is about 11 minutes too long. This seems a small error, but it amounts to one day about every 130 years. This error became increasingly serious for the Catholic Church, concerned as it was that the dates of major religious events like Easter be set

October   **1582**

Sunday	Monday	Tuesday	Wednesday	Thursday	Friday	Saturday
	1	2	3	4	15	16
17	18	19	20	21	22	23
24	25	26	27	28	29	30
31						

Figure 3.10.   The calendar for the month of October 1582, when the Gregorian calendar was adopted. Ten days were eliminated to bring the calendar in agreement with the seasons.

properly. On the advice of the astronomer Clavius, Pope Gregory XIII in 1582 instituted the system we use today, the **Gregorian calendar**. To bring the calendrical and seasonal years into agreement, ten days were eliminated, and the day after October 4, 1582, became October 15 (see Figure 3.10). Then, to better approximate the length of the solar year, century years were dropped as leap years (in the Julian calendar all those divisible by four) when not also divisible by 400. Thus the years 1700, 1800, and 1900, leap years on the Julian calendar, were not leap years on the Gregorian, but the year 2000 will be. This shortens the calendrical year by three days or 72 hours every 400 years, which compensates quite nicely for the 75 hours gained every 400 years in the Julian calendar. More precisely, the length of the Gregorian year is 365 days, 5 hours, 48 minutes, 8 seconds—only 38 seconds shorter than the tropical year. A few thousand years must pass before our calendar will be even one day out of phase with the seasons. Finally, the year was taken to begin on January 1, rather than sometime in March, as had usually been the custom.

As you would expect, Catholic countries adopted the Gregorian calendar first. Britain and its empire, including the American colonies, did not change until 1752. According to Mrs. Washington, her son George was born on February 11, 1732; only with the new calendar does his birthday fall on the 22nd. Czarist Russia never did adopt the Gregorian calendar. The revolutionary government did, however; so the anniversary of the "October" Revolution was celebrated in November in the Soviet Union.

Astronomy and Astrology

Not surprisingly, given the uncertain world in which they lived and the meager state of their learning, people of antiquity felt they were surrounded by many invisible and mysterious powers that influenced all aspects of their lives. It was crucial for them to avoid the wrath of these spirits, to placate them, to try to discover what they might do next. The search for signs, omens, and charms became important. Since the ancients thought they lived in a small universe, these spirits, these gods, must be nearby, many of them in the sky. So not only were omens sought in rainbows, earthquakes, flights of birds, and the innards of sacrificial animals, they were to be sought in the sky as well. There are references to omens in the heavens as early as about 2500 B.C.E.

Obviously, it was the unusual celestial events that attracted attention. The regular, unchanging stars were of little interest except as a backdrop for celestial activity. But the Sun and especially the Moon and the planets, moving independently of the stars, constantly changing their positions with respect to each other, sometimes reversing their motions, appearing in different constellations singly or occasionally in groups, rising and setting at different times, must hold the secrets of the spirits. For the same reasons, eclipses, comets, and meteors were also thought to be of great significance.

The influence of the Sun on the Earth is obvious. It is a small step to imagine that when the Sun was in the direction of a particular "sign" (constellation) of the zodiac, its power would be modified. For example, Aquarius (the Water Bearer) may have been so named because when the Sun was in this sign, the season was generally rainy. Thus Aquarius somehow modified in soggy ways the influence of any object that passed through it. (Make sure you understand the circularity of this reasoning!) It is easy to imagine how particularly striking events like eclipses or the appearance of comets were thought to indicate especially significant events, though what those might be was open to a wide range of speculation. Also, the ancients could entertain the possibility of relations between wildly disparate phenomena, since they had no idea of the physical causes of these phenomena. Thus, the idea of astrology—that our lives are somehow influenced by celestial objects—did not seem at all unreasonable to them; we have no such excuse.

This relation became even more plausible after the Babylonians made their gods visible and perhaps predictable by associating them with celestial objects. For example, Shamash became the Sun, Sin the Moon, Ishtar and Marduk (the city-god of Babylon) became the planets we now call Venus and Jupiter, respectively, etc. (see Figure 3.11). Astronomical observations were thereby closely associated with religious practices and with attempts to discover the wills of the gods, and so became an important societal function. Since the gods each had special characteristics and personalities, these came to be associated with corresponding influences on terrestrial phenomena, strengthening or weakening as the god-star was rising or setting. The observation and interpretation of celestial omens were crucial because these god-stars concerned themselves with matters of great significance, like the destiny of empires and the fates of rulers. Ordinary individuals were of no interest to them.

The Babylonians were assiduous observers, motivated by the needs of the calendar but especially by those of astrology, which required much more extensive observations. Astrology required that every unusual celestial or terrestrial event had to be probed for its possible significance. The Babylonians kept detailed records of all sorts of natural phenomena for hundreds of years, and searched these for regularities, for any cycles of recurring phenomena that might be present. (Recall the 19-year lunar cycle described earlier.) In this way, they were able to predict certain aspects of planetary motions and lunar eclipses well, but not solar eclipses, since these can be seen only rarely from a given place on Earth. They also made the first star catalogs and denoted the positions of stars within the constellations in which they were found.

Having little understanding of the causes of most natural events, they did not, for example, distinguish between what we would now call meteorology and astronomy. Instead, they attempted to find regularities and relations among all these phenomena. Though useful information may be gained by recording all such natural events and searching for possible connections, it is important to note that such collections of data by themselves do not constitute science. These data are not attached to any conceptual framework, nor do they illustrate any general ideas that relate them and make them comprehensible in terms of a few broadly applicable concepts. Nonetheless, astrology

Figure 3.11. A sculpture of Ashurnarsipal II (reigned 883–859 B.C.E.) showing important symbols. From the king's head, counterclockwise: helmet with horns—Ashur (supreme); winged disk—Shamash (Sun); crescent within circle—Sin (Moon); wiggly fork—Adad (thunderbolt of the storm god); star within circle—Ishtar (Venus).

was a powerful motivation for the study of the sky and continued to be so for many centuries.

It was not until about the fourth century B.C.E. that the Babylonian astrological tradition became a vital force in Greek astronomy. By this time astrology had evolved, and celestial events came to be considered as the *causes* of events rather than primarily as signs or omens of events. Relationships between celestial objects as gods and certain events on the Earth were extended to causal relationships between celestial objects in general and earthly events.[3] Now the lives of ordinary people and not just the destiny of states became subject to planetary influences. Astrology became a kind of fortune-telling carried out according to a large number of arbitrary rules, conventions, and calculations. And with the Earth centrally located within the firmament, it seemed perfectly natural that we should be the object of all this celestial attention.

---

[3] That the Moon (*luna* in Latin) has long been associated with abnormal behavior is demonstrated by the word "lunatic" and by any number of monster movies.

Astrology was widely practiced and accepted through the seventeenth century, though becoming more and more divorced from astronomy. Because of their training and often as a requirement for their positions, many astronomers practiced astrology throughout this period, though sometimes with a rather skeptical eye.

Despite its utter silliness as a causal agent in our lives, astrology continues to be practiced, even in our supposedly enlightened age. Rather than telling us anything about the future, its practice provides continuing testimony to our deep wish for certainty in what many believe to be a capricious world, a wish so profound that otherwise rational people readily deny their intelligence to embrace this celestial hocus-pocus. If it is of any interest at all today, it is in its historical and psychological aspects.

The Days of the Week

Celestial objects also came to be related to the days of the week. In all likelihood our seven-day week is a consequence of there being seven "planets": the Sun and Moon, and the five naked-eye planets. In contrast to the lunar month and the solar year, the week has no foundation in natural phenomena. In fact, many societies, the ancient Greeks, for example, had no week. For much of their history the Romans had an eight-day week that only in the fourth century C.E. was changed to seven days under Christian influence. The Old Testament account of the creation (six days plus a seventh for rest) was also a factor. As a result, seven has become a special number for us; we talk of the seven ages of man (for example in Shakespeare's *As You Like It*), the seven deadly sins, the seven wonders of the world, and seven as a "lucky" number.

The practice of naming the days after the seven "planets" in this way apparently originated with the Babylonians. We follow that practice, but the names of our days come to us from Norse and Celtic mythology and language: Sun-day, Moon-day, Tiu (Mars)-day, Woden (Mercury)-day, Thor (Jupiter)-day, Freya (Venus)-day, and Saturn-day. Those of you who know French or Italian will recognize that Romance languages show the connections between the days of the week and the names we use for the planets more clearly than does English.

Early Models of the World

A basic need we all have, and which must have been particularly acute for our ancient ancestors, is to feel at home in what often appears to be a hostile environment. People have always made models of the world; that is, they have attempted to account for the most important features of the world in terms of familiar objects and processes. In other words, they have tried to bring the world under some sort of intellectual understanding. These models were generally accompanied by descriptions of how the world came to exist, descriptions that often became intertwined with religious beliefs.[4]

The earliest world models were myths or allegories and had little to do with astronomy except in an incidental way. The Egyptians quite naturally pictured the world as long and narrow; through it the Nile flowed. The goddess of the sky arched herself over the solid Earth, and let the boat of the Sun glide over her back. When the boat sank beneath the Earth, entering the realm of the dead, night fell. After it traveled through this region, it reappeared, beginning its daily journey again (see Figure 3.12). The Babylonians imagined a disk-shaped Earth surrounded by the seas, beyond which

---

[4] In their astronomical contexts we lump world models and ideas of origin together in the term "cosmology," though strictly speaking cosmology applies only to structural models; "cosmogony" has to do with origins.

Figure 3.12. The Sun boat travels over the back of the Egyptian goddess of the sky who is arched over the Earth.

the celestial sphere descended. To some of the early Greeks, the sky was two curved shields surrounded by fire that, shining steadily through holes in the shields or flaming suddenly, was responsible for the stars and planets or for lightning, respectively. Shining through the crack between the shields, the fire appeared as the Milky Way. Surrounding the whole was the sea, which touched the shields at a great distance.

These myths, charming as they may be, have little to do with the origins of astronomy in particular or of science in general. They suggest, however, the primitive context of attitudes and beliefs out of which the precursors of science did arise. These precursors are to be found in more abstract concepts, in ideas concerning the nature of the world. As with so much in Western culture, these come down to us from the Greeks. We will look at some of these concepts in the next chapter.

Terms to Know

Cardinal points; solstice; heliacal rising; azimuth and altitude; sidereal and synodic period; tropical year; Julian calendar; Gregorian calendar.

Ideas to Understand

Why our ancestors studied the sky; simple ways of finding north; how to keep time; how to find the length of the day, the lunar month, and the year; how early calendars were made and how our present calendar works; origin of astrology; the need for models of the world.

Simple Observing Projects

When they looked up at the night sky, people of ancient times saw more than most of us when we do so today. The difference is not the numbers of stars, obviously, or even smog and light pollution (though they don't help), but simply a difference in awareness. Most of you have looked out at the night sky and seen the stars and planets, but have you seen them move across the sky in the course of a night, or noted the motions of the Moon or the planets against the stars? Long ago people were familiar with this celestial activity; often it played a significant role in their lives.

The purpose of these projects is to help you gain at least a little familiarity with the sky. With the growth of cities, it is becoming increasingly difficult to see the night sky in all of its beauty, but I hope you can find a place from which you can see the brightest objects, at least.

When you make your observations, record immediately and carefully what you see; don't rely on your memory. Unless you are very lucky, you will discover one of the chief frustrations of the astronomer—clouds. You can't do anything about them except to take advantage of every clear night to make your observations.

1. Find a convenient place from which you have a good view of the east, south, and west. Many of your observations will be made soon after sunset; keep this in mind in choosing your observing station. Carefully make a sketch of features on your horizon—trees, houses, chimneys—from east to south to west.

Depending on the time of year and your geographic location, your instructor will tell you which constellation is best suited for you to observe; it might be Orion, for example. Soon after sunset, find that constellation and sketch it on your horizon drawing, showing its position as carefully as you can. Note the time. About an hour later, draw it on a second copy of your horizon sketch, again noting the time. Repeat this until you go to bed.

(a) In which direction does the constellation move? Why does the position of your constellation change during the time you observed it?

(b) Considering only what you have observed, can you think of any reason to suppose that the sky is fixed and that the Earth is rotating?

(c) Would you be able to observe this constellation six months from now? Why or why not?

2. On a star chart provided by your instructor, plot the Moon's appearance (that is, its phase) and its position with respect to the stars. Follow the Moon's motion for at least a week; record the time and date for each of the positions of the Moon you plot. Draw a line on the star chart that best represents the path of the Moon against the stars. This path is very nearly the same as the celestial track called the _____. Why isn't it exactly the same?

3. Plot the position of Mars, Jupiter, or Saturn on your star chart. You should follow these objects for several weeks, marking their positions at intervals of a few days. Draw a line on the chart that best represents the path of the planet. Compare the average angular motion (degrees per day) of the Moon and the planet. Which moves faster? By about how much?

If where you live the sky is too bright to plot the motions of the Moon or planets against the stars, record the position of the Moon or planet at the same time each night on your horizon drawing. In this case, why is it important for you to make your observation at very nearly the same time each night? (Remember the result of Project 1.)

4. The naked-eye star Polaris is within about a degree of the north celestial pole. Find Polaris early in the evening. Carefully draw the Big and Little Dippers (Ursa Major and Ursa Minor) and indicate Polaris (α Ursae Minoris). Suppose you were at the North Pole. Where in the sky would Polaris be with respect to your horizon? Now imagine yourself at the equator; where would Polaris be? What is the relation between the latitude of your location and the smallest angle between the horizon and the position of Polaris? A few hours after making your first sketch, make another. How have the positions of the Big and Little Dippers changed?

5. Since people are all built to about the same proportions, we can easily estimate angles in the following way: stretch your arm straight out in front of your face; the width of your thumb subtends about one degree (that is, the line of sight from one side of your thumb to the other is about one degree); the width of your fist (thumb to little finger) is about 10°; the width of your outstretched hand, thumb to little finger, is about 20°.

(a) Convince yourself that the diameter of the full Moon is about 0.5°.

(b) How far apart are the pointer stars of the Big Dipper (the two stars at the end of the bowl that point to Polaris)?

(c) What is the altitude of Polaris above the northern horizon? How does this compare with the latitude of your location? How should it compare?

Here are a few daytime projects for you to do.

6. Carefully sketch the western horizon from your observing station (or eastern horizon, if you like to get up early). At intervals of about a week for about two months, indicate the position on your horizon diagram of the setting (or rising) Sun. Label each position with the date and the time of sunset (sunrise).

How has the setting (rising) point of the Sun changed over the time you observed it? Has it moved northward or southward? Were the days getting longer or shorter during this time?

7. In a south-facing window, position a narrow rod (a ruler, yardstick, etc.) so that it stands approximately vertically. Mark on a piece of paper the end point of the shadow as the Sun crosses your meridian.

(a) How accurately can you decide the time of the shortest shadow (that is, local noon)?

(b) Compare the lengths of the shortest shadow cast ten days apart.

(c) How near to 24 hours is the time interval between two successive meridian crossings you can measure in this way? How would you improve the accuracy of your measurements?

8. Explain, with a diagram of the Sun, Earth, and Moon, why the lunar (synodic) month is longer than the sidereal month.

9. How many years must pass before the Gregorian calendar is 24 hours out of phase with the seasons?

Suggestions for Further Reading

Below are a few of the many books and articles that describe some of the results from the new field of archaeoastronomy.

Aveni, Anthony F., *Skywatchers of Ancient Mexico*. Austin: University of Texas Press, 1980.
    Not only gives an account of some of the early astronomical observations and observatories in what is now Mexico, but also describes and explains in some detail the various celestial phenomena available to the naked-eye observer.

Burl, A., "The Recumbent Stone Circles of Scotland," *Scientific American*, **245**, p. 66, December 1981.

Hadingham, E., *Early Man and the Cosmos*. New York: Walker and Co., 1984. An introduction to archaeoastronomy.

Hawkins, G., *Stonehenge Decoded*. Garden City, NY: Doubleday and Company, Inc., 1965. The book that revived popular interest in the astronomy of Stonehenge.

Krupp, E. C. (ed.), *In Search of Ancient Astronomies*. Garden City, NY: Doubleday and Company, Inc., 1977. A collection of interesting articles on aspects of archaeoastronomy in the Near East, North and Central America, Britain, and France. An introductory chapter describes relevant celestial motions.

Krupp, E. C., *Echoes of the Ancient Skies*. New York: Harper and Row Publishers, 1983. A good discussion of archaeoastronomy.

Williamson, R., *Living the Sky: The Cosmos of the American Indian*. New York: Houghton Mifflin, 1984.

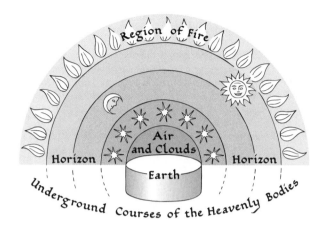

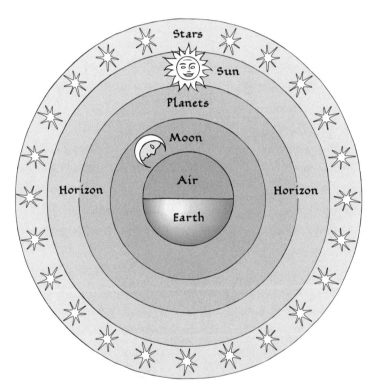

(upper) Anaximander (ca. 611–547 B.C.E.) assumed that the inhabited Earth was the flat top of a cylinder that was surrounded by transparent shells containing celestial bodies at different distances. Note that the stars were made nearer the Earth than the Moon, even though the latter occults stars. The shells extended somewhat below the horizon to account for the appearance and disappearance of celestial objects as they rose and set.

(lower) About a century after Anaximander's death, Leucippus held that the Earth was a hemisphere capped by a hemisphere of air. Concentric transparent spheres supported the Moon, Sun, planets, and stars.

CHAPTER 4

Early Greek Astronomy

The Idea of an Orderly Universe

A crucial notion necessary for the development of science was first put forward by a few Greek thinkers around the seventh century B.C.E. (Figure 4.1 shows the Greek world.) This was the remarkable idea that there was a basic reality, an underlying order to the universe. There was no compelling evidence that this was so, but they were convinced that somehow or other the universe must make sense. As far as we know, they were the first people who thought it possible even to ask, "What is the fundamental nature of the world around us? How is the universe organized?" Just the assumption that such an order exists is liberating. With it we no longer need feel quite so much at the mercy of forces that seem completely beyond our understanding. A person could feel more confident in such a world, a world in which order prevailed, especially if the basis of that order could be discovered. And these Greeks did believe that it was possible for ordinary mortals to discover that underlying basis by observation and reasoning. This was essentially an act of faith, but it was and is an obvious and necessary prerequisite to science.

Note the huge difference between this way of looking at nature and that of the Babylonians! The latter hoped to find simply that an event *A* might be associated with event *B*. Each case was approached separately, with no notion that there might be some fundamental basis by which many phenomena could be related. Note also that the Greek view implies a particular relation with nature. We must try to stand apart from it and look at it as objectively as possible, rather than being such an integral part of it that we cannot separate ourselves from it.

So the Greeks asked, "What is basic to understanding the form, the structure, and the changes we see in the universe? What are the fundamental properties, the important characteristics of the world that make it what it is?" Some people claimed that underlying the nature of things was matter, that is, substances, or perhaps a single substance in many forms. Some thought that atoms—very tiny, indivisible particles—formed the basis of things. Others, however, thought that mere material substances were inadequate and that the basic principles must be mathematical, either arithmetic or geometric.

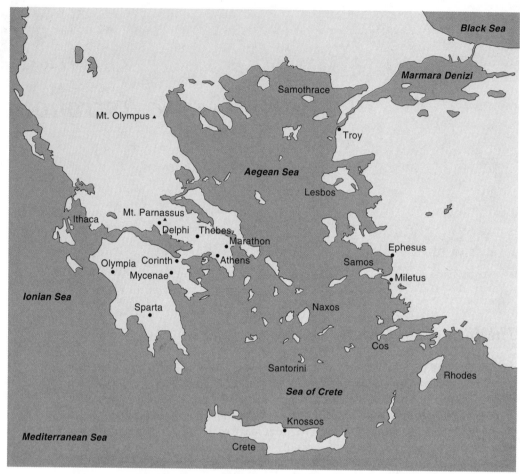

Figure 4.1. The Aegean world at the time of classical Greece. Troy, Knossos, and Mycenae were old ruins at this time, but are shown for orientation purposes.

Matter as the Basic Reality
Water

An example of the idea that matter was basic was given in the seventh century B.C.E. by Thales, who is considered to be the traditional founder of Greek philosophy, but about whom practically nothing is certainly known. Later writers said that Thales thought that all things were made of water in various forms. This idea has nothing to do with modern chemistry with its compounds and reactions. Rather, it is more directly connected with the experiences of everyday life. The Greeks knew that much of the Earth is covered with water, and that water falls from the sky as rain, sleet, or snow. As snow it is soft and can feel dry; as ice it is hard. Thus water, which can be transformed into liquid, solid, and gaseous forms, encompasses many of the polar characteristics of things around us—softness, hardness, dryness, wetness. Silt is deposited by rivers such as the Nile, so earth (dirt) is built up from water, apparently. Furthermore, it is obvious that life would be impossible without water to drink or without the fluids that course through our bodies. The whole Earth was thought to float on water, and earthquakes were thought to be caused by water waves. Significantly, however, the god of the sea, Poseidon, plays no role in Thales' speculations. Suggesting that many natural phenomena might be explained by a fundamental substance is Thales' great contribution. That he was unable to say just how water could be transformed somehow into the incredible diversity of things around us is relatively unimportant. What is significant is his daring idea that the universe is knowable.

Earth, Air, Fire, and Water

Two centuries later, another philosopher, Empedocles of Akragasin Sicily, held that four "fundamental" substances were necessary to account for all things: earth (as in soil), air, fire, and water.[1] A piece of green wood, for example, could be thought of as being made up of all four: the unburnt wood feels damp and so contains water; when burned, fire and smoke (air) are released, and ash (earth) remains. The four **elements** were also thought of as representing states of matter as well as basic substances, so that change could be expressed in terms of transformations of one element to another. For example, when boiled, water is transformed to air, and when frozen, to earth. These ideas lasted for more than 2,000 years; we will return to them later.

Atoms

A quite different approach was suggested by Democritus (born about 460 B.C.E.), who taught that matter could not be endlessly subdivided, but consisted of ultimate, invisible particles—atoms (from *a-tomon* meaning "non-cuttable," that is, indivisible). There existed only one kind of atom, which, however, could be arranged in many different ways, thereby producing the variety of substances we see around us. Note that he was suggesting that the world looked very different on a small scale than it does in our everyday experience. The modern, very rough parallel to this idea is familiar to everyone, though in its own time the notion of explaining the world by means of atoms was not very successful.

Mathematics as Basic

Other Greeks, however, found unsatisfactory the suggestions that basic reality could be understood in terms of primary substances. They felt that such an approach was too limited and did not explain enough. For them a mathematical basis was more fundamental. Ultimate reality was to be found in mathematical relations, relations that when taken together produced a pleasing harmonious whole. In following this path the Greeks invented abstract mathematics, that is, mathematics that is not derived from practical concerns, but stands apart from the immediate world. This was a new approach, quite in contrast to what had gone before, which is exemplified in the origin of the word **geometry**; it comes from the Greek words *geo* and *metron*, having to do with Earth and measurement, respectively. The Babylonians, for example, had discovered a good deal of algebra and geometry, but these were tied directly to their practical needs, such as measuring the boundaries and areas of plots of land. The Babylonians knew of the 3-4-5 right triangle, but they did not generalize this to what we now call the Pythagorean theorem (see Figure 4.2), which relates the hypotenuse, c, of any right triangle with the other two sides: $c^2 = a^2 + b^2$.

---

[1] The Chinese defined five basic elements: earth, fire, water, wood, and metal. These along with the Sun and the Moon they attached to their names of days.

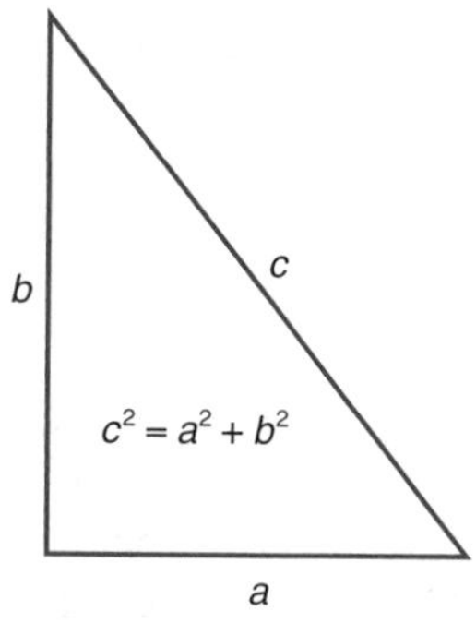

Figure 4.2. The well-known Pythagorean theorem: in a right triangle (two sides of which form a 90° angle) the square of the hypotenuse equals the sum of the squares of the other two sides.

$$c^2 = a^2 + b^2$$

By contrast, some Greeks looked on mathematics as a pathway to eternal truths. After much effort and many false starts, these Greeks discovered a crucially important way of thinking. They found that, starting from a few basic definitions and assumptions, they could work out by strict deductive reasoning (reasoning from the general to the particular) all of the consequences implicit in the framework defined by these definitions and assumptions. Euclidean (plane) geometry, perhaps the crowning achievement of Greek mathematics, is a good example of this method. **Axioms**—assumptions—are defined, and then properties of geometric figures (**theorems**) are proven within the set of assumptions. Although such mathematical systems are internally consistent, it is important to note that they may not be based on nature. We take it for granted now that they are often useful tools for describing nature, but it is not really obvious that they should be. In fact, in a way it is remarkable that the physical universe can be described in terms of mathematical systems not directly based on it!

Numbers and the Pythagoreans

Pythagoras (born about 580 B.C.E. on the island of Samos; see Figure 4.1) is an example of one who wanted to explain things in mathematical terms, in particular, in arithmetic terms. He was a philosopher and the spiritual leader of a religious community that he founded in southern Italy and that lasted some 200 years.[2] This is the same Pythagoras for whom the theorem ($a^2 + b^2 = c^2$) is named, because he or one of his followers discovered the general proof. He was impressed by the discovery that sounds produced by plucking strings (those of a harp, for example) sounded harmonious when the lengths of the strings were related to each other by whole-number ratios. A whole vibrating string produces a tone; divide it in half, and it sounds the tone that is an octave above the root tone, that is, with twice the pitch; divide it into thirds, and it produces a tone an octave and a fifth above the original tone. These notes, when played together, sound pleasing to the ear, harmonious, not discordant. Musical intervals in the Greek scales, the Pythagoreans discovered, could be represented by the ratios of the small whole numbers one, two, three, and four.[3] These, when taken together, add up to ten, a number to which they gave great significance. Ten was important, in fact, the perfect number, because if one represents a point, two a line (two points define a line), three a surface, and four a volume, then their sum, ten, encompasses all forms! Other numbers came to have a more earthy significance. For example, 224, written on food and swallowed, was supposed to be an aphrodisiac! These

---

[2] Among the rules of the community were prohibitions against eating beans, picking up something that has fallen, and walking on highways.

[3] This is probably why the study of simple fractions (for example, 3/4 + 1/3 = ?) was included in "music" until well into the Middle Ages.

examples suggest how numbers and their relationships seemed to pervade all aspects of nature for the Pythagoreans.

The Pythagoreans also knew a little astronomy: for example, that the Big Dipper is circumpolar in northern latitudes, but not when seen from farther south. They correctly interpreted this as evidence for the sphericity of the Earth. This had the important consequence of explaining how different climates arise, as we have seen. They also recognized that the Moon does not shine by its own light, and understood how eclipses occur.

The Importance of Spheres

For several reasons, the "purest" geometrical form, the sphere, was fundamental to the structure of the universe, Pythagoreans thought. For one thing, the sky looks spherical; the celestial sphere was an image of the terrestrial sphere. For another, the stars move across the sky in circles, as seen most clearly in the paths of circumpolar stars. This was the beginning of the importance of spheres in ancient astronomy (even though at that time it is unlikely that most people accepted the idea that the Earth was spherical, if they wondered about it all).

The Pythagoreans conceived of the universe as consisting of spheres that moved each of the then-recognized "planets"; these, you recall, included the Sun and Moon as well as the five naked-eye planets. There was also a sphere in which the stars were embedded, and another outside of that which moved all the rest about a near-central Earth once every 24 hours. The planets' spheres slipped backward at rates sufficient to account for their motions against the stars. Now, all of these added up to only nine spheres, but ten was the Pythagoreans' "perfect" number; so apparently they reasoned that there must be yet another sphere doing something useful in their cosmos. Therefore they postulated what they called the "counter-Earth," which was always opposite the "central fire" from the Earth and consequently (and conveniently) never seen. It and the Earth were on this tenth sphere. (The Pythagoreans were by no means the only, or last, people to require Nature to fit their theories; the practice is not unheard of even today!)

As with most of Pythagorean doctrine, it is not clear what they meant by the "central fire"; it was certainly not the Sun, which had already been given its own sphere. In some versions of this cosmology the outermost sphere was surrounded by fire; it might have been thought appropriately symmetric (or harmonious) for fire to be at the center of the system as well. In any case, there are a couple of interesting notions in this muddle: the model was spherically symmetric and emphasized circular motions around a common center. In addition, they put the Earth in motion about the mysterious central fire. That the Earth moved was not generally accepted, however. The Pythagoreans also emphasized the distinction between the planets and the stars and, following the Babylonians, considered celestial objects to be divine.

Harmony of the Spheres

We mentioned earlier that the Pythagoreans discovered the relations between small whole numbers and what the ear perceives as pleasing harmonies. They thought that these relations were universal and should apply *everywhere*. For example, the ratios of the different speeds of the shells that moved the celestial objects across the sky, or possibly the ratios of the shells' diameters, should be in the ratios of small numbers, like 2:1 or 3:2. In this way was born the idea of the **harmony of the spheres**, an idea Kepler seized upon some 2,000 years later. In moving around the Earth, the planets pro-

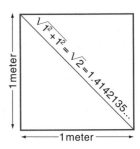

Figure 4.3. The diagonal of a square is an irrational number, that is, one that cannot be expressed as a ratio of two integers; it is an infinite decimal with no repeating pattern of digits. Note that it doesn't matter what length is taken for the sides of the square; the diagonal can always be factored to some number times the square root of two, which is irrational.

duced tones proportional to their speeds, the slower-moving object making lower tones. They all moved in such a way that these tones, when taken together, produced pleasing, harmonious sounds. These harmonies were supposed to be physical manifestations of spiritual truths. But how is it that we don't hear this celestial symphony? Well, since we have heard it from birth, we aren't consciously aware of it and so don't hear it at all.

This idea, that small numbers and their ratios represent fundamental aspects of nature, received a severe blow when it was discovered that the diagonal of the square (see Figure 4.3) cannot be expressed as a simple fraction of the length of its sides. (If a square is one meter on a side, then the length of its diagonal is just the square root of two or 1.4142 . . . meters; this is a number that cannot be expressed as a ratio of two whole numbers.) This came as a great shock to the early Greeks, a shock that is expressed by the name they gave to such quantities, namely, "irrational" numbers. If ratios of small numbers did not work for something as simple as this, then why should one expect them to describe the planetary spheres? What happens to the idea of the harmony of the spheres?

Perhaps the idea of an arithmetic basis of the universe seems strange, mystical, or even primitive. This notion keeps popping up in various guises even today, however. For example, eminent scientists have wondered why some of the physical constants— the charge on the electron, the mass of the proton, or the gravitational constant, for example—have the values they do. These physicists have tried to find relations between these numbers, which they feel to be fundamental to an understanding of the universe. (Such speculations are kept in bounds, however, by the necessity that they agree with the facts of experiments.) Thus the Pythagorean idea still has its faint echoes in modern science.

Geometry as Basic

If the idea of a universe based on numbers did not work, perhaps one with a *geometric* foundation might be more fundamental and capable of being made specific. Maybe structure or shape is the key. Again, this idea has modern overtones. The French painter Cezanne, one of the precursors of Cubism in art, wrote in the late nineteenth century, "Everything in nature is modeled on the sphere, the cone, and the cylinder. One must teach oneself to base one's painting on these simple figures, then one can accomplish anything one likes." A recent scientific example of the importance of structure is the discovery of the geometric shape—the double helix—of the DNA molecule shown in Figure 4.4.

The Greek philosopher Plato (428–348 B.C.E.) was one of those who felt that form (that is, geometry) was a key to understanding the universe. It was said that over the

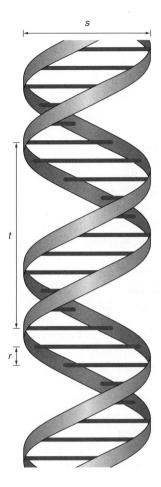

Figure 4.4. Imagine a ladder twisted uniformly about a central axis so that the rungs of the ladder are always perpendicular to the axis. The sides of the ladder are always the same distance (s) apart, the spacing of the rungs (r) is constant, and the pitch of the twist (t) does not change. Such a structure represents the DNA double helix molecule.

entrance to Plato's academy appeared the message "Let no one enter here who knows no geometry" (note, not arithmetic). Euclid (*ca.* 330–275 B.C.E.), who systematized plane geometry in the form we learn today, was a disciple of Plato.

Plato taught that nature could be understood only if we aren't misled by the ever-changing world around us, which we perceive through our imperfect and unreliable senses. Instead, we must push beyond appearances to fundamental and constant notions, which form the true basis of our experience. A trivial example: we can't draw an absolutely perfect circle, but we can agree that such exists, think about it, and all know what we mean by the term. In this sense reality is beyond our immediate everyday experience.

Uniform Circular Motion

Now one of these fundamental ideals by which the world was to be understood, Plato thought, was that of **uniform circular motion**, that is, circular motion at constant speed. This idea sounds strange to us, but consider: a circle has neither beginning nor end; so circular motion can go on forever, never changing, always the same distance from the center. This perfect motion is clearly exemplified in nature by the stars, which unendingly move in circular paths, always at the same speed, constant in all their aspects. Motions of the planets and of the Sun and Earth were more difficult to understand, however, because of the nonuniform and retrograde motions of the first and the nonuniform motions of the latter two. The Greeks argued that these apparent disordered motions are deceptive; these motions must in reality be based on an ordered, perfect system of uniform circular motion. The same general principle of uniform circular motion must somehow apply to them also. This idea took hold and set the ground rule for the basic problem of astronomy up through the time of Copernicus: *construct a geometric representation of celestial motions using only combinations of uniform circular motions about the central, fixed Earth.* So great was the appeal and strength of this idea that for 2,000 years astronomers worked within this framework, giving little thought to the possibility that it might be wrong.

Since the sphere is the "purest" form—any point on it is equivalent to any other point, it looks the same from any direction, it is the simplest of closed figures—the stars and planets must be attached to spheres, each of which rotates at a constant rate. Clearly (to the ancients), the solid, stationary Earth must be the unmoving center of the system. Whatever their aesthetic or philosophical virtues, circles do have the important practical advantage of being easy to manipulate. It takes only a point (to fix the center) and a length from the center (to fix the radius) to define completely a circle or a sphere.

Although the notion of a moving Earth had been suggested, it flew in the face of ordinary experience and was not accepted. Hence, uniform circular motion around a fixed, central Earth became an integral part of the intellectual baggage carried by most thinkers for two millennia. Until the time of Copernicus and Kepler, they provided the framework—or prison—within which all attempts to understand the universe had been made.

Before we smile at the naiveté of those who held these ideas, we should remember that we live our own lives within a forest of constructs, unspoken assumptions, and attitudes that are so ingrained in us that most often we aren't even aware of them. Such structures are generally necessary and useful, and most of the time serve us well. Occasionally, however, their consequences become questionable or even intolerable; a particularly bold thinker then replaces one or more of these basic assumptions, and in extreme cases we might say that a revolution—scientific, political, artistic, religious, etc.—has taken place. If as a result of college you begin to become aware of these implicit attitudes, to think about them, and to evaluate them during the rest of your life, you will have received a handsome return on your investment of time and money.

Before proceeding further, let us quickly recall the primary features that models of the solar system had to account for:

(1) the daily motions of the Sun, Moon, planets, and stars from east to west;
(2) the much slower motions of the Sun, Moon, and planets from west to east with respect to the stars;
(3) the occasional retrograde motions of the planets from east to west, which interrupt the motions noted in (2); and
(4) the increased brightness of planets during retrograde motion.

Plato himself described an eight-sphere model, but it was far too primitive to account, even in a general way, for retrograde motions. It was useful, however, as an example of the imaginative leap by which one can stand apart from the world and look at it as a whole rather than remaining imprisoned within it.

Eudoxus' Model

Eudoxus (about 406–350 B.C.E.), a colleague and perhaps a student of Plato, made the first more-or-less successful model using these ideas. Obviously, neither retrograde nor simple nonuniform motions can be represented with just one uniformly moving sphere per planet. Eudoxus' system required a total of 27 spheres, all centered at the same point, the Earth. Each sphere could have a different axis and a different rate and sense of rotation. Four spheres were used for each of the five planets, three each for the Sun and Moon, and one for the stars. The poles of each sphere were attached to the surface of the one surrounding it.

Its general approach can be seen as follows (see Figure 4.5). The outermost

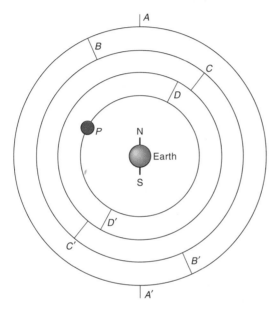

Figure 4.5. Eudoxus' scheme for representing the motion of a planet. The outer sphere, which carries the stars and drives the other spheres, rotates from east to west once in 24 hours. The second sphere rotates from west to east once in the planet's year. Its equator is in the ecliptic. The third and fourth spheres, inclined to each other, account for retrograde motion. The planet, P, is attached to the equator of the fourth sphere.

sphere, which drives all the others, rotates from east to west once in 24 hours and provides all the daily motions. Its poles lie on a north-south line. The second sphere, its axis inclined to the first so that its equator is in the ecliptic, rotates once west to east in the planet's year (technically, its sidereal year; see Chapter 7). The poles of the third sphere are located on the equator of the second, and the poles of the fourth sphere are inclined to those of the third at an angle different for each planet. The third and fourth spheres, rotating at equal rates but in opposite senses, account for retrograde motions. The planet itself is attached to the equator of the innermost sphere closest to the stationary, central Earth. Each planet requires a separate set of constructions, and so it is not a unified system in Aristotle's sense (see below).

Impressive geometric skill is required to work out the motions of each sphere, which, when taken with all the others for that object, result in something qualitatively approximating the daily, yearly, and retrograde motions of the planets. This system works moderately well for Mercury, Jupiter, and Saturn, but not for Mars and Venus. Furthermore, it cannot account for the changing brightnesses of the planets nor the differences in the length of the seasons, since all the spheres have a common center.

As far as is known, no predictions of planetary positions were ever made using Eudoxus' model, which is just as well, since its positional accuracy is no better than five degrees or so. (This is the angular distance between the pointer stars of the Big Dipper—see Figure 3.1—or ten lunar diameters.) It should be noted, however, that most of the Greeks (and even some of their successors until the time of Tycho Brahe in the sixteenth century) did not see a strong need to represent observations much more accurately than this. Often they were satisfied with qualitative agreement. The requirement of the highest possible precision in both theory and observation is relatively recent, and is a hallmark of modern science.

Aristotle's Cosmological Ideas

It is not clear if Eudoxus believed that the spheres on which celestial objects were attached were physically real or not. Others did take them seriously, however, and wondered how they were attached to each other, what kept them moving, and of what they were made. That is, they wondered about the *physics* of the system. Aristotle (384–322 B.C.E.) was one of these. (He was also the archetype of the professional student, entering Plato's Academy at age 17 and staying there for twenty years. He went on to found a school in Athens that rivaled that of his teacher.) Aristotle was interested in nearly everything and wrote about nearly everything. He attempted to pull together all the knowledge of his time into a coherent, unified whole. First in the Graeco-Roman world, then in Islam, and later in the West the influence of his ideas was enormous.

Like his predecessors, Aristotle wanted to discover the general principles of nature. To him, however, this involved both organic and inorganic bodies. He approached nature more from the viewpoint of a biologist or naturalist than of a mathematician. In fact, Aristotle felt that mathematics had no place in his physics, since the latter was concerned with change or movement, whereas mathematics dealt with the unchanging world. He was impressed by the differences between the constantly changing Earth, where things were born and died, grew and decayed, and the unchanging stars, which were always predictable, always the same year after year. Also he was struck by the differences in the "natural" motions on Earth and those in the sky—vertical and limited in the former, circular and endless in the latter.

Aristotle distinguished two ways by which changes could take place. Changes could be natural, like the ripening of olives, or forced, like the chopping down of the olive tree. Applying these ideas to motion, Aristotle thought the upward surge of fire to be natural, but the upward motion of a rock to be forced. On Earth, natural motions were vertical—flames move up, rocks fall down; and they were limited—fire reaches its natural level in the atmosphere and stops, rocks hit the ground and stop. By contrast, in the sky far above the Earth, natural motions were circular and unending, as exemplified by the stars. According to these ideas, uniform circular motion was as much to be expected of stars in the sky as was the vertical motion of a falling rock on Earth. Such motions were an essential part of the very nature of stars and rocks.

Furthermore, changes on the Earth took place when any of the four elements—earth, air, fire, and water—were out of their natural places. Water flowed downhill until it came to rest in its natural place, surrounding the Earth. Steam moved upward, coming to rest only when it had found its natural place among the clouds. Fire, the lightest element, escaped from burning wood and rose from the Earth, stopping only

Figure 4.6. The upward motions of fire to a region below the Moon, of steam to the clouds, and the downward motions of rain to a lake on Earth's surface and of a stone to the bottom of the lake are all examples of natural motion, straight line and limited. The stars, far beyond the Moon, move along circular, unlimited paths, in a natural motion but of a fundamentally different kind from that near the Earth. The stone thrown upward is an example of a forced (unnatural) motion.

when it surrounded the atmosphere (see Figure 4.6). In contrast to things on the Earth, the stars and planets did not change, because the substance of which they were made was already in its proper place. This substance, obviously different from the four earthly elements, must be yet another, a fifth element or essence, a *quintessence* that was the perfect, unchanging, immortal form of matter.

Aristotle applied these ideas to Eudoxus' model, but took them a step further by trying to conceive of a workable mechanism made of real transparent spheres, all physically connected, at least in principle. The whole system was driven by the outermost sphere, the prime mover, as a consequence of its own natural motion. By means of some fifty-odd nested and linked spheres, Aristotle attempted to account for all the observed motions in the sky. The Earth was at the center, because, being made of the heaviest element, its natural place was "down" as far as possible. In a spherical system that meant it had to be at the center. The changing, mortal Earth was separated from the unchanging, perfect heavens at the Moon's sphere.

Several features of this picture should be noted. Two completely different kinds of physics—theories of motion and substance—are required, one for the Earth and one for the skies. Since nothing beyond the lunar sphere could change, comets, which appeared for a short time and then disappeared, must be some sort of fiery condensation moving entirely within the Earth's atmosphere. With the constancy of the stars explained as part of their nature, it was necessary to map their positions only once and for all. There was little reason for further observations of the stars, since, according to these ideas, they could never change. This view of the sky was very nearly closed and complete, and did not lead to new questions and predictions. Today such a picture would be considered fundamentally deficient, because one way we judge competing scientific theories is by their ability to predict new consequences. These provide crucial checks on a theory's validity. If a prediction turns out to be incorrect, the theory is discarded or at least modified.

Aristotle's system was of little use for predicting the future positions of the planets, but it did provide a qualitative description of the universe that seemed to account for many of its features. This world picture, along with an overlay of Christian ideas and symbolism, was the standard model taught and studied everywhere in the West until only a few hundred years ago. It provided the setting within which the drama of man and nature unfolded (see Figure 4.7).[4]

I hope that this chapter has given you some notion of how our ancestors tried to understand the world. They insisted that nature was rational and knowable. They developed ideas that, though appearing naive by our standards, went far beyond the simple myths held by most people at that time. These ideas contained some of the seeds from which modern science grew.

Another point is appropriate here. The number of people who were involved in developing these concepts was astonishingly small—a minuscule fraction of the population of even the small Greek city-states. There was nothing like the large universities, industrial and government laboratories, and think tanks that play such a prominent role in our lives today. Few people in antiquity even knew of the existence of these thinkers or of their theories. Yet these ideas have played a major role in shaping our way of thinking, our way of looking at the world down to the present day; this in turn has affected the lives of most people on Earth. Ideas can be more powerful than armies.

It would be a mistake to think that Greek astronomy was entirely qualitative. Several quantitative measurements were made, some of which were surprisingly accurate. We will describe some of these in the next chapter.

Terms to Know

Harmony of the spheres; uniform circular motion.

---

[4] In its most sophisticated form, which flourished in the fifteenth century, the standard model included Ptolemy's planetary mechanisms, to be described in the next chapter.

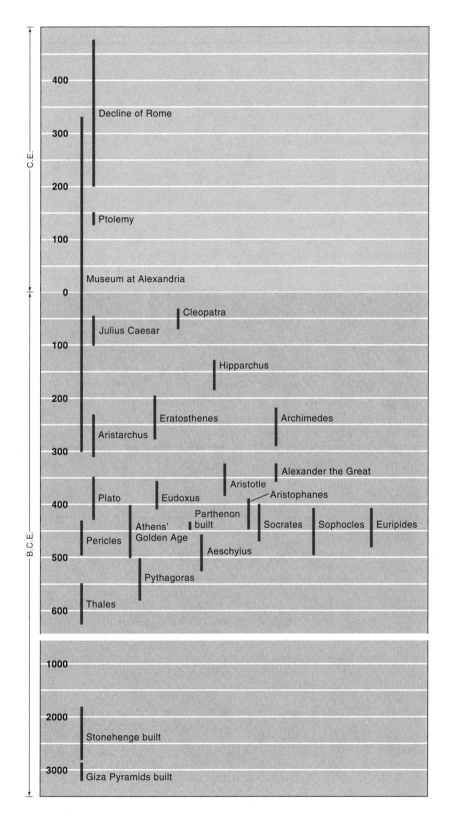

Figure 4.7. A historical timeline, showing people and events during the early centuries of astronomy.

Ideas to Understand

Greek attempts to describe the universe in terms of general principles; in terms of substances and mathematics; how spheres became important in astronomy; the harmony of the spheres; the origin of uniform circular motion; Eudoxus' model; Aristotle's system of terrestrial and celestial physics; his view of the astronomical universe.

Questions

1. List all the events in the natural world that you can think of that suggest that there is some sort of order or regularity in the universe.

2. (a) In attempting to define the fundamental nature of the world, the ancient Greek philosophers took two general approaches: matter and mathematics. Describe examples of each. If you had been a Greek philosopher, which of these approaches do you think would have appealed to you the most? Why?

(b) Do you think that a "model" of the universe should be a coherent, accurate representation of it in all details, or do you feel that it need account for only all the observational data, even if it could not work as a single piece of clockwork? Explain.

3. According to Aristotle's ideas, how far up in the sky did a meteor shower occur? Explain your answer.

4. How does a square compare with a circle as being a "perfect" form in the Greek sense?

5. What attitudes did the Greeks develop that were necessary for the idea of science? According to comments in this chapter, in what ways were these attitudes deficient?

Suggestions for Further Reading

Clagett, Marshall, *Greek Science in Antiquity*. New York: Collier Books, 1963. Describes Greek medicine, biology, physics, and mathematics as well as astronomy.
Dicks, D. R., *Early Greek Astronomy to Aristotle*. Ithaca: Cornell University Press, 1985. An authoritative description of the development of early astronomy.

The following books include discussions relevant to Chapter 4, but are also broader in scope, as indicated.

Chapters 4–6

Lindberg, D., *The Beginnings of Western Science*. Chicago: The University of Chicago Press, 1992. An authoritative and readable account covering the period from 600 B.C.E. to 1450.

Chapters 4–7

Dreyer, J. L. E., *A History of Astronomy from Thales to Kepler*. Mineola, NY: Dover Publications, Inc., 1953. An inexpensive reprint of a standard book on the subject.

Chapters 4–8

Berry, A., *A Short History of Astronomy*. Mineola, NY: Dover Publications, Inc., 1961. Originally published in 1898, this is still a useful history of astronomy from the beginnings through the nineteenth century.
Pannekoek, A., *A History of Astronomy*. London: George Allen and Unwin, Ltd., 1961. This history covers the subject up through the 1920s or so.
Toulmin, S. and Goodfield, J., *The Fabric of the Heavens: The Development of Astronomy and Dynamics*. New York: Harper and Row, 1965. A concise and well-written account of the development of astronomy through Newton's time.

Size of Star

Size of Earth's Orbit

Size of Jupiter's Orbit

Hubble Gallery. An ultraviolet image of the enormous red star, Betelgeuse, in the constellation of Orion. The central bright spot is at least 2000 K hotter than the surrounding surface. This is the best image yet made of the surface of a star other than the Sun.

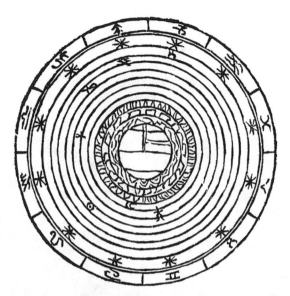

Pera est corpus solidum vna tantum superficie contentum.
ⓒSpere celestes decem sunt.ⓒ Prima est firmamentum seu
circumferentia.ⓒSecunda orbis signorum.ⓒTertia orbis
stellatus.ⓒQuarta saturni.ⓒQuinta iouis.ⓒSexta mar//
tis.ⓒSeptima solis.ⓒOctaua veneris.ⓒ Nona mercurij.
ⓒDecima lune.ⓒOrbis supremus seu spera a virtute prime
cause quem imobilis est mouetur que mota mouet omnes alias: ꝛ planete
mouentur contra ipsum. Iste orbis dicitur magnus ꝛ est rectus capatior ꝛ ve
locior omnibus alijs: ꝛ eos inter se comprehendit ꝛ reuoluitur in die ꝛ no//
cte reuolutione. 360.partium cum orbe signorum: ꝛ reuoluit secum orbem
stellarum fixarum ab oriente in occidentem. Et orbis illarum reuoluit de
ipso ad orientem in. 100. annis vnum gradum secundum. Ptbolomeum.
Deinde alij orbes secundum quantitatem constrictionis sue ꝛ amplitudinis: ꝛ
propter ipsum sunt dies ꝛ nox: ꝛ tempora diuersa veris ꝛ estatis autuni ꝛ bie/
mis ꝛ ipse permutat septem planetas: ꝛ terra est fixa in medio ipsius: quod
si non esset nunꝗ equarentur dies ꝛ nox: ꝛ est non stellatus ꝛ dicunt aliqui ꝗ
ipse est spiritualis.

A iiii

(translation) The sphere is a solid body bounded by only one surface. §There are ten celestial spheres. §The first is the firmament or circumference. §The second is the orb of the signs. §The third is the stellar orb. §The fourth is that of Saturn. §The fifth is that of Jupiter. §The sixth is that of Mars. §The seventh is that of the Sun. §The eighth is that of Venus. §The ninth is that of Mercury. §The tenth is the lunar [sphere]. §The highest orb or sphere is moved by virtue of the immovable first cause, and that motion moves all the others; and the planets are moved contrary to it. That orb is called the great [orb], and is appropriately more capacious and faster than all the others; and it contains them within itself and, along with the orb of the signs, revolves 360 degrees in the course of a day and night; and the orb of the fixed stars revolves with it from east to west. The orb of those [i.e. the stellar orb and the orb of the signs] revolves eastward one degree per century, according to Ptolemy. Finally, the other orbs [follow] according to their smallness or largeness; and owing to it [the "great" orb] there are day and night and the varying lengths of spring, summer, autumn, and winter; and it truly moves the seven planets. The earth is fixed in the center of it; and if it were not, then day and night would never be equal. And [the "great" sphere] is starless, and some say that it is spiritual.

—The *Compilatio* of Leopold of Austria, reprinted in 1520;
Leopold was a thirteenth-century astrologer

CHAPTER 5

Quantitative Greek Astronomy

Though located in Egypt, the city of Alexandria was a center of Greek culture. It was founded in 330 B.C.E.; about 40 years later a museum was established there that became an important institution of learning. In fact, with one notable exception, all the important astronomers of the next several centuries worked there. It was dedicated to the Muses (hence our word "museum"): the goddesses of poetry, music, dance, history, drama, and astronomy, nine in all. State-supported, it functioned more as a research institute than as what we would think of as a museum. Its library was the largest of antiquity, and its fate mirrored the general decline of Hellenistic culture; it was damaged by Caesar's Roman occupation, later by religious fanatics during disorders in the early Christian era, and finally by the conquering Muslims in the seventh century.

As a result of the conquests of Alexander the Great in 356–323 B.C.E. (see Figure 5.1), the Greeks gradually became aware of Babylonian astronomy. Partly because of this, they paid more attention to careful observational work. Many ingenious astronomical measurements were made by people associated with the Museum. These measurements often gave surprisingly accurate results, for example, the diameter of the Earth and the Moon's distance from the Earth. Even when the results were not very accurate, the difficulty was usually with the observational techniques available to the astronomers, rather than with their basic methods. Let's consider some of these measurements.

The Relative Distances and Sizes of the Sun, Moon, and Earth

Aristarchus of Samos (310–230 B.C.E.) attempted to find the relative distances of the Sun and the Moon. He did this by measuring, as precisely as his method permitted, the angle as seen at the Earth between the direction to the Moon and the direction to the Sun at the time of the quarter Moon (see Figure 5.2). The quarter Moon is important, because at that moment the angle *SME* is exactly 90°, by definition. Thus, if the angle *MES* can be found, then the shape of the triangle *SME* is known, and so the ratio of the Earth-Moon distance (*EM*) to the Earth-Sun distance (*ES*) is determined.

Measuring the angle *MES* directly is difficult (*MS* and *M'S* are very nearly parallel); so Aristarchus devised a clever indirect method. Like all other astronomers of the day, he thought that the Sun was relatively nearby, and so expected the time for the Moon to go from *M'* to *M* to be measurably shorter than the time for it to go from *M* to *M'*. This time difference, which Aristarchus thought was one day, could then be

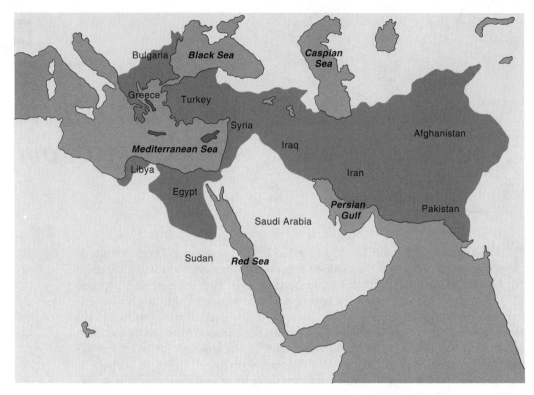

Figure 5.1. Alexander the Great's empire. Its maximum east-west extent was more than 5000 kilometers, stretching from Greece to India. At one time more than 50 cities and towns bore Alexander's name. The names of modern countries are given for orientation.

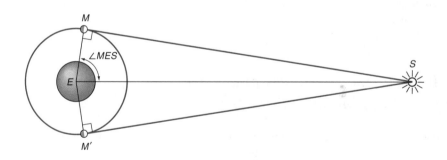

Figure 5.2. Aristarchus' method for finding the relative distances of the Sun and the Moon from the Earth. This requires finding the angle *MES* at exactly the time of the quarter Moon, because at that time the angle *SME* is 90°. With these two angles known, the shape of the triangle *EMS* is known, giving the ratio of *EM* to *ES*. Though this method is correct in principle, it is difficult to carry out, because the angle *MES* differs from 90° by only about 10 arc minutes; that is, the Sun is much farther away than Aristarchus (and any of his contemporaries) could imagine; in fact, light rays from the Sun are very nearly parallel at the Moon.

translated into the angle *MES*. In this way Aristarchus calculated that the Sun was about 20 times farther away than the Moon. The true value, however, is about 20 times greater yet! His result was so inaccurate because, in reality, the Sun is much farther away from the Earth than Figure 5.2 suggests; the line *ES* should be about 400 times longer than *EM*! In fact, the angle *MES* differs from 90° by less than 10 arc minutes. Measuring such a small angle requires a precision far greater than Aristarchus could achieve. (Remember, these were naked-eye observations.) An additional source of error was Aristarchus' assumptions that the Moon's orbit is circular and its motion is uniform, neither of which is correct. Given his optimistic assumptions, however, his method was correct in principle and was the first measurement of distances of celestial objects in the history of astronomy.

His attempt to find the relative sizes of the Earth, Sun, and Moon was more successful. It was known that lunar eclipses occur when the Moon moves into the Earth's shadow. By timing how long it took for the Moon to travel through the Earth's shadow and comparing that with the time required by the Moon to move a distance equal to its diameter, Aristarchus found that the shadow was about 8/3 the diameter of the Moon. (A crude estimate could be made by comparing the curvature of the Earth's shadow on the Moon with the curvature of the Moon itself.) He also knew that the angular diameters of the Sun and the Moon were about the same, one-half degree, and he thought that the Sun was only 20 times farther away from the Earth than the Moon.

Aristarchus could then have produced something like Figure 5.3. First, the solid lines are drawn, making two angles of 0.5° at the Earth, *E*. The Sun, *S*, is then placed 20 times farther from the Earth than is the Moon, *M*. (Note that the figure is not drawn to scale—the angle is much greater than 0.5° and the distance *ES* is not 20 times *EM*.) Next, the dashed lines are drawn from the Sun to the Moon in such a way that *DD'* is 8/3 times the diameter of the Moon. Finally, the dashed Earth is drawn in, centered at *E*, tangent to the dashed lines (since the Earth cast the shadow with diameter *DD'* at the Moon). The relative sizes of the Moon, Earth, and Sun could then be measured from a properly scaled figure.

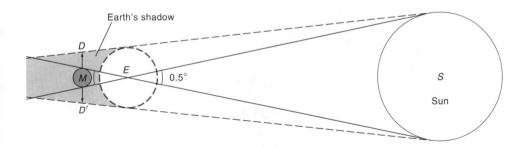

Figure 5.3. Aristarchus' method for finding the relative diameters of the Earth, Sun, and Moon. Eclipse observations tell us that the angular diameter of the Sun and Moon are both about 1/2°, and that the Earth's shadow is about 8/3 the Moon's diameter. Thinking that the Sun was only about 20 times farther away than the Moon, Aristarchus could have made this diagram. The solid lines are drawn making an angle of 1/2° (for clarity the angle actually shown is much greater than this), the dashed lines are drawn so that *DD'* is 8/3 the Moon's diameter, and finally the Earth is drawn in as shown. The relative sizes of the three bodies follow from the diagram.

Though it is known that Aristarchus used the method just described to find the relative sizes of Sun, Earth, and Moon, it appears that he himself did not make any of the individual measurements he needed. In fact, the values he used were incorrect; for example, he took the angular diameter of the Sun and the Moon to be 2°!

Even though Aristarchus' answers were wrong (but his methods were sound), they did indicate that the Sun was much larger than the Earth. Perhaps it was this result that led him to the belief that the Sun, not the Earth, was the center of the solar system. However that may be, this revolutionary idea ran counter to the knowledge and understanding of the time, was not accepted, and died.

The Size of the Earth

Eratosthenes (276–196 B.C.E.), for a time the director of the Museum and tutor to the son of the Egyptian ruler, Ptolemy III, was primarily a geographer. He calculated the absolute size of the Earth by a simple but clever astronomical method. To understand his idea you must remember that the Sun is so far away compared to the size of the Earth that all the rays of the Sun strike the Earth along nearly parallel lines. Put in another way, the direction to the Sun is very nearly the same as seen from any place on Earth. In Figure 5.4, Z is the zenith at Alexandria, A, and Z' is the zenith at Syene, S (the site of the present Aswan dam), nearly due south of Alexandria. Eratosthenes knew that at noon on June 21 (the summer solstice), the Sun is at the zenith at Syene (the Sun illuminated the bottom of a deep vertical pit then). At the same time in Alexandria, however, the Sun was 7.2° south of the zenith. He could then write the following proportionality:

$$\frac{\text{Syene to Alexandria}}{\text{(circumference of Earth)}} = \frac{(7.2)}{360} = \frac{1}{50}.$$

That is, the Earth's circumference is about 50 times the distance from Syene to Alexandria.

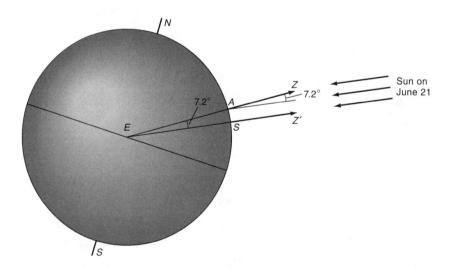

Figure 5.4. Eratosthenes' method of finding the diameter of the Earth. At noon on the summer solstice, the sun is at the zenith in Syene (S), but about 7.2° of the zenith in Alexandria (A). Thus the angle AES is also 7.2° and with the distance AS known, the circumference and hence the diameter of the Earth can be found.

This method is completely correct in principle, but just how accurate Eratosthenes' result was depends on two factors. How well did he know the distance from Alexandria to Syene? Did he correct for the fact that Syene is not exactly due south of Alexandria? We aren't certain of the answer to either question. The unit of length used was the *stadium*. If by this was meant the ordinary Greek stadium used for the games at Olympia and elsewhere, then the result was too large by about 20 percent. If, as some scholars believe, the unit of length was about 0.1 mile (a unit commonly used then in measuring long distances), Eratosthenes' answer was very nearly correct. In any case, the circumference of the Earth could be converted to its diameter by using the formula $C = 2\pi R$, where R is the radius of the Earth.

Later, this confusion about the unit of length gave rise to even more discrepant estimates for the size of the Earth. About 100 years after Eratosthenes, another measurement was made by the same method, but using a star instead of the Sun. The maximum altitude of the bright star Canopus as seen on the Mediterranean island of Rhodes (see Figure 4.1) and at Alexandria was measured. Again, uncertainty about the size of the unit of measurement led to two different values for the diameter of the Earth, but the error was compounded because somehow the smaller value became attached to the smaller unit of length. It was this value, too small by about 25 percent, that was quoted in the second century C.E. by the astronomer Ptolemy in his book on geography.

Incidentally, the uncertainty at the time of Columbus was not whether the Earth was flat or round—it was known to be round—but rather how large it was, and how far eastward in longitude Asia extended. The fourteenth-century Venetian traveler Marco Polo had estimated that the land mass from Spain eastward to China extended for 225° in longitude, much larger than the 180° given by Ptolemy, and very much larger than the actual value of 130°. To strengthen his arguments for financial support for his voyage to the Orient, Columbus assumed that both the small value of the Earth's diameter given by Ptolemy and the maximum eastward extension of Asia given by Polo were correct. He could then claim that a ship could carry enough provisions for such a voyage. Since Columbus' estimates put the easternmost part of Asia only about 2,500 miles west of the Canary Islands (at about Bermuda!), he and his crew were indeed fortunate that the Americas were in the way.

The Work of Hipparchus

As impressive as were the achievements of these early Greeks, they were surpassed by those of Hipparchus, often regarded as the greatest of all the ancient astronomers. He was not associated with the Museum at Alexandria, but worked around 150 B.C.E. on the island of Rhodes, which was about 600 km north of Alexandria, and was then something of a rival of Alexandria in its cultural and intellectual life. Hipparchus was an assiduous observer, making measurements to the greatest precision allowed by his instruments. For example, he cataloged the positions of about 850 stars, and divided them into six brightness groups, or **magnitudes**. This is the basis of the astronomical magnitude system still in use today. By comparing his measured star positions with those made earlier by Babylonian observers, he discovered that the direction of these stars with respect to the vernal equinox had changed. (We now know that this precession of the equinox, as it is called, is caused by the slow change in the direction of the Earth's axis of rotation.) He measured the length of the year to within a few minutes, and the length of the synodic month (that is, one full Moon to the next) to within one second! His abilities as an observer were matched by his mathematical skills. To make his calculations more easily and accurately without having to make carefully drawn geometric constructions, he developed some of the methods of trigonometry.

Precession

Let us consider briefly Hipparchus' discovery of **precession**. The Sun's direction on the first day of spring (when the Sun appears to cross the celestial equator moving south to north) has long been a convenient reference point against which to measure other directions in the sky. When Hipparchus compared his own measurements of the positions of several stars relative to the vernal equinox with the positions of the same stars measured about 150 years earlier, he found that all had moved. He realized that he could account for the motions of all the stars if only one direction had changed, namely the direction to the vernal equinox. But how could this change? Hipparchus reasoned that if the axis about which the stellar sphere rotates slowly changed its direction, the orientation of the celestial equator (always 90° from the celestial poles) must also change. But, since the intersection of the ecliptic and equatorial planes defines the direction of the vernal equinox, this direction must change also. Hence, the positions of stars measured with respect to the direction of the vernal equinox would change. (We would say that it is the direction of the Earth's axis of rotation that changes. The spinning Earth must slowly be wobbling on its axis, just like a gyroscope.)

The motion is indeed leisurely, the axis moving only one degree in 72 years. Thus it takes about 26,000 years for the celestial poles to complete one circuit, always staying about 23.5° from the ecliptic pole (see Figure 5.5). Instead of the Sun being in the constellation of Aries on the first day of spring, as it was when the zodiacal signs were established, it is now in Pisces, and in several hundred years will be in the constellation of Aquarius when spring begins.[1]

---

[1] So-called Sun signs in popular-press astrology columns have nothing to do with the constellations of the same name. If you were born on March 27, for example, your "Sun sign" is still said to be Aries, even though, because of precession, the Sun is now in Pisces on that date.

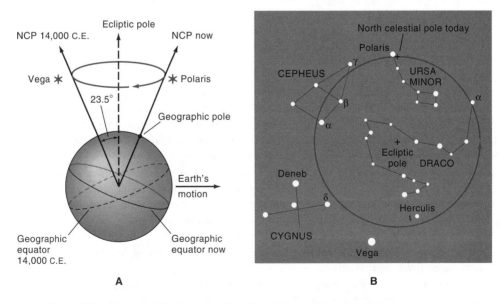

A   **B**

Figure 5.5. Diagram (A) shows the direction of the Earth's axis of rotation (and along with it, the equator) at two orientations about 12,000 years apart. In (B) the circle, centered on the ecliptic pole (i.e., the direction perpendicular to the plane of the ecliptic), is the path of the north celestial pole in the sky over the 26,000-year period of precession.

Because of precession, Polaris is the pole star only temporarily. When the pyramids were built 5,000 years ago, the pole star was in Draco; in about 12,000 years the north celestial pole will be fairly near the bright star Vega. Note how the small effect of precession can nonetheless be detected by naked-eye observations if these are carried out over a time span long enough that the effect, being cumulative, becomes large. Hipparchus could not explain the physical cause of precession; this had to await Newton and his invention of the theory of gravity.

The Earth's Motion

The actual orbit of the Earth around the Sun is not quite circular but elliptical. The Earth moves along this orbit with slightly varying speed, somewhat faster when closer to the Sun than when farther away. As a consequence, the Earth takes about six days longer to go from the vernal to the autumnal equinox than from the autumnal to the vernal equinox. Thus the northern hemisphere summer is slightly longer than its winter. The Moon moves in a similar nonuniform fashion around the Earth. Therefore, as seen from the Earth, neither the Sun nor the Moon moves along its apparently circular path at an exactly constant rate; that is, neither moves through exactly the same angular distance from one day to the next. Hipparchus accurately represented each of these nonuniform motions by means of a circle in which the Earth was not precisely at the center, but was slightly displaced from it; this construction is called an **eccentric circle** (see Figure 5.6). The Sun was imagined to move along the circle at constant speed (remember, uniform circular motion!), but as seen from E', the off-center Earth, it *appeared* to travel more rapidly at B than at A. This construction can represent quite well the small nonuniform motions of the Sun and the Moon, and with it Hipparchus could predict lunar eclipses to within one hour.

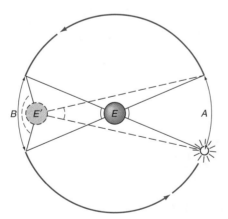

Figure 5.6. The motion of the Sun as seen from a central Earth, E, is uniform; that is, if the Sun took six weeks, say, to move along arc A, it would move along any other arc of equal angular extent (such as arc B) in the same time. Next imagine the solar motion as seen from an off-center Earth, E'. Now arc A subtends a considerably smaller angle than it did before and arc B a much larger angle. Thus when moving uniformly along A, the Sun would appear to move fewer degrees than it did before in the six weeks required to traverse the arc, and many more degrees than before in the six weeks when moving along arc B. The motion of the Sun would appear to be nonuniform, even though its "real" motion was uniform. (The Earth's eccentric position at E' is greatly exaggerated for clarity.)

Ptolemy the Synthesizer

Little astronomy of importance was accomplished in the 300 years after Hipparchus. It is only with Ptolemy (no known connection with the Egyptian ruling dynasty of Ptolemies, the last of whom was Cleopatra), who worked in or near Alexandria in the middle of the second century C.E., that we again encounter first-rate astronomical work. Ptolemy was one of the great synthesizers of classical times, ranking with Euclid, whose systematic presentation of plane geometry eclipsed the work of his predecessors. Ptolemy wrote a book describing world geography as it was then understood, as well as a treatise on astrology, both of which became the standard works on these subjects for many centuries. It is, however, as the author of a compendium of astronomy known to us as the *Almagest* that Ptolemy is best remembered today. (Ptolemy referred to his book as the "great composition," which when translated into Arabic was upgraded to the "greatest" or "Al Magisti." This finally became *Almagest* when it was assimilated into Latin literature.) In this work (for which astrology was an important motivation), Ptolemy gathered together all the astronomy known to him, and synthesized it into a cosmological system that attempted to account for all the available data. In fact, most Greek astronomy is known to us through the *Almagest*.

Distance to the Moon

Ptolemy devised an extremely clever way of measuring the distance of the Earth from the Moon. This method depends on the fact that the Moon is not very far from the Earth (only about thirty times the Earth's diameter); so the directions to the Moon as seen simultaneously from two places on Earth are quite different. (This is in contrast to the situation for the Sun, which is nearly 400 times farther away from the Earth than is the Moon.) Figure 5.7 shows the Earth and the Moon as seen from above the North Pole. An observer at *A* sees the Moon on his meridian at the same time that the observer at *B* sees it make an angle *MBC* to his meridian. This determines angle *MBE*, since the sum of the two equals 180°. Now angle *AEB* is just the difference in longitude between the places *A* and *B*, which was known, and *EB* is the radius of the Earth, also known. Thus with two angles and the included side of the triangle *MEB* known, Ptolemy could solve for either of the other two sides of the triangle, in particular for *EM*, the Earth-Moon distance.

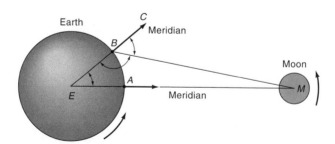

Figure 5.7. Ptolemy's method for finding the Earth-Moon distance. The diagram shows the Earth and Moon as viewed from above the Earth's North Pole. Suppose that when the Moon is crossing observer *A*'s meridian, observer *B* finds that the Moon makes an angle *MBC* with her meridian. Angle *AEB* is just the difference in longitude of the two observers, and angle *EBM* is just 180° minus the measured angle *MBC*. Thus the shape of the triangle *MEB* is known, and hence the ratio of *EB* to *EM*.

We have glossed over one point, however. How can the two observers be certain that they are making their observations at the same time? Ptolemy neatly solved this difficulty by letting the rotation of the celestial sphere carry the Moon from A to B, thereby enabling him to make both observations himself. (We would say that the rotation of the Earth carried him from A to B.) He had to make a small correction for the amount of the Moon's eastward motion during the time required for the Earth to move through the angle AEB, but that was not difficult. In this way Ptolemy found that EM was 59 times greater than EA, or in modern terms about 378,000 km, very close to the correct value of 384,000 km.

Ptolemy's System

In writing the *Almagest*, Ptolemy organized astronomy into a useful, working system. In effect, he did for astronomy what Euclid did for geometry. This effort was so successful that the Ptolemaic system remained the standard until the time of Copernicus some 1,400 years later.

What did Ptolemy "know"? First, his basic assumptions:

- Celestial motions are made up of combinations of uniform circular motion; and
- the Earth is spherical and stationary, nearly at the center of the universe, and small by comparison with it.

What observational data did he have to work with?

- The relative diameters of the Sun, Moon, and Earth, and the distances to the Sun and the Moon;
- the causes of lunar and solar eclipses, and how to predict the times of their occurrence, at least approximately;
- the positions of the planets and of the brighter stars to about one degree;
- the length of the year to a few minutes and the length of the lunar month to about one second; and
- various "details" like precession and irregularities in the Moon's motion.

What were the main observational features that a theory or model of the universe must account for?

- The daily motions of all celestial objects from east to west;
- the slow eastward motion of the Sun and the faster eastward motion of the Moon against the background stars; the nonuniform motions of the Sun and Moon;
- the generally eastward motions of the planets and their occasional westward (retrograde) motions; that planets are brightest during retrograde motion;
- that planets are restricted to the zodiac; that Mercury and Venus are never more than 28° and 47°, respectively, from the Sun, whereas the other planets can be up to 180° from it; and
- various minor motions, of the Moon, for example.

Ptolemy's Geometric Constructions.   How did Ptolemy attempt to represent nonuniform celestial motions? What geometric constructions did he use? The simplest was the eccentric circle, used so effectively by Hipparchus. A more complicated construction was the deferent-epicycle system shown in Figure 5.8A. The large circle is the **deferent** on which is attached a smaller circle, the **epicycle**, to which a planet is fixed. Both circles rotate uniformly, but at rates that may differ from each other, in the same or opposite sense.

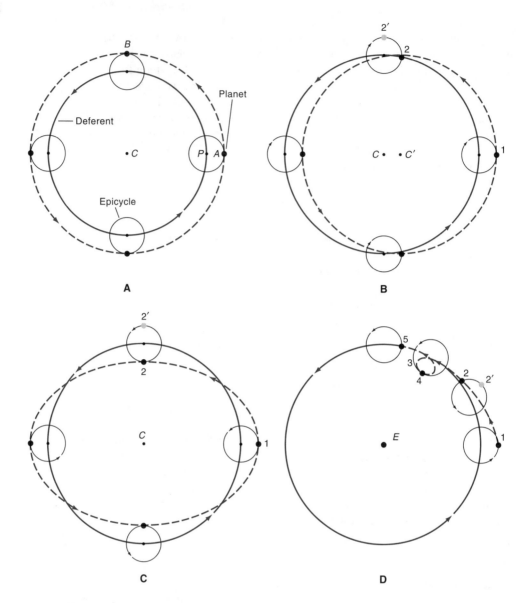

Figure 5.8. These diagrams show a few of the sorts of motions that can be achieved with the deferent-epicycle system. In (A) a larger circle is made—not very interesting; in (B) an eccentric circle results; in (C) an elliptical orbit is made; in (D) retrograde motion is represented. See the text for a detailed explanation.

Let us see what can be done with this construction. If the epicycle is fixed on the deferent and does not rotate about *P*, all that happens as the deferent rotates is that the planet moves along a circle having a radius different from that of the deferent; obviously, we don't need such a fancy construction to produce this simple result.

Next, let the epicycle rotate about its center once in the clockwise sense during the time the deferent rotates once in the opposite sense (Figure 5.8B). Without the epicyclic rotation, a quarter-turn of the deferent would take the planet from 1 to 2′, but since the epicycle also has rotated one quarter-turn, the planet actually is at position 2. When we trace out the motion of the planet in this manner, we will find that the resulting path is circular. The center of the circle, however, is displaced from the center of the deferent, *C*, to *C′*; that is, an eccentric circle is produced. Hipparchus knew of this construction and could have used it to represent the small uniform motions of the Sun and Moon. He chose not to do so, for the very good reason that the simpler eccentric circle worked.

Now let the epicycle rotate twice as rapidly as the deferent in either the same or the opposite sense. A careful look at Figure 5.8C, shows that the resulting path is elliptical. By adding more epicycles on epicycles, orbits of any shape—even squares—may be produced. The technique is a powerful one.

Finally, let the epicycle rotate in the same sense as the deferent, but at three times its rate. The resulting motion is shown in Figure 5.8D. As seen from the Earth, *E*, the planet's motion from position 1 to 2 is eastward, that is, direct. As it moves from 2 to 3 it would appear to slow down against the stars (since its motion is increasingly toward the Earth and not across the sky), and at position 4 it would appear to be moving in a westerly (retrograde) direction. Next, it would appear to slow down again, stop, and finally resume its eastward motion, completing its retrograde loop. See what has happened. Nonuniform and retrograde motion have been produced by two uniform circular motions, just as was required by the basic requirement guiding the representation of celestial motions. But there is even more: the planet, being nearer to the Earth during its retrograde loop, would appear to be brighter then, just as was observed! This result must have been regarded as a big plus for the deferent-epicycle system. Furthermore, by changing the rate of rotation of the epicycle and the size of its radius, Ptolemy could produce as many periods of retrograde motion per orbit and as large retrograde loops, respectively, as were necessary to fit the observations.

The Equant.   Ptolemy added a curious wrinkle of his own to the deferent-epicycle system. This was the **equant**, *Q*, shown in Figure 5.9. The equant was located on the opposite side of the center of the deferent from the Earth. It was as seen from *Q* that the angular motion of the center of the epicycle was to be uniform. Uniform angular motion was to occur not as seen from the center of the deferent, nor as seen from the Earth, but as seen from this arbitrary equant point. To a true believer in uniform circular motion, this would be questionable at best, heretical at worst. Evidently, however, Ptolemy was not the strictest adherent to that doctrine. By introducing the equant,

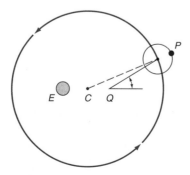

Figure 5.9.   Ptolemy's equant. The planet *P* is observed from the Earth, *E*. The center of the epicycle moves around *C*, but it moves uniformly as seen from the equant, *Q*.

he could account for both the size of the retrograde loops and the variable velocity with which the center of the epicycle had to move. It resulted in much more accurate predictions of planetary positions along the ecliptic than were otherwise possible. Ptolemy must have felt that the practical advantages of the equant outweighed any theoretical objections to it.

What did Ptolemy's System Look Like?   To account for the maximum elongations of Mercury and Venus, the centers of their epicycles always had to be on the Earth-Sun line; that is, the period of their deferents had to be one year. Their maximum elongations were then set by the radii of their epicycles, as shown in Figure 5.10. Next,

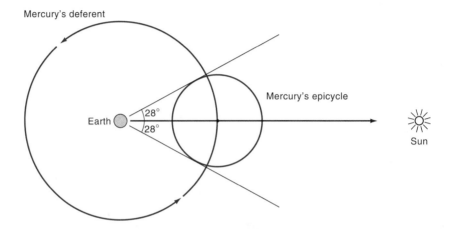

Figure 5.10.   The diameter of the epicycle of Mercury relative to the diameter of its deferent must be such that its maximum elongation (its maximum angular distance east or west of the Sun) is 28°.

the radii of the epicycles of Mars, Jupiter, and Saturn had to be always parallel to the Earth-Sun line (so the period of the epicycle is one year) to produce the correct number of retrograde loops during each orbital revolution (see Figure 5.11). Note that although the Earth and not the Sun was near the center of Ptolemy's system, the Sun played a special role in it, because, as we have just seen, all the planets were in a special relation to the Earth-Sun direction. Thus, Ptolemy's Earth-centered system itself gave a strong hint as to the centrality—both literally and figuratively—of the Sun. Notice also that in the Ptolemaic system the phases of Mercury and Venus could be only new or crescent. Since planetary phases cannot be distinguished by the naked eye, this became important only when Galileo turned his telescope to the skies. Then, as we shall see in Chapter 7, it became crucial.

Ptolemy did not know the distances to any of the planets; so, like his predecessors, he made the not-unreasonable assumption that the slower the object, the farther away it was. Thus, located at increasing distances from the Earth were the Moon, Mercury, Venus, the Sun, Mars, Jupiter, Saturn, and the stars (see Figure 5.12 and Color Plate C1).

An important feature of the *Almagest* was that it included the mathematical techniques and tables necessary to carry out the calculations of planetary positions, times

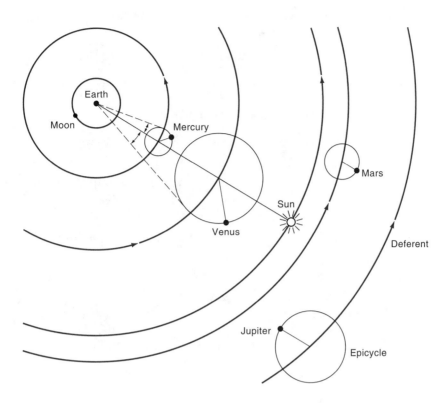

Figure 5.11. A much-simplified version of Ptolemy's planetary system, showing only the major epicycles. The Earth, Sun, and centers of the epicycles of Mercury and Venus must all be on the same straight line. The sizes of their epicycles are determined by their maximum elongations, which are actually much larger than shown here. The radii from the centers of their epicycles to each of the outer planets must always be parallel to the Earth-Sun line.

of lunar eclipses, etc. It was a practical, "do it yourself" book for the astronomers—and astrologers—of the day. On the average, the accuracy of this system was only about 5°, or the angular separation of the pointer stars in the bowl of the Big Dipper. That this was not nearly as accurate as the observations did not greatly concern most astronomers of antiquity, who, in general, seemed to be satisfied with qualitative agreement between theory and observations.

Finally, it should be pointed out that this "system" of Ptolemy's, though made up of spheres in uniform circular motion about a fixed Earth, was not at all the sort Aristotle had in mind. It was not a coherent unit that, in principle at least, could have actually been built and made to function like some great machine. Instead, it was a collection of geometric models, often more than one for a given object, each of which accounted for a different aspect of the planet's motion. Probably Ptolemy was not at all concerned with the reality of his geometric constructions. He simply wanted a set of geometric devices that could give fairly accurate representations of celestial motions. In this he largely succeeded.

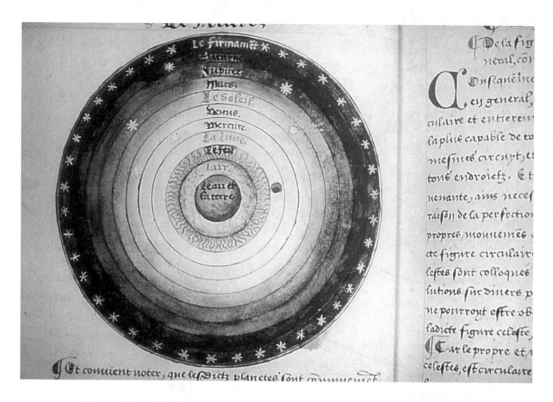

Figure 5.12. This illustration of the geocentric universe is from a French manuscript of 1549. It simply shows a general arrangement without any of the geometric constructions required to account for the various motions. Note that the "natural" locations of the four elements (earth, water, air, and fire) within the lunar sphere are shown as well as the locations of the Sun, Moon, planets, and stars. Plate C1 gives this figure in its original color.

The Greeks as Scientists

Were these Greeks scientists? Probably not in the modern sense. True, they took the first necessary steps in believing that the universe was ordered and knowable. They felt that natural phenomena should be accounted for in rational ways and not by invoking supernatural forces. Thus it was possible for them to construct theories. These attitudes are necessary for science, but in themselves don't go far enough. For example, they felt that if it was possible to give more than one explanation for a particular phenomenon, why, all the better; lightning might be this, or it might be that, or something else yet again. The idea of a single correct explanation was not firmly embedded in their attitude toward nature.

Though the Greeks did compare their theories with nature and did make quantitative measurements, in general they seemed content with qualitative rather than quantitative agreement, as we noted earlier. For our taste, their theories were not sufficiently tied to the observed and measured facts of nature. For example, modern science would require that Ptolemy's system predict planetary positions as accurately as the measurements allow. We have seen that it did not, nor does this seem to have been of great concern to astronomers of the time. We insist that theories must be tested rigorously against nature, that they must agree in all their particulars with what we take to be physical reality. Without this, theories have no strong limits, no hard boundaries to keep them from becoming little more than speculation. We shall see a particularly striking example of the force of this requirement when we discuss Kepler's work in Chapter 7.

Terms to Know

Precession of the equinox; eccentric circle; deferent; epicycle; equant.

Ideas to Understand

How the Greeks found the distances and sizes of the Earth, Moon, and Sun; Ptolemy's assumptions, celestial motions he attempted to account for, the uses of the eccentric circle, deferent and epicycle, equant; virtues and defects of his geocentric system; Greek steps toward science.

Questions

1. (a) What is implicitly assumed about the shape of the Earth in Eratosthenes' method for finding its size?

(b) Suppose two cities, due north-south from each other, are separated by 10 degrees of latitude. On a day when the Sun is at the zenith of the southern city, how many degrees is the Sun from the zenith of the northern city at the same instant? Explain your answer with a sketch. If the two cities are 2,000 km apart, what is the diameter of the planet? Are the cities located on the Earth?

2. Ptolemy set the order of the planets out from the Earth according to their rate of motion against the stars.

(a) What is the significance of the distance of the Sun's orbit from the central Earth?

(b) Why was it necessary for the epicyclic centers of Mercury and Venus to be fixed on the line between the Sun and the Earth?

3. Make a sketch of the Ptolemaic solar system showing the central Earth, the deferent and epicycle of Venus, and the Sun. Show that this arrangement limits the phases of Venus. Where would Venus be on its epicycle to show its various phases? Where on its epicycle and at what phase would Venus appear to be smallest? largest?

4. Why doesn't the Moon show retrograde motion?

Suggestions for Further Reading

See also the books listed at the end of Chapter 4.

Van Helden, A., *Measuring the Universe: Cosmic Dimensions from Aristarchus to Halley*. Chicago: The University of Chicago Press, 1986. An interesting history of how distances within the solar system were measured from earliest times to about 1700.

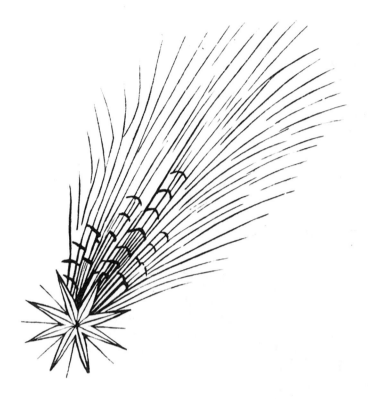

A dramatic drawing of Halley's comet in its apparition of 684, as given in the Nuremberg Chronicles of 1493. A description of the disasters accompanying the comet was also given.

"If the Lord Almighty had consulted me before embarking on Creation, I should have recommended something simpler."

—Alfonso X (The Wise), (1252–1284), after hearing an explanation of the Ptolemaic system

CHAPTER 6

The Medieval Interlude

The Middle Ages lasted nearly 1,000 years and encompassed the rise and fall of kingdoms and of social and economic systems, both secular and religious, as well as the development of various technologies and movements in art and scholarship. Generalizations about such a span of history are dangerous, and opposing examples can always be found. The difficulty is compounded when, as in this text, we can devote only a few pages to this period. Since we are interested here only in the course of astronomy during these centuries, however, perhaps the comments sketched below have some validity.

The Decline of Astronomy in the West

A striking indication of the fundamental sterility of astronomy in the Middle Ages is that, had Ptolemy been magically transported to the sixteenth century, he would have been completely comfortable with the problems addressed by contemporary astronomers and the conceptual framework within which they worked. There are several reasons that little progress was made during this period. First of all, in the fifth century the Roman Empire began to fall apart. Western society became fragmented and often lawless, communities were isolated, trade declined, and economic life became precarious. Societies were barely able to meet basic needs, much less support activities, such as astronomy, which had little immediate benefit.

Also, with the rise of Christianity the attitude toward science and philosophy changed. Christians, quite naturally, placed the greatest emphasis on Scripture. For some scholars concerned with explaining and proving the truth of Scripture, Greek science and philosophy simply were not particularly relevant. In addition, many early Christians felt it important to separate themselves from anything having to do with the taint of the old religions, including the pagan learning associated with them. They placed little value on the study of the natural world, and discarded some of what had been learned. Theirs was not the only or even the dominant view, however.

Other early Christian scholars valued the old philosophy, especially that of Plato, whose views on the immortality of the soul and the finiteness of the universe had become those held by the Church. In adapting his philosophy to their view of the world, they helped preserve the Greek learning. In fact, the Church, feeling that science might serve theology, preserved much ancient learning at a time when the secular society was not at all interested in it. Beginning with Augustine (354–430), the attitude took hold that Christians should take what was useful from the pagan study of nature rather than reject it outright. The sciences should not dominate or be in conflict with the Church,

Figure 6.1. The Bayeux tapestry is an embroidery 230 feet long and 20 inches wide giving an account in 72 scenes of the Norman conquest of England in 1066. It was made some years after the event and was probably commissioned by the Bishop of Bayeux, the half brother of William the Conqueror. The panel above depicts the consternation of the English at the appearance of Halley's comet.

but become its servant, the handmaiden of religion. Apparently most Western scholars adopted this approach up through the time of Copernicus, not because of any coercion, but because it seemed reasonable to them.

In many monasteries old texts were carefully preserved and copied, though generally little effort was made to study and expand upon them. Nevertheless, as time went on the knowledge of Greek slowly faded. Fewer and fewer scholars knew how to read the language and so were unable to read the Greek manuscripts even had they desired to do so. From about 500 to 1100 there is no indication that any Latin translations were made of the Greek works.

As always, superstition flourished, particularly astrology. The Church had difficulty with astrology and could not let it go unchallenged. For example, Augustine allowed that the stars might in some general way influence nature—the rainy season occurred because certain stars were out, for example. He could not, however, admit that the stars had any direct influence on human life, or else what became of the idea of our having wills free to choose between good and evil, and what happened to God's power over the universe? Augustine argued strongly against the fortune-telling sort of astrology, citing twins conceived at the same instant and born very nearly at the same times, yet having completely different personalities and lives.

Nevertheless, much of the astronomy done in the West during this time was motivated by the needs of astrology. Unusual events in the sky were noted and interpreted by astrologers. The best known of these is the reference in the Bayeux tapestry (Figure 6.1) to what we now call Halley's comet, which, appearing in the spring of 1066, was later associated with Harold's defeat at the hands of William the Conqueror that October. (Had the battle gone against William, the comet would doubtless have been connected with *his* defeat.) Astrologers also associated unusual celestial happenings with events that had already taken place, thereby doubling the opportunities to justify the

Figure 6.2.   Decoration on the facade of a sixteenth-century house, originally owned, it is said, by an astrologer, in the town of Bazas in southwest France.

faith of the astrological true believer. Planetary **conjunctions**—two or more planets in nearly the same line of sight (and so appearing to be close together in the sky)—were also taken to have great significance.

Our language provides many reminders of the pervasiveness of astrology, for example, "disaster"—against (counter to) the stars, and "exorbitant"—out of orbit, unusual. Shakespeare's writing is full of such references: from *King Lear*, "The stars above us govern our condition," or "These late eclipses in the Sun and Moon portend no good to us"; from *Julius Caesar*, "The fault, dear Brutus, is not in our stars, but in ourselves that we are underlings"; or "When beggars die, there are no comets seen; the heavens themselves blaze forth the death of princes" to cite only a few.

Thus poor economic conditions, the relatively passive attitude of the Church toward science, and the constriction of learning all contributed to the decline of science. For several centuries in the West, astronomical theory and observations were largely limited to timekeeping, to navigation, to setting the dates of the various Church holy days, and to the needs of astrology (Figure 6.2).

Islam and Science

Astronomy in the Greek tradition was carried on in Islam, however. In just a century the followers of Muhammad (about 570–632) conquered all the Middle East to India, as well as North Africa and most of Spain, as shown in Figure 6.3. After their victories they became tolerant of diverse ideas and attitudes, and conditions for intellectual life improved in their domains. Starting around 750 and for the next 250 years, the caliphs of Baghdad became patrons of science and the city a center of learning. By the eleventh century the library of the caliph of Cairo contained roughly 150,000 volumes; by contrast, a Western monastery considered itself fortunate to have 150.

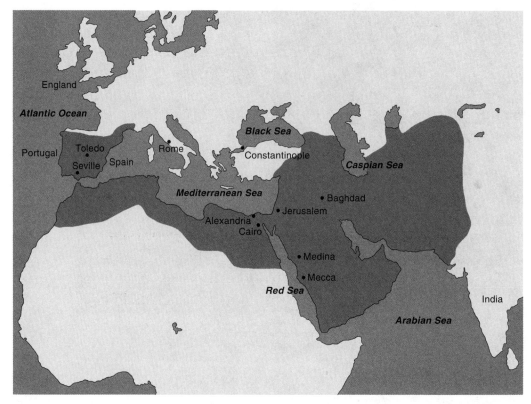

Figure 6.3. The Islamic world in 750. Of the territories shown, all are still Muslim except Spain and Portugal.

An important factor in the development of Islamic science was the old Greek writings that the Arabs found in the lands they conquered. Within just a few decades after 750 the major Greek scientific works were translated into Arabic; by the end of the tenth century essentially all the known Greek manuscripts had been translated. The atmosphere of tolerance was such that this work was done by Christians, Jews, and pagans, as well as by Islamic scholars.

Science was supported because of its usefulness to rulers—medicine, for example, and astronomy in its relation to astrology. Astrology, in fact, did double duty in being used not only to discover the destinies of caliphs, but also to decide, in part, on their medical treatment.

The religious requirements of Islam also were a powerful impetus to the study of astronomy. For example, Islam adopted (and generally still uses) a strictly lunar calendar, which, however, begins not at the time of the new Moon, but at the first sighting of the crescent Moon just after sunset. Calculating when this occurs requires fairly complex geometry. Since the lunar year is about 354 days long, the months of the Islamic year cycle through the seasons in about 33 years. Mosques had to be oriented toward Mecca, and Muslims were to pray facing in that direction. Furthermore, timekeeping was required to properly set the five times for daily prayer.

In attacking such problems Islamic scholars developed mathematics, especially trigonometry and spherical geometry, far surpassing the Greeks. They learned of the sine trigonometric function from India, and invented the other five, the cosine, tangent, etc. They also derived some of the relations among these trigonometric func-

98

tions, such as the law of sines. With these developments it was far easier to solve the geometric problems of astronomy. The Arabs also used a system of numbers, including the concept of zero, which they had acquired from India, which in turn they transmitted to the West, and which we use today. These Arabic numerals, as they are called, are far easier to calculate with than are Roman numerals. Arab texts describing these new mathematical developments as well as various summaries and commentaries on the *Almagest* eventually found their way in Latin translation to the West, where they became standard works for centuries.

Islamic astronomers refined many of the basic astronomical constants, for example, the length of the year and the eccentricity and inclination of the Sun's (really the Earth's) orbit, data that were later used in the West. Although they worked within the Ptolemaic tradition, they questioned some of the constructs of Ptolemy. The equant (the point around which the angular motion was uniform), for example, was felt not to conform to the requirement of uniform circular motion. Even epicycles were questioned, and attempts were made to construct a system that did not require them.

From our point of view, however, the Muslims' most important contribution was that they preserved much of the Greek learning, and then, beginning around the year 1000, became the means by which it was retransmitted to the West. Cities near the boundaries between Islamic and Christian domains, such as Toledo in Spain, became centers of a "translation industry" (from Arabic to Latin) and Arabic words such as *zenith*, *nadir*, *alchemy*, *algebra*, and *algorithm* entered our language, along with star names such as Algol, Aldebaran, Alcor, Vega, Deneb, and Betelgeuse. One of the most prominent of these translators—and surely one of the most industrious—was Gerard of Cremona (ca. 1114–1187), who, unable to find a Latin translation of the *Almagest* in Italy, came to Toledo in search of one. He stayed there and eventually put into Latin all of Aristotle, Plato, Euclid, and Ptolemy, about 100 major works in all, including a dozen on astronomy and 17 on mathematics.

The richness, diversity, and power of the Greek writings made a deep impression on Western scholars, and soon they were attempting to accommodate them to the Christian tradition. Thomas Aquinas (1225–1274) was a key figure in this effort (called Scholasticism), and he showed how much of Aristotle's thought could be integrated into Christianity. This amalgam became the foundation for natural science in the West.

Universities Appear

In the late Middle Ages universities developed, growing out of the cathedral schools that had been established to train clerics. The charter of the University of Paris dates from 1200, that of Oxford from a few years later. These associations of scholars formed at the same time as did the guilds of craftsmen and commerce. Though secular patronage of science was to be found in some royal courts, in Paris and the Holy Roman Empire, for example, the Church-associated universities provided the greatest opportunities for scholars and became the primary institutional home for science. Nowadays universities are such a common feature in our lives that we take them for granted. You should remember, however, that they were an essentially new kind of institution, international in orientation, that enabled relatively large numbers of students to acquire systematic training in various subjects, including science. Universities have had a profound effect on our lives.

To a much greater degree than they do now, universities, with their common traditions and curricula, formed a subculture within the larger society. They were also quasi-independent entities, and to some extent not subject to the local authorities, primarily because of the economic leverage they exercised.[1] Not surprisingly this occasionally caused friction between townspeople and students. Problems between town and gown are nothing new.

Printing

Although long known in Asia, printing using movable type was independently invented in Europe in the 1440s, probably by Johann Gutenberg (Figure 6.4). By 1476 a printing press was in operation in England, and by 1539 in Spanish Mexico. The ability to print at a relatively small cost large numbers of copies of pamphlets, books, documents, etc., had an enormous impact throughout society; it is perhaps the key development in the entire cultural history of the West. No longer were the Greek works laboriously copied by hand; they could now be printed in quantity.[2] Personal and institutional libraries expanded from a few volumes to hundreds and thousands. Scientific works could be circulated widely, thus helping to establish a community of scholars with common interests and knowledge of the work of their colleagues. One of the earliest publications was an astronomical almanac printed in 1447, about a decade before Gutenberg's famous Bible.

Figure 6.4. Gutenberg's printing press, one of the most important inventions in the Western world.

---

[1] The "Latin Quarter" of Paris refers to the early university area, where Latin was spoken.

[2] Paper, of course, was a critical element in this development. It originated in China early in the second century, reached India by the sixth century and Baghdad by 793. Its use spread westward across North Africa and then to Spain in 1150, to Italy in about 1270, Germany in 1390, and England in 1494. The first paper made in North America was in Philadelphia in 1690.

If, by the beginning of the sixteenth century, the same problems of astronomy addressed by Ptolemy remained, the circumstances under which they could be attacked had changed drastically. Much sharper mathematical tools provided by the Muslims were now in use, as well as their improved values for many of the astronomical constants that entered Ptolemy's system. Arabic commentaries on the *Almagest* as well as their modifications of the theory were studied by astronomers in the West. The universities provided the means whereby learning could be passed on to others as well as broadened and deepened. The printing press made volumes once difficult to find widely accessible, hastening the spread of new learning. The stage was set for new ideas, different approaches to old problems.

The Late Medieval World

The late medieval world of Europe became what might seem to us to be a crazy mixture of Christianity, Greek and local pagan ideas, and astrology. Looking at this world from a perspective of a millennium, we tend to ascribe to it more of a unity, more of a cohesiveness than is probably justified. Many different theories, ideas, and attitudes were expressed by scholars, often in response to the newly acquired Greek learning. Nonetheless, it is fair to say that the European medieval world was not as fragmented, as disconnected within itself as our own is. At various times medieval scholars made connections between the social order and the religious order or between theories of personality and theories of matter, for example. Things were generally related, directly and indirectly, so that one could not make a fundamental change in one part of the structure without affecting all the rest.

Consider briefly the views of a well-educated man, the poet Dante Alighieri (1265–1321), which he expressed in his *Divine Comedy*. Like Aristotle, he thought that change occurred on Earth because the four elements—earth, air, fire, and water—were out of their proper places; that man's corruptible body was made up of these elements, but that his soul was made up of unchanging, perfect matter like that constituting the heavens. In Dante's cosmology there were nine major spheres—five for the planets and one each for the Sun, Moon, stars, and the prime mover, which made everything go (Figure 6.5). These were echoed in his description of Paradise. Moving upward from the Earth, one passed through circles that became increasingly perfect until ultimate perfection was reached in heaven, at the tenth sphere. With a central Earth surrounded by heaven, man was always under the eye of God. Hell was located at the center of the Earth, and with its nine circles of greater and greater sinners it was a perverse reflection of the heavenly spheres. Man lived between hellish corruption and heavenly perfection, and his soul could go in either direction.

These ideas had their echoes throughout the medieval structure. For example, the social structure with an absolute ruler at the top and serfs at the bottom could be claimed to be merely an earthly manifestation of the natural order. Corresponding to the four elements were the four temperaments (personality characteristics) of optimism, pessimism, calmness, and hot temper; the four bodily fluids having to do with health, which were blood, phlegm, yellow bile, and black bile; and the four qualities of matter. Thus, the particular mixture of the four elements and four temperaments determined what sort of person you were, while the proportions of the four bodily fluids determined your health. Shakespeare reflects this notion in *Julius Caesar*: "His life was gentle and the elements so mixed in him that Nature might stand up and say to all the world, 'This was a man.'"

A brief consideration of the implications of an infinite universe will give some idea of the interconnectedness of this world. If the universe were infinite, then there could

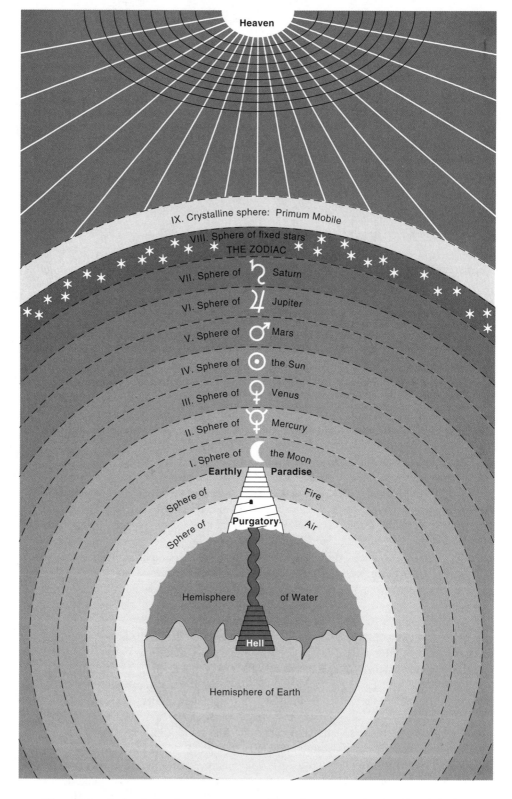

Heaven

IX. Crystalline sphere: Primum Mobile

VIII. Sphere of fixed stars
THE ZODIAC

VII. Sphere of ♄ Saturn

VI. Sphere of ♃ Jupiter

V. Sphere of ♂ Mars

IV. Sphere of ☉ the Sun

III. Sphere of ♀ Venus

II. Sphere of ☿ Mercury

I. Sphere of ☽ the Moon

Earthly Paradise

Sphere of   Fire

Purgatory   Air

Sphere of

Hemisphere   of Water

Hell

Hemisphere of Earth

Figure 6.5. A sketch showing some aspects of Dante's cosmos as he described it in the *Divine Comedy*. Satan has fallen (literally and figuratively) as far as it is possible, to the deepest part of Hell at the center of the Earth. Note that the Earth is divided into two parts, a hemisphere of land and one of water, reflecting the geographical knowledge of the time. The circles in the sky are echoed by those in Hell and Purgatory.

be no center, so what happens to the idea of "up" and "down"? What would be the natural directions of motion of the elements? If there is no center, why should the four elements be collected at just one point, so why could there not be other worlds? What then happens to the uniqueness of the Earth, created for man, always to be under the eye of God? If the Earth were just another planet, then other planets must also be changeable, corruptible, so what happens to the perfection of the celestial regions? Where is heaven, where is God?

This medieval construction, in many ways sturdy and complete, was also fragile—in principle if not in fact—in that the crumbling of a stone or two might start the whole edifice tumbling down. What actually happened, however, was that the medieval worldview came apart bit by bit during a rather long span of time. For example, the doctrine of the four elements long outlived the idea of a central, unmoving Earth. We shall see the role astronomy played in this process.

Terms to Know

Planetary conjunction.

Ideas to Understand

Why astronomy declined during the "Dark Ages"; why the Muslims supported astonomy; their contributions to science; importance of universities and printing to the development of science; the unity of the medieval world; the "impossibility" of an infinite universe.

Questions

1. Why was there little scientific activity during much of the Middle Ages?

2. Can you think of any invention that has had as profound an effect on society as the printing press? Justify your answer.

3. For what reasons was it difficult for scholars in the Middle Ages to accept the possibility of an infinite universe?

Suggestions for Further Reading

See also the books listed at the end of chapter 4.

Gingerich, O., "Islamic Astronomy," *Scientific American*, **254**, p. 74, April 1986.
Grant, E., *Physical Science in the Middle Ages*. New York: John Wiley and Sons, Inc., 1971. A short book giving an overview of various aspects of medieval science including the "translation industry" and the rise of universities.
Kunitzsch, P., "How We Got Our 'Arabic' Star Names," *Sky & Telescope*, **65**, p. 20, January 1983.
Nicolson, M., *Science and Imagination*. Ithaca, NY: Cornell University Press, 1956. The poetic response to the discoveries of the telescope, the microscope, and new astronomy.
North, J., "The Astrolabe," *Scientific American*, **230**, p. 96, January 1974. A description of how the most widely used astronomical instrument of the Middle Ages was made and used.
North, J., *Chaucer's Universe*. Oxford: Oxford University Press, 1988.
Olson, D. and Jasinski, L., "Chaucer and the Moon's Speed," *Sky & Telescope*, p. 376, April 1989. Discussion of one of the many astronomical allusions in the *Canterbury Tales*.

omnia (infinita in potentiâ) permeantes actu : id quod aliter à me non potuit exprimi, quam per continuam seriem Notarum intermedia-

Saturnus   Jupiter   Mars ferè   Terra

Venus   Mercurius   Hic locum habet etiam

CAP. VI

rum. Venus ferè manet in unisono non æquans tensionis amplitu-
dine vel minimum ex concinnis intervallis.

Atqui signatura duarum in communi Systemate Clavium, & for-
matio sceleti Octavæ, per comprehensionem certi intervalli concinni,

Kepler's search for universal harmonies led him to a musical representation of the orbital speeds of the planets. Venus' orbit is nearly circular, so that its speed (and tones) is constant, whereas Mercury's orbital speed and its representative tones vary considerably.

"The aims of scientific thought are to see the general in the particular and the eternal in the transitory."

—Alfred North Whitehead (1861–1947); British philosopher

This is also a concise description of Galileo's genius.

CHAPTER 7

The New Astronomy

Nicolaus Copernicus (1473–1543; see Figure 7.1) received a thorough education, first spending three years at the university in Krakow in his native Poland, and then about ten years in various Italian universities. He studied theology, law, and medicine, as well as mathematics and astronomy. Before finishing his studies at Krakow, he had been appointed to the lifelong position of a canon in the Church. This was a lay administrative post, usually requiring only light duties (and made possible Copernicus' long academic sojourn in Italy). Copernicus' uncle, a bishop who was a good friend of the king of Poland, secured this choice appointment for him. Copernicus served as his uncle's secretary until the latter's death in 1512, after which he moved to Frauenberg, on the shores of the Baltic Sea. For the next thirty years Copernicus lived a busy life as an administrator, as an economist (he reformed the local system of weights and measures, for example), as a practicing physician, as a military man (he organized a successful defense against an attack by the Knights of the Teutonic Order), and as an astronomer, working on his planetary theory.

What Motivated Copernicus?

Aesthetic Reasons

Copernicus' dissatisfaction with the Ptolemaic theory did not stem from a preconceived notion that the Sun, and not the Earth, must be at the center of our system. Rather, he was troubled for more subtle reasons. Like Aristotle, Copernicus felt that a satisfactory representation of the solar system should be coherent and physically plausible, not requiring a different construction for each phenomenon, as Ptolemy's system did. In his *Commentariolus* (*The Little Commentary*), Copernicus made his feelings about this quite clear: "It is as though, in his pictures, an artist were to bring together hands, feet, head, and other limbs from quite different models, each part being admirably drawn in itself, but without any common relation to a single body: since they would in no way match one another, the result would be a monster rather than a man." In other words, Copernicus had *aesthetic* objections to Ptolemy's system: to him it was an ugly theory. This is a very important point. It may seem strange, but a scientific theory, supposedly so "objective" in all its aspects, must also pass the fuzzy, subjective test of beauty or elegance to be completely acceptable. Scientists will seek alternatives to an ugly theory, even if it seems to account for the experimental or observational data. That scientific beauty is just as hard to define as beauty in any other sphere of our experience does not diminish its power as a discriminant. It was so important to Copernicus that it ultimately drove him to throw out 2,000 years of tradition, that of the stationary Earth.

105

Figure 7.1. Nicolaus Copernicus (1473–1543), a Pole who was the first to present a detailed description of a universe that had at its center the Sun and not the Earth. Though at first few accepted the idea that the Earth was not stationary and at the center of the universe, this model soon became enormously influential.

Purifying Ptolemy

In addition, Copernicus felt that Ptolemy, in his invention of the equant, was not playing according to the rules of the game. For him, as for Islamic astronomers before him, the equant violated the principle of uniform circular motion. He wrote in the *Commentariolus*:

> Yet the planetary theories of Ptolemy and most other astronomers, although consistent with the numerical data . . . present no small difficulty. For these theories were not adequate unless one also thought up certain equants: it then seemed that the planets moved with uniform velocity neither on their deferent circles nor around the centers of their epicycles. Such a system appears neither sufficiently absolute nor sufficiently attractive to the mind.
>
> Being aware of these defects, I spent much time considering whether one might perhaps find a more reasonable arrangement of circles, from which every apparent inequality could be calculated, and in which every element would move uniformly about its own center, as the rule of absolute motion requires.

During his studies, Copernicus had heard of Aristarchus' suggestion that the Sun was at the center of the solar system. How much this may have influenced him is not clear. In any case, in struggling with ways to overcome his objections to the Ptolemaic system, Copernicus found that with an Earth moving around the Sun he could construct a model that conceptually, at least, was physically coherent and did not require the equant.

The Copernican System

Assumptions

In 1512 Copernicus circulated in manuscript his *Commentariolus*, which gave a qualitative description of his ideas. His main assumptions can be given as follows:

(1) the Sun, not the Earth, is very nearly at the center of the universe;
(2) the Earth rotates on its axis once in approximately 24 hours. Along with the other planets, the Earth revolves around the Sun in a circular orbit;
(3) the distance from the Earth to the Sun must be very small compared to the distances of the stars from the Sun.

Features of the System

Several consequences of these assumptions follow immediately. The apparent daily motions of celestial objects result from the Earth's axial rotation, whereas the Sun's apparent yearly motion is in reality simply a reflection of the Earth's revolution around the Sun. Though these inferences are indeed correct, you must remember that they did great violence to the common sense of the time. Opponents claimed that if the Earth rotated to the east, then clouds, the wind, and projectiles would be left behind and always appear to move to the west. Furthermore, they argued, a rotating Earth would fly apart. To the first objection Copernicus could reply that everything, including the air, was carried along by the rotating Earth; to the second he could point out how much more difficult it would be for the celestial spheres to stay in one piece, since, being so much larger than the Earth, they must move more rapidly. But his critics could counter by reminding Copernicus that the celestial spheres were not made of ordinary matter, and so his argument lost its force.

A moving Earth, however, produced a straightforward explanation of retrograde motion. It is simply a consequence of the motions both of the planets *and* of the Earth. It occurs whenever the Earth laps (overtakes) an outer, more slowly moving planet as shown in Figure 7.2, or is itself lapped by a faster-moving inner planet. No special

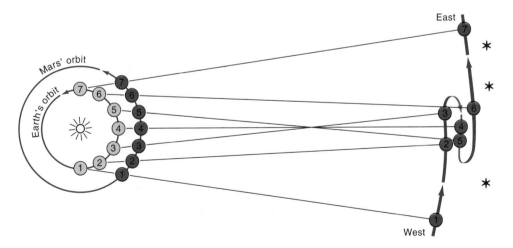

Figure 7.2. Retrograde motion according to Copernicus' Sun-centered system. The positions of Earth and Mars are given at monthly intervals. With the Earth just another planet in motion around the central Sun, retrograde motion is seen to be simply the consequence of the Earth overtaking the slower-moving Mars.

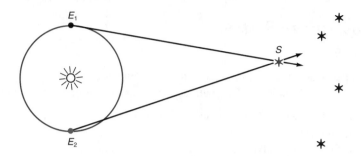

Figure 7.3. When the Earth is at E_1, the direction to the star S would be along E_1S; six months later the Earth is at E_2 and the direction to the star would be E_2S. This effect is called stellar parallax and is enormously exaggerated in the sketch (that is, the stars are so far away that rays of light reaching the Earth are essentially parallel). Stellar parallax is hopelessly beyond detection with the naked eye and was an argument against a moving Earth.

constructions, no epicycles are required. Retrograde motion is just a consequence of observing a moving planet from a moving Earth. Notice also how the increase in brightness of a planet during retrograde motion is accounted for.

The third assumption (that the stars are very far away) was necessary because, if the stars were nearby but at various distances, we would have a **parallax** effect: the directions of the nearer stars would appear to change with respect to the more distant stars, as the Earth revolved around the Sun (see Figure 7.3). We don't see this with our naked eye; so the stars must be very distant, which is indeed the true situation (remember the ping-pong-ball stars separated by hundreds of miles). Opponents of a moving Earth, however, turned the argument around, and said that since we don't see stars changing their directions, a Sun-centered system could not be correct. Note that, because of parallax, stars in the Ptolemaic system had to be far away as compared to the Earth's diameter; in Copernicus' system, however, they had to be far away as compared to the diameter of the Earth's orbit! The Copernican universe had to be much larger than Ptolemy's.

In Copernicus' system the maximum elongations of Mercury and Venus simply become consequences of the sizes of their orbits. Furthermore, from these elongations, one can easily calculate the sizes of the orbits of these two planets in terms of that of the Earth, as shown in Figure 7.4. That is, the scale of the inner portion of the solar

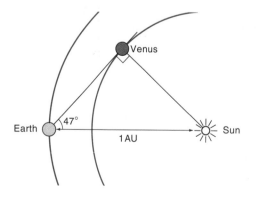

Figure 7.4. The scale (relative distances from the Sun) of the part of the planetary system closer to the Sun than the Earth is easy to establish. The maximum elongation of Venus, for example, is shown in the diagram. Recall that maximum elongation is the greatest angular distance from the Sun that Venus can achieve, about 47°. This means that the orbit of Venus must just touch the line Earth-Venus as shown. You can verify that the Sun-Venus distance is about 0.7 AU. A similar construction can be made for Mercury.

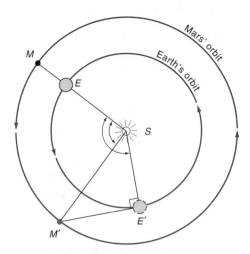

Figure 7.5. Measuring the scale of the outer part of the planetary system is only a bit more involved than finding the relative distances to Venus and Mercury. Begin with the planet (in this case, Mars) in opposition, that is, along the line *SEM*. About five months later Mars will be at quadrature; *SE'M'* form a right angle. Knowing the number of days it has taken Mars to go from opposition to quadrature along with the sidereal periods of Earth and Mars, the angles *ESE'* and *ESM'* can be calculated (*ESE'* is approximately the number of days times 1° per day). Subtracting the two angles gives the angle *E'SM'*, and so the shape of the triangle *E'SM'* is determined. The ratio of the distance of Mars, *SM'*, to the distance of the Earth, *SE'* (1 AU) follows.

system comes directly out of this model. In addition, it is only a little more difficult to calculate the relative sizes of the orbits of the outer planets (see Figure 7.5). Table 7.1 shows how successful Copernicus was in calculating the scale of the solar system.[1] The largest error (for Saturn) is only 4 percent.

This is a striking feature of Copernicus' universe. For the first time, the sizes of the planetary orbits are linked together in a necessary and unambiguous way. Recall that in the Ptolemaic scheme several different constructions were required for every planet, each model representing a different aspect of the planet's motion. With Copernicus, however, a single arrangement of all the planets enables all the orbital sizes to be calculated. In this sense it is a true system rather than just a collection of individual models.

Notice also that in the Copernican system Venus and Mercury should show nearly all phases (that is, look like crescents in some part of their orbits, and be full or nearly full in other parts). By contrast, in the geocentric system they could appear to be only either new or crescent.

This qualitative picture is appealing in its conceptual simplicity. It required, however, a basic, fundamental reorientation in looking at the world, one that seemed to violate ordinary common sense[2] and experience: namely, the Earth is not central and stationary, but is in constant motion. Then, just as now, nothing in our direct expe-

---

[1] Note that the Sun-planetary distances are only *relative*: Jupiter's distance from the Sun is 5.2 times that of the Earth's. To convert these distances to kilometers, we need to know the distance in kilometers of any planet from the Sun.

[2] Einstein is said to have remarked that common sense is the collection of prejudices we have acquired by age 16.

Table 7.1.   The scale of the solar system according to Copernicus

Mean orbital radius in terms of Earth's orbit		
Planet	Copernicus	Modern value
Mercury	0.376	0.387
Venus	0.719	0.723
Earth	1.00	1.00
Mars	1.52	1.52
Jupiter	5.22	5.20
Saturn	9.17	9.54

rience forces this notion on us. Furthermore, it is important to remember that Copernicus was unable to give any *proof* that the Earth in fact moved. Nothing in his theory forced anyone to believe the crazy idea that this solid, stable Earth was actually spinning its way around the Sun. Nevertheless, Copernicus' work was read with interest by some astronomers and even by a few churchmen, who urged him to publish a quantitative account of his model in which all the mathematical details would be worked out. This he did, but only about 30 years later. Tradition has it that he saw the first copy of his *De Revolutionibus Orbium Coelestium* (*On the Revolutions of the Celestial Spheres*) on his deathbed.

Problems with Copernicus' System

As willing as Copernicus was to adopt radical ideas, he could not break out of the conceptual scheme of uniform circular motion. Consequently, though he did not use the equant, he employed the techniques of deferents and epicycles, (As mentioned earlier, a practical reason for the persistence of uniform circular motion was that a circle is easy to work with, requiring only two points for its definition—one at the center and one on the circumference.)

In contrast to Ptolemy's system, epicycles were not fundamental for Copernicus; they were not needed, for example, to account for retrograde motion. Instead they were more in the nature of relatively small corrections to the basic model. Nonetheless, Copernicus' fully worked out model was very complicated, and required about as many circles as Ptolemy's. Furthermore, it did not predict planetary positions any more accurately than did the older system, because of Copernicus' continuing adherence to uniform circular motion. Also, being no observer himself, Copernicus (like some theoreticians even today!) accepted as fact all the observational data that came his way. He "explained" some motions that just don't exist, for example, that the inclination of the planets' orbital planes oscillated, or that the eccentricity of the Earth's orbit varied.

Late in his life he apparently realized that he had erred in several ways and that his system was, therefore, deficient. He also knew that he would be ridiculed for his idea that the Earth moved, having already been made fun of in a locally performed skit. Nonetheless, he firmly believed in his model's basic correctness. He would have completely disagreed with the sentiment expressed by his colleague, Andreas Osiander (a Lutheran clergyman), who was seeing *De Revolutionibus* through publication. Osiander wrote a preface (without Copernicus' knowledge), saying that the system should be taken simply as an aid in carrying out the mathematical calculations, rather than as representing the real, physical state of the world. (Incidentally, it was Kepler who, about 50 years later, first realized that Osiander and not Copernicus had written the preface.)

Reception of Copernicus' System

In a sense, Copernicus was not a revolutionary, but a conservative who attempted to cleanse the Ptolemaic system of its departures from Platonic ideals. Restoring the idea of uniform circular motion to its central position in physics was more important in his mind than assuming the centrality of the Earth in the physical universe. See how important the circular-motion assumption had become! Apparently Copernicus did not see himself as a revolutionary, nor did the Church immediately see his ideas as radical. In fact, Copernicus' book attracted little attention at first, probably because of its technical nature. *De Revolutionibus* was put on the Index of forbidden books only in 1616,

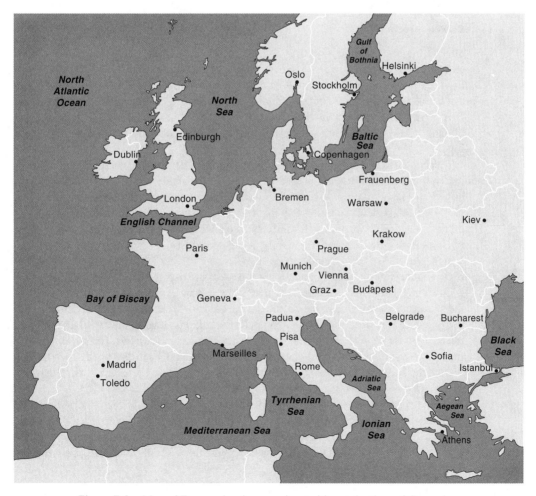

Figure 7.6. Map of Europe showing prominent cities at the time of Copernicus.

73 years after its publication, and then for only a dozen or so passages that referred to the moving Earth more as fact than as hypothesis. Nevertheless, it remained on the Index for two centuries. Nor were many Protestant churchmen any friendlier to the new system of Copernicus. Martin Luther made his feelings clear: "An upstart astrologer, this fool wishes to reverse the entire science of astronomy; but sacred Scripture tells us that Joshua commanded the Sun to stand still, and not the Earth." This comment was probably prompted also by Copernicus' affront to common sense.

Few scholars took Copernicus seriously at first, though his work was soon taught in many Lutheran universities; even most astronomers who used his more convenient planetary tables did not take as literally true the system on which they were based. It was not until about 1600 that his ideas began to be accepted by a few scholars, some of whom saw their disquieting implications. For example, if the stars' daily motions were simply a consequence of the Earth's rotation, why need they be fixed to a sphere? Why could they not be distributed throughout space at varying distances? Why could not the universe be infinite?—with the consequences for Aristotelian physics and Christian theology described previously in Chapter 6.

Tycho Brahe
Early Work

Several factors combined to bring about these changes in attitude. One was the work of Tycho Brahe (1546–1601—see Figure 7.7). A member of a prominent Danish family, Tycho was kidnapped, adopted, and educated by his childless rich uncle. Tycho was a colorful and irascible fellow who in his youth even fought a duel with another student over who was the better mathematician. Although it is not known if he upheld his "honor," we do know that as a consequence of the duel he had to wear a metallic nose for the rest of his days.

When he was 13, he witnessed a partial eclipse of the Sun, in itself not a particularly impressive event. Its predictability, however, made a profound impression on him. Later he wrote that it seemed to him "divine that men could know the motions of the celestial bodies so accurately that they could long before foretell their places and relative positions." Though his family had a career in law planned for him, his interest in astronomy persisted and was finally accepted by them.

In November 1572 he noticed a "new" star—a *nova stella*[3]—where one had not been visible before (see Figure 7.8). We now know that this sudden appearance of a bright "new star" resulted from an incredibly violent explosion, which blew most of the star apart. According to the physics of the day, however, this object, appearing suddenly, could not be in the unchanging celestial regions beyond the Moon. Many surmised it must be a slow-moving comet without a tail. However, Tycho (along with other astronomers of the day) could not detect any parallax of the nova, even though that of the Moon was easily measured; hence it must be at a greater distance than the Moon, in the supposedly constant region of the heavens.

Five years later he presented another bit of evidence that ultimately cast doubt on traditional views, when he showed that the bright comet of 1577 was at least six times as far away as the Moon. Interestingly, it was a full decade and many more comet observations later that Tycho himself realized that the crystalline spheres carrying the

---

[3] Hence our term **nova** for a star that quickly increases in brightness, sometimes by enormous factors, then slowly declines. This nova, appearing three months after the St. Bartholomew's Day massacre of French Protestants, was taken by some as an evil sign of it.

Figure 7.7. Tycho Brahe (1546–1601), a Danish nobleman who spent 20 years at his observatory on the island of Hven making the most accurate measurements possible without a telescope, of the positions of planets and stars. Without observations of such accuracy, Kepler would not have been able to discover the laws of planetary motion.

Figure 7.8. Bayer's atlas of the sky, published in 1603—two years after Tycho's death—was based on the latter's star catalog. The brightest object shown in the constellation of Cassiopeia is Tycho's nova of 1572, though it had long since faded to obscurity.

114

Figure 7.9. The Uraniborg, Tycho's observatory and residence on the 2,000-acre island of Hven. The square enclosure was about 78 meters on a side. The structure at the right corner of the square was the servants' quarters; that on the left housed Tycho's printing press.

planets could not be solid, otherwise they would be shattered as comets passed through the solar system. As we will see, this was important for the development of his model of the solar system.

His book *De Nova Stella* (*On the New Star*), which described his work on the nova, brought Tycho considerable fame and, in particular, the attention of King Frederick of Denmark. Frederick, wanting to keep Tycho for Denmark, gave him the feudal estate on the island of Hven, located between Copenhagen and Hamlet's Elsinore castle. Here he set up an observatory to be supported by rents and taxes from the forty or so farms located on the 2,000-acre island. It has been estimated that Tycho received about $1.5 million to support his twenty years of work on the island. Government support of research is nothing new. He lived on Hven in the grand manner customary to the nobility (see Figure 7.9). In addition to an elaborate residence and observatory, he set up a paper mill and printing press, a game preserve, fish ponds, and a prison for peasants foolish enough to object to his high-handed ways. Frederick's successor eventually did object to Tycho's manner, however, and in 1597 Tycho left Denmark to become the Imperial Mathematician to Emperor Rudolph II in Prague.

Tycho's Planetary Observations

Not having to suffer the consequences of his quirky personality, we can easily overlook it and concentrate on his observational work, which was superb. In the atmosphere of Aristotelian science, more qualitative than quantitative, a need for observations of the highest possible accuracy was not generally perceived. Furthermore, with the basic celestial geometry of circles, it was not felt necessary to make measurements of a planet's position over its entire orbit, because a circular orbit was defined by only two observations. Like some of his teachers, however, Tycho was not of this mind. Over a period of two decades, he built about two dozen instruments, many with significant improvements of his own design, and hired assistants to help operate them (Figure 7.10). He made observations as often as he could, and attempted to understand the sources and evaluate the magnitudes of the errors in his measurements. All of this sounds commonplace today, but Tycho was the first to carry out such a program over a long period of time and did so with incomparable skill and persistence. During his twenty years at Hven, he amassed a large amount of data on the positions of the planets and stars. His best planetary measurements had uncertainties of only one arc minute, as did the positions of the 777 stars in his catalog. Some of the reference stars against which the other stars' positions were determined were measured many times and had positional errors as small as 25 arc seconds, or about 1/60 the diameter of the Moon! In brief, these measurements were the best that can be made without a telescope.

Figure 7.10. The mural quadrant, perhaps the most famous of Tycho's instruments. The wall was accurately oriented along a north-south line and a brass plate, carefully marked off so that angles as small as one minute of arc could be read, was mounted on it. As a star crossed the meridian, its altitude was measured by sighting it through the slit in the outer wall.

Tycho's Attitude toward Copernicus

Tycho was aware of Copernicus' ideas, but did not believe them for a very good reason: he was unable to confirm one of their consequences, namely, the apparent change in the direction of stars as the Earth revolved around the Sun (see Figure 7.3). Tycho's observations indicated to him that the stars must be more than 3,700 times farther away than the Sun, which he, and nearly everyone else at that time, thought was absurd. How could the universe possibly be so enormous? The maximum change in direction to even the nearest star (other than the Sun) as observed from two opposite points in the Earth's orbit is only about 1.5 arc seconds. This is much smaller than the naked eye is able to detect, so Tycho's instruments were utterly incapable of measuring this so-called stellar parallax. In fact, about 250 years passed before it was first detected! Tycho simply could not accept the possibility that stars were so far away that his instruments could not measure their parallax. This also confirmed his distaste for the idea of a moving Earth. In addition, he was influenced by the medieval notion that God (or nature) does nothing in vain; if the stars were so very far away, what possible use could there be for all that space? None; therefore, there was no such space, the stars were relatively nearby, and the Earth did not move, but was at the center of the universe.

Tycho's System

Apparently Tycho felt that the Ptolemaic system was essentially correct but, like Copernicus, he found the equants objectionable and tried various schemes by which they could be eliminated. The path that he took in developing his system is complex and need not concern us. An important factor for him, however, was giving up on the idea of planets attached to solid spheres, since in his system he had to have Mars' orbit intersect the Sun's. In 1588 he published his mixed system, shown in Figure 7.11. It was Copernican in that the planets revolved around the Sun, but at the same time Aristotelian in that the Sun, with its planetary family, revolved around the stationary Earth.

Though it agreed with the observational evidence of the time, this model achieved only a modest popularity, and that primarily after Galileo's difficulties with Catholic authorities several decades later, because Tycho's system had all the geometric advantages of the Copernican system without its attendant theological pitfalls. Tycho recognized that he was not enough of a mathematician to work out the details of his model; he invited the young German astronomer-mathematician, Johannes Kepler, to join him in Prague and use his observational data to develop the new system.

Kepler

In many ways, the German Johannes Kepler (1571–1630) is one of the most fascinating characters of this period, standing as he does with one foot firmly planted in the medieval past, the other in the scientific future (see Figure 7.12).[4] He suffered from many of the miseries endemic to the times—poor health, disruptions to his life because of the devastating Thirty Years' War, and the plague of "witch" persecutions that infected Europe for 200 years. Kepler barely prevented his mother from being burned as a witch, as her sister had been before her. (The New World was not immune to this dis-

---

[4] Hence the title *The Watershed* for the separately published section on Kepler's life taken from Arthur Koestler's larger work, *The Sleepwalkers*. One need not accept Koestler's ideas of scientific discovery embodied in the latter title to enjoy thoroughly his well-written account of the lives of Copernicus, Tycho, Galileo, and Kepler.

Figure 7.11. In 1588 Tycho published a book on the comet of 1577, which he showed could not be within the lunar sphere, as Aristotelians maintained. In it he also described his mixed system of the world, shown here, in which the Sun revolved around the central and stationary Earth (Ptolemaic), but the other planets revolved around the Sun (Copernican). Note that Mars' sphere intersects that of the Sun. Tycho could not detect any stellar parallax as the Earth revolved around the Sun so, not knowing the actual distances of the stars, he believed the Earth to be motionless.

Figure 7.12. Johannes Kepler (1571–1630), the German astronomer-mathematician who used Tycho's observations to discover the three laws of planetary motion.

ease; recall the Salem witchcraft trials in 1692.) For much of his life he earned his living by teaching mathematics and astronomy, but since professors' earnings then depended on the number of students they could attract, Kepler, a poor lecturer, was usually in financial difficulties. He supplemented his income by casting horoscopes, thinking that astrology might be useful in describing the overall path a person's life would take, but little more. He kept score of his predictions and was not impressed by their accuracy.

Kepler's Attitudes and Motivations

Kepler was both a Copernican and a Pythagorean. Not only was he convinced that the Earth was just another planet revolving around the Sun, but he also wanted to know why there were six and only (!) six planets, why they were spaced as they were, and why they moved as they did. In this he was infected for his entire life by a species of numerology nearly as virulent as any that struck the Pythagoreans. During one of his lectures (at the Protestant school in Graz), he wrote, the answer came to him. There are five and only five regular solids (that is, solid figures having all faces of the same shape). These are the pyramid with four faces, each an equilateral triangle; the cube, each of its six faces a square; the octahedron, with eight equilateral triangles; the dodecahedron, with twelve pentagons; and the icosahedron, with twenty equilateral triangles (see Figure 7.13). His idea was that planetary orbits (which he took to be contained within thin spherical shells) must be inscribed within one of these figures, just touching its faces, or circumscribed around one of these solids, just touching its edges (see Figure 7.14). Kepler had the orbits encompassed by thin spherical shells because at this time he thought that orbits were nearly, but not exactly, circular. Thus, he thought that the five regular solids could account for both the number of the planets and their spacing. He associated this idea with the ideals of universal harmonies; it obsessed and guided his work for the rest of his life.

Today we attribute no particular significance to the number of planets in the solar system; it is simply an accident of the formation process. In Kepler's time, however, it was not obvious what a theory of the solar system should explain. That is, which of the many characteristics of celestial objects were fundamental and should be accounted for by theory? What needed explaining? Kepler thought that in addition to their number, the sizes and motions of the planets were significant. His attempt to account for the first was sterile and led nowhere, but his investigation of planetary motions helped to create modern science. Focusing one's attention on the key phenomena to be explained is a crucial first step in constructing a successful theory.[5]

Kepler worked long before the conventions governing the writing of scientific papers were established. He writes about his work with its false starts, hopes, failures, and triumphs in a very personal and detailed manner. By contrast, modern scientific papers are at best bland in style and sometimes brief to the point of near-unintelligibility, partly because of the enormous increase in the amount of scientific research, which forces the scientific journals to require papers to be as short as possible. To get any idea from reading a modern paper of how research is done you must know the shorthand: for example, "It can easily be shown that . . ." sometimes means that it took the author two months of head-wracking labor to prove the statement or equation that follows; or "for the purpose of this paper, we will assume the following . . ." often means that a more physically realistic problem was beyond the

---

[5] By comparison with a contemporary account that the number of planets was somehow connected with the number of "windows" on the face—two eyes, two ears, and two nostrils—Kepler's explanation seems rather sophisticated.

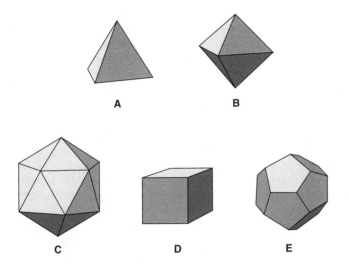

Figure 7.13. The Platonic solids. Only five three-dimensional solids can be constructed having all faces identical. Three are made from equilateral triangles. (A) the tetrahedron (four triangles), (B) the octahedron (eight triangles), and (C) the icosahedron (twenty triangles); the cube (D) is made from six squares and the dodecahedron (E) from twelve pentagons.

Figure 7.14. Kepler wondered why there were only six planets (Uranus, Neptune, and Pluto had not yet been discovered) and thought that the five Platonic solids gave the answer. Each one determined the spacing between an inscribed and circumscribed sphere that contained a planet's orbit; five spacings, six orbits! This figure is from his book *Mysterium Cosmographicum*, published in 1597.

author's abilities. This shorthand is probably necessary, but it's too bad that modern scientific papers have had most of life's juices squeezed out of them. In contrast, the actual scientist must be full of life; science is always done for *human* reasons.

Kepler's Insistence on a Physical Cause

Surprisingly, Kepler's idea that the planetary spacings might be determined by the regular solids works moderately well (see Table 7.2). Had he been entirely a man of his own time, he would have been satisfied to stop with this result, and would today scarcely merit a footnote in the history of astronomy. Instead, he insisted that his theory should agree with the observations to within their accuracy. He demanded quantitative, not just qualitative agreement. In giving expression to this attitude, he took the crucial step necessary for science. Nor was this just a theoretical idea that Kepler advocated for others; his vigorously applying it to his own work cost him months and years of the most tedious labor.

As a Copernican, Kepler realized that since the stellar sphere was not in rotation, the traditional driving force for the universe had disappeared. He showed his modernity once again by insisting that the physical cause for planetary motions must be associated with the Sun. This idea was at least partly the result of his discovery that the planets' orbital planes all intersect in the Sun. Copernicus, still under the spell of circular geometry, had put the center of his system at the center of the Earth's orbit, a geometric point. Kepler could not accept this and placed the center of his system in the Sun, since in some physical way, he thought, the Sun controlled the solar system. At the risk of oversimplification, it can be said that Copernicus' was a geometric system, but Kepler's goal was a physical system. Somehow, Kepler thought, the Sun exerted a continuously acting force that weakens with increasing distance from the Sun. That the force weakened was suggested, for example, by Jupiter: though its orbit is only about five times larger than Earth's, it requires nearly 12 years for one round trip; so it moves much more slowly than does the Earth. Magnetic phenomena, described at this time by Gilbert in England, were much in vogue, and Kepler speculated that perhaps the Sun swept a "magnetic broom" around the solar system, causing the planets to be pushed along in their orbits. Whatever the source of the force, Kepler was so convinced of its reality and its necessity that his 1609 book, in which he published his first two laws of planetary motion, was titled in part *A New Astronomy Based on Causation, or A Physics of the Sky*.

Table 7.2. Kepler's spacings of the planets using the Platonic solids[a]

Planets	Polyhedron	Ratio of radii of inscribed and circumscribed spheres (Kepler)	Ratio of mean orbital radii of neighboring planets (Copernicus[b])
Mercury/Venus	Octahedron	0.577	0.523
Venus/Earth	Icosahedron	0.795	0.714
Earth/Mars	Dodecahedron	0.795	0.658
Mars/Jupiter	Tetrahedron	0.333	0.291
Jupiter/Saturn	Cube	0.577	0.569

[a] This is a somewhat simplified version of what Kepler did, but it illustrates his idea.
[b] See Table 7.1.

Kepler's Work with Tycho's Data

From the beginning of their short relationship, Kepler and Tycho had difficulty in getting along. Their differences in social standing and character were even greater than their quarter-century gap in ages. Tycho was as stingy with his observational data as he was with Kepler's salary, and gave far too little of both for Kepler's taste. For his part, Kepler realized that Tycho's observations were unique and was anxious to get his hands on all of them. Kepler was not interested in Tycho's model, which he didn't believe was correct, but wanted to find his own celestial harmonies. So when, in 1601, Tycho died, Kepler quickly gathered up Tycho's data before his heirs could claim them.

In his work on Mars (suggested by Tycho as being crucial, because its orbit departs more from a circle than do the orbits of any of the outer planets), Kepler began with the assumption already noted: though the Sun is not at the geometric center of a planet's orbit, it exerts a controlling tangential force on the planet's orbital motion. Since the planet's motion is not centered on the Sun, however, but rather on the center of its orbit, Kepler thought that there must be another force acting on the planet as well. As a consequence of these notions, Kepler discarded the idea of uniform motion, since it seemed reasonable that the Sun's controlling force should be greater when nearer the planet; so the planet would move more rapidly when it was nearer the Sun than when it was more distant. After all, a planet far from the Sun like Jupiter moves more slowly than Mercury, which is near the Sun, so why couldn't the speed of an individual planet vary as its solar distance varied? Note how Kepler's ideas were guided by physical notions, and not just geometric ones.

Using Tycho's observations, Kepler also quickly exorcised one of Copernicus' demons, his oscillating planetary orbits, showing that the planes of the orbits remained fixed in space.

Though he gave up on uniform motion from the beginning, for a long time he kept the notion of circular orbits and even the equant. In fact, Kepler first set himself the problem of finding the position of the Sun and equant with respect to the center of Mars' circular orbit, the size of the orbit, and the orientation of the Sun-center-equant line with respect to the stars (Figure 7.15). Kepler had first thought this would be an

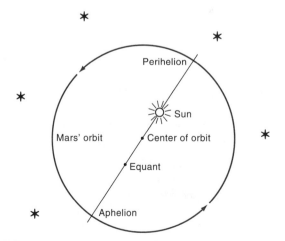

Figure 7.15. The quantities Kepler at first wanted to determine in the orbit of Mars: the position of the Sun and equant with respect to the center of the orbit and the orientation of the orbit in space.

easy task, but it dragged on for years. Finally, after more than 70 tedious attempts (made geometrically, with compass and ruler) he was able to fit four of Tycho's observations of Mars to the circular orbit within the completely unprecedented accuracy of 2 arc minutes. "Thou seest now, diligent reader, that the hypothesis based on this method not only satisfies the four positions on which it was based, but also correctly represents within 2 minutes, all the other observations . . . ," he wrote in his *New Astronomy*.

As an additional test, however, he took two more observations and found to his dismay that his orbit missed them by 8 arc minutes! Even though this was far better agreement than anyone had achieved before, Kepler knew that the observations were more accurate than 8 arc minutes, and felt that such a discrepancy was impermissible. Consequently, he concluded that his orbit must be wrong, and 900 pages of calculations must be thrown out. "Who would have thought it possible? This hypothesis, which so closely agrees with the observed oppositions, is nevertheless false. . . . If I had believed that we could ignore these 8 minutes, I would have patched up my hypothesis accordingly. But since it was not permissible to ignore them, those 8 minutes point the road to a complete reformation of astronomy: they have become the building material for a large part of this work." Why did Kepler insist on the primacy of observations? Probably because he was convinced that the universe was governed by physical causes and not only by geometric principles. It must be possible to find the true, physical system of the world, to discover just how the universe is built. It was not possible simply to tinker with his model, adjusting it slightly here, modifying it a bit there. For the system to reflect the observations in all their accuracy, a new approach *must* be taken.

Kepler's Laws of Planetary Motion

His new approach was radical; it was to give up—reluctantly—on circular orbits, thereby discarding both parts of the 2,000-year-old idea of uniform circular motion. But if the orbits were not circular, what shape were they? He first tried egg-shaped orbits—one end larger than the other—using ellipses as computing aids. After many attempts, he eventually found a mathematical formula that represented Mars' orbit well, but he did not recognize it as being that of an ellipse until after another year's work. He then wrote in *A New Astronomy*,

> Why should I mince my words? The truth of Nature, which I had rejected and chased away, returned by stealth through the back door, disguising itself to be accepted. That is to say, I laid [the original equation] aside, and fell back on ellipses, believing that this was quite a different hypothesis, whereas the two, as I shall prove in the next chapter, are one and the same. . . . I thought and searched, until I went nearly mad, for a reason why the planet preferred an elliptical orbit [to mine]. . . . . Ah, what a foolish bird I have been!

Thus, in 1609, after eight years of concentrated effort and persistence, he published the first two of the three laws that bear his name. These laws state the shape of planetary orbits—ellipses—and the speed with which planets move along them.

Ten years later, he published what he thought was the capstone of his life's work, a description of the harmonies pervading the world. The first two sections of his 1619 book, *The Harmony of the World*, described mathematical harmonies; the third, musical harmonies; the fourth, those in astrology; and the fifth, those in astronomy. For the latter he found, by trial and error, what he thought to be his greatest achievement:

that the ratios of the planets' maximum and minimum angular velocity as seen from the center of their orbits (not from the Sun) gave the musical harmonies. For example, for Saturn the ratio is 5/4. Buried in all this numerology was a relation between the length of time a planet takes to travel once around its orbit and the size of that orbit. This relation, which we call Kepler's third law (or, in recognition of its origin, his "harmonic" law) is what we find important, not his other harmonies.

We will first state Kepler's three laws of planetary motion briefly and then discuss each in some detail. The laws are:

(1) the orbits of the planets are ellipses, with the Sun at one focus (we will describe the properties of an ellipse and the definition of its focus shortly);
(2) the line joining the Sun and planet sweeps out equal areas in equal times;
(3) the square of a plant's period (that is, the time to complete one circuit of its orbit) is proportional to the cube of the semimajor axis[6] of its orbit; more succinctly, $P^2 \propto a^3$.

Finally, after two millennia, uniform circular motion was buried, and a completely new way of looking at the motions of planets was proposed. With his laws, Kepler could calculate planetary positions, especially that of Mars, to a much greater accuracy than had been possible with the Copernican system. Even so, many astronomers could not bring themselves to accept this "crazy" new system, and believed as the clergyman David Fabricius did, who wrote to Kepler: "With your ellipse, you abolish the uniformity and circularity of the movements, which seems to me more absurd than I am prepared to admit. . . . If you would simply retain uniform circular motion and justify your elliptical orbit by a new epicycle, I would regard your work as being of more value." Even Galileo, generally hospitable to new ideas (especially his own), continued to hold to uniform circular motion. He never even acknowledged Kepler's work in his own astronomical writings, much less accepted it.

In defense of Kepler's contemporaries, we should remember that, strictly speaking, Kepler, like Copernicus, *proved* nothing. Especially for those who could not accept the possibility of a moving Earth, or give up the guiding principle of uniform circular motion, his laws could be taken as just another conceptual scheme by which planetary positions could be calculated (albeit to an undreamed-of accuracy) rather than as representing physical reality, as Kepler was convinced they did. In any case, they confirmed brilliantly the old Greek assumption that the universe was orderly, and that this order could be discovered. We now see them as the first modern physical laws, precisely stated generalizations derived from and agreeing with observations, freed from restrictive, artificial notions such as uniform circular motion. Though Kepler could not describe them, he thought that somehow *forces* between the Sun and the planets governed the motions of the latter through space, not crystalline spheres. Thus his laws reject a large part of Aristotelian physics. They exemplify the two salient features of modern science: the faith that the universe is orderly and knowable, so that one can build theories; and the requirement that such theories be tested against and agree with nature.

Kepler's laws, distilled from observations, stood alone, however, and at this point were not part of any wider context, nor could they be seen as consequences of a powerful, general theory of motion and gravity. Kepler could not manage this final step toward modern science, the integration of his laws into a broad, physical theory. For this we must wait until Newton.

---

[6] The semimajor axis of a circle is its radius; see the next section.

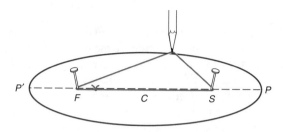

Figure 7.16. How to draw an ellipse: attach a string to two points, F and S, and put a pencil in the loop; keep the string tight and move the pencil about F and S.

Properties of Ellipses. Before considering Kepler's laws in more detail, we must first define a few quantities that characterize ellipses. **Ellipses** can be drawn by looping a string around two points, F and S (in Figure 7.16), putting a pencil in the loop of the string, and, keeping the string tight, moving the pencil about F and S. Since the length of the string doesn't change, the distance from F to any point on the ellipse plus the distance from S to that point is always the same. The two points F and S are called the **foci** of the ellipse; when an ellipse represents a planet's orbit, the Sun is at one of the foci. Ellipses have two axes of symmetry, the longer one of which (PP') is called the **major axis**, the shorter, the **minor axis**. Half of the major axis is called the **semimajor** axis, CP or CP', and is generally denoted by a. The **eccentricity**, e, of the ellipse is defined as $e = CS/CP$. Notice that as F and S get closer and closer to C, e becomes smaller and smaller, and the major and minor axes become nearly the same size. In the limit, when F and S are at C, $e = 0$, and the ellipse becomes a circle. In other words, a **circle** is just a special case of the more general ellipse. As F and S move away from the center and approach P' and P, respectively, the ellipse becomes increasingly elongated until, when S is at P (and F at P'), $e = 1$. The ellipse has become a straight line. Thus the *shape* of the ellipse is given by its eccentricity, e (between 0 and 1), but its *size* is given by the semimajor axis, a. It is easy to show that **aphelion**, the maximum distance from the Sun that a planet in an elliptical orbit can be, is $r_A = a(1 + e)$; similarly, **perihelion**, the smallest Sun-planet distance, is given by $r_P = a(1 - e)$. The average Sun-planet distance is just $(r_P + r_A)/2 = a$. Note that when $e = 0$ (so the orbit is a circle), $r_A = r_P = a$.

Synodic and Sidereal Period. We must deal with two kinds of periods of planetary revolution, the sidereal period and the synodic period. The Copernican system, treating the Earth as just another planet revolving around the Sun, requires both. We encountered these two kinds of periods in Chapter 3, but they are important enough to warrant additional explanation. The **sidereal period** is simply the length of time a planet takes to go once around the Sun as seen against the stars. It is the period with respect to the stars, which are so far away that they provide a fixed reference frame against which to measure the motion. But we make observations of moving planets from a moving Earth, not from the distant stars. What we in fact measure directly is the **synodic period**, the length of time required for the same arrangement (**configuration** is the astronomical term) of Sun, planet, and Earth to repeat itself.

Consider the Earth, a planet, and the Sun all lined up at E, P, and S, respectively (Figure 7.17). The stars are "fixed" in the figure. The planet P is closer to the Sun than is the Earth, so it travels in its orbit more rapidly than does the Earth. It makes one complete trip around its orbit back to P (in a time equal to its sidereal period) while

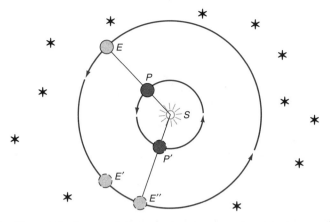

Figure 7.17. Planet *P* is shown at inferior conjunction (*EPS* in a straight line); after one sidereal year the planet has returned to position *P*, and the Earth has traveled only to *E'*. By the time the planet has moved to *P'*, Earth has traveled to *E''*, when the planet is again at inferior conjunction. The time required between successive configurations (in this case from one inferior conjunction to the next) is the planet's synodic period.

the Earth has moved only to *E'*. For the same straight-line configuration of planet-Earth-Sun to repeat itself, *P* must "catch up" with the Earth, which it will do at *P'* by the time the Earth is at *E''*. We call the time required for the planet to go from one such configuration at *P* to the next same configuration at *P'* the synodic period of *P* as seen from the Earth, or, simply, the synodic period of *P*. A little thought will convince you that this is also the apparent period of revolution of the planet with respect to the Sun as seen from the Earth. (In astronomical usage **revolution** refers to motion around an external body, whereas **rotation** refers to turning about an axis fixed within the body itself; the Earth *revolves* around the Sun in a year, but *rotates* once every 24 hours.) With an inner planet as shown in Figure 7.17, the synodic period of *P* is longer than its sidereal period, but shorter than the Earth's sidereal period.

For a slower-moving planet beyond the Earth (Mars, for example), it is the Earth that catches up to it and repeats the same configuration that prevailed at the starting point. The synodic periods of planets farther and farther out from the Sun than the Earth progressively approach the sidereal period of the Earth, that is, one year. (Do you understand why this is so? Remember, the more distant a planet is from the Sun, the more slowly it moves in its orbit around the Sun.)

We measure the synodic period of Mars, but we want to find the more fundamental quantity, its sidereal period. It turns out to be easy to derive it. Consider: in one day, the Earth goes 1/365 of its way around the Sun, while Mars goes $1/P_{\text{sidereal}}$ of its way. Therefore, each day the Earth gains on Mars by $1/365 - 1/P_{\text{sidereal}}$. But if P_{synodic} is the number of days required for the Earth to catch up with Mars, then $1/P_{\text{synodic}}$ is just the amount of catch-up each day. Therefore,

$$\frac{1}{365} - \frac{1}{P_{\text{sidereal}}} = \frac{1}{P_{\text{synodic}}}$$

or, solving for $1/P_{\text{sidereal}}$,

$$\frac{1}{P_{\text{sidereal}}} = \frac{1}{365} - \frac{1}{P_{\text{synodic}}}.$$

Now, the synodic period of Mars is 780 days, so

$$\frac{1}{P_{\text{sidereal}}} = \frac{1}{365} - \frac{1}{780}$$

or

$$\frac{1}{P_{\text{sidereal}}} = 0.00274 - 0.00128 = 0.00146 ,$$

and

$$P_{\text{sidereal}} = \frac{1}{0.00146} = 685 \text{ days .}^{[7]}$$

If the periods are given in Earth years instead of days, the expression is even simpler:

$$\frac{1}{P_{\text{sidereal}}} = 1 - \frac{1}{P_{\text{synodic}}} .$$

For Mercury and Venus, which move more rapidly than the Earth, the formula becomes

$$\frac{1}{P_{\text{sidereal}}} = \frac{1}{365} + \frac{1}{P_{\text{synodic}}} ;$$

or in years,

$$\frac{1}{P_{\text{sidereal}}} = 1 + \frac{1}{P_{\text{synodic}}} .$$

Kepler's First Law.   How did Kepler discover that Mars' orbit is an ellipse? First, Kepler used Tycho's observations of Mars to find its synodic period. Then, using the formula just given, he calculated Mars' sidereal period, 687 days. Next, suppose that when the Earth is at some arbitrary point in its orbit, say E (see Figure 7.18), the direction to Mars is given by the arrow. At this point all Kepler knew was that Mars was somewhere along the direction of the arrow, but he did not know just where. Now after 687 days, that is, one sidereal period of Mars later, Mars will be back at exactly the same place in its orbit. The Earth, however, will be at E', having made not quite two trips around the Sun in that time. From the Earth at E', Mars appears to be in the direction given by the second arrow. Clearly Mars must be at the intersection of the two arrows, and so Kepler knew one point on the orbit of Mars. By repeating this process for several more pairs of observations separated by 687 days, he built up Mars' orbit. (See the importance of long series of observations!) Thus he knew the size and shape of Mars' orbit in terms of the Earth's. The shape of the latter orbit, nearly circular, he had calculated from analysis of observations of the apparent motion of the Sun with respect to the stars. He then showed that Mars' orbit (and by inference that of other planets) was an ellipse and not a circle. So Kepler's first law states that planetary orbits are ellipses, with the Sun at one focus of each ellipse, and not at the center as perhaps might have been expected.

Kepler's Second Law.   The second law describes how a planet moves around its orbit, namely: the straight line joining a planet and the Sun sweeps over equal areas in

---

[7] Mars' sidereal period is actually 687 days. The difference of 0.3 percent is caused by round-off errors in the input values.

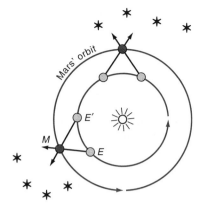

Figure 7.18. Kepler's method for calculating the orbit of Mars. When the Earth is at E, Mars is in the direction EM. After 687 days (one sidereal period), Mars is back at M, but the Earth is at E', and Mars is in the direction $E'M$. Mars must be at the intersection of the two arrows EM and $E'M$. With many pairs of observations separated by the sidereal period, the orbit of Mars can be defined.

equal intervals of time. In moving from P to P' in a month, say, the line from the Sun to the planet sweeps out the shaded area in Figure 7.19; in another month it moves from Q to Q' in such a way that the second shaded area equals the first. As the planet moves closer to the Sun, it travels faster. It moves fastest at P and slowest at A. The planet's motion is not only noncircular, it is nonuniform as well! With this law it is a relatively simple matter to calculate the position of a planet at any time. In fact, Kepler himself used it to extrapolate back 1,600 years (!) to investigate the possibility that the Christmas "star" might have been two or more planets in conjunction, that is, in very nearly the same direction in the sky. He suggested that a conjunction of Jupiter and Saturn that took place in 7 B.C.E. might have been the "star."

Kepler's Third Law. Kepler's third law, which gives the relation between the time, P, it takes a planet to go around an orbit of a given size, a, can be written as $P^2 = ka^3$. Let's write this for two planets, denoted by 1 and 2, and divide the expression for planet 1 by that for planet 2:

$$\frac{P_1^2}{P_2^2} = \frac{a_1^3}{a_2^3}.$$

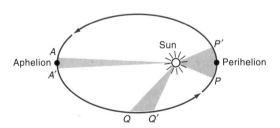

Figure 7.19. If the Sun-planet line sweeps out a given area in say, one month, Kepler showed that during one month in any other part of its orbit the Sun-planet line will sweep out the same area. As the planet gets closer to the Sun, the faster it moves; it travels only from A to A' at aphelion, but moves from P to P' at perihelion.

Let's take the Earth for planet 2, so that $P_2 = 1$ year and $a_2 = 1$ Astronomical Unit (recall that the AU is the average Earth-Sun distance). Then

$$P^2(\text{years}) = a^3(\text{AU}) \, .$$

We know that the sidereal period of Mars is 687 days or 1.88 years; it is easy to calculate the semimajor axis of its orbit:

$$P^2 = (1.88)^2 = 3.54 = a^3 \, ,$$

so

$$a = (3.54)^{1/3} = 1.52 \text{ AU}$$

(because $1.52 \times 1.52 \times 1.52 = 3.54$). Note that this law gives the *relative* sizes of the orbits, but not their *absolute* sizes (in miles or kilometers, for example). We just found that for Mars, $a = 1.52$ AU, or 1.52 times the semimajor axis of the Earth's orbit. Finding the absolute size of the Earth-Sun distance, that is, the length of the Astronomical Unit, is a separate problem, which was solved with reasonable accuracy only decades later. Notice also that the orbital period depends only on the *size* of the orbit and not on its shape. As long as two orbits have semimajor axes of equal length, they will have the same orbital periods, even if one is a circle and the other is cigar-shaped!

The Beauty of Kepler's Laws.   Kepler's laws are simple, easy to use, and equally applicable to all the planets in our solar system. Imagine the difficulties in computing the positions of planets using deferents and epicycles. Also, remember that several different constructions are required for each planet in the Ptolemaic system. Compare that with using Kepler's laws, which apply equally to all the planets. Here are good examples of ugliness and elegance in science. Even if the Ptolemaic system were capable of the same accuracy as Kepler's, who today would hesitate to prefer the latter?

Galileo and the Telescope
Emergence of the Telescope

The Ptolemaic universe was dealt another serious blow by technology, by the invention of that most important of astronomical instruments, the telescope. Its origins are somewhat obscure. Magnifying glasses and then eyeglasses appeared by the end of the thirteenth century, developed largely to help the elderly overcome farsightedness. By 1500 eyeglasses with concave and convex lenses[8] were in common use, but it was another century before lenses strong enough for use in a telescope were available. In the fall of 1608 several people in Holland claimed the invention of an optical device that made distant objects appear nearer. Within a few months, not only had the news of this invention spread, but the gadgets themselves were for sale in Holland, Germany, and France.

By the spring of 1609, the Italian Galileo Galilei (1564–1642)[9] was convinced that the rumors concerning this device were based on fact and should be taken seriously (see Figure 7.20). He discovered the optical principles involved, taught himself how

---

[8] The cross-sectional shape of the lens, thicker at the middle than at the edge, apparently reminded people of a lentil bean (*lens* in Latin); hence the name.

[9] Note that Galileo was born in the same year as Shakespeare, and died in the year of Newton's birth.

Figure 7.20. Galileo Galilei (1564–1642), an Italian, whose work on the physics of motion made him the founder of experimental physics. He was the first person to use a telescope to systematically observe the sky, and his discoveries, published in 1610 in the *Sidereal Messenger*, created a sensation throughout Europe.

to grind and polish lenses, and was soon making the best instruments in Europe. By the end of the year he turned his most powerful telescope toward the sky. Thus began a new era not only in astronomy but in all of science. For the first time in history, the range of our physical senses was extended. No longer were we limited by what we can perceive directly with our own eyes. New worlds were quickly revealed, first in the sky by the telescope, and a few years later in a drop of water, when the microscope was developed. It is nearly impossible for us to conceive of living in a world without such instruments and the knowledge they have brought us; imagine their impact on a thoughtful person in the seventeenth century.

Incidentally, Galileo's construction of a telescope is a good example of how once it is known that a particular gadget has been invented, that it can be made, it will be quickly duplicated elsewhere. That is, the knowledge of the mere existence of a device, be it a better mousetrap or an atomic bomb, is often its most important "secret," not the details of its construction.

Galileo was well prepared for a career in science. His father, a cultured man who wrote about and composed music, some of which is occasionally played today, saw to it that his son received a good education. The young Galileo studied mathematics and mechanics at the university in his home town of Pisa, in central Italy, and on his own. He spent most of his life as a mathematician at the University of Padua (near Venice) and as mathematician to the grand duke of Tuscany. Galileo was primarily what we today would call a physicist. In fact, he is considered the founder of experimental physics. His astronomical work was secondary to his main interests. Though educated in the Aristotelian tradition and trained in Ptolemaic astronomy, he eventually became a confirmed Copernican, but not a Keplerian; he still believed in uniform circular motion. For Galileo, the central problem in astronomy was to prove that the Earth really moved, to show that a Sun-centered system was physically real, not just a model convenient for calculating planetary positions.

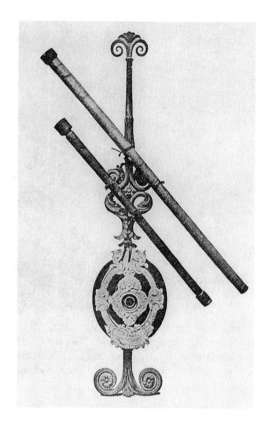

Figure 7.21. Galileo's telescopes. His best instrument had an objective lens about one inch in diameter and a magnifying power of about 30 times.

Though Galileo was not the very first person to use a telescope to study the heavens, he was the first to do so systematically and to publish his results.[10] His fourth telescope (Figure 7.21), with its one-inch-diameter lens and a magnifying power of about 30 times, was surely one of the most important telescopes in the history of astronomy. (Probably its closest competitors are Herschel's 18-inch and the Mount Wilson 100-inch telescopes, both of which we will hear more about later.)

In the spring of 1610 he published his observations in a pamphlet, *Sidereus Nuncius (Star Messenger)*, which quickly created a sensation in Europe. It was written in Italian rather than Latin and was not a technical treatise; hence it could be read and understood by large numbers of people. Understandably, its impact on the popular mind was far greater than that of Kepler's work. Within weeks news of his discoveries had reached much of western Europe. By autumn independent verification of some of his observations had been made. The telescope, this small bit of technology, revolutionized astronomy, eventually expanding the observable universe to extents far beyond the comprehension and even the imagination of Galileo and his contemporaries. Galileo's small book made telescopes and astronomy fashionable among the reading public and sparked the beginning of what we now call amateur astronomy, perhaps the first scientific hobby.

Galileo's Telescopic Discoveries

It is not hard to understand why the *Sidereus Nuncius* produced such an uproar. New worlds were made visible, worlds never even dreamed of. Consider the telescopic observations that Galileo described (see Figure 7.22). The Moon had craters, valleys, and

---

[10] The Englishman Thomas Harriot had carefully examined the Moon with a telescope in the summer of 1609. He did not publish his lunar maps, however.

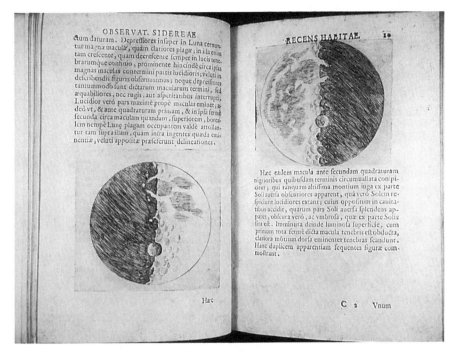

Figure 7.22. Two of Galileo's sketches of the Moon, showing craters and other rough surface features. These observations made it difficult to claim—as did Aristotle's cosmology—that as a heavenly body, the Moon was a perfectly smooth sphere.

mountains, the heights of which Galileo calculated from the lengths of the shadows they cast. The Moon's "landscape" had a distinctly Earth-like appearance, and was not that of a smooth, uniform, "perfect" celestial object! He even compared a particular lunar feature with a region in Germany, making it clear that the Moon's appearance did not differ fundamentally from that of the Earth. Furthermore, Saturn had "ears" (Galileo's small telescope could not resolve Saturn's rings clearly), so evidently celestial objects were not always spherical, the perfect shape.

He also showed from the geometry of the situation that a portion of the Moon that would otherwise be dark was faintly illuminated by sunlight reflected from the Earth, by Earthshine. For example, when the Moon is new, the telescope shows that it is not completely dark. This indicated that the Earth shines like the Moon, as perhaps do the other planets.[11]

Most astonishing, Jupiter had four moons, which were not left behind as Jupiter moved, but continued to revolve around the planet (Figure 7.23). Galileo could not explain why these satellites continued to orbit a moving Jupiter, but the fact was that they did. This countered the argument that a moving Earth would quickly lose its Moon. It allowed the possibility of many centers of motion, not just one—the Earth— as Aristotelian theory would have it.

Countless stars became visible through his telescope. Where one was seen before, ten or a hundred were visible now. Galileo saw that the Milky Way itself consisted of a myriad of stars that blended together to form the faint band of light we see with the naked eye. This indicated clearly that the ancients did not know everything. Further-

---

[11] At this time it was not clear if the planets were self-luminous or not.

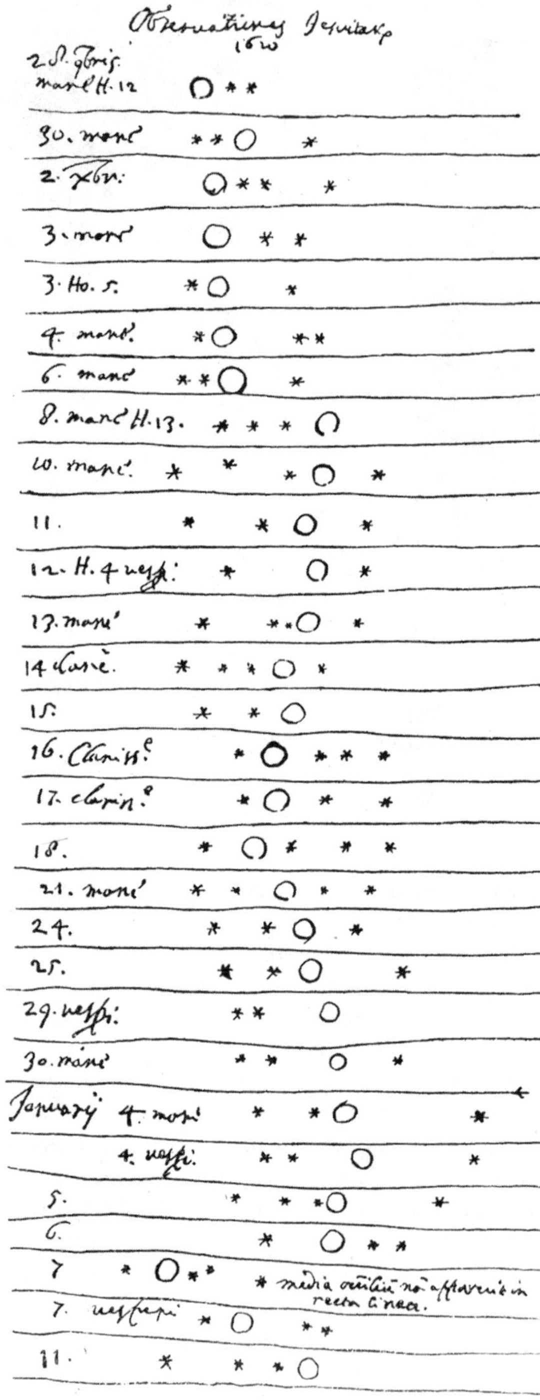

Figure 7.23. These sketches of the positions of Jupiter's satellites, sometimes attributed to Galileo, were actually made by Jesuits in Rome and later copied by Galileo. The first sketch was made on the morning of November 28, 1610, and the last shown was on January 11, 1611. Only eight months after Galileo's pamphlet was published, one of its most significant discoveries was confirmed.

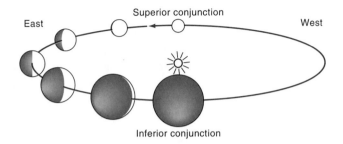

Figure 7.24. Venus' orbit is within the Earth's; so the sketch shows how Venus appears to us during half its orbit. Note the phases and the substantial change in its apparent diameter. We see a full Venus when it is at superior conjunction, but it is not brightest then, because it is at its maximum distance from Earth, so its apparent diameter is smallest. Venus is brightest about 36 days before and after inferior conjunction.

more, if, as contemporary Christianity taught, the world was created just for us, how could you explain objects seen only by means of telescopes? Planets' diameters were magnified by Galileo's telescope, but stars were not, so perhaps they were fundamentally different. Perhaps Copernicus was right: that stars showed no parallax because they were so far away.

Remember that the naked eye cannot detect any of the phases of Venus. The telescope enables their observation to be made, however, and in 1612 Galileo announced that Venus showed full and gibbous as well as new and crescent phases (Figure 7.24); so it, like the Moon, must be moving around the Sun and shining by reflected light. The Ptolemaic model allowed new and crescent phases only; so here, finally, was direct proof that this 1,500-year-old system was wrong. This did not prove, however, that Copernicus was right; Tycho's model also predicted the observed phases. Nonetheless, this discovery strengthened Galileo's pre-telescopic belief that Copernicus' system was the correct one. Also, he showed that even the Sun's surface was not "perfect," but was blemished by spots. In addition, Galileo used these spots to show that the Sun rotated on its axis, and in the same sense as the rest of the solar system, just as Copernicus claimed the Earth did.

The Reception of Galileo's Discoveries

These sensational discoveries met with a mixed reception. Jesuit astronomers confirmed what Galileo had reported in the *Sidereus Nuncius* and honored him. Poems were written in praise of Galileo and his discoveries. These discoveries were so newsworthy that when ambassadors in Italy first heard of them they wrote to their sovereigns. Some people, however, simply refused to believe their eyes. This is not too surprising, perhaps, since these early telescopes were of poor quality and difficult to use. They were hand-held and had a small field of view, so that it was no mean trick simply to find an object, much less to keep the telescope pointing at it. Furthermore, few people understood the physical principles by which telescopes worked, and thought they might produce optical illusions. Perhaps all these wonders were produced by the telescope itself. Others applied the old medieval argument that since we can't see these objects with our naked eyes, they can have no influence on us; so they must be useless, and, therefore, don't exist.

Kepler had no telescope of his own, but nevertheless was convinced that Galileo's observations were correct, and wrote a short pamphlet to that effect. A few months later, Kepler was able to borrow a telescope for a short time and confirmed the motions of Jupiter's moons. The published description of his observations was the first public confirmation of Galileo's discovery and, coming from such a well-known source, helped to turn the tide in Galileo's favor. (It was also the first public use of **satellite**, from a Latin word meaning "attendant," to describe astronomical moons.)

Galileo's Difficulties with the Church

Perhaps as a consequence of his astounding discoveries as well as because of his argumentative and aggressive personality, Galileo offended many of his contemporaries. Today he often seems less than generous in his treatment of his opponents. He was a talented and clever polemicist, and rather than simply proving his opponents wrong, he delighted in making them appear ridiculous as well.

I mention this to suggest that his difficulties with the Catholic Church need not have occurred. The Church's position was that since there was no real proof that Copernicus was right, his system could be taught as a hypothesis, but not as actual fact. After he was told to do so in 1616, Galileo followed this restriction for several years. In 1623 Cardinal Barberini, a cultivated man, became Pope Urban VIII, and received Galileo on several occasions. This emboldened Galileo to publish in 1632 a *Dialog Concerning the Two Chief World Systems* between a supporter of Copernicus and an advocate of Ptolemy's system. In this work, he presented the Copernican system as fact, citing his incorrect theory of tides as proof of the Earth's motion. Worse, it was clear that Galileo associated the not-too-bright proponent of the Ptolemaic system in the dialog with the Pope's position on the matter. The book was clearly written in Italian instead of Latin (then the language of scholarship), which meant that many people were reading it. A scientific work written in everyday language was distasteful to seventeenth-century scholars, just as scientists today look with disfavor on one of their number who pushes unfashionable or discredited ideas in the popular press. Remember, too, that at this time the Church was under attack by the ideas of the Protestant Reformation; it felt its authority had to be maintained. Yet here was Galileo insisting that the "Book of Nature" could be made to yield truths about the world that called into question the truth of the Book of Scripture. Consequently, Galileo was brought before the Inquisition, made to recant, and spent the rest of his days under house arrest.

Neither side behaved very well in this business. Galileo tricked the censor into allowing his book to be published. For its part, the Church forged a document it used in the trial. Galileo may look like a "martyr to science" to us now, but in Galileo's own time (when the notions of freedom of speech and expression were generally suspect), the situation was not so clear.

Ironically, in consistently ignoring Kepler's laws of motion and their success in predicting planetary positions, Galileo turned his back on the nearest thing to a proof that the Earth moved that was then available. This, however, would have required him to give up Copernican circular motion, which he was not willing to do, even though he could not give any physical arguments in support of circular orbits. Galileo may also have found Kepler's fuzzy, even mystical, ideas of the universal harmonies not to his taste and so discounted the planetary laws along with the mysticism.

Incidentally, the first direct proof of the Earth's revolution around the Sun did not come until 1729 (Figure 7.25); that of the Earth's rotation on its axis not until 1851 (Figure 7.26).

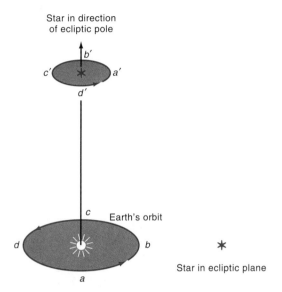

Figure 7.25. In attempting to detect stellar parallax, the English astronomer Bradley found a large annual displacement of the position of α Draconis. These shifts were not in the directions required by parallax, however, and Bradley realized that they were a consequence of the motion of the Earth and the finite velocity of light. A windless rain provides a familiar analogy. If you stand still, the drops fall straight down, but if you walk, they seem to be coming toward you at an angle; the faster you walk, the greater the angle. Similarly, the Earth's motion makes a star's position appear to be displaced in the direction of that motion. A star at the ecliptic pole would appear to move in a circle over the course of the year. When the Earth was at *a*, the star would appear to be at *a'*; when it was at *b*, the star would appear to be at *b'*, etc. The radius of the circle (called the constant of aberration), about 20.5 arc seconds, is simply the ratio of the Earth's velocity to that of light, so it is the same for all stars, and is much greater than even the largest stellar parallax. You should convince yourself that a star in the ecliptic plane would appear to move in a straight line, and one between the pole and plane of the ecliptic in an ellipse with semimajor axis equal to 20.5 arc seconds.

The Old World Starts to Crumble

We have seen that Kepler's laws, by eliminating the assumption of uniform circular motion, gave a powerful assist to the idea of Earth moving around a central Sun. Galileo's telescope showed that the Ptolemaic system must be incorrect, and that much of classical astronomy must be wrong as well. With the sharp distinction between Earth and the heavens gone, so also disappeared the connections between the physical universe and the social order. Where then was the model of perfection to which we should aspire? What happened to the "natural" relation between prince and subject, father and son? That the old world was disintegrating was painfully clear to people like the English cleric and poet John Donne (1572–1631), who wrote in 1611, in *The First Anniversary*,

> And new philosophy calls all in doubt,
> The Element of fire is quite put out;
> The Sun is lost, and the Earth, and no man's wit
> Can direct him where to look for it.

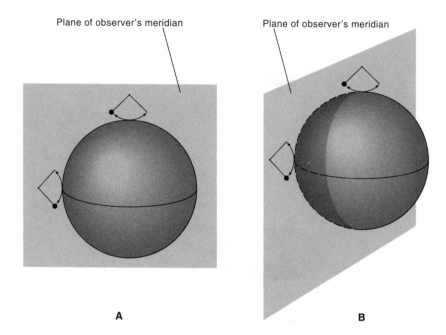

Plane of observer's meridian

Plane of observer's meridian

A

B

Figure 7.26. The French physicist, Foucault, set up a long pendulum in the Pantheon in Paris. A pin in the bottom of the weight made a mark in sand each time the pendulum swung. It soon became clear that its apparent direction of swing was changing; actually, the Earth was rotating underneath the pendulum. The effect is easily visualized when the pendulum is at the pole and the Earth rotates beneath it at 15° per hour. At (A) the observer's meridian and the plane of the pendulum's swing are in the plane of the paper. Several hours later, when the observer's meridian has rotated to (B), the pendulum at the pole is still swinging in its original plane. At the equator, however, the plane of the pendulum's swing always moves with the meridian, no motion is detected, and the experiment fails. In between the pole and equator, the length of time for one rotation is greater than 24 hours. At Chicago, for example, it would take a little more than 36 hours.

> And fully men confess that this world's spent,
> When in the Planets, and the firmament
> They seek so many new; then see that this
> Is crumbled out again to his Atomies.
> 'Tis all in pieces, all coherence gone;
> All just supply, and all Relation.

René Descartes (1596–1650), the French soldier of fortune and philosopher, is a good example of the response to this new world, to the new spirit. He wanted to throw out everything we thought we knew, begin afresh, and ask, "What do we really know?" The only certainty was doubt. Since so much of what our senses told us turned out to be incorrect, he put little trust in them and made a clear distinction between man and nature. For Descartes it was the awareness of ourselves and not of the world around us that served as his starting point: "I think; therefore, I am." I know I exist because I know I am thinking—not looking at something outside of myself, but thinking. Pure reasoning would enable him to build not only a new philosophy, he maintained, but a new all-encompassing theory of physics in which all the modern ideas, including planetary motions, would find a natural home. Though it is forgotten now, his physical sys-

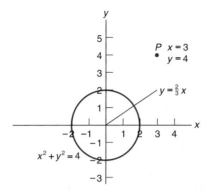

Figure 7.27. An example of a Cartesian coordinate system. Point *P* has the coordinates *x* = 3 and *y* = 4; the straight line is represented by *y* = 2/3*x*; the circle by the equation $x^2 + y^2 = 4$. Geometric constructions are converted to expressions that can then be manipulated by algebraic means.

tem of vortices, by which planets supposedly moved, so dominated much of European thought for most of the seventeenth century that Newton went to great pains to prove it wrong.

Descartes has had a longer influence through his mathematical ideas. In fact, he felt that only mathematics gave security against doubt. In particular, he showed how geometric figures could be represented and manipulated algebraically, inventing what we call analytic geometry. We still speak of Cartesian coordinates (see Figure 7.27).

He also formulated the concept of **momentum**, the product of an object's mass and velocity. Imagine two billiard balls striking each other. Before the impact, each had a certain momentum (which is a vector quantity, since velocity has direction as well as magnitude); after the impact, the balls go off with different velocities in different directions. Descartes asserted that the total momentum before the impact must equal the total afterward; that is, the total momentum is unchanged or *conserved* in the encounter. This is always true unless there is an outside force acting on the billiard balls. (Strictly speaking, there is such a force in our example—friction between the ball and the billiard table. This is relatively small, however, so we will ignore it.)

Note that nothing is said about the details of the interaction. All that is asserted is that the total momentum doesn't change. This is a very powerful concept. It enables you to understand what goes on without having to consider any of the particulars of the situation. This is just one example of many so-called **conservation laws** in physics. Conservation of energy is perhaps a more familiar concept. A ball is dropped from a height. Initially its velocity is zero, so its energy of motion is zero, but it has a certain potential energy by virtue of its height above the ground. Partway to the ground it has energy of motion, but less potential energy, since it is not as high up as it was initially. Potential energy has been converted to energy of motion, but the total energy remains the same; energy is conserved (again in the absence of any outside influences). We will use these ideas many times throughout this book. As we shall see in the next chapter, Newton made use of the idea of momentum.

It was Newton who succeeded in creating the new physics that satisfied the requirements of modern knowledge. In doing this so successfully, he has had, for the last 250 years, an influence and authority comparable only to Aristotle's in earlier days. We will examine Newton's ideas in the next chapter.

138

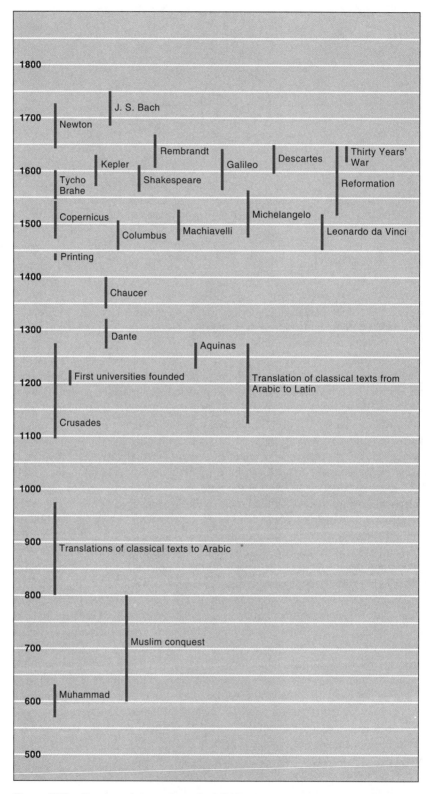

Figure 7.28. People and events from the Middle Ages to the time of Newton.

Terms to Know

Sidereal and synodic periods; ellipse; focus; major and minor axes; semimajor axis; eccentricity; aphelion and perihelion; configuration; momentum.

Ideas to Understand

Copernicus' reasons for proposing a heliocentric system; how it explains daily and yearly motions, retrograde motion, maximum elongations of Mercury and Venus; its predictions about the phases of Mercury and Venus, its virtues and defects. Tycho's observations that cast doubt on the Ptolemaic system; why Tycho rejected Copernicus' system; why his planetary observations were so important. Kepler's motivations; his laws and how they accounted for the observed motions of the planets; the old and the new in Kepler's attitude toward his work. Galileo's telescopic observations, what they showed; the response to the "new" astronomy. Conservation of momentum.

Questions

1. (a) Why did most astronomers at first reject Copernicus' planetary theory?

(b) In what ways were Copernicus' astronomical ideas both conservative and revolutionary?

(c) Explain why Copernicus' universe had to be much larger than Ptolemy's.

2. (a) What was Kepler trying to achieve with his Platonic solids? How did this approach relate to Greek attempts to understand the cosmos?

(b) In what sense can Kepler be thought of as having one foot in the medieval world and one foot in the modern world? Illustrate your answer with examples.

3. (a) Make a sketch of the Sun-centered solar system, showing the central Sun and the orbits of Venus and the Earth. What phases does Venus show? At what phase would it appear to be smallest? largest? Could Venus show a quarter phase?

(b) What was the significance of the discovery of the phases of Venus for the question of the motion of the Earth? for the question of the centrality of the Earth?

4. Explain in a few words the significance of each of these discoveries made by Galileo with his telescope:

(a) Jupiter's moons;

(b) Earthshine; and

(c) mountains on the Moon.

5. (a) Is there any restriction on the relative positions of the Sun, Earth, and Mars (on an epicycle) when the latter is in retrograde motion in a geocentric system? In Copernicus' system?

(b) What phases does Mars show?

6. The synodic period of Mercury is about 0.32 years.

(a) Calculate Mercury's sidereal period.

(b) Calculate the mean distance of Mercury from the Sun. In what unit of distance is your answer?

7. Look up the synodic periods of Uranus, Neptune, and Pluto. Why do they approach the sidereal period of the Earth?

8. (a) If a planet had a sidereal period of 3 years, what would its synodic period be?

(b) If a planet had a synodic period of 3 years, what would its sidereal period be?

(c) If a planet had a semimajor axis of 3 AU, what would its sidereal period be?

(d) If a planet had a sidereal period of 3 years, what would its semimajor axis be?

9. Do full Moons occur at intervals of one synodic period or one sidereal period of the Moon around the Earth? Explain your answer with a sketch.

10. (a) Make an accurate drawing of the relative sizes of the orbits of Mercury, Venus, and Earth using the greatest elongations of the first two; assume circular orbits. What are the semimajor axes in AU of the orbits of Mercury and Venus?

(b) Which planet, Mercury or Venus, can be seen for a longer time after sunset, or before sunrise? Explain your answer.

11. Comets occupy the outermost fringes of the solar system. If the orbit of a particular comet has a semimajor axis of 10,000 AU, how long will it take to make a complete trip around its orbit? Where in its orbit will it be moving most slowly? most rapidly?

12. The semimajor axis of Jupiter's orbit is about 5 AU.

(a) Show that it takes Jupiter about 11.2 years to make one trip around the Sun.

(b) About how many times will Jupiter appear in retrograde motion during that time? (Hint: calculate the synodic period of Jupiter.)

13. The leftmost sketch below (Figure 7.29) represents the Earth's orbit. The other three orbits are drawn to the same scale, with the position of the Sun indicated by the usual symbol. One of the three orbits is a possible planetary orbit with a period of two years.

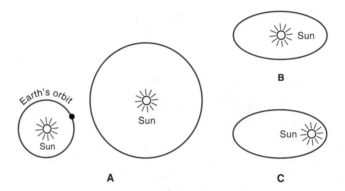

Figure 7.29. Figure for problem 13.

(a) Which is the correct orbit? Explain why. For each of the wrong orbits, explain why you know it is wrong.

(b) On the correct orbit, put a dot labeled F at the point the planet is moving the fastest; S, the slowest.

14. Suppose an asteroid's closest approach to the Sun is 2 AU, and its greatest distance from the Sun is 4 AU. Sketch its orbit. How large is its semimajor axis? What is its period? What is the eccentricity of its orbit?

15. The semimajor axis of the orbit of Mars is 1.52 AU. At its closest approach Mars is 1.38 AU from the Sun. Calculate the eccentricity of Mars' orbit.

16. Show that the perihelion distance of a planet moving in an orbit of semimajor axis a and eccentricity e is $r_P = a(1 + e)$.

17. (a) The Earth's orbit around the Sun is slightly elliptical. When the angular diameter of the Sun appears smallest, will the Sun appear to be moving most slowly or most rapidly with respect to the stars? Explain your answer.

(b) Make a sketch of the relative positions of the Sun, Earth, and Mars when Mars' angular diameter appears to us to be largest. How is Mars moving during this time?

18. Using Kepler's laws, explain why a planet in a very eccentric orbit would have large extremes of orbital velocities.

19. Imagine an observer on Mars. What would the Martian measure be for the synodic period of Jupiter (in Earth years)?

20. Show how the new Moon is illuminated by Earthshine. How would you convince someone that the light is not from the Moon itself, that is, that the Moon is not self-luminous?

Suggestions for Further Reading

See also the books listed at the end of Chapter 4.

Adamczewski, J., *Nicolaus Copernicus and His Epoch*. Philadelphia: Copernicus Society of America, 1973. A well-illustrated account of his life and times.

Byard, M., "Poetic Responses to the Copernican Revolution," *Scientific American*, **234**, p. 121, June 1977.

Christianson, J., "The Celestial Palace of Tycho Brahe," *Scientific American*, **204**, p. 118, February 1961.

Galilei, G. *Sidereus Nuncius or The Sidereal Messenger*. Translated and with an introduction and notes by A. Van Helden. Chicago: University of Chicago Press, 1989. An exciting book, even today.

Gingerich, O., "Johannes Kepler and the Rudolphine Tables," *Sky & Telescope*, **42**, p. 328, 1971.

————"Copernicus and Tycho," *Scientific American*, **229**, p. 86, December 1973.

————"Tycho Brahe and the Great Comet of 1577," *Sky & Telescope*, **54**, p. 452, 1977.

————"The Galileo Affair," *Scientific American*, **247**, p. 133, August 1982.

————*The Eye of Heaven; Ptolemy, Copernicus, Kepler*. New York: American Institute of Physics, 1993. A collection of 25 articles on the three astronomers listed.

Heninger, S., *The Cosmographical Glass; Renaissance Diagrams of the Universe*. San Marino, CA: The Huntington Library, 1977. An interesting survey of the images depicting the universe taken from books published before 1700.

Koestler, A., *The Sleepwalkers*. New York: Grosset and Dunlap, 1963. Though many will not agree with Koestler's ideas on scientific creativity, this is nonetheless an enjoyable and well-written account of the period from Copernicus through Newton.

————*The Watershed*. Garden City, NY: Doubleday and Co., 1960. This is the section on Kepler taken from *The Sleepwalkers*.

Kuhn, T., *The Copernican Revolution*. Cambridge, MA: Harvard University Press, 1957. An excellent account.

Thoren, V., *The Lord of Uraniborg*. Cambridge: Cambridge University Press, 1990. An authoritative account of the life and times of Tycho Brahe.

————"New Light on Tycho's Instruments," *Journal for the History of Astronomy*, **4**, p. 25, 1973. An illustrated description of Tycho's instruments.

Wilson, C., "How Did Kepler Discover His First Two Laws?" *Scientific American*, **226**, p. 93, March 1972. Interesting insights on Kepler's way of working.

"One had to be a Newton to notice that the moon is falling, when everyone sees that it doesn't fall."

—Paul Valery (1871–1945); French poet

"I don't know what I may seem to the world, but, as to myself I seem to have been only like a boy playing on the sea-shore and diverting myself in now and then finding a smoother pebble or a prettier shell than ordinary, whilst the great ocean of truth lay all undiscovered before me."

"If I have seen farther than other men, it is because I stood on the shoulders of giants."

—Sir Isaac Newton (1642–1727)*

---

* The second of Newton's remarks above is not original, but has been traced back to the twelfth century by R. Merton, in his wonderful book *On the Shoulders of Giants*.

CHAPTER 8

The Newtonian Synthesis

Less than a year after Galileo died, Isaac Newton (1642–1727; see Figure 8.1) was born. His father, a farmer, had died before Newton's birth, and he was brought up by his mother and grandmother. After it became clear that farming was not for him, Newton entered Cambridge University when he was 19. In 1665–66, when the bubonic plague was sweeping over England, Cambridge was closed, and Newton returned home to the village of Woolsthorpe. Here he passed the time by discovering the

Figure 8.1. Isaac Newton (1642–1727) at age 46, two years after the publication of his *Principia*. No other scientist's work has had more influence on the modern world than Newton's.

binomial theorem, beginning work on the calculus, experimenting with light, showing that white light was composed of all colors, and beginning the work that led to the idea of gravity. Not a bad way to spend a break from college! Within another three years he had developed his corpuscular (particle) theory of light, constructed the first reflecting telescope, and published his work on the calculus.

Newton's genius was quickly recognized. At the age of 27 he became the Lucasian Professor of Mathematics at Cambridge, where he remained for 27 years. He was a solitary and extremely reserved person, and disliked controversy of any sort—in this

Figure 8.2.   Edmund Halley (1656–1742) played a key role in persuading Newton to write the *Principia*. In addition, he both facilitated and financed its publication. He is also known for his work on comets and in particular for recognizing that the comets of 1531, 1607, and 1682 were the same object, now known as Halley's comet.

about as different from Galileo as one can imagine. In his early fifties he suffered episodes of virtual insanity, which some have attributed to mercury poisoning brought on by his nearly 30 years of experiments in alchemy. Whatever the cause of his health problems, his scientific creativity declined after 1693. In 1696 he left academia and moved to London, becoming Warden, and four years later Master, of the Mint. He was an effective administrator and introduced many reforms. Queen Anne knighted him in 1705.

In the 1680s, Edmund Halley (of comet fame; see Figure 8.2) and several other scientists in London had been trying to understand how a force from the Sun could govern the planets' motions. The Dutch scientist Christian Huygens had already shown that circular orbits could be understood in terms of a Sun-planet force that varied as $1/R^2$. (We read this as "inversely as the square of the distance.") But planetary orbits are elliptical, and the London scientists were unable to prove that a $1/R^2$ force would result in elliptical orbits. When Halley asked Newton about this, he answered that he had proven gravity's inverse-square dependence on distance a long time earlier. Halley urged Newton to publish his results, and in fact provided the money to do so. Newton finally completed this work three years later, in 1687. His book, *Philosophiae Naturalis Principia Mathematica* or *The Mathematical Principles of Natural Philosophy* (Figure 8.3), is one of the most important of the modern world. In it Newton sets forth a unified, comprehensive system of definitions and laws by which the motions of objects can be described. It includes the laws of motion, which form the basis of modern dynamics, and the law of universal gravity, which describes the action of what we now know to be one of the fundamental forces of nature. This work demonstrated clearly the power of modern physical science, especially as it is expressed mathematically.

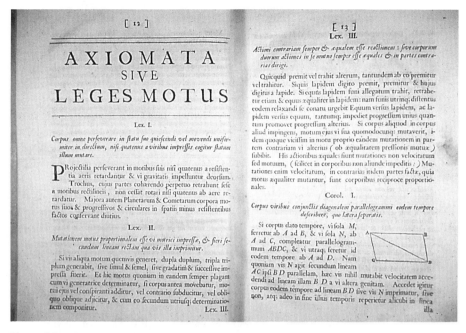

Figure 8.3. Two pages of the *Principia* in which the three laws of motion are stated. Note that it is written in Latin, the scholarly language of the time.

Newton showed how to calculate the mass of the Sun and the masses of the Earth and other planets, and showed how Kepler's laws were just particular examples of the new general system of dynamics. He predicted, for example, the small flattening of the Earth due to its rotation and the variation of one's weight with latitude, and explained quantitatively the causes of the tides, the precession of the equinoxes, etc. It soon became clear that the fundamental problems that astronomy had struggled with for two millennia were finally solved, and solved brilliantly. As we shall see, the *Principia* (as it is usually called) is often taken as the beginning of modern science. Indeed, Halley felt that Newton's work was one of the pinnacles of human achievement, comparable only with the development of society, the establishment of agriculture, and the invention of wine and of writing!

Newton's Theory of Motion

From late antiquity through the Middle Ages, there were many points of view about dynamics, that is, how bodies move under the action of forces. To put Newton's system in sharper relief, let us look at some ideas that at first glance might seem plausible. For example, if one object is to exert a force on another, it must be in contact with it; the direction of motion is in the direction of the force, and unless that force is continually exerted, the motion will eventually stop. In other words, if you want to move a heavy stone straight ahead, you must put your hands on it and push straight ahead; when you stop pushing the stone will stop moving. Put this way these ideas sound reasonable, even obvious. As generalizations concerning the nature of force and motion, however, they are all wrong.

Kepler, Descartes, and Galileo all had thought about the question of motion and contributed various ideas. Kepler thought that a force must emanate from the Sun to cause the planets to move in elliptical orbits. But what was the nature of this force, and how could it be transmitted over the empty space between the Sun and the planets? Galileo, however, followed Aristotle and Copernicus, and held that objects "naturally" moved in circles, so that no solar force was necessary. The concept of inertia in one form or another was generally accepted. **Inertia** is the property of mass by which it resists changes in its motion: a bowling ball rolling down an alley is not easy to stop; its inertia is large. Galileo's idea of inertial motion was tied to the surface of the spherical Earth, a sort of "circular inertia." Descartes correctly maintained that inertia caused objects to continue to move in straight lines, not along circular paths. But planets certainly do not move in straight lines; so something more must be involved. Thus there was considerable confusion in the ideas then current.

Newton's achievement was to combine many contemporary ideas with his own concepts to make a coherent, consistent framework on which the theory of *both* celestial and terrestrial motions could be constructed. He formulated three laws of motion, three generalizations having an enormously wide range of application from the Earth to the heavens.

Before describing these laws, however, we should define a few terms. It is useful to distinguish two kinds of quantities: **vectors** and **scalars**. The latter is just a number along with its associated units, such as 3 meters or 4 kilograms. To specify a vector, however, requires a number and an associated *direction*. The *speed* of an object, 25 km/hr, is a scalar, but 25 km/hr toward the northeast is its *velocity*, a vector. Speed is the amount or magnitude of the velocity. Consider a toy train moving with uniform motion around a circular track; its speed along the track, say, 5 km/hr, is constant, but its velocity is always changing, since the direction of motion of the train is always changing.

Another important concept to understand is that of **acceleration**. Acceleration is the rate (in time) at which the velocity of an object changes, that is, how rapidly its velocity changes, either increasing or decreasing. The performance figure given in automobile ads—"from 0 to 60 miles per hour in 10 seconds"—is just the acceleration of the car. Numerically, it is the change of velocity divided by the length of time during which the velocity change occurs; thus its units are (cm/sec)/sec, or (km/hr)/sec, etc.

Is acceleration a vector or scalar quantity? Dividing a vector (change in velocity) by a scalar (time) results in another vector (acceleration), changed in magnitude but not in direction. That is, dividing by the scalar time does not change the direction of the velocity. Thus, acceleration is a vector having the same direction as the *change* in velocity.

Consider a toy train motionless on a straight track: when you start it up and increase its velocity to 5 km/hr, it is accelerating; when it moves at a constant velocity (5 km/hr in a straight line), it has 0 acceleration, because neither its speed nor its direction is changing; when you bring it to a stop, it again accelerates (or "decelerates" if you prefer). In this example of straight-line motion, the acceleration is a result of a change in speed only. When the train moves at constant speed around a circular track, it is constantly accelerating, because its direction is always changing, even though its speed along the track is not. With these definitions in mind, let us return to Newton's laws of motion.

The First Law

An object at rest remains at rest, and an object in motion remains in uniform motion in a straight line unless acted on by a force.

The first part of the statement is Newton's statement of the idea of inertia. The second part says what a force does: it *changes* motion. It is not, "no force, no motion," but rather "no force, no *change* in motion." Change in motion is the characteristic consequence of a force. A ball set rolling on a very large smooth surface is constantly subject to the small frictional force between the ball and the surface; this slowly changes its motion until the ball finally comes to a stop. Without this frictional force (and with an infinitely large surface) what would cause the ball to stop? Nothing! It would roll forever!

Since a change in motion is an acceleration, which is a vector, force, which produces the acceleration, must also be a vector. An alternative formulation of this idea can be given in terms of Descartes' "quantity of motion" or momentum. (Recall that this is the product of the mass of a body and its straight-line velocity, or mV.) The first law then becomes: the momentum of an object is constant unless a force acts on it. The ball mentioned above has momentum mV. Without the frictional force, its momentum would not change, and it would continue to move in a straight line with velocity V.

The Second Law

The rate of change of an object's motion is directly proportional to the force acting on it and is in the direction of that force.

This is essentially a quantitative definition of a force. According to Newton, the greater the acceleration, the greater must be the force acting; that is, $a \propto F$. If you want to double the acceleration of an object, you must double the force exerted upon it. (Unfortunately, it is customary to denote both the semimajor axis of an ellipse and the acceleration of an object by the letter a; the context in which they are used should keep you from confusing the two, however.)

This proportionality relation can't be the whole story, however, since it makes no distinction between the force required to push a marble and that to push a bowling ball. So, on what else does the acceleration depend? How do we change this statement of proportionality into one of equality, that is, into an equation? Obviously, the force needed to produce any detectable change of motion in the bowling ball is much larger than that required for the marble, because the mass of the ball—the total amount of matter it contains (a scalar quantity)—is so much larger than that of the marble. Newton understood that the acceleration resulting from the action of a given force is smaller for a more-massive object; we say that the acceleration is *inversely* proportional to the object's mass. Thus, to completely describe the action of a force, he wrote, $a = 1/m \times F$, or, as it is more commonly written,

$$F = ma.$$

Remember that when a scalar (such as mass) multiplies a vector (such as acceleration), the direction of the product (the force) is still that of the acceleration. Thus, the force is in the direction of the acceleration, that is, in the direction in which the motion is changed, not in the direction of the motion itself.

A few examples should help to clarify these ideas. Suppose your vehicle has stalled, and you want to push it out of traffic. To do so you must change its velocity from 0 to, say, 2 kilometers per hour. Since you want to change its motion, that is, accelerate it, you must exert a force on it. Clearly, the force required will be greater if you are driving a truck instead of a compact car. Similarly, it will be harder to stop the motion of the truck than that of the compact car. Furthermore, when you are speeding the truck up, you must push in the direction opposite to that in which you would push when trying to slow it down. That is, the force that you exert is in the direction of the change in motion (the acceleration), rather than in the direction of the motion itself.

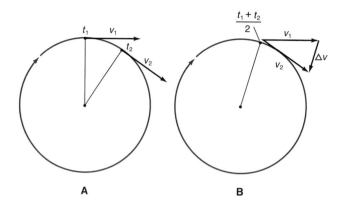

Figure 8.4. The velocity of a train at two times, t_1 and t_2, is shown in (A). The average change in velocity during the time interval is given in (B) and is ΔV, and is the average acceleration. Note that ΔV is pointing down, toward the center. If the time interval were taken shorter and shorter, the acceleration ΔV would point ultimately directly at the center. Thus, a force toward the center acts on the train.

Next, consider our example of a toy train moving with constant speed on a circular track. Though its *speed* is constant, its *velocity* is always changing direction. Consequently, the train must experience an acceleration and, therefore, a force, both being constant in magnitude but always changing direction. In what direction does that force act? The situation is shown in Figure 8.4, where the two, equally long arrows V_1 and V_2 indicate the direction of the train's velocity at two moments, t_1 and t_2, close together in time. The difference in the direction of the two velocity vectors is ΔV and is the average acceleration during the time between points 1 and 2. To find the exact direction of the force, however, we want the direction of the instantaneous acceleration at any point, not an averaged value between two points. So let the time interval between the two points become smaller and smaller. As we do so the two velocities become more and more nearly parallel, and ΔV becomes more and more nearly perpendicular to them both. In the limit, when the difference in time between the two points approaches 0, the direction of ΔV (and so of the force) becomes more and more nearly the direction to the center of the track. So for circular motion a force must be acting toward the center of the circle, that is, along a radius in a direction always *perpendicular* (at 90°) to the instantaneous velocity. Hence this force does not increase the speed of the train, but only changes its direction. What is providing the force on the train? It is the track, constantly pushing the train toward the center. If the track were suddenly to vanish, the train would move in a straight line (Newton's First Law) along the direction in which it was moving at the moment when the force disappeared.

The above description of finding the instantaneous acceleration is an example of a major problem faced by seventeenth-century scientists. They had to deal with quantities, like velocity, which were always changing. For example, Kepler had shown that planets move around the Sun with varying velocities; closer to home, a pendulum swings back and forth with varying velocity. For velocities that are constantly changing, the notion of an average velocity is of limited usefulness, and certainly won't represent the velocity at a particular instant. To find the instantaneous velocity of a car, for example, we can imagine carrying out the following series of calculations. First, find the average velocity during, say, 10 seconds, centered on the instant in question. Then squeeze the time interval down to 1 second, and calculate the average velocity during that interval. Repeat the process again and again, with smaller and smaller time intervals, all centered on the instant at which you wish to know the velocity. As the time intervals approach 0, the velocities will approach the particular value that is the instantaneous velocity you want. Graphical methods of handling such problems are cumbersome (as you no doubt noticed when you tried to understand the force acting on the toy train, described above).

To handle these problems more easily, Newton and some of his contemporaries began to develop mathematical techniques that enable formulae to be derived by which instantaneous velocities or accelerations, for example, could be calculated. These techniques, which we now call **calculus**, are extremely powerful, and have an enormously wide range of application. They allow us to find the instantaneous rate of change of any quantity with respect to another. The quantities need not be only distance or velocity with respect to time, but can also be the Earth's atmospheric pressure with respect to altitude above the Earth, the change of the volume of a sphere with respect to its radius, etc. The power of calculus helped to convince scientists that the language of nature is mathematics. It is still the single most important branch of mathematics used by astronomers and other physical scientists.

The Third Law

When any two objects interact, the force exerted by the first object on the second is equal and opposite to the force exerted by the second on the first. An equivalent statement is that *for every action (force), there is an equal and opposite reaction; that is, forces are mutual and act in pairs.*

This law is deceptively simple. A change in motion results only from the interaction of two bodies (which does not necessarily mean direct contact), both of which have their motion changed as a result. When you push a bowling ball, you exert a force on it; the ball in turn exerts an equal force in the opposite direction on you. Both your motion and the bowling ball's motion are changed. Since your mass is much greater than the mass of the bowling ball, your acceleration is much less than that of the ball. To convince yourself that forces act in pairs, imagine trying to push a car while you are wearing roller skates!

Note that any interaction involves both the active and the reactive aspects. Even though one may be more apparent than the other (it is impossible to detect the recoil of the Earth caused when you walk on it, for example), one is not more important than the other. They are simply the two aspects of the single interaction.

Unlike the first two laws, which were "in the air," that is, had already been discussed by several scientists, the third law is entirely Newton's own. It played a key role in the development of the idea of universal gravitation.

Why Were Greek Theories of Motion So Different?

Two important points are illustrated by these laws of motion. Though the facts of nature available to the Greeks and to seventeenth- and eighteenth-century Europeans were the same, their ways of looking at them were not. Aristotle's ideas were closely tied to everyday experience, to a world pretty much taken at face value. It was not idealized to any extent in the Platonic sense of getting to the "basic realities" underlying our experience. For example, unless you understand that the fluttering motion of a falling leaf is produced by air resistance, and mentally separate that from its downward motion, you aren't likely to realize that the leaf and a rock are both falling to the ground under the action of the same force, the one known as gravity.

By contrast, Galileo extrapolated the results of his experiments on the motions of falling objects to situations that were beyond his ability to achieve in practice. For example, he imagined a frictionless interaction between a ball and the inclined plane on which it was rolling. He did this in order to isolate and understand what he took to be the essentials of the experiment. In this sense his approach was like Plato's: he wanted to go beyond mere appearances and get to the essence of things, to the "reality" basic to the phenomenon in question. Our modern ideas of motion, then, are consequences of a different way of looking at the phenomena, of getting to the fundamental factors involved, rather than any change in the basic facts of experience themselves.

The second point is that one of the most difficult things to do in science is to create the concepts, the tools we need that are most powerful, most useful for describing the phenomena at hand. Such tools are never obvious and simply ours for the taking from a generous nature. We must abstract them from nature; we must *invent* them. They become the means by which we develop new ways of looking at the world. In the preceding sections we have discussed several of these concepts, most of which are probably so familiar to you that you take them for granted, like velocity and acceleration. Yet consider: is it really sensible to divide, one by the other, two such unlike and unrelated quantities as distance and time in order to define velocity? It didn't seem sensible to the Greeks, nor would dividing three apples by four oranges make much sense to us. Then we compound the confusion by again dividing the result, velocity, by time, to define acceleration. The concepts of velocity, acceleration, inertia, force, and gravity don't jump out at us from nature; they must be invented. Only those concepts that prove to be useful are kept. For example, we have defined the quantities $V = d/t$, and $a = V/t = d/t^2$; we could also define $a/t = d/t^3$. The first two quantities—velocity and the change in velocity with time (or acceleration)—are useful. The change in acceleration with time is not a very useful quantity, and we rarely talk about it. The concepts we keep are those that are found to be useful tools with which to build a coherent and unified way of dealing with the physical circumstances of nature.

The Laws of Motion Applied to Gravity
An Experiment With Falling Objects

How do these general laws of motion and force, which were formulated by Newton, apply to the particular force of gravity? Imagine the following experiment (Figure 8.5). Drop a marble in a vacuum from a height and time its fall. You would find that at the end of 1 second, it has dropped 490 cm, and its speed is 980 cm/sec; at the end of 2 seconds, the total distance fallen is 1,960 cm, and its speed is 1,960 cm/sec; at the end of 3 seconds, it has fallen 4,410 cm, and it is moving with a speed of 2,940 cm/sec, or about 70 mph. If measured in air and not in a vacuum, the speed would be somewhat less than this because of the resisting force exerted by the air. Taking this force into account, and eliminating it as being extraneous to the basic experiment, is a good example of the abstraction from the "real" world of experience mentioned earlier.

Let's see what we can learn from this experiment. Notice that during each second the speed increases by 980 cm/sec, so the acceleration is 980 (cm/sec)/sec throughout. Repeat the experiment with a bowling ball instead of a marble, and you would get the same result. The velocity of the bowling ball and the distance traveled by it at the end of each second would be the same as for the marble. Thus the marble and bowling ball hit the ground at the same time. This is a very important point: the gravitational acceleration is the *same* for all falling bodies, regardless of their mass. (This is, of

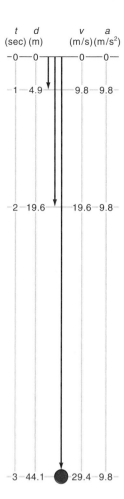

t	d	v	a
(sec)	(m)	(m/s)	(m/s²)
0	0	0	0
1	4.9	9.8	9.8
2	19.6	19.6	9.8
3	44.1	29.4	9.8

Figure 8.5. If you dropped a marble from a height, and measured the distance fallen and its velocity at the end of 1, 2, and 3 seconds, you would get the results seen here.

course, the result of the famous Leaning Tower of Pisa experiment in which Galileo supposedly showed that two objects of different mass would fall together. Actually, Galileo did not perform such an experiment.) We conclude, then, that the force causing the object to fall must be proportional to the mass of the object. The greater the falling mass, the greater the force acting on it, otherwise the bowling ball and the marble would not hit the ground at the same time.

By Newton's third law (forces occur in equal and opposite pairs), some other body must be interacting with the falling marble; this can only be the Earth. Thus, the magnitude (or strength) of the gravitational force must be proportional to the mass of both interacting objects: $F \propto Mm$. The direction of the force must be along the line joining the two interacting objects. Since, according to the third law, bodies interact in pairs, the Earth must experience the same force upward as it exerts downward on the marble. However, the mass of the Earth is much, much greater than that of the marble, and so by the second law, the more massive Earth undergoes a much smaller acceleration upward than does the ball downward; it is, in fact, quite imperceptible.

Newton's Law of Gravity

Does the force between the Earth and the falling ball depend on anything else? Kepler showed that the planets move more slowly the farther away they are from the Sun; so how does the force depend on the separation of the two interacting masses? Newton showed that Kepler's first and third laws require the magnitude of the force to decrease with the square of the distance between the two masses, that is, $F \propto 1/R^2$. If the separation of the objects is doubled, the force acting between them is decreased to one quarter of its original value; if tripled, to one ninth its original value, etc. Thus, the force between the Earth and the falling object, and by generalization, the force between *any* two objects of mass M and m (Jupiter and one of its moons, for example), is given by $F \propto Mm/R^2$, where R is the distance separating the two. We can write this as an equation by introducing the gravitational constant, G, which is just a number with appropriate units, a universal constant denoting the strength of the gravitational force:

$$F = \frac{GMm}{R^2}.$$

Suppose M and m are each 1-gram masses separated by 1 centimeter; then G is the gravitational force between the two masses, expressed in units of grams, centimeters, and seconds. The direction of the force and the resultant acceleration will always be understood to be along the line joining the two masses.

Newton's idea that gravity is a *universal* force, acting between *all* masses *everywhere*—not just between the legendary falling apple and the Earth, or between the Earth and the Moon—is an enormous extrapolation. It amply testifies to his genius. As mentioned earlier, gravity is one of the four fundamental forces in nature. Compared to the other three, gravity is a very weak force intrinsically, but because we live on the surface of a massive object, it is the dominant external physical force in our lives.

Considering that it describes one of the basic forces of nature, the above formula is remarkably simple. Familiarity may have dulled some of its strangeness, however. As intrinsically weak as it is, the gravity acting between the Sun and the Earth is enormous because the masses involved are so large. In fact, if this force were pulling on a steel cable, the cable would have to be roughly half the diameter of the Earth to keep from breaking! Yet this enormous force acts over empty space; no medium, no substance, carries it.

Note that Newton's law of gravity describes *what* happens, but not *how* or *why* it happens; no fundamental cause is given for gravity. Newton invented the concept, showed how it depended on mass and distance, but said nothing about the origin of gravity, about why nature behaves this way. He could think of no mechanism to explain gravity's working. Furthermore, nothing is said about how such a force is transmitted. Newton (and generations of physicists after him) was troubled by this problem, but could give no solution to it. In fact, Newton wrote (to the Reverend Richard Bentley in 1693), "That gravity should be innate, inherent, and essential to matter, so that one body may act upon another, at a distance through a vacuum . . . is to me so great an absurdity, that I believe no man who has in philosophical matters a competent faculty for thinking, can ever fall into it." Also, if all matter attracts all other matter, why do we see it spread throughout space and not collected together in one place? We shall return to both of these questions later.

A Closer Look at the Law of Gravity

Now let us look at Newton's laws in a little more detail. First of all, what should we take for R, the distance between the two attracting masses? Spherical objects, like marbles and stars, act gravitationally as if all their mass were concentrated at a point at their center (see Figure 8.6). Thus the appropriate distance to take is the distance between the *centers* of the two spherical masses. Newton was pretty sure that this was the case, but it took him a long time to prove it. In fact, this was one of the reasons he delayed so long in publishing his work.

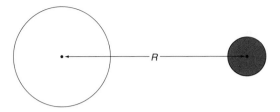

Figure 8.6. For the purposes of gravity, the mass of a spherical object acts as if it were concentrated at a point at the center of the object. Most planets and stars are very nearly spherical; so the appropriate distance, R, to take in calculating the gravitational force between them is the center-to-center distance.

That both masses experience the same force can be illustrated by imagining them to be connected by a stretched spring, which represents the gravitational force between them (Figure 8.7). Clearly, the spring can't pull on one mass with a greater force than on the other; so both are subject to the same force (the third law). Both respond (according to the second law) with accelerations inversely proportional to their masses, that is, $F = ma$, or $a = F/m$; the larger mass experiences the smaller acceleration resulting from the action of the given force, F.

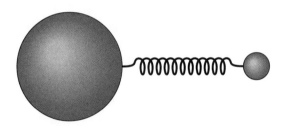

Figure 8.7. The force between two balls of different masses connected by a spring is the same on each. One end of the spring cannot pull harder than the other end. Similarly, the gravitational force between two masses acts equally on both.

The Force Exerted by the Earth

Now consider a marble falling freely near the surface of the Earth. Only one force acts on it—gravity:[1]

$$F = G \frac{M_\oplus m_m}{R_\oplus^2} ,$$

---

[1] In the next few pages you will be faced with several equations. Don't panic! There's nothing mysterious about the math we are using. It is simply a convenient shorthand. Instead of writing the formula for Newton's law of gravity, we could write it out in English: "the force between any two masses is directly proportional to the product of the masses and inversely proportional to the square of the distance between them." But this is very cumbersome and inconvenient compared to writing simply the formula.

Furthermore, the manipulations we will make amount to little more than canceling identical quantities appearing in the numerator and denominator of an expression, or on each side of an equation.

where $M_\oplus$ and m_m are the mass of the Earth and the marble, respectively, and $R_\oplus$ is the Earth's radius. Now, Newton's second law, $F = ma$, which is true for any kind of force, can be applied to this gravitational situation:

$$F = G\frac{M_\oplus m_m}{R_\oplus^2} = m_m a_m \ ;$$

a_m is the acceleration of the marble. The force and acceleration are understood to act along the direction between the centers of the marble and the Earth. Since m_m is tiny compared to $M_\oplus$, we say the marble falls, not the Earth; nonetheless, an infinitesimal acceleration is experienced by the Earth. Since m_m occurs on both sides of the equation, it cancels out, and so we have for the marble's acceleration,

$$a_m = \frac{GM_\oplus}{R_\oplus^2} \ .$$

See what this equation says: the acceleration of the marble depends not on its own mass (m_m does not appear in the formula), but only on the Earth's mass and radius. This is just what is found by the experiment described earlier—the falling bowling ball and marble hit the Earth together, since their acceleration depends not on their mass, but only on the mass of the Earth.

The Mass of the Earth.   Since the value of the **gravitational acceleration** at the Earth's surface is important to us, we give it a special symbol, g. From laboratory measurements its value is found to be 980 (cm/sec)/sec. So, for the marble and the bowling ball,

$$g = \frac{GM_\oplus}{R_\oplus^2} = \frac{980 \ (\text{cm/sec})}{\text{sec}} \ .$$

This simple equation is very powerful; it gives us a way to find the mass of the Earth! G can be measured in laboratory experiments, $R_\oplus$ can be found by Eratosthenes' method; so we can solve for the remaining unknown, $M_\oplus$, the mass of the Earth.

We call the gravitational force acting on an object its **weight**, w. In general, $F = ma$, but when referring to the gravitational force on the Earth's surface, we can write $w = mg$: the force acting on an object at the Earth's surface, that is, its weight, equals its mass times the acceleration of gravity at the Earth's surface. When you are on top of a mountain your weight is slightly decreased, because you are a little farther from the center of the Earth, but your mass is unchanged.

The Motion of the Moon.   Now let's leave the Earth and see if we can understand how the Moon moves in its orbit. According to Newton's first law, the Moon would fly out of its orbit along a tangent to the orbit and travel in a straight line, unless it was being acted on by a force (see Figure 8.8); so there must be a force keeping the Moon in its orbit. The two interacting objects (the third law) must be the Moon and the Earth; so the force acting between them must be along the line joining them. The direction of that force is not in the direction of the Moon's motion, but in the direction of its acceleration, that is, toward the Earth. The force acting on the Moon (and equally on the Earth) is

$$F = G\frac{M_\oplus m_\mathbb{C}}{R_{\oplus\text{-}\mathbb{C}}^2} = m_\mathbb{C} a_\mathbb{C}; \text{ so } a_\mathbb{C} = G\frac{M_\oplus}{R_{\oplus\text{-}\mathbb{C}}^2} \ .$$

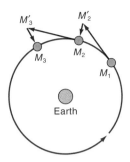

Figure 8.8. If it were not for the force of gravity acting between the Earth and the Moon, the latter would fly off into space along $M_1M'_2$, for example. Instead, the Moon is constantly falling toward the Earth as indicated. Actually, the Earth and Moon are both "falling" toward each other, but since we are on the Earth, we attribute all the motion to the Moon.

$R_{\oplus\text{-}\mathbb{C}}$ is just the Earth-Moon distance.

The Moon falls toward the Earth with an acceleration depending only on the mass of the Earth and on the Moon's distance from the center of the Earth, just as was true for the falling marble. In other words, Newton predicts that the *same* physics applies to the celestial realm as to the terrestrial! Let's test this crucial idea and see if it is true.

The Earth-Moon distance is about 60 times greater than the Earth's radius; so the acceleration produced by the Earth's gravitational force at the distance of the Moon is only $1/(60)^2 = 1/3600$ that which it produces on us here on the Earth. Instead of falling 490 cm by the end of the first second because of its interaction with the Earth, the Moon falls only $490/3600 = 0.136$ cm. In that same second it travels about 103,000 cm in a direction perpendicular to its direction of fall. It's not too hard to show that, small as it is, this tiny "bending" of the Moon's path by 0.136 cm every 103,000 cm is just the right amount to produce the Moon's orbit.[2] The Moon is continually falling toward the Earth, but never gets any closer to it. This sounds strange, but what is happening is that the Moon is falling away from the straight line path it would follow (because of its inertia) if no gravitational force were acting. So, as a consequence of gravity, it always remains the same distance from the center of the Earth. The constantly falling Moon remains securely trapped in its orbit. Newton was right. The same physics that accounts for the motion of the falling marble on the Earth accounts for the orbit of the Moon in the sky!

Actually, the first time Newton tried to check his ideas about the Moon's motion by making the kind of calculation we have outlined here, he found that the numbers didn't work out. Six years later, when a better value for the Earth-Moon distance was measured, he tried again and found that his calculations gave exactly the right answer.

Your Weight in Space

How much would you weigh out in space at a distance from the Earth equal to the Moon's distance, but not on the Moon itself? Since you would then be 60 times farther from the center of the Earth than when you are on its surface, the acceleration and the force (your weight) that you would feel would decrease by a factor of $1/(60)^2$ or

---

[2] Note that in this discussion we are assuming that the Moon's orbit is circular, whereas in fact it is slightly elliptical. Thus the distance the Moon falls each second is not quite constant. The idea of the argument is still correct, however.

1/3600. If you weighed 180 lbs on Earth you would weigh only 180/3600 = 1/20 lb out there. This is an effective—but costly—method of weight loss (but not of mass loss!).

Let's calculate how much you would weigh on the Moon itself. On Earth you weigh

$$w_\oplus = mg, \text{ so } w_\oplus = \frac{M_\oplus m}{R_\oplus^2} G, \tag{A}$$

whereas on the Moon you would weigh

$$w_\mathbb{C} = ma_\mathbb{C} \text{ and } w_\mathbb{C} = \frac{M_\mathbb{C} m}{R_\mathbb{C}^2} G. \tag{B}$$

Let's find out what you would weigh on the Moon as compared to what you weigh on Earth by dividing equation (B) by (A). Then

$$\frac{w_\mathbb{C}}{w_\oplus} = \frac{G\,(M_\mathbb{C}m)/R_\mathbb{C}^2}{G\,(M_\oplus m)/R_\oplus^2} = \frac{M_\mathbb{C} R_\oplus^2}{M_\oplus R_\mathbb{C}^2}.$$

Now,

$$\frac{M_\mathbb{C}}{M_\oplus} = \frac{1}{81} \text{ and } \frac{R_\oplus}{R_\mathbb{C}} = \frac{4}{1},$$

so

$$\frac{R_\oplus^2}{R_\mathbb{C}^2} = 16 \text{ and } \frac{w_\mathbb{C}}{w_\oplus} = \frac{16}{81} \approx 0.20.$$

Thus, on the Moon you would weigh about 20 percent of what you weigh on Earth. A hundred-pound person would weigh only twenty pounds, and could easily jump to Olympic heights and distances.

Newton's Generalizations of Kepler's Laws

Newton showed that Kepler's laws were just special examples of the much broader consequences of the laws of motion and gravitation applied to two interacting masses. He showed that the paths of such objects need not be only ellipses, but could also be parabolas or hyperbolas. They are ellipses if the two objects are moving rather slowly around each other and so are permanently bound together by gravitation, like the Earth and the Sun. If, however, they are moving rapidly enough that they can either just barely escape or easily escape from each other, then the path of one relative to the other will be a parabola or a hyperbola, respectively.

Also, Newton showed that Kepler's second law (equal areas in equal times) is a consequence of the gravitational force being directed along the line joining the planet and the Sun. Not only gravity, but any force between two objects acting in this way will result in the motion described by the law of areas. A more physical interpretation of Kepler's second law is that it is a consequence of the conservation (constancy) of angular momentum. In Chapter 7 we described the idea of linear (straight-line) momentum that is given by the product of a mass, m, and its straight-line velocity, V. Unless a force acts on the mass, its linear momentum, mV, remains constant; this is one way of stating Newton's first law.

It is extremely useful to define an analogous concept for **angular momentum** (that is, the momentum of an object not in straight-line motion). Consider an object moving

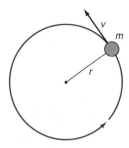

Figure 8.9. The angular momentum of the mass m moving with velocity V along a path of radius r is $L = mVr$. Unless some outside force acts on m, its angular momentum remains constant.

in a circular orbit with radius r (Figure 8.9): the product of the object's mass and its velocity, V, about an axis a distance r away is constant unless a force acts on the system. That is, the angular momentum, $L = mVr$, is conserved (does not change) in the absence of a force. This is one of the most important conservation laws in physics, and since astronomical objects revolve and rotate (the Earth revolves around the Sun and rotates on its axis, for example), it has wide applicability in astronomy.

Let's consider for a moment how the conservation of angular momentum applies to an object rotating about an axis within itself. A familiar example is the ice skater spinning slowly with arms outstretched (Figure 8.10). As she brings her arms to her body, her spin increases. The decrease in the distance of the mass of her arms from her rotation axis must be compensated for by a corresponding increase in V, because the product mVr must remain constant unless a force is acting. (Actually, there is a frictional force between the skate blades and the ice, which would eventually stop her spinning, but it is relatively small and we neglect it here.)

The application of this concept to a planet orbiting the Sun is straightforward. At a distance r from the Sun a planet moves with a velocity V. Its angular momentum is just mVr, and the product remains constant because no outside force is acting on the

Figure 8.10. A spinning skater is a familiar example of the constancy of angular momentum. As she brings her arms and leg to her body, she spins up, because r has decreased; so V must increase.

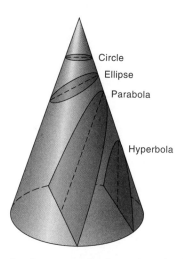

Figure 8.11. The intersection of a plane perpendicular to the axis of a cone is a circle; when the plane is at an angle to the axis but still cutting the cone, the intersection is an ellipse; both these figures are closed. A parabola results when the plane cuts the cone parallel to the side of the cone; if the plane cuts the cone at an even smaller angle, a hyperbola is produced. The latter two curves are open; the two sides of a parabola become parallel to each other at infinite distance, whereas the two ends of a hyperbola keep on diverging.

Sun-planet system. As the planet moves closer to the Sun on its elliptical orbit, r decreases, so V must increase to keep the angular momentum the same. This results in just the motion Kepler found.

Be sure you understand the concept of angular momentum and the circumstances under which it remains constant or conserved. Any time two or more objects are interacting as an isolated system (so that the outside forces acting on the system are zero or very small), the principle of the conservation of angular momentum applies. It applies to two galaxies moving around each other just as well as to planets moving around the Sun. We will make use of this concept time and again in the rest of this book.

As mentioned above, Newton showed that any two gravitationally interacting objects—Sun and planet, star and star—escape from each other in such a way that Kepler's second law holds for parabolic and hyperbolic paths as well as for elliptical orbits. More succinctly, the orbits of gravitationally interacting masses are **conic sections** for which the law of areas holds (see Figure 8.11). It is interesting that all gravitational orbits are just sections of a single geometric figure, the cone. A medieval astronomer might have seized upon this as indicative of a geometric basis to the cosmos. We shall see later, however, that there is a much more profound geometric foundation to the universe than this.

What is Newton's more general expression corresponding to Kepler's third law?

$$P^2 = \left[\frac{4\pi^2}{G(M_1 + M_2)}\right]a^3. \tag{C}$$

An alternative form of this equation is

$$(M_1 + M_2)P^2 = \left(\frac{4\pi^2}{G}\right)a^3, \tag{C'}$$

where P and a (now the semimajor axis) have their customary meaning, and M_1 and M_2 are the masses of *any* two objects moving in a closed, elliptical orbit around each other. Note that P^2 is on one side of equation (C) and a^3 is on the other side; these are just the quantities in Kepler's third law. The part in the brackets is new, however; in particular, the masses of the two objects appear. The two masses could be a planet and its Moon, the Sun and a planet, two stars, or even two galaxies, as long as they are gravitationally bound to each other. By measuring P and a of the two orbiting objects, we can get the sum of the two masses, whatever the objects might be. So with this formula we can measure the masses of planets, stars, and galaxies! Note how much more powerful than Kepler's is Newton's result, since it is derived from general physical principles rather than from a particular set of data.

Why Kepler's Third Law Works

What is the relation between Newton's generalization of Kepler's third law above and Kepler's form of it? To answer this question, let's write equation (C') twice—once for planet A moving around the Sun, and again for another planet B orbiting the Sun and divide the first expression by the second:

$$\frac{(M_\odot + M_A)\, P_A^2}{(M_\odot + M_B)\, P_B^2} = \frac{(4\pi^2/G)\, a_A^3}{(4\pi^2/G)\, a_B^3}.$$

The $(4\pi^2/G)$ terms cancel out, and we are left with

$$\frac{(M_\odot + M_A)\, P_A^2}{(M_\odot + M_B)\, P_B^2} = \frac{a_A^3}{a_B^3}.$$

Since planetary masses are so small compared to that of the Sun, it is a very good approximation to put $(M_\odot + M_A) \approx M_\odot$ and $(M_\odot + M_B) \approx M_\odot$; a flea does not add much to the mass of an elephant. So we are left with

$$\frac{M_\odot\, P_A^2}{M_\odot\, P_B^2} = \frac{a_A^3}{a_B^3} \quad \text{so} \quad \frac{P_A^2}{P_B^2} = \frac{a_A^3}{a_B^3},$$

which is what Kepler found! Thus we discover that Kepler's third law works because the mass of any planet in our solar system is negligible compared to the mass of the Sun. Recall that if P is in years and a in AU, then Kepler's law can be written as $P^2 = a^3$.

As we stated above, Newton's generalization of this law applies not only to solar-system objects orbiting each other, but to *any* two objects gravitationally bound to each other. So let's write this law for any two mutually orbiting masses, M_1 and M_2 (they could be two stars in orbit around each other), and once again for the Earth-Sun system, and divide the two expressions:

$$\frac{(M_1 + M_2) P_{1\text{-}2}^2}{(M_\odot + M_\oplus) P_{\odot\text{-}\oplus}^2} = \frac{(4\pi^2/G)\, a_{1\text{-}2}^3}{(4\pi^2/G)\, a_{\odot\text{-}\oplus}^3};$$

since $M_\oplus \ll M_\odot$,

$$\frac{(M_1 + M_2) P_{1\text{-}2}^2}{M_\odot P_{\odot\text{-}\oplus}^2} = \frac{a_{1\text{-}2}^3}{a_{\odot\text{-}\oplus}^3}.$$

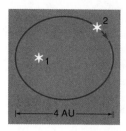

Figure 8.12. The orbit of star 2 around star 1. The major axis of the orbit is 4 AU.

If we express the mass in units of the solar mass, P in years, and a in AU, then all the quantities in the denominator in the preceding equation equal 1 solar mass or 1 year or 1 AU, and

$$(M_1 + M_2)P^2 = a^3. \tag{D}$$

See what we have done. The equation in (C′) requires that masses, the period, and the semimajor axis be given in grams, seconds, and centimeters, respectively (or, if you like, kilograms, seconds, and meters). The gravitational constant G must be given in the same set of units. In equation (D), however, by taking M in solar masses, P in years, and a in AU, we made $(4\pi^2/G)$ equal to 1. Equations (C′) and (D) have the same physical content; they say the same thing. They are simply expressed in different units.

Let's work a simple example using formula (D): suppose two stars are in orbit around each other with a period of 2 years and an average separation of 2 AU, as shown in Figure 8.12. What is the sum of their masses?

$$(M_1 + M_2)2^2 = 2^3 \;;\; \text{so} \;\; (M_1 + M_2) = \frac{8}{4} = 2M_\odot.$$

To relate solar masses to physical mass (grams, for example), we must use equation (C) to find the mass of the Sun. Apply the formula to the Sun-Earth system, remembering that the mass of the Earth is negligibly small compared with that of the Sun. P, a, and G must, of course, all be in the same system of units, like seconds, grams, and centimeters. The result is that 1 solar mass is very nearly 2×10^{33} grams, or about 330,000 times greater than the Earth's.

Note that we cannot find the mass of some other planet by substituting the mass of the Sun just found into equation (C), along with the values of P and a appropriate to the planet in question. This is because the mass of any planet is only a tiny fraction of the mass of the Sun. Even Jupiter, the most massive planet, is less than 0.001 the mass of the Sun. (Subtracting the mass of an elephant from the combined mass of the elephant and the flea is not a good way to measure the mass of the flea!) Instead, use the motion of a planet's satellite around its parent body to find the latter's mass; that is, substitute the period of the satellite's orbit and its semimajor axis a into equation (C). Now it is the satellite's mass, which is so much smaller than that of the planet, that the sum of the two essentially equals the mass of the planet alone.

Planets with no satellites, like Venus and Mercury, present a more difficult situation. We must first calculate the orbit of some object—a comet, for example—when it is so far away from Venus, say, that its motion is unaffected by the planet. Then the difference of that orbit from the observed motion of the comet as it passes Venus gives the effect, the **perturbation**, of Venus on the comet's motion (see Figure 8.13). From the magnitude of the perturbation, the mass of the planet can be deduced. Today this can be done with much greater accuracy using space probes, whose orbits can be accurately tracked as they pass very near to Venus or Mercury.

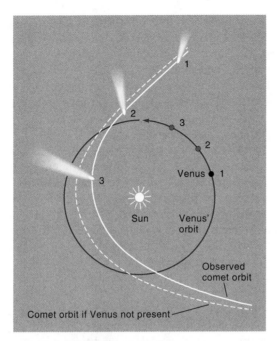

Figure 8.13. The dashed line shows the orbit a comet might follow in the inner part of the solar system as derived from observations of the comet's motion when it was at a much greater distance from the Sun, undisturbed by Mercury or Venus. The solid line shows how this orbit is modified by the gravitational influence of Venus.

The Motion of a Satellite in Orbit

In his *Principia*, Newton has a diagram similar to that shown in Figure 8.14. If a projectile is fired from the top of the mountain with a relatively slow speed in the horizontal direction (that is, in a direction perpendicular to the direction to the center of the Earth), it would soon fall to the Earth. If the force of gravity somehow were not

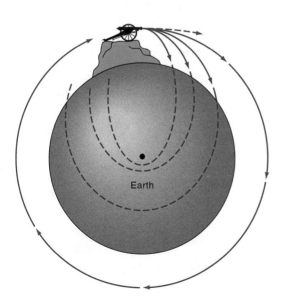

Figure 8.14. A figure adapted from Newton's *Principia*. Paths of a projectile fired from a mountain are shown. Orbits are sketched that the projectile would follow if the Earth were replaced by a point mass.

acting, however, the projectile would fly off the Earth along a straight line. The path of the projectile is the resultant of two motions: that imparted horizontally by the explosion of the gunpowder, and the accelerating motion toward the center of the Earth.

If its speed is increased, the projectile would fall a greater distance from the mountain. The paths followed by the projectiles before they hit the Earth are sections of Keplerian ellipses, with one focus at the center of the Earth and the other near the mountain top. (If all the Earth's mass was actually located at a point at its center, the projectiles would move along elliptical paths shown by the dotted lines in the figure.) When the horizontal speed is increased to about 8 km/sec, the projectile's speed is large enough that it would move around the Earth in a circular path, returning to its starting point each orbit. In this case the projectile's speed is such that it falls toward the Earth at just the rate that returns it to its starting point. Further speed increases would result in elliptical orbits of increasing size. When the speed was about 11 km/sec, the projectile would escape from the Earth along a parabolic path, never to return. That is, the **escape velocity** from the Earth is 11 km/sec. With still-greater initial velocities, the projectile would leave the Earth along a hyperbolic track.

In all the examples given above, the projectile is subject to the same acceleration toward the center of the Earth. The mass of the Earth is not changing; so the projectile is always falling to the Earth at the same rate. What is different in each case is the horizontal speed with which it is fired. We call the position, direction, and speed of the projectile as it leaves the cannon the **initial conditions**; they determine the path the projectile will follow when it is fired from the mountain. Note that the direction as well as the magnitude of the initial velocity is important.

Now consider a satellite moving in a circular orbit around the Earth. Let's find an expression for the speed with which it is traveling. The time it takes to make one trip around the Earth is just the distance it travels divided by its speed, or $t = d/V$. We can rewrite this in terms of the period and semimajor axis (now just the radius of the circular orbit) as follows:

$$t = P = \frac{2\pi a}{V} \; ;$$

squaring both sides,

$$P^2 = \frac{4\pi^2 a^2}{V^2} \; .$$

If we substitute $(4\pi^2 a^2)/V^2$ for P^2 in equation (C′) above, we have

$$\frac{(M_1 + M_2)(4\pi^2 a^2)}{V^2} = \frac{4\pi^2 a^3}{G} \; ;$$

so

$$\frac{(M_1 + M_2)}{V^2} = \frac{a}{G} \; ,$$

or

$$V = \sqrt{\frac{G(M_1 + M_2)}{a}} \; . \qquad \text{(E)}$$

Since a is in the denominator, the larger the radius of the orbit of the satellite, the smaller is its speed; this is just what we would expect.

What is the speed required to put an artificial satellite in a low-altitude circular orbit around the Earth? If M_1 is the mass of the Earth, then the mass of the satellite M_2 is completely negligible compared to M_1; also, $a = R_\oplus$, and so we have

$$V = \sqrt{\frac{GM_\oplus}{R_\oplus}}.$$

With this formula we can calculate that the speed of a satellite in a low Earth orbit is indeed about 8 km/sec. Furthermore (although we will not show it), it turns out that the speed required for a satellite to barely escape (in a parabolic orbit) from its primary is just $\sqrt{2}$ times its circular velocity. Thus for our near-Earth satellite, the escape velocity is $\sqrt{2} \times 8$ km/sec = 1.41×8 or about 11 km/sec, as mentioned above.

Suppose the orbit is not circular, but elliptical—what then? In that case the speed of a satellite at any point a distance r from the focus of the ellipse is given by

$$V = \sqrt{G(M_1 + M_2)\left(\frac{2}{r} - \frac{1}{a}\right)}. \tag{F}$$

Note that in a circular orbit $r = a$ so that equation (F) reduces to equation (E). We shall use these formulae in Chapter 23 when we plan a trip to Mars.

An astronaut in an Earth-orbiting spaceship is weightless, but not because he or she is somehow beyond the reach of gravity. Rather, our astronaut is falling toward the Earth at just the same rate as the spaceship. Hence, no muscular effort (which produces the perception of weight) is needed to keep the astronaut from crumpling to the floor. Weightlessness is easy to achieve, at least momentarily. Sky divers, for example, are weightless until their parachutes open.

Some Aspects of Newton's Laws

Comets

Newton's theory of gravity was the first theory of modern science. It provided a conceptual framework within which all the observational data then available fit exactly. With the laws of motion and of gravitation, the fundamental problem of astronomy (and a lot more!) for the previous 2,000 years was finally solved. The motions of celestial objects were understood and could be calculated to an accuracy limited only by that of the observational data.

Comets, for example, were now understandable as material objects subject to gravitational forces like any other masses. Their appearance and disappearance was simply a consequence of their large, elliptical orbits, which take them far out in the solar system before they return to visibility again near the Sun. This view was in stark contrast to the ideas held by some of the leading scientists of the time. For example, only a few decades earlier Galileo maintained that comets were not celestial objects orbiting the Sun, but rather clouds of atmospheric vapors that moved straight up from the Earth. When a cloud was out of the Earth's shadow, it would be illuminated by sunlight and we would see it as a comet. Now, however, there was no need to imagine them as vapors within the lunar sphere, or as mysterious objects, portents of disaster. Instead, they were objects firmly encompassed within the new physics.

This view was completely vindicated after Halley's successful suggestion that one comet, at least, was periodic. (By "periodic" is meant that it returns again and again to the vicinity of the Earth.) He noted the similarities of the orbits of the comets of 1531, 1607, and 1682, suspected that these were simply three different appearances of the same comet, and correctly predicted that it would reappear in 1758–1759.[3] It was then shown that the orbit of Halley's comet could be described accurately using Newton's laws.

Discovery of Neptune

Another example: in 1781 the first non-naked-eye planet, Uranus, was discovered by William Herschel (who at first thought it was a comet). An accurate orbit for it was soon established. By 1840, the difference between the calculated (predicted) and observed position of Uranus in its orbit amounted to more than 1 arc minute, an unacceptably large error (see Figure 8.15). This discrepancy was taken to be an indication that Uranus' motion was being perturbed by some other, as-yet-unknown, object. Two people carried out the laborious calculations required to find out the position of the disturbing object—a young, little-known Englishman, J. C. Adams, and the well-known French mathematician, U. J. Leverrier. As sometimes happens in such cases, Adams' work was largely ignored, and his request for a search of the part of the sky where he predicted the new object to be was not carried out by British astronomers. Astronomers at the Berlin Observatory, however, had just finished charting that part of the sky, and in 1846 quickly responded to a request from Leverrier to look for the unknown object. Within a few minutes, they found it—Neptune—a little less than 1° from its predicted position! So not only could Newton's laws accurately reproduce the motions of known objects, they were of such power that they could predict the existence of planets heretofore unsuspected.[4]

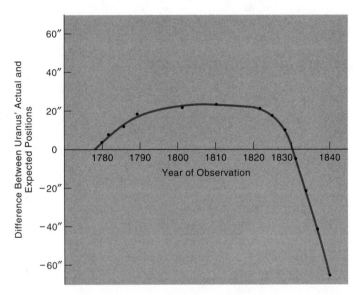

Figure 8.15. The discrepancy in the motion of Uranus; that is, the difference in the actual position of Uranus and the expected position based on calculations of its orbit from earlier observations. By 1840, the unexplained difference amounted to more than 1 minute of arc.

---

[3] Mark Twain, born in 1835, the year of the comet's next appearance, predicted that just as he came in with Halley's comet, he would go out with it; and he did in 1910.

[4] In January 1613, while observing Jupiter, Galileo noted the position of two "stars" in a line a few minutes of arc from Jupiter. In 1980 it was shown that one of these objects was indeed a star, but that the second one must have been Neptune, noted but unrecognized by Galileo 243 years before its "discovery."

Beyond the Solar System

Not only do Newton's laws work in the solar system, they also apply to the stars. For example, the motions of two stars gravitationally bound in orbits around each other, so-called binary stars, follow Kepler's laws. We gave an example above in which the masses of two orbiting stars were calculated. The existence of globular clusters implies the action of strong attractive forces between the stars. Though these clusters are too far away to allow the motions of individual member stars to be followed, their velocities are consistent with the action of Newton's gravity. The simple law of gravity does indeed apply throughout space.

Comments on Newton's Laws

Note the difference between Kepler's and Newton's laws: the former were derived by trial and error; there was no underlying basis, no well-defined physical context from which they arose. Newton organized various ideas, some his own, some from others, into a simple conceptual framework from which Kepler's laws fell out in a natural way. In his system there was no distinction between celestial and terrestrial physics: the *same* laws applied *everywhere*. These laws could be made to yield the position of Jupiter ten years in the future as well as the causes of today's tides; the wobble of the Earth's rotation axis as well as the time and place of visibility of the next solar eclipse. Furthermore, Newton's work clearly showed the power of mathematics in describing nature (or at least the physical world). The law of gravity could not be taken as just a convenient device for computing planetary motions. So many different phenomena could be accounted for so accurately by the mathematical expressions of Newton's laws that there could be no doubt as to the validity of the mathematical physics that he established.

Newton's work was the first—and still perhaps the best—example of the modern enterprise of physical science, namely, the search for the relations between various physical phenomena, with the aim of finding the smallest set of assumed concepts (like gravity) by which the greatest number and variety of phenomena may be accounted for. To be fully convincing, the theory must also predict new effects that must agree with observations as accurately as the data allow.

Note that Newton's laws describe only how the gravitating universe behaves; the "why" of gravity is not addressed. Most scientists since Newton have been satisfied with this approach, or at least accept it. Questions as to why masses exhibit gravity or why electrical charges interact are generally considered beyond the scope of science. If the ultimate aim of science is less ambitious than that envisioned by pre-Newtonian philosophers, it is also more successful. In fact, one can argue that success has come as a consequence of adopting a somewhat limited view of the aim of science.

It should not be hard to imagine the enormous impact made by Newton's laws and all the new physical understanding that resulted. The universe was popularly pictured as a giant piece of clockwork, requiring God only to set it up; forever after it ticked according to Newton's laws. Mechanical models to explain physical phenomena became popular, and in fact we still use them whenever possible. Science became fashionable; people felt that it provided a royal road leading inevitably to a better world, that "progress" was well-nigh guaranteed. It was one of the important factors in the rise of eighteenth-century optimistic rationalism. All this is well exemplified in Alexander Pope's lines:

> Nature and nature's laws lay hid in night.
> God said, "Let Newton be!" and all was light!

After Newton, astronomy progressed along two different paths. The one followed by most eighteenth- and nineteenth-century astronomers involved working out the mathematical consequences in the solar system of Newton's work, and obtaining the increasingly accurate observations of positions and motions of planets and satellites they required. The second path, the one we shall follow in Part III, took a few astronomers on telescopic journeys far beyond the solar system to see what was out there.

Until about the time of the second World War, cosmology developed very nearly independently of the rest of astronomy. To follow its more recent growth, however, requires an understanding of the nature of stars and the matter in between them, as well as some of the theoretical foundations of modern astronomy. In addition, our understanding of stars and stellar evolution has been one of the most spectacular astronomical success stories of the last half-century, and is worth pursuing for its own sake. This is what we will do in Part II of this book.

With two relatively minor exceptions,[5] *all* the information we have concerning the universe beyond our solar system comes from an analysis of light collected by telescopes. Consequently, we shall begin the next section by discussing briefly the characteristics of light or more generally, electromagnetic radiation, and then see how telescopes—and the instruments astronomers put on them—work. With these tools in hand, we will be able to understand a considerable portion of the life history of stars.

Terms to Know

Velocity, speed, acceleration, mass, inertia, angular momentum, force, weight; vectors and scalars; conic sections, escape velocity; perturbation.

Ideas to Understand

Newton's three laws of motion and of gravity; how they describe the motion of a stone falling to the Earth, the orbit of the Moon, or the orbit of an artificial satellite around the Earth. How to calculate the mass of the Earth, Venus, and the Sun, and your weight on Mars. Why Kepler's laws work; Newton's generalizations of Kepler's laws. How Newton's laws embody the characteristics of modern scientific laws; how scientific laws are abstracted from nature.

Questions

1. In the first eight chapters of this book there are nearly twenty comments concerning the nature and the characteristics of science.

(a) After identifying these, write a short essay on the important features of modern science.

(b) With these in mind, discuss in what ways and to what extent political science, economics, and clinical (i.e., bedside) medicine are and are not "scientific."

2. In what ways does the Ptolemaic system not meet the requirements of a modern scientific theory?

3. If the same force is applied to object A with mass M and to object B with mass $3M$, which object will experience the greater acceleration? by how many times more? Which of Newton's laws of motion is relevant here?

---

[5] The two exceptions are cosmic rays—very high energy particles that are produced in our galaxy—and neutrinos—neutral, massless particles detected from the Sun and a recent supernova.

4. (a) If Galileo had dropped a lead ball and a wooden ball, both the same size, from the tower of Pisa, both would have hit the ground at the same time. Explain in words why both balls experience the same acceleration.

(b) Explain why astronauts are weightless when in orbit around the Earth. Explain why, if an astronaut were to let go of the same lead and wooden balls, she wouldn't see either fall to the "floor" of the spaceship (i.e., the side of the ship toward the Earth).

5. The Earth and the Moon are bound together gravitationally. Explain your answers to the questions below.

(a) Compare the force exerted by the Earth on the Moon with that of the Moon on the Earth. Which of Newton's laws of motion is relevant here?

(b) Compare the acceleration experienced by the Earth because of the Moon with that experienced by the Moon because of the Earth. Which of Newton's laws of motion is most relevant here?

6. You should work this problem, as with many problems throughout this book, by taking ratios (see Appendix B).

(a) The diameter of Mars is about half that of the Earth. How many times greater is the surface area of the Earth than that of Mars? How many times greater is the volume of the Earth than that of Mars?

(b) The mass of Mars is about 1/10 that of the Earth's, and its radius is about half the Earth's. Would you weigh more or less on Mars than you do on Earth? By what factor?

7. (a) Suppose the mass of the Earth were equal to that of the Sun. How long would it take the Earth to make one trip around the Sun? (Assume the size of the Earth's orbit is unchanged.)

(b) Suppose the mass of Jupiter were equal to that of the Sun. How long would it take Jupiter to make one trip around the Sun?

Compare your results with the orbital periods of the Earth and Jupiter. Would Kepler's third law have worked if the planets had been massive? Explain.

8. Which of the planets—Venus, Mars, Jupiter, and Saturn—exert the largest gravitational force on the Earth at its closest approach to the Earth? Give your answers in terms of the Mars–Earth interaction (i.e., take the ratio of planet–Earth interaction to that of Mars–Earth). Which planet exerts the smallest gravitational force? (You will need the relative masses of the planets and their orbital sizes. Assume all orbits are circular.)

9. If on Earth you weigh 100 pounds, how much would you weigh on the surface of a planet with twice the mass and twice the radius of the Earth? How would your mass compare on the two planets?

10. Suppose the Sun were nine times as massive as it now is, and the Earth's orbit were unchanged. Would the year be longer or shorter? By about how many times? Explain your answers.

11. Dione, a satellite of Saturn, orbits Saturn at almost the same distance as our Moon orbits the Earth. The sidereal period of Dione is much shorter, however, only 2.7 days (i.e., 0.1 of our Moon's sidereal period). What is the mass of Saturn relative to that of the Earth (use ratios!)?

12. Two small objects of equal mass, *a* and *b*, follow orbits *A* and *B*, respectively, around the Sun (see Figure 8.16). Both objects happen to be at point *P* where their orbits nearly touch. The semimajor axis of orbit *B* is three times that of orbit *A*. Briefly explain your answers to the following.

(a) Which object has the longer orbital period?
(b) Which has the more eccentric orbit?
(c) Compare the force exerted by the Sun on *a* and on *b* at point *P*.
(d) Compare the accelerations of *a* and *b* at *P*.
(e) Compare the speeds of *a* and *b* at *P*. (Use equation E.)

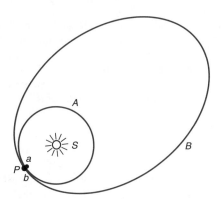

Figure 8.16. Sketch for Question 12.

13. We know the sidereal periods and semimajor axes of the orbits of Jupiter's satellites.

(a) Would we find P^2 (years) $= a^3$ (AU)? Explain.

(b) What useful information about Jupiter or its satellites could we derive from the values of P and a? How?

14. If the Moon were five times farther away from the Earth, how much would the force it exerts on the Earth be reduced? About how long would its sidereal period be, in months? (You do not need to know the numerical value of G to answer these questions. Just compare the "new" situation to the present one.)

15. A rocket burning chemical fuel is to be injected into orbit from the Earth. We wish to maintain a constant acceleration for 4 minutes during ascent. Does the applied force have to remain constant, increase, or decrease during this period? Explain your answer using Newton's laws of motion.

16. (a) If you drop a stone from a cliff, after 3 seconds it will be moving at nearly 30 m/sec. Show how to calculate this result from what we know about acceleration of gravity near the Earth's surface.

(b) Now throw the stone horizontally from the cliff with a speed of 10 m/sec. Three seconds later what is its horizontal speed? (Remember Newton's first law!) What is its vertical speed? Explain briefly.

Suggestions for Further Reading

Christianson, G., "Newton's *Principia*: A Retrospective," *Sky & Telescope*, **74**, p. 18, 1987. An interesting account of how the *Principia* came to be written and published.

Cohen, I., *The Birth of a New Physics*. Garden City, NY: Doubleday and Co., Inc., 1960. This short, well-written book gives some of the Greek background, then picks up the story with Copernicus, but its emphasis is on the development of Newtonian physics.

————*The Newtonian Revolution*. Cambridge: Cambridge University Press, 1980.

————"Newton's Discovery of Gravity," *Scientific American*, **244**, p. 167, March 1981.

Feynman, R., *The Character of Physical Law*. Cambridge, MA: The MIT Press, 1965. Six popular lectures given by a well-known physicist. Topics discussed include gravity, the relation of math to physics, and conservation principles.

Gamow, G., *Gravity*. Garden City, NY: Doubleday and Co., Inc., 1962. The nature of gravity, especially the work of Newton, but also that of Galileo and Einstein, written by an eminent physicist who was also a skillful popular writer.

Nicolson, M., *Newton Demands the Muse: Newton's Opticks and the Eighteenth Century Poets*. Princeton: Princeton University Press, 1966. Some of the literary consequences of Newton's work on optics.

Westfall, R., *The Life of Isaac Newton*. Cambridge: Cambridge University Press, 1993. The non-technical version of a longer biography written by an authority on Newton.

Hubble Gallery. A portion of M16 is shown here. Stars are forming in the finger-like protrusions in the cloud of molecular hydrogen. Each "nob" is a few times larger than our solar system.

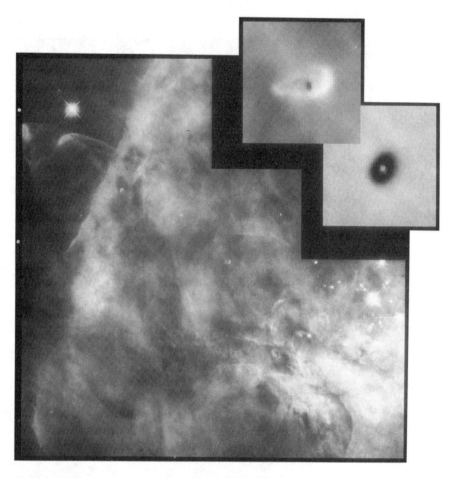

Hubble Gallery. A portion of the Orion nebula, a large cloud of gas and dust about 1500 light-years away. The small dark and bright blobs across the center of the image are "cocoons" of gas and dust in which stars are forming. Close-up views of two of these objects, both a few times larger than our solar systems, are shown in the insets.

PART II

The Life and Death of Stars

"When it comes to atoms, language can be used only as in poetry. The poet, too, is not nearly so concerned with describing facts as with creating images."

—J. Bronowski, 1975; British scientist

"The atom of modern physics can be symbolized only through a [mathematical equation]. All its qualities are inferential; no material properties can be directly attributed to it. That is to say, any picture of the atom that our imagination is able to invent is for that very reason defective. An understanding of the atomic world in the primary sensuous fashion . . . is impossible."

—Werner Heisenberg, 1945; one of the founders of the new physics

"The electron is not as simple as it looks."

—Sir William Bragg (1890–1971); British physicist

CHAPTER 9

Radiation and Atoms

We have finished giving an account of the development of early ideas concerning the motions of solar-system objects and their organization or arrangement. These ideas are necessarily tied up with the beginnings of modern astronomy, which we also described. In this, the second part of the book, we will consider what is known about the structure and evolution of stars, including how stars are formed, their energy sources, how they change as they run out of energy, and finally how they die.

The content of the last sentence above is straightforward enough, but that should not mislead you about just how remarkable a statement it is. Remember that stars are far away and, with the exception of the Sun, appear to us only as points of light. We know them only through the radiation they emit. As late as the middle of the last century, most people thought that learning the chemical composition of even the Sun would be forever beyond our capability. But now we know not only what the Sun and many other stars are made of, but also a great deal about their internal structure and how that changes during their lifetimes. We understand the source of the enormous energy stars produce, how long they will live, and their ultimate fate.

This is the result of many factors, such as the development of the telescope as well as the various kinds of instruments that analyze and record the light collected by the telescope. The spectrograph is particularly important in this connection, since it enables us to "dissect" the spectrum of starlight. Though Newton showed in 1666 that ordinary white light is made up of all the colors, our empirical understanding of light advanced little until the nineteenth century. In 1814 the German Joseph Fraunhofer observed the solar spectrum and mapped more than 300 of its features. We still use some of the notation he established to designate particular features in the spectrum, for example, the yellow sodium "D" lines. As the century progressed, instruments were improved that produced better spectra of the Sun, and by the 1870s instruments sensitive enough to record the spectra of bright stars photographically were in use. Astronomers found that spectra of stars could be arranged in such a way that spectral features changed gradually from one group to the next; that is, they classified the stellar spectra to distinguish one from the next.

It would have been impossible to interpret all the new observational data, however, without a corresponding development in physical theory, especially the theory of atoms. The physical understanding of stellar spectra began in 1913, with the work of the Danish physicist Neils Bohr (1885–1962). This was a major step along the road to the development of modern physics, which has provided the theoretical basis for understanding not only the stars, but other astronomical objects as well. In the following chapters we will describe some of these developments.

Since very nearly all the observational data we can acquire in astronomy come to us in the form of electromagnetic radiation, we must first describe some of its general properties. We must also learn about **continuum radiation**—a spectrum with all wavelengths present, the characteristics of which depend on the temperature alone. In addition, we will describe **line radiation**—the radiation absorbed or emitted by atoms at specific wavelengths. It is by the analysis of both of these kinds of radiation from stars and nebulae that we have been able to measure the physical properties of these objects—their temperatures, densities, pressures, chemical composition, etc.

Characteristics of Electromagnetic Radiation
Electromagnetic Waves

Light is said to be an **electromagnetic wave**. What does this mean? A wave, like one in water, is a periodic (or regularly repeating) disturbance that carries energy. Undisturbed, the surface of a lake is flat and smooth. When the wind blows, waves are formed, their energy acquired from the wind. A piece of wood bobbing up and down derives its energy of motion from the wave. Though we can't see a light wave directly, the energy it carries stimulates receptors in the retinas of our eyes, producing the sensation of light.

That light is an electromagnetic wave means that many of its properties can be understood in terms of a periodic disturbance that has both electric and magnetic properties. This means, for example, that the wave interacts with (that is, exerts a force on) electrical charges. Figure 9.1 represents a snapshot of a wave passing over a row of electrons; the arrows indicate the magnitude and direction of the electromagnetic force the wave exerts on the charges.

Electromagnetic radiation does not require a medium to be transmitted; it can travel through empty space (otherwise there would be no astronomy!). When a light wave passes through a vacuum there is, of course, no motion of material particles, only the regularly increasing and decreasing strength of an electric and magnetic field. By an electric "field" we simply mean a region of space in which an electric charge would experience a force. Similarly, a gravitational field is a region of space in which a mass experiences a gravitational force. The magnetic field surrounding the Earth causes a compass needle to point north.

The electric and magnetic aspects of an electromagnetic wave can be pictured as two separate waves both traveling in the same direction, but whose planes of oscillation are at right angles (at 90°) to each other; see Figure 9.2. If the wave passes by an electrically charged particle like an electron, the electron is set in motion, moving up and down in a direction *perpendicular* to the direction of travel of the wave. A par-

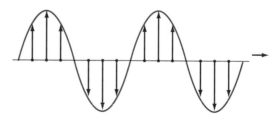

Figure 9.1. The arrows show the size and direction of the force acting on electrons as an electromagnetic wave passes over them. Note that as the wave moves to the right, the electrons simply bob up and down vertically in response to the changing force acting on them.

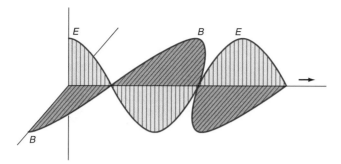

Figure 9.2. As its name implies, an electromagnetic wave has two components—one electrical (*E* in the diagram) and one magnetic (*B*). Their planes of vibration are perpendicular to each other. The wave exerts both electrical and magnetic forces on the region of space it passes over.

ticle that is electrically neutral would experience no force at all as the wave went by, since it would have no electrical interaction with the wave.

Sound is also a wave phenomenon, but one that is quite different from light. Sound requires a medium for its transmission—for us most commonly that medium is air; for whales it is water. Molecules of air are put into a back-and-forth motion, in the *same* direction as the direction of travel of the wave, by the energy of the sound wave. The oscillating molecules strike our eardrums, which send signals to the brain we interpret as sound.

In this book we will use the term **light** as a convenient shorthand to denote electromagnetic radiation in general. Thus the term includes radiation to which our eyes are not sensitive, such as x-rays. Electromagnetic radiation that our eyes detect we will call **optical radiation** or the **optical spectrum**.

Wavelength, Amplitude, Speed, and Polarization

Any wave, be it light, sound, or water, can be characterized by its wavelength, its amplitude, and its speed. In addition, an electromagnetic wave can be characterized by the direction of a force on a stationary electric charge. As shown in Figure 9.3, the **wavelength** is just the distance between successive peaks or troughs of the wave; it is usually denoted by the Greek letter λ (lambda). The **amplitude** is the height of the

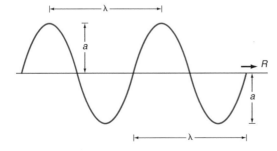

Figure 9.3. The wavelength of a wave is the distance between two successive corresponding points: from one peak to the next, or one trough to the next, etc. Its amplitude, *a*, is a measure of the energy of the wave. The direction along which it travels is denoted by the ray *R*.

wave; the greater the amplitude, the more energy carried by the wave. (Compare the energy of a gentle wave lapping on a lake shore with a ten-foot-high ocean wave in the surf.) The **speed** of the wave is the rate at which its energy—the "disturbance"—travels. It is not the speed of any particles that may be responding to the action of the wave. The direction of travel of the wave is usually specified by a direction we call a **ray**.

A useful concept concerns the number of waves—the number of peaks or troughs—that pass a given point in one second. For a given wave speed, the shorter the length of the wave, the greater will be the number of waves passing a point each second. We call the number of waves passing by each second the **frequency**, f, of the wave. If ten waves each with a length of two centimeters pass a point in one second, then $10 \times 2 = 20$ centimeters of wave will have gone by; so its speed is 20 cm/sec.[1] In general,

$$\lambda f = V .$$

The speed of an electromagnetic wave in empty space (usually denoted by c) is 3×10^{10} cm/sec (or 3×10^5 km/sec or 186,000 miles/sec). So for electromagnetic waves in empty space we write

$$\lambda f = c .$$

The speed of light is one of the fundamental constants of nature, the same at all places and at all times, as far as we know. Furthermore, theory as well as countless numbers of experiments show that it is the *maximum* velocity that a material object can attain or at which energy can be transmitted.[2] Whenever a light wave passes through a medium of any kind, like glass, for example, its speed is less than c, never more. In air the speed is decreased so slightly that for nearly all purposes one can still use c, the vacuum speed.

Radiation from an ordinary light bulb consists of innumerable waves traveling out from the bulb, their electric planes of oscillation oriented in all directions, randomly. (In discussing polarization it is customary to consider the oscillations of the electric portion of the wave, not the magnetic.) Such radiation is said to be **nonpolarized**. If more than the average number of the electric waves vibrate in one particular plane of oscillation, however, the radiation is said to be **polarized** (see Figure 9.4). The degree of polarization depends on just what fraction of the waves are vibrating in a particular plane; if they all are in a given plane, then the radiation is completely polarized. Most astronomical sources emit radiation that is unpolarized or only very slightly so, but some objects emit strongly polarized light. Sunlight bouncing off surfaces on the Earth is partially polarized. Polaroid sunglasses allow vibrations in only one plane to pass through to your eyes, thus eliminating much reflected light and reducing the brightness of the optical radiation you see. We shall see in later chapters that polarization can give us important information concerning the process by which radiation is generated.

The Electromagnetic Spectrum

The wavelength is a convenient way to characterize one aspect of electromagnetic radiation. For example, blue light has a wavelength of about 4500 Å (pronounced Ångström, after the nineteenth-century Swedish spectroscopist; 1 Å $= 10^{-8}$ cm); the wave-

---

[1] Note that the units are consistent: the unit of λ is a length, say centimeters; that of f is "per second" as in 10 waves per second. Their product is cm $\times$ 1/sec = cm/sec, which is the unit of speed. Note that you can multiply or divide units just as you do algebraic quantities.

[2] Speeds greater than c, so beloved of science fiction writers, are not science but fiction.

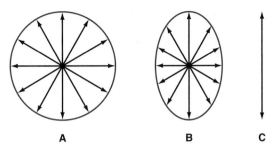

Figure 9.4. Sketch (A) shows the planes of vibration of many waves moving directly toward you (out of the page). This beam of radiation is completely unpolarized, that is, the amplitudes of the waves are the same in every direction; there is no preferred plane of vibration; all—vertical and horizontal—are represented equally. In (B), however, the amplitudes of the waves are greater in the vertical planes than in the horizontal, and the light beam is partially polarized. Sketch (C) shows completely polarized radiation; all the electric parts of the waves vibrate in the vertical plane.

length of red light is about 6500 Å. In other words, wavelength specifies the color of optical radiation. There's a lot more to electromagnetic radiation than just **visible** light, however. Wavelength regions shorter than the visible include γ (**gamma**)**-rays**, **x-rays**, and **ultraviolet** light; **infrared** radiation and **radio** waves have longer wavelengths than visible radiation (see Figure 9.5). These kinds of radiation are all the same phenomenon, and differ from each other only by their wavelength or frequency (the list just given is in order of increasing wavelength). It is important to realize that these names—ultraviolet, radio waves, etc.—are used for convenience only, and do not imply any natural divisions or sharp boundaries in the spectrum.

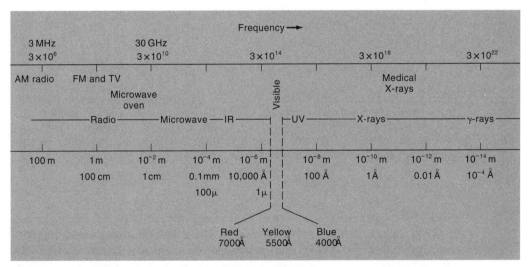

Figure 9.5. A representation of the electromagnetic spectrum. The various regions of the spectrum (radio, ultraviolet, etc.) are given names for convenience only. Note how narrow the strip of spectrum is to which our eyes are sensitive. In our daily lives, however, we make use of a broad range of the electromagnetic spectrum: medical and dental x-rays, visible light, infrared heat lamps, microwave ovens, AM and FM radio, and television. Notice the various units of wavelength shown in the lower part of the figure.

The dial on your radio gives the frequency of the radio waves in the number of cycles (waves) per second, or **Hertz**, after the nineteenth-century German physicist Heinrich Hertz, who first investigated the properties of electromagnetic radiation. "One thousand" on your AM dial is to be read as 1000 kHz (kiloHertz) or $f = 10^6$ Hz; one million waves pass a given point each second. This corresponds to a wavelength of $\lambda = c/f = (3 \times 10^{10})/10^6 = 3 \times 10^4$ cm or 300 m. By contrast, the wavelength of the x-rays your dentist uses is about 0.5 Å or 5×10^{-9} cm.

Note that our eyes are sensitive to an extremely narrow range of wavelengths, and what we call ordinary visible light is only a tiny portion of the entire electromagnetic spectrum. In that sense, we are very nearly blind. The explosive growth of astronomy since World War II has been caused largely by new kinds of ground- and space-based telescopes and detectors, which enable us to observe the radiation emitted by celestial objects over very nearly the whole electromagnetic spectrum.

How the Intensity of Light Falls Off with Distance

Imagine a very small light bulb shining at the center of an empty sphere of radius R (Figure 9.6). A certain amount of radiation from the bulb falls on each square centi-

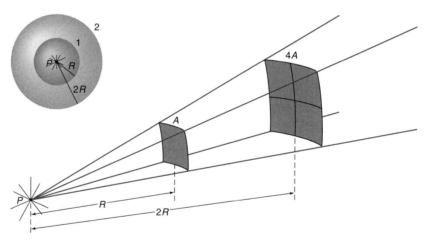

Figure 9.6. Light from a point source at P radiates out in all directions. When the light strikes sphere 1, it has spread over an area of $4\pi R^2$. When it reaches sphere 2, which has twice the radius of sphere 1, it has spread over an area of $4\pi(2R)^2 = 4\pi(4R^2) = 16\pi R^2$ or 4 times the area of sphere 1. Hence at twice the distance, the amount of energy falling on each unit of area has decreased by a factor of 4.

meter of the sphere's interior surface. Now double the radius of the sphere; its area ($= 4\pi R^2$) is increased by four times (2×2). The light bulb is unchanged, and none of the light is absorbed, so the same total amount of energy from the bulb is now spread over four times as much area as before; hence only one-fourth as much energy falls on each square centimeter. Double the sphere's radius again, and the energy per square centimeter will be only one-sixteenth the original amount. In other words, the intensity of radiation falls off inversely as the square of the distance from the source, that is, as $1/R^2$. (Note that this is the same dependence on distance as is true for the force of gravity.)

Thermal Radiation
The Continuous Spectrum

What general sorts of processes can electromagnetic radiation undergo? It can be *transmitted* (through a medium like glass or water as well as through empty space); it can be *reflected* (which is how the Moon and planets shine, and how we see each other); it can be *absorbed* (which is how foods are cooked and how we get sunburned); and it can be *emitted* (like light from stars or the glow from a red-hot electric burner). The first three processes are not strongly affected by the temperature. As long as you don't boil the water or vaporize the Moon, the transmission of the water and reflectivity of the Moon don't change much with temperature. The fourth process, however, the emission of radiation, depends strongly on the temperature: the higher the temperature of an object, the shorter the dominant wavelength radiated, and the greater the total amount of energy emitted. Our bodies emit radiation (roughly as much energy as a 100-watt bulb) as well as reflect and absorb it, but since our body temperature is relatively low, the little energy we do emit is in the far infrared. This radiation can be detected by infrared-sensitive night-vision devices, but not by the unaided eye.

The Concept of Temperature

Temperature is a familiar notion—though perhaps a slightly fuzzy one—that we use daily. Because it is such an important concept in astronomy, you should be certain that you understand how the term is defined in science.

The molecules in the air surrounding you are not stationary, but are rapidly moving this way and that as they constantly collide with their neighbors. Their **kinetic energy**—their energy of motion—is given by the formula

$$K.E. = \frac{1}{2}mV^2,$$

where m is the mass of the atom or molecule and V is its velocity. As they collide, their energy tends to become equally distributed, the less energetic particles gaining from the more energetic ones. Because the air molecules undergo many collisions each second, they all quickly acquire the same average kinetic energy. Temperature is a measure of that energy; the greater the average kinetic energy of the particles of a gas, the higher its temperature.

It is important to realize that electrons, atoms, and molecules in a volume of gas of a given temperature (that is, with the same average kinetic energy) will not have the same speed. In order for a molecule of O_2 to have the same kinetic energy as a molecule of CO_2, it will have to move more rapidly than the heavier molecule. Similarly, an electron, being the least-massive particle, will be moving faster than any other constituents of the gas, so that, on average, it has the same kinetic energy as any of the other particles.

In this country the Fahrenheit scale is in everyday use, but in science—and in this book—either the centigrade or the Kelvin scale is used. If you are not familiar with them, you should read Appendix B where they are described.

Blackbody Radiation

Now consider the following situation: construct a well-insulated hollow box, heat it uniformly, and keep it at a constant temperature (Figure 9.7). Make the inside walls

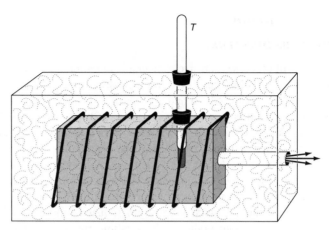

Figure 9.7. This shows schematically how a blackbody might be made. The inner hollow box, with its interior walls painted a dull black, is uniformly heated by an electrical current passing through resistive wire (like in a toaster). A thermometer, *T*, measures the temperature in the box. This box is placed in a larger container, where it is surrounded by insulation. A tiny hole in the inner box enables radiation to escape to the outside as shown. The spectrum of this radiation will be like that shown in Figure 9.8.

of the box such that they neither transmit nor reflect any radiation, but instead absorb all the energy striking them. This could be done by painting the walls a dull black or by coating them with soot, for example. Because all the radiation is absorbed, the wall temperature would go up unless this energy were re-emitted. Since the temperature inside the cavity of the box is kept constant, however, as much energy must be emitted by the walls as is absorbed. If there were an imbalance, the temperature would change. Such an object is called a **blackbody radiator** (black because no light is reflected by the cavity walls). The radiation inside the box (which is constantly being absorbed and re-emitted by the walls) is said to be in **thermal equilibrium** with the walls; that is, the temperature of the walls and of the radiation are the same.

Cut a tiny opening in the cavity, so that only a negligible fraction of the radiation escapes, and analyze the spectrum of that light. We would find that the characteristics of the radiation depend only on the *temperature* of the cavity, not on its composition or shape. Furthermore, these radiation characteristics can be completely described mathematically from general theoretical considerations.

The spectrum of blackbody radiation is shown in Figure 9.8. The amount of energy at each wavelength is plotted against wavelength for radiation emitted by blackbodies at three different temperatures, T_3 being the hottest. Notice several things about these energy curves:

(1) Each is continuous; that is, there are no gaps. The emitted energy changes smoothly from one wavelength to the next.

(2) The spectrum of the highest-temperature blackbody (T_3) reaches a peak at the shortest wavelength (λ_3), and the peak shifts to longer wavelengths as the temperature decreases.

(3) The hotter object emits more energy at every wavelength than does the cooler one. Therefore, the total energy emitted over the whole spectrum is greatest for the hottest blackbody and least for the coolest. All of this corresponds to our experience: as an object is heated, its color changes from dull red, to orange, to yellow (that is, to shorter wavelengths), and it becomes brighter (it emits more energy).

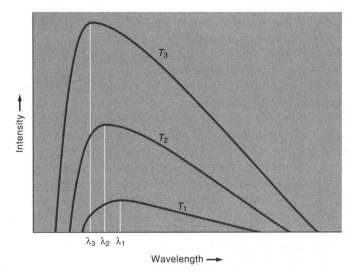

Figure 9.8. The spectrum of blackbody radiation at three temperatures, T_3 being the hottest. Note that as the temperature is increased, the wavelength at which the greatest amount of energy is radiated shifts toward shorter wavelengths, and that the area under the curves (which represents the total amount of energy radiated) rapidly increases.

The formulae representing the last two properties of blackbody radiation are quite simple. The relation between the temperature and the peak wavelength of the spectrum is given by

$$\lambda_{max} T = \text{constant} .$$

Here λ_{max} is the wavelength at which the maximum energy is radiated (the wavelength corresponding to the highest point on the energy curve for a given temperature; it is denoted by λ_1, λ_2, λ_3 in the figure). Since the product of the two quantities λ_{max} and T remains constant, if one increases, the other must decrease. Consequently, as T increases, λ_{max} decreases; that is, the wavelength at which the maximum amount of radiation is emitted becomes shorter. If λ is in Ångströms, T in degrees Kelvin, the constant in the equation is 29,000,000 Å K.

The formula that describes how an object becomes brighter as it gets hotter is also simple:

$$E = \sigma T^4 .$$

Here E is the total energy per square centimeter radiated per second over *all* wavelengths by a blackbody at temperature T (σ—the Greek letter sigma—is just a constant).[3] The important point is the way the energy depends on T: if T is doubled, the total energy radiated (which is related to the brightness) increases by $2^4 = 2 \times 2 \times 2 \times 2 = 16$ times; if T is tripled, the energy radiated increases by 3^4 or 81 times.

Note that these formulae say nothing about the shape of the blackbody itself, or even its composition. It doesn't make any difference if it is made of iron, copper, or uranium; the emitted radiation depends only on the temperature of the blackbody. This is a remarkable fact! The radiation in the cavity arises from the absorption and emission by the many tiny atomic oscillators—vibrating systems—in the walls. At any given temperature these oscillators shuffle the available energy among themselves

[3] The value of this constant is 5.67×10^{-5} in cgs units.

through collisions so that overall they attain the most probable distribution of energy. The blackbody spectrum is just a consequence of this most probable distribution of energy among the oscillators. In turn this distribution depends only on the average energy of the oscillators, that is, only on their temperature.

A formula that reproduced the spectrum of a blackbody—how much energy is emitted at every wavelength—was much more difficult to derive than the two equations just given. Before 1900, physicists had made the perfectly natural assumption that the oscillators (which we now know to be electrons) emitting and absorbing radiation were in no way restricted in terms of the exact amount of energy they could have, within the limits implied by their temperature. Working within this framework, however, physicists were unable to find a mathematical formula that represented the spectrum of blackbody radiation. Finally, in 1900, the German physicist Max Planck (1858–1947) succeeded, but only by making the radical (and puzzling!) assumption that the oscillators could not take on just any random value of the energy. Instead, a formula giving the correct shape of the spectrum could be derived only if the oscillators' energies were restricted to certain well-defined, discrete values that depend on the frequency of the light emitted. That is, heat could be converted to light only in tiny, individual chunks of energy. The energies are said to be **quantized**. From this apparently innocuous beginning grew the profound revolution in science called the quantum theory, one aspect of which is the wave-particle duality we describe below.

Photons

Five years later, Albert Einstein (1879–1955) became convinced that something much more fundamental was going on than Planck's arbitrarily imposed discrete energy limitations on electron oscillations. Instead, he showed that it was in the very nature of light to be quantized, that is, that light itself acted as if it were made up of particles or packets of energy; these were christened **photons**. Each photon is an indivisible unit: we can talk about 385 photons, but not about 385.7. Einstein found that each photon carries an energy that is proportional to the frequency of the radiation;[4] the higher the frequency (and shorter the wavelength), the greater the photon's energy. Its energy is given by the simple formula

$$E = hf,$$

where h is another important physical constant, everywhere the same. It plays as fundamental a role in modern physics as does the speed of light, and is called **Planck's constant**. It gives the conversion factor between the frequency and the energy of photons.

X-ray photons have shorter wavelengths, higher frequencies, and so greater energy, than ordinary light, and hence greater penetrating power. The greater the number of photons in a beam of radiation, the brighter is its intensity. A 100-watt bulb produces much more light—many more photons each second—than a 5-watt bulb.

Wave or Particle?

Understand the situation: in many circumstances it is impossible to describe electromagnetic radiation as a wave—it just doesn't work—and we have to think of it as particles. Note the big difference between the two models: we think of a wave as spread

---

[4] Note the mixture of concepts: frequency is associated with a wave; yet it is being applied to something which we tend to think of as a particle.

out in space and a photon as being sharply localized in space; yet both are applied to electromagnetic radiation. Obviously, models based on everyday experience, our familiar waves or particles, may not work well in the microscopic world.

Not only does light sometimes behave like a wave and in other circumstances like a particle, but what we ordinarily think of as a particle—an electron, for example—sometimes exhibits wave properties. That is, in some situations it is necessary to associate a wavelength with an electron. These are aspects of the so-called wave-particle duality in our representation of physical phenomena on the microscopic scale. The recognition of this duality forms one of the foundations of the revolution in physics, which began around the turn of the twentieth century. Furthermore, this duality exemplifies well one of the characteristics of modern science; namely, it describes how light (in this case) *acts*, how it behaves, sometimes like a wave, in other circumstances like a particle, but nobody says what it "really is." In fact, this is considered a meaningless question in modern physics just as is asking the same question about gravity.

Even though a star doesn't look much like a blackbody (its hot surface is not enclosed but radiates out to cold, interstellar space), we shall see that these laws are useful in representing the radiation characteristics of stars, which is, of course, why we have described blackbody radiation.

Atomic Structure and Line Spectra
Constituents of Atoms

As we have seen, blackbody radiation depends only on its temperature, not on any details like the composition of the blackbody. That is, it is a function of a macroscopic (large-scale) property, not of any microscopic ones. Now let's look at radiation that depends explicitly on microscopic—small-scale—structure, namely that of atoms. We will begin with a brief review of their constituents.

Atoms are made of **electrons**, denoted by e^-, which carry a negative electrical charge; positively charged **protons**, p; and electrically neutral **neutrons**, n (see Table 9.1).[5]

Table 9.1. Charge and mass of the electron, proton, and neutron

Particle	Charge	Mass
Electron (e^-)	-1	9.1091×10^{-28} grams
Proton (p)	$+1$	1.67262×10^{-24} "
Neutron (n)	0	1.67493×10^{-24} "

The terms "negative" and "positive" charge were introduced in 1749 by Benjamin Franklin in his descriptions of his experiments with electricity. He believed that electricity was a fluid that could be excessive (positive) or deficient (negative) in an object. Note that the neutron and the proton are of very nearly equal mass, and that each is about 1,840 times more massive than the electron.

---

[5] There also exists an analogous family of antimatter, which consists of negatively charged protons called antiprotons; positively charged electrons, called antielectrons or, more commonly, positrons; etc.

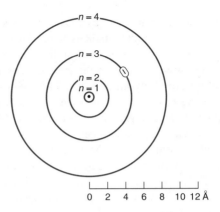

Figure 9.9. A simple model of the hydrogen atom. The dot at the center is the positively charged proton and the circles represent the allowed orbits of the electron (which is shown in the $n = 3$ orbit). The scale shows the radii of the orbits in Ångström units ($1 \text{ Å} = 10^{-8}$ cm). Although the orbits are to scale, the proton is not; in reality it is only about 10^{-13} cm in diameter or 1/100,000 the size of the atom!

The simplest atom of all is hydrogen. It consists of one proton (the nucleus of the atom) and one electron, which can be thought of as being in orbit around the proton, as sketched in Figure 9.9. When seen as a unit, the total electrical charge of the atom is zero, since the positive charge of the proton cancels the negative charge of the electron; we say that it is electrically neutral.

Note that the diameter of the atom is only a few Ångströms, which, however, is about 100,000 times larger than the nucleus when we consider it as a particle. Matter is mostly empty space! Atoms are so small, however, that under ordinary circumstances we are unable to detect this lumpiness or granularity. Instead, matter appears to us to be smooth and continuous. (This is a simple example of how our view of nature is affected by our size.)

The nucleus of ordinary helium contains two protons and two neutrons. In the neutral atom two electrons "circulate" around the nucleus. A neutral atom of common carbon has six protons and six neutrons in its nucleus, and is surrounded by six electrons. The charge on the nucleus is six, and its mass is about twelve times that of hydrogen. The number of protons in the nucleus (its **atomic number**) determines what species— hydrogen, helium, lithium, carbon, etc.—the atom is. A nucleus with six protons is still carbon even if it has only five, or four, or no electrons around it.

The number of protons plus the number of neutrons is called the **atomic weight** of the atom. Thus the atomic weight of hydrogen is one, and that of carbon is twelve. If a nucleus has six protons but seven neutrons, we say it is another **isotope** of carbon. It is carbon because its nucleus has six protons, but it is not common carbon because it has seven, rather than six, neutrons.

What Holds an Atom Together?

An atom as a single entity, that is, as a nucleus surrounded by its orbiting electrons, is held together because of the **electromagnetic force** of attraction between the electrons and the positively charged nuclear protons. (Recall that unlike charges attract and like charges repel each other.) In some ways this force acts like gravity, since it depends directly on the product of the interacting charges and inversely as the square of the distance between them. Hence, the electromagnetic force exerted by a charge goes to zero only at an infinite distance from it. It differs from gravity in that the electromagnetic force can be repulsive as well as attractive. In addition, electromagnetism is enormously stronger than gravity; the electrical force between an electron and a proton is about 10^{39} times stronger than the gravitational force between them! Gravity plays no role whatsoever in atomic structure.

But we have two problems here. Given the strong electromagnetic force between the nucleus and its surrounding electrons, why don't atoms collapse; that is, why don't the positively charged nucleus and the negative electrons coalesce? This question will be addressed in the next section. Second, why doesn't an atomic nucleus, with its positive charges all crammed together in a tiny volume, explode apart? After all, the six protons making up a carbon nucleus all repel each other through the electromagnetic force.

Let's first consider what overcomes this force and holds the nucleus together. The answer is provided by yet another kind of interaction between particles, the so-called **strong nuclear force**. It attracts two protons, for example, with a force about 100 times stronger than that with which electromagnetism repels them. Unlike the latter, however, the domain ruled over by the nuclear force is very small, only about 10^{-13} cm in diameter, or roughly the size of the nucleus. At greater distances, the strength of this force rapidly goes to zero. The strong nuclear force is indifferent to electrical charge: it is just as effective in gluing together protons and neutrons, or two neutrons, as it is in binding two protons. Furthermore, the strong force between two nuclear particles becomes *repulsive* when they are about 0.4×10^{-13} cm apart. As a consequence, an atomic nucleus doesn't collapse on itself. Thus, the nuclear force maintains the stability, the very existence, of nuclei; they neither explode nor collapse. Without it there would be no nuclei, hence no atoms, hence none of the matter we see around us, including ourselves.

For the sake of completeness let's mention the last of the known interactions, the **weak nuclear force**. It governs the spontaneous emission of particles like electrons from nuclei (radioactivity) and the decay of the neutron. Within a nucleus the neutron is stable, but outside it is unstable, decaying (breaking up) into a proton, an electron, and an antineutrino in about ten minutes (see Chapter 13).

As far as we know, these four—the strong and weak nuclear forces, electromagnetism, and gravitation—are the only fundamental forces in nature.[6] Together they account for all the known phenomena and the incredible variety of objects we see in the universe. Every kind of object, every kind of process, involves interactions described by one or more of these forces. We will return to this point on various occasions in the remainder of the text and in Chapter 21 in particular. Next, however, let's turn to the question of why the electrons surrounding the nucleus don't fall into it.

The Energy Levels of Atoms

For convenience we may say that the electrons orbit the nucleus, but the analogy between the atom and the solar system is very poor. First, the quantum picture of nature tells us that it is wrong to think of a microscopic particle like an electron as being similar to a tiny marble following a well-defined orbit. Instead it is more accurate to think

---

[6] We will see later that the electromagnetic and weak nuclear forces have been shown to be different aspects of a single "electroweak" force; strictly speaking, we should speak of three fundamental forces.

of it as able to follow many slightly different orbits, which, however, are not all equally probable. Put in another way, we can think of the electron as being slightly smeared or distributed around the nucleus at various distances from the atom with some of these distributions being much more likely than others. These more probable locations we will call **orbitals** to distinguish them from the notion of a sharply defined planetary orbit.

There is a second major difference between the structure of the solar system and the atom. In contrast to the planets, which in principle could have orbits of any size, it turns out that an electron in an atom is most likely to be found around the nucleus only in orbitals of certain sizes, each of which corresponds to a certain energy. We say that the orbitals are quantized. Furthermore, the electron does not coalesce with the positively charged nucleus despite the strong electrical force of attraction between the two. Instead, there is an orbital of *minimum* radius that represents the closest approach of the electron to the nucleus. More physically, it represents the lowest energy state of the atom; the electron never "falls into" the nucleus. This orbital is called the **ground level** or **ground state**. Left to itself, an electron will stay in the ground level. To move an electron from an inner to an outer orbital requires energy, just as it requires energy to move an artificial satellite from a smaller orbit around the Earth to a larger one. Suppose a photon with energy $E = hf$ (corresponding to a particular frequency or wavelength), equal to the energy difference between an inner and outer orbital, interacts with or "hits" an atom. The photon can be absorbed all at once, and its energy will "lift" the electron to the outer orbital. When the photon is absorbed it disappears, since all its energy has been taken up by the electron in the atom. A photon whose energy does not correspond to the exact difference in energy between any allowed orbitals (or energy states) will not be absorbed. The absorption is an all-or-nothing event. As a consequence, an atom absorbs light of only certain discrete wavelengths.

If an electron jumps from an outer to an inner orbital, it gives up energy in the form of a photon of just the energy difference between the two orbitals. In other words, it emits light of the same wavelengths it can absorb. It is convenient to illustrate the different electron energies for an atom by an **energy-level diagram** in which the energy levels correspond to the different orbitals. A sample energy-level diagram is given in Figure 9.10. The vertical separations between the energy levels are proportional to their energy differences. The $n = 1$ level is the ground state, and the $n = 2, 3$, etc., levels are higher-energy or **excited** states. Transitions 1 and 2 represent absorption (a photon is absorbed, and its energy used to "kick" the electron from the ground level to the higher-energy, excited levels); transitions 3 and 4 represent the emission of photons as the electron jumps from a higher to a lower level. Although we speak, crudely, of the electron jumping from one level to another, it is really the atom as a whole that is gaining or losing energy.

Each chemical element has a unique set of energy levels, and therefore a unique set of wavelengths it can absorb or emit. These wavelengths have been measured in the laboratory. We can identify the radiating atoms in a star as hydrogen, helium, carbon, etc., by measuring the wavelengths at which they emit or absorb energy, and comparing these wavelengths with those measured in the lab. In this way we can determine the chemical composition of distant stars just by analyzing their radiation.

If enough energy is supplied to an atom, either by photons of a high-enough frequency or by energetic collisions with other particles, an electron can be completely removed from the nucleus. In this case, the atom is no longer electrically neutral, and we say that it has been **ionized**, has become an **ion**. (This is somewhat analogous to a satellite being given enough energy to escape from the solar system.) Transition 5 in Figure 9.10 represents the ionization of the atom.

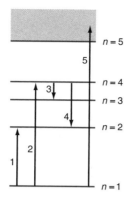

Figure 9.10. The energy-level diagram of an atom. The ground state (the lowest energy state) is labeled $n = 1$, and energies greater than $n = 5$ ionize the atom. Transitions 1 and 2 represent the absorption of energy by the atom, raising it to higher energy levels, whereas it loses energy in transitions 3 and 4. Transition 5 represents the ionization of the atom.

When an atom is ionized, the escaped electron is not on a particular energy level, but has made a transition to the **continuum**. It can take on any of a continuous (unquantized) range of energies, because it is no longer bound to the atom. It moves freely through the gas, colliding with other particles until it is captured, usually by a positively charged ion. As long as an ion is left with at least one orbiting electron, the ion will have a set of energy levels and can produce a spectrum. This spectrum will, however, be different from that produced by the neutral atom or by ions of the same atom but of different states of ionization. Thus, the various spectral features can tell us not only which chemical elements are producing them, but also the states of ionization of those elements. (A little jargon: the spectrum of neutral carbon can be denoted as C I as well as by C; once-ionized carbon, C^+, is C II; three-times-ionized carbon, C^{+++}, is given as C IV; etc.)

The Spectrum of Hydrogen

Hydrogen is both the simplest atom and the most abundant element in the universe; so let's take a look at its spectrum. When its electron is in the ground state, for example (see Figure 9.11), it can be kicked to a higher level, say, $n = 2$, by absorbing a photon of just the right energy. The photon loses all its energy in the process and so disappears. Alternatively, the hydrogen atom can gain the energy corresponding to the $n = 2$ level by colliding with a passing electron. The energy lost by the electron is gained by the atom. In either case, if nothing further happens to it within about 10^{-8} seconds, the electron will fall from the $n = 2$ level back to the ground state. As it does so it emits a photon with a wavelength equal to the energy difference between the ground state and the $n = 2$ level (the first excited state). If, however, before the electron has a chance to fall back to the ground level, it absorbs another photon of the right energy, it can be kicked to a still higher level, absorbing the photon in the process. Thus the atom can absorb photons from any level.

We call the set of transitions in hydrogen from $1 \rightarrow 2$, $1 \rightarrow 3$, $1 \rightarrow 4$, etc., the **Lyman series**—Lyman-α, Lyman-β, etc., respectively—of absorption lines. (You will see in the next chapter why they are called "lines.") As you can see from the fig-

188

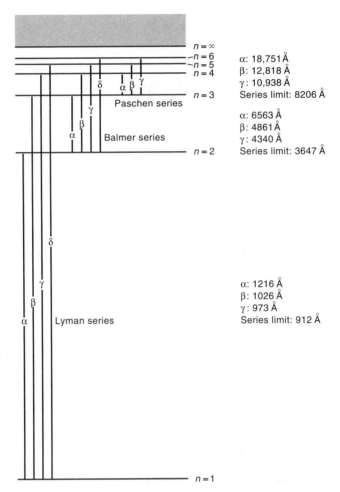

Figure 9.11. The energy level of the hydrogen atom, drawn to scale. The transitions associated with the ground state ($n = 1$) all fall in the ultraviolet region of the spectrum, unobservable from the ground. The transitions of the Balmer series, having the $n = 2$ level in common, all fall in the visible part of the spectrum, from the red to the blue. The transitions of the Paschen series are in the near-infrared part of the spectrum. Note that the higher energy levels get closer and closer together, until they coalesce and form the continuum.

ure, the change in energy for these transitions is relatively large, so that the wavelengths of the corresponding photons are in the ultraviolet region of the spectrum. The Lyman-α transition ($n = 1$ to $n = 2$) requires a photon with a wavelength of 1216 Å, Lyman-β (the ground state to $n = 3$) a photon of wavelength 1026 Å, etc. Note that since the energy difference is greater in the latter transition than in the former, the wavelength of Lyman-β is shorter than that of Lyman-α. Transitions in the opposite direction ($2 \to 1$, for example) correspond to the Lyman lines in emission. Since ultraviolet radiation is absorbed by the Earth's atmosphere, in order to measure the Lyman lines we must fly telescopes and detectors to high altitudes in rockets or on orbiting satellites.

Similarly, transitions arising from (or all ending on) the $n = 2$ level of hydrogen give rise to the so-called **Balmer series** in absorption (or emission, respectively).

These are particularly important because, in contrast to the Lyman lines, which are absorbed by the Earth's atmosphere, the Balmer lines occur in the easily observed visual region of the spectrum. Balmer-α (more usually called Hα) arises from the $n = 2$ to $n = 3$ transition and corresponds to a photon with wavelength of 6563 Å (red); Hβ corresponds to 4861 Å (blue), etc. The Balmer lines appear in the spectra of a wide variety of stars.

Note that the higher-energy levels of hydrogen become closer and closer together, so that successive lines of the Lyman or Balmer series differ from each other by smaller and smaller increments of energy and wavelength. The energy levels finally merge into the continuum; for the Lyman and Balmer series this occurs at energies corresponding to about 912 Å and 3647 Å, respectively. These wavelengths are called the **series limits**. Photons of wavelengths shorter than 3647 will ionize hydrogen from the $n = 2$ state (the Balmer level). If a ground-state hydrogen atom absorbs a 930 Å photon, a high-level Lyman line may result; if it absorbs a 900 Å photon, however, the atom will be ionized.

What determines which series of spectral lines will dominate the spectrum of a volume of gas? If the energy of the gas atoms is small, most of the atoms will be in the ground state, and few will be in the higher states from which the Balmer and other series arise. As the energy available to the gas increases, however, photon or electron collisions will cause the higher-energy levels to be populated. Temperature, you will remember, is a measure of the kinetic energies of the atoms, ions, and electrons of the gas. Hence it is the temperature that is the primary determinant of the level of ionization of the gas as well as the relative populations of the excited energy states; it dictates which series will be strongest. We will return to this point in Chapter 11.

Kirchhoff's Laws

Spectra can also yield useful qualitative physical information, which is embodied in **Kirchhoff's laws** (see Figure 9.12 and Plate C2). These are three empirical laws discovered in the mid-nineteenth century, which we can now understand with the aid of atomic theory. The laws are

    (1) a hot solid or liquid or hot gas at high pressure emits a continuous spectrum;
    (2) a hot gas at low pressure has an emission-line spectrum;
    (3) a gas illuminated by a hot continuous spectrum shows absorption lines.

The first law is exemplified by the filament (a hot solid) in a light bulb, which is heated to incandescence when an electrical current is passed through it (see Figure 9.12A). Its spectrum is entirely continuous with neither absorption nor emission lines. In this sense it is like a blackbody, though the shape of the spectrum only approximates that of a blackbody. We will see later that for a gas to emit a continuous spectrum, it must be opaque (very absorbing) at the wavelengths of interest. To be opaque, relatively high gas pressures are required, hence the second part of the statement of the first law.

An example of the second law is given by sodium-vapor street lights. High-temperature, low-pressure sodium gas radiates two strong emission lines in the yellow-orange region of the spectrum (5890 and 5896 Å), which produce the characteristic color of such lights (Figure 9.12B). This light is produced because sodium atoms are kicked to an excited level by the energy of an electric current; in the transition back to the ground state, the atoms give up their excitation energy by emitting yellow photons.

Finally, the third law can be demonstrated by shining the light from the hot tungsten filament (producing the continuous radiation) through cooler sodium vapor (Fig-

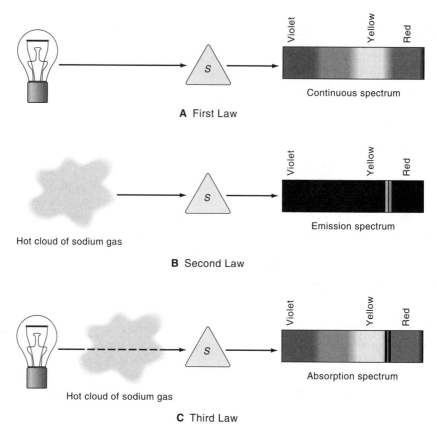

Figure 9.12. A schematic representation of Kirchhoff's laws. The hot filament in a light bulb, A, produces a continuous spectrum, that is, all wavelengths are present (the first law). In B, a cloud of hot gas emits an emission line spectrum (the second law). In this case the gas is sodium, so the most prominent lines are two in the yellow-orange region of the spectrum. A gas cloud absorbs photons, C, from a hotter continuous spectrum producing absorption lines at the same wavelengths at which it would otherwise emit them (the third law). (The *S* enclosed by a triangle simply represents a spectrograph.)

ure 9.12C). The result is a continuous spectrum with two absorption lines in the yellow-orange, at exactly the same wavelengths at which the emission lines had appeared. This occurs because the sodium atoms now are excited by incident photons from the hot filament that then disappear from the beam, resulting in the absorption lines. Note that both the emission and the absorption processes described here take place between the same energy levels. The photons that are re-emitted when the excited atoms return to the ground state are radiated in all directions; only a tiny fraction will find themselves in the original beam. We will see that Kirchhoff's laws provide important insights into the nature of astronomical objects.

The Shell Structure of Atoms

We have described the simplest atom, hydrogen, with its one electron making transitions between energy levels. What about heavier atoms, which have many electrons surrounding the nucleus? Are all these electrons in the same lowest-energy level when

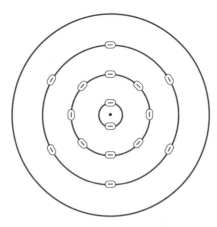

Figure 9.13. The shell structure of sulfur in its lowest-energy level. The nucleus, containing 16 protons and, in ordinary sulfur, 16 neutrons, is the black dot at the center. Two electrons occupy the first shell, eight the second, and six the third. The first two shells are completely filled; no more electrons can occupy them.

the atom is in the ground state, and do they all participate in these transitions? Let's consider a carbon atom. In its neutral form it has six electrons orbiting the nucleus. Two of these are in the $n = 1$ level, but the other four are in the $n = 2$ level. It is simply a fact of nature that in its lowest-energy state, no atom can accommodate more than two electrons in the state (or shell) of lowest energy. The others must be in the next higher shell. Furthermore, in heavier atoms no more than eight electrons can occupy the $n = 2$ level; others must go to the next higher level. For example, a neutral sulfur atom (atomic number of 16) has, when it is in its ground state, two electrons in the first shell, eight in the second, and six in the third (see Figure 9.13).

This so-called **shell structure** of the atom is understood by means of the **Pauli Exclusion Principle**,[7] which asserts that no two electrons in a given atom can have the same energy and angular momentum. We have spoken of only one of these quantities, the size of the most probable electron distribution (the orbital) that represents the energy of the atom and that we denoted by $n = 1, 2$, etc. A complete description of an atom requires additional parameters, however. These include the shape of each electron's orbital—circular or elliptical—which can be associated with the angular momentum of the electron revolving around the nucleus. Particles like protons and electrons can also be thought of as rotating or spinning in a clockwise or counterclockwise sense; they have angular momentum because of this spin. According to quantum theory, all three of these quantities—the size and shape of the orbital and the particle spin—are quantized, that is, can take on only certain discrete values. When all the appropriate quantities are taken into account, it turns out that only two electrons can occupy the $n = 1$ level, because a third electron would have the same values of these quantities as one of the other two. Similarly, no more than eight electrons occupy the second level, eighteen the third, etc. It is important to realize that the third and subsequent electrons *must* occupy the $n = 2$ and higher levels even when the temperature is very low. There is simply no place for them in the lower-energy levels. Only in this way can the arrangement of the electrons in shells around the nucleus be understood.

---

[7] Wolfgang Pauli (1900–1958), an Austrian physicist who made major contributions to quantum theory.

The Pauli Exclusion Principle also explains why atoms in ordinary matter can't be squeezed together to arbitrarily large densities. Identical electrons in two neighboring atoms cannot overlap, so the atoms must remain some minimum distance apart. The Exclusion Principle has important applications in the description of matter in very dense objects such as white dwarfs and neutron stars, as we shall see in a later chapter.

In an atom with more than one electron, it is usually one of the electrons in the highest-energy shell of the ground state that performs transitions to states of higher energy and produces the pattern of spectral lines characteristic of the atom. The other electrons remain undisturbed in their lowest-energy states.

A Few Last Comments

Some of the ideas of quantum theory described in this chapter appear to be weird: particles that act like waves; waves having particle-like characteristics; photons carrying only discrete chunks of energy; electrons in atoms not being pulled into the positively charged nucleus and allowed to have only certain energies; and so on. It is perfectly natural and sometimes useful for us to try to picture these ideas, to make models that represent at least some aspects of what is going on. To do so they, of course, must be expressed in terms with which we are familiar, namely, our own experience. But our experience takes place in a world whose scales of size, time, and mass are utterly different from those in the atomic and sub-atomic realm. Furthermore, the force that we experience most directly, that of gravity, plays no role in the microscopic world we have been describing. Our intuition is of little use here. So should we be surprised when our attempts to picture this world in terms directly meaningful to us result in a certain amount of confusion and inconsistencies? Why should models that might be appropriate representations of reality at one scale be applicable on a completely different scale? We get into trouble by taking them too seriously. This may not be a satisfactory situation, but it's the way things are.

Now we can see why a mathematical approach is so crucial for physical science. The power of a mathematical description of nature is that it enables us to transcend the limitations of our experience, of our imagination shaped by that experience. In a sense, the "reality" of the microscopic world we have been trying to describe resides in the mathematical equations of quantum theory: the equations, not models, describe how that world works, and do so wonderfully well. Quantum theory is probably the most successful structure in science, accounting for an incredibly broad range of phenomena to high, sometimes incredible, precision. For example, quantum theory predicts a value for a certain quantity that agrees with the measured value to one part in 10^{16}! The "strange" concepts are forced upon us; we know of no other way to describe the microscopic world.

With this account of the properties of electromagnetic radiation and atomic structure as background, let us now turn our attention to the tools that astronomers use to collect and analyze this radiation.

Terms to Know

Electromagnetic radiation, optical radiation, wavelength, amplitude, frequency, Hertz, polarization; gamma- and x-rays; ultraviolet, visible, infrared, and radio radiation. Photons, Planck's constant, blackbody radiation, continuous radiation, line radiation. Electrons, protons, neutrons, nucleus, atoms, atomic number, atomic weight, isotope, orbital, energy-level diagram, ground level, excited level, ionization; absorption, emission, Balmer series, Lyman series, series limit. Shell structure, Exclusion Principle.

Ideas to Understand

The distinction between continuous and line spectra, blackbody radiation laws, thermal equilibrium. The structure of atoms; how they produce absorption and emission lines; Kirchhoff's laws; the Pauli Exclusion Principle and the shell structure of atoms.

Questions

1. (a) How much more energetic is a photon with a wavelength of 2000 Å than one of 20 cm? Can either of these photons be detected with the naked eye? In what region of the spectrum does the 2000 Å photon lie? the 20-cm photon?

(b) The ordinary AM radio broadcast band includes frequencies from 540 kHz to 1600 kHz (kilo = 1,000, Hz = Hertz or cycles/sec; thus 540 kHz = 540 × 1000 = 5.4×10^5 Hz). Your radio dial probably simply has numbers 5, 6, 7, . . . , 16 to represent these frequencies. Suppose you are tuned to a station at 9, i.e., 900 kHz. What is the wavelength in centimeters of the radio waves transmitted by that station?

2. (a) Our eyes are most sensitive to light having a wavelength of about 5500 Å. What is the frequency of such radiation? What is the speed through empty space of this radiation in km/sec?

(b) How many waves of yellow light (5500 Å) are required to span 1 cm?

(c) Dental x-rays are taken with radiation having a wavelength of about 0.5 Å. What is the frequency of this radiation? How many times more energy does one of these x-ray photons carry than does a photon of 5500 Å radiation?

3. Briefly explain your answers to the following. Which is greater:
(a) the speed in a vacuum of an x-ray photon or of a radio photon?
(b) the frequency of an x-ray photon or of a radio photon?
(c) the energy of an x-ray photon or of a radio photon?
(d) Considering a beam of x-rays and of radio energy as each a wave, can you say which wave has the greater amplitude?

4. Venus and Jupiter are about 0.7 and 5.0 AU, respectively, from the Sun. What is the ratio of the solar energy received at each of the two planets to that received at the Earth, per unit area and unit time?

5. The supernova discovered in 1987 in the Large Magellanic Cloud (a nearby satellite galaxy of the Milky Way) was nearly as bright at maximum light as Jupiter. Jupiter is about 5 AU away from us, whereas the supernova is about 10^{10} AU away. If the supernova were only 5 AU away, how many times brighter than Jupiter would it appear to be?

6. Consider two blackbody furnaces of the same size. The temperature of furnace *A* is 2,000 K and that of furnace *B* is 4,000 K.
(a) What is the wavelength at which the peak of the spectrum occurs in *A*? In *B*?
(b) The total energy radiated by *A* is how many times greater than that radiated by *B*?

7. Explain why curves representing the radiation from blackbodies at different temperatures can't cross.

8. Our bodies have a temperature of about 310 K. At what wavelength do we radiate the greatest amount of energy? In what region of the spectrum is this?

9. The effective temperature of the Sun is about 5,700 K.

(a) At about what wavelength (in Ångströms) does the Sun radiate the most energy?

(b) The temperature of Betelgeuse (a red star in Orion) is about one-half that of the Sun. At what wavelength does it radiate the largest amount of energy?

(c) Betelgeuse radiates about 10,000 times more energy than the Sun, even though the Sun is hotter. How can this be?

10. Some hot stars have temperatures of about 50,000 K. Calculate the wavelength at which their spectrum is a maximum. Why is it advantageous to observe these stars by using telescopes in space?

11. The energy-level diagram (Figure 9.14) shows various transitions denoted by the labeled arrows. Which represents:

(a) emission of the longest wavelength Lyman photon?

(b) absorption of the shortest wavelength photon?

(c) absorption from the ground state?

(d) absorption of a Balmer photon?

(e) ionization?

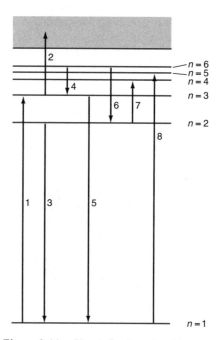

Figure 9.14. Sketch for Question 11.

12. A certain atom has energy levels as shown in Figure 9.15. The energy difference between $n = 1$ and $n = 2$ is exactly half that between $n = 2$ and $n = 3$. In the Sun, the line produced when the atom makes a transition between $n = 1$ and $n = 2$ is at 6000 Å. Assume that the temperature of the Sun is 6,000 K.

(a) What is the wavelength of the transition between $n = 1$ and $n = 3$ in the Sun? Explain.

(b) What is the wavelength of a transition between $n = 1$ and $n = 2$ in a star with a temperature of 3,000 K? Explain.

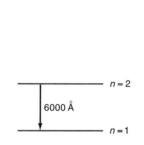

Figure 9.15.   Sketch for Question 12.

13. What are the differences between an atom, an isotope, an ion, and a molecule?

14. In what ways is the solar system a poor model of the atom?

15. How many neutrons and protons are there in ^{24}Mg, ^{56}Fe, ^{235}U, and ^{238}U? (The atomic numbers of Mg, Fe, and U are 12, 26, and 92, respectively.)

16. Can an atom in its ground (lowest-energy) state emit a photon? Absorb a photon? Explain.

Suggestions for Further Reading

Augensen, H. and Woodbury, J., "The Electromagnetic Spectrum," *Astronomy*, **10**, p. 6, June 1982.

Bova, B., *The Beauty of Light*. New York: John Wiley and Sons, 1988.

Cline, B., *Men Who Made a New Physics*. New York: Signet, 1965. A non-technical account of the development of quantum theory.

Hey, T. and Walters, P., *The Quantum Universe*. Cambridge: Cambridge University Press, 1987. A well-illustrated book describing various aspects of quantum theory.

Pagels, H., *The Cosmic Code: Quantum Physics as the Language of Nature*. New York: Simon and Schuster, 1982. Written by a physicist, a popular account of the ideas of the new physics.

Rowan-Robinson, M., *Cosmic Landscape: Voyages Back Along the Photon's Track*. Oxford: Oxford University Press, 1980. Well-written popular account of what the various spectral regions tell us about the universe.

G. W. Ritchey (1864–1945), a talented and remarkably far-sighted optician and telescope designer. Though he was responsible for the successful 60-inch Mount Wilson reflector, most of his subsequent work was completely ignored until recently. Now his optical design for increasing the field of view of reflectors is standard (the Ritchey-Chrétien system, which is used, for example, in the Hubble Space Telescope), and many of his ideas on sharpening telescopic images are being incorporated in the new-generation telescopes under construction.

CHAPTER **10**

Telescopes

Optical Telescopes

The telescope is the basic astronomical instrument. Without it the last 300 years of astronomical exploration and discovery would have been impossible, and our knowledge of the universe would have remained rudimentary; this book would have ended at Part I. Telescopes are simple devices, at least in principle, and it is easy to understand how they work and what they do. Also, in this chapter we will consider the instruments and detectors that are mounted on telescopes to measure various properties of the radiation they collect.

General Characteristics of Telescopes

A telescope forms an image of a distant object, which can then be examined by eye, photographed, or otherwise measured or recorded. Consider a glass lens collecting radiation from a star, as shown in Figure 10.1. The lens forms an image as a consequence

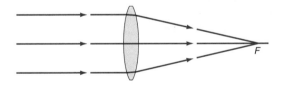

Figure 10.1. A lens in cross-section (a "side view") receiving parallel light from a star; the light is brought to a focus at *F*.

of two factors: the index of refraction of glass, and the curvature of the lens. The **index of refraction**, n, of glass, for instance, is the ratio of the speed of light in a vacuum, c, to the speed of light in the glass, V, or $n = c/V$. For glass, n is about 1.5, so that the speed is reduced by about one-third. When light rays traveling in air strike the surface of glass, their direction is changed, and the angle they make with the perpendicular to the surface is decreased, as shown in Figure 10.2. That a stick appears bent when stuck into water is just a consequence of the fact that the speed of light in water is less than it is in air. By properly curving the two surfaces of a lens, incident light rays can be brought to a focus, because at each point on the surface of the lens, the light rays are bent toward the perpendicular to the surface at that point. Conversely, when the light emerges from the glass into air, it is bent away from the perpendicular.

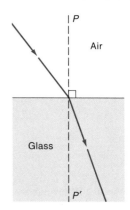

Figure 10.2. A ray of light traveling from air to glass is bent toward the perpendicular *PP'* to the surface because light travels more slowly in glass than in air.

The result of these two actions is that an image can be formed. Figure 10.3 shows how this works for a simplified lens made up of two prisms.

Astronomical objects such as stars are so far away that light rays coming from them strike the Earth and our telescopes traveling along parallel paths (see Figure 10.4).[1] A lens called the **objective** (shown in cross-section in the figure) brings this light to a **focus** at *F*; that is, it forms an image at *F* of the star. (The image of an extended object like a planet or galaxy is formed in the **focal plane**.) This image is small; so an **eyepiece** is used to magnify it, enabling the eye to see it better. In essence, this is all that a telescope consists of—an objective to form an image, and an eyepiece with which to examine it. The objective lens is characterized by two numbers: its diameter (called **aperture** in astronomy), and its **focal length**, *f*, which is the distance from the lens at which parallel light is brought to a focus. The ratio of the focal length to the aperture is called the ***f*-number** or *f*/ratio, and for telescopic objective lenses is typically about *f*/15. (The photography buff will recognize this nomenclature as the same used for adjustable-focus cameras.) A focal ratio of *f*/15 means that the distance from the lens to the focal point is 15 times greater than the diameter of the lens. The second lens, the eyepiece, is a high-quality magnifying glass, and generally has a very short focal length.

The function of the telescope tube is simply to keep the objective and eyepiece properly aligned with respect to each other, and to provide a means for mounting the instrument on its pedestal so that it can be pointed to any part of the sky. Since celestial

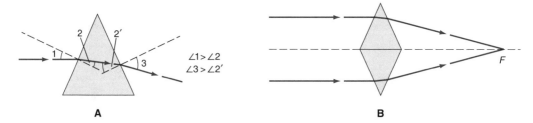

A   **B**

Figure 10.3. Sketch A shows how a ray of light is bent as it enters and leaves a glass prism. The ray strikes the prism making an angle (1) with the perpendicular to the surface; in the glass it is bent toward the perpendicular and makes the angle (2); as it leaves the prism it is bent away from the perpendicular at angle (3). Sketch B shows two prisms acting like a lens.

---

[1] Perhaps this point will become clearer if we point out that regardless of where you are on Earth, the direction to a given star is the same.

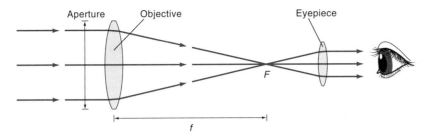

Figure 10.4. A schematic diagram of a telescope. The image of a star at F formed by the objective is examined by the eye looking through the eyepiece (essentially a magnifying glass). To form a parallel bundle of light (which is easy for the eye to focus on), the eyepiece is placed its own focal length away from F.

objects are constantly moving across the sky, the telescope should be mounted so that it can follow them easily. This is usually done by building the instrument so that it can rotate about two axes that are at right angles to each other (see Figure 10.5). One of these, the **polar axis**, is made parallel to the Earth's axis of rotation, so that a single motion about this axis at the same rate as the Earth's rotation enables the telescope to follow the stars as they move across the sky.

One of the major differences between amateur and professional telescopes, apart from the aperture of the objective, is that the mounting of a professional telescope is very sturdy; pointings can be maintained to a high precision. If you buy a telescope, look for one with as rigid a mounting as you can afford. It will make the telescope's use much easier, and add a lot to your enjoyment of it.

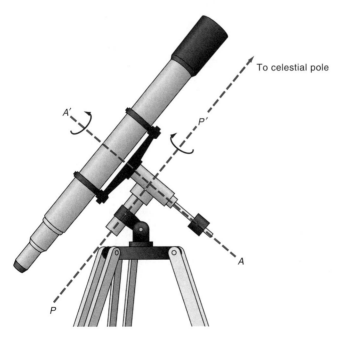

Figure 10.5. By rotations around the two axes AA' and PP', the telescope can be pointed to any point of the sky. PP' is in the direction of the celestial pole so that a continuous motion around this axis can compensate for the daily motion of the sky. This is called an **equatorial mounting**.

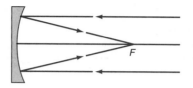

Figure 10.6. A mirror can bring parallel light from a distant object to a focus. A telescope in which a mirror is the objective is called a reflector.

There are two general kinds of telescopes: **refractors**, in which the incoming light passes through an objective lens, and as a consequence its direction is bent (or "refracted"); and **reflectors**, in which the direction of a light ray is changed upon reflection by the objective, in this case, a curved mirror (see Figure 10.6).

Refracting Telescopes

As we have just noted, a telescope with a lens as its objective is called a refractor. It was the only kind of telescope in use for the first half-century after Galileo turned his toward the sky. It suffers from two basic disadvantages, however: **chromatic aberration** and a limit to the diameter of its objective. The former refers to the fact that light of different colors, say, red and blue, is not equally refracted or "bent" by a lens. Instead, the blue light comes to a focus nearer to the lens than does the red (see Figure 10.7). In a medium such as glass, the speed of red light is slowed down somewhat less

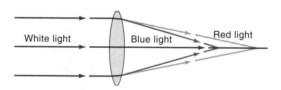

Figure 10.7. Since shorter-wavelength radiation is slowed down more than that of longer wavelengths when it passes through glass, a lens refracts (bends) blue light more than red light. Hence violet and blue light is brought to focus at a point nearer to the lens than is orange and red light. This effect is called chromatic aberration.

than that of blue. That is, chromatic aberration is a consequence of the fact that the index of refraction of glass, or any refractive medium, varies with wavelength. (A rainbow has its origin in this phenomenon, as shown in Figure 10.8). Thus, a focused red image will be surrounded by a blue halo, and vice versa. A fuzzy compromise can be achieved by focusing the eyepiece between the two sharp images (Figure 10.9). For many years the only way around this problem was to use a very long focal-length objective lens ($f/100$ or even $f/200$!) so that the different colored images were well separated and would not interfere with each other. Very large separations of objective and eyepiece present an obvious difficulty, however, which resulted in some strange and cumbersome telescopes, as shown in Figure 10.10.

Incidentally, our eyes, consisting of single lenses, suffer from a certain amount of chromatic aberration. We aren't aware of it because the eye-brain combination elim-

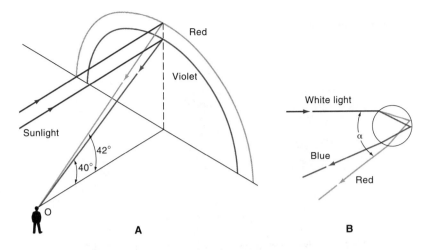

Figure 10.8. A rainbow (A) is just the spectrum of sunlight produced by water droplets acting like tiny prisms. Droplets disperse white light back to the observer at O most intensely at an angle of 42° for red light, 40° for violet light, and intermediate angles for green and yellow, producing the characteristic two-degree-wide band of the rainbow. When white light strikes a droplet (B), blue light is dispersed more than red light as it enters and leaves the drop. As a result, red light forms the upper edge of the bow, violet the lower.

inates most of it. This is similar to the adjustment made when someone wears glasses equipped with prisms that invert everything. The brain soon compensates and reinverts the images.

Chromatic aberration was a problem from the beginning, but the limit to the diameter of a refracting lens did not become a problem until the nineteenth century. The difficulty is an obvious one: since light must pass through it, a lens can be supported only at its edge. As the lens becomes larger, a point is reached at which the strength of the glass is insufficient to maintain the precise curvature (**figure** is the astronomical term) required for the lens to form sharp images. The lens will sag under its own weight and produce a distorted image. Worse yet, the amount of sag—and hence image distortion—depends on whether the telescope is pointing toward the horizon, straight up, or some direction in between. As a consequence of this effect, the largest astronomical lens is only 40 inches in diameter.

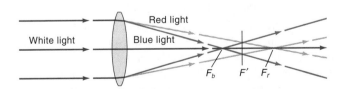

Figure 10.9. With chromatic aberration, the image of a star at F_b would have a blue center and a reddish halo; the opposite would be the case at F_r. A compromise consists of making the focus at F' where the image, though not sharp, is the smallest.

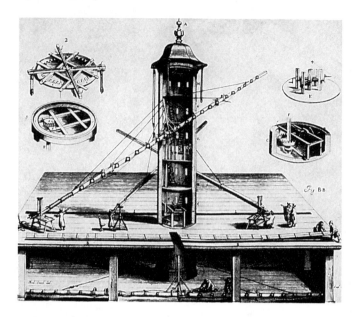

Figure 10.10.   An observatory designed by Jan Hevelius in the 1670s. The very long focus lengths were required to reduce the effects of chromatic and spherical aberration produced by the lenses of the time. The top of the tower was to revolve, allowing the telescope to track a star. Hevelius commented that this observatory would not be ". . . one for a private person to undertake, but for some great nobleman possessing ample room, money and above all, enthusiasm. . . ." The observatory was never built.

Reflecting Telescopes

To overcome the problem of chromatic aberration, Newton invented the reflecting telescope in 1668. The objective mirror of this instrument had a diameter of only a little over an inch, and the overall length of the telescope was about six inches. A mirror reflects a ray of white light (containing all optical wavelengths) as a single ray. That is, all wavelengths are reflected equally so all wavelengths are brought to the same focus; since reflected light does not pass through the glass, there is no chromatic aberration. Furthermore, a mirror can be supported on its back as well as at its edges, so that it is much easier to maintain the proper figure regardless of the mirror's size or its orientation. Thus reflecting telescopes do not suffer from either of the two major difficulties of refractors.

One of the functions of the objective mirror in a reflector, like that of the objective lens in a refractor, is to bring light from a distant object to a sharp focus. To do so the mirror is carefully ground and polished to a particular shape, usually a paraboloid (the shape formed by rotating a parabola about its axis of symmetry). Like a lens, a mirror is characterized by its aperture and its focal length. Now, however, the final image can be formed at several different locations, as shown in Figure 10.11. When no other optical element is in the light path, an image is formed at the **prime focus**. This is impractical for small telescopes, since putting your eye or a camera at this focus would block out too much of the incoming light. However, a flat mirror can be mounted near the prime focus, as shown, reflecting the light to a point outside the tube, where the image can be examined or recorded. This is the type of telescope built by Newton; such an arrangement is called a **Newtonian focus**. Alternatively, a mirror with a curved sur-

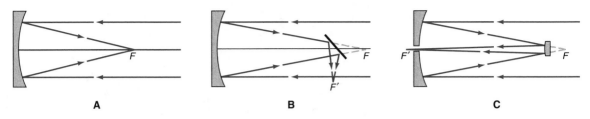

Figure 10.11. When the aperture of a reflector is small, putting your eye or an instrument at the prime focus (A) blocks out too much light. Newton put a small flat mirror in the converging beam from the primary (B) so that the focus is completely outside the tube at F'. Alternatively, a curved mirror can send the light back through a hole in the primary, placing the focus behind the objective (C) at the so-called Cassegrain focus.

face can be placed in the converging beam and the light reflected back through a hole in the objective. The image is then formed at the so-called **Cassegrain focus**, named after a French telescope designer.

In addition to those already mentioned, reflectors have additional advantages: since the light is reflected off the mirror, only one surface need be ground and polished to the highly precise shape (typically to within 10^{-5} centimeters) required to form sharp images. Also, the glass need not be completely free of internal bubbles or other small blemishes, since the light does not pass through it. Furthermore, for a given aperture, reflectors are shorter than refractors, since the primary mirror has a focal ratio of only about $f/5$ or even less; in fact, large astronomical mirrors can now be constructed with focal ratios of $f/1.5$. Thus, the mechanical structure holding the primary and secondary mirrors is shorter and less massive. This in turn means that a less massive mounting and smaller dome are required than would be needed for a refractor of the same aperture. Since the mounting and dome are among the costliest elements of a large telescope, these are all significant advantages for reflectors.

Not until the beginning of the twentieth century, however, did large reflectors dominate observational astronomy, because one critical piece of technology was missing: how to make a mirror reflect most of the light striking it. Until the middle of the nineteenth century, the best solution to this problem was to make the mirror not out of glass, but out of speculum metal, an alloy of copper and tin, mostly, along with a pinch of arsenic. This metal is extremely brittle, and great care must be taken in grinding and polishing it, but its reflectivity for visible light can be as high as 60 percent (that is, it reflects 60 percent of the light striking it). Unfortunately, a freshly polished surface begins to oxidize immediately, thereby reducing its reflectivity. After a time the mirror must be removed from the telescope and repolished. These difficulties, along with the discovery of how to reduce chromatic aberration in lenses,[2] were sufficient to maintain the supremacy of refractors throughout most of the nineteenth century.

Nonetheless, attempts were made to build large reflectors, the most notable of which was by Lord Rosse (William Parson) in Ireland. In 1845 he put into operation a 60-inch[3] speculum-metal reflector, "The Leviathan of Parsontown," by far the largest

---

[2] Instead of being a single element, the objective is made of two or more lenses having different refractive properties, enabling chromatic aberration to be corrected for two or more colors. Obviously, this at least doubles the number of surfaces to be figured. It also means, however, that refractors could be made much shorter than the earliest monsters, typically about $f/15$.

[3] A telescope is usually referred to by the diameter of its objective; thus Lord Rosse's telescope had an objective 60 inches in diameter.

Figure 10.12.   The 60-inch reflecting telescope built in Ireland by Lord Rosse in 1845. The "Leviathan of Parsontown," as it was called, was indeed impressive. Its mirror weighed 4 tons, the masonry walls were 72 feet long and 56 feet high, and the telescope tube was 56 feet long. It could be pointed to any altitude from the horizon through the zenith to the pole, but only about 15 degrees east or west of the meridian. Since the telescope could not track a star, the observer, at the Newtonian focus, would simply watch objects drift through the field of view.

telescope constructed up to that time (Figure 10.12). It was in a very poor climate for astronomy and was extremely difficult to use, however, and never really fulfilled the hopes of its builder. We will mention its most notable discovery in Chapter 17.

A Technique for Coating Glass Mirrors

In 1851 at the Crystal Palace Exhibition in London, vases were displayed that had been given a coating of silver by a chemical process. Five years later the first silvered-glass reflecting telescopes were made. Slivered-glass mirrors have many advantages over speculum mirrors. It is much easier to polish a glass mirror to the required figure than it is one of speculum, and a freshly coated silver surface reflects about 50 percent more light than a fresh speculum surface. Like speculum, silver begins to tarnish almost immediately, but it is a relatively simple matter to wash away the old coating and replace it with a new one. No polishing or refiguring of the glass surface is necessary. Note that the function of the glass is simply to support the thin layer of silver in the proper shape. With this technical development, large reflectors soon dominated astronomy. Thus, a simple bit of technology, first developed for aesthetic and commercial purposes, had a profound effect on astronomy. It is a good example of the influence of technology in science. Incidentally, household mirrors are still made by this silvering process.

Since the 1930s, astronomical mirrors have been coated by evaporating a thin layer of aluminum onto the figured glass; the aluminum has an even higher reflectivity than silver. Furthermore, the aluminum layer can be covered by evaporating a transparent coating on top of it. This seals off the aluminum from the air, thereby preventing it from oxidizing and allowing the mirror to be used for a longer time before it has to be cleaned or re-coated.

Large Telescopes

The largest and last (1897) major refractor constructed was the 40-inch at Yerkes Observatory in Williams Bay, Wisconsin (Figure 10.13). About a decade later the 60-inch reflector was put into operation at Mt. Wilson in southern California; a few years later the 100-inch reflector was built on the same site (Figure 10.14). Not only were large reflectors being constructed, but in marked contrast to nearly all previous observatories, they were being located in climates good for astronomical observing rather than at sites chosen for their easy access. By the early part of this century, the combination of large telescopes and clear skies established the preeminence of U.S. observational astronomy. The rise of U.S. astronomy (as well as that of most of our major art collections and museums) was largely a result of the huge fortunes created in this country after the Civil War. Astronomy became among the most richly endowed of the sciences.

Figure 10.13. The 40-inch refractor of the University of Chicago's Yerkes Observatory, located in Williams Bay, Wisconsin. The world's largest refracting telescope, it began operation in 1897 and is still in regular use. Its tube is about 64 feet long, the dome is 90 feet in diameter, and the floor can be moved up or down to make it easy for the observer to reach the eyepiece. (Note that a reflecting telescope of equal aperture can fit inside a dome about one-third as large.)

Figure 10.14. The 100-inch telescope on Mt. Wilson in southern California is mounted within a rectangular, rotating structure that comprises the polar axis. This mounting makes it impossible for the telescope to reach a small region of sky around the north celestial pole. The 100-inch began operation in 1918, and with it some of the most important astronomical discoveries of this century were made, including the nature of the spiral nebulae and the expansion of the universe, both of which we will describe later in this book.

For many years the largest reflector in the U.S. has been the 5-meter[4] at Mt. Palomar in California (Figure 10.15), which was exceeded in aperture only by the 6-meter reflector in the former Soviet Union. The largest telescopes constructed until recently have been about 4 meters in aperture. Now, however, several much larger telescopes are under construction. The largest of these is the Caltech/University of California Keck 10-meter reflector (Figure 10.16). It is located on Mauna Kea in Hawaii, and is now fully operational. A second Keck telescope, located next to the first and essentially identical to it, is under construction.

The Keck telescope is a good example of new techniques that are being developed for making large mirrors. Why are new approaches necessary? The problem is that the old method of making ever-larger mirrors by simply scaling up their diameters and thicknesses has reached its limit. Very large mirrors would be too massive. For example, to construct an 8-meter mirror by scaling it up from a conventional 4-meter (that weighs about 15 tons) would result in a mirror weighing about 120 tons! It would be extremely difficult to maintain its figure as it changed its orientation with respect to gravity.

Two general approaches are being taken to make large but light mirrors (Figure 10.17). The Keck telescope exemplifies one of these. Instead of being made of one giant piece of glass, the primary mirror consists of 36 individual hexagonal mirrors,

[4] For several decades this telescope was referred to as the "200-inch." Recently, however, it has become customary to give telescope apertures in meters; so the 200-inch has become the 5-meter.

Figure 10.15. For many decades after its completion in 1950, the Palomar 200-inch (5-meter) telescope was the largest in the world. Within a few years, however, it may not rank among the top ten, such is the extent of telescope building going on in the 1990s. The 200-inch was the final achievement of a remarkable astronomical entrepreneur, George Ellery Hale, who was also responsible for the construction of the Yerkes 40-inch, as well as the 60-inch and 100-inch telescopes on Mt. Wilson.

Figure 10.16. The first 10-meter Keck telescope on Mauna Kea in Hawaii began operation in 1993. It is not only the largest of the new breed of telescopes, it and its twin (the dark dome still under construction) are the only fully pointable telescopes in which the primary mirror is made up of many (36) individually mounted mirrors rather than a single piece of glass. The two telescopes were built with funds donated by the W. M. Keck Foundation to the California Institute of Technology and will be shared by astronomers at Caltech and in the University of California system, which provides for operations. Under construction at the right is the building for the 8.2-meter Japanese telescope, Subaru. In the distance is the 3.6-meter Canada–France–Hawaii telescope building.

each 1.8 meters across and only about 7 centimeters thick. The shape of each thin mirror must be maintained as the telescope points to different parts of the sky, and the alignment of all the mirrors must be carefully computer controlled to enable them to act together like a single mirror. This control system as well as the manufacture of all these segments posed many problems, but these have been overcome. An advantage of this construction technique is that in principle it can be used to manufacture even larger telescopes in the future.

Other large telescopes are also being planned (see Table 10.1). All of these use the large single-mirror approach, but also use one of two new manufacturing techniques that result in a lighter mirror requiring less glass and a less-massive support system. In one approach (Figure 10.17), the front and back of a large mirror are made of two thin sheets of glass (about 25 mm thick) held together by a lightweight honeycomb structure of glass less than a meter thick. The mold for such a mirror containing the molten glass is spun as the glass cools. This produces a front surface having the shape of a paraboloid, so that much less glass must be removed to obtain the final figure. An 8-meter mirror made in this manner weighs no more than a conventional 4-meter.

The second approach uses a very thin mirror shaped like a meniscus; that is, the front and back sides of the mirror are both curved parallel to each other (Figure 10.17). An 8-meter telescope of this design would be only about 20 centimeters thick. Such a thin mirror is not very strong, and so requires a set of "pistons" to push on the back of the mirror so that it always retains its shape as it is pointed in different directions with respect to gravity. An 8-meter meniscus mirror will be mounted in each of four side-by-side telescopes under construction by the European Southern Observatory located in the Chilean Andes. The Subaru and Gemini projects are also using this type of mirror.

Another important feature of the new telescopes is that provisions are being made to keep the temperature of the primary equal to that of the surrounding air. The temperature in a dome tends to warm up in the daytime and cool off at night. Left to itself, a massive mirror cannot adjust quickly to temperature variations, and so is never at the same temperature as the air. As we will see when we discuss the effects of the at-

Table 10.1. New-generation telescopes

Telescope	Aperture	Location	Owner
Keck I and II	10-meter	Hawaii	Caltech/UCal
Very Large Telescope	Four 8-meter	Chile	European Southern Observatory
Gemini	Two 8-meter	Hawaii, Chile	U.S., U.K., Canada, Argentina, Brazil, and Chile
Subaru	8.2-meter	Hawaii	Japan
MMT	6.5-meter	Arizona	Smithsonian, Arizona
Magellan	6.5-meter	Chile	Carnegie Institution, Arizona
Hobby-Eberly (fixed primary mirror)	8-meter	Texas	University of Texas, Pennsylvania State University, Universities of Munich and Gottingen, Stanford University

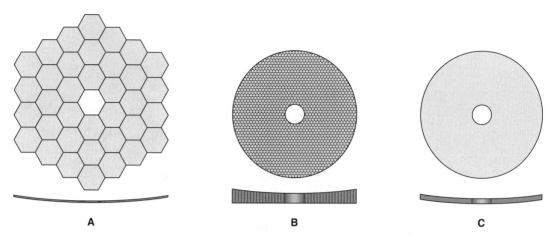

Figure 10.17. Approximate scale diagrams of the (A) 36-segment 10-meter mirror for the Keck telescope, (B) an 8-meter honeycomb mirror, and (C) an 8-meter meniscus mirror. Note that the cross-sections of the mirrors are also approximately to scale; the Keck mirrors are 7.5 centimeters thick, the honeycomb is about one meter thick at the edge, and the entire meniscus mirror is 20 centimeters thick. The weight of glass in each mirror is about 14.5 tons, 19 tons, and 26 tons, respectively. Each type of mirror has its advantages and disadvantages.

mosphere, this temperature difference degrades the quality of the images produced by the mirror.

It is instructive to compare the 5-m Palomar and 10-m Keck telescopes. The weight of the glass in both primary mirrors is very nearly the same, about 14.5 tons. Because the Keck primary has the very fast focal ratio of 1.75, the length of its "tube" and therefore the diameter of the dome is only 37 meters, 7 meters smaller than the Palomar dome!

These new telescopes are being constructed on so-called alt-azimuth mountings rather than on the traditional equatorial mounting. In the former, they are rotated about a vertical axis to any point on the horizon (azimuth) and also rotated around a horizontal axis to point to any angular altitude above the horizon. This type of mounting is much stronger mechanically than an equatorial mounting, but it has been used only recently on optical telescopes because it requires accurate tracking around both altitude and azimuth axes, rather than around a single (polar) axis of the equatorial mount. Tracking about the two axes, under computer control, can now be done with sufficient precision.

This new generation of giant telescopes should produce much new exciting data in the coming years.

What Do Telescopes Do?

The most important function of a telescope is to collect radiation. The eye is a reasonably good lens, but with a maximum aperture of about ¼ inch, or 6 mm, it can collect only a tiny fraction of the radiation gathered by a telescope of even modest aperture, say, 1 meter. (Imagine a small tin can and a large bucket side by side in the rain. In a given length of time the bucket will collect much more rain than will the tin can, as shown in Figure 10.18.) The amount collected is proportional to the area of the aperture, πr^2, where r is the radius of the objective lens or primary mirror. Thus, to calculate how much more light a 1-meter telescope can collect than the eye, we have

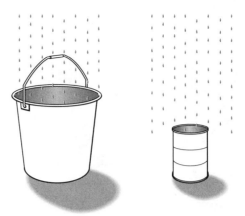

Figure 10.18. A large bucket and a small tin can are side by side in the rain. The bucket, with its large collecting area, will contain more water after a given time than the tin can. The same applies to telescopes. One with a large aperture will collect more photons/sec than one with a small aperture.

$$\frac{\text{Area (telescope)}}{\text{Area (eye)}} = \frac{\pi(50)^2}{\pi(0.3)^2} = \frac{2{,}500}{0.09} = 27{,}780.$$

The 1-meter telescope collects about 28,000 times as much light as the eye! Putting it another way, using such a telescope to gather light and then putting that light into our eye enables us to see objects as faint as 1/28,000 of those we can see with the unaided eye. The Keck 10-meter telescope will collect 100 times more light than the 1-meter, or 2,800,000 times as much as the eye.

A telescope can produce a magnified image of an object in the sense that it can increase its apparent *angular* size. If an object that has an angular diameter of 10 arc seconds is made to appear to be 1,000 arc seconds in size, it has been **magnified** 100 times. The magnification depends on the ratio of the focal lengths of the eyepiece and objective: $m = f_{\text{obj}}/f_{\text{eyep}}$. Since the focal length of the objective is fixed, the magnification is changed by changing the eyepiece; the shorter the focal length of the eyepiece, the higher the magnification. Several factors limit the useful magnification, however. One is that the greater the magnification is, the fainter the image becomes, since the same amount of light (collected by the objective) is spread over a larger and larger area in the focal plane. Another has to do with the unsteadiness of the Earth's atmosphere, which we shall discuss below. A third factor affecting the magnification is the limit set by the ability of a telescope to resolve detail, which we will discuss next.

An important function of a telescope is to reveal detail in, or to "resolve," extended objects such as the surfaces of planets or nebulae. (Without very elaborate computer analysis of the images, which is just now becoming possible, stars other than the Sun always appear as points of light.) This resolution or **resolving power** is usually measured as the smallest angular separation of two features that can just be separated, for example, neighboring craters on the Moon.

One of the properties of waves (including electromagnetic waves) is that their direction is bent (**diffracted** is the technical term) when they encounter an obstacle. (Note that this is not the bending we call refraction that occurs when light passes from one medium to another.) Loudspeakers radiate sound waves over a somewhat larger angle than they would without diffraction. The obstacle in this case is the edge of the speaker cone or of the speaker enclosure. For telescopes, the edge of the objective lens or mirror is such an obstacle. The light is diffracted slightly as it passes by this obstacle, the amount depending on the ratio of the wavelength of the light to the size of the obstacle; the longer the wavelength, the greater the diffraction, for a given aperture telescope. Diffraction thus sets a fundamental limit to the ability of an optical system to produce sharp images.

Assuming the telescope optics were of perfect quality, we have seen that their resolving power depends on just two quantities: the wavelength of the radiation by which the object is being observed, and the diameter of the telescope objective. In particular, the resolution θ is given by $\theta \propto \lambda/d$, or, in arc seconds,

$$\theta = \frac{(2.5 \times 10^5)\lambda}{d} \, ,$$

where λ and the aperture d are in the same units of length, for example, centimeters or meters. Thus, for a given telescope the resolution improves as shorter wavelengths are used for the observation; for a given wavelength, it improves if the objective of the telescope is larger. In both cases, θ, the *minimum* detectable angular separation, becomes smaller. For visible light ($\lambda = 5500$ Å) this equation becomes, to a good approximation, $\theta = 14/d$, where the aperture, d, is in cm and θ is in seconds of arc. According to this formula, our 1-meter telescope would enable resolutions of about 0.1 arc sec to be achieved. This sets a theoretical limit to the useful magnification of a telescope, since the detail one can see is set by its resolution. Magnification beyond this limit does not enable more detail to be seen.

The Effects of the Atmosphere

Unfortunately, an even more severe limit on performance attainable by ground-based telescopes is set by the blanket of air through which radiation must travel. Our atmosphere is not a perfectly smooth, uniform, and stable gaseous layer. Instead, it consists of small blobs of gas, some of lower density than average, some of greater density, rising and falling, respectively, all the time being blown across our line of sight by winds. The blobs slightly refract the light by different amounts, continually changing the direction from which it appears to come, and as a result, smears out the observable detail. We call this astronomical **seeing**; it is most obviously seen as the twinkling of starlight. Figure 10.19 shows the effect of poor seeing.

The quality of seeing varies considerably with the particular location on a given site as well as from one observing site to another and with changes in the weather at a given location. At a good site in the southwest U.S. or in Hawaii, 1 second of arc seeing occurs often; occasionally it may be as good as 0.5 seconds of arc or so. In the midwest, the eastern U.S., and throughout most of Europe, however, rarely does the atmosphere allow even 1 or 2 seconds of arc seeing. Thus the effective resolution of a ground-based telescope is set by the quality of the atmosphere and not by the size of its aperture. Until now, there has been no point in making the optics of a large ground-based telescope as nearly perfect as possible, because its resolution will be set by the seeing, not by the optics.

Efforts are underway that should change this, however. Telescope enclosures are now designed and located at sites in ways that minimize the part of the seeing produced near the telescope. Seeing produced by temperature differences between the mirror and surrounding air is minimized by carefully ventilating the new low-mass mirrors in the daytime and nighttime both. This turns out to be very important, since a temperature difference of only 1°C can produce a blur of about 0.4 arc seconds. In addition, ex-

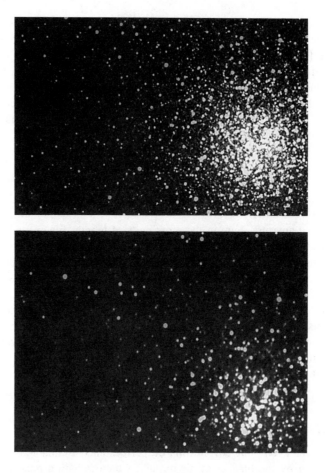

Figure 10.19. The effect of atmospheric seeing. Both images are of the same field of the globular cluster M13, taken with the 36-inch refractor of Lick Observatory. Note the improvement in resolution and, in particular, the larger number of faint stars visible in the upper image when the viewing was better than it was when the lower photograph was taken.

periments are underway with the aim to compensate directly for seeing fluctuations. For example, attempts are being made to design optical mechanisms that sense the distortions of the incoming light waves and then compensate for them by appropriately distorting an optical element in the light path many times each second. Because of the small-size scale of the atmospheric fluctuations, however, they can be compensated for over small fields of view only, perhaps 5 arc seconds in diameter. Though these efforts at adaptive optics (as such techniques are called) are promising, it will likely be several more years before successful image-compensation systems are in routine operation. Nonetheless, the new generation of telescopes should provide the best images so far obtained from the ground.

Telescopes in space, where there is no atmosphere, don't have problems with seeing. There the angular resolution obtainable depends directly on the size and quality of the optics and the stability provided by the pointing system. The spectacular images of planets taken on the Voyager missions were made by high-quality but small-aperture telescopes. The high resolution was obtained by taking the instruments very near the planets, an impractical solution for stars and galaxies, needless to say.

The atmosphere is an astronomical nuisance in another way: some of its constituents act like a filter, and let only a relatively narrow range of wavelengths reach the Earth's surface. In the infrared, the absorbers are carbon dioxide and water vapor; and oxygen, especially as ozone (O_3), absorbs ultraviolet radiation. (It is a good thing for us that the UV is absorbed; otherwise we would all be much more at risk for skin cancer. This is why the so-called ozone hole in the Earth's atmosphere is causing concern.) Of the whole electromagnetic spectrum, only that portion from about 3000 Å to about 20000 Å (the near ultraviolet to the near infrared), along with the very-short-wavelength γ-rays and the radio band longward of about 1 millimeter, can be observed from the ground (see Figure 10.20). Even within these spectral regions the atmosphere

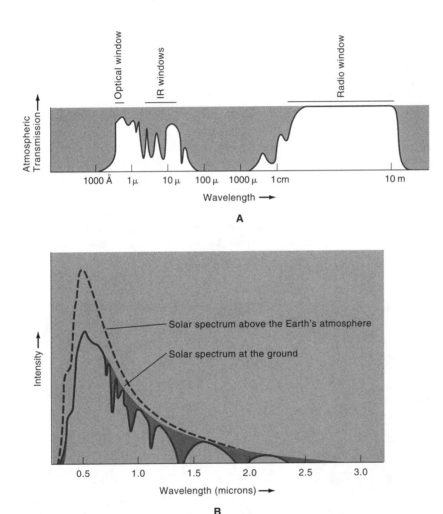

Figure 10.20. The major "windows"—optical, infrared, and radio—in the Earth's atmosphere through which one can observe celestial objects from the ground are shown schematically in (A). Note that the atmosphere does not transmit radiation in the x-ray, ultraviolet, and far-infrared spectral regions (the shaded areas). (B) shows how the solar spectrum from about 3,000 Å to 3 microns (30,000 Å) is absorbed by particular atmospheric molecules, water vapor, and carbon dioxide, primarily. In addition to this molecular absorption over specific wavelength bands, the Earth's atmosphere produces a general extinction that is strongest in the ultraviolet and weakens toward the red. It is significant even in the optical window, where only about three-quarters of the incident radiation, shown by the dashed curve, reaches the ground.

is not completely transparent, and allowances must be made for the radiation that is absorbed to a lesser or greater degree. Outside of these wavelength regions, the atmosphere is essentially opaque.

Some hot stars, for example, radiate a hundred times as much energy in the ultraviolet as in the visible. Some galaxies radiate most of their energy in the far infrared. Limited to the optical region of the spectrum, we are blind to much of the energy radiated by celestial objects. We are in the position of a listener whose hearing is limited to much less than an octave trying to make sense of a symphony. In order to observe the rest of the spectrum, we must use radio telescopes and put our instruments above the atmosphere, in high-flying aircraft, rockets, or satellites. We will describe some of the techniques of space astronomy later in this chapter and again in Chapter 23.

Radio Telescopes

The Beginnings of Radio Astronomy

Only a few years after the discovery of radio waves in 1888, it was suggested that celestial objects might be a source of such radiation, and there were even one or two attempts to detect radio waves. It was only in 1933, however, that Karl Jansky, then a young engineer working for the Bell Telephone Laboratories in New Jersey, presented a paper describing the discovery of "Electrical Disturbances Apparently of Extraterrestrial Origin." Relatively short wavelengths were becoming useful for long-distance communications, and Jansky was investigating sources of radio static or interference at a wavelength of 14.6 meters. After several years of observations with a crude antenna (Figure 10.21), he was able to show that a faint "hiss-type static" came from a

Figure 10.21. Jansky's antenna reconstructed at the National Radio Astronomy Observatory at Green Bank, West Virginia. The wooden structure, about 30 meters across and 4 meters high, supports the antenna (the black metal rods). The structure was mounted on four Model-T car wheels that enabled it to be rotated on the circular track visible below the wheels. The instrument was sensitive to radio waves of 14.6-meter wavelength.

Figure 10.22. The 10-meter dish built by Grote Reber has been reconstructed at the National Radio Astronomy Observatory. It was the first antenna built solely for radio astronomy.

fixed direction in the sky, which he identified with the Milky Way. This was not as easy as it may sound, because the directivity (or angular resolution) of his antenna was extremely poor. Though this discovery created a brief stir in the press, only one or two astronomers showed any active interest in continuing this work: lack of knowledge of radio techniques, of money (this was during the Great Depression), and of understanding of the significance of these first crude observations—all contributed to this lack of interest. Nor was Jansky himself able to devote much more time to his discovery, although he was aware of its importance.

This work was taken seriously by Canadian-born Grote Reber, an enterprising engineer and radio ham. In fact, Reber, who lived in Wheaton, Illinois, built a 31-foot-diameter radio dish in his backyard during the summer of 1937. The wooden frame was made mostly of two-by-fours, and the surface of the dish was of galvanized iron curved into a parabolic shape (Figure 10.22). It took Reber nearly two years of experimentation before he was able to construct a receiver sensitive enough to detect radiation from the Milky Way, at a wavelength of 2 meters. (The radio power from all celestial sources beyond the solar system amounts to only about 4 watts over the entire surface of the Earth! Many radio stations broadcast with 50,000 watts of power.) He published preliminary results in 1940, and by the late forties was publishing radio maps of the sky available to him from Wheaton. For nearly a decade, Reber was the only radio astronomer in the world!

Radio astronomy has had an enormous influence on astronomy, not only by revealing for the first time exotic objects such as pulsars, quasars, and radio galaxies, but also because it showed how fruitful it could be to observe celestial objects at wavelengths outside the conventional visible region of the spectrum. The entire visible spectrum extends over only a narrow band, less than a factor of 2 in wavelength. By contrast, the longest wavelengths used in radio astronomy are more than a thousand times longer than the shortest wavelengths.

Single Dishes

Because of the rapid advances made in radar technology during World War II, radio astronomy developed rapidly in the post-war era. Many radar engineers returned home to apply their newly developed knowledge and skills to observing the radio sky. Military radar units first transmitted and then detected radio waves after they bounced off incoming enemy aircraft. Some of these radars were converted to radio telescopes simply by using them to receive radiation from celestial objects.

In its simplest form, a radio telescope consists of a large objective that reflects the radiation it collects to its focus, where a detector—a sensitive radio receiver—is mounted (Figure 10.23). The paraboloidal surface of the objective (usually called a **dish**) is formed by metal plates or by a metallic mesh. Incidentally, because radio dishes are such large structures, they were, from the beginning, almost always constructed on alt-azimuth mounts for reasons of mechanical stability, as are the current crop of large optical telescopes. The "optical" properties of these radio telescopes are just the same as those we described earlier for ordinary optical reflectors. In particular, the resolution of a single dish is given by the same formula as we presented earlier in this chapter. Recall that the angular resolution of a telescope (the smallest resolvable angular separation) is proportional to the ratio of the wavelength of the radiation being collected to the diameter of the collector, that is, λ/d.

The wavelength of visible light is of the order of a few hundred-thousandths of a centimeter, but radio wavelengths range from millimeters to meters. Consequently, to have a resolution of, say, 1 arc second at a wavelength of 1 meter, a radio telescope would have to have an aperture of 200 km! By contrast, the largest steerable radio telescopes have apertures about equal to the length of a football field. Operating at a wavelength of 1 meter, such an instrument has a resolution of approximately the diameter of the Moon. Under these circumstances the direction to a given radio source can be specified only as being someplace within an area of the sky equal to that of the full

Figure 10.23. The 100-meter radio telescope near Bonn, Germany. This is the largest steerable radio telescope in the world. To reduce its weight, the outer part of the dish is made of a metal mesh, rather than solid panels. Receivers are located at the prime focus (the apex of the grid structure) and in the cylinder at the Cassegrain focus. Note the fast *f*-ratio of the dish; the focal length is less than the diameter of the dish.

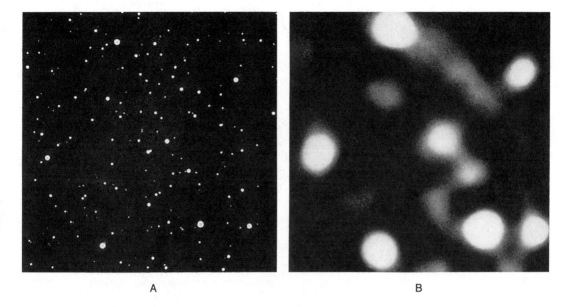

A   B

Figure 10.24. A simulated field of radio sources (A); how that same field would be "seen" by a single radio dish (B) 100 meters in diameter operating at 21 centimeters, giving a resolution of 7 arc minutes. Note that only the brightest objects would be detected and that many of them would appear to be connected.

Moon. With such poor resolution it is impossible to say anything about the detailed structure of a radio source, and even identifying a visible object with the radio object is very difficult. This situation can be improved by various means, however.

The resolution of a given telescope can be improved by using it at shorter wavelengths (see Figure 10.24). If the football-field-size radio dish mentioned earlier could be used to collect 1-centimeter wavelength radiation instead of 1-meter radiation, its resolution would be a hundred times better; instead of about 40 arc minutes, its resolution would be 0.4 arc minutes or about 25 arc seconds. There are two major difficulties in doing this, however. The surface of the dish must be made much smoother to efficiently collect and bring to a focus the shorter-wavelength radiation. This is a difficult and costly process. In addition, the radio receiver at the focus of the dish must, of course, be able to detect and measure the short-wavelength energy. In the early days of radio astronomy technical difficulties prevented this; fortunately, these problems have been solved. So the largest radio telescopes operating at centimeter wavelengths can achieve resolutions perhaps only 20 times poorer than those attainable at visible wavelengths.

Radio Interferometers

Through the development of interferometry, however, it is now possible for radio telescopes to have resolutions far better than any telescope sensitive to visible light. **Interferometry** is a technique whereby several small telescopes are operated together in such a way that they have the equivalent resolution of a much larger radio dish. For example, instead of building an impossibly costly single dish 10 kilometers in diameter, suppose we build two 50-meter dishes and place them 10 kilometers apart. We can then point them at the same object and combine the radiation collected by the two

217

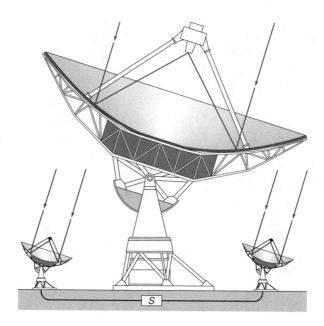

Figure 10.25. When the signals they receive from a celestial object are properly combined at S, two small dishes placed 10 kilometers apart can have the same resolving power as a giant dish 10 kilometers across.

dishes exactly as if the individual telescopes were two small parts of a 10-kilometer single dish (see Figure 10.25). Analysis of the resulting radio signal can be made to give a resolution equivalent to that from the 10-kilometer dish. Obviously, the total amount of radiation collected, and so the strength of the signal, will be much less than what a 10-kilometer dish would collect, but the large effective diameter of the system with its high angular resolution is preserved.

The Very Large Array (Figure 10.26) near Socorro, New Mexico (operated by the National Radio Astronomy Observatory), is a modern example of an interferometer. It consists of an array of 27 dishes, each 25 meters in diameter. These telescopes can be placed in various positions along a Y-shaped railroad track about 20 kilometers long. With this instrument, resolutions as high as about 0.06 arc seconds at a wavelength of 1 centimeter can be achieved.

Even higher angular resolutions have been achieved by combining the signals from several dishes spaced an ocean apart, for example, in Britain and on the east coast of the U.S. Such telescopes are separated by too great a distance to be linked directly by cable. Instead, the radio radiation collected by each telescope is recorded along with extremely accurate time signals generated by atomic clocks. By using the clock "ticks" to determine the exact time, the signals from the radio telescopes can be matched very precisely, combined, and analyzed as if they had been made by an instrument 5,000 kilometers in diameter. This technique, called **Very Long Baseline Interferometry** (or VLBI), can give angular resolutions as high as 0.001 arc seconds, roughly a thousand times better than that attainable in the visible region of the spectrum. It is a relatively slow process to make measurements in this way, however. A new instrument, the **Very Long Baseline Array** (or VLBA) will consist of ten 25-meter dishes located across the U.S., as shown in Figure 10.27. It is to be completely devoted to VLBI work, so that when it begins operating in the mid-1990s, radio astronomers will be getting a steady flow of very high-resolution data.

Figure 10.26. The Very Large Array, located in New Mexico, consists of 27 radio dishes, each 25 meters in aperture. By arranging these at various separations along the Y-shaped railroad track, different angular resolutions can be achieved.

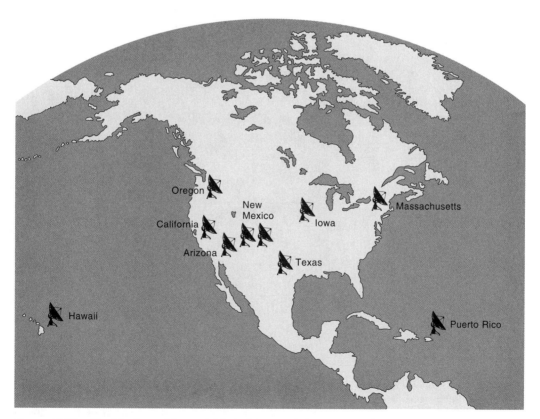

Figure 10.27. The locations of the ten VLBA dishes, each 25 meters in diameter. This array will make observations at wavelengths from 1 meter to a few millimeters with better angular resolution than has been achieved in any other spectral region.

Space Telescopes

Early Work

Though science fiction writers had based some of their stories on space flight, for example, Jules Verne in *From the Earth to the Moon* (1865), the earliest technical work on space flight was begun by Russians, most notably Konstantin Tsiolkovsky (1857–1935). In 1903 he published his *Exploration of Cosmic Space with Reactive Devices* (i.e., rockets). He originated the idea of multi-stage rockets, and speculated about Earth-orbiting colonies. In 1926 in the U.S., Robert Goddard built and flew the first liquid-fuel rocket (Figure 10.28). At the same time in Germany a group of enthusiasts began experiments that led to the World War II V-2 rocket used to bomb London and targets in Europe.

Perhaps the first suggestion that existing rockets be used for scientific purposes was made by E.O. Hurlburt, who proposed using Goddard's rockets to carry a spectrograph to record the Sun's ultraviolet radiation that is absorbed by the Earth's atmosphere. Nothing came of this, but just after the end of World War II, when the leading German V-2 engineers were preparing to resume their work at White Sands, New Mexico, he suggested using captured V-2 rockets for this purpose. These rockets could

Figure 10.28. Dr. Robert H. Goddard, the U.S. rocket pioneer, with the world's first liquid-fuel rocket. It was successfully flown on March 16, 1926 at Auburn, Massachusetts, when it burned for a little over 2 seconds and rose 41 feet. Fuel-gasoline (in the lower of the two cylinders at the bottom) and the oxidizer-liquid oxygen (in the larger diameter cylinder just above the gas tank) were carried by tubes to the ignition chamber at the top of the structure, where they were combined, ignited, and ejected through the nozzle (the flared cylinder). At the time, liquid oxygen was an exotic and dangerous substance. It must be colder than about −180°C (−300°F) if it is not to boil.

Figure 10.29.   The so-called Aerobee-Hi sounding rocket was the workhorse of early space-age science. It was capable of carrying a payload of perhaps 70 kilograms to an altitude of 150 kilometers or more, giving telescopes a view lasting 5 minutes or so above most of the Earth's atmosphere before it fell back to the ground. In the U.S., most of these were launched from the White Sands Missile Range in New Mexico.

reach altitudes as high as about 100 miles before falling back to Earth, allowing a few minutes of data-taking above most of the atmosphere. In February 1946, Richard Tousey at the Naval Research Laboratory in Washington, D.C., began constructing a spectrograph designed to analyze ultraviolet light from the Sun, to be mounted on a V-2. The first V-2 was flown in the U.S. on April 6, 1946; just two months later, the fourth V-2 was flown carrying Tousey's spectrograph to an altitude of 70 miles, but the instrument and the film on which the spectrum was recorded were destroyed on landing, and no data were obtained. Less than four months later, Tousey mounted a second spectrograph on the fin of a V-2, which was separated from the rocket and parachuted safely to Earth. His success with this flight marks the beginning of space astronomy.

Note that two instruments were built and flown in less than a year, a remarkably short time in any case and in marked contrast to the situation today. This is, of course, primarily a consequence of the incomparably more sophisticated instruments and rockets flown nowadays. To a much lesser but still significant extent, however, it was also due to the informal and relatively bureaucracy-free atmosphere in which the work was done nearly half a century ago. Note also that, as happened with radio astronomy, the technology of space astronomy developed quickly only as a result of wartime needs.

V-2 flights ended in 1952, when more than 60 had been launched, after which the U.S. developed its own small rockets with capabilities similar to those of the V-2. By the mid-1950s small instruments on Aerobee-Hi rockets (Figure 10.29) were obtaining measurements of the ultraviolet brightnesses of the Sun and stars; in the early 1960s stellar x-ray astronomy was born, when instruments on Aerobees detected the first x-ray source.

Orbiting Observatories

The pace of development speeded rapidly after the launch of the Soviet Earth-orbiting Sputnik in 1957, and both the Soviets and the U.S. began extensive programs of space science. In astronomy the Soviets concentrated on the solar system and, for example, took the first images of the far side of the Moon in 1959. The U.S. undertook a broad program of space astronomy, launching in 1962 the first of a series of Earth-orbiting observatories to study the Sun, as well as a series of lunar probes. In 1968 the Orbiting Astronomical Observatory (Figure 10.30), the first successful stellar space observatory, was launched. The following year, the U.S. landed the first men on the Moon. The U.S. planetary program will be described in Part IV of this book.

Space observatories are commanded by radio from the ground to point to a particular object such as a star, and observe it for an appropriate length of time with a particular telescope and photometer or spectrograph (see below) by which the desired wavelength band is isolated and measured. The data are stored on board the observatory until they are transmitted to a ground station. These observatories can operate for months or years, rather than for only a few minutes like the instruments on the small Aerobee-Hi sounding rockets. They have most of the capabilities of ground-based observatories, but are operated remotely, obviously. Interestingly, some ground-based observatories are beginning to be operated in a similar fashion, with the observer many miles away from the dome.

Figure 10.30. The Orbiting Astronomical Observatory 2, the first successful space observatory for stellar (as opposed to solar) astronomy, was launched on December 7, 1968. The spacecraft was 3 meters long and 2.1 meters across, weighed about 1,500 kilograms, including a 450-kilogram payload. Its purpose was to measure the ultraviolet radiation (about 1200 to 3200 Å) emitted by celestial objects. For more than 4 years, the small OAO reflecting telescopes (the largest had an aperture of only 40 centimeters) collected UV photometric and spectrophotometric data of hundreds of planets, stars, and galaxies. It will remain in its circular 800-kilometer-altitude orbit for a few hundred years before finally breaking up as it re-enters Earth's atmosphere.

Since the 1960s a wide variety of Earth-orbiting observatories have been flown by the European Space Agency, and recently by Japan, as well as by the former Soviet Union and the U.S. (Figures 10.31–10.32). These have been designed to observe celestial objects in regions of the spectrum that can't be observed from the ground—especially the x-ray, ultraviolet, and infrared. The ability to measure all the radiation emitted by an object and not just that in the visible has revolutionized astronomy.

Note that the space observatories so far described have been designed to take advantage of the circumstance that at high altitudes the Earth's atmosphere is too thin to filter out incoming radiation. But we have seen that the atmosphere produces another hindrance to astronomy—seeing—that limits the detail possible from the ground. Above the atmosphere the resolution is limited only by the size and quality of the telescope optics and by the stability of the pointing system. We are all familiar with this last point: if you move your camera while taking a photograph, it will be blurred.

Manufacturing near-perfect mirrors and extremely stable pointing systems is not only costly, but presents severe technical difficulties as well. Consequently, it is only

Figure 10.31. In 1978, the U.S.-European International Ultraviolet Explorer was launched into a 24-hour orbit. This small satellite carries a 45-centimeter aperture telescope that feeds two spectrographs sensitive to ultraviolet radiation from 1150 to 3200 Å. For 16 hours each day it is operated in real time from a ground station at Goddard Space Flight Center in Maryland, and 8 hours a day from a ground station in Spain. Its high orbital altitude means that the Earth blocks out very little of the sky and so the IUE can observe throughout most of its 24-hour orbit. Remarkably, it was still operating in 1995. In terms of the numbers of scientific publications based on IUE data, this small telescope has been the most productive in the world for much of its life.

224

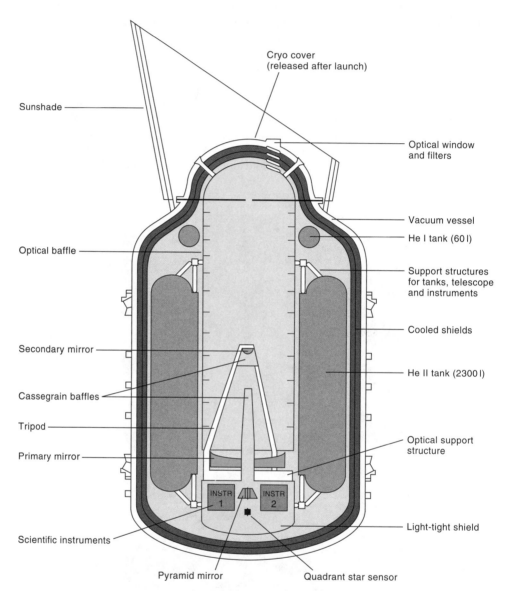

Sunshade

Cryo cover
(released after launch)

Optical window
and filters

Vacuum vessel

He I tank (60 l)

Optical baffle

Support structures
for tanks, telescope
and instruments

Cooled shields

Secondary mirror

He II tank (2300 l)

Cassegrain baffles

Tripod

Optical support
structure

Primary mirror

INSTR 1

INSTR 2

Light-tight shield

Scientific instruments

Pyramid mirror

Quadrant star sensor

Figure 10.32. A drawing of the European Infrared Space Observatory (ISO), launched in November 1995. The satellite is 5.3 meters high and 2.3 meters wide, and weighs about 2,400 kilograms at launch. Its 60-centimeter telescope feeds instruments (a camera and spectrographs) sensitive in the range 3–200 microns (1 micron = 10,000 Å). Since it is detecting heat radiation, the telescope and detectors must be cooled so that their radiation does not swamp that from celestial objects. Liquid helium in tanks surrounding the telescope is used to cool the telescope and instruments to 3–4 Kelvin.

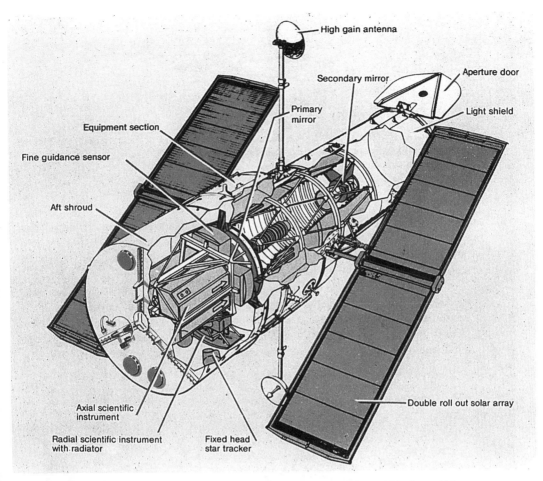

High gain antenna

Secondary mirror

Aperture door

Primary mirror

Light shield

Equipment section

Fine guidance sensor

Aft shroud

Axial scientific instrument

Radial scientific instrument with radiator

Fixed head star tracker

Double roll out solar array

Figure 10.33. A cutaway drawing of the U.S.-European Space Agency Hubble Space Telescope showing some of the major subsystems. The spacecraft is 13 meters long and 4.2 meters across at its widest; its total weight is 11,600 kilograms. The aperture of the primary mirror is 2.4 meters. Its field of view is shared by two cameras, two spectrographs, and a photometer, which together weigh 1,490 kilograms. The instruments provide sensitivity from about 1200 Å through the visual region of the spectrum. Energy to operate the spacecraft is derived from the nearly 50,000 solar cells mounted on the two solar array "wings."

recently that attempts have been made to achieve the high resolution possible from space, most notably with the Hubble Space Telescope (Figure 10.33). Unfortunately neither its optics nor its pointing stability were of the quality that had been expected. The cores of the images of stars produced by the telescope are sharp—less than 0.1 arc second—but contain only about 0.15 of the total energy of the image. The rest is contained in a "skirt" about 2 arc seconds in diameter surrounding the core, because the 2.4-meter objective mirror suffers from **spherical aberration**; light reflected by the outer part of the mirror is not brought to the same focus as light from the inner part (see Figure 10.34). These extremely poor images resulted from the excessive removal of only 0.0002 inches of glass from the outer part of the mirror. This is a large error when compared to the tolerances within which opticians work. In addition, the spacecraft was not as stable as expected, because unequal heating of the two sides of the supporting structure during transitions from day to night caused the large solar arrays to oscillate, resulting in a corresponding motion of the telescope. As a result of these

225

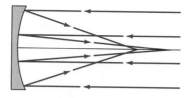

Figure 10.34. Parallel rays of light reflected from the outer part of a spherical mirror will be brought to a focus nearer to the mirror than rays reflected from the inner part of the mirror. This defect is called spherical aberration.

two defects, the performance of all of the instruments on HST—two cameras, two spectrographs, and a photometer—were compromised in various ways. For the first three years of its life, HST acquired data that, though not of the quality hoped for, were still better than obtained previously.

In December 1993, an ambitious servicing mission using the space shuttle was successfully completed. After capturing the HST and placing it in the cargo bay of the shuttle, astronauts replaced the solar arrays and several boxes of malfunctioning equipment (Figure 10.35). They also removed the primary camera and installed a new one fitted with optics that corrected the spherical aberration of the primary mirror. Finally,

Figure 10.35. This photograph was taken as the December 1993 servicing mission of the Hubble Space Telescope was drawing to a close. Two astronauts can be seen, one in the shuttle bay, the other riding the shuttle's arm to the upper part of the HST.

Figure 10.36. The three images are of the same field centered on a bright star in the Large Magellanic Cloud (a small satellite galaxy of the Milky Way). The image on the left was taken from the ground when the seeing was very good, about 0.6 arc seconds. The image has been magnified considerably, showing the individual square picture elements (pixels) of the detector, a CCD. The middle image, taken with WFPC-1, shows the effects of the spherical aberration of the primary mirror of the HST. The core of the image is a little smaller than 0.1 arc seconds, but it has a skirt more than 2 arc seconds in diameter. During the servicing mission, the camera was replaced with one having optics that corrected the spherical aberration. It produced the picture on the right. Note the large number of faint stars that appear, since their light is now concentrated into a sharp image that is visible above the sky and instrumental background.

they replaced the photometer with a box containing corrective optics for the two spectrographs and the second camera. Although these procedures compromised various aspects of the observatory performance (for example, the viewing fields of the cameras are considerably smaller now), the image quality is as good as was originally hoped (see Figure 10.36).

Detecting and Analyzing Electromagnetic Radiation
Information Carried by Radiation

An interesting bit of trivia is that all the energy ever collected from objects outside the solar system—stars, nebulae, galaxies—by all the optical telescopes of the world amounts to no more than what would power a flashlight for a second or two! Most of the observational basis for our knowledge of the universe rests on this tiny bit of energy. Rather than making you skeptical of our world picture, this fact should drive home how much information can be carried by so little energy.

What can we learn by analyzing the electromagnetic radiation from a star? In fundamental terms, the list is short: we can measure the radiation's intensity (brightness) as it depends on (1) direction, (2) wavelength, and (3) plane of polarization. Furthermore, all three of these quantities may change with time; so we may want to repeat the measurements many times.

By (1) above we mean the direction along which the radiation travels to us, that is, the direction to the star from which the radiation comes. For some objects, this direction may be seen to change with time, indicating that the object is moving. In its simplest terms, (2) answers the question: is the star red or blue; does it emit more red light or more blue light? More sophisticated observations enable us to measure the brightness at every point in the star's spectrum, Ångström by Ångström. Finally, (3) tells us whether or not the electric waves from the star show any preference for oscillating in one particular plane. If so, the light is said to be polarized. If the brightness, spectrum, or polarization of a celestial object changes with time, we say that it is a variable star or variable galaxy. Throughout the rest of the book we shall encounter many different examples of each of these measurements.

We can now make measurements of the intensity of a source as it depends on direction, wavelength, or polarization with varying precision throughout the whole electromagnetic spectrum, from γ-rays to the radio region. Such measurements form the observational basis by which astronomers try to describe and understand the universe.

There is an important point behind our consistent use of "observed" or "observation" when referring to astronomical measurements. Unlike physics, chemistry, or biology, astronomy is *not* an experimental science. The objects of astronomy—planets, stars, galaxies—cannot be created in or brought to the laboratory and then experimented on there. Even many of the physical conditions encountered in astronomy, such as the high temperatures or the very large and very small densities, cannot be approached in a useful way in the laboratory. Instead of doing experiments, we make observations[5] and squeeze as much information as possible from the radiation collected by our telescopes. We try to infer the physical nature of stars and galaxies by analyzing the radiation they emit, using the laws and techniques of physics. This is why it is so important to observe as much as possible of the entire electromagnetic spectrum. Otherwise, our information is incomplete.

Detectors

The Eye. Let us next consider briefly the means by which we detect and analyze the light collected by optical telescopes, that is, those used to collect visible light. The eye (Figure 10.37) was, of course, the first detector used, but it has serious limitations as an astronomical device. First of all, its wavelength sensitivity at low brightness levels (night vision), extending from about 4000 Å to 6300 Å with a peak sensitivity at 5100

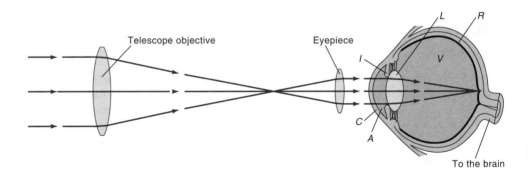

Figure 10.37. The eye was the first astronomical detector. Light from the eyepiece of the telescope passes first through the cornea, C, which contains the aqueous humor, A; then through the lens, L; through the vitreous humor, V; and finally to the retina, R, where an electrical signal is produced and sent to the brain. The aqueous and vitreous material and the lens all have somewhat different indices of refraction, and so the eye can correct reasonably well for chromatic aberration. The focal length of the lens is about 20 millimeters, but we adjust the shape of the lens (and hence its focal length) to bring objects at different distances to a focus on the retina. The iris, I, can change the aperture of the lens to adjust for varying light levels; in very faint light, the iris can open to a diameter of about 8 millimeters.

---

[5] The term "observations" comes from the time when the eye was the only detector and astronomers spent their time really looking through their telescopes. Now that they have far more sensitive detectors with which to make measurements, astronomers spend very little time "observing."

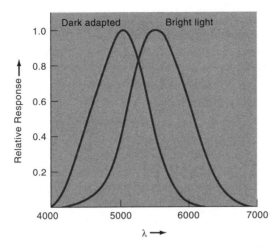

Figure 10.38. The sensitivity of the eye under bright light and when dark adapted. Note that the latter is shifted about 500 Å to the blue. Since the relative sensitivities are given, both curves peak at 1.0; in actuality, the dark-adapted eye is about three times more sensitive than the eye seeing bright light.

Å (Figure 10.38) is much narrower than even the atmosphere's optical "window." Second, just looking doesn't result in a permanent record of what was seen. Either we have to make a sketch of the image or hope our memory is adequate (which, at 3:00 A.M., it rarely is). Finally, although the eye is quite sensitive, it can't add up the photons it receives to produce a brighter image; it is not an "integrating" device. Stare as long as you like at a light; it won't appear any brighter. The effective integration time of our eyes is less than 1/20 second (this is why the frame rate in motion pictures is 24 per second; we can't perceive any break from one frame to the next). By contrast, if we want to take a photograph in faint light, we can increase the exposure time and collect enough photons to produce a detectable image on the film.

Photography. The application of photography to astronomy began in the 1840s. This was an enormously important development, because a photographic plate or film has many of the virtues the eye lacks. Up to a point, the photographic plate can respond to more and more photons by producing a darker and darker image on the negative. Also, it provides an objective, permanent record. Although at first the wavelength sensitivity of an emulsion was limited to the blue region of the spectrum, various photographic emulsions have been developed that are sensitive to other regions of the spectrum. These emulsions, combined with appropriate filters, enable photographs to be made in the light of about any wavelength band desired, from the ultraviolet to the near infrared. Finally, the grains that make up an emulsion are very small, only a few thousandths of a millimeter in diameter, so that there are an enormous number of image elements, or **pixels**, on a photographic plate. Consequently, emulsions can record all the detail in the telescopic image. The major disadvantages of film are its nonlinear response to light (see below) and its still relatively poor sensitivity. Though today's emulsions are hundreds of thousands of times "faster" (more sensitive) than the early plates, only about one photon in every hundred affects a grain—is detected—in a modern emulsion.

The simplest way to take an astronomical photograph is to put a piece of film or a photographic plate at the focus of a telescope and guide carefully so that no image smearing occurs. Generally, however, a camera with high-quality optical elements is mounted at the focus of the telescope. (Remember that the Earth's rotation makes the celestial object continuously change its direction with respect to the Earth's surface. The application of photography to astronomy provided a considerable impetus for the development of stable telescope mountings and good drive mechanisms.)

The Photoelectric Detector.   From about the 1920s on, another kind of detector played an increasingly important role in astronomy. It was based on the **photoelectric effect**, in which photons, striking the surfaces of certain kinds of metals such as cesium, can kick electrons out from those surfaces. (This is one of those effects that requires the "particle" or photon nature of light for its description; it cannot be accounted for if light is described as a wave.) The resulting flow of electrons constitutes a weak electrical current, which can be amplified and detected by electrical means. Thus a detector based on this phenomenon converts a beam of light into an electric current. By the 1940s, the photoelectric effect had been incorporated in the **photomultiplier** tube (Figure 10.39), a photoelectric detector in which great amplification or multiplication (typically a million times) of the weak photoelectric current takes place within the device itself. The resulting signal is thereby much easier to amplify and record.

This kind of detector has two important advantages over photography. First, it is much more sensitive than the photographic plate. About one photon in five is detected, that is, will cause an electron to be ejected from the sensitive surface. Second, the electrical signal produced (the number of electrons ejected) is directly proportional to the number of photons striking the detector. If we double the intensity of the light, we double the resulting signal; that is, the response of the detector is linear. This is not true with photographic emulsions, which respond to light in a much more complicated way. Thus, the brightnesses of celestial objects can be measured much more accurately by photoelectric than by photographic techniques. Its major disadvantage is that the photomultiplier is a "point" detector—it has only one sensitive element instead of the millions of pixels (the grains) on a photographic plate.

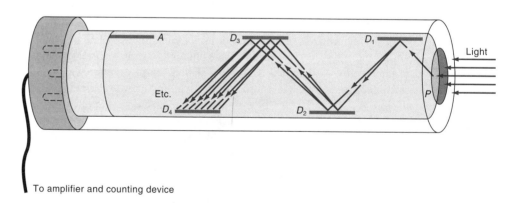

To amplifier and counting device

Figure 10.39.   Schematic diagram of a photomultiplier. Five photons strike the light-sensitive coating on the inside of the evacuated glass tube and release one electron (a typical efficiency for this kind of detector). A positive voltage attracts the negative electron to the first dynode, D_1, with sufficient energy to cause it to release three or four electrons (for clarity, only two electrons are shown). These electrons are attracted to D_2, where they each release three or four more, etc. The process is repeated eight or ten more times with the result that a pulse comprising as many as a million electrons is collected at the anode, A, after which it is amplified and recorded.

Figure 10.40. A charge-coupled device (CCD). A detector for astronomical use might consist of an array of 2000-by-2000 light-sensitive pixels (4 million in all), each 20 microns in size. The left-hand image of Figure 10.36 shows individual pixels. An electrical signal representing the charge released in each pixel by incident photons is stored in a computer for further processing.

The CCD.   Today's solid-state detectors such as the so-called **charge-coupled devices** (or CCDs), however, combine the advantages of high sensitivity and linear (one-to-one) response to the incident light on each of many small, sensitive elements arranged over a substantial area (Figure 10.40). Each pixel can detect about one of every two photons that strikes it, which means that the efficiency of the CCD is approaching the maximum possible. In addition, a single CCD has a broad spectral response, from the blue to the near infrared. One of the latest of these detectors has about four million separate pixels, each of which is only a few times larger than a grain clump in an emulsion. Consequently, the resolution of the device—its ability to detect detail—is nearly as good as that of a photographic plate. Thus, the best aspects of both photoelectric and photographic detectors are combined in the CCD. Finally, since each CCD pixel produces a number representing the brightness of the image at that point, the image is "digitized." This means that the numbers representing the image can be read into a computer and easily stored for later processing, for example, removing an unwanted background. Modern digital computers are now an integral part of the observing astronomer's toolbox.

The CCD has become the dominant astronomical optical detector, rapidly replacing photographic emulsions. Its major disadvantage, which it shares with much modern astronomical equipment, is that to achieve maximum sensitivity to the extremely low light levels encountered in astronomy, it requires sophisticated electronics and highly trained people to maintain the performance of the detector. By contrast, CCDs intended to operate at ordinary light levels, as in today's hand-held video cameras, are straightforward to operate. Nevertheless, these new efficient solid-state detectors, in instruments at the focal plane of the giant telescopes being constructed, will give astronomers their best view of the optical and near-infrared universe yet.

Instruments

The Photometer.   What kinds of instruments do we mount on telescopes to analyze the light they collect? We have already mentioned the camera. Another simple but ex-

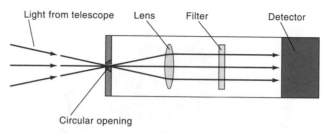

Figure 10.41. The essentials of a photometer. Light collected by the telescope is focused onto a small opening, made parallel by a lens, passed through a filter and then onto the detector. The small opening allows light from only the object of interest to pass through. The filter, often of colored glass, passes light of a relatively small range of colors and absorbs the rest. The detector might be a photomultiplier tube or a CCD.

tremely useful instrument is the **photometer**. As its name implies, it measures light, in particular, its brightness. More information about a star or galaxy can be derived if the photometer measures how bright the object is in several different regions of the spectrum, for example, in the red, yellow, and blue. Usually glass filters are used to pass a particular band of wavelengths while blocking the rest. In essence, a photometer consists of one or more filters that can be inserted into the light beam focused by the objective, and a lens to focus the light onto a detector, which could be a photomultiplier or a CCD, for example (Figure 10.41). The filters, blue or yellow glass, for example, allow us to measure the relative amounts of blue and yellow light emitted by the star. Usually these filters transmit a broad band of wavelengths: a blue filter may pass radiation from about 4000 Å to 5000 Å, and the yellow filter may be transparent from 5000 Å to 6000 Å. Using such broad chunks of the spectrum enables us to measure the colors of faint stars, but we will see that it also throws away a lot of information that the light could give us.

The Spectrograph. The instrument that can give detailed information about the spectrum of radiation collected by the telescope is the **spectrograph**. Such instruments were first applied to astronomy in the mid-nineteenth century. Not surprisingly, they were inefficient and difficult to use. By the beginning of our century, however, they had become very useful devices, and soon took their place, along with cameras and photometers, as one of the basic astronomical instruments.

A spectrograph breaks up light into its component wavelengths and spreads them apart; we say it disperses the light. There is a direct correspondence between the position or distance along the spectrum and the wavelength (see Figure 10.42). In fact,

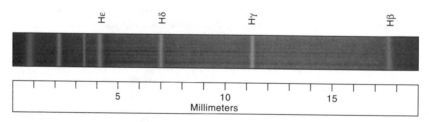

Figure 10.42. A much-enlarged spectrum of an A-type star along with a ruler. Note the small size of the original spectrum.

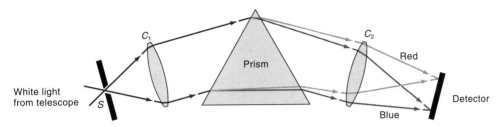

Figure 10.43. The essential elements of a spectrograph. Light from the telescope passes through a slit, S, is made parallel (collimated) by lenses (or mirrors) called the collimator, C_1, is dispersed by the prism, and focused by the camera optics, C_2, onto the detector. The latter used to be a photographic plate, but today is nearly always a solid-state detector such as a CCD. (Also, the dispersing element is more likely to be a diffraction grating than a prism, but the latter is easier to show.)

one of the characteristics of a spectrograph is how many Ångströms of spectrum are included within 1 millimeter of it. A typical dispersion might be 100 Å/mm; with this dispersion, 1000 Å of spectrum would extend over 10 millimeters. Light can be dispersed either by a **prism** or by a **diffraction grating**. A prism spreads out light into its constituent wavelengths because the index of refraction of the glass and, hence the amount of refraction of the light, varies with wavelength. A diffraction grating consists of very many fine grooves—typically several hundred per millimeter—usually ruled on a flat mirror. These grooves diffract the light because they act as obstacles around which the light is bent. Thus, two phenomena we encountered earlier, chromatic aberration and diffraction, presented then as difficulties, now appear as the means by which an extremely important instrument, the spectrograph, is made to work.

Figure 10.43 shows a prism spectrograph in which the light, passing through a narrow entrance slit at the focus of the telescope, is made parallel before it passes through the prism and is dispersed. A second lens then focuses the resulting spectrum onto a detector, which today would be a CCD. The spectrum is just a series of images of the slit made in the wavelengths of light incident on the slit. If the radiation consists of one color, then only one image—an emission line—results. When the brightness of the source at a given wavelength is low compared with that of adjacent wavelengths, a dark, narrow gap in the spectrum is produced. As we have seen, we call such a feature an absorption line ("line" because it is an image of the long, narrow spectrograph entrance slit).

Ordinarily the light from just one object at a time—a star or a galaxy, for example—enters the slit of a spectrograph, where its spectrum is produced and recorded. In the late 1980s a technique was developed that enables a spectrograph to produce the spectra of many objects simultaneously. Suppose you want to obtain the spectra of several stars in a cluster that, of course, must all be within the field of view of the telescope. You map out carefully the position of each star in the focal plane of your telescope and then place a thin optical fiber at each position (see Figure 10.44). The fibers might be held in place by a specially constructed metal plate or perhaps by a series of arms. Each fiber carries the light from one star to the spectrograph, where all the fibers are arranged in a line along the entrance slit. The spectrograph then produces a spectrum of each star in the usual manner. Multi-object spectrographs able to measure simultaneously the spectra of as many as 100 objects are in operation, with a corresponding increase in their efficiency.

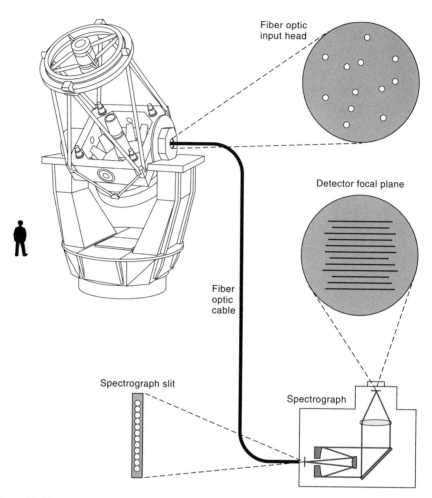

Figure 10.44. A schematic diagram of a multi-object spectrograph. Very thin optical fibers placed in the focal plane of the telescope carry the light of the stars or galaxies being observed to a spectrograph where the fibers are arranged along the entrance slit. Light from each fiber produces a spectrum that is then recorded in the usual manner. In this way, many spectra can be produced and recorded simultaneously, with a corresponding increase in observing efficiency.

The Doppler Effect

Next let us see how, by measuring accurately the wavelengths of these spectral lines, we can find the relative line-of-sight velocity between the source and the observer. Most of us have experienced the **Doppler effect** in sound, the drop in pitch of the horn of a rapidly moving car just as it speeds by you. As it approaches you the horn's pitch is slightly higher than it is when the car is stationary, and as it recedes from you it is slightly lower. The important factor is the ratio of the **radial velocity** (the velocity directly toward or away from you) to the velocity of sound. The greater the radial velocity, the greater the change in pitch. A schematic representation of what is going on is given in Figure 10.45, where the waves emitted by the moving source when it is at positions 1, 2, 3, and 4 are shown. An observer at *A* will see more waves passing by than one at *B*. Hence *A* measures a higher pitch, or frequency, than does *B*.

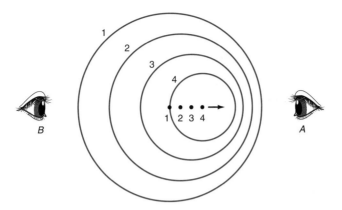

Figure 10.45. Imagine a light source moving to the right through positions 1, 2, 3, and 4. The light emitted when it was at 1 has expanded to the sphere labeled 1, the light emitted when it was at 2 has spread a somewhat smaller distance to sphere 2, etc. In a given time, observer *A* will see more waves pass by than observer *B*; that is, *A* will measure a higher frequency than he would if the light weren't moving, but *B* will measure a lower frequency. Hence to *B* the light will be blue-shifted and to *A* it will be red-shifted.

The same effect takes place in light. Its magnitude depends on the same factor as it does for sound: the ratio of the radial velocity of the source and the observer to the velocity of light. Because the speed of light is very large compared to speeds in our everyday experience, we do not ordinarily see the Doppler effect in light. The color of a car does not become bluish (higher frequency) as it speeds towards us! Astronomical objects, however, can have very large velocities. Stars in the solar neighborhood typically have speeds of 10 to 40 kilometers per second, and galaxies can have speeds approaching large fractions of that of light. These give rise to large Doppler shifts in the spectra of such objects. If a star is moving away from us, its whole spectrum is shifted to the red (lower frequency or longer wavelengths) by an amount proportional to its speed; we say it is red-shifted. Conversely, a star's spectrum is blue-shifted when it approaches us.

As long as we can measure accurately the wavelength of the spectral lines emitted by the source, we can find its radial velocity by using the simple formula

$$\frac{\lambda_{obs} - \lambda_{lab}}{\lambda_{lab}} = \frac{V}{c},$$

where λ_{lab} is the "laboratory" wavelength (the wavelength of the spectral line when source and observer are at rest with respect to each other), λ_{obs} is the measured wavelength of the line from the astronomical source; $(\lambda_{obs} - \lambda_{lab})$ is the amount by which the line is shifted because of the radial velocity of the object; c is the speed of light; and V is the radial velocity of the object that we wish to find.

A simple example will help make this more concrete. Suppose a star is moving away from us with a velocity of 30 km/sec. Then $V/c = 30/300,000 = 1/10,000 = 10^{-4}$. Thus, to detect this redshift the wavelength must be measured with an accuracy of 1 part in 10,000 or 0.01 percent! The shift to the red of a spectral line that has a rest wavelength of 5000 Å would be only 0.5 Å to 5000.5 Å.

We shall see later that this simple formula must be modified when we want to calculate the radial velocity of objects moving at large fractions of the velocity of light.

Astrophysics

Probably more than any other astronomical instrument, the spectrograph has been responsible for the development of **astrophysics** (as opposed to astronomy). Classical astronomy is usually (though not always) thought of as being the study of the gross properties of celestial objects—their brightnesses, distances, motions, and distribution in space. By contrast, astrophysics has to do with determining the physical properties of celestial objects—their temperatures, masses, densities, chemical compositions, etc. Astrophysics requires not only the appropriate observations (obtained largely by the spectrograph), but also the physical theories by which these observations may be interpreted. Without these theories of the structure of atoms and of nuclei, of electromagnetic radiation, and of special relativity, for example, astrophysics would not have been possible. We will have to consider several of these physical theories at appropriate points in this book.

Next, however, let us apply what we have so far learned to the spectra of stars.

Terms to Know

Index of refraction; refractor, reflector, aperture, objective, *f*-number; eyepiece, focal point, chromatic aberration, spherical aberration, dispersion; diffraction, resolving power, seeing. Newtonian focus, Cassegrain focus, prime focus, light-gathering power, equatorial mounting and alt-azimuth mounting. Radio dish, VLA, VLBA, VLBI, CCD, photoelectric detector, photometer, spectrograph; Doppler effect, radial velocity.

Ideas to Understand

What telescopes do and how they work. Why all large optical telescopes are reflectors. Properties of electromagnetic radiation. Important astronomical instruments and detectors, what they do and how they work; why the resolution of a single radio dish is poor; how a radio interferometer works. The Doppler effect and what it measures. The distinction between astronomy and astrophysics.

Questions

1. (a) The largest optical telescope in the continental U.S. is the 200-inch-diameter (5-meter) reflector on Mt. Palomar in southern California. The objective of Galileo's telescope was a lens having a 1-inch diameter. How many times more light does the Palomar telescope gather than did Galileo's instrument?

(b) The Keck telescope, the largest optical telescope in the world, has a 10-meter-diameter objective. How many times more light does it gather than the Palomar reflector?

2. To give the best performance, which has to be more accurately made, the parabolic surface of an optical telescope or that of a radio telescope? Why?

3. Why are all large telescopes reflectors rather than refractors?

4. What advantages would there be in placing an optical telescope on the surface of the Moon? A radio telescope on the far side of the Moon?

5. Would it be a good idea to build a huge telescope with perfect optics on the Earth in order to observe very small surface details on the Moon?

6. (a) What is the theoretical resolution of a 10-meter optical telescope at 5500 Å? Compare this with typical seeing of the atmosphere.

(b) Similarly, calculate the theoretical resolution of a 10-meter telescope at 10 microns, and compare that with atmospheric seeing.

7. Describe two advantages of placing telescopes into space.

8. Compare the properties of the photographic emulsion, the photoelectric detector, and the CCD.

9. The radial velocity of the bright star of the summer sky, Vega, is about 14 km/sec toward us.

(a) What fraction of the speed of light (300,000 km/sec) is Vega's velocity?

(b) By how many Ångströms would the wavelength of 5000 Å light be shifted? of 10,000 Å light?

(c) To the red or the blue?

10. Why are radio interferometers built?

11. Explain carefully how, by using the Doppler effect, you could show that the Earth rotates on its axis and revolves around the Sun.

12. Astronomy is primarily an observational, rather than experimental, science. Explain how you would characterize chemistry, meteorology, geology, and political science.

Suggestions for Further Reading

Books and articles on the history of the telescope:

Bell, L., *The Telescope*. New York: Dover Publications, Inc., 1981. A reprint of a 1922 book that discusses many aspects of telescopes—history, glass, mountings, eyepieces, etc.

King, G., *The History of the Telescope*. New York: Dover Publications, Inc., 1979. A reprint of the 1955 book. A fascinating and authoritative account of the development of the telescope.

Wolff, S., "The Search for Aperture: A Selective History of Telescopes," *Mercury*, **14**, p. 139, 1985.

Woodbury, D., *The Glass Giant of Palomar*. New York: Dodd, Mead and Co., 1939. An interesting account of the construction of what was then the world's largest telescope.

Various aspects of optical and IR telescopes:

Bruning, D., "Three Nights on Kitt Peak," *Astronomy*, **20**, p. 38, April 1992. An account of an astronomer's observing run at the national optical observatory.

Gillett, F., Gatley, I., and Hollenbach, D., "Infrared Astronomy Takes Center Stage," *Sky & Telescope*, **82**, p. 148, 1991. An overview of the prospects for ground- and space-based infrared astronomy.

Janesick, J. and Blouke, M., "Sky on a Chip: The Fabulous CCD," *Sky & Telescope*, **74**, p. 238, 1987. Now the standard detector for optical astronomy.

Strom, S., "New Frontiers in Ground-Based Optical Astronomy," *Sky & Telescope*, **82**, p. 18, 1991. Prospects for the decade.

Many large telescopes have recently been built or are under construction. Some of these are described in the articles below.

Bunge, R., "Dawn of a New Era: Big Scopes," *Astronomy*, **21**, p. 48, August 1993. A nicely illustrated description of the new generation of ground-based telescopes.

Fischer, D., "A Telescope for Tomorrow," *Sky & Telescope*, **78**, p. 248, 1989. A description of the European Southern Observatory 3.6-meter telescope that incorporates many of the techniques being applied to the new generation of telescopes.

Keel, W., "Galaxies Through a Red Giant," *Sky & Telescope*, **83**, p. 626, 1992. Describes recent improvements made to the Russian 6-meter telescope.

Sinnot, R., "The Keck Telescope's Giant Eye," *Sky & Telescope*, **80**, p. 15, 1990. A well-illustrated description of the largest ground-based telescope in the world.

Sinnot, R. and Nyren, K., "Sky & Telescope's Guide to the World's Largest Telescopes," *Sky & Telescope*, **86**, p. 27, 1993.

Old and new radio telescopes are described in these books and articles:

Kellermann, K., "Radio Astronomy: The Next Decade," *Sky & Telescope*, **82**, p. 247, 1991.

Shore, L., "VLA: The Telescope that Never Sleeps," *Astronomy*, **15**, p. 15, August 1987. The VLA and how it is used.

Spradley, J., "The First True Radio Telescope," *Sky & Telescope*, **76**, p. 28, 1988. Grote Reber and his telescope.

Sullivan, W. (ed.), *The Early Years of Radio Astronomy: Reflections Fifty Years After Jansky's Discovery*. Cambridge: Cambridge University Press, 1984. A collection of several historical articles, many written by the pioneers in this field.

Some technical and political aspects of space astronomy:

Blair, W. and Gull, T., "Astro: Observatory in a Shuttle," *Sky & Telescope*, **79**, p. 591, 1990. The shuttle as a platform for astronomy.

Carroll, M., "Cheap Shots," *Astronomy*, **21**, p. 38, August 1993. Possibilities for space science in an era of declining funding.

McDougall, W., *The Heavens and the Earth: A Political History of the Space Age*. New York: Basic Books, Inc., 1985. How politics shaped space science and technology.

Smith, R., *The Space Telescope: A Study of NASA, Science, Technology, and Politics*. Cambridge: Cambridge University Press, 1989. A prize-winning account of the selling and construction of the most expensive telescope ever built.

Tucker, W. and Tucker, K., *The Cosmic Inquirers: Modern Telescopes and Their Makers*. Cambridge, MA: Harvard University Press, 1986. Description of the planning and construction of major new instruments in radio, x-ray, gamma-ray, infrared, and ultraviolet astronomy, with emphasis on the astronomers involved.

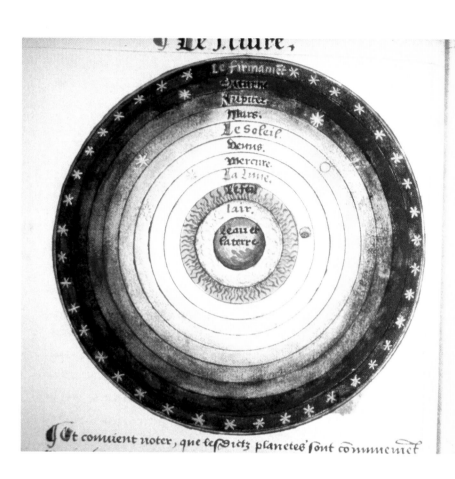

Plate C1

This illustration of the geocentric universe, shown here in its original color, is from a French manuscript of 1549. It simply shows a general arrangement without any of the geometric constructions required to account for the various motions. Shown are the "natural" locations of the four elements (earth, water, air, and fire) within the lunar sphere as well as the locations of the Sun, Moon, planets, and stars.

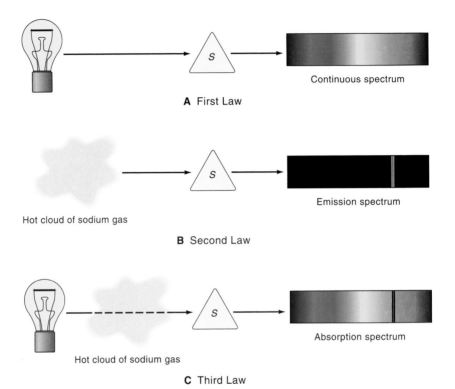

A First Law

Continuous spectrum

Hot cloud of sodium gas

B Second Law

Emission spectrum

Hot cloud of sodium gas

C Third Law

Absorption spectrum

Plate C2

A schematic representation of Kirchhoff's laws. The hot filament in a light bulb, A, produces a continuous spectrum; that is, all wavelengths are present (the first law). In B, a cloud of hot gas emits an emission line spectrum (the second law). In this case the gas is sodium, so the most prominent lines are two in the yellow-orange region of the spectrum. A gas cloud absorbs photons, C, from a hotter continuous spectrum producing absorption lines at the same wavelengths at which it would otherwise emit them (the third law).

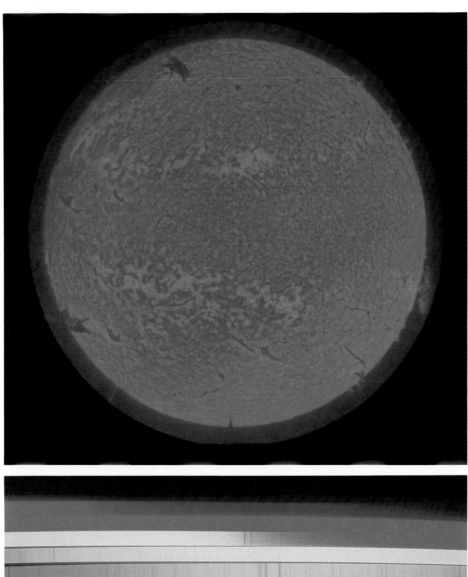

Plate C3

(upper) An image of the Sun made in the light of the H-α line. Note the prominences seen at the limb of the Sun, and the dark filaments (prominences seen in projection against the disk). The lighter regions are plages, areas of magnetic activity. Compare with Figure 11.22.

(lower) A portion of the spectrum of the Sun with much higher spectral resolution than that in Figure 11.4, for example.

Plate C4

(upper) The Trifid nebula showing the red emission nebula and the blue reflection nebula. See Figure 14.7.

(middle) The Ring nebula. Compare with Figure 16.3.

(lower) The Crab nebula. The relatively smooth blue synchrotron radiation is clearly distinguished from the pinkish filaments of the emission nebulosity. Compare with Figure 16.15.

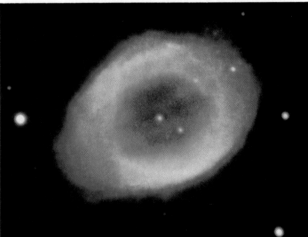

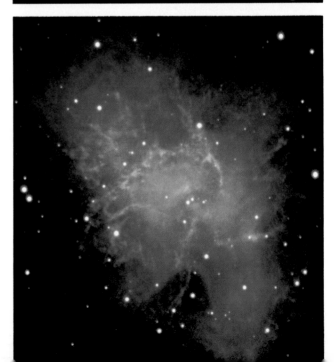

Plate C5

(upper) The bright spiral galaxy M83. The blue regions indicate hot, young stars, the red spots are H II regions, and the dark patches are obscuring dust.

(lower) Centaurus A (NGC 5128), the result of the collision of an elliptical and a spiral galaxy. The old, yellowish stars of the elliptical are crossed by a broad and disrupted band of dust, a remnant of the spiral. Young, blue stars can be seen along the edges and end of the dust band. Compare with Figure 20.4.

EARTH

MARS

MERCURY

MOON

IO

EUROPA

GANYMEDE

CALLISTO

VENUS

TITAN

Plate C6 *(facing page)*

(upper) Images of the terrestrial planets, the Moon, Galilean satellites, and Saturn's Titan, all to scale.

(lower) A Viking view of the surface of Mars covered with frost.

EARTH

Plate C7

Jupiter, Saturn, Uranus, and Neptune compared with the Earth, all to scale. Though the colors here are too vivid (as they are with the terrestrial planets in Plate C6), they do point up the difference in color between brown and orange Jupiter and blue Neptune.

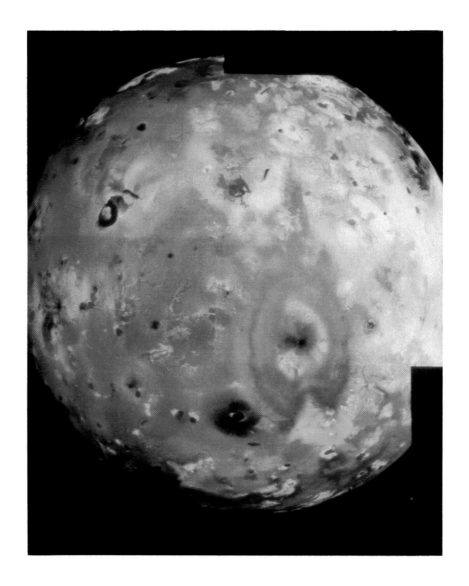

Plate C8

Io, a satellite of Jupiter and the most active volcanic object
in the solar system. See Figure 24.39.

Hubble Gallery. Image clarity of Mars is limited only by the properties of the camera when the latter is above the Earth's atmosphere, as was the case here. Mars was at opposition, about 100 million kilometers from Earth, when this image was taken by the HST. Note the volcanic peak above the cloud deck on the western (left) limb.

"We understand the possibility of determining their shapes, their distances, their sizes and motions, whereas never, by any means, will we be able to study their chemical compositions. . . ."

"I persist in the opinion that every notion of the true mean temperature of the stars will necessarily always be concealed from us."

—Auguste Comte, 1835; French philosopher

Twinkle, twinkle little star,
I don't wonder what you are,
For by spectoscopic ken
I know that you are hydrogen.

—Anonymous

CHAPTER 11

Reading Stellar Spectra

Recall that the general subject of this part of the book is the evolution of stars. A necessary prelude to understanding their life history is to have a good idea of the physical state of stars; that is, how much energy they radiate, how large they are, how massive, how hot, and of what chemical elements they are composed. This information can be obtained only by analyzing starlight, using all the observational and theoretical tools of astrophysics available to us. Let us begin by taking a look at the spectra of stars and the information they contain.

Classification of Stellar Spectra

The overwhelming majority of stars show absorption lines superposed on a continuous spectrum (see Figure 11.1). The spectra of some stars are rather simple, with only a few absorption lines; other stars show complex spectra with many lines produced by both atoms and molecules. When faced with such a variety of spectra, astronomers did just what biologists had done before them—they attempted to classify the spectra according to their appearance, to arrange them in groups containing similar members. A few attempts had been made to classify spectra at the telescope by eye before photography was applied to astronomical spectroscopy, but these were of necessity very primitive. Only the spectra of the very brightest stars could be seen and these only faintly; furthermore, they were smeared by seeing and were difficult to sketch. One of the earliest of these efforts was made in 1798 by William Herschel (much more about Herschel in Chapter 17). He placed a prism at the eyepiece of his telescope and examined the spectra produced by a few of the brightest stars. He did not pursue this work, however, because of the difficulties just mentioned and because he had no way of interpreting his observations.

By the 1880s, however, the techniques of photography and spectroscopy had both advanced sufficiently to allow large-scale programs of stellar spectroscopy to be undertaken. The most ambitious of these was carried out at Harvard College Observatory, through a gift from the widow of a pioneer in astronomical photography, Henry Draper, a medical doctor. Using a telescope equipped with an objective prism (see Figure 11.2), Harvard astronomers photographed and classified the spectra of more than a quarter of a million stars.

Photographic emulsions at that time were sensitive only in the blue, so spectral features from about 3800 Å to 5000 Å were used in the classification system. At first, spectra were divided into groups according to the strength of their hydrogen lines, class A stars having the strongest lines, class B the next strongest, and so on. With the hy-

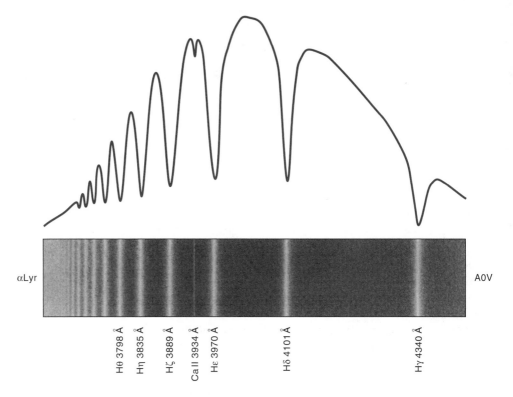

αLyr   A0V

Hθ 3798 Å Hη 3835 Å Hζ 3889 Å Ca II 3934 Å Hε 3970 Å   Hδ 4101 Å   Hγ 4340 Å

Figure 11.1. Vega, the brightest star in the summer sky, has a relatively simple spectrum, that is, not many absorption features superposed on the star's continuum. As is customary in astronomy, this spectrum, and all the others in this book, is a negative print rather than the more familiar positive of our snapshots. Consequently, dark regions in the prints are actually bright, and the lighter absorption lines are dark. Above the spectrum is a tracing showing the relative energy in the continuum and lines. Note that with the exception of the relatively weak line of ionized calcium, all the absorption lines are members of the Balmer series of hydrogen.

drogen lines ordered in a sequence continuously declining in strength, other spectral features did not change smoothly from class to class, however. Ultimately the various spectral classes were arranged so that *all* spectral features changed smoothly and continuously along the sequence (Figure 11.3). This requirement, along with several types being merged with others or dropped, resulted in the final spectral classes being denoted by O, B, A, F, G, K, and M (see Figure 11.4; compare with Plate C3). The number of classes of spectra that can be distinguished even at moderately low spectral resolution is quite large. Consequently, these major groups were further divided, usually into ten subclasses, such as A0, A1, A2, . . . , A9, F0, F1, . . . , F9, G0, etc.

It is important to realize that this classification is entirely empirical; that is, it is based on only the *appearance* of spectra taken with a particular spectroscope. It is not shaped by any preconceived notions of what, if anything, the spectral sequence might signify physically. Again, this resembles the classification technique used in biology. In essence, then, the classification system works this way. Stars whose spectra illustrate each of the spectral types that the classification system is able to distinguish are taken as standards; they *define* the various classes. Any star with a spectrum identical to one of the standards is then a member of that particular class. In practice, certain spectral lines—or more usually, the ratios of the strengths of certain features—are

A 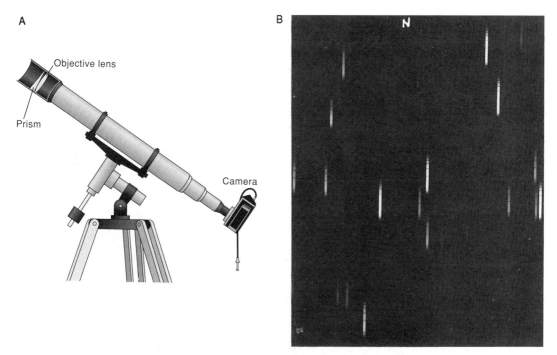 B

N

Figure 11.2. A prism placed in front of the objective lens of a telescope (A) produces a spectrum of every star in the field of view (B). Spectra of many stars suitable for classification purposes can be produced with just one exposure. If the seeing is poor, however, the spectra will not be sharp.

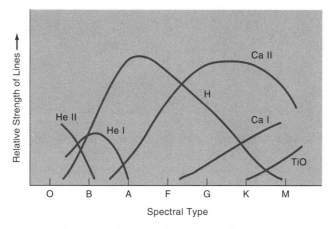

Figure 11.3. The spectral sequence is arranged so that all of the spectral features change smoothly from one class to the next. Hydrogen lines, for example, weak in O-type stars, strengthen through the B-types, reach a maximum at about type A2, then slowly become weaker through the F-, G-, and K-type stars. The sketch also indicates, for example, that at maximum, neutral helium lines are much weaker than the hydrogen lines at their maximum.

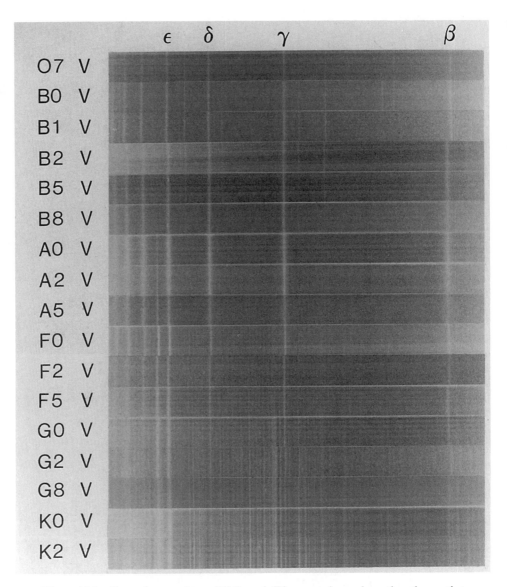

Figure 11.4. Stars of spectral type O7 through K2 arranged one above the other so that their spectral lines are aligned. Note how the strengths of the lines change smoothly from one spectral type to another as indicated in Figure 11.3. Also, note how many more lines there are in G- and K-type spectra than in the spectra at the top of the figure. The Balmer lines of hydrogen can be seen over most of the spectral sequence.

found to be good indicators of spectral type. These line ratios are then used to classify the star. In this way, extraneous effects, such as a spectrum under- or overexposed compared to the standard, are removed.

Notice that only a glance is required to distinguish the spectra of stars at the "OB" end of the spectral sequence from those in the middle "FG" range or at the "KM" end. The number of spectral lines increases markedly along the spectral sequence. Much practice is required to use a classification system in all its detail, but the eye is good at recognizing intensity ratios and other spectral patterns. Consequently, experienced astronomers can classify spectra rapidly and accurately.

A little jargon is useful here. Stars whose spectra place them toward the left end (O, B) of the spectral sequence are said to be **early-type** stars; those toward the right end (K, M) are called **late-type** stars. In addition, we speak of "early B-type stars"— B0, B1, B2—and "late B-type stars"—B7, B8, B9—and similarly for other spectral classes. This nomenclature had its origin in old (and incorrect) ideas of stellar evolution.

The classification system in use today is one developed by W. W. Morgan and P. C. Keenan of Yerkes and Perkins Observatories, respectively, and is a major revision and refinement of the Harvard system. This so-called MK system is based on spectra of much better quality than those available earlier. Various spectral criteria (usually the ratios of strengths of neighboring absorption lines) were found that not only allow accurate classification of the spectral types of stars, but also enable a measure of their intrinsic luminosity to be made. For example, a spectral type indicator from A0 to F5 is the ratio of the strengths of the Ca II line (3933 Å) to that of Hδ (4101 Å), which changes steadily over this part of the spectral sequence. We shall return to this important classification system later, but first let us consider what stellar spectra tell us.

Interpretation of Stellar Spectra
How Absorption and Emission Lines Are Formed

First of all, why are most of the features in stellar spectra *absorption* lines rather than emission lines? A qualitative answer to this question can be found in Kirchhoff's laws of spectroscopy, which we described in Chapter 9. In this picture a continuous spectrum is formed by the hotter, higher-pressure gas near the surface of the star. As these continuum photons travel outward, some of them are absorbed by atoms in the cooler, overlying, low-density gases of the star's atmosphere. The layer of the star where the continuum is produced is called the **photosphere** (the light sphere), whereas the cooler regions in which absorption lines are formed is called the **reversing layer**. The dark absorption lines are a "reversal" of the bright continuum.

We now know that it is an oversimplification to imagine two such well-defined regions in a star's atmosphere. In fact, both the absorption lines and the continuum are formed throughout the photosphere. This outermost region of a star extends over a narrow range of depths, only several hundred kilometers thick. Since typical stellar radii are on the order of hundreds of thousands of kilometers, a star's atmosphere is the thinnest skin of the onion. Its maximum extent is set by the depth into which we can "see"; that is, the depth from which photons can directly escape from the star without undergoing any further absorptions and re-emissions.

The photospheric layers at which particular spectral lines and the continuum are formed depend on the physical conditions in the stellar atmosphere, such as temperature and density, and on atomic parameters of the different elements, such as the probability that a photon of a given wavelength will be absorbed. For example, the probability that a photon in a strong absorption line will be absorbed as it travels out is very large. Consequently, such photons will have a good chance to escape only from the uppermost layers of the atmosphere, from which they don't have to travel through much material. Since the strength of the various features of a spectrum depends on the physical conditions of the gas, it is possible to derive the temperature, density, and chemical composition of the gas from the appearance of its spectrum.

The spectra of a small fraction of stars show emission as well as absorption lines. Kirchhoff's laws also suggest how these stellar emission lines come about. Suppose the star's atmosphere is surrounded by a low-density gas that extends a great distance outward from the star instead of closely surrounding it (see Figure 11.5). Then our line of sight to the star will include not only the photosphere where the continuum and absorption lines are formed, but also the extended low-density cloud layer alone. Since this hot gas is not seen against a hotter background, it produces emission lines only (Kirchhoff's second law; see Chapter 9). Such extended atmospheres are relatively rare, so most stars show only absorption-line spectra. Nevertheless, many interesting types of stars do have emission lines, and we will refer to this physical picture later in describing some of them.

In addition to information about the structure of the outer layers of stars, spectra tell us other things. Though it may not be obvious at a casual glance, different stars have different colors—red, orange, yellow, blue-white, etc. If on a clear night you are away from bright lights, and you allow your eyes to become dark-adapted, you will be able to notice the differences in the colors of stars.[1] Early-type stars appear to be blue or bluish-white, whereas late-type stars are yellow or red. If you interpreted this intuitively as meaning that early-type stars are hotter than those of later types, you would be correct.

This can also be seen by looking at the spectral features prominent in each class, shown schematically in Figure 11.3 and in the spectra of Figure 11.4. For example, in O-type stars, ionized helium lines are stronger than those of neutral helium, whereas in B-type stars the opposite is the case. Hotter temperatures (greater energy) are required to produce higher stages of ionization of a given element. Thus the atmospheres of O-type stars must be hotter than those of B-type. Similarly, A- and F-type stars must be hotter than G- and K-type stars, because in the former, lines of some ionized metals are present, but these have disappeared from the G and K spectral types, where instead lines of the neutral metals appear. Another indication of this same general trend is provided by the appearance of spectral features produced by molecules in K- and M-type stars. The electrical force binding atoms together to form molecules is quite small, generally smaller than the force with which an electron is bound to an atom. Hence, the appearance of molecules in late-type spectra indicates that the temperatures of such stars must be low, otherwise the molecules would be dissociated (broken apart) into their constituent atoms.

The Hydrogen Balmer Lines

Since hydrogen lines are so prominent in the spectra of many stars, let us look at their behavior along the spectral sequence in light of the comments above. From Figures 11.3 and 11.4 we see that the Balmer lines are weak in the O-type stars, strengthen in the B-types, reach maximum prominence at about A2, then slowly decline toward later spectral types. This behavior can be understood in the following way. In the hottest stars, an appreciable fraction of hydrogen is ionized, few hydrogen atoms exist in the $n = 2$ level to absorb photons, and hence the Balmer lines are weak. As the temperature drops, more hydrogen is neutral, some of these atoms are in the $n = 2$ state from which Balmer photons are absorbed, and the lines become stronger. Among stars at the cool end of the spectral sequence, all the hydrogen must be neutral, but nevertheless the lines are weak, because the temperature is so low that only a small fraction of hydrogen atoms acquire the energy necessary to put them into the $n = 2$ state from which they can absorb Balmer photons. As the temperature is raised, however, more of the hydrogen is raised to the $n = 2$ level, and the lines become stronger. At about

---

[1] Our color vision at low illuminations is poor, because the eye's color receptors, the cones, function best at high illuminations.

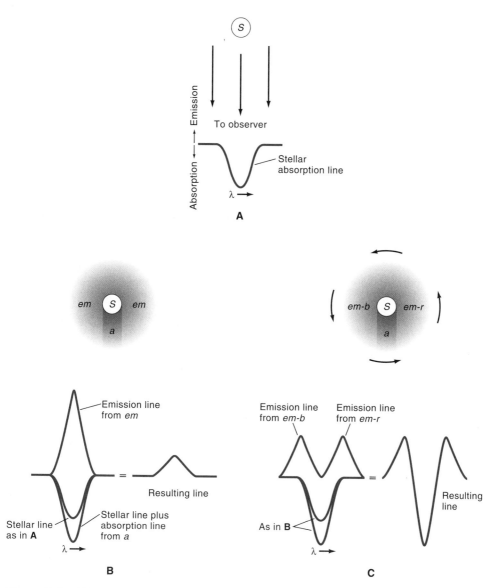

Figure 11.5. (A) A star *S* radiates a continuous spectrum interrupted by absorption lines (one of which is shown) formed in the photosphere of the star. (B) The star is now surrounded by a large outer atmosphere of warm gas. That part of the cloud (marked *a*) between the star and the observer is seen against the star and adds to the absorption, deepening the stellar absorption lines (Kirchhoff's third law, see Chapter 9). At the same time, that part of the cloud marked *em* is not seen against a hotter continuum source and so produces emission lines in accordance with Kirchhoff's second law. Depending on the depth of the photospheric absorption line and the density and temperature of the surrounding cloud, the latter might partially fill in the absorption, producing a shallow line, or it might more than fill it, resulting in an emission line as shown in the sketch. (C) Now suppose the outer cloud is rotating as shown. The absorption line is formed as before, unshifted by the Doppler effect because its motion is across the line of sight. Because of the Doppler effect produced by their rotation, the regions of the cloud marked *em-r* and *em-b* are producing red-shifted and blue-shifted lines, respectively. The result is a narrow absorption line bordered on each side by emission. All of these phenomena are observed in stars.

spectral class A2 the temperature is high enough that an appreciable fraction of hydrogen atoms are in the first excited state, absorbing Balmer photons, but not so high that too many hydrogen atoms have become ionized. In other words, at A2 the temperature is such that the maximum fraction of neutral hydrogen atoms exist in the second level. It is the interplay of both the ionization and excitation of hydrogen that produces the observed maximum in the line strengths.

The Temperatures of Stars

So we have two fairly easily measured quantities—spectral type and color—that are both related to the temperature of the star. But what is the exact relationship? Much effort, both theoretical and observational, has been expended to establish as precisely as possible the temperatures corresponding to the various spectral types. It is not an easy job.

A star's photosphere cannot be characterized by a single temperature. Instead, an atmosphere encompasses a range of temperatures, from the hottest, deepest layers from which photons can escape to the coolest, outermost layers. Some spectral features arise primarily from the hotter layers; others are formed in cooler regions under different conditions of excitation and ionization. Consequently, the idea of *the* temperature of a star may seem ambiguous. Nevertheless, the atmosphere of a star can be usefully characterized by its **effective temperature**. This is the temperature of a blackbody that radiates from each unit area (square centimeter, for example) the same *total* energy per second as a unit area of the star in question does. By total energy we mean the energy over the entire spectrum. This does not imply that the spectrum of the star looks like that of a blackbody. In fact, the shapes of blackbody spectra are rather poor approximations to the spectra of some stars. So the effective temperature of a star tells us the total energy radiated each second by each unit area of surface of the star, but it says nothing about the shape of the spectrum radiated. Figure 11.6 shows two spectra, A and B, which have the same effective temperature as the blackbody spectrum shown in C.

The total energy emitted by a star from its entire surface and over the whole spectrum is just the product of the surface area of the star, $4\pi r^2$ (the surface area of a sphere with radius r), and the amount of energy radiated by each square centimeter, σT^4; that is, $L = 4\pi r^2 \sigma T^4$. This quantity is called the **bolometric luminosity**, because a bolometer is a detector capable of measuring energy from all wavelengths. The energy radiated by the Sun over the whole spectrum, its bolometric luminosity, is 3.96×10^{33}

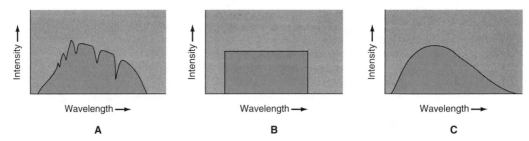

Figure 11.6. The shaded areas in (A), (B), and (C) are all equal and represent the energy radiated per second per unit area by a star, by some (strange!) object, and by a blackbody of temperature *T*, respectively. Hence the star and object also have effective temperatures equal to *T*.

Table 11.1. *Effective temperatures of stars*

Spectral type	$T_e(K)$	Spectral type	$T_e(K)$
O5	40,000 K	G0	6,050
B0	28,000	G2 (Sun)	5,750
B5	15,500	G5	5,500
A0	9,900	K0	4,900
A5	8,500	K5	4,150
F0	7,400	M0	3,500
F5	6,600	M5	2,800
		M8	2,400

ergs/sec.[2] Since we know the Sun's radius, we can use the formula above to find its effective temperature. It turns out to be 5,750 K.

Hot stars radiate most of their energy in the ultraviolet, which is absorbed by the Earth's atmosphere (see Figure 11.7). Hence measurements of the energy radiated by stars in this spectral region must be made with telescopes carried above the Earth's atmosphere in rockets or satellites. Such observations are difficult, but much progress has been made in the last few decades in establishing the effective temperature scale. The values given in Table 11.1 are probably fairly accurate.

Chemical Composition of the Stars

Several factors determine how strong absorption lines are in the spectra of stars. For a particular line to be seen in the spectrum of a star, there must, of course, be atoms of the chemical element in question in the star's atmosphere. But that is not enough. These atoms must also be present in the right stage of ionization and in the appropriate state of excitation (the $n = 2$, 3, or whatever level) to produce absorption lines. Thus the absence of spectral lines from a certain element does not necessarily mean that that element is not present in the star's atmosphere. Instead, it could be there, but the temperature may not be appropriate for the stage of ionization or excitation required to form the line, or you may be observing in the wrong region of the spectrum. Thus O-type stars are not necessarily rich in helium and deficient in calcium; similarly, G-type stars are not necessarily full of calcium and empty of helium. By the 1930s, in fact, careful analysis of stellar spectra showed that the spectral sequence could be understood nearly entirely as a *temperature* sequence,[3] with temperatures decreasing from O- to M-types, all different spectral types having about the *same* chemical composition. That is, the proportions of the various elements—H, He, C, N, etc.—are pretty much the same from star to star.

The relative abundances of the more plentiful chemical elements of this standard stellar recipe can be given as follows. The simplest element, hydrogen, makes up about 75 percent of the total mass; helium, the next simplest atom, constitutes about 23 per-

---

[2] An erg is a unit of energy in the cgs system of units. A 100-watt bulb produces 10^9 ergs of heat and light every second.

[3] In the next chapter we will see that atmospheric pressure plays a minor role in determining the appearance of a spectrum.

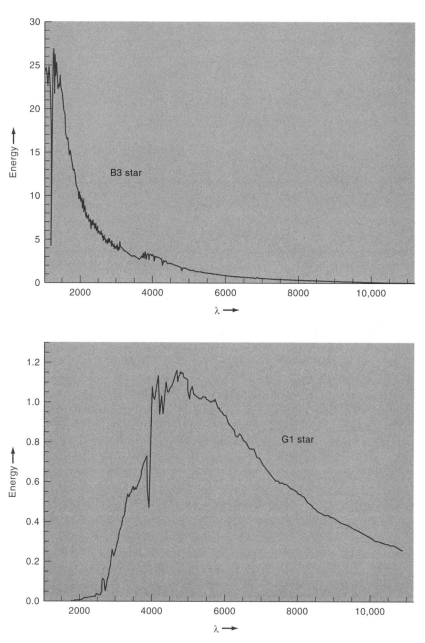

Figure 11.7. The relative energy distributions of a B3 and G1 star. The vertical scales are such that the two stars would appear to be about equally bright in the visual region of the spectrum. Note that the G1 star radiates essentially no energy shortward of 2000 A, whereas the B3 star spectrum is very bright there and rising. Since the Earth's atmosphere absorbs radiation shortward of about 3000 Å, we can observe only a small portion of the total energy emitted by a hot star.

Table 11.2. Relative abundances by number of some of the chemical elements

Element	Relative abundance	Element	Relative abundance
Hydrogen	1,000	Neon	0.1
Helium	100	Magnesium	0.04
Carbon	0.4	Silicon	0.04
Nitrogen	0.1	Iron	0.03
Oxygen	0.8	Sulphur	0.02

cent of the total. All of the other elements, which astronomers call (misleadingly) the **metals**, make up only about 2 percent of the total mass of a star or gas cloud. Prominent among the metals are carbon, nitrogen, oxygen, and iron. The relative abundances by *number* of the more common elements are given in Table 11.2. This standard composition mix is called the **cosmic abundances**. Despite the rather grand name, we must remember that it is based on a detailed analysis of only the Earth's crust, a few meteorites, the Sun, and a handful of stars and gaseous nebulae, along with an approximate chemical analysis of a few hundred other stars and nebulae. Furthermore, the vast majority of these objects are located within our galaxy. We have comparatively little direct knowledge about the chemical composition of other galaxies. What we do know, however, supports the notion that their composition is the same as that of our galaxy.

It turns out that some stars show departures from these abundances in both the total fraction of the elements lumped together as the metals (in Table 11.2, everything from carbon on), and the relative abundances of the elements within the metals. Nevertheless, the realization that there does exist a "standard" mix of chemical elements against which abundance departures can be measured was perhaps the most significant advance in astrophysics in the first half of this century. It plays a crucial role in areas as diverse as the early history of the universe and the last stages of a star's life.

The most obvious example of a wide departure from the norm of cosmic abundances is the Earth itself, where hydrogen and helium are conspicuously rare. This anomaly can be understood in terms of the origin of the solar system, which we will discuss in Part IV.

Other Information Provided by Spectra

We have described atomic energy levels as if they were extremely sharp so that transitions between them give rise to spectral lines of a precise energy or wavelength. This is not true in a real gas, however. Several factors act to broaden energy levels, to smear them, so to speak, so that transitions between two levels can occur with a wavelength range corresponding to the amount of smearing (see Figure 11.8). Among these factors are the temperature and pressure of the gas forming the lines. Other circumstances also influence the line strength and shape (called the line **profile**), for example, the abundance of the element in question, whether the star has a strong wind or not, how fast the star is rotating, etc. If we understand how line profiles are affected by these quantities, we can turn the situation around and use the observed profiles to determine these quantities. That is, line profiles are powerful diagnostics of physical conditions in stars and nebulae. Let us consider the factors that determine spectral line profiles.

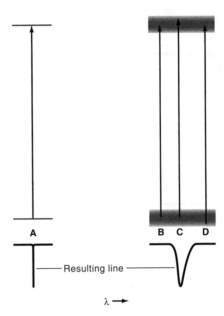

Figure 11.8. The energy levels of atoms are not infinitely sharp as shown in (A), but are smeared by various processes. This is indicated on the diagram by the shading of the energy levels. Transitions are most likely to occur where the shading is darkest (B), corresponding to the levels in (A), but could also occur with smaller probability at (C) or (D), for example. The result is that a spectral line is spread over a small range of wavelengths; that is, it is broader than it would be otherwise.

Thermal Broadening. Recall that temperature is a measure of the kinetic energy (energy of motion) of the particles in a gas; the higher the temperature, the greater their average kinetic energy. We can write this as

$$T \propto mV^2, \text{ or } V \propto \left(\frac{T}{m}\right)^{1/2},$$

because the kinetic energy of an atom of mass m is simply $1/2mV^2$, where V is its speed. Since the kinetic energy depends on the mass m of the gas particles as well as on their speed, then in a gas of hydrogen and helium all at the same temperature, the latter will move more slowly than the former because a helium atom is four times as massive as a hydrogen atom. In fact, the average speed of the helium atoms will be one-half that of hydrogen. (Make sure you understand why this is so.) Similarly, at a temperature of 6,000 K (roughly that of the Sun's atmosphere), a hydrogen atom has a speed of about 10 kilometers/second; in the 24,000 K atmosphere of a B-type star, its speed would be 20 kilometers/second because the speed increases as the square root of the temperature; temperature four times higher produces twice the average speed.

So what does this have to do with line profiles? We observe atoms in a gas moving with radial velocities from 0 kilometers/second (across our line of sight) all the way up to the maximum allowed by the temperature. It is easy to see that this results in a corresponding range in the amounts by which wavelengths of emitted photons are shifted to shorter or longer wavelengths by the Doppler effect. (A little thought should convince you that absorbed photons will show the identical effect.) Thus a spectral line formed by atoms in a hot gas is broader in wavelength than it would be if the gas were cold. This process is called **thermal broadening**.

An iron atom has a weight about 56 times that of hydrogen, so in the Sun (where hydrogen has a thermal speed of 10 kilometers/second) its average speed would be $(1/56)^{1/2} \times 10 = 0.13 \times 10 = 1.3$ kilometers/second. By the Doppler formula,

$$\frac{\Delta\lambda}{\lambda} = \frac{V}{c} = \frac{1.3}{(3 \times 10^5)} = 4.33 \times 10^{-6}.$$

So an iron line with a wavelength of say, 6000 Å, will not be extremely narrow, but will have a width due to thermal broadening of

$$\Delta\lambda = 6000 \times 4.33 \times 10^{-6} = 0.026 \text{ Å}$$

in a star like the Sun.

Pressure Broadening. When atoms collide, they experience additional electrical forces produced by the electrons surrounding both nuclei. During the short time of the collision these forces momentarily distort the electron orbits, further smearing their energy levels. During the collision, photons can be absorbed having a range of wavelengths shortward and longward of the wavelength of the undisturbed transition. Clearly, the greater the number of particles in a given volume of gas and the higher their temperature, the larger will be the number of collisions each second. The greater the rate of collisions, the greater is the probability that a photon will be absorbed during a collision resulting in this line smearing. Now the product of the number of particles in a given volume and their temperature is just the pressure, that is, $P \propto NT$. Consequently, this process is called **pressure broadening**.

Some atoms are extremely sensitive to this effect. The most striking example is provided by hydrogen in A-type stars where collisions so smear the energy levels that Balmer lines can be several Ångströms wide, rather than the fraction of an Ångström produced by thermal broadening. Since the broadening of the line depends on the gas pressure, and the spectral type gives us the temperature of the star, measurement of the broadening allows the number of particles per unit volume, N, to be found. We will see particularly striking examples of this effect in the next chapter.

Abundance Effects. Imagine that you are in a laboratory, examining the absorption lines produced in the continuous spectrum of a hot source by a cooler gas contained in a bottle. (This is the same arrangement used to demonstrate one of Kirchhoff's laws in Chapter 9.) Obviously, if there are no sodium atoms in the bottle, no yellow sodium lines will be produced. If you add only a little sodium to the gas, their lines will be weak, but as you add more sodium, the line profiles will deepen and broaden (assuming of course that the temperature is favorable for producing the lines). At first this increase in line strength proceeds just as you would expect: double the number of sodium atoms and their line strengths are doubled. In this regime, where there is a one-to-one relation between the abundance of the gas and the line strength, it is relatively easy to use the line profile to measure the abundance of the sodium.

This behavior does not continue indefinitely, however; instead, the yellow lines, now quite prominent, are said to become saturated. This means that a significant increase in the abundance of sodium produces only a small increase in their total absorption. You can understand this qualitatively by imagining that there are more than enough sodium atoms to absorb all of the appropriate incident photons. Relating the absorption-line strength to the abundance becomes more complicated, but it can still be done though usually not as reliably as when the lines are not saturated.

The profile of a stellar absorption or emission line is the result of many processes acting simultaneously, so the total effect can be quite complicated. However, it is fairly well understood how the processes discussed so far influence the shapes of stellar lines, as well as how other relevant factors operate on the atomic scale (like magnetic fields). Consequently, we can deduce with reasonable accuracy the physical conditions, such as the temperature and density of the gas where the lines are formed, as well as chemical composition. Line profiles also give us information concerning gross properties of stars, such as their rotation and the outflow of matter in winds. Let us consider those next.

Stellar Winds.   In various phases of their lives, several types of stars constantly lose mass back to interstellar space. High-luminosity stars produce the most pronounced of these so-called **stellar winds**, primarily by the strong outward force they exert through radiation pressure acting on the star's outer atmosphere. That is, the enormous numbers of energetic photons produced in such stars interact with the atoms and ions in the atmosphere, driving them out at velocities of as much as a few thousand kilometers per second, and at mass-loss rates as high as one solar mass in one hundred thousand years.

We detect such winds from the **P Cygni** spectral-line profiles they produce (named after the prototype object that defines the phenomenon), an example of which is shown in Figure 11.9. Radiation from the rapidly expanding gas surrounding the star produces broad emission lines as a consequence of the high velocities of the gas toward and away from us (*em-b* and *em-r*, respectively, in the figure). The gas approaching us (marked *a-b* in the figure) is seen against the star. Consequently this gas will form a blue-shifted absorption line, as indicated. The amount of energy absorbed is related to the amount of material in the wind, and the shortest wavelength point of the absorption feature gives the maximum rate at which the matter flows out.

Stellar Rotation.   The shapes—profiles—of spectral lines also tell us that stars rotate on their axes. Imagine a rapidly rotating star with its axis perpendicular to our line of sight (see Figure 11.10). Half the star's hemisphere facing us is moving toward us, and half is receding from us. Radiation from the part of the star that is moving across our line of sight will show no Doppler shift, but radiation from the limbs (the edges) will show the maximum blue and red shifts. The resulting absorption line, then, will consist of the sum of contributions formed by radiation having Doppler shifts ranging from zero to the maximum amount set by the star's rotation velocity. Compared to a nonrotating star, the rotating star will have broad and shallow absorption lines. Since the faster the rotation, the wider the line, measuring the width of the absorption line gives us the rotational velocity.

The total amount of energy absorbed by a given line is the same in a nonrotating as in a rotating star of the same spectral type; in the latter the line is simply spread out more and therefore shallower. There is a complication, however. Suppose the star is oriented so that its rotational axis points straight at us. Then regardless of how fast the star is spinning, its absorption lines would not be broadened because all the rotational motion is *across* our line of sight. We can measure only that part of the rotational mo-

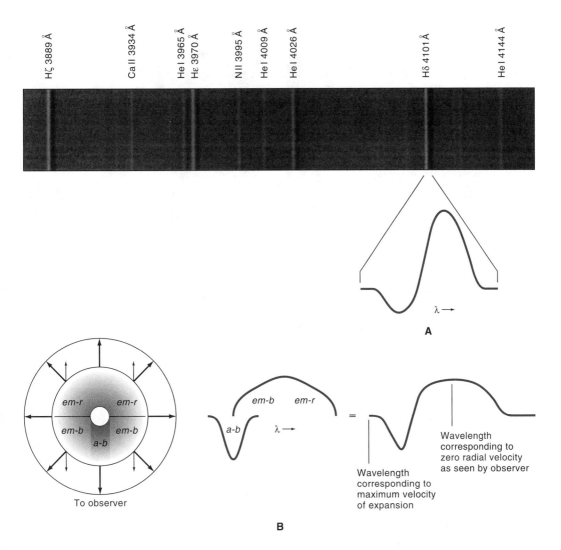

Figure 11.9. (A) A portion of the spectrum of the B-type star, P Cygni, is shown. Note that both absorption and emission lines are present. (Remember that the spectrum is a negative print so that absorption lines are bright and emission lines are dark.) Beneath the spectrum is a sketch (expanded in scale for clarity) of a so-called P Cygni line profile that indicates that the star has a strong stellar wind. (B) How P Cygni lines are formed: The cloud surrounding the star is expanding away from it so that every part of the cloud has a particular velocity toward or away from the observer; the shaded region, *a-b*, produces a blue-shifted absorption line, and the regions marked *em-b* and *em-r* produce blue and red-shifted emissions as shown. The result is the schematic line profile. Note that no stellar photospheric absorption line is shown. In these stars, the lines formed in the expanding wind are usually of higher ionization than those in the star.

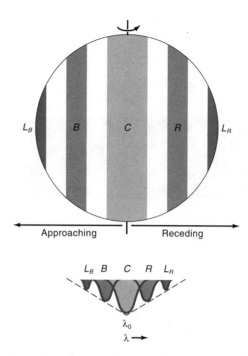

Figure 11.10. A star is rotating as shown. The part of the star marked *C* moves primarily across our line of sight. The radiation it emits is not Doppler shifted and it produces an absorption line centered at the rest wavelength, λ_0. Radiation emitted from *B* and *R* is somewhat blue- and red-shifted, respectively, while that from L_B and L_R produces the largest shifts. The absorption lines produced by these stellar strips (along with others between them) result in the broadened absorption line indicated by the dashed line. The faster the star rotates, the broader will be its absorption lines, allowing the rotational velocity to be determined from the shape of the absorption line.

tion that is projected *along* our line of sight. Given measurements of enough stars, however, we can calculate the average rotational velocity of each spectral type, assuming the star's axes are randomly oriented.

Table 11.3 gives the average rotational velocities for various spectral types. We will discuss the trend shown in Table 11.3 in later chapters.

Table 11.3. *Average stellar rotational velocities*

Spectral type	Avg. velocity	Spectral type	Avg. velocity
O5	180 km/sec	F0	100 km/sec
B0	200	F5	30
B5	230	G0	4
A0	190	K	1
A5	150	M	1

A Detailed Look at the Surface of an Ordinary Star: The Sun

Thus far, the kind of information given to us by the spectra of stars has to do with their overall properties, such as their effective temperature, rotation speed, or chemical composition. This is so because we can collect light only from a star's entire facing hemisphere, and not from one small part or another of it. The Sun is the only star near enough to us for us to have extensive and detailed knowledge of its surface. In this section we want to describe briefly some of the surface features of the Sun, as revealed primarily by spectra, but also by other means. The Sun is a *very* ordinary kind of star, so its properties are probably representative of those of a vast number of stars.

The Outer Regions of the Sun

The Photosphere.   We have seen that the outer layer of a star is called its photosphere, from which photons escape directly to space. This "skin," which is what we see when we look at the Sun, is only a few hundred kilometers thick. Direct photographs of the Sun show that the photosphere is not featureless, but consists of a myriad of small bright cells called **granules**, shown in Figure 11.11. These range from a few hundred to a thousand kilometers in diameter, are about 100 K warmer than the darker intergranule region, rise with velocities of about 2 km/sec, and maintain their identity for 15 to 20 minutes. They are the tops of **convection cells** rising up from the interior. The high-resolution spectrum in Figure 11.12 shows the Doppler shifts in the absorption lines, produced by the rising and falling convection cells, from which their ve-

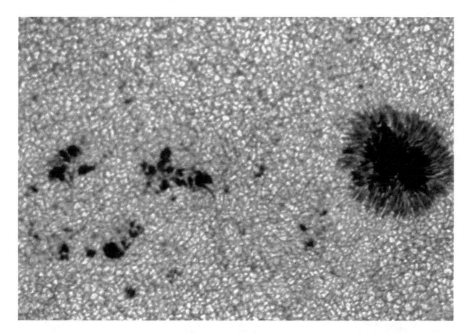

Figure 11.11.   A photograph of a small area of the Sun, showing a sunspot group and the photospheric granulation. The granules are rising columns of gas, up to about 1,000 kilometers in diameter and 50 to 100 K warmer than the darker intergranular regions where cooler gas is sinking.

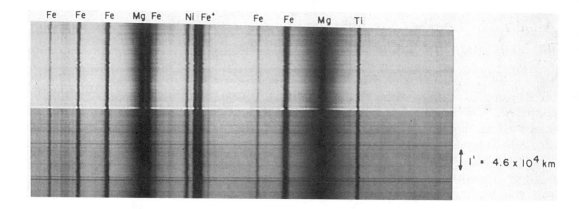

Fe Fe Fe Mg Fe Ni Fe* Fe Fe Mg Ti

1' = 4.6 x 10⁴ km

Figure 11.12. About 10 Å of two high dispersion spectrograms of the Sun; the two prominent lines of singly ionized iron nearest the center are at 5168.91 Å and 5169.0 Å. The two spectra were made with the slit on the same part of the Sun, but with different exposure times. Note that the edges of the absorption lines are jagged, not smooth, indicating that small gaseous elements (the granules) along the slit have different radial velocities. Also, a marked "wiggle" in the lines of the shorter exposure is not present in the longer one.

locities can be measured. Granules are thought to play an important role in carrying energy to the chromosphere and corona of the Sun.

The Chromosphere and Corona. Extending out from the photosphere for a few thousand kilometers is a transparent and much more tenuous region called the **chromosphere** (Figure 11.13). Its name, the "color sphere," is derived from its reddish appearance seen during solar eclipses. An even more rarified **corona** (Figure 11.14) surrounds the chromosphere, and extends outward for millions of kilometers. Finally, the entire inner portion of the solar system is bathed in a very low density stream of coronal particles, electrons and protons mostly, but with a few helium nuclei and heavier ions, called the **solar wind**. The structure of the Sun is sketched in Figure 11.15.

Since the temperature from the center of the Sun out toward the photosphere falls steadily, we might expect the temperature of the chromosphere and corona to continue this trend, so that the latter would be the coolest region of all. This is not the case, however. The temperature decrease stops and reverses in the chromosphere, and begins to increase rapidly, so that coronal temperatures reach one to two million degrees!

How do we know this? Until about the turn of this century, the outer atmosphere of the Sun could be observed only during infrequent solar eclipses, so knowledge of the chromosphere and corona was slow in coming. Since then, however, instruments like the **coronograph** have been developed, which enable us to observe the brighter portions of the Sun's outer layers at any time. In this device a disk of just the right diameter placed in the focal plane of a telescope occults—blocks out—the light from the Sun's photosphere. This allows one to observe the much fainter chromosphere and corona. All the optical elements of the instrument must be carefully polished to keep scattered light to a minimum. In addition, coronographs are usually operated on mountaintops, to reduce light-scattering by the Earth's atmosphere. With instruments such as this, the chromosphere and corona can be photographed directly and also investigated spectroscopically.

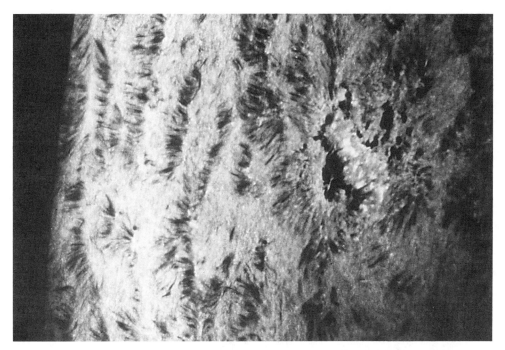

Figure 11.13. A photograph in the red Balmer line of hydrogen of the chromosphere. The dark filamentary structures are spicules, jets of gas that form the wavy top of the chromosphere. On average, they rise about 5,000 kilometers above the photosphere and last for about 10 minutes before subsiding. They mark the boundaries of magnetic regions.

Figure 11.14. The Sun's corona, photographed during the total eclipse of February 16, 1980, in Hyderabad, India.

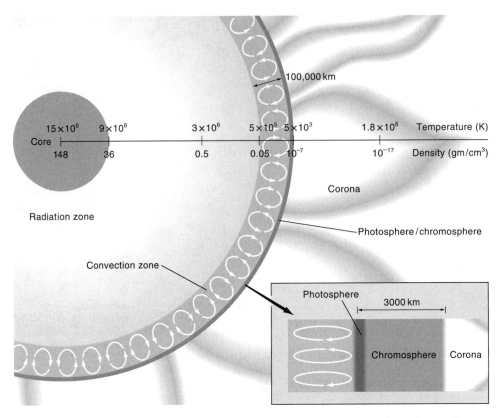

Figure 11.15. A sketch of the various regions of the Sun shown approximately to scale except that the corona can extend outward several solar radii. Note that the chromosphere, the transition region between the photosphere and corona, is very narrow.

Spectroscopic observations show that nearly all the photospheric absorption lines become emission lines in the chromosphere. All become weaker with height, but at different rates, as a result of the decreasing density and increasing temperature with height in the chromosphere. The characteristic reddish color of the chromosphere is produced by Hα (H-alpha) in emission. Lines of many other elements are present, of course, including those of helium, which were first discovered in the solar chromospheric spectrum, hence its name (from Helios, the Greek word for the Sun). Analysis of these spectra reveals that the temperature at first continues to decrease outward from the photosphere, reaching a minimum of about 4,300 K, then rapidly increases to about 100,000 K, after which it merges into the million-degree corona.

The High Temperature of the Corona. Evidence for the very high coronal temperature comes from several sources. One clue is the presence of emission lines from highly ionized ions, such as Fe XV, Ca XIV. (Figure 11.16). To remove 14 electrons from an atom of iron or 13 from a calcium atom requires a great deal of energy, which means a high temperature. Another piece of evidence comes from a faint continuous spectrum seen in the inner corona (within about two solar radii). Because of the high ionization, the corona contains large numbers of free electrons, which reflect the photospheric spectrum. Even though the photosphere shows numerous absorption lines,

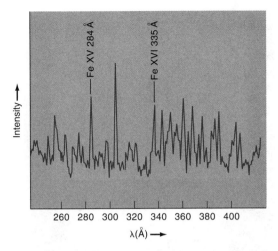

Figure 11.16. A portion of the far ultraviolet spectrum of the Sun, taken from an Orbiting Solar Observatory. Note the two high ionization iron coronal lines at 284 Å (Fe XV) and 335 Å (Fe XVI). That so many electrons have been stripped off iron atoms indicates that the coronal temperature is very high.

the reflected coronal spectrum does not. In order that these lines, some of which are an Ångström or more wide, be washed out, the scattering electrons must produce large random Doppler shifts. This requires high speeds, which means high temperatures, again one or two million degrees.

Why aren't we cooked by these high temperatures? The answer can be inferred by direct observation: the corona is extremely faint. We can't see the corona unless the photosphere is blocked out by the eclipsing Moon or by an occulting disk in a coronograph. This is so because, although the corona is roughly half as bright as the full Moon, it is very faint compared with the Sun's photosphere; in the visual region of the spectrum it is only about a millionth as bright. Therefore the density of the corona must be very small, no more than about 10^{-8} that of the upper photosphere. Evidence supporting the low density comes from another quarter, namely, the so-called forbidden lines observed in the corona. We will see in Chapter 14 that forbidden lines can appear only when the gas density is small. The low density (equivalent to a very good laboratory vacuum) means that the heat content of the corona is extremely small, and so it contributes a negligible amount of energy to that radiated by the photosphere.

The mechanism by which the chromosphere and corona are heated is not well understood. The rate at which energy is lost by the faint, low-density corona is not large, so it does not take much energy to heat it to a high temperature. Complex interactions of ionized matter moving through the Sun's magnetic field cause mechanical (acoustic) energy to be transported by the rising convection cells in the photosphere and from there to the outer layers, where it is deposited and heats the chromospheric and coronal gas.

Since the corona is so hot, it radiates x-rays as well as visible and ultraviolet light. The x-rays come primarily from blobs of gas trapped in the corona's knotted magnetic field. **Coronal holes** (Figure 11.17), regions of weak emission and low gas density, exist between these blobs, and it is through these holes that the solar wind blows. The existence of this wind was first inferred in the 1950s as an explanation for the long-known fact that comet tails point away from the Sun. (Kepler, in fact, had speculated that some repulsive force from the Sun was responsible for this effect.) A decade later

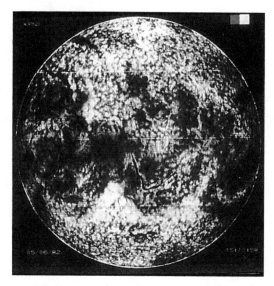

Figure 11.17. The Sun as it appears in the light of a chromospheric helium line radiating at 10,830 Å. The dark areas are holes in the corona through which flow the particles of the solar wind.

direct measurements by space probes confirmed that electrons and protons stream out from the Sun with an average velocity and density of about 400 km/sec and about 5 ions/cm^3, respectively. Although this plasma interacting with the cometary tail gas is sufficient to push it away from the Sun, the solar wind is a very light breeze indeed compared with the winds produced by highly luminous stars. Nevertheless, the solar wind is responsible for various effects here on Earth, such as radio-communication blackouts.

Solar Activity

A wide variety of phenomena are included under the general heading of **solar activity**, including plages, flares, prominences, and sunspots. They are all associated with a 22-year **magnetic cycle**, which, in a way not yet fully understood, seems to be the driving force behind the various manifestations of solar activity. The general, overall magnetic field of the Sun is rather weak, only a few gauss.[4] Much stronger but localized fields occur in **sunspots**, however, ranging from about 100 to 4,000 gauss (Figure 11.18). They are the most obvious manifestation of solar activity.

Sunspots. These phenomena appear dark because they are as much as 1,500 K cooler than the surrounding photosphere. We know this because their spectra show absorption lines that can arise only in gas having a temperature of 4,000 to 4,500 K. Hot as this is, compared with their surroundings, spots are giant refrigerators. The strong magnetic field associated with a spot must play a role in cooling it, perhaps by reducing convection, and therefore the heat flow, within the spot. An individual spot or a spot group might persist for as little as a few hours or as long as several months. They

---

[4] Don't worry about the unit of magnetic field strength, the gauss; just note that the Earth's field is a few tenths of a gauss, and that the field strength of a bar magnet you might have seen in a science class is about 1,000 gauss.

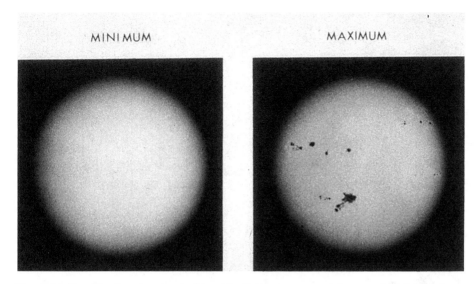

MINIMUM   MAXIMUM

Figure 11.18. The Sun photographed in white light at sunspot minimum, when the Sun has no spots, and at maximum, when several spot groups are visible simultaneously. They are aligned parallel to the solar equator. A detailed view of a sunspot is shown on Figure 11.11, where the dark umbra and the lighter penumbral region can be seen. Well-developed spots like this one have diameters several times that of the Earth.

do not move appreciably on the surface, so they can be used as "markers" by which the Sun's rotation can be measured (and in fact was first measured by Galileo). Careful measurements of the spots' motions as they are carried around by the Sun shows that the Sun does not rotate like a solid body (which, of course, it isn't), but rather rotates differentially. That is, at the equator the Sun rotates once in about 25.8 days; at higher latitudes the rotation period is longer, for example, about 28 days at latitude 40 degrees north and south of the equator. This differential rotation, acting on the photospheric magnetic field, is thought to play a key role in solar activity (Figure 11.19).

The numbers of spots seen at any one time in the northern and southern solar hemispheres increase and decrease in a fairly regular cycle, from only a few (or, rarely, even 0) at the minimum of the cycle, to as many as 200 at cycle maximum. Spot number differs greatly from one cycle to the next, as does associated solar activity.

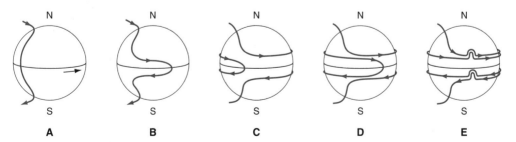

   A   B   C   D   E

Figure 11.19. The Sun, rotating more rapidly at the equator than at higher latitudes, causes the magnetic field to be stretched out and strengthened on either side of the equator. Convective motions constantly buffet and stress the field. In effect, the kinetic energy of convection is locally transformed to magnetic energy until such a patch of field breaks through the photosphere, when we see it as a pair of sunspots. The energy contained in the magnetic field is analogous to that stored in a tightly wrapped and twisted rubber band.

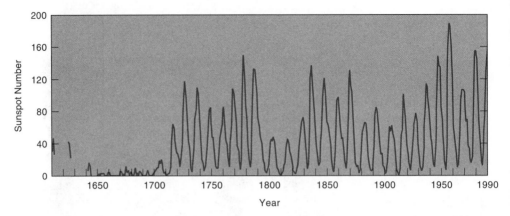

Figure 11.20. Sunspot numbers since early in the 1600s. Note the absence of spots during the second half of the seventeenth century.

The **sunspot** or **activity cycle** lasts about 11.1 years on the average (Figure 11.20), but has been as short as 8 and as long as 16 years. The magnetic behavior of spots, however, requires an average of 22 years to complete its cycle. Often spots appear in pairs, and if in the northern hemisphere the leading spot has a north magnetic polarity, then the following spot has a south polarity (see Figure 11.21). In the opposite hemisphere the magnetic arrangement of such a pair would be reversed—the leading spot would have a south magnetic polarity, the following a north polarity. After some 11 years, when the next activity cycle begins, the leading spots in the northern and southern hemispheres would now have south and north polarities, respectively, so that two spot-number cycles are required before the magnetic cycle repeats itself. The weak general magnetic field of the Sun also reverses itself every 11 years. The causes of these phenomena are not well understood.

Plages, Flares, and Prominences. As we saw earlier in this chapter, different parts of an absorption line are formed at different heights in a stellar atmosphere. For example, the core of a line is formed higher in the atmosphere than the wings because the absorption at line center is greater (and so the probability of a photon being able

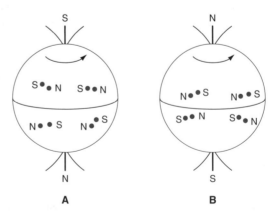

Figure 11.21. If in a particular 11-year-cycle, the leading (in the sense of solar rotation) spot of a spot pair in one hemisphere has a north magnetic polarity, and the following spot a south polarity, then the opposite is the case in the other hemisphere. Eleven years later, in the next cycle, the spot fields are reversed. Thus the magnetic cycle is 22 years. The very weak general magnetic field of the Sun also reverses itself in a 22-year cycle.

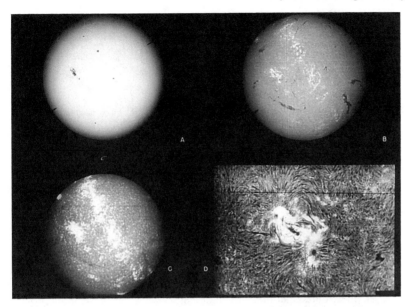

Figure 11.22. (A), (B), and (C) are photographs of the solar disk taken on the same day in white light, red light of Hα, and blue light of Ca II, respectively. White light shows the Sun's photosphere, Hα the layer of the chromosphere just above the photosphere, and Ca II the upper chromosphere. Note the increasingly coarse structure as the height increases. (D) is the area around a sunspot seen in red light (and therefore above the spot). The white areas in (B), (C), and (D) are plages (from the French for "beach" or white area) that indicate regions where strong magnetic fields are concentrated. Plages indicate where solar activity, including flares, are likely to occur.

to escape is smaller) than it is in the wings. Hence, a photograph of the Sun taken in the light from the core of Hα records radiation from a higher layer than one taken in light from the wing of the line. Similarly, a very strong absorption line will be formed higher in the atmosphere than a weak line, everything else being equal. The Ca II line at 3934 Å is stronger than Hα so a solar image taken with light from the center of the Ca II line reveals a higher stratum than does an Hα image. Hα and Ca II images enable us to see the lower and upper chromosphere, respectively. Both images are produced by radiation coming from higher layers than one made in the light of the visible continuum where the average absorption is relatively small; such a white-light image gives us a view of the base of the atmosphere, the photosphere.

How are such photographs actually made? The simplest way is to put a special filter in the light path of the camera that passes only a very narrow band of radiation. The filter can be "tuned" to be at the core or somewhat off the core of a strong line such as the red Balmer line. In this way, photographs of chromospheric and coronal features can be made. Figure 11.22 shows the Sun in white light, Hα, and Ca II (compare with Plate C3).

Associated with strong local magnetic fields are several phenomena: bright areas called **plages** and **flares**, which occur primarily in the chromosphere, and **prominences**, which are based in the chromosphere and extend into the corona. Plages (Figure 11.22) are bright regions surrounding sunspots, but often extending over much larger areas of the solar disk than that occupied by the spots. Plages indicate the emergence of a magnetic field from beneath the photosphere into the chromosphere and corona, which ultimately produces the whole array of solar activity. Appearing before a spot and persisting after the spot has disintegrated, plages are more fundamental in-

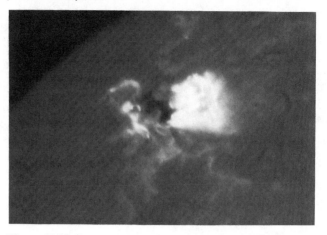

Figure 11.23.   A very bright flare that occurred in March 1989.

dicators of solar activity than are sunspots. They are most easily seen in the light of Hα or Ca II lines, but sometimes are sufficiently intense to be seen against the photosphere in white light. They don't represent concentrations of hydrogen or calcium; instead, they are hot, dense regions where the conditions for their excitation are favorable.

Flares are the most energetic events that take place in the exterior of the Sun (Figure 11.23), the largest releasing an amount of energy equivalent to one or two billion megaton hydrogen bombs! They are most commonly seen in the light of Hα as a rapid brightening of a small part of a plage. A given flare lasts only a short time, rising to maximum brightness in a few minutes and disappearing after an hour or so (Figure 11.24). During this period, intense ultraviolet, x-ray, gamma-ray, and radio radiations are emitted, as well as charged particles. The radiation is both thermal and nonthermal in nature (see Chapter 16). The former indicates that gas in the flare is heated to temperatures as high as 2×10^7 K, and the latter that high-energy electrons are emitted. These eruptions are generally associated with sunspot groups, several flares often occurring over and over again among the same spots.

Some of the most visually striking phenomena are the solar prominences, a few examples of which are shown in Figures 11.25–11.27. They have different origins and

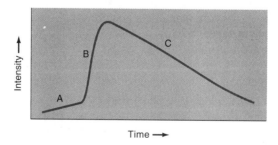

Figure 11.24.   How the energy output of a solar flare changes with time. Stresses in the magnetic field slowly build up and release small amounts of energy (A), when suddenly all the energy stored in the field is released (B). As a result, large numbers of electrons and protons are accelerated to high energy. Some of their energy goes into the production of electromagnetic radiation spanning the wavelength region from γ-rays to radio during the few minutes of this phase. The particles also heat the surrounding photosphere and chromosphere to very high temperatures, which produce thermal radiation in the flare's decay phase (C).

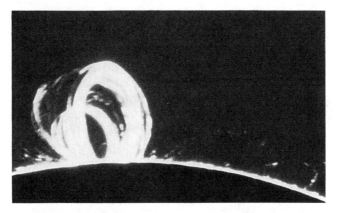

Figure 11.25. The Sun's magnetic field forces the hot, ionized gas of this prominence into a loop form. Prominences of this type are associated with flares, and, like them, last only minutes to an hour or so.

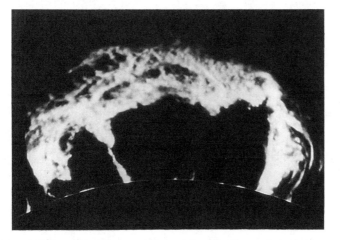

Figure 11.26. A spectacular prominence formed from cooling and sinking coronal material guided by the magnetic field to form the structure seen. Such prominences can last days or weeks.

Figure 11.27. This eruptive prominence was photographed from Skylab in 1973 in the light of an ionized helium line at 304 Å, indicating that its temperature is several times hotter than most prominences. It began from a quiescent prominence not far above the Sun, and after only 20 minutes, it had risen about 600,000 kilometers! Though the prominence has a light, filmy appearance, its mass is on the order of a billion tons! Its form is affected by the magnetic field lines torn from the chromosphere in the eruption.

structures, but do have in common cool temperatures (usually)—less than 10,000 K—and a density greater than the coronal gas into which they extend. Like a rock in water, they are not buoyant. Consequently, stable, non-eruptive prominences must somehow be supported to keep them from sinking, or they must be continuously replenished to exist. Like everything else connected with solar activity, it is the magnetic field that plays a key role. By interacting with the ionized prominence gas it can support it, for example, in a loop-like structure (Figure 11.25), or guide cooling and sinking coronal material downward to form another type of prominence (Figure 11.26).

The shape of the corona varies somewhat during the activity cycle. At maximum the corona extends about equally around the Sun, but at minimum it is considerably elongated near the solar equator and flattened near the poles.

The Sun as a Variable Star.   In later chapters we will encounter stars that fluctuate in brightness by amounts ranging from a few percent up to factors of two or three. Our everyday experience tells us that the Sun, fortunately for us, is not that kind of variable star. But we have seen that its magnetic characteristics, along with many associated phenomena, vary in a fairly regular pattern. From the ground it has been impossible to detect any variability in the amount of radiation emitted by the Sun on timescales longer than those of transient events such as flares. The measurement errors mask any small variations that might occur. With detectors in satellites well above the obscuring effects of the Earth's atmosphere, however, fairly reliable measurements of the intensity of solar radiation can now be made. These measurements have shown that the total amount of energy emitted can vary by as much as 0.2 percent during periods of weeks, that is, during the month-long rotation period of the Sun. These variations result primarily from the changing number and sizes of sunspots visible on the Sun at any given time. The sunspots, being cooler than the photosphere, emit less energy; the larger the number of spots, the greater the decrease in the emitted solar energy.

Longer-term variations are even more difficult to measure, but it appears that the Sun's brightness changes by about 0.1 percent during a sunspot cycle, becoming brighter as the number of spots increases, that is, as solar activity increases. Apparently, the increasing number of bright faculae (which radiate more than their surroundings) more than compensate for the increasing number of spots (which radiate less than their surroundings). None of these variations is large enough to affect the Earth's temperature.

There are other aspects of solar activity that do affect the Earth, however, especially the charged particles emitted by flares. As they interact with Earth's upper atmosphere, they cause disturbances in the magnetic field as well as radio communication problems. Protons colliding with the atoms and molecules of the Earth's upper atmosphere excite them, producing the characteristic green and red colors, both due to oxygen, of the aurora borealis and aurora australis (the northern and southern lights) shown in Figure 11.28. High-energy particles emitted by flares can pose a danger to anyone flying above the Earth's atmosphere.

Some scientists claim that there is evidence, admittedly weak, for a relation between solar activity and climate, but this remains dubious at best. An example perhaps is given by the so-called **Maunder minimum** in the sunspot number (named after the British astronomer who noted it in 1890). From about 1650 to 1715 there were far fewer spots than were observed before or since.[5] At several times during this period (see Figure 11.20) up to a decade passed when no spot appeared anywhere on the Sun!

[5] The reign of Louis XIV of France, the Sun King, coincided nearly exactly with the Maunder minimum!

Figure 11.28. An auroral display photographed from the South Pole in 1984.

Associated phenomena such as aurorae were also rare; in fact, during one 37-year span not one aurora was reported anywhere on Earth. On the simplest picture this would suggest that Earth should have received slightly more solar energy than normal during this time. In Europe, however, climate records indicate the opposite; the period is known as the Little Ice Age. Dutch paintings of the time, for example, show people ice skating on canals that today usually don't freeze. Neither the reason for the marked decrease in sunspot activity nor its possible effect on Earth is understood.

Variations analogous to those occurring in the Sun have been discovered in other F-, G-, and K-type stars. For example, many stars show emission lines of ionized calcium just as are observed in the Sun's chromosphere. As in the Sun, the strengths of these lines vary over the years of their activity cycle. Observations of the changing brightnesses of some stars as they rotate have been interpreted as being due to one or more very large starspots.

Not all solar variability is associated with the activity cycle. In the 1960s, astronomers discovered from radial-velocity measurements of solar absorption lines that the Sun's surface is made up of a myriad of small patches, each independently pulsating—rising and falling—in a cycle lasting about 5 minutes. The change in velocity is measured only in meters/second or even centimeters/second, so the measurements are difficult. At first it was thought that these motions were random, one patch independent of another. In the 1970s, it was realized that what was going on is more complex and much more interesting than that: the entire Sun is oscillating under the action of millions of different sound waves, each with its own period and velocity. Each of these waves is regular, but together they interfere with each other and produce the chaotic appearance observed. In order to discover the regularities behind the apparent disorder, observations lasting for long periods of time are required. One of the more spectacular of these attempts was made by astronomers at the South Pole during its summer who measured these oscillations in the Sun continuously for seven days!

Now, the speed of a sound wave depends on the temperature, density, and composition of the gas through which it travels. A sound wave in the Sun is refracted— bent—in response to the changing physical conditions in the gas as it travels inward and outward. A given wave yields information about the layer of the Sun between the

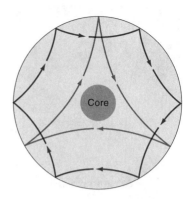

Figure 11.29. The paths in the Sun of sound waves of two
different wavelengths are shown. The speed of sound inside
the Sun increases with depth, causing the waves to be refracted.
When they encounter the solar surface they are reflected.

surface and the depth at which it is bent back toward the surface (see Figure 11.29).
If the most significant of the oscillations can be measured, we will have the means to
probe the internal structure of the Sun. Measurements of tiny changes in the spectrum
at the solar surface can tell us about the deep interior! This technique is analogous to
our use of earthquake—seismic—waves to give us information about the interior of
the Earth, and so is called **helioseismology**.

This is such a promising way of learning about the solar interior that the U.S.
is establishing a worldwide network of six specially designed telescopes to observe
these oscillations continuously. The GONG project (Global Oscillation Network
Group) should be producing data by the mid-1990s. The European Space Agency is
preparing an instrument to be placed on an Earth-orbiting satellite to make similar
measurements.

Analogous oscillations have been found in stars, for example, A-type stars that
have peculiar chemical compositions. Since stars are so much fainter than the Sun, the
potential yield of information is not as great. Nevertheless, initial observations are en-
couraging, and important results are beginning to be obtained.

With this background in stellar spectra, we are ready to look at the physical properties of stars, by which we mean not only their effective temperatures, but also their
diameters, masses, luminosities, and chemical compositions. This information is conveniently and concisely contained in the Hertzsprung-Russell diagram, to which the
next chapter is devoted.

Terms to Know

Early- and late-type spectra, the spectral sequence, photosphere, reversing layer, cosmic
abundances; color, effective temperature, and bolometric luminosity of a star; stellar
wind. Granules, convection cells, chromosphere, corona, coronal holes; solar activity,
magnetic cycle, sunspots, plages, flares, prominences; the Maunder minimum;
helioseismology.

Ideas to Understand

How spectra are classified; the significance of the spectral sequence; why hydrogen line
strengths peak at about A2; why molecules appear in late-type stars; why stellar spectra
have such a wide variety of appearance even though their chemical compositions are all
about the same. How the temperature changes in the Sun's outer atmosphere, and how we
know; phenomena associated with solar activity; the role of the magnetic field.

Questions

1. Why are hydrogen lines weak in O-type stars? in M-type stars?

2. Describe the evidence that the spectral sequence is a temperature sequence.

3. What factors cause spectral lines to be broader in wavelength than they would otherwise be?

4. Explain why molecules are not seen in the spectra of B-type stars.

5. Two stars, A and B, have the same radius; the luminosity of Star A is 16 times that of Star B. If the effective temperature of Star B is 3,200 K, what is the effective temperature of Star A? Show how you get your answer.

6. If two stars have the same spectral type, but one is 100 times more luminous than the other, which star has the larger radius? by how many times? Explain your answers.

7. Two stars have the same radius and effective temperature; the apparent brightness of Star A is 9 times that of Star B. Which is nearer, and by how many times? Explain.

8. Suppose you obtain the spectra of two stars (Stars 1 and 2), and find that the spectrum of Star 1 peaks at about 8700 Å, that of Star 2 at 2900 Å.
    (a) What are the approximate effective temperatures of each star?
    (b) About what is the spectral type of each star?
    (c) Which spectral lines are strongest in each star?

9. Is the rotational velocity calculated from the spectrum of a star its true rotational velocity or not? If not, just what is measured?

10. Suppose a particular A0 star has a rotational velocity of 200 km/sec and a radius of 2.5 solar radii. The Sun's radius is about 7×10^{10} cm. How long does it take for the A-star to rotate once?

11. What are sunspots? Why are they darker than their surroundings?

12. What are some of the characteristics of solar activity?

13. Explain the difference between the 22-year magnetic cycle and the 11-year activity cycle.

14. From Figure 11.20 find when the sunspot cycle lasted 8 years; when it lasted 16 years.

15. What evidence indicates that the solar corona is hot?

16. The hydrogen-alpha line (Hα) is seen as an absorption line when we observe the center of the Sun's disk, but as an emission line when we observe the chromosphere at the edge of the Sun. Explain why.

Suggestions for Further Reading

Delancy, M., "The Case of the Missing Sunspots," *Astronomy*, **9**, p. 66, February 1981.
Golub, L., "Heating the Sun's Million Degree Corona," *Astronomy*, **21**, p. 26, May 1993.
Kaler, J., *Stars and Their Spectra*. Cambridge: Cambridge University Press, 1989. Much of this appeared in several articles published in *Sky & Telescope*, beginning in February 1986.
LoPresto, J., "Looking Inside the Sun," *Astronomy*, **17**, p. 20, March 1989. What helioseismology (solar sound waves) tells us about the structure of the Sun.
O'Leary, B., "The Stormy Sun," *Sky & Telescope*, **60**, p. 199, 1980.
Noyes, R., *The Sun, Our Star*. Cambridge, MA: Harvard University Press, 1982.
Sneden, C., "Reading the Colors of the Stars," *Astronomy*, **17**, p. 36, April 1989. Information from stellar spectra.
Sampson, R., "Fire in the Sky," *Astronomy*, **20**, p. 38, March 1992. Activity on the Sun and aurorae on Earth.
Stahl, P., "Prominences," *Astronomy*, **11**, p. 66, January 1983.
Wolfson, R., "The Active Solar Corona," *Scientific American*, **248**, p. 104, February 1983.

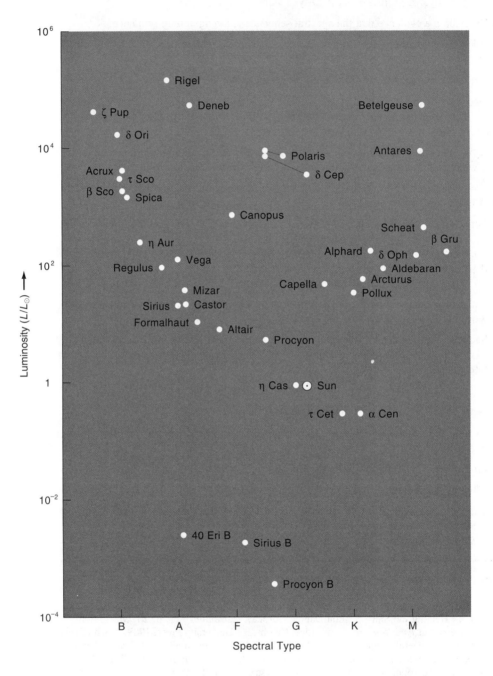

The positions of some of the brighter naked-eye stars on the Hertzsprung-Russell diagram. Note the range in brightness and spectral type of the two Cepheid variables, Polaris and δ Cephei.

CHAPTER 12

The Hertzsprung-Russell Diagram

Some Properties of Stars

After the preliminary discussions of the last two chapters, we are ready to begin our study of the properties of stars. As a first step, however, it is useful to devote a few pages to establish the context of the stellar world, in particular the sizes and distances of stars, as well as the groups into which they are organized.

Stellar Distances

Recall that in describing distances in the solar system we spoke in terms of astronomical units; Jupiter is 5.2 AU from the Sun, for example, and Pluto is about 40 AU out. Remember also that if a ping-pong-ball-size Sun is in the end zone of a football field, Pluto would be in the other end zone. With all sizes and distances reduced in this same way, we saw in Chapter 2 that we would have to travel about 350 miles to reach the nearest star, α Centauri.

Because distances between stars are so large compared with the AU, a larger unit of length is more meaningful and convenient for interstellar distances, namely, the **light year**. As the name implies, it is the distance that light, traveling at a speed of 300,000 km/sec, moves in one year (= 31,500,000 sec) or 9.45×10^{12} km. On this scale, the nearest star is about 4.3 light years away. That is, it takes light from α Centauri 4.3 years to reach us. By comparison, the light-travel time from the Sun to the Earth is only 8 minutes, and from the Sun to Pluto is about 5 hours and 20 minutes. (Make sure you understand that even though we are talking about travel time here, the light year is a unit of distance.) Notice what this implies. We know celestial objects only by the radiation they emit, most familiarly, visible light. That radiation can travel to us no faster than 300,000 km/sec. It follows, therefore, that our latest information about an object is always out of date by the light-travel time from the object to us. Thus we know that the Sun was shining 8 minutes ago, but we will have to wait another 8 minutes to see if it is still shining "now." Similarly, we know only that α Centauri was shining 4.3 years ago. In contrast to distances here on Earth, astronomical distances are so huge that light-travel times are significant. This means that as we observe increasingly distant objects, we are of necessity looking back into time, seeing these objects as they were at earlier epochs, when they emitted the light now reaching us. Built into astronomical observations is a one-way time machine enabling us to look back into the astronomical past, but, alas, not into the future.

Clusters of Stars

When you look at the sky on a clear, dark night, you can see perhaps 2,000 to 3,000 points of light that you perceive as single stars. Appearances can be deceiving in the sky just as on Earth, however. In reality roughly half these points of light are two stars so close together compared to their distance from us that the eye cannot see them as separate objects. This is one of the characteristic features of stars, namely, their tendency to occur in groups bound together by the force of gravity. The simplest such group is a pair of stars, called a **double star**. (See Figure 12.1 for a famous example

Figure 12.1.   The two stars of the visual binary, 61 Cygni, are gravitationally bound and revolve around each other with a period of 653 years. The separation of the K5 and K7 stars is 24 arcseconds. The trigonometric parallax of this system, 0.294 arcseconds, was one of the first to be measured, in 1838. These stars are visible to the naked eye at a dark site.

of a double star.) Three or more stars occur in **multiple systems** (Figure 12.2), and hundreds of stars congregate in loose collections called **open** (or **galactic**) **clusters** (see Figures 12.3 and 12.4). By contrast, **globular clusters** are usually much tighter, fairly symmetric collections of tens or even hundreds of thousands of stars (see Figure 12.5).

Now, on the average in the vicinity of our Sun, single stars represented by ping-pong balls are separated from each other by about 500 kilometers. Hence, it is completely implausible that a group of stars such as an open cluster could have formed by a process of one-by-one capture of passing stars. The stars are just too far apart. In other regions of our galaxy, the central bulge, for example, the average separation of stars is much smaller (about one-tenth that near the Sun), but even there this argument still holds. Thus it is clear that not only the stars making up a multiple system such as α Centauri, but also those in open clusters such as the Pleiades and those in globular clusters such as M3, must have always been together. This implies that they must have formed together at about the same time.

Figure 12.2.   The star θ Orionis is actually four stars, the so-called Trapezium, located in the heart of the Orion nebula. These O- and early B-type stars provide most of the energy to heat the nebula. The photograph is a short exposure so that most of the nebula is not visible. (The nebula is shown in Figure 14.3.)

A tightly packed group of stars such as a globular cluster has remained as a cluster for billions of years because of the strong gravitational attraction each cluster member exerts on the others. Indeed, globular clusters are among the oldest objects in the universe. On the other hand, an open cluster, much more loosely packed, with correspondingly weaker mutual gravitational forces, may slowly lose its members as they "evaporate" away from the group over a period of perhaps a billion years.

Figure 12.3.   The Pleiades (sometimes called the Seven Sisters) is certainly the best-known open cluster in the northern sky. For example, it is mentioned in Chinese records of the third millennium B.C.E, in Homer's *Odyssey*, and in the Bible. The photograph shows the brightest members of the cluster and the nebulosity surrounding them. The cluster is approximately two degrees in diameter (four times that of the full Moon), is about 400 light years away, contains several hundred stars, and is young—less than 100 million years old.

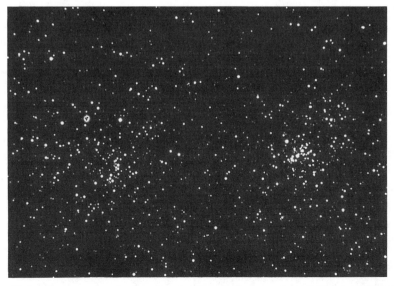

Figure 12.4.   The two open clusters, h and χ Persei. They are about 7,000 and 8,000 light years away, respectively, each is about the diameter of the Moon, and both are less than 10 million years old. By astronomical standards, they were born yesterday.

Figure 12.5.   The globular cluster 47 Tucanae, visible only in the southern hemisphere, is one of the brightest in the sky. It is about 16,000 light years away. Globular clusters are much more tightly packed with stars than open clusters, often containing as many as a million stars. A typical diameter of a globular cluster is about 30 light years. Think of what the night sky would look like if we were located inside such a cluster!

Sizes of Stars

So far we have spoken of the diameter of only one star, the Sun. It is a very ordinary star, not at all unusual in any of its aspects (except for the crucial one of being the source of the energy that makes life on Earth possible!). Most stars have diameters smaller than that of the Sun; considerably fewer have diameters larger than the Sun's. Sirius, the brightest[1] star in the sky other than the Sun (see Figure 12.6), has a diameter about twice that of the latter; the two brightest stars in the constellation of Orion, Rigel and Betelgeuse, have diameters about 80 and 800 times, respectively, that of the Sun (see Figure 12.7). Betelgeuse is truly enormous; at the Sun's location in the solar system it would extend out well beyond the orbit of Mars! Such gigantic stars are extremely rare, however; far more numerous are stars smaller than the Sun, stars whose diameters range down to perhaps one-tenth that of the Sun, or about Jupiter's diameter. Dead stars are even smaller; white dwarfs are about the size of Earth, and neutron stars are roughly city-size. Black holes, the most compact objects in the stellar graveyard, require a special definition of their diameter, as we shall see in Chapter 16.

Brightness of Stars

The amount of energy a star radiates, either over the whole spectrum or within some well-defined wavelength band, is certainly one of its fundamental properties. The amount of energy the star is generating tells us the rate at which it is using up its fuel. This in turn is related to how long it will be able to live, at least in that condition. What we see when we look at a starry sky, however, is how bright a star appears to be. A

---

[1] We use "brightest" here to mean how a star appears to us from Earth, that is, its apparent brightness. In reality, other stars are radiating a lot more energy per second into space. We call the total energy output of a star or galaxy its **luminosity**.

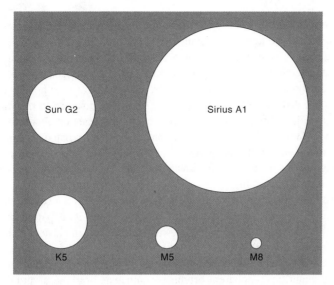

Figure 12.6. The relative diameters of the Sun and four other main-sequence stars. Note that the later the spectral type, the smaller the diameter of the star.

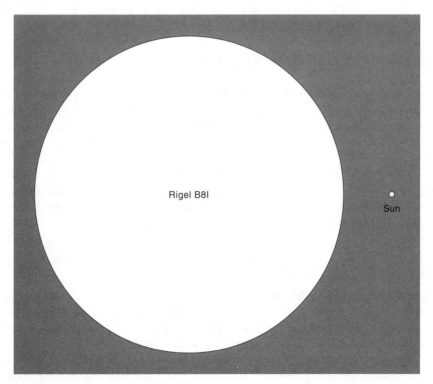

Figure 12.7. The relative diameters of the Sun and Rigel, an early-type star in the constellation of Orion.

star's **apparent brightness** is easy to measure, but by itself is not a very useful quantity. Clearly, unless we know its distance, we can't tell if a star appears to be bright simply because it is nearby (the extreme example of which is the Sun), or if, even though it is at a great distance, it appears to be fairly bright because it is *intrinsically* very bright. In more familiar terms, a nearby 25-watt light may appear to be just as bright as a more-distant 100-watt light. If we know the distance to the star, however, we can calculate how bright it really is, that is, its **intrinsic** or **absolute brightness**. This quantity, from which the distance factor has been eliminated, is fundamental. It tells us how much energy the star radiates away and hence how much it is generating. Intrinsic brightness, or a quantity related to it, is the vertical coordinate of a Hertzsprung-Russell diagram.

Now, we know that the brightness of a light decreases as the square of its distance from us; double your distance from a 100-watt bulb, and its apparent brightness decreases to only one quarter of what it was before. Its intrinsic brightness hasn't changed, however; it is still a 100-watt light. Obviously, a star's apparent and intrinsic brightness are mathematically related through its distance from us. If we know any two of these three quantities—apparent brightness, intrinsic brightness, and distance—we can calculate the third. In practice we can always measure the apparent brightness, so if by some means we measure the star's distance, we can then calculate its intrinsic brightness. Conversely, if somehow we know the star's intrinsic brightness, this, with its measured apparent brightness, allows us to calculate its distance.

Distance determination is not only one of astronomy's most basic problems, but also one of its most difficult. Before discussing the Hertzsprung-Russell diagram, let's consider some of the techniques used to find distances to the stars.

Measuring Distances to Stars

As we have mentioned, many attempts had been made to detect the change in the direction to a star as viewed from opposite sides of the Earth's orbit. Despite this large baseline, the stars are so far away that the change in direction is very small, too small for any but the most sensitive instrumental techniques. Recall that Tycho's inability to measure this effect caused him to propose his own mixed model of the solar system, part Sun-centered and part Earth-centered. Finally, in 1837–38, three observers working independently succeeded in detecting this so-called **trigonometric parallax**, and so the distances, to three nearby stars—α Centauri, Vega, and 61 Cygni. Since this method is still fundamental, let us look at it in some detail.

Trigonometric Parallax

Let us first consider a slightly oversimplified situation: that the Sun and all other stars do not move with respect to each other, while the Earth orbits the Sun. Suppose a star S is located in the direction of the ecliptic plane (see Figure 12.8); when observed from E_1 it would appear to be in the direction E_1S as measured against the fainter and presumably more distant stars; six months later its direction would appear to have moved to the direction along E_2S. We have assumed that the star itself hasn't changed direction; instead it is being observed from the ends of a baseline that is 2 AU long. Extend your arm and look at your thumb first with one eye, then the other. Your thumb will appear to move back and forth against the background because it is viewed from the opposite ends of a baseline a few inches long. This is analogous to the situation for trigonometric parallax.

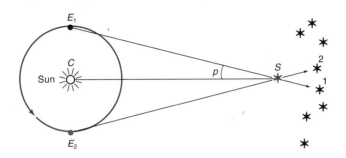

Figure 12.8. The star S, when observed from Earth at E_1, is in the direction of the background star 1. Six months later, when star S is observed from the Earth at E_2, S is in the direction of background star 2. This apparent change in direction to the star results from its being observed from the two ends of a baseline 2 AU long.

Arbitrarily, the parallax of a star is defined to be the angle p, that is, half the total possible change in direction to the star. This is the angular change in direction that would be measured from C to E_1 or C to E_2 as seen from the star S. A convenient distance unit in this situation can be defined as follows. Imagine yourself traveling out from the solar system along CS, measuring as you go the angular separation of the Sun and the Earth when the latter is at E_1 or E_2. Travel outward until at a distance X this angle decreases to only 1 second of arc. At that point, one AU is said to "subtend" an angle of 1 arc second; the corresponding distance, Sun to X, is called one **parsec**—*par* for parallax, and *sec* for second. As you continue traveling out, 1 AU subtends a smaller and smaller angle. Travel ten times farther away, and the parallax angle would be only 0.1 arc seconds, and the distance would be 10 parsecs (abbreviated pc). The distance in parsecs is the reciprocal of (one over) the parallax angle in seconds of arc, or

$$d(\text{pc}) = \frac{1}{p(\text{arc seconds})}.$$

It is easy to show that 1 pc = 3×10^{18} cm = 206,265 AU = 3.26 light years (see Figure 12.9). The parallax of α Centauri is only 0.76", so d = 1/0.76 = 1.3 pc = 4.2 l.y. or about 270,000 times the Sun-Earth distance.[2] And this is the star nearest the Sun! Note the enormous change in the scale of distances as we go from the solar system out to the stars. When one remembers that this tiny angle was first measured by eye at the telescope, it is not surprising that it was so long in being detected.

An interesting question is the minimum number of observations needed, in principle at least, to measure a parallax. If, as we have so far assumed, the stars were really "fixed," then only two measurements, taken six months apart, would be required. Stars in the ecliptic plane would simply oscillate back and forth in a line, taking a year to make a complete round trip; stars in the direction to the ecliptic pole would appear to move in circles; between the ecliptic and the poles they would appear to move in ellipses (see Figure 12.10). Stars aren't fixed in space, however, but are all moving. Despite their enormous distances from us, the motions of the nearer and faster stars are detectable. Hence, a minimum of three observations are required—when the Earth is at E_1, at E_2, then back at E_1 again (see Figure 12.10). The difference between the two

---

[2] Up to now we have been using light year for distances outside the solar system. Parsec, however, is the conventional distant unit in astronomy, which we will also use from now on.

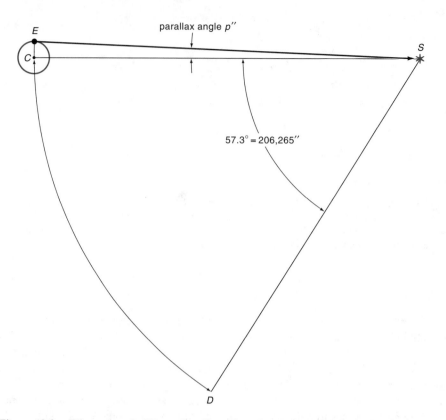

Figure 12.9. When a length *CD* equal to the radius of a circle, *CS*, is laid off along its circumference (that is, *CS* = *CD*), then the angle (called the **radian**) subtended by *CD* is 57.3 degrees or 206,265 seconds of arc, as shown. We can then write the proportionality: *CD/CE* = 206,265/*p*. Because *CD* = *CS* = *d*, the distance to the star, and *CE* = 1 AU, and if we define a new unit of distance, 206,265 AU (the parsec), then we have *d* = 1/*p*, where *d* is in parsecs and *p* is in seconds of arc.

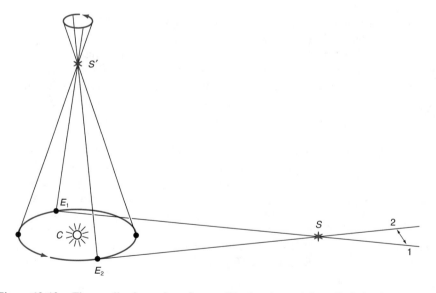

Figure 12.10. The parallactic motion of a star *S* in the plane of the ecliptic is along the line 1-2, whereas the parallactic path of the star *S'* at the ecliptic pole is a circle.

measurements at E_1 gives the motion of the star itself, unaffected by any parallactic displacement. In practice, however, many more observations are taken, often extending over 20 or 30 years, in an attempt to improve the accuracy of the result.

The Motions of Stars

Let us look at stellar motions in more detail, remembering that motions across the sky are detectable only for the nearest stars. Figure 12.11 shows a nearby star moving from A to A', as seen from the Sun (so we have eliminated any parallactic effect caused by the Earth's motion around the Sun). The change of direction to the star per year as it

Figure 12.11. The proper motion of a star as its direction changes from *SA* to *SB* is given by the angle μ. Its velocity in space is given by V_S, which has two components, V_R along the line of sight and V_T across the line of sight. Note that even a nearby star with a large space velocity could have a nondetectable proper motion if V_T were small.

moves from A to A' is denoted by μ (the Greek letter *mu*) and is called the star's **proper motion**, motion that is "proper to" the star itself; it is usually given in seconds of arc per year. Few stars have proper motions larger than one second of arc per year; so the angle μ in the figure is enormously exaggerated. As time goes by, the star moves a greater distance; by comparing its positions on photographs taken a few decades apart, one can detect small yearly motions.

The star's velocity producing the displacement AA' is called its **space velocity**, V_s, given in km/sec. This motion can be broken into two parts: one, AB, is a consequence of the star's **tangential velocity**, V_T, *across* the line of sight from the Sun. It is tangent to the plane of the sky; hence the name. It is this velocity component that produces the change of direction to the star we call its proper motion. The other component of motion, BA', is the result of the star's velocity *along* the line of sight, or its radial velocity, V_R. The three velocities are related by the Pythagorean theorem, $V_s^2 = V_T^2 + V_R^2$.

Notice that a given proper motion can be produced by a nearby star moving slowly, or a more distant star moving more rapidly. In Figure 12.12 stars A and B have the same proper motion, μ, but the space velocity of B is greater than that of A [or $V_s(B) > V_s(A)$] and the tangential velocity of B is greater than that of A [or $V_T(B) > V_T(A)$]. A moment's reflection should convince you that the tangential velocity is pro-

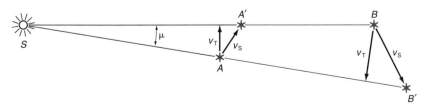

Figure 12.12. Stars A and B have the same proper motion, but since B is farther away than A, its tangential velocity must be greater than A's.

portional to the product of the proper motion and distance: $V_T \propto \mu d$; in commonly used units it turns out that

$$V_T = 4.74\mu d,$$

where μ is in arc seconds per year, d is in parsecs, and V_T is in km/sec. In general, if a star has a large proper motion, it is probably located fairly near to the Sun, which is useful information. To find its more physically meaningful space velocity, however, we must also know the distance to the star (to get V_T) and its spectrum (to get its radial velocity V_R from the Doppler shift). Radial velocities and proper motions have been measured for thousands of stars, but accurate distances are much more difficult to come by. Consequently, relatively few stellar space velocities are known.

The Limits of Trigonometric Parallaxes

No physical measurement is complete unless its error or uncertainty is given (an idea Tycho rigorously applied to his own observations). Obviously, we would give much more weight to a measurement that was accurate to within 1 percent as compared to one that was good to only 20 percent. As we have seen, trigonometric parallaxes are very small, difficult to measure, and therefore, subject to significant—and sobering— errors. With considerable effort, one can measure a parallax with an error of about $\pm 0.005''$. The most modern techniques enable parallaxes to be measured with an error of about $\pm 0.003''$, but relatively few parallaxes have so far been measured this precisely.

Suppose the measured parallax is $0.01''$; with an error of $\pm 0.005''$, the measurement would be given as $p = 0.01'' \pm 0.005''$. It would be misleading to think that the star is $d = 1/0.01 = 100$ pc away, however. Instead, all that we can say is that there is a 70 percent chance that the parallax is between 0.01 ± 0.005 (that is, *between* $0.01 + 0.005 = 0.015''$ and $0.01 - 0.005 = 0.005''$), and a 30 percent chance that it falls *outside* these limits. Hence it is only about twice as probable that the star is between 67 ($0.015''$) and 200 ($0.005''$) parsecs away, as that it is nearer than 67 or farther than 200 parsecs (see Figure 12.13). In other words, we really don't know very much about this star's distance. Trigonometric parallaxes are generally too inaccurate to give useful distances for stars that are a mere 100 parsecs away. Because the center of our galaxy is about 8,000 parsecs distant, it is obvious that the parallax method is useful only for objects located in a tiny local region of our galaxy. We shall take as a practical useful limit for trig parallaxes about 30 pc (about 100 l.y.). Fortunately, many hundreds of stars are known to be located within this distance, so that we know their intrinsic brightnesses fairly accurately. Interestingly, the high angular resolution achieved by radio interferometers (see Chapter 10) has allowed the measurement of the trig parallaxes of a handful of radio objects at distances of a few hundred parsecs.

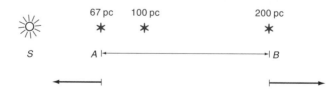

Figure 12.13. A representation of the range of distances from the Sun S at which a star having a parallax of $0.01'' \pm 0.005''$ could be located. Its nominal distance is 100 pc, but actually all we can say is that it is about twice as likely that it is located somewhere between A and B than that it is outside those limits.

In 1989 the European Space Agency launched the first satellite devoted to the measurement of stellar positions. Appropriately named Hipparcos, the satellite did not achieve the intended 24-hour orbit, because of a failure of the upper-stage rocket. Nevertheless, it appears that most of its goals will be achieved, which are to measure the positions, proper motions, and parallaxes of about 100,000 stars to an accuracy of about 0.002 arcsec. This is much better than has been possible up to now, because a space instrument does not suffer from limitations, such as seeing effects and telescope flexure, which plague ground-based positional measurements. In addition, a single instrument will measure a huge number of stars, thereby eliminating errors that arise when observations from several different telescopes are combined. When the Hipparcos data catalogs become available in the late 1990s, positional astronomy should be on much firmer ground.

Method of Similar Objects

In order to measure distances of objects more than about 30 pc away, we must use various indirect techniques, the most common group of which we call the **method of similar objects**. The idea is simple enough. Suppose we want to find the distance to a certain star we'll call S. We then find a nearby star that looks like the same kind of star as S, and whose distance—and so whose intrinsic brightness—we already know. We then assume that the intrinsic brightness of S is the same as that of the nearby star; in other words, we again invoke the idea of the uniformity of nature. Then, with the measured apparent brightness of S and its assumed intrinsic brightness, we calculate the distance to S.

The crucial step in all this is justifying our assumption that the two stars really are sufficiently alike that they have the same intrinsic brightness. The best way is to see if their spectra are essentially identical. If this is so, we can fairly safely assume that the two stars are similar in other respects, in particular in their intrinsic brightness. This assumes that the distant star's brightness is not dimmed by absorption by interstellar material, or if it is, that we have corrected for the effect. (We will see later that interstellar absorption makes starlight redder than it would otherwise be—a B-type star might have an F-type color; hence the effect is readily detectable.)

This "bootstrap" method can be applied to clusters of stars as well as to single objects, with a great increase in the return on our investment of effort. Find several stars in an open cluster, for example, that we convince ourselves are identical in nature to those whose distance and, therefore, intrinsic brightness we already know. Assuming these brightnesses for the corresponding cluster stars gives us their distances, which we can then average to obtain a more reliable distance for all the cluster stars. In general the cluster size will be small compared to its distance, so that all its members are essentially equally distant from us (just as everyone in Chicago is about the same distance from the Washington monument). In this way we now have the distances and, therefore, the intrinsic brightnesses of all the cluster members, some of which may be types of stars not represented in our original list of trigonometric parallaxes. We can then use these stars to find distances to others, and so on.

Though simple in principle, this technique is not without its pitfalls. We must be careful that the supposed cluster stars really are members of the cluster, and not foreground objects that just happen to be in the same direction as the cluster. This can be checked by seeing if the proper motions and radial velocities of the stars in question are about equal to the average for the cluster as a whole. As always, we have to do everything possible to assure ourselves that the "known" and "unknown" stars are really the same kind with the same properties.

The Hertzsprung-Russell Diagram
Plotting an HR Diagram

Now that we have become familiar with some of the methods used to measure stellar distances and therefore intrinsic stellar brightnesses, we are ready to tackle the subject of this chapter. All of the most important stellar parameters—temperature, luminosity, mass, and radius—can be inferred, at least in an approximate way, from the position of a star on the **Hertzsprung-Russell** diagram (conveniently abbreviated as the HR diagram). Such a diagram, named for the two early twentieth-century Danish and American astronomers who devised it (E. Hertzsprung and H. N. Russell, respectively), is easily the most useful way we have of representing stellar data.

In an HR diagram some measure of a star's energy output is plotted against a measure or index of its temperature (see Figure 12.14). For the former we will use the energy output of stars in the visible region of the spectrum in units of the Sun's output:[3] $15 L_\odot$ means that the star radiates 15 times more energy in the visual region of the spectrum than the Sun does; for a measure of temperature we have spectral type, color, or effective temperature described in the last chapter. Now, suppose we plot the intrinsic brightnesses of a representative sample of stars against their effective temperatures (or what is equivalent, against their colors or their spectral types) as shown in Figure 12.14. (Notice that, contrary to general practice, the temperature, represented by spectral type, decreases to the right along the horizontal axis, rather than increases.)

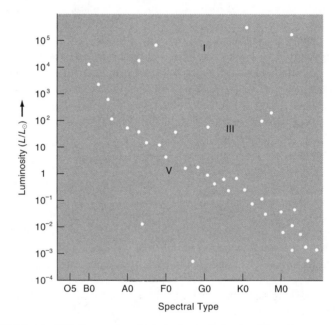

Figure 12.14. An HR diagram in which the intrinsic luminosities of a representative sample of stars are plotted against their spectral types. Note that the stars are not distributed randomly over the diagram.

---

[3] It is often more meaningful and convenient to put many stellar parameters—especially luminosity, radius, and mass—in terms of those of the Sun, rather than in ergs/sec, centimeters, and grams, respectively. These values are accurately known for the Sun; they fall conveniently at about the midrange of the values exhibited by stars. Thus, in this and subsequent chapters we shall speak of a star with a mass of 10 solar masses (ten times the mass of the Sun), or a luminosity of 50 solar luminosities, etc.

The first—and most important—thing to notice is that points representing the stars are not scattered randomly across the diagram. Instead, most stars fall along a relatively narrow band running diagonally across the diagram. A few are sprinkled across the upper part (in areas marked III and I), and some others are found in the lower left-hand side of the diagram. A V-shaped area called the **Hertzsprung gap** and falling in the upper center of the diagram is, largely but not entirely, devoid of stars. That the points are not randomly distributed over the HR diagram means that stars cannot have every possible combination of temperature (spectral type) and energy output. This in itself is an extremely important result; it must be telling us something significant about how stars are constructed. In fact, understanding the physical processes and properties of stars that cause them to fall on one part or another of the HR diagram is just another way of describing the study of the structure and evolution of stars.

The Main Sequence.   The band (marked V) running diagonally across the HR diagram contains the vast majority of stars. For this reason it is called the **main sequence**. Most stars have only the combination of temperature (spectral type) and luminosity (or radius, because $L \propto R^2 T^4$) indicated by the main sequence. Figure 12.15 shows the approximate values of various stellar parameters along the main sequence. We see that in going down the sequence from upper left to lower right, stars become systematically cooler, less luminous, less massive, and smaller in radius. Furthermore, the relative numbers of stars generally increase as you go down the main sequence.

An interesting question concerns the range of physical characteristics such as radius, mass, or temperature that stars can have. From the diagram we see that the hottest stars have effective temperatures about 20 times greater than the coolest, and the largest stars have radii (found, for example, from their temperature and luminosity) about 100 times greater than the smallest. These values are somewhat arbitrary, because it is

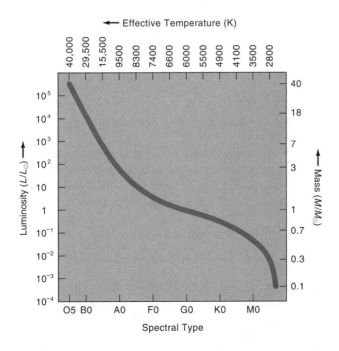

Figure 12.15.   An HR diagram giving values of luminosity, temperature, and mass along the main sequence. For example, an A0 star has an effective temperature of about 9,500 K, and a mass of about 3 solar masses, and is nearly 100 times more luminous than the Sun.

impossible to say exactly how hot is the hottest star or how cool is the coolest, etc. The ranges of parameters given in this chapter, however, include the vast majority of stars.

From the HR diagram we see that the luminosities of main-sequence stars range over a factor of about one billion! This huge range results because the luminosity of a star depends on both the square of its radius and the fourth power of its temperature. Let us compare the luminosity of an O-type star with that of a late M-type star; remember that $L = 4\pi R^2 \sigma T_e^4$, so that (because the $4\pi\sigma$'s cancel)

$$\frac{L_{\text{O-star}}}{L_{\text{M-star}}} = \frac{R_{\text{O-star}}^2 \, T_{\text{O-star}}^4}{R_{\text{M-star}}^2 \, T_{\text{M-star}}^4} = (100)^2 (20)^4 = 10^4 (16 \times 10^4) = 1.6 \times 10^9,$$

since from above, $(R_{\text{O}}/R_{\text{M}})$ is about 100, and $(T_{\text{O}}/T_{\text{M}})$ is about 20 (that is, the radius of the O-type star is 100 times greater than that of the M-star, and the temperature of the O-star is 20 times greater than the M-star's).

Giants and Supergiants.   Next consider the stars along sequences III or I. G2-stars (having the same effective temperature as the Sun but on sequences III or I), are about 50 and 5,000 times, respectively, more luminous than the main-sequence Sun. Now, since the three stars we are considering all have about the same effective temperature, each unit of area on each star must emit the same amount of energy. Therefore, to be more luminous, the G2-star on sequence III must be larger in area than the Sun, and that on sequence I must be even larger yet. If we call main-sequence stars **dwarfs**, then it is appropriate to call stars on sequence III **giants**, and the larger stars along sequence I, **supergiants**.

Let us calculate how much larger than the Sun a G2 supergiant must be. Remember that $L = 4\pi R^2 \sigma T_e^4$, and since the two stars have the same temperature and the $4\pi\sigma$'s also cancel, we can write just as we did above,

$$\frac{L_{\text{SG}}}{L_\odot} = \frac{R_{\text{SG}}^2 \, T_{\text{SG}}^4}{R_\odot^2 \, T_\odot^4} = \frac{R_{\text{SG}}^2}{R_\odot^2} = \frac{5000}{1},$$

so

$$R_{\text{SG}} = \sqrt{5000}\, R_\odot \approx 71 R_\odot .$$

If such a star were located where our Sun is, it would extend out nearly to the orbit of Mercury!

Supergiants with the same luminosities but later and later spectral types have larger and larger radii, because each unit of area of a supergiant with a lower effective temperature radiates less energy than a hotter one. But if the luminosities of the supergiants are about the same, the surface areas and hence the radii of the cooler ones must be larger than those of the hotter ones. Indeed, the latest-type (coolest) supergiants are so enormous that if one replaced the Sun it would extend out beyond the orbit of the Earth. We would be inside the star! Figures 12.6 and 12.7 show the relative diameters of various stars.

Luminosity Classification.   We saw that the primary factor that determines the physical state of a stellar atmosphere and the appearance of its spectrum is its temperature, but the pressure of the gas also plays a role, albeit a secondary one. Pressure differs from star to star because of differences in the acceleration of gravity in the atmosphere of the star. Just as you would expect, the greater the acceleration (the "pull") of gravity, the greater the gas pressure, other things being equal. The differences in

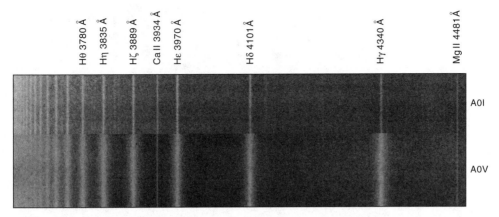

Figure 12.16. The spectra of two A0-type stars. The upper spectrum is that of a supergiant, the lower of a main-sequence star. Note the striking difference in the widths of the hydrogen lines. Also notice that the Mg II line at 4481 Å does not change appreciably; in contrast to the hydrogen lines, it is not luminosity-sensitive.

atmospheric gravity between dwarf stars is relatively small, and so pressure effects are not marked along the main sequence. The situation is quite different among the giants and supergiants than among the dwarfs, however.

Consider two A0 stars, one on the main sequence, the other a supergiant; their spectra are shown in Figure 12.16. The mass of the supergiant is somewhat larger than that of the dwarf, but its radius is very much larger. Hence, in the atmosphere of the supergiant, the gravity (which as you recall is proportional to M/R^2) is much smaller than in the main-sequence star. In fact, it is about 0.01 or even 0.001 of the main-sequence value. Consequently, the pressure of the atmospheric gas in the A0 supergiant will be much lower than that of the gas in the atmosphere of the main-sequence star.

This gives rise to a marked effect in the spectra of the two stars: the widths of the hydrogen lines are much narrower in the supergiant than in the dwarf. Hydrogen line widths are determined in part by the collisions hydrogen atoms undergo as they absorb and emit radiation—the greater the number of collisions, the greater the line widths. But at a given temperature, the greater the pressure, the greater the number of collisions. Hence the Balmer lines in the main-sequence star are systematically stronger than those in the supergiant.

To summarize, the two stars have certain spectral features (or more likely, ratios of line strengths) that are identical; that is why they are both classified as A0. In addition, however, the hydrogen lines are narrower in the supergiant than in the dwarf, so one can discriminate further between these two stars. This discriminant is called the **luminosity class**, because it is related to the intrinsic luminosity of the star.

Pressure differences give rise to another effect. Consider a volume of gas at some temperature: the larger the pressure of the gas, the greater the number of particles in the volume (after all, no particles, no pressure); also, the greater the number of particles—electrons, ions, and atoms—in that volume, the larger the probability that an ion will capture an electron, thereby reducing the state of ionization of the gas. Hence, at a given temperature, a higher atmospheric pressure will tend to reduce the degree of ionization of the gas. This can give rise to a fairly subtle effect in the spectrum of a star. Within a given temperature range of the spectral sequence, the strength of a particular absorption line might be especially sensitive to the degree of ionization of the

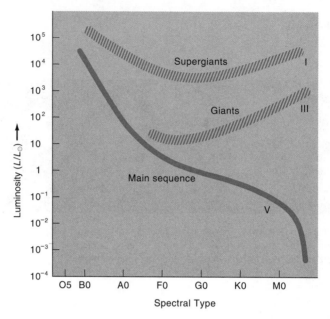

Figure 12.17. An HR diagram showing the positions of luminosity classes I, III, and V, or supergiants, giants, and the main sequence, respectively.

gas. Consequently, the appearance of that feature could be detectably different in the spectrum of a dwarf as compared to its strength in the spectrum of a supergiant. Such ionization effects exist throughout most of the HR diagram and, along with the pressure effects described above, make possible the additional refinement of luminosity class in the MK classification system.

Morgan and Keenan found that five **luminosity classes** could be distinguished on their spectra. They denoted them by I, II, . . . , V. (Actually the most luminous stars can be further subdivided into Ia and Ib and even finer subdivisions, but usually we will lump them all together.) Three of these so-called luminosity classes are shown on the HR diagram of Figure 12.17, where I refers to supergiants, III to giants, and V to main-sequence stars. Luminosity classes II and IV fall between I and III, and III and V, respectively, as you would expect. Thus, the complete MK classification of a star might be A5III or B3V; the Sun is a G2V star.

It is important to understand that a given luminosity class does *not* imply one single value of the intrinsic luminosity for all stars of the class. Instead, it means that the star belongs to a particular luminosity *sequence* in which individual luminosities might take a wide range of values. This is particularly true for luminosity class V.

Spectroscopic Parallaxes.   To the extent that the HR diagram is well calibrated, that is, to the extent that we know the intrinsic luminosities of the various kinds of giants, supergiants, etc., then we have another method for finding the distances to stars. All we need do is classify the star on the MK system (spectral type and luminosity class, which together determine its position on the HR diagram), and then read off its intrinsic luminosity from the diagram. That, along with its apparent brightness, enables us to calculate its distance. Distances found in this way, called **spectroscopic parallaxes** ("spectroscopic" because the star's spectrum is needed and "parallax" in the general sense of a distance), are sometimes rather uncertain because of the intrinsic

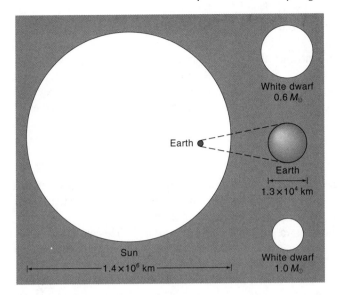

Figure 12.18. The relative diameters of the Sun, the Earth, and white dwarfs of two different masses. Note that the more massive a white dwarf is, the smaller (not larger!) it is.

spread in the luminosities of various kinds of stars. For example, the spread of luminosities of main-sequence stars is rather small at A0, but it is large among the late-
type giants. Hence spectroscopic parallaxes for the latter will not be as reliable as for
the former. Nonetheless, the technique has proven to be extremely useful.

White Dwarfs.   The last major group of stars is located in the lower left-hand portion of the HR diagram. These stars are hot, but have very low luminosities and hence
must be small in radius. Because of these characteristics, they are called **white dwarf**
stars. Logically this name should refer to early-type main-sequence stars, but for historical reasons the term is used to describe hot, low-luminosity objects instead.

Let us calculate just how small white dwarfs are. Consider a main-sequence
A-type star and a white dwarf of the same effective temperature, but having a luminosity only about 10^{-5} that of the main-sequence star. Just as we have done before in
this chapter, we can write (remembering that the temperatures as well as the constants
cancel),

$$\frac{L_{wd}}{L_A} = \frac{R^2_{wd}}{R^2_A} \ ;$$

so

$$R_{wd} = \sqrt{10^{-5}} \, R_A \approx 3 \times 10^{-3} R_A.$$

Since the radius of an A-type star is about three times that of the Sun, that is, $R_A \approx 3R_\odot$,

$$R_{wd} \approx (9 \times 10^{-3}) R_\odot \approx (10^{-2}) R_\odot \ ,$$

or just about the radius of the Earth! Figure 12.18 shows the diameters of a few white
dwarfs compared with the Sun and Earth. Since the white dwarf's radius is about

$10^{-2}R_\odot$, its volume must be only about $10^{-6}V_\odot$ (since volumes are proportional to R^3). Thus a 1-solar-mass white dwarf will have an average density (given by mass/volume) about one million times that of the Sun! Now, the densest matter on Earth, osmium, has a density only 22.5 times that of water, or about 16 times the average density of the Sun (which is 1.4 times that of water). Clearly, white-dwarf material must be very strange indeed. In Chapter 16 we will describe the physical properties of these stars.

Measuring Stellar Masses

Kinds of Binary Stars.   We've just mentioned the mass of a white dwarf. How are stellar masses determined? They are found from binary-star systems; so let's first consider how we know a star is really a double star.

Most double stars are too far away from us to be seen as separate objects; they appear only as a single point of light. Nonetheless, such apparently single stars can be shown to be binary systems in either of two ways. Spectra of an apparently single point of light taken every few hours or every few days will often show a periodic shift of the spectral lines of the object (see Figure 12.19). The lines might be shifted first to the blue, then back to no displacement, then to the red, back to no displacement, repeating the cycle over and over again. Clearly, this must be the result of the Doppler effect with the star first moving toward us (blue-shifted spectrum), then across our line of sight (no displacement), then away from us, regularly repeating this sequence. In some systems we see two sets of spectral lines displaced in the opposite sense, indicating that when one star is approaching us, the other is moving away; in other systems only one spectrum is visible. Even if the companion star is too faint to produce a spectrum that we can see, its presence can be inferred from the regular motion of the visible star. Such binary stars, detected by the periodic shift of their spectra, are called **spectroscopic binaries**.

Other binary systems are oriented in such a way with respect to our line of sight that one star periodically eclipses the other; this causes the total light from the system to vary in a regular way (Figure 12.20). The graphical representation of the changing

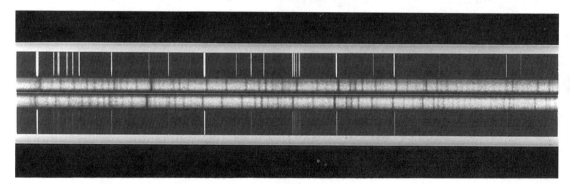

Figure 12.19.   This bright star, α^1 Geminorum, is a spectroscopic binary with a period of 2.9 days. A part of its early (and somewhat peculiar) A-type absorption line spectrum is shown at two different phases; above and below those spectra is an emission line comparison spectrum. The spectrum of only one star of the α^1 Geminorum pair is visible, its lines shifting redward and blueward as the star orbits its invisible companion. Incidentally, α^1 Geminorum is also a member of a visual binary system, having a period of 420 years. The companion, now 2 arcseconds away, is also an early A-type main-sequence star and a spectroscopic binary. Together these stars are known as Castor (of Castor and Pollux fame). An M-dwarf star, 73 arcseconds away, is gravitationally bound to the four stars comprising Castor, and it too is a spectroscopic binary. Thus, this is a system of six stars!

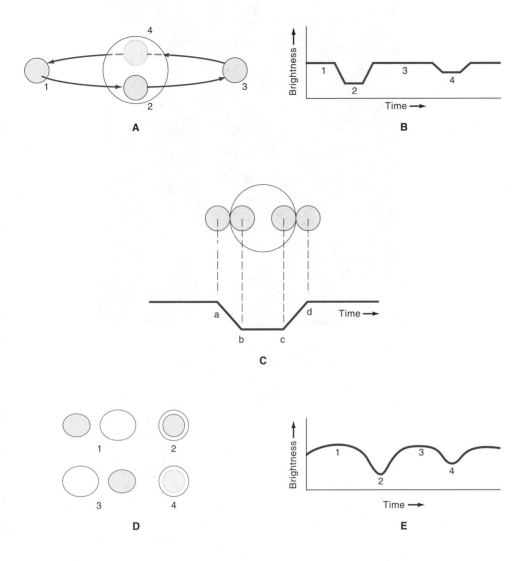

Figure 12.20. (A) A small, cool star is shown in orbit around its companion, a large, hot star. The light curve of this eclipsing binary system is given in (B). Outside of eclipse at 1 and 3, the light from the system is constant and is the sum of the radiation from both stars. At 2, when the smaller star passes in front of the larger one, some of the light from the hotter star is blocked and the brightness of the system decreases. At 4, the larger star eclipses the smaller one; the light loss is less than in 2, because the cooler star radiates less energy per unit area than the hotter one. (C) A careful look at the eclipse light curve shows that in moving from a to b and from c to d the smaller star moves its own diameter, whereas in moving from a to c (or b to d) it moves the diameter of the larger star. Hence, the ratio of the time from a to b to that from a to c is the same as the ratio of the diameter of the smaller star to that of the larger (assuming that the star moves from a to d with a uniform speed). That is, $t_{ab}/t_{ac} = R_s/R_1$. (D) This is the same sort of binary system as is shown in (A), with the difference that the two stars are so close together that they gravitationally distort each other's shape from spheres to ellipsoids. Such stars generally keep the same face toward each other as they revolve. (Imagine the centers of the two stars joined by a rigid bar, with the axis of rotation perpendicular to the bar.) Because the amount of stellar surface facing us changes constantly, the amount of light we measure is never constant, even outside of eclipse, producing the light curve shown in (E).

291

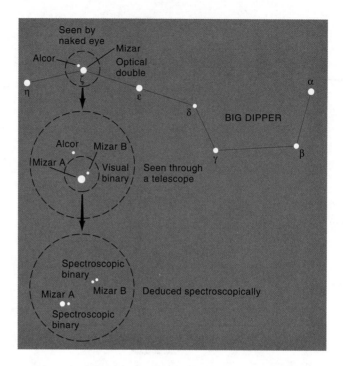

Figure 12.21. Mizar, one of the stars forming the handle of the Big Dipper, is an optical double with Alcor (that is, they are not gravitationally connected, but just along the same line of sight) located about 3.8 arcminutes away. Alcor is a spectroscopic binary. Mizar is a visual binary having a gravitationally bound companion about 14 arcseconds away. Alcor and Mizar are easily separated by the naked eye, but Mizar requires a telescope to be seen as two stars. Each component of Mizar is itself a spectroscopic binary; in fact, Mizar A was the first spectroscopic binary discovered, in 1889. So the apparently single star Mizar is in reality four stars!

brightness of any star or stellar system is called its **light curve**. Analysis of the light curve of these objects (called **eclipsing binaries**, not surprisingly) enables us to deduce the stars' sizes and sometimes even their shapes, if rotation, for example, distorts them from the usual spherical figure. It is remarkable that such information can be extracted from the variations in brightness of a single point of light!

A relatively small number of double stars are sufficiently separated from each other and at the same time near enough to us so that both stars can be seen in a telescope; these are called **visual binaries** (see Figure 12.21). Since the two stars are widely separated, the period of revolution of each around their common center of mass (see below) is always long, many years in fact. By contrast, eclipsing binaries can have periods as short as hours.

Calculating the Masses of Stars in a Visual Binary System. If we measure the motion of one star with respect to the other (usually the fainter with respect to the brighter; see Figure 12.22),[4] after many years we can deduce the shape of the orbit and how long it takes for one star to make a complete trip around the other. That is, we can find the orbital ellipticity and period of the system. (A little thought should convince you that we would get the same answers if we measured the motion of the

---

[4] Imagine yourself on the brighter star watching the fainter star revolve around you.

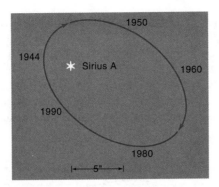

Figure 12.22. The observed 50-year orbit of Sirius B (a white dwarf) with respect to Sirius A (an A-type star). Note that Sirius A is not at the focus of the elliptical orbit, because the plane of the orbit is inclined to the plane of the sky, and what we observe is the projection of the true orbit on the sky. The orbital inclination is the angle that tips the observed orbit back to the true orbit, which has Sirius A at its focus.

brighter star around the fainter. After all, a given double-star system has only one orbital shape and one period.) Knowing the shape of the orbit gives the size of its semimajor axis, but only in angular measure—seconds of arc. To convert this to AU, we must know the distance to the system, which, as you must now realize, may be difficult to measure. Generally speaking, however, if a binary system is close enough to us that we can see both stars, it is close enough that a trigonometric parallax can be measured. Let's assume that this is the situation, and that an accurate trigonometric parallax can be measured. Then we can find the semimajor axis in AU, and using Newton's generalization of Kepler's third law (Chapter 8), we find the sum of the masses of the two stars: $(M_1 + M_2)P^2 = a^3$. This is useful information, but what we really want is the mass of each star. If we can measure the motion of each star around the common **center of mass** of the system, then we can find the ratio of the two masses, that is, the value of M_1/M_2. With the sum and the ratio of the two masses, we can find the individual masses, because we have two equations for the two unknowns, M_1 and M_2.

So the problem is to find the mass ratio. Suppose you and a friend are on a seesaw (Figure 12.23). If you both weigh the same, then the seesaw will balance when you are both the same distance, a, from the bar on which the seesaw pivots. If you weigh less than your friend, however, you will have to sit farther from the pivot point to keep the seesaw in balance. If you weighed half as much as your friend, you would have to sit twice as far away. In general, you would have to sit in such a way that $M_1a_1 = M_2a_2$,

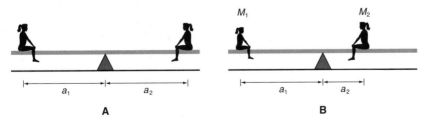

Figure 12.23. (A) If you and your seesaw partner weigh the same, then to balance the seesaw you would both be at the same distance, $a_1 = a_2$, from the pivot point. If one of you, M_2, is twice as heavy as the other, M_1, the lighter one will have to be twice as far from the pivot point as the heavier, so that $M_1a_1 = M_2a_2$.

where M_1 and a_1 are your mass and distance, respectively, from the balance point; quantities with subscript "2" refer to your friend. Now, both stars of a double-star system move around their common center of mass (analogous to the seesaw's pivot point) such that at every point in their motion around each other, $M_1a_1 = M_2a_2$, or, equivalently, $(M_1/M_2) = (a_2/a_1)$. If we can measure the proper motions of both stars (which will be in wavy rather than straight lines because of the orbital motions of the two stars) and measure a_1 and a_2, we can find the mass ratio of the two stars (see Figure 12.24).

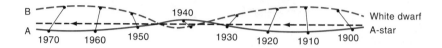

Figure 12.24. The wavy motion of each component of Sirius across the sky. The heavy dashed line is the path of the center of mass of the binary system (the "balance point") of the two stars; it moves in a straight line, as Newton's first law requires. The two stars move so that the ratio of their distances from the center of mass is always in inverse ratio to their masses, that is, the white dwarf is always 2.2 times farther away from the center of mass than is the A-type star.

Note that though the distance of each star from the center of mass constantly changes, the ratio of any pair of these distances is constant and is equal to the ratio of their masses.

Let us apply these ideas to the double-star Sirius, which consists of an A1V star and a white dwarf. The parallax of Sirius is $p = 0.375$ arcsec, so that its distance $d = 1/p = 1/0.375 = 2.67$ pc. The semimajor axis of its orbit has been measured to be 7.7 arcsec, which, given the distance just found, corresponds to 20 AU (see Appendix A). The period of the binary system is 50 years. Now,

$$(M_A + M_{wd})P^2 = a^3 \; ;$$

so

$$(M_A + M_{wd}) = \frac{20^3}{50^2} = \frac{8,000}{2,500}$$

and

$$(M_A + M_{wd}) = 3.2M_\odot.$$

From the motion of Sirius around its center of mass, the mass ratio of the two stars has been found to be $M_A/M_{wd} = 2.2$; that is, the less-massive star is always 2.2 times farther away from the center of mass than is the more-massive star. Knowing the sum and the ratio of the masses enables us to find the masses separately:

$$M_A = 2.2M_\odot, \text{ and } M_{wd} = 1M_\odot.$$

The Mass-Luminosity Relation. Using techniques such as these, the masses of perhaps 150 stars have been determined reasonably accurately. On the right-hand side of the HR diagram in Figure 12.15 the masses of main-sequence stars are given. Note that they are systematically smaller as the stars are fainter. An O-type star is about 400 times more massive than a late M-type star, but it is about 10^9 times more luminous.

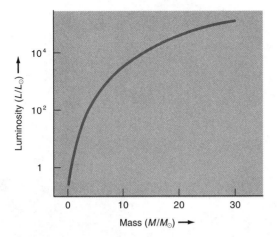

Figure 12.25. The relation between mass and luminosity for main-sequence stars. Note how a small change in mass corresponds to a very large range in luminosity.

The O-type star generates much more energy per gram of its mass than does the M star. For main-sequence stars there exists a relation between a star's mass and its luminosity, which we can write approximately as $L \propto M^{3.5}$ (you can check that $(400)^{3.5} = (400)^3(400)^{1/2} \approx 10^9$). This is called the **mass-luminosity relation** (Figure 12.25), and it can be derived theoretically from general physical principles. We shall soon see its usefulness. It is significant that many giants do not follow such a relation between mass and luminosity. This indicates that their structure is basically different from that of main-sequence stars.

Intrinsically Variable Stars

We have seen that light from a star may vary because a companion star eclipses it. Other stars vary in light even though they are single and have no eclipsing companion; that is, they are intrinsically variable. In some, the variability is strictly periodic, and their light curves repeat in every detail over and over again. Other types of variables do not behave in so predictable a fashion, and are only semiregular in their variability (see Figure 12.26). Some stars regularly change their brightness by a factor of two or more, whereas others vary by only a percent or less.

Intrinsically variable stars are found throughout most of the HR diagram. A particularly interesting group is found along the so-called Cepheid instability strip shown in Figure 12.27, and named after its most prominent members. (The prototype of Cepheid variables is δ Cephei, the fourth brightest star in the constellation of Cepheus.) Any star within this strip will be variable. This instability strip includes stars of an extremely wide range of luminosities—supergiants (the Cepheids themselves), near-main-sequence objects, and even white dwarfs.

Both the radius and the atmospheric temperature of a Cepheid vary as its brightness changes. When the star is brightest its temperature is the hottest, but its radius is near its minimum size (see Figure 12.28). The cause of the light variability in a Cepheid is a consequence of the way energy is periodically bottled up and released in its lower atmosphere, not of any change in the rate at which energy is generated by the star; that remains constant. We shall see later that Cepheids play a crucial role in establishing the distances to galaxies.

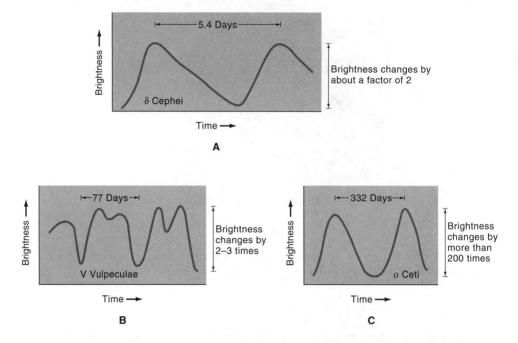

Figure 12.26. (A) The light curve of the prototype classical Cepheid variable. Cepheids are F- and G-type supergiants with periods ranging from 1 to 50 days and brightness variations from a factor of about 1.2 to more than 5 times. (B) RV Tauri stars are G- and K-type supergiants with periods of 30 to 150 days; they show double-humped brightness variations of about 2 to more than 10 times. Their maxima and minima may change from one cycle to the next. (C) The long-period variable, o Ceti or Mira, has a very large amplitude, so it is not too surprising that it was the first variable star of any type to be discovered, in 1596.

Figure 12.27. The distribution of some of the types of intrinsically variable stars over the HR diagram. Many RR Lyrae stars, A- and F-type giants, have light curves and amplitudes similar to those of classical Cepheids (see Figure 12.26), but with much shorter periods ranging from about 4 hours to 1 day. RR Lyrae itself has a period of 13.6 hours. Delta Scuti stars bridge the gap between the RR Lyraes and the main sequence. They generally have shorter periods and smaller amplitudes than RR Lyrae stars; their periods range from less than 1 hour to about 5 hours, and the amplitude of light variation from a factor of about 1.2 to less than 1.01. ZZ Ceti stars are white dwarfs with periods from 1 to about 15 minutes and amplitudes of only a few percent. They are probably the most numerous variables in the neighborhood of the Sun, followed in frequency by the δ Scuti stars. Long-period variables (see Figure 12.26) are M-type giants and supergiants with variable maxima and minima.

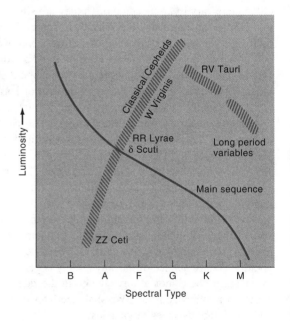

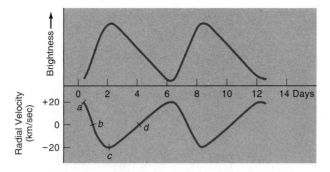

Figure 12.28. The variation in brightness and radial velocity of a Cepheid. At *a*, the star is contracting most rapidly; at *b* it is smallest; at *c* it is expanding most rapidly; at *d* it has expanded to its maximum diameter. By combining the radial velocity measurements with the length of time in which they occur, it can be shown that the radius of the star expands and contracts by about 10 percent over a brightness cycle. Except for the ZZ Ceti white dwarfs, all of the stars discussed in this section pulsate in a similar manner.

Relative Numbers of Different Kinds of Stars

Most stars are located on the main sequence, and their relative numbers generally increase as you go down the sequence to intrinsically fainter stars. There is a curious exception to this rule: from A0 to about A4, the numbers of stars decrease before increasing again. As far as we can tell, this is a real effect, but its cause is uncertain.

Giants are much less common than main-sequence stars, but more abundant than supergiants, which are very rare indeed. It is impossible to establish the true relative numbers of stellar types by counting all the stars in a given volume of space. To contain even a few of the rare supergiants, the volume of space must be so large that most of the intrinsically faint stars will be too far away to be detected by even the largest telescopes. The difficulty is well exemplified by comparing a list of the nearest stars (within about 5 parsecs of the Sun) with the apparently brightest stars (which includes stars over 400 parsecs away) given in Appendixes D and E. Among the former are main-sequence and white-dwarf stars only; the earliest type star is Sirius (A1V), and there is only one F-type star, two of G-type, four of K-type; the rest are M-type stars. All these are plotted on the HR diagram in Figure 12.29.

The list of the apparently brightest stars includes only three stars that are also among the nearest and nine other main-sequence stars; the rest are giants and supergiants. The volume of space that contains the twenty brightest stars is more than 500,000 times larger than the volume containing the twenty nearest stars. The vast majority of naked-eye stars are upper main-sequence objects, giants, or supergiants. High-luminosity stars make up for their rarity by being visible over long distances.

Given an accurately calibrated HR diagram, note how much information is implicitly contained in the spectral classification of a star: accurate indications of its color and temperature, and moderately precise measures of its intrinsic luminosity, radius, atmospheric density, distance, and mass. It is little wonder that the HR diagram is fundamental to stellar astronomy. We will refer to the HR diagram time and time again when we discuss the evolution of stars.

298

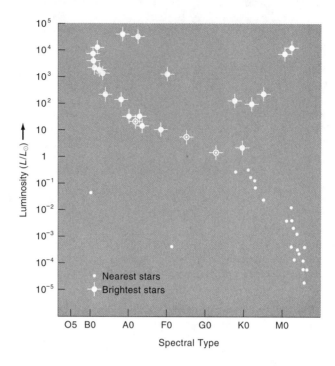

Figure 12.29. An HR diagram of the 25 nearst stars and the 25 brightest. Note that except for two white dwarfs the nearest stars are all on the main sequence and, except for Sirius, are all F-type or later. By contrast, nearly half of the brightest stars are above the main sequence. Only three stars are in both groups.

Terms to Know

Trigonometric parallax, proper motion, tangential velocity, space velocity, parsec, light year; apparent and intrinsic brightness; HR diagram, main-sequence stars, giants, supergiants, and white dwarfs; Hertzsprung gap; luminosity class; spectroscopic parallax. Open cluster, globular cluster; visual binary, spectroscopic binary, eclipsing binary, light curve, center of mass, mass ratio; mass-luminosity relation. Cepheid instability strip.

Ideas to Understand

Accuracy of trig parallaxes; method of similar objects. The significance and uses of the HR diagram; the ranges of stellar properties; how stellar masses can be measured; the mass-luminosity relation; how a spectroscopic parallax is derived.

Questions

1. (a) The trigonometric parallax of the bright star Altair is 0.20 seconds of arc. What is its distance in parsecs? in light years?

(b) Suppose the Earth's orbit were the size of Jupiter's (about 5 AU instead of 1 AU). What would Altair's parallax be then?

2. Antares, a bright red star in the summer sky, has a measured parallax of 0.02 seconds of arc. If the error in this measurement is ±0.005 arcsec, what can you say about the distance to this star from its parallax? (How near might Antares be? how far away? with what probability?)

3. (a) Two stars, *A* and *B*, appear to be equally bright, but star *A* is 25 times brighter intrinsically than star *B*. Which star is farther away? Why? How many times farther away?

(b) Suppose both stars move across the sky with velocities of 20 km/sec. Which star has the larger proper motion? Why? If the proper motion, μ, of star *B* is 1 arc second/year, how large is the proper motion of star *A*?

4. Could a star with a large proper motion have zero radial velocity? zero tangential velocity? zero space velocity? Explain your answers.

5. Vega, one of the first stars to have its parallax measured, has a parallax of 0.12 seconds of arc. How far away is it in parsecs? Its proper motion is 0.35 seconds of arc per year. What is its tangential velocity? Its radial velocity is 14 km/sec toward us. What is its space velocity? Draw a sketch showing the Sun, and the proper motion and all the velocities of Vega.

6. A certain star has the same spectrum as the Sun. Its apparent brightness is only 10^{-12} that of the Sun. How far away in parsecs is the star? What is the parallax in seconds of arc of this star? (There are 2×10^5 AU in one parsec.)

7. (a) The Sun has an effective temperature of about 5,700 K. At about what wavelength (in Ångströms) does the Sun radiate the most energy?

(b) The effective temperature of the supergiant Betelgeuse is about half that of the Sun's. At what wavelength does it radiate the maximum amount of energy? Compare the colors of the Sun and of Betelgeuse.

(c) Even though the Sun is hotter, Betelgeuse radiates about 10,000 times as much energy as the Sun. How can this be? Calculate the radius of Betelgeuse in terms of that of the Sun.

8. The two apparently brightest stars in the sky are the Sun and Sirius (neither is very bright intrinsically, but both—especially the Sun!—are nearby). Both are main-sequence stars, and the effective temperature of Sirius is about 10,000 K. The intrinsic luminosity of Sirius is about 25 times that of the Sun.

(a) Which star has the earlier spectral type? Why?

(b) Which star is bluer in color? Why?

(c) How many times more energy/cm² does Sirius radiate than the Sun? (For this part take the effective temperature of the Sun to be 5,000 K.)

(d) How many times larger in radius is Sirius than the Sun?

(e) What is the approximate spectral type and mass of Sirius? (Use the HR diagram given in this chapter.)

9. In this (and similar) exercises, first calculate the effect of the mass of the object on your weight, then the effect of the radius, then combine the two (see Appendix B).

If you weigh 100 pounds on the surface of the Earth, how much would you weigh on the "surface" of

(a) the Sun (mass of Sun = 3×10^5 Earth masses, radius of Sun = 100 Earth radii);

(b) a supergiant (mass = 5 solar masses, radius = 100 solar radii);

(c) a white dwarf (mass = 1 solar mass, radius = 1 Earth radius). Recall that $g = GM/R^2$; remember, take ratios!

(d) What do your answers have to do with pressure effects in the spectra of these stars?

10. Suppose the following stars all have the same apparent brightness. Which is farthest away? Which is nearest? Why?

(a) G2V (b) B3III (c) F0V (d) B0I (e) white dwarf

11. Three stars, a main-sequence M-type, a giant M-type, and a supergiant M-type, all have the same apparent brightness and the same proper motion. Explain briefly your answers to the following:

(a) Which star has the largest diameter?

(b) Are these stars bluer or redder than the Sun?

(c) Which star is nearest to us?

(d) Which star is moving with the greatest velocity (km/sec) across the sky?

12. (a) Why are the periods of visual binaries likely to be very long, many years, in fact, whereas spectroscopic binaries generally have short periods?

(b) Would you expect binary systems that are both visual and spectroscopic systems to be common or rare? Why?

13. A particular visual binary system consists of a red and a blue star; the orbital period is 4 years, and the semimajor axis of the orbit of one star with respect to the other is 4 AU. What is the sum of the masses of the two stars? If they are both on the main sequence, which is the more massive?

14. Refer to the radial velocity curve in Figure 12.28. Explain how you know that the star is contracting most rapidly at a, why it is smallest at b, why it is expanding most rapidly at c, and why it has reached its maximum diameter at d.

15. Binary system 1 consists of two B0 stars, system 2 consists of a B9 and an F8, and system 3 consists of two G5 stars. All six stars are on the main sequence. If the average distance between the two members of each pair is the same in all three binary systems, which system has the shortest period? Why?

16. A binary system consists of two G2V stars. What is the period of the system if the semimajor axis of the orbit of one about the other is

(a) 0.5 AU?

(b) 2 AU?

17. The sketch (Figure 12.30) shows an HR diagram with five stars indicated (points A–E) along with the main sequence.

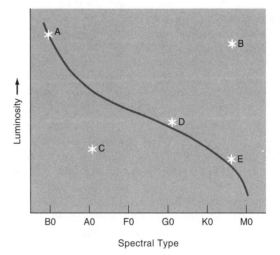

Figure 12.30. The HR diagram for Question 17.

Luminosity →

Spectral Type

B0   A0   F0   G0   K0   M0

   (a) The largest star is?
   (b) The smallest star is?
   (c) The hottest star is?
   (d) The star most like the Sun is?
   (e) The late-type supergiant is?
   (f) The star with strong lines of ionized helium is?
   (g) The star with strong lines of ionized calcium is?

18. The spectral types and luminosities (in terms of the Sun's luminosity) are given for each of the stars below. Identify each as belonging on the main-sequence, giant, supergiant, or white-dwarf branches.
   (a) Sirius (A1, $L=25$)
   (b) Arcturus (K2, $L=100$)
   (c) Polaris (F2, $L=6000$)
   (d) Spica (B1, $L=2000$)
   (e) Barnard's star (M5, $L=0.005$)
   (f) van Mannen's star (A, $L=0.01$)

19. A stellar spectrum is obtained in which there are prominent absorption lines of both ionized helium and molecules. How could this happen?

Suggestions for Further Reading

Getts, J., "Decoding the Hertzsprung-Russell Diagram," *Astronomy*, **11**, p. 16, 1983. History and interpretation of the HR diagram.
Gingerich, O., "The Search for Russell's Original Diagram," *Sky & Telescope*, **63**, p. 36, 1982.
Philip, A. and Green, L., "Henry Norris Russell and the H-R Diagram," *Sky & Telescope*, **55**, p. 306, 1978.
————"The H-R Diagram as an Astronomical Tool," *Sky & Telescope*, **55**, p. 395, 1978.

The following two articles by Kaler give a good overview of most of the kinds of stars on the HR diagram.

Kaler, J., "The Brightest Stars in the Galaxy," *Astronomy*, **19**, p. 30, May 1991.
————"The Faintest Stars," *Astronomy*, **19**, p. 26, August 1991.

Many particularly fascinating stars are described in the following:

Moore, P., *Astronomers' Stars*. London: Routledge and Kegan Paul, 1987.

"Astrophysics has changed all our illusions—in other words, our very soul."

—Anatole France (1844–1924); French novelist

". . . it is reasonable to hope that in a not too distant future we shall be competent to understand so simple a thing as a star."

"We do not argue with the critic who urges that the stars are not hot enough for this process (the conversion of hydrogen to helium); we tell him to go and find a hotter place.*"*

—Sir Arthur Eddington, 1926; a distinguished British astrophysicist

Only a few years later a development in the new quantum physics led to the understanding of the source of a star's energy.

CHAPTER 13

Stellar Evolution: Physical Ideas

Although significant work was done earlier, the modern beginnings of stellar astrophysics can be taken to begin (rather arbitrarily) around 1915, when the British astrophysicist A. S. Eddington began his theoretical studies of stellar interiors. That is, he used the laws of physics, including the newly emerging quantum theories of atoms and radiation, to deduce what the properties—temperature and density, for example—of the interior of a star must be like. He was able to do this even though he did not know what the source of a star's energy was, what causes it to shine.

By 1939, through the work of several scientists, but especially that of the German-American physicist H. A. Bethe, one of the major nuclear processes by which stars generate their enormous energy was identified. We will describe these processes in this chapter. After World War II, both observational and theoretical astrophysical studies flourished, especially when electronic computers became generally available in the 1950s. As we shall see, calculations of mathematical models that represent the physical conditions throughout a star are quite complicated and were extremely tedious to carry out using mechanical calculating machines. Electronic computers have made possible investigations of stellar processes and conditions that would otherwise have been impossible; they are responsible for much of our present understanding of stellar structure and evolution. (Like anything else, computers can be misused, however. One of the common ways this happens in science is that it is often easier—and more fun— to compute models than to think; as a consequence we may, perhaps, lose some physical understanding of what is going on, as well as waste computer time.)

With some knowledge of the physical properties of stars acquired in the last two chapters, we can tackle the question of how stars form, live, and die. In this chapter we will introduce the subject of stellar evolution by describing the physical factors that determine the structure of stars and how that structure changes with time.

General Considerations

How Long Do Stars Shine?

First of all, why must stars evolve? Why don't they remain unchanged forever? Because they emit enormous amounts of energy, the source of which must sooner or later run dry. Consequently they must eventually change. For example, it would take about 5 million years for the U.S. to generate enough electrical energy to equal the amount of energy that the Sun (not a particularly bright star) radiates in 1 second. Since the Sun's main-sequence lifetime is about 10 billion years, its energy production is incredible by Earth standards. An O-type star radiates energy at a rate 10,000 times greater than the Sun, but its mass, and hence its fuel supply, is only about 50 times greater. Clearly, its lifetime in that highly luminous state must be much shorter than the Sun's. Two factors are involved: the size of the fuel supply, and the rate at which it is being used up. Consider first the amount of fuel only; the greater the mass of a star, the larger its fuel supply, and hence the longer its life, t. We can write this mathematically as follows: $t \propto M$. Now consider the role played by luminosity; the greater the luminosity of a star, the faster it uses up its fuel supply, and so the shorter its life. That is, $t \propto 1/L$. In general, then, we can write that the lifetime of a star is proportional to M/L. In particular, the main-sequence lifetime of an O-type star compared to that of the Sun's is

$$ t \propto \frac{M_O/M_\odot}{L_O/L_\odot} = \frac{50}{10^4} = 5 \times 10^{-3} \text{ as long.} $$

On the other hand, an M-type dwarf, with a luminosity only 0.001 the Sun's and a mass of 0.1 solar masses, will have a lifetime a hundred times longer than that of the Sun. In general, the more massive the star (or the earlier the main-sequence spectral type), the shorter will be its main-sequence lifetime. This fact is expressed in the mass-luminosity relation, $L \propto M^{3.5}$, which shows that the energy radiated by main-sequence stars increases much more rapidly than their mass (and hence than their fuel supply). Double the mass of a star, and its luminosity increases by $2 \times 2 \times 2 \times \sqrt{2} = 11.3$ times. This gives one clue to why the numbers of stars increase so rapidly as you go down the main sequence: the lower-mass stars live longer.

We can combine the two equations above to get a convenient approximation for the main-sequence lifetime of a star as it depends on its mass. That is, $t \propto M/L$ and $L \propto M^{3.5}$, so $t \propto M/M^{3.5} = 1/M^{2.5}$. The main-sequence life of the Sun is about 10 billion years, so if the masses are expressed in solar masses, we can write

$$ t = \frac{10^{10}}{M_\odot^{2.5}} = \frac{10^{10}}{\sqrt{M_\odot}\, M_\odot^2} \text{ years.} $$

For example, a $25M_\odot$ star has a main-sequence life of about

$$ t = \frac{10^{10}}{\sqrt{25}\,(25)^2} = \frac{10^{10}}{5 \times 625} = 3.2 \times 10^6 \text{ years.} $$

What do observations tell us about stellar lifetimes? Our historical record of the sky is of little use here, since it goes back only about 5,000 years, and indicates that the bright, visible stars have remained unchanged over this period. Such a timescale is hopelessly short for purposes of stellar evolution. The Earth's biological and geological records are of much greater help. Analyses of fossil remains show that the temperature of the Earth has not changed drastically—the oceans have neither boiled nor frozen—for about the last 3 billion years, so the Sun cannot have changed much during that period of time. The geological record takes us even further back in time. By measuring the radioactive decay products of uranium in rocks, we can find the length of time since they solidified (see Chapter 23). The oldest Earth rocks so far discovered solidified about 3.9×10^9 years ago, nearly the same as the oldest Moon rocks returned to Earth by the Apollo astronauts. We will see later that the Sun and the planets formed at the same time. Hence it is safe to say that the Earth and the Sun must be at least 4 billion years old, a long time even by astronomical standards.

Incidentally, it is only in the past 100 years or so that time has become an important concern in astronomy. This was first prompted by the need to find energy sources capable of maintaining the Sun's lifetime for geological and biological time-scales, which were then recognized as being very long.

Stellar Energy Sources

The energy-generation process that could keep the Sun shining at about its present energy output for several billion years was long a puzzle. Chemical energy is completely inadequate. Were it composed entirely of coal along with enough oxygen to burn it, the Sun could produce its present energy output for only about 2,000 years. More exotic chemical fuels could not increase this by more than a few times. In the nineteenth century it was suggested that meteors—debris of the solar system—bombarding and heating the Sun could produce enough energy to power it. The amount of material falling in to release this much energy, however, would increase the Sun's mass by about 1 percent every 300,000 years. As a consequence, the Earth's orbital period would decrease by about two seconds per year, an easily observable effect. No such change is seen.

Another process suggested in the last century by the British physicist Lord Kelvin[1] is more promising. Recall that the temperature of a gas is just a measure of the kinetic energy of its particles. So, as a volume of gas such as a star radiates away energy, it cools, and as it cools its pressure decreases. As its pressure decreases it is no longer able to maintain its volume against its own gravity. Hence, the gas is compressed, which causes it to heat up, and so it radiates away more energy, which causes it to cool, and so on (see Figure 13.1). In essence, gravitational potential energy is converted to heat. This process, called **gravitational contraction**, would enable the Sun to shine at its present luminosity for about 20,000,000 years. During this time the Sun would shrink by about 20 meters every year. This is an appreciable span of time, but during the latter part of the nineteenth century, when it became clear that Darwin's theory of biological evolution required very long time intervals, and geologists realized that sedimentary deposits took a long time to accumulate, a timescale measured in only millions of years became unacceptable. Hence this process could not be the major source of energy for the Sun. As we shall see, however, there are phases in stars' lives during which gravitational contraction does play a significant role.

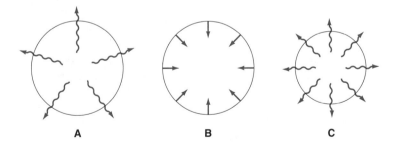

    A   B   C

Figure 13.1. Schematic representation of gravitational contraction. The star radiates away energy (A) and so cools. Consequently the interior pressure drops, and gravity, always acting, is able to compress the star (B), which then heats up and radiates away more energy (C), allowing the star to be further compressed, heated, etc.

---

[1] His mundane name was William Thomson (1824–1907). He is the Kelvin of the temperature scale.

The basic source of stellar energy remained a mystery until the early part of this century, when Einstein showed that one of the consequences of his special theory of relativity (see Chapter 19) was that energy and matter are two aspects of the same thing, that they are fundamentally equivalent. The energy equivalence, E, of matter, m, is given by Einstein's famous equation, $E = mc^2$, (c being the velocity of light). For the Sun to maintain its present energy output, $4.2 = 10^{12}$ grams of matter must disappear and be converted completely to energy each second.[2] Though by our standards this is a huge amount of matter, it is totally negligible compared with the mass of the Sun. This process could keep the Sun shining for many billions of years. By the 1920s many astronomers were convinced that the conversion of matter to energy must somehow power the Sun, but the detailed nuclear process by which this conversion took place remained unknown until 1939.

It is important to understand the fundamental difference between chemical and nuclear reactions. In essence, all of chemistry involves nothing more than various manifestations of the electromagnetic force between electrons and ions. Molecules are atoms bound together by ions or by their outer electrons, for example, by sharing electrons between atoms. Chemical reactions (such as oxidation or burning) occur as these electrons combine atoms into molecules different from those that began the reaction. Not much energy is involved in these rearrangements of the outer electrons; after all, most reactions can be strongly influenced by adding just a little energy, like the heat from a gas flame, to the test tube containing the reactants. Chemical reactions (including those producing violent explosions) are hopelessly inadequate in accounting for stellar energy.

Recall that because the protons and neutrons within a nucleus are roughly 100,000 times closer together than the electrons and nuclei of atoms, the electromagnetic force between the former is enormously greater than that between the latter. Yet the strong nuclear force is so powerful that it overcomes this repulsion and binds the protons and neutrons into stable nuclei. Consequently, processes resulting in changes of the particles within a nucleus will involve much more energy than that produced in chemical reactions—enough in fact to power the stars.

Studying the structure of stars presents some obvious difficulties, because the lifetimes of stars are incomparably longer than those of astronomers. We can't detect directly evolutionary changes in any individual main-sequence stars. Trying to discover how stars evolve by direct observation is roughly equivalent to a Martian attempting to understand the evolutionary relationships of people—babies, adolescents, adults, tall people, short people—with only a few seconds look at the multiracial population of a large city.

To make any progress, our observations must be guided, informed, and interpreted by **models of stars**. Clearly, we can't construct a real model of a star in the laboratory (that is, a massive hot ball of gas), but we can make mathematical, "paper" models. These are the results of lengthy calculations that give the run of temperature, density, energy generation rate, and chemical composition with depth in the star. (Since most stars are spherical, their properties vary only with the depth in the star.) These quantities are calculated in a way that is consistent with the relevant physical laws and processes at each point in the star. For example, at every point in the star, the gas law must be satisfied; that is, the relation between the pressure P, density N (the number of particles per unit volume), and temperature T *must* be $P = NkT$ (see the next section). Series of models can be computed that predict how stars will change with time.

---

[2] The mass equivalent of the Sun's luminosity of 3.8×10^{33} ergs/sec is $m = E/c^2$ or $m = (3.8 \times 10^{33})/(3 \times 10^{10})^2 = 4.2 \times 10^{12}$ grams/sec, or about 4.6 million tons of matter converted to energy each second!

We will see that these paper models can be related to and checked against observations of actual stars.

In general, mathematical models focus our attention on what we think are the important aspects of a particular physical situation, and highlight what we know and don't know. They can organize our knowledge (or ignorance!), and bring a degree of coherence to what would otherwise be a muddled situation.

To construct models of stars we must apply the appropriate laws of physics to large masses of gas. In doing this we are assuming, as always, that the laws of physics derived here on Earth are applicable everywhere. Fortunately, stars are relatively simple entities, and their general structure and evolution can be understood using only a few straightforward physical ideas, which we will now describe.

Understanding the Interior of a Star

The Forces on a Star

What forces act on a star? The hot gas of a star exerts a pressure that tends to expand it, whereas the mass of the star gravitationally attracts itself, which tends to compress the gas (see Figure 13.2). Ordinarily, these two forces are in exact balance at every point in the star; the star neither expands nor contracts, and we say it is in **mechanical equilibrium**. This is certainly true for the Sun, as well as for the vast majority of stars in the sky. A mathematical equation can be written that expresses this equilibrium condition. (Even pulsating stars such as Cepheid variables are essentially in mechanical equilibrium. Though they expand and contract periodically, averaged over a long time they do not change, because both the expansion and the contraction are limited.) Note that gravitational contraction, described earlier in this chapter, is a situation in which gravity wins; the star is not in equilibrium, but instead continues to contract.

The matter making up most stars is in a gaseous state, which is the simplest form of matter to describe mathematically. In a physics or chemistry class you may have encountered Charles' law or Boyle's law; these represent some aspects of the simple be-

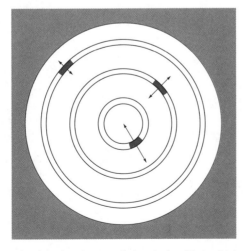

Figure 13.2. A star can be thought of as consisting (onion-like) of a large number of thin shells, only three of which are shown here. At every depth in the star, outward-acting pressure and inward-acting gravity are in exact balance. There is no net motion of mass inward or outward, and the star is in mechanical equilibrium.

havior of a gas. Here we will write the **gas law** in its more general form, $P = NkT$: P and T are the pressure and temperature, respectively; N is the number of gas particles—atoms, molecules, electrons—per cubic centimeter; and k is a number called Boltzmann's constant. It is a physical constant like c or G, independent of temperature, density, etc. The gas law is said to be an **equation of state**, because it relates the variables P, N, and T, which describe the state of the gas.

Consider a closed flask containing a given amount of gas. If you double the temperature of the gas, then the product NkT doubles, which means that the pressure doubles. If you put more gas into the flask, so that N increases, then again the pressure increases in direct proportion to the increase in N. What could be simpler?

Because of this simplicity, we know much more about the interiors of the distant stars than we do about the interiors of the nearby planets (probably including even the Earth). This is because stars are gaseous throughout, and the equation of state of a gas is much simpler than the corresponding formulae describing the behavior of the solids and liquids that compose planetary interiors.

Chemical Composition: Opacity

Next, we must know the chemical composition of the material of the star. This is important in three respects. First, the number of particles per unit volume, N, in the gas law (above) depends on the composition of the gas. If, for example, the gas is pure hydrogen, then each ionized atom contributes two particles, a proton and an electron; in a carbon gas, however, each fully ionized carbon atom contributes seven particles, six electrons and one nucleus. So we must know the composition in order to calculate N. Second, the composition is a factor in determining by what process and at what rate the star generates energy. Finally, it also is a factor in determining how this energy makes its way through the interior of the star to the surface where it is radiated away. Let us consider this latter aspect first.

In order that energy be driven out from the central region of the star, where it is generated, to the surface, from which it escapes, the center of the star must be hotter than the surface. This is simply an example of the law that heat can flow only from regions of higher temperature to regions of lower temperature. The heat flows at a rate determined by the temperature drop per centimeter distance outward. This temperature drop per centimeter is called the **temperature gradient**. The steeper the temperature gradient, the more heat will flow, other things being equal. Now, the temperature gradient required to drive out a given amount of heat energy depends on the resistance of the stellar material to that flow of energy. This resistance is called its **opacity**; this simply means how opaque the gas is or how easily photons pass through it. As its name implies, opacity is determined by how efficiently the material of the star absorbs radiation or scatters it (changes its direction), thereby preventing or impeding its escape to the surface. Thus opacity depends on the chemical composition and physical state of the star—what it is made of, and how hot and how dense it is. That is, the opacity depends on the effectiveness of the various processes by which radiation loses energy when it interacts with matter; these in turn are determined by the physical state of the matter.

This situation is analogous to the flow of water through a pipe, as shown in Figure 13.3. The rate of the flow depends on the pressure difference between the two ends of the pipe (corresponding to the temperature gradient in a star). It also depends on the resistance of the pipe to the flow of water, that is, the pipe's diameter and how smooth its walls are (corresponding to the opacity). The greater the pressure difference, the

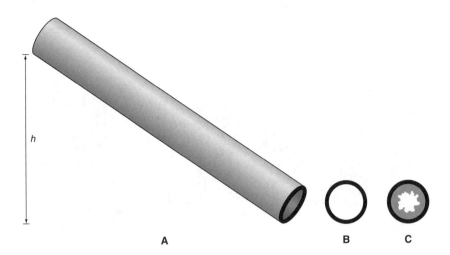

Figure 13.3.   (A) The steeper the pipe (the larger *h* is), the greater will be the flow rate of water through it. Also, a clean pipe (B) will allow a more rapid flow rate than one that has been encrusted with minerals (C).

faster the flow for a given diameter of pipe. The greater the temperature gradient, the more energy flows out for a given opacity.

The opacity of air to visible light is very small; we have no difficulty seeing each other. The water vapor contained in air is opaque to various wavelengths in the red and infrared, however, and oxygen absorbs ultraviolet radiation. Thus the opacity of a gas depends not only on its composition and physical state, but also on the wavelength of radiation passing through the gas.

Why is the edge (the limb) of a star like the Sun so sharp and well defined (see Figure 11.18)? After all, the Sun is a sphere of gas, not a solid, and we might expect to see it slowly fade out at increasing distance from its center. That the limb is so pronounced is a consequence of the high opacity of the gas in the outer layers of the Sun. If air were as opaque, per gram, as the gas in the outer layers of the Sun, we would be able to see only a few meters, as in a very heavy fog. Because of this large opacity, photons from deep within the solar atmosphere cannot travel very far before they are absorbed or scattered. They don't have a direct path to the surface. The photons that can make a straight-line path to the surface, that is, the photons by which we see the Sun, can escape only from the very outermost layers, just a few hundred kilometers thick. This is an insignificant fraction of the Sun's radius, and so the Sun's limb is sharply defined.

Convection.   When the opacity of the stellar material is small, heat flows outward with little difficulty. The energy is carried to the surface by radiation, by photons. The larger the opacity is, however, the larger is the temperature gradient required to drive a given amount of heat energy out. If the temperature gradient becomes large enough, a blob of gas accidentally pushed upward slightly might find itself surrounded by appreciably colder, denser gas; so it would continue to rise, just as a piece of wood, less dense than water, rises toward the surface of a pond. The warm blob of gas would rise until it reached a region of the star where the opacity is low and radiation—photons—can again carry off the energy contained within the blob. The star as a whole is still

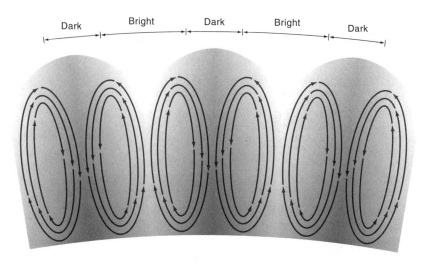

Dark　　　　Bright　　　　Dark　　　　Bright　　　　Dark

Figure 13.4. Warmer gas rises up through the star until it cools and falls back. As much gas rises as falls, so there is no net change in the structure of the star. Despite the motion, the star is still in mechanical equilibrium. When convection reaches the top of the photosphere, as in the Sun, the rising gas is seen as bright granules, and the cooler gas falling inward is seen as the darker intergranular region (see Figure 11.11).

in equilibrium, neither expanding nor contracting. Hence, cooler blobs, having lost much of their energy in their outward journey, must sink back toward the center of the star at a rate that just balances the outward-moving hotter blobs. In this way circulation currents are set up in the star, hot matter rising, carrying energy outward, and cool matter sinking (see Figure 13.4); this process is called **convection**. In the situation just described, convection occurs because the opacity is so high (and so the temperature gradient required to drive radiation out becomes so large) that mass motions become a more effective method of transporting energy than radiation. (Consider a pot of oatmeal on a stove: when the heat is low, the temperature gradient is small and no convection takes place; raise the temperature of the burner, and the oatmeal "bubbles" by convection.)

A somewhat more general way of looking at convection is to note that when there is too much energy to be carried through a zone of a star by radiation, by photons, convection will take place. As we have just described, such a situation occurs when the opacity in a particular layer of a star is very large. Convection also takes place when the energy-generation rate is so large that radiation simply cannot carry all the energy out. The outer layers of a red giant star are convective because the opacity there is very large. The centers of massive stars are convective because too much energy is generated there for it to be transported by radiation. That a star is convective in some zone does not necessarily mean it is convective throughout, however. In the inner regions of the red giant and in the outer layers of a massive star, energy is carried by radiation (see Figure 13.5).

Another important aspect of convection is that it thoroughly mixes the stellar material in the convective region. By contrast, layers in which the energy is carried by radiation are relatively stable and do not get stirred up. As a consequence, systematic nonuniform chemical compositions, that is, compositions that change with depth in the star, or **composition gradients**, can result in these layers. As we shall see, this effect also has important consequences for the structure of a star.

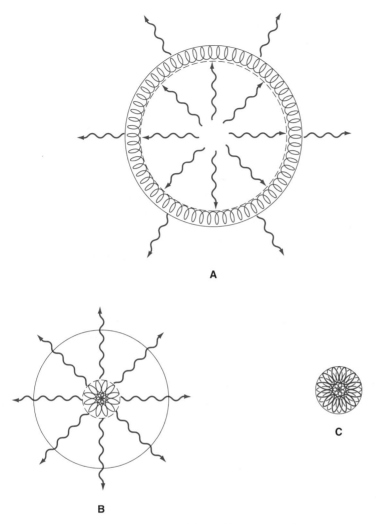

Figure 13.5. How energy is carried in different kinds of stars (not drawn to scale). A red giant (A) is convective in its outer regions, whereas an upper main-sequence star (B) is convective in its core, but radiative outside of it. Stars much less massive than the Sun are convective throughout (C). However, energy is always ultimately carried out of the star by the radiation that enables us to see it.

Though we understand the physical principles underlying convection, relating these to the detailed structure of a star has proven to be extremely difficult. Indeed, a good theoretical description of how convective energy is transported is probably the major deficiency in our understanding of stellar structure.

In addition to radiation and convection, there is a third way by which energy can be transported in a star, namely, by **conduction**. This is familiar to all of us; for example, it is the means by which the handle of a metal spoon is heated in a hot cup of coffee. What happens is that atoms of the spoon that are at higher temperatures collide with their cooler (and hence more slowly moving) neighbors, thereby transfering some of the energy of motion to them. Although conduction does not play a role in the transport of heat in ordinary stellar matter, we will see that it becomes important when the matter reaches high density, as in white dwarf stars, for example.

Chemical Composition: Energy Generation

The other way in which the star's composition affects its structure is through the series of nuclear reactions by which matter is converted to energy in its core. Obviously, if a particular kind of nucleus necessary for a reaction is not present, that reaction will not take place. Let us examine the most important of these energy-producing nuclear reactions.

Energy is generated in stars primarily through a series of nuclear reactions, the most important of which involves the transformation of four nuclei of hydrogen into one helium nucleus. The total mass of the product of the reaction (the helium) is slightly less (by 0.7 percent) than the mass of the particles (the protons) entering the reaction. This small mass is released as energy according to Einstein's relation $E = mc^2$. Even though the vast majority (99.3 percent) of the matter in these reactions is transformed from hydrogen into helium rather than converted to energy, the process provides an ample energy supply for stars.

What happens in this reaction is that the helium nucleus resulting from the combination of the protons is in a lower-energy state—more tightly bound—than are the individual particles, which originally weren't bound together at all. The energy holding any nucleus together is called its **binding energy**. The **mass defect** of a nucleus is the small mass difference between the components of a nucleus taken separately and the resulting nucleus taken as a whole. The energy binding the nucleus is just the energy equivalent of the mass defect; in the reaction just described, the binding energy of a helium nucleus is its mass defect in the amount given by $E = mc^2$. Interestingly, the relation of the mass defect to the binding energy was one of the first experimental demonstrations of Einstein's prediction of the equivalence of mass and energy.

The particles taking part in these reactions are the positively charged nuclei of hydrogen and helium. In order to interact, that is, for the very short range (about 10^{-13} cm) nuclear force to take hold, these particles must come extremely close to each other. But, since the interacting particles are both positively charged and like charges repel, this does not occur easily. The two nuclei must have sufficient energy to overcome the repulsive electrical force between them. This means that the rate at which they interact depends critically on their energy of motion, that is, on their temperature. Only the central cores of stars have the required high temperatures.

In any case, the net result of the process is

$$4\,^1\text{H} \rightarrow\,^4\text{He} + 2e^+ + \text{energy},$$

where e^+ denotes a positively charged electron, or positron. (Recall that this is the antimatter analog of the electron; see Chapter 9). The electrons surrounding the nuclei have been stripped away (the atoms are completely ionized) because the density and temperature at the centers of stars are very high (about 150 times that of water and 15,000,000 K, respectively, in the Sun). The above reaction occurs in stages in which the reacting particles interact with each other only two at a time. This is so because even at stellar densities, nuclei are very far apart compared with their sizes. Hence it is unlikely that at any instant more than two particles could approach each other closely enough to interact.

Proton-Proton Chain.   Let's look at the energy generation processes in a little more detail. Two series of reactions convert hydrogen to helium in stars: the **proton-proton**

chain and the **carbon** (or **CNO**) **cycle**. First the former:

$$(1) \qquad ^1\text{H} + {}^1\text{H} \rightarrow {}^2\text{H} + e^+ + \gamma + \nu$$

$$(2) \qquad ^1\text{H} + {}^2\text{H} \rightarrow {}^3\text{He} + \gamma$$

$$(3) \qquad ^3\text{He} + {}^3\text{He} \rightarrow {}^4\text{He} + {}^1\text{H} + {}^1\text{H}$$

Remember that when we write the symbol for a chemical element in the nuclear reactions above, we mean the nucleus of that element, because the atoms are completely ionized. For hydrogen, this means a proton; for helium it means a nucleus of two protons and two neutrons, also called an **alpha** (α) **particle**. ^{2}H, consisting of a proton and a neutron (n), is an isotope of hydrogen called **deuterium**; ^{3}He is an isotope of helium. An ordinary electron is denoted by e^-. By comparison with ordinary electrons, positrons (e^+) are rare in nature. Furthermore, when a positron and electron encounter each other, both are very quickly annihilated and completely converted to pure energy. Thus, the e^+ in (1) contributes to the total energy released. **Neutrinos** are denoted by ν (the Greek letter *nu*). Neutrinos are weird particles that have no charge, and probably no mass, but carry momentum and energy. In contrast with the protons, electrons, and neutrons, neutrinos hardly react with matter; the overwhelming fraction of them pass right through the Sun and the Earth as if these objects weren't there. Countless numbers zip through us each second. (In fact, it has been estimated that each cubic centimeter of the universe contains about 150 neutrinos.) Since they don't interact with the atoms and molecules of our bodies, they have no effect whatsoever on us. Finally, recall that a γ-ray is just a very high frequency photon. It carries most of the energy released in these reactions.

You should note several things in the three steps of the proton-proton chain given above.

(a) The first step of this process demonstrates the action of three of the four fundamental forces of nature (recall the discussion in Chapter 9). To interact, the hydrogen nuclei must have sufficient energy to overcome the electromagnetic force of repulsion between them. The strong nuclear force binds together a proton and neutron to form deuterium, the product of the interaction. Before that happens, however, one of the protons must be converted to a neutron by means of the weak interaction. Of the basic forces, only gravity plays no direct role in these interactions; it does, however, create the conditions (the hot stellar core) necessary for the reactions to occur.

(b) Electrical charges must balance. For example, in (1) there are two positive charges on the left-hand side (the two protons) and two on the right (^{1}H and e^+).

(c) Furthermore, in a gross way, the masses must balance. In (1) there are two units of mass on the left (^{1}H + ^{1}H) and two on the right (^{1}H + n). Remember that the electron has only about 1/1835 the mass of the proton or neutron, and so can be neglected when we use only whole numbers to represent the masses. To calculate the amount of mass lost to energy, however, we must use the exact masses of the reacting particles.

(d) Since reactions (1) and (2) together produce one ^{3}He, and (3) requires two ^{3}He's, reactions (1) and (2) must take place twice for each reaction (3). Thus, six protons enter into the proton-proton chain, and one alpha particle and two protons result. This gives a net conversion of four protons to one alpha particle plus two positive electrons (because charges must balance).

(e) Finally, you should realize that the above reactions are not equally likely to occur; that is, they do not all take place at the same rate. At a given temperature, reaction (2) will proceed at a rate nearly 10^{17} times faster than that of reaction (1)!

Reaction (1) sets the overall rate for the conversion of hydrogen to helium by the proton-proton chain. This step of the process proceeds at such a glacial pace because it involves not only the strong force, but also the weak nuclear force, which acts relatively slowly. (Recall that the decay of the neutron through the weak interaction requires ten minutes, an eternity on the timescales of the strong interactions.) That is, the slow weak interaction must take place during the very short time that two protons are flying close by each other, a most unlikely occurrence.

How much energy is liberated in this set of reactions? The mass of the four protons (6.6904×10^{-24}) grams is slightly greater than the mass of the products (6.6421×10^{-24}) grams. The difference or mass defect, 0.0483×10^{-24} grams, disappears as mass and is converted to energy. (About 2 percent of this energy is carried away by neutrinos.) Note that this represents only 0.0073 of the mass of the four protons, so that this process is only 0.7 percent efficient. Even the conversion of this tiny mass, however, produces a substantial amount of energy,

$$E = mc^2 = (0.0483 \times 10^{-24})(3 \times 10^{10})^2 = 4.3 \times 10^{-5} \text{ ergs,}$$

which is the energy of ten million blue photons. After about five billion years of producing energy through this process, the Sun's mass has decreased by only a few hundredths of 1 percent.

The CNO Cycle.   The other primary process by which hydrogen is converted to helium, the carbon cycle, produces energy in fundamentally the same way. A small amount of matter disappears to energy with essentially the same efficiency as in the proton-proton chain (here 6 percent of the energy is lost to neutrinos). However, though carbon takes part in the reaction, it appears again at the end, able to participate in another round of reactions; it acts like a catalyst. The steps of the cycle are as follows:

$$^{12}\text{C} + {}^{1}\text{H} \rightarrow {}^{13}\text{N} + \gamma$$

$$^{13}\text{N} \rightarrow {}^{13}\text{C} + e^+ + \nu$$

$$^{13}\text{C} + {}^{1}\text{H} \rightarrow {}^{14}\text{N} + \gamma$$

$$^{14}\text{N} + {}^{1}\text{H} \rightarrow {}^{15}\text{O} + \gamma$$

$$^{15}\text{O} \rightarrow {}^{15}\text{N} + e^+ + \nu$$

$$^{15}\text{N} + {}^{1}\text{H} \rightarrow {}^{12}\text{C} + {}^{4}\text{He}$$

Note that ^{13}N and ^{15}O are naturally unstable, and decay in 15 and 3 minutes, respectively, by emitting a neutrino through the weak interaction. After this time, the decay products, ^{13}C and ^{15}N, capture a proton and carry the process forward. In contrast to the situation in the proton-proton chain, the weak interaction in the carbon cycle does not slow the overall process.

Obviously, if the star had no carbon, the CNO cycle could not take place. Even the small fraction of carbon in most stars, however (where, you recall, the abundances by mass are about 75 percent H, 23 percent He, and about 2 percent everything else, including carbon), is sufficient to make the carbon cycle work. The rate at which this process takes place depends on a very high power of the temperature, about T^{18}, whereas the corresponding rate for the proton-proton chain goes at about T^4.

The Core's Safety Valve.   Now we can understand why stars are so constant, so stable, generally changing their properties only very slowly with time. The reason is that the core, the center, of a star where the star's energy is produced acts like a **safety valve**. For example, suppose that for some reason a star's energy-generation rate increased: the temperature and pressure of the core would increase and, according to the gas law, it would expand. But this expansion would cool the core, causing the energy-generation rate to decrease. As a consequence, the temperature and pressure would decrease, and the core would contract to its proper equilibrium value. This safety valve is very sensitive, because the energy-generation rate depends on high powers of the temperature; a small change in temperature produces a huge change in the rate of energy production. For example, a 1 percent change in the temperature can result in a 20 percent change in energy production by the CNO process. In nearly all phases of its life, a star's equilibrium is maintained by this mechanism. We shall soon see, however, some special instances when this safety valve fails.

P-P or CNO?   What determines which reaction, the proton-proton chain or the CNO cycle, will be the dominant process in a given star (assuming, of course, the presence of carbon in the star)? The decisive factor is the temperature, as can be seen in the following way. The nuclear force is strongly attractive, but it acts over very short distances only (about 10^{-13} cm). In order to interact, therefore, the nuclear particles must approach each other very closely. But it is hard for the two particles (for example, the two protons in the first step of the p-p chain) to get close enough together, because their positive charges strongly repel each other. The ability of the two nuclei to get close enough together so that they can react, however, increases enormously with only a modest increase in their energy. For two protons to interact, the temperature must be high, at least 10,000,000 K.

Now, the temperature required for the first step of the carbon cycle is higher than that required for any step of the proton-proton chain, because a carbon nucleus has six positive charges, and the repulsive force is correspondingly larger than that associated with only two protons. Therefore in stars with relatively cool cores the carbon cycle will not take place, but the proton-proton chain will.

It so happens that the Sun is about at the dividing line, producing about 85 percent of its energy by the proton-proton chain, the rest by the carbon cycle. Main-sequence stars hotter than the Sun generate their energy primarily by the carbon cycle; the Sun and main-sequence stars cooler than the Sun produce energy primarily by the proton-proton chain. It is important to keep in mind that the high temperatures we are talking about are those prevailing at and near the center of stars. Only there are temperatures and densities great enough to support the conversion of hydrogen to helium.

For about 80 to 90 percent of their energy-producing lives, stars are powered by the conversion of hydrogen to helium, for two reasons. First, hydrogen is by far the most abundant element in the universe, so stars have the most of this kind of fuel. Second, the energy released in hydrogen burning is large. This can be seen in Figure 13.6, where the binding energy of a nucleus divided by its mass (that is, the average binding energy per unit mass of a nucleus) is plotted against the nuclear mass. Notice that in going from a mass number of 1 to mass 4 (hydrogen to helium), the binding energy increases especially rapidly. This means that the mass defect is large, and a correspondingly large amount of energy is released in the process. No other jump in the binding-energy curve is so large; that is, no other fusion of lighter to heavier nuclei produces so much energy.

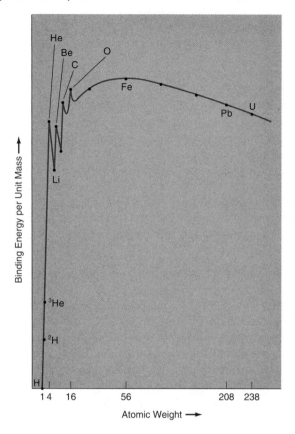

Figure 13.6.   The binding energy per unit mass of a nucleus plotted against atomic weight. Energy will be released as nuclei are transformed to heavier and heavier species until the peak of the curve is reached. Note that ^{56}Fe is the most strongly bound nucleus, so that reactions involving iron require energy rather than release it. Reactions transmuting H to ^{4}He release more than three-quarters of the total energy available, which is why a star lives the largest fraction of its life while converting hydrogen to helium.

The conversion of hydrogen to helium is the energy source for main-sequence stars. In fact, we will see in Chapter 15 that the main sequence is simply the position of stars in the HR diagram at which core hydrogen burning occurs. Thus the longest part of a star's life is lived on the main sequence, and this is why most stars are found there on the HR diagram.

Limits to Stellar Masses.   We now can understand why there are limits to masses of stars, why 1,000-solar-mass objects or 0.001-solar-mass objects cannot become stars. As we pointed out earlier, most stars are in mechanical equilibrium between their own gravity and gas pressure. In very massive (and, therefore, luminous) stars, however, **radiation pressure** becomes significant. Radiation pressure is the pressure exerted by photons as they interact with matter. In most circumstances radiation pressure is small, but since it depends on the fourth power of the temperature, it can add substantially to the gas pressure in the hottest and most luminous stars. At some value of the luminosity, the star's total pressure (radiation plus gas) becomes large enough to overcome gravity and the star becomes unstable and starts to lose mass. It might seem that increasing the star's mass to increase its gravitational force could make the star stable again. Instead, that just makes the situation worse. Remember that for main sequence stars, $L \propto M^{3.5}$, and so doubling the mass means increasing the luminosity by about 10 to 15 times. As a consequence of this large increase in the number of energetic photons, the radiation pressure becomes even more important and easily overcomes the additional gravitational force; the star becomes even more unstable. From considerations of this sort, it appears that the maximum mass a star can have is about 200 solar masses. Not surprisingly, such objects are extremely rare.

We can also see why main-sequence stars cannot form with arbitrarily small masses. A blob of gas less than about 0.08 solar masses is just not massive enough for gravity to squeeze its core to the high densities and pressures required to trigger the fusion of hydrogen to helium. Jupiter, at 0.001 solar masses, is far below this limit. There is no reason why objects with less than 0.08 solar masses can't form and gravitationally pull themselves together, of course, but they won't "burn" hydrogen; they won't become stars.

A small amount of nuclear energy is produced well before a star's core becomes hot enough for the fusion of hydrogen to helium to occur. When the core temperature of a contracting star reaches about 1,000,000 degrees, several nuclear reactions take place involving the light elements, which are destroyed in the process. In particular, protons interact with deuterium (one proton and one neutron) as well as the nuclei of lithium, beryllium, and boron, converting them all to helium. In these reactions, protons are able to overcome the repulsive force exerted by as many as five protons (in boron). We have seen that it takes a temperature of ten million degrees for two protons to interact in the first step of the proton-proton chain, so it seems odd that reactions involving protons and the light elements can occur at temperatures of only a few million degrees. It is instructive to understand why this is so. Recall that the first step of the proton-proton chain requires the emission of a positron. This process is governed by the weak nuclear force, in which interactions are inherently improbable and so proceed at a slow rate unless the temperature is high. By contrast, the nuclear burning of the light elements mentioned above (lithium, for example) involves only the strong nuclear force and so can proceed rapidly even at relatively low temperatures. These reactions liberate little energy and so play no significant role in the evolution of ordinary stars; they simply slow the contraction slightly.

In low-mass objects whose interiors will never become hot enough for hydrogen burning to occur, however, these reactions provide the only means (other than gravitational contraction) by which they can produce energy. For a short time such objects will glow faintly with a dull red color. Rather quickly they will exhaust their meager energy supply, cool, and simply fade away.

Nuclear Bombs. The conversion of hydrogen to helium is what goes on in a hydrogen or "fusion" bomb; hydrogen is fused into helium (actually, more reactive isotopes of hydrogen—deuterium and tritium [^{3}H]—are used). This is quite different from what happens in an atomic or "fission" bomb, in which heavy elements, such as ^{235}U, are split (fissioned) into lighter ones such as ^{137}Ba, with the release of much energy, though less than in the much more powerful fusion bombs. A hydrogen bomb is made by surrounding the fusion material with fissionable elements. When the latter explode, the temperature is raised sufficiently for the fusion of hydrogen to helium to take place, releasing devastating amounts of energy.

We can understand where the fission energy comes from by referring again to the binding-energy curve in Figure 13.6. When uranium is split into lighter nuclei, the binding energy per nucleon (per nuclear particle) increases—the products of the reaction are higher on the binding-energy curve than is the original nucleus. The increase, however, is not nearly as much as it is in a hydrogen bomb, where the change in binding energy is very large.

Nuclear Power.   Nuclear-power stations generate their energy through the fission process. Many countries, including the U.S., are spending large amounts of money to develop techniques so that the fusion process could be used for power generation. An advantage of fusion power is that the fuel it would use, deuterium (one proton and one neutron), is much more plentiful than the uranium or plutonium used in fission reactors. However, keeping large amounts of matter at temperatures of ten million degrees for long periods of time so that the fusion process can occur continuously and in a controlled manner is no small trick. It has not yet been adequately demonstrated in the laboratory, much less in a working power plant. Note that a fusion power source is quite different from a fusion bomb, where the idea is to fuse hydrogen to helium as quickly as possible to produce a very rapid, one-time release of energy. Let's return now to the controlled fusion process in stars.

Energy Generation after Hydrogen Burning.   When most of the hydrogen in the core of a star has been converted to helium, the hydrogen-to-helium energy-generation rate will fall off. Without sufficient energy from hydrogen "burning" to maintain itself against the ever-present inward force of gravity, the core must contract. This gravitational contraction produces energy, part of which escapes to the surface, where it is radiated away, and part of which remains in the core, increasing the core temperature. The larger the star's mass is, the greater will be the core compression by the weight of the overlying matter, and the greater the consequent increase in temperature. If the compression is sufficient to cause the core temperature and density to reach about 100,000,000 K and about 1,000 gm/cm^3, respectively, then helium "burning"—the fusion of helium to carbon—will begin. The high temperature is required in order that the strong electrical force of repulsion between the alpha particles can be overcome. The reactions are

$$^4\text{He} + {}^4\text{He} \rightarrow {}^8\text{Be},$$

followed by

$$^4\text{He} + {}^8\text{Be} \rightarrow {}^{12}\text{C}.$$

Schematically the process can be represented as

$$3\,^4\text{He} \rightarrow {}^{12}\text{C} + \text{energy}.$$

Not surprisingly, this is called the **triple-alpha process**. See what's happening: the proton-proton chain or the carbon cycle converts hydrogen to helium until the core hydrogen is used up. Then, if by the subsequent gravitational contraction the core temperature becomes hot enough, the triple-alpha process converts some of the helium to carbon. The ash of one process has become the fuel for the next.

Incidentally, the beryllium produced in the first step of this process is unstable and will decay back into two helium nuclei in the very short time of about 10^{-16} seconds. Unless the beryllium nucleus captures an alpha particle during this fleeting instant, the triple-alpha process will not go to completion and no energy will be produced. On the average only a tiny fraction of the core material will consist of beryllium. In fact, when the triple-alpha process is operating, there is one beryllium nucleus for about a billion helium nuclei. Nonetheless, this is sufficient to support the conversion of helium to carbon if the core density is very large, since then the particles are close enough together that the relatively few beryllium nuclei have a chance to interact with alpha particles.

Energy is produced by the triple-alpha process because the mass of the three alpha particles is slightly greater than that of the carbon nucleus; it is this mass difference that is converted to energy. As you can see from Figure 13.6, going from helium to carbon increases the binding energy, but by a small amount compared to the fusion of hydrogen to helium. The triple-alpha process produces only a little more than one-fourth as much energy as the hydrogen-to-helium reaction, but by it a star like the Sun will produce about 10 percent of the total amount of energy emitted during its lifetime. Subsequent reactions produce even less energy, because the binding-energy curve flattens out.

When in its turn the helium in the hot core begins to be depleted, the rate at which the triple-alpha process proceeds will decrease. If the core is sufficiently massive, gravitational contraction will again raise the temperature until it becomes hot enough for the next reactions to take place. At a temperature of about 2×10^8 K, a series of alpha-particle captures begins with **carbon burning**:

$$^{12}C + {}^4He \rightarrow {}^{16}O + \gamma.$$

Subsequent captures of alpha particles produce ^{20}Ne and ^{24}Mg.

Other carbon-burning reactions can occur at temperatures of about 6×10^8 K, for example:

$$^{12}C + {}^{12}C \rightarrow {}^{23}Na + {}^1H, \text{ or}$$

$$\rightarrow {}^{20}Ne + \alpha, \text{ or}$$

$$\rightarrow {}^{24}Mg + \gamma.$$

If temperatures of about 1.5×10^9 K are reached, **oxygen-burning** reactions occur, such as

$$^{16}O + {}^{16}O \rightarrow {}^{32}S + \gamma, \text{ or}$$

$$\rightarrow {}^{31}P + {}^1H, \text{ or}$$

$$\rightarrow {}^{28}Si + \alpha.$$

At higher temperatures the reactions become even more complex. Some of the previously formed nuclei are broken down into lighter nuclei, along with protons, neutrons, and alpha particles. These latter particles then take part in reactions to produce still heavier nuclei, for example, the iron group (iron, nickel, chromium, manganese, and cobalt). The number of possible reactions is very large indeed.

Now, since iron is at the peak of the binding-energy curve, interactions involving iron cannot produce energy, since the product of such reactions must have a lower binding energy per nucleon than iron. These reactions require energy rather than produce it. So the iron group is the end of the line for energy production, in that it cannot serve as a nuclear fuel. At this point (and often well before) even a massive star is in trouble. We will see what happens in such situations in Chapter 16.

It is important to remember that the vast majority of stars never get beyond the hydrogen- or helium-burning phases. Their masses are just too small to compress their cores enough to raise them to the high temperatures required by the later reactions.

Stellar Models

How do we put all these ideas together to construct a mathematical model of a star? We begin with an assumed chemical composition and mass of the star we want to model. We use equations that represent mathematically the general characteristics of a stable star:

(1) the condition of mechanical equilibrium, that is, that the star is neither significantly expanding nor contracting;
(2) the condition of thermal (temperature) equilibrium, that is, that as much energy escapes from the star as it generates, so that no part of the star heats up or cools down; and
(3) the temperature gradient—the rate at which temperature drops with distance from the center of the star—which depends on the mode of energy transport (by radiation or convection).

Note that these are rather general: for instance, nothing has been said about any particular energy-generation process. So we need information, in mathematical form, that describes the behavior of stellar matter:

(1) the perfect gas law, since the stellar material is gaseous throughout;
(2) how opaque to the transport of radiation the stellar material is; and
(3) the rate at which energy is generated (the luminosity) by the processes appropriate to the conditions in the interior of the star.

Recall that these last three all depend on the chemical composition of the star.

To construct the model, we imagine the star to be divided into a large number of thin layers, like an onion. Starting at the center of the star, we guess what the values of temperature, pressure, etc., are there and, using the mathematical relations described above, we calculate what the physical properties of the next layer out must be. With these quantities, we compute the properties of the next layer. This process is continued until we reach the surface of the star, that is, the radius at which the pressure becomes zero. This results in a table of values of pressure, temperature, density, and energy production in a large number of shells from the center of the star to its surface. At each depth in the star all these values must satisfy (that is, be consistent with) the mathematical relations we have just listed. If they aren't, the model must be adjusted until a consistent solution throughout the star is established.

The model is checked by comparing its mass, radius, and luminosity with those of a real star, and if all observable parameters agree, we have reason to believe that the model is at least a good approximation to a real star. In particular, the run of temperature, density, pressure, etc., with depth, which we can't observe directly, must be about right. The observable quantities predicted by the best stellar models do in fact agree so well with observations that the view they provide us of the interiors of stars must be very nearly correct.

Models are important in science. Ideally, one can derive their physical properties in detail and make exact statements or predictions concerning their characteristics or behavior. In a sense, models are experiments done mathematically with a computer rather than in the laboratory. If the model of, say, the interior of a star predicts various parameters (the effective temperature and radius, for example) that agree with the observed quantities, then the properties of the model—the interior—can be ascribed to the real star. In this way we get information that would otherwise be unknowable. Unfortunately, things rarely work out that neatly. The difficulty is that most models are idealizations, simplifications of the real situation, because the latter is too complicated to be described in every detail. Thus most models are at best approximations to the real physical circumstances, and their results and predictions are, therefore, also approximate. The crucial questions are, how well does a model represent reality? Have all the important features been properly modeled? Usually it is questions of this sort that lie at the root of arguments among scientists concerning, for example, the safety of nuclear-power plants or the exact threat to the Earth's climate posed by the greenhouse effect and the rate at which it might occur. For now these situations are beyond our ability to model precisely; hence predictions of what might happen inevitably turn on the adequacy of the imperfect model being used to generate those predictions.

A Model of the Sun's Interior. Let us look at a stellar model (given in Table 13.1) meant to represent the present-day Sun. (The "Fraction of radius" of 0.00 means the center of the Sun; it is 1 at the Sun's surface.) Notice how the energy generation is concentrated toward the center; nearly all the Sun's luminosity is created within a volume

Table 13.1. A model of the interior of the Sun

Fraction of radius	T(K) (millions)	Density (gm/cm^{-3})	Fraction of luminosity	Fraction of H (by wt)	Fraction of mass
0.00	15.0	148	0.00	0.38	0.00
0.10	12.5	86	0.45	0.59	0.08
0.20	9.0	36	0.94	0.71	0.33
0.40	4.9	4	1.00	0.73	0.79
0.60	3.1	0.5	1.00	0.73	0.95
0.80	0.8	0.02	1.00	0.73	1.00
1.00	0.006	3×10^{-7}	1.00	0.73	1.00

having a radius of only 20 percent of the Sun; this means that only the central 1 percent of the Sun's total volume is directly involved in energy production, because the high temperature and density required for the fusion of hydrogen to helium exist only at and very near the center of the star.

Even though the density of solar material at the center is nearly 150 times that of water (more than 11 times the density of lead!), the matter is still a gas because its temperature is so high. Finally, note how the hydrogen content in the core of the model has been depleted by its conversion to helium. The outer part of the Sun, however, is unaffected and retains the chemical composition with which it was formed and that originally made up the whole star (fraction of hydrogen by weight of 0.73), because during the main-sequence phase of a star's life, matter in the core does not in general mix with the matter outside the core. It is the slow change in the chemical composition of the core that is responsible for the gradual change in the structure of a star and hence its evolution.

In general, none of the products of nucleosynthesis is mixed up to the surface until the very advanced stages of evolution. Consequently, the surface layers of most stars have the same chemical composition with which they were formed. It is this happy circumstance that has enabled us to discover how the composition of matter from which stars formed has changed over the age of the galaxy.

Evolving Models.   Most of this discussion has had to do with stars early in their lives, but stars use up their fuel and slowly change with time. To calculate a series of models representing the evolution of a star over time, we begin with a model computed as described above. Since its energy-generation rate (its luminosity) is known, we can calculate how its core chemical composition changes after a certain number of years; that is, how much hydrogen is converted to helium. This time should be short compared to the time required for any major changes to take place in the star's structure; for a star like the Sun, this may be about 300,000,000 years. We then calculate a new model with a core that now is helium-enriched at the expense of hydrogen as a result of 300,000,000 years of hydrogen burning. This model will have a slightly different luminosity, density, and temperature distribution with depth than the first model. We next calculate a third model based on the second one, but with new core abundances to reflect the additional helium enrichment after another time step of a few hundred million years or so, and so on.

Because computer time is expensive, we want to make the time steps from one model to the next as large as possible. Time steps that are too large, however, may fail to represent important changes in the star and thus give quite misleading results. Generally speaking, a star evolves more and more rapidly as it becomes older, and hence more models are required to represent its evolution. In some particularly rapid phases, for example, during the so-called helium flash or during a supernova outburst (see Chapter 16), conditions change so rapidly that models spaced only minutes or even seconds apart must be calculated to represent these changes properly!

As we have emphasized, since stars spend the overwhelming fraction of their lives undergoing extremely slow changes, we cannot compare the predictions of evolutionary models with the observed behavior of just one star. Rather, we must compare evolutionary theory with observations of stars belonging to clusters. Before seeing how this is done, we shall first describe the characteristics of the material between the stars from which stars form.

Terms to Know

Gravitational contraction, mechanical equilibrium; gas law, opacity, convection, conduction, temperature gradient; binding energy, mass defect; proton-proton chain, carbon cycle, triple-alpha process, safety valve; deuterium, positron, alpha particle, neutrino, radiation pressure.

Ideas to Understand

Why stars must evolve; how a star holds together; how energy is transported in a star; energy sources of stars, the difference between chemical and nuclear reactions, understanding energy generation through the binding-energy curve; the role of the fundamental forces in nuclear burning, the role of gravitational contraction in stellar evolution, late energy-generation processes; how a star's safety valve works. Why there are limits to stellar masses. How stellar models are computed.

Questions

1. (a) Explain why a B0 star lives a shorter time on the main sequence than a K0 star.
   (b) How could you convince someone that the Sun is not hollow?

2. Sirius is twice as massive as the Sun. How would you expect the temperature and pressure at the center of Sirius to compare with those at the center of the Sun? Explain briefly. Which star should generate more energy by thermonuclear reactions? Why?

3. The Sun radiates 4×10^{33} ergs/sec.

(a) How much of the Sun's mass is converted to energy each year? ($E = mc^2$, where E is the energy in ergs produced by the conversion of m grams of matter; $c = 3 \times 10^{10}$ cm/sec, the velocity of light; 1 year $= 3 \times 10^7$ seconds.)

(b) How much of the Sun's mass is converted to energy in its 10^{10} years of main-sequence life? How many Earth masses is this equivalent to? (Mass of the Earth $= 6 \times 10^{27}$ gm.)

4. If most of the Sun's energy is generated within 20 percent of its center, show that this represents only about 1 percent of the Sun's total volume.

5. (a) Explain why increasing the temperature at the center of a star will cause the energy-generation rate to increase. Why would an increase in density also increase the energy output of the star?

(b) Suppose the temperature in the core of the Sun dropped slightly. How would the safety-valve mechanism work to restore the Sun's luminosity?

6. Consider a B-type main-sequence star that has four times the surface temperature, five times the radius, and ten times the mass of the Sun.

(a) How many times greater than the Sun's is the B-type star's total energy output?

(b) If the Sun and the B-type star have available the same fraction of their mass for hydrogen burning, and the sun has a main-sequence lifetime of ten billion years, how long is the main-sequence life of the B-type star?

7. How many neutrons and protons are in each of the following nuclei?

(a) ^{3}He

(b) ^{13}C

(c) ^{8}Be

8. The masses of four atomic nuclei are given below:

H: 1.67262×10^{-24} gm   He: 6.64463×10^{-24} gm   C: 1.99209×10^{-23} gm
O: 2.65527×10^{-23} gm

(a) Find the mass defect that occurs in the fusion of hydrogen to helium.

(b) Find the mass defect that occurs in the triple-alpha process.

(c) Find the mass defect that occurs in the synthesis of oxygen from carbon and helium.

(d) Which of these processes releases the most energy? the least? Explain.

9. The energy-generation rate in a star depends sensitively on the core temperature. Use this fact to explain why a relation between a star's mass and its luminosity should exist, and why it is not surprising that $L \propto M^{3.5}$ rather than just $L \propto M$.

10. The Earth's atmosphere is in hydrostatic equilibrium, just as a star is. Explain what this means and how hydrostatic equilibrium is established.

Suggestions for Further Reading

All of the many elementary astronomy texts available on the market describe the subject of this chapter in varying degrees of complexity and completeness. A few more specialized books that might be useful are listed below.

Aller, L., *Atoms, Stars and Nebulae*. Cambridge: Cambridge University Press, 3rd edition, 1991. Chapter 7 discusses energy generation in stars.

Goldsmith, D. and Owen, T., *The Search for Life in the Universe*. Reading, MA: Addison-Wesley Publishing Company, 2nd edition, 1992. Chapter 5 gives a good overview of energy generation.

Kippenhahn, R., *100 Billion Suns*. New York: Basic Books, Inc., Publishers, 1983. Chapters 3 and 4 discuss stellar energy generation and models.

Meadows, A., *Stellar Evolution*. Oxford: Pergamon Press, 2nd edition, 1978. This is a concise description of stellar evolution. Chapter 3 is relevant here.

The Lagoon nebula (M8) is one of the few gaseous nebulae visible to the naked eye. See the summer star chart (3) for its location. Seen in a dark sky with binoculars it is about 40 arc minutes across and has associated with it an open cluster of stars. M8 is in the plane of the galaxy and about 6 degrees northeast of the center of the Milky Way, but at a distance of 5,000 light-years, it is only about one-fifth of the way there. A few degrees to the southeast of M8 is the Great Sagittarius Star Cloud, some of which is associated with the center of the galaxy. Incidentally, the Lagoon nebula is just 1 degree south of the southernmost point of the ecliptic.

CHAPTER 14

The Interstellar Medium: Birthplace of the Stars

We now know something about the physical characteristics of stars, their structure, and their energy sources. But where do stars come from? Where are they formed? And how are they formed? The vast space between the stars is not entirely empty, but is occupied by an extremely low-density gas and tiny solid particles referred to as **dust**. In some places this gas and dust collect in huge clouds sometimes called **nebulae**. Stars form from these clouds of gas and dust. We know this because the youngest, most recently formed stars are nearly always still associated with such nebulae. It is these bright nebulae and the young stars that have recently formed from them that define the spiral arms as seen in Figure 14.1. As we would expect, the abundances of the elements in the nebulae are the same as in their stars, with hydrogen and helium making up all but a few percent of the total. At various stages of their evolution, stars

Figure 14.1. A spiral galaxy in Ursa Major known as M81. The prominent arms are delineated by young, hot stars and the gas clouds they heat.

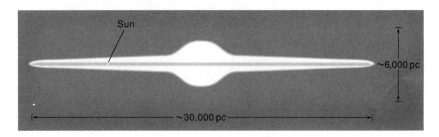

Figure 14.2. A schematic "side view" of our galaxy. Much of the interstellar matter is concentrated in a thin layer in the galactic plane. The rest of it is distributed in a much larger volume extending some 3,000 parsecs above and below the galactic plane.

expel matter back into the interstellar medium; some of this matter results from nuclear processes inside stars. Also, the medium is stirred up by stellar winds, and in particular by supernovae blast waves; so it is a dynamic entity, its structure and composition changing with time as it participates in the continuing cycle of stellar birth and death.

The material of the interstellar medium is not uniformly distributed throughout our galaxy. About half of it—made up of clouds—is found in a thin layer extending only about 100 parsecs above and below the plane of the Milky Way (see Table 14.1 and Figure 14.2). Material between these clouds makes up the other half, and is in a much thicker layer extending as far as 3,000 pc above and below the galactic plane. You can think of it as the "atmosphere" of the galaxy. The "filling fraction" given in Table 14.1 is just the fraction of the volume of space occupied by each component of the interstellar medium. Note that clouds occupy only about 2 percent of the interstellar volume.

Although the total mass of the interstellar medium is a few percent of that of our galaxy, it occupies such a vast volume of space that its average density is very small. Though there are huge variations in the density of various components of the medium, a value of one atom in each cubic centimeter is about in the middle of the range of densities given in Table 14.1. By contrast, the air we breathe has about 10^{19} particles per cubic centimeter. In a very good "vacuum" that we might create for laboratory work, about 10^9 particles still remain in each cubic centimeter. It may help bring the

Table 14.1. *Components of the interstellar medium*

		T(K)	Density in plane (cm^{-3})	Filling factor (%)[a]	Average height above plane (pc)
Clouds	H$_2$	15	200	0.1	75
	H II	8,000	15	0.1	100
	H I	120	25	2	100
Intercloud	warm H I	8,000	0.3	40	500
	warm H II	8,000	0.15	20	1,000
	hot H II	10^6	2×10^{-3}	40	3,000

[a]"Filling factor" is the fraction of the volume of the galaxy filled by the component in question.

point home by noting that it would take only about 1,000 grams of gas to fill a volume equal to that of the Earth with one atom in each cubic centimeter. By any standards, the interstellar medium is extremely tenuous indeed.

So why is an interstellar cloud like the Orion nebula (Figure 14.3) such a spectacular sight? Because it is so huge, about 10 parsecs across. Even though each cubic

Figure 14.3. The Orion nebula is a large cloud of hot gas in the central part of the sword of Orion. If your eyesight is good, it can be seen with the naked eye; it is easily visible with binoculars. It contains approximately 300 solar masses of gas and dust and is 1,500 light years away. It is heated primarily by the Trapezium stars (see Figure 12.2). The nebula is a "blister" on the front side of a giant molecular cloud, as we will see later.

centimeter of the Orion nebula contains on average only about 100 atoms and molecules of gas, in a column 1 square centimeter in cross-section but 10 parsecs long (= 3×10^{19} cm) there will be 3×10^{21} particles in our line of sight. Thus, we see a beautiful, glowing interstellar cloud.

Our understanding of the interstellar medium has benefited enormously from the technological advances that enable us to observe radiation from this material not only in the optical, but also in the x-ray, ultraviolet, infrared, and radio regions of the spectrum. Generally speaking, the x-ray and ultraviolet data give us information pertaining to the high-temperature regions of the medium, and the infrared and radio data tell us about the cold clouds that are often hidden from us by obscuring dust.

Interstellar Clouds

As indicated in Table 14.1, it is useful to speak of the cloud and intercloud regions. The former consists of three kinds of clouds, denoted by their predominant constituent. Remember, however, that their overall chemical compositions are all about the same:

(1) warm clouds of ionized hydrogen (often called "H II regions");
(2) cold clouds of neutral hydrogen (H I regions); and
(3) cold clouds of molecules.

You should not think of these as independent entities. Instead it is more appropriate to regard them as parts of a single, very patchy, clumpy, nonuniform structure. Cold (about 15 K) molecular clouds are the densest of all, and are the site of star formation. When O-type stars are formed, they cause the surrounding gas to become ionized and to radiate, producing H II regions. Neutral gas (H I regions) often surrounds H II regions. Small solid particles, dust grains, are always associated with gas, sometimes in very dense, dark clouds. Finally, all these structures are embedded in a low-density intercloud region that occupies a large volume of our galaxy. Let us consider some of the properties of these clouds.

H II Regions

The Recombination Spectrum. **H II regions** are certainly the most visually impressive structures in the interstellar medium and are among the most beautiful objects in nature. A spectacular example is the Rosette nebula (Figure 14.4), which is visible even with a relatively small telescope.

The temperatures of H II regions, about 8,000 K, are maintained by the ultraviolet radiation emitted by O-type stars formed within the nebulae. These stars are very hot, and so emit much of their energy shortward of the Lyman limit, that is, at wavelengths shorter than 912 Å. Such photons have enough energy to ionize the hydrogen in the cloud (hence the name, H II region). Sooner or later, however, a given proton captures a free electron, usually into a high-energy level. The electron then loses its energy by cascading through successively lower energy states to the ground level (Figure 14.5). The resulting emission line spectrum is called a **recombination spectrum**, because it is produced by the electron recombining with the ion. We see the energy lost by the electron as emission lines, for example, the Balmer series of hydrogen in the visual region of the spectrum. In effect, the nebula converts high-energy ultraviolet photons (which ionized the hydrogen atom) to low-energy visual photons (which are emitted when the electron is recaptured).

Figure 14.4. The Rosette nebula, an H II cloud 3,000 light years distant in the constellation of Monoceros. Strong winds from a cluster of stars formed in the center of the nebula about one million years ago have blown away the surrounding gas, leaving a cavity at the center of the nebula.

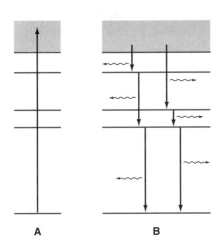

A   B

Figure 14.5. In (A), a high-energy photon is absorbed by an atom and is ionized. Eventually, the atom captures an electron (it "recombines") that cascades down through the energy levels, emitting photons (B). Two possible paths by which an electron reaches the ground level are shown.

The size of an H II region depends on the amount of gas it contains. If there is more than enough gas to absorb all the ultraviolet photons, then the size of the nebula is limited by the temperature of the central star (or stars), because that determines the number of Lyman continuum photons available to ionize the gas. If so, the H II region would be surrounded by un-ionized, neutral gas, a so-called H I region. If, on the other hand, there is not much gas surrounding the hot stars, not all the far-ultraviolet photons will be absorbed, and the size of the H II region will be limited by the amount of gas present.

Forbidden Lines. Two emission lines in the green region of the spectrum (4959 Å and 5007 Å, shown in Figure 14.6) are of particular interest. Although these are usually the strongest lines in the visible spectrum of H II regions, for many years they could not be identified with any known spectral lines or chemical element. It was even speculated that perhaps an as-yet-unidentified element, given the name "nebulium," was responsible for these lines. Finally, in 1927 it was shown that these green lines arise from so-called metastable states of the common element oxygen in its doubly ionized state.

Why was it so dificult to identify these lines? As you know, the lifetime of an electron in an ordinary excited atomic level is only about 10^{-8} seconds. After this time the atom spontaneously decays—makes a transition—to lower levels. Some atoms, however, have excited levels in which an electron can stay for enormously longer times—seconds or even hours—before it falls to a lower level. Such levels are called **metastable** and the radiation they emit is said to be **forbidden**. Since the radiation does

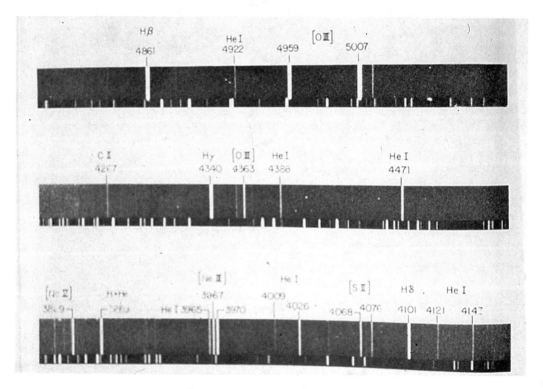

Figure 14.6. An 800-Å portion of the recombination spectrum of emission lines formed in the Orion nebula. A comparison spectrum is shown below each strip of nebular spectrum. Note the forbidden lines—for example, those of twice-ionized oxygen at 4959 Å and 5007 Å.

occur, obviously it is not absolutely forbidden, but the term indicates that it is far less likely to be emitted than that from ordinary, non-metastable levels. In fact, forbidden radiation is not observed in ordinary laboratory spectra.

That these lines are common in nebulae but difficult to produce in laboratory sources is directly related to the physical conditions in nebulae. As we saw earlier, the temperatures of nebulae are quite high, and so their free electrons move with great velocity, about 600 km/sec. When one happens to collide with a twice-ionized oxygen ion, it can kick one of the ion's electrons to a metastable state. It will remain in this excited metastable level for about 100 seconds (10^{10} times longer than an electron in an ordinary excited level does) before eventually falling to the ground state, giving up its energy in a photon of 4959 Å or 5007 Å. We don't see these spectral lines in a laboratory source, because the density in the best laboratory vacuum is much greater than that in a nebula. Consequently, if an oxygen atom has been excited to a metastable level, it will collide with another particle or with the walls of the container *before* the oxygen can radiate. It will give up its energy and leave the metastable state through this collisional process rather than by emitting a photon. It is the high temperature of the nebula that enables electrons to collisionally excite oxygen ions, and it is the low density (and no walls) that allows enough time to elapse for spontaneous transitions from metastable states to the ground level to occur. As with many astronomical objects, a nebula provides us with physical conditions difficult to achieve on Earth, a laboratory in space where unusual conditions prevail.

The probability that the forbidden-line radiation will be reabsorbed is very low, so it escapes. This is one of the ways by which the nebula loses energy and maintains its constant temperature despite the continuing inflow of energy from the associated O-type stars.

H I Regions

Reflection Nebulae and Dust.   If the forbidden lines of oxygen are particularly bright, an H II region will appear to be greenish in color, but if H-alpha and the nitrogen lines on either side of H-alpha dominate, the nebula will be red. Some bright nebulae, however, are blue in color (see Figure 14.7). This simple observation tells us that these nebulae are shining by a process different from that which causes H II regions to shine. Suppose the stars that illuminate a nebula happen to be of later spectral type than about B1. Being much cooler than O-type stars, such stars radiate few photons capable of ionizing hydrogen, and so no recombination spectrum, no emission lines, are produced in the nebula. It is not an H II region. Since the hydrogen is neutral and not ionized, the cloud is said to be an **H I region**. Its color is blue because blue starlight is reflected (or scattered) more efficiently than red starlight by the dust associated with the gas within the nebula. Such objects, called **reflection** nebulae, are revealed by the presence of the dust they contain, not the gas. Even if no gas at all were present, we would see the light reflected by the dust cloud. Though the color of the nebula is blue, the embedded star will be redder than it would otherwise be; we say that it is reddened. (Can you explain how this happens? See Figure 14.8.) This reflection interpretation is clinched by the spectrum of such a nebula: it shows the same absorption lines that appear in the spectrum of the illuminating star. The dust simply reflects (scatters) the light of the illuminating star. When a stellar photon is scattered (rather than absorbed), its direction is changed but its wavelength is usually little affected.

Suppose circumstances are such that an H I region does not appear as a reflection nebula (which is the case most of the time); can it be detected? Possibly; if a bright

Figure 14.7. The Trifid nebula is 3,500 light years away in the constellation of Sagittarius. The larger and brighter nebulosity is an H II region; the smaller cloud is a reflection nebula. The difference between the two can be seen in the color photograph of the nebula (see Plate C4). The Hα emission line produces most of the red color of the H II region, and the dust within the smaller nebula reflects the blue light of the embedded stars.

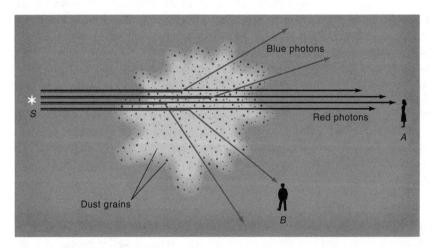

Figure 14.8. Red photons from star *S* are much more likely to get through the nebula to the observer at *A* than are blue photons. The latter are more likely to be scattered out of the direction to *A* by the dust grains within the nebula. Hence the star will appear to be redder and fainter than it actually is. By the same token, an observer at *B* will receive more blue than red light, giving the reflection nebula its characteristic blue color.

star happens to be behind the cloud and shines through it, then superposed on the stellar spectrum might be absorption lines that are much narrower than those produced in the star (see Figure 14.9). Where are these lines formed? How do we know that these lines are associated with an intervening cloud and not the star? That the lines are narrow means that their broadening by the random velocities of atoms must be small. This implies that they arise in a gas at very low temperature, about 100 K in fact, in marked contrast to the broad line widths that are produced in a hot stellar atmosphere. In addition, these absorption lines arise from atoms in their ground states or from levels very near to them. Collisions are simply not energetic enough to raise atoms to high excited levels. Again, this indicates that, in contrast to the star, the gas is cold. The Balmer lines of hydrogen, which arise from the $n = 2$ level, are not produced in such clouds. Finally, unless the star and cloud just happened to be physically associated, their radial velocities (derived from the Doppler effect) are likely to be quite different. All this can mean only one thing: the two sets of lines must arise in different sources, namely, a hot stellar photosphere and a cold gas cloud.

Note that the interstellar medium provides good examples of two of Kirchhoff's laws: a hot gas produces emission lines (H II regions); and as we have just seen, absorption lines are produced in the spectrum of a hotter object when its radiation passes through a cooler cloud (H I region). This last is a powerful way to probe the interstellar medium. Imagine a bright star at a great distance from us. Its light will most likely pass through several cold clouds, each one of which is revealed by the set of narrow absorption lines (all with the radial velocity of that cloud) it impresses on the stellar spectrum. Thus one can detect discrete clouds in a particular line of sight, measure their radial velocities, and say something about their physical conditions like temperature and composition. Finding their distances is not so easy; nevertheless, this technique has been used to establish the existence of a tenuous halo of gas out to distances of a few thousand parsecs above and below the plane of our galaxy.

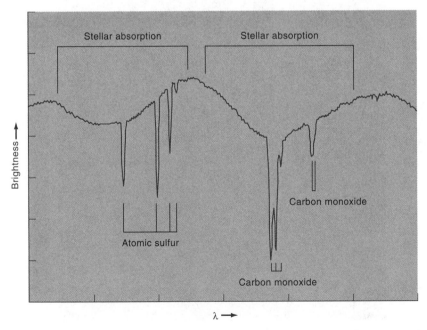

Figure 14.9.   Narrow atomic and molecular absorption lines are superposed on two much broader stellar-absorption features. The narrow lines are formed in cold clouds of interstellar gas when light from a background star passes through them on the way to the Earth.

Why is an H I cloud so cold? Because, apart from producing these narrow absorption lines, the gas is transparent to the starlight; that is, the cloud absorbs little energy and so it remains cold. Light from the star does interact with the dust that is always associated with gas, however. Dust produces two primary effects on starlight: it makes it both fainter and redder than it otherwise would be. Dramatic examples of the dimming effect are provided by the dark nebulae, one of which is shown in Figure 14.10. Here, lanes of dust block out most of the light from the more distant stars. This effect of dust is also prominent in H II regions like the Trifid nebula (Figure 14.7).

A B-type star embedded in or behind a dust cloud appears fainter than it would if the dust were not present, and it has the color of a later-type star, for example, one of G-type. Its line spectrum is unchanged, however, and will reveal that it is really a B-type star. This allows a correction to be made for both its reddening and its dimming. The amount of reddening is just the difference between the observed and the intrinsic color of the star, the latter of which is known from its spectral type. Since the amounts of reddening and dimming are correlated (the more dust, the redder and fainter the star), a correction can also be made to the apparent brightness of the star

Figure 14.10. When there are no stars near a cloud of gas and dust, neither a diffuse nebula nor a reflection nebula is seen. By blocking the light of stars behind the nebula, the dust is revealed.

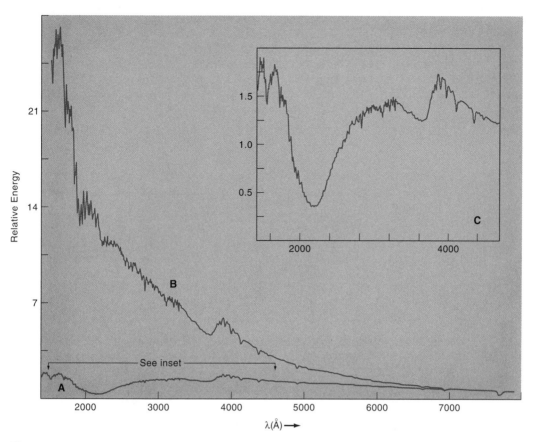

Figure 14.11. The spectrum in (A) is that of a B2 III star as it is observed, heavily dimmed by the interstellar dust that happens to be between us and the star. Spectrum (B) shows what the energy curve for this star really is; that is, it is spectrum (A) corrected for the extinction effects of the dust. (The relative magnitude of this correction is shown in Figure 14.13.) Note that in the visual part of the spectrum, this B-type star would appear to have the color of a much cooler, later-type star. In the visible region, the dust cuts the brightness of this particular star in half, and the dimming effect of the dust increases at shorter and shorter wavelengths. In the far ultraviolet the brightness is reduced to only about one-tenth of its true value. Inset (C) is the spectrum on an expanded scale from about 1600 Å to 4500 Å in order to show how the dust is particularly effective in absorbing light over a broad spectral region centered at about 2175 Å.

(see Figure 14.11). Note how distances to stars can be overestimated if the dust extinction effect is neglected.

A secondary effect of the scattering process is that it polarizes starlight slightly. That is, the scattered electric waves are not randomly oriented, but show a slight tendency to oscillate more in one plane than another (see Chapter 9). This polarization is the result of two factors. The first is that the dust grains themselves are not perfectly spherical in shape, but are slightly elongated. The second factor is that a weak magnetic field, only about 10^{-5} as strong as the Earth's, pervades the disk of our galaxy and tends to align the dust grains so that their longer axes are more or less parallel. As a consequence, they scatter starlight a little more effectively along the direction of the long axis than perpendicular to it, thereby polarizing it. In fact, the existence of this magnetic field was first inferred from the polarization of starlight.

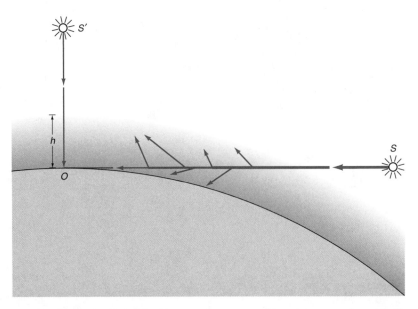

Figure 14.12. Molecules in the Earth's atmosphere scatter blue photons much more efficiently than red, so less blue light gets to the observer at O when the Sun is setting (S) and it appears red. When the Sun is high in the sky (at S'), the path length through the atmosphere, h, is not nearly as long as the path length when the Sun is setting, so the light is little affected.

It may seem surprising that, in contrast to interstellar dust, interstellar gas does not affect starlight; after all, scattering by air molecules in Earth's atmosphere produces our red sunsets. The latter happen in the following way. Because of their very small size, air molecules scatter blue photons out of an incident beam of sunlight much more efficiently than they do red photons (see Figure 14.12). Consequently, many more of the latter reach our eyes than do the former and so the setting Sun appears red. Since the path length of sunlight through the atmosphere, and therefore the probability of scattering by an air molecule, is much smaller at noon than at sunset, the color of the midday Sun is little affected.

Sometimes sunsets will be especially spectacular or the Sun will appear red even when it is not setting. The diameters of ordinary atmospheric dust particles are too large to affect the color of the Sun, so these phenomena arise on the relatively rare occasions when tiny particles (not molecules) are in the air, for example, mist or fine volcanic dust. If the water droplets or dust particles are very small (say 3000 Å in diameter), then like atmospheric molecules, they will scatter blue light much more efficiently than red light, and produce the unusual effects mentioned.

The same molecular scattering process that causes our red sunsets is also responsible for the blue color of the sky. Because blue photons are much more likely to be scattered than those with a longer wavelength, the former are scattered to us from all over the sky, whereas the latter are much more likely to pass through the atmosphere. If it were not for our atmosphere, no radiation would be scattered and, except in the direction of the Sun, the sky would be black.

So why can we say that interstellar gas does not affect starlight in a similar manner? There simply is not enough of it along any line of sight. As we look out from the Earth toward the Sun, a column of air one square centimeter in cross-section, extending all the way through our atmosphere, contains about one thousand grams of ma-

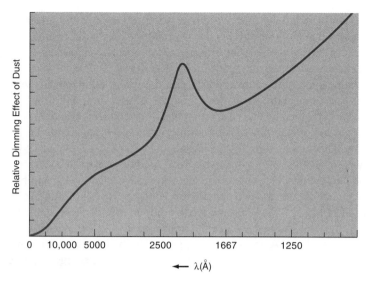

Figure 14.13. The relative dimming effect of interstellar dust as it depends on wavelength; that is, how much more one wavelength is affected than another. (The strange wavelength scale arises for reasons that need not concern us.) The actual amount of the dimming depends on how much dust is in the line of sight. This can be estimated by comparing the red color of the star to its intrinsic color. The steady, nearly straight line, increase in the dimming of light to shorter wavelengths is interrupted only by the extinction "bump" at about 2175 Å. We saw the effect of this feature in Figure 14.11.

terial. A similar column extending *all the way across the galaxy* contains only about 0.2 grams of matter(!), a few percent of which is dust, while the rest is gas. Now the 1,000 grams/cm² of air do not decrease the brightness of sunlight by any appreciable amount,[1] but the 0.2 grams/cm² of matter across the galaxy markedly dim and redden starlight. Since sunlight is transmitted very well through clear gas-filled skies, the incomparably smaller number of interstellar gas atoms and molecules along a sight line cannot be responsible for dimming starlight (or for making it redder, either). It is interstellar dust particles that both dim and redden starlight, not interstellar gas.[2]

What kinds of objects could make incoming radiation fainter? Interstellar bricks or baseballs would certainly do so, but since such objects are so much larger than the wavelengths of visual radiation, they would decrease the brightness of blue and red light equally. We have seen, however, that red light is much less dimmed than blue light by the interstellar dust. It turns out that particles of dust, which are much larger than air molecules but about the size of the wavelength of ultraviolet light, can both dim and redden light very effectively. In the visual and infrared regions of the spectrum the efficiency of this scattering goes as about 1/wavelength; that is, the longer the wavelength of light, the less it is scattered by interstellar dust (see Figure 14.13). Consequently, a given star will appear to be much less dimmed and reddened by dust when observed at longer wavelengths than when observed at shorter wavelengths. For example, if there is so much dust along a particular line of sight that only one-millionth of visible light can pass through it, fully one-quarter of infrared radiation at 2 microns

---

[1] The Sun appears to be about as bright at sea level as it does when seen from an airplane at 35,000 feet.

[2] Like the Sun, rising and setting stars are reddened by molecular scattering in Earth's atmosphere. This has to be corrected for before the effects of interstellar dust are considered.

(20,000 Å) will get through. Dust becomes so transparent in the infrared and radio spectral regions that at these wavelengths we can "see" right through dust clouds. Radio waves easily reach us from the galactic center and beyond, whereas visible light from the center is absorbed and scattered by dust long before it reaches us.

The composition of interstellar dust is still uncertain, but it appears that primarily compounds of carbon, nitrogen, silicon, oxygen, magnesium, and iron form the dust grains. This is supported by the observation that these elements are often depleted from interstellar gas as compared with their cosmic abundances; apparently they are tied up in the grains. Dust grains form by condensing out of a cool gas (about 1,000 K) when its density is from 10^9 to 10^{10} atoms/cm^3. Such conditions exist in a variety of objects, such as the outer atmospheres of cool giants and supergiants, the expanding envelopes of planetary nebulae, novae, and supernovae (Chapter 16), and in the star-formation process itself (Chapter 15). What the relative importance of these sources may be is not yet entirely clear, however. Once the grains are formed, they can grow as atoms strike and stick to them. At the same time, grains can be heated by starlight and evaporate or be destroyed by collisions with other grains or with energetic particles.

21-cm Radiation.   As we have seen, reflection nebulae are revealed by their dust, not by any associated gas. In general gas and dust exist together, however, so it is most likely that gas exists in reflection and other cold nebulae. The most abundant element is hydrogen; is it possible to detect it when it is cold and un-ionized? It happens that there is a powerful technique for doing so.

The ground level of hydrogen is not a single level, but is actually split into two levels, which differ only slightly in energy. The ground state is split as a result of the spin of the proton and electron in the hydrogen atom. As we saw in Chapter 9, particles such as protons and electrons can be thought of as spinning, as having a very small angular momentum. These spins can be represented as either clockwise or counterclockwise, represented by an arrow pointing either "up" or "down." The energy of a hydrogen atom when both the electron and proton are spinning in a parallel direction is a tiny bit greater than when the spins are in antiparallel directions (see Figure 14.14). In fact, the energy difference between the two levels is roughly 1/300,000 the energy difference that produces H-alpha. The wavelength of the photon that is emitted when the spin flips from parallel to antiparallel is, therefore, correspondingly longer, about **21 cm**. The upper energy level of the hydrogen ground state (spins parallel) is an extreme example of a metastable level. Its lifetime is about ten million years! Only in the vast, extremely low-density regions of interstellar space can hydrogen atoms exist undisturbed in this state long enough to emit a 21-cm photon.

The 21-cm radiation emitted by neutral hydrogen is in the radio region of the spectrum; by 1951 radio instrumentation necessary to detect these photons had been developed. Because 21 cm is such a long wavelength, incomparably longer than the size of interstellar particles, it is not at all affected by interstellar dust. Consequently, it reaches our telescopes from all parts of the galaxy where neutral hydrogen is found, for example, in spiral arms. This has enabled us to map out the distribution in our galaxy and in other galaxies of the most abundant chemical element. In our galaxy, about 3×10^9 solar masses of neutral hydrogen exist in the interstellar medium.

Molecular Clouds

By the 1940s, absorption line spectra in the visible region had revealed the existence of only three molecules in the interstellar medium—CH, CH$^+$, and CN. The number of these molecules in a given volume of space is relatively small, perhaps 10^{-7} the

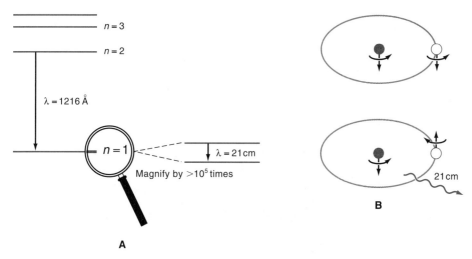

Figure 14.14. The ground level of the hydrogen atom is not single, but is very slightly split into two levels (A) because of the spin of the proton and electron. These two particles can be thought of as spinning in either a clockwise or counterclockwise direction, represented by the direction of the arrows in (B). When both particles are spinning in the same sense (arrows parallel), their energy is slightly greater than when they are spinning in opposite senses (one arrow "up," the other "down"). So when the spins flip from parallel to antiparallel, a low-energy photon is emitted; its wavelength is 21 centimeters, nearly two million times longer than Lyman-alpha at 1216 Å.

number of hydrogen atoms. Since the density of the gas is so very low, the probability of individual atoms encountering each other in a way that allows them to combine and form a molecule was thought to be extremely small. Furthermore, appreciable numbers of stellar photons pass through the medium with sufficient energy to dissociate molecules into their constituent atoms. Also, in higher-density regions collisions would destroy the molecules. All in all, it seemed extremely unlikely that many molecules could survive in such a harsh environment.

In 1963, however, radio radiation at 18 cm was detected and identified with the OH molecule; five years later, 1.25-cm radiation emitted by NH_3 (ammonia) was discovered. Since then more than 60 molecules have been found, all except H_2 having been detected by ground-based radio telescopes. (Absorption lines from H_2 have been observed in the far ultraviolet by instruments carried above the Earth's atmosphere on rockets and satellites.) Most of the constituent atoms are the common elements—hydrogen, carbon, nitrogen, oxygen, sulfur, silicon—but there are also molecules with deuterium (heavy hydrogen) as a component. Some of these molecules are fairly complex; for example, ethyl alcohol has nine atoms: C_2H_5OH. (It has been estimated that the Orion clouds contain about 10^{27} fifths of alcohol!) The largest molecule so far discovered has 13 atoms, $HC_{11}N$. The poisonous gases HCN and CO have also been found. The discovery of these molecules has provided us with a new way of studying the interstellar medium, and a new field, interstellar chemistry, has developed. The processes responsible for the formation and destruction of molecules give us much information about the conditions of the gas and dust between the stars.

Despite the discovery of large varieties and numbers of molecules, the arguments given above that molecules should be rare are not all wrong. They do apply to the general interstellar medium. Astronomers did not realize, however, that high-density clouds could exist surrounded by protective layers of dust that effectively shield molecules from the destructive ultraviolet radiation emitted by hot stars. Furthermore, typical cloud temperatures are only 10 K, so that collisions are not sufficiently energetic to destroy the molecules. Such an environment allows molecules to form in profusion. These **molecular clouds** have populations of molecules that are hundreds of millions times greater than those in ordinary nebulae. Though the average density in a molecular cloud is a few hundred hydrogen molecules per cubic centimeter, cloud cores have much greater densities. Densities as large as 10^5 to 10^6 molecules per cubic centimeter, and temperatures of 2,000 to 3,000 K have been inferred in the cores. It is here that star formation is taking place. The high temperatures are produced by rather violent shock waves created by outflowing gas from embedded, recently formed stars.

Sometimes these dense molecular clouds are sources of extremely intense radiation in the radio frequency lines of OH and H_2O. The source diameters are so small (just a few AU) and the radiation so intense that if the latter were produced by a thermal process, incredible temperatures—on the order of 10^{14} K!—would be implied. Clearly, such temperatures in the interstellar medium are absurd, so the assumption that the radiation is a consequence of high temperature must be wrong. The observational evidence and the chain of theoretical arguments are both rather complex and still not completely understood. Suffice it to say that these sources are naturally occurring **masers**,[3] sources in which conditions are wildly nonthermal, that is, they cannot be accounted for by processes that depend purely on temperature. In particular, excited levels of the OH and H_2O molecules are enormously overpopulated compared to the number that would be produced in gas that was in thermal equilibrium. When all these molecules are stimulated to release their energy, they produce an extremely intense stream of radiation. These cosmic masers are often coincident in position with bright infrared sources that are probably **protostars**, that is, stars in the process of formation. It is the radiation from these objects that selectively "pumps" huge numbers of electrons to the excited levels of OH and H_2O, from which they emit the intense maser radiation (see Figure 14.15).

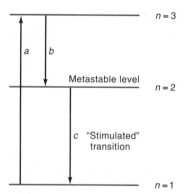

Figure 14.15. Principle of a maser: molecules are excited to the $n = 3$ level by collision or by absorbing a photon, a. After a short time, photon b is emitted, and the molecules fall to the $n = 2$ metastable level. The transition $2 \rightarrow 1$ is forbidden, so many molecules remain in this excited state for a relatively long time. Whenever a photon with the energy corresponding to the $2 \rightarrow 1$ transition interacts with a molecule, it can stimulate that transition to occur, which in turn stimulates other molecules to emit the $2 \rightarrow 1$ photon, producing an intense cascade of these photons.

---

[3] You have no doubt heard the term "laser," denoting a light source that produces an extremely bright, parallel beam of light of very nearly one wavelength. If laser light were produced by a thermal process (like that of a blackbody), its temperature would have to be enormous. "Maser" describes the analogous process in the microwave (very short radio wavelengths) region of the spectrum.

The chemical composition of a given mass of the interstellar medium is pretty much like that of the Sun if all the material in the mass is included, and not just a particular component of it, such as molecular clouds or dust. As we have noted, the gas phase in clouds often shows abundance deficiencies (compared with the cosmic values) of certain elements such as iron and magnesium. These elements most likely have become part of the dust grains associated with the gas. These grains must be taken into account when one estimates the chemical abundances in the entire mass of the medium. Similarly, some elements may appear to be underabundant in their atomic form because they are tied up in the molecules that form the molecular clouds found in some gaseous nebulae.

As would be expected, H_2 is by far the most abundant molecule in the medium. It probably forms by the combination of hydrogen atoms on the surfaces of dust grains. Since most other molecules are thought to need H_2 for their formation, this is a crucial step. Here is another role played by interstellar dust: no dust, no molecular clouds; no molecular clouds, no star formation. The mass of hydrogen in molecular form is about equal to that in atomic form, a few billion solar masses. Carbon monoxide (CO), though less abundant than H_2 by factors of 10^{-4} or 10^{-5}, is still the next most abundant molecular species. It is a good tracer of the extent of cold molecular clouds. Some of these, called **giant molecular clouds**, are enormous—50 to 75 parsecs in diameter with up to one million solar masses of gas! They are the largest known structures within our galaxy (see Figure 15.2), and probably grow by the collisions and merging of smaller clouds. Embedded within the giant clouds are small, denser clouds where star formation is occurring or has recently occurred, because very young clusters of stars are often found nearby. Most stars now being formed in our galaxy come from molecular clouds. Apparently it is the activity produced by these newly formed stars that eventually disrupts molecular clouds and limits their mass to about a million solar masses.

The Intercloud Region

In addition to the discrete clouds we have just described, an equal mass of matter, but at very low density, is distributed over a huge volume of the galaxy. This low-density gas extends far above and below the relatively thin layer of clouds; nonetheless, it is traditional to call it the **intercloud region**.

Just as we discussed three kinds of clouds, we can identify three different regions of the intercloud component (see Table 14.1):

(1) warm regions of neutral hydrogen;
(2) warm regions of ionized hydrogen; and
(3) hot regions of ionized hydrogen.

These regions were discovered only relatively recently, roughly in the decade of 1965 to 1975, and our understanding of them is poor. Though the warm neutral gas was the first component of the intercloud material to be identified by the 21-cm radiation it emits, little is known about it. It extends about 500 parsecs above and below the plane, and makes up about half the interstellar neutral hydrogen. Its relationship to and its interaction with the rest of the gas in the medium is unknown.

The layer of ionized hydrogen is about twice as thick as that of neutral hydrogen, and occupies about half the volume of either the neutral or the hot gas. It is not yet certain what causes it to be ionized.

The hot ionized gas has the lowest density of the three intercloud components, but its temperature is high, from 10^4 to 10^6 K. We know this because, by means of telescopes on Earth-orbiting satellites, ultraviolet absorption lines of C IV, N V, and O VI arising in this low-density gas have been observed. It takes a lot of energy to strip five electrons from each oxygen atom, so the temperature must be very high, several hundred thousand degrees, in fact. Even low-energy x-rays are produced in this hot, tenuous material.

The gas is heated to these high temperatures by supernova explosions. As you will learn in Chapter 16, some stars end their lives as a supernova, a cataclysmic explosion that tears the star apart. This outburst sends a blast wave traveling through the interstellar medium, heating it to temperatures of millions of degrees. Given the number of galactic supernovae (one or two each century) and the amount of energy released by each outburst, it appears that any point in the galactic disk is subjected to this heating about every two million years. Since it takes much longer than this for the low-density material to cool off, supernova explosions are easily able to maintain the gas at a high temperature. The energy injected into the medium by supernovae also stirs up the gas, causing the extent of the galactic disk "above" and "below" the plane to be greater than it would otherwise be.

The high-temperature gas also provides an explanation of why cold clouds like H I regions can exist for long times. Generally, the gravitational force of a cloud on itself is insufficient to keep it from expanding, dispersing, and eventually disappearing as a discrete entity. But many cold clouds do exist, because, even though the density of the intercloud gas is very low, its temperature is so high that the pressure it exerts ($P \propto NT$) is significant. The hot gas, surrounding the denser but cold clouds, exerts enough pressure to keep the H I region from expanding and merging into the interstellar medium. Thus the existence of most cold clouds depends on the very hot intercloud gas. Interestingly, to account for the stability of the cold clouds, the existence of this hot gas was predicted about 17 years before it was discovered.

We are now ready to discuss some of our ideas about how stars are formed from the interstellar medium. We will do so in the next chapter, as well as give an account of the evolution of stars throughout the greater part of their lives.

Terms to Know

H I and H II regions, recombination spectrum, forbidden lines and metastable states, 21-cm radiation; reflection nebula, interstellar dust, molecular cloud, giant molecular cloud, the intercloud regions, cosmic maser.

Ideas to Understand

The organization and distribution of interstellar matter; the ways different parts of it are heated. Why metastable lines are observed in the interstellar medium, but not in the laboratory. How dust reddens and dims starlight and how it is corrected for. Origin of the 21-cm line. Why molecules can exist in the interstellar medium; why cold H I clouds can exist.

Questions

1. Although hydrogen is the most abundant element, no Balmer lines are seen in absorption in the interstellar medium. Explain why.

2. Do you think "nebulium" lines are observed in the Sun's photosphere? Explain your answer.

3. Emission nebulae (like Orion) are generally reddish or greenish in color; yet reflection nebulae can be blue, yellow, or red. Why is this so?

4. How is 21-cm radiation produced?

5. Why isn't 21-cm radiation emitted by the Orion nebula?

6. (a) What causes most of the hydrogen in a bright nebula to be ionized and only a small fraction to remain neutral?

(b) When a hydrogen ion captures an electron into the $n = 3$ level, it may then emit the red (6563 Å) line or instead emit another line. In what spectral region—UV, visible, infrared—is this other line? How do you know?

(c) Can an ion that captures an electron emit both the Balmer red lines (6563 Å) and the blue line (4861 Å), one after the other? Explain.

7. (a) How do we know that interstellar space contains dust particles?

(b) How do we know that these particles are not the size of basketballs?

(c) Why do radio waves pass through dust clouds but optical light does not?

8. Explain why gas clouds such as H I regions don't slowly expand and disappear.

9. Give two reasons why the rare O-type stars are better probes of the interstellar medium than the much more common M dwarfs.

Suggestions for Further Reading

Aller, L., *Atoms, Stars and Nebulae*. Cambridge: Cambridge University Press, 3rd edition, 1991. Chapter 10 is on the interstellar medium and gaseous nebulae.

Churchwell, E. and Anderson, K., "The Anatomy of a Nebula," *Astronomy*, **13**, p. 66, June 1985.

Gingerich, O., "Robert Trumpler and the Dustiness of Space," *Sky & Telescope*, **253**, p. 213, 1985.

Kanipe, R., "Inside Orion's Stellar Nursery," *Astronomy*, **17**, p. 40, August 1989.

Shore, L. and Shore, S., "The Chaotic Material Between the Stars," *Astronomy*, **16**, p. 6, January 1988.

Spitzer, L., *Searching Between the Stars*. New Haven: Yale University Press, 1982. A popular account of the interstellar medium written by an expert in the field.

Verschuur, G., "Interstellar Molecules," *Sky & Telescope*, **83**, p. 493, 1992. A good overview and table of molecules detected.

———*Interstellar Matters*. New York: Springer Verlag, 1989. Articles on various aspects of the interstellar medium.

Wynn-Williams, G., *The Fullness of Space: Nebulae, Stardust and the Interstellar Medium*. Cambridge: Cambridge University Press, 1992.

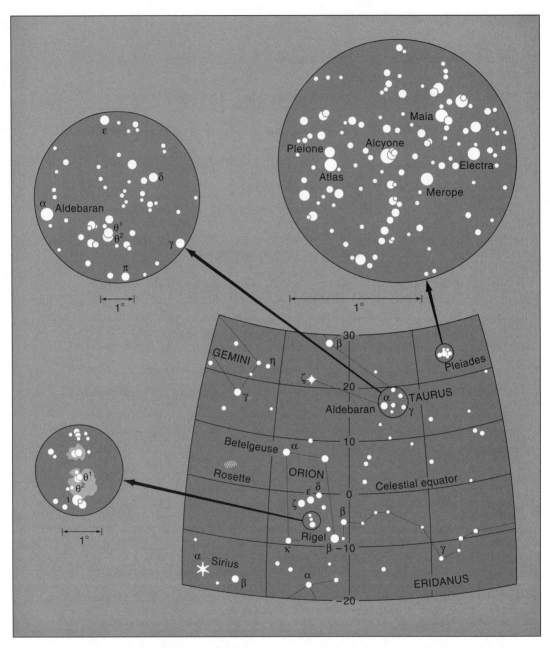

Charts of the Pleiades, the Hyades, and of the area around the Orion nebula showing stars visible in binoculars or small telescopes. The Hyades is the second nearest open cluster, only about 150 light-years away. The V-shape outlined by the brightest stars is the core of the cluster, but its members extend over more than 20 degrees on the sky. Aldebaran, the most prominent star in the area, is a foreground star, not a cluster member. The Pleiades cluster is about 400 light-years distant and its brightest members are hot, bluish stars. The Orion area is full of interesting objects, the best known of which is M42, the nebula in the sword of Orion. It can be seen with the naked eye from a dark site (though it was recorded as a nebula only in 1610, after the invention of the telescope), and with binoculars it is a spectacular object. The two objects θ^1 and θ^2 are multiple stars; θ^1 (the four stars that form the Trapezium) is part of a group of very recently formed stars. In fact, the Orion nebula region is the site of active star formation even now. Thus, ages of stars in this area of the sky range from the now-forming to a few million years old, to about 60 million years old, to roughly 1 billion years old in Orion, the Pleiades, and the Hyades, respectively.

CHAPTER 15

Stellar Evolution: The Early Phases

In this chapter we will trace the evolution of stars from their prenatal contraction phase in interstellar clouds, to their birth and youth on the main sequence, and finally to their middle age as giants. Let us begin at the beginning and first consider the formation of stars.

Stars are formed from the interstellar medium, which, per gram of material, is extremely opaque, as we have seen. As a consequence, formation is the most poorly understood part of a star's life, partly because during this phase the observational data are most obscure, literally as well as figuratively. A main-sequence star or a giant or supergiant is there for all to see. Even white dwarfs, though intrinsically faint, can be studied in some detail, as can some aspects of neutron stars. But stars in formation—**protostars**—are often so buried in dust that no visual radiation can escape, and even their near-infrared radiation may be appreciably dimmed by the surrounding dust.

Enormous changes in their diameter and density are required for protostars to contract from cloud to stellar dimensions. Protostars also show varieties of erratic behavior, some of which are a consequence of their interaction with the material from which they have formed. In addition there are many theoretical difficulties remaining: for example, just how is the formation of a protostar triggered in the first place? Consequently, we will give here only the sketchiest of outlines of the star-formation process.

Evolution to the Main Sequence
How a Cloud Collapses

It can be shown that an ideal, completely uniform, constant-density interstellar gas is inherently unstable. Under its own gravitational force, the medium eventually will break up into higher-density clouds surrounded by a lower-density medium. As a cloud contracts, its internal temperature will increase. Hence its pressure increases and tends to balance its inward gravitational force. Only if the contracting cloud is very massive and the initial internal temperature is low will the cloud be able to overcome its internal pressure and continue to contract. Hence this process strongly favors the production of very massive clouds, rather than those of stellar mass.

In a real cloud, however, the situation is much more complicated. There are local factors, most of which are not well understood, that can inhibit or promote the formation of stars. The processes that help in the formation of stars can be divided into two types: those that provide an inward-acting force in addition to gravitation; and those that cool the contracting cloud, thereby enabling the contraction to be maintained.

Let's consider the cooling process first. Dust, mixed in with the gas, plays the crucial role here. Dust grains in the cloud are heated slightly by collisions with the gas. As a consequence, they emit photons in the far infrared, a region of the spectrum in

345

which the gas is transparent. In this way energy of motion (kinetic energy) of the gas atoms and molecules is transformed to radiation and lost to the cloud; in other words, the gas is cooled. This cooling process is so effective that in the early phases of proto-star formation the cloud collapses as if it were in free-fall, that is, at the rate set only by the local gravitational acceleration. The contraction is not effectively impeded by internal heating.

Triggering Collapse

Several processes external to the cloud can help compress it. Newly formed stars may themselves trigger the formation of others by a variety of mechanisms. For example, massive stars formed in a molecular cloud emit such huge amounts of radiation that the pressure of this radiation tends to compress nearby portions of the cloud. When the mass of the compressed gas exceeds some critical value, it will collapse by its own gravitation and break into many blobs, which ultimately become protostars. Later, when these stars are on the main sequence, their radiation pressure may compress nearby gas and another group of stars will be formed.

Another phenomenon that can speed up star formation involves the action of stellar winds (see Chapter 11). Matter streaming out from stars at velocities of up to a few thousand kilometers per second carries a lot of momentum, and so can compress the nearby interstellar medium, thereby improving the chances for star formation.

Young O- and B-type stars, recently condensed out of a molecular cloud, will ionize the surrounding medium and form expanding H II regions. A shock front driven into the molecular cloud ahead of the H II region will compress it. In the thin layer between the shock and expanding H II region, more stars will form. In turn these might trigger the collapse of more material, giving rise to several newly formed star groups arranged by age (see Figure 15.1).

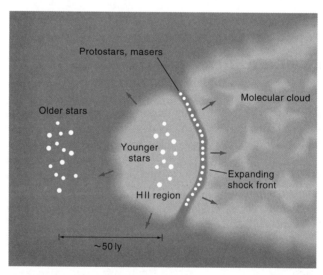

Figure 15.1. Hot stars formed recently from a molecular cloud produce strong winds that compress the molecular cloud along a relatively thin layer called a shock front. Radiation from these hot stars heats and ionizes the gas behind the shock. Gas in the thin layer between the ionization and shock fronts is still cold and neutral, but compressed to a density perhaps 100 times greater than the rest of the cloud. When the density is high enough, the layer breaks up into protostellar blobs pulling themselves together by their own gravitation. When this new group of stars is formed it repeats the process.

An example of these processes in which star formation leads to more star formation is seen in Orion, where four groups of stars are found roughly in a line. As one proceeds from north to south, the age of the groups decreases from 12 to 2 million years (see Figure 15.2), strongly suggestive of the influence of one episode of star formation on the next. The Orion nebula is on the surface of the molecular cloud, expanding into it.

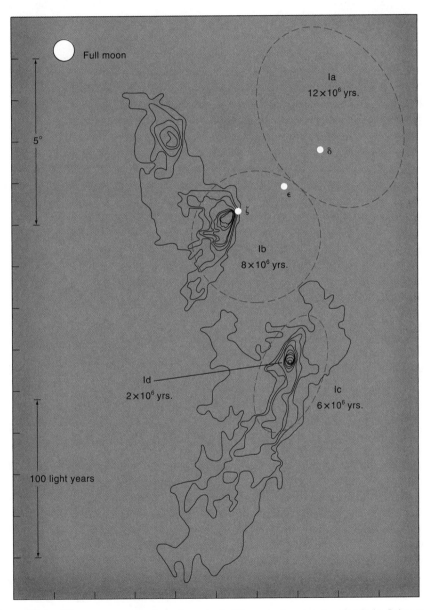

Figure 15.2. The contour maps show the extent of two giant molecular clouds in Orion as delineated by the carbon monoxide gas they contain. They are enormous, as indicated by the various scales on the left side of the diagram. The objects labeled ζ, ε, and δ are the belt stars in Orion. The dashed lines enclose groups of very young stars; their average ages are given on the diagram. The fourth and youngest group is centered on the Orion nebula, which is on the edge and expanding into the cloud. This map suggests that star formation occurs on the boundary of a molecular cloud that is slowly eaten away by the formation process itself.

Figure 15.3.   M3 is a globular cluster about 35,000 light years away. It contains an unusually large number of variable stars.

Single Stars or Clusters?

It has been estimated that no more than a percent or so of the mass of the typical molecular cloud is converted to stars; the star-formation process makes very inefficient use of the available mass. An interesting consequence of this is that a group of stars formed in a cloud most often will not be gravitationally bound together into a cluster, because not enough mass will be left in the star group to bind them gravitationally against their energies of motion. After ten to one hundred million years, these stars will have moved away from each other and merged into the general stellar background, no longer distinguishable as a group. Such temporary groups are called **OB associations**, after the spectral types of their most prominent members.

We do observe many groups of stars that appear to be bound, however: open clusters like the Hyades and the Pleiades, and globular clusters like M3 (see Figure 15.3). How they are formed is still poorly understood. Theoretical work indicates that three factors play a role: the initial density of the cloud, the efficiency by which gas is turned to stars, and the length of time over which the gas is dispersed. It is possible to find combinations of these parameters that result in a gravitationally bound cluster. It appears that to form such a cluster, 20 to 50 percent of the mass of a cloud must be turned into stars. Since in typical cases it is on the order of 1 percent, we can understand why clusters of stars are relatively rare.

Evolution of a Protostar

In any case, we have something like the following picture. A slowly rotating mass of gas embedded within a molecular cloud begins to contract, possibly set off by one or more of the mechanisms just described (see Figure 15.4). The mass of the cloud de-

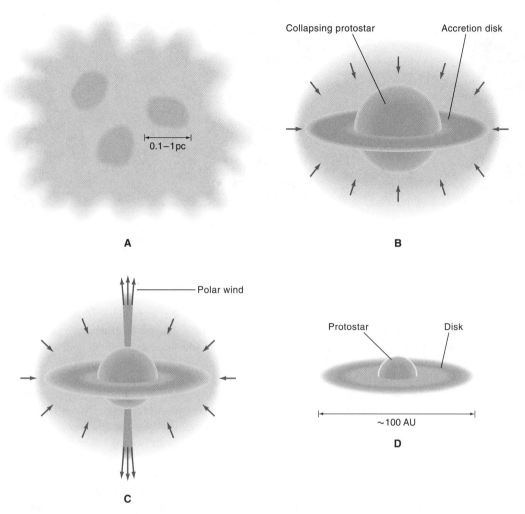

Figure 15.4. Within a molecular cloud, matter gathers into slowly rotating flattening cores (A) that eventually form rapidly collapsing protostars surrounded by a large accretion disk (B). Matter continues to fall onto the protostar and disk and the system radiates in the infrared. The protostar becomes hot enough to burn deuterium, and a polar outflowing wind is developed, but matter still accretes onto the disk (C). Finally, the accretion phase ends, leaving a protostar surrounded by a disk (D).

pends on the initial conditions of mass, density, and temperature in the original cloud; it may fragment into several blobs. Typically these blobs would have a temperature of about 20 K, a density of 10^4 to 10^5 particles/cc, a diameter of about 0.1 parsec, and a total mass of a few tenths to about ten solar masses. The collapse of a blob is fastest at its center, where it is densest, so that the core falls away from the surrounding dust and gas.

As material falls inward toward the core, its angular momentum remains constant and so its rotation about the center of the cloud speeds up. This makes it more difficult for the cloud to contract in its equatorial plane than at the poles. As a result, the outer part of the cloud tends to be flattened, and eventually an **accretion disk**[1] forms around

---

[1] Disks are common in astronomy; they are found around planets (Saturn's rings, e.g.), protostars, and stars, and as a major component of spiral galaxies.

Figure 15.5. These two Herbig-Haro objects formed when a jet from a young star (obscured by dust) near the center of this HST image slammed into interstellar matter. The distance across the image is about one light year. See also the Figure on p. 418.

the contracting protostar. Both the central cloud and the surrounding disk are fed by infalling gas and dust from the surrounding envelope. The gravitational energy lost by the infalling matter heats the dust in the disk and protostar, making them both luminous in the infrared. The environment is too dusty for visible light to escape, however.

Eventually, the interior of the protostar becomes hot enough to "burn" deuterium ($D + p \rightarrow {}^{3}He$). This causes the star to become convective and eventually develop a strong stellar wind. Because of the disk, the wind blows preferentially out from the polar regions of the star. It is this wind that is responsible for the jets sometimes observed in protostars. So in this phase, material flows both inward (increasingly onto the disk) and outward (from the poles of the protostar). Eventually the surrounding material is used up, and the inflow stops. Visible radiation from the protostar is now able to escape, and we might see the object as a **T Tauri** star. These objects have masses between 0.2 and 3 times that of the Sun, and are still contracting toward the main sequence. The radius of the disk might be as large as 100 AU (solar-system size), contain 0.1 to about 1 solar mass of gas and dust, and radiate about as much energy as the protostar.

At the ends of some of the beamed outflows from the T Tauri stars are nebulae, so-called **Herbig-Haro objects** (Figure 15.5), which apparently are energized by the rapidly outflowing streams of matter. These objects are generally knotty in appearance, and are moving away from the associated T Tauri star with velocities of up to 500 km/sec. In addition, H_2O and OH masers are often associated with the outflows. Enough objects with these outflows have been found to suggest that they probably are an evolutionary phase of all stars having masses of a few solar masses or less.

After becoming visible, T Tauri stars can be placed on the HR diagram above the lower (later spectral types) main sequence. Their subsequent evolution follows the so-called **Hayashi tracks** (see Figure 15.6), named after the Japanese astronomer who investigated this phase of evolution. Stars less massive than about 0.5 solar mass approach the main sequence along nearly vertical paths corresponding to a temperature of about 3,500 K. The tracks in the HR diagram of stars more massive than about 0.5 solar mass, however, make a turn to the left and follow a horizontal line to the main sequence, as shown on Figure 15.6. The turn occurs when the interior temperature becomes hot enough to decrease the opacity sufficiently for radiation to be capable of carrying the thermal energy outward. At this point, the star is no longer convective

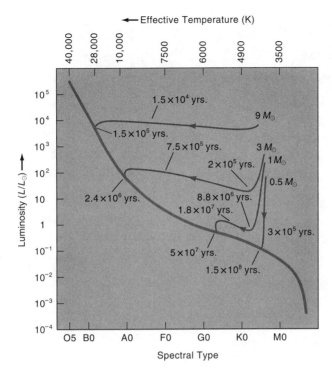

Figure 15.6. Pre-main-sequence evolutionary tracks for several stellar masses. The times on the tracks begin when the star first appears on the Hayashi track, that is, when the protostar's effective temperature is about 3,500 K.

throughout. Horizontal motion to the left in the HR diagram (like vertical motion downward) again implies contraction, since the effective temperature increases, but the luminosity remains constant.

When Contraction Stops

Contraction of the protostar finally stops when core temperatures and densities become high enough to allow the conversion of hydrogen to helium to take hold. The resulting energy generation enables the star's internal pressure to balance its gravitational force. The star is then in equilibrium on the main sequence. In other words, the main sequence is the band of points in the HR diagram at which stars of various masses derive all their luminosity from hydrogen burning.

If a massive object reaching the main sequence as an O-type star is still surrounded by a sufficiently high density cloud of gas and dust, we may see it as a compact H II region. Associated with this H II region may be an OH maser, emitting large amounts of radiation at 18 cm. Radiation pressure and the wind from the O-type star will continue to blow the H II region away. When its density has fallen below about 10^6 cm^{-3}, the OH maser will fade out. Eventually we will see the star surrounded by an ordinary rather than by a compact H II region.

The infrared-sensitive instruments on IRAS discovered a few A-type stars that are surrounded by solar-system-size disks of dust, radiating in the infrared. One example

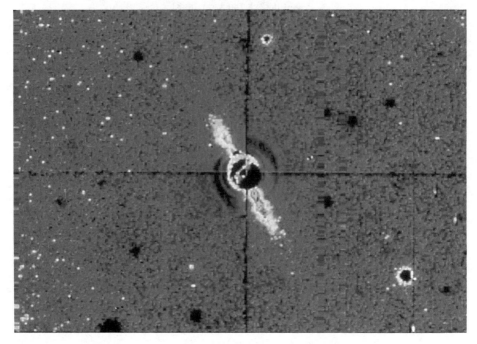

Figure 15.7. This is an optical image of the disk around the star β Pictoris. The disk was discovered by infrared-sensitive detectors on the IRAS spacecraft. What the evolutionary status of the disk may be is uncertain.

is β Pictoris, shown in Figure 15.7. These disks are presumably the dusty remains of the disks of gas and dust that formed during the early accretion phases in the parent molecular cloud. It is speculated that massive disks may form a companion star—a binary system—whereas low-mass disks, like that of β Pictoris, may evolve into planets. We will return to this possibility in Part IV.

Stellar Rotation

Though something like this very rough picture of star formation may take place, several major difficulties (as well as most of the details) remain to be clarified. Rotation is one of these major problems. A large interstellar cloud is not completely still, but has internal motions of perhaps 1 km/sec. That is, different blobs of the cloud move at these velocities with respect to each other. Because it is extremely unlikely that these motions completely cancel each other, the cloud has a very small net spin, or more accurately, a net angular momentum, which, recall, is the product of mass, the velocity of the mass, and its distance from the axis about which it is rotating.

It is important to realize that a cloud must shrink by an enormous factor in order to form stars, very roughly by the factor by which a skyscraper would shrink to a tiny grain of sand! As the cloud contracts, the radius of the cloud decreases, so its rotational velocity increases. Since the cloud must shrink by a huge factor to become even a protostar, the spin rate would increase so much that the cloud could no longer contract, and no star would be formed. But stars do form, so this can't be the whole story. What is missing? What we have neglected thus far is the fact that clouds don't exist in isolation, but are magnetically threaded to surrounding clouds by the weak galactic magnetic field. As they move, these magnetic threads carry ionized gas along with them. This causes most of the rotation developed by a contracting cloud to be transferred via the magnetic lines of force to neighboring clouds, which take up some of the angular momentum. Consequently, the original cloud can shrink enough to form a star, though it will in general still be rotating rapidly. It is believed that early-type stars, which have equatorial velocities as given in Table 11.3, undergo this process.

A glance at Table 11.3 shows, however, that later-type stars have very little rotation. The Sun, rather than rotating once in a day or two as does an early-type star, takes about a month to do so. A much more efficient process must be operating to so markedly slow down these stars, and various mechanisms have been suggested to do this. For example, angular momentum may be transferred to planets forming along with the star, or it may be transferred by a strong stellar wind (during the T Tauri stage) interacting with an external magnetic field tied to surrounding material.

The Zero-Age Main Sequence

Even though much is not understood, we have a general idea, at least, of how stars form from the interstellar medium. As they contract toward the main sequence, their internal temperatures and densities increase. When the core temperature becomes high enough that nuclear burning liberates sufficient energy to completely halt the gravitational contraction, the star has reached the main sequence. To emphasize this point, we call the band in the HR diagram along which stars stop contracting the **zero-age main sequence**, or **ZAMS** for short. In other words, we are defining the age of a star in terms of its energy source: at zero age it has begun to derive all its energy from the transformation of hydrogen to helium. The position of a star on the main sequence is determined by its mass; the more massive the star, the higher on the main sequence is its equilibrium position.

Main-Sequence Lifetimes

Regardless of its mass, a star's greatest source of energy is the fusion of hydrogen; thus its main-sequence phase lasts the largest *fraction* of its life. The actual length of time spent by a star on the main sequence decreases for more massive stars; the more massive the star, the more rapid its evolution. This makes sense because the more massive the star, the higher its central temperature must be, so that pressure can support the mass. But the higher the central temperature, the faster the rate at which hydrogen is converted to helium (remember how rapidly the rate of nuclear energy generation increases as the temperature is increased; see Chapter 13).

This relation between stellar mass and evolution rate is true not only for stars on the main-sequence phase, but for all phases of a star's life, and can be taken as a fundamental rule of stellar evolution: *the rate of evolution through any phase of a star's life is proportional to the mass of the star*. Table 15.1 gives the number of years spent on the main sequence by stars of various masses.

Figure 15.8 gives the HR diagram for a young cluster. Each point represents an individual star, and the band indicates the zero-age main sequence. The fundamental rule just mentioned provides an interpretation of this diagram: the cluster is so young that only its more massive stars have reached the main sequence. Low-mass stars, evolving more slowly, are still in their contraction phase toward the main sequence. Many of these display the T Tauri characteristics described earlier.

***Table 15.1.   Main-sequence lifetimes of stars of
various masses***

Mass ($M_\odot$)	Lifetime (yrs)	Mass ($M_\odot$)	Lifetime (yrs)
15	1.0×10^7	1.5	2×10^9
9	2.1×10^7	1.25	4×10^9
5	6.4×10^7	1.0	1×10^{10}
3	2.2×10^8	0.5	3×10^{10}
2.25	5.3×10^8	0.25	7×10^{10}

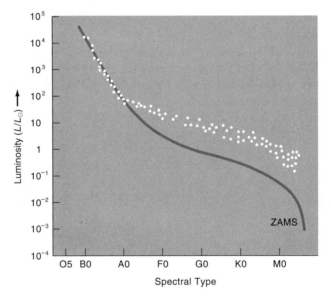

Figure 15.8.   Only the more massive stars in a young cluster will have reached the zero-age main sequence in a few million years. Less massive stars are still contracting to the main sequence.

Evolution off the Main Sequence

Massive Stars

After a time, the lowest-mass stars in Figure 15.8 will finally reach the main sequence, where they will stay for many billions of years. The massive stars at the upper end of the main sequence, however, will evolve rapidly in only millions of years. Soon, the most massive star will begin to evolve off the main sequence, leaving it at the so-called **turn-off point**. The position on the main sequence of the turn-off point for a particular star is directly related to the mass of that star: the greater the mass is, the higher on the main sequence is the turn-off point.

Let's consider the evolution of such a massive star, for example, one of 9 solar masses. The path it follows in the HR diagram in Figure 15.9 is given by the line that

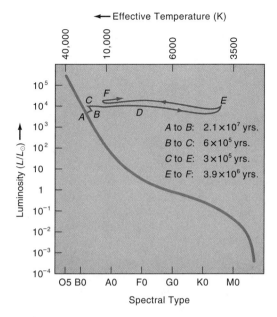

Figure 15.9. The post-main-sequence evolution of a 9-solar-mass star. See the text for an account of the various evolutionary phases.

represents the positions taken by the star as it ages.[2] Leaving the main sequence at its turn-off point and slowly moving toward slightly higher luminosities and lower temperatures on the HR diagram (toward *B*), the star expands slightly. This evolution is a consequence of the changing chemical composition in the core, that is, the steadily increasing abundance of helium in the core produced by the carbon cycle.

When nearly all the core hydrogen has been transformed to helium, the entire star contracts, releasing gravitational energy that maintains the star's luminosity. On the HR diagram the star moves slightly to the left, toward higher effective temperature (to *C*). The temperatures in the central regions of the star are now so high that hydrogen burning can take place outside the hot, inert helium core in a thin, surrounding shell. The core contracts because the little nuclear-energy generation going on there does not balance its gravity. The energy released by gravitational contraction does two things: part of it heats the core, and part of it is radiated away, adding to the energy produced in the hydrogen-shell burning region. As a result, the outer portion of the star expands and cools. The star's energy output decreases slightly because much of the internal energy of the star is used up in doing the mechanical work of expanding its outer layers. On the HR diagram the star moves rapidly to the right (*C* toward *D*).

This evolution is ultimately stopped when the core becomes sufficiently hot and dense for helium burning to begin by the triple-alpha process (at *E*). The star is then a red giant, or if more massive, a red supergiant. The triple-alpha process converting helium to carbon produces much less total energy than does the fusion of hydrogen to helium. Consequently, the giant or supergiant phases of massive stars are much shorter than their main-sequence lifetimes, lasting on the order of 10 percent of the latter.

Subsequent evolution of massive stars is very complicated, going through many twists and turns in the HR diagram. Note that a supergiant may pass through the in-

---

[2] Make sure you understand that the "motions" we are describing on the HR diagram represent changes in the star's effective temperature and luminosity, and not the motion of the star in space.

stability region in the HR diagram more than once, becoming a Cepheid variable each time. We can also understand why giants and especially supergiants are so rare: they evolve from massive upper-main-sequence stars, which are formed rarely compared to objects having masses like the Sun; in addition, their lifetimes as giants or supergiants are short.

Intermediate-Mass Stars

The near-main-sequence evolution of stars with smaller masses, say, from about 0.5 to 4 solar masses, is roughly like that of massive stars, just described. In their red-giant phases, however, these stars show a rapid and marked increase in luminosity (see Figure 15.10). This occurs because the electrons in their hydrogen-depleted cores are degenerate. We have already encountered the phenomenon of degeneracy when we discussed atomic structure in Chapter 9. We saw there that electrons can be located around the nucleus of an atom along certain "orbits" only, and that only a limited number of electrons can occupy a given path at any time. Otherwise, two electrons might have identical energies, orbital angular momenta, and spins, a circumstance that does not occur in nature and that is embodied in the Pauli Exclusion Principle.

Similar ideas apply to electrons that are not bound to any atom but are free. Within a small volume, no two free electrons may have the same energy or take the same path. When densities are low (as in all the gases we deal with in everyday circumstances), plenty of unoccupied energy states exist, and the motions of the electrons are not at all affected by the Exclusion Principle. When the density reaches about 10^6 gm/cc, however, the allowed lower-energy trajectories begin to be filled, and the electrons are forced (according to the Exclusion Principle) to occupy higher-energy states than would otherwise be the case. This has nothing to do with their temperature (just as the

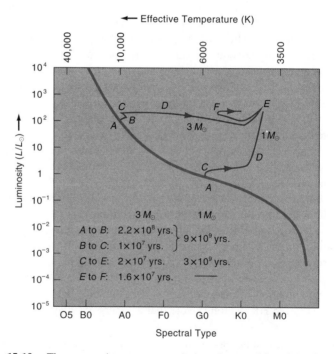

Figure 15.10.   The post-main-sequence evolution of stars of 3 and 1 solar masses.

third through sixth electrons of a carbon atom *must* go to the $n = 2$ level regardless of the temperature, because the $n = 1$ shell is already filled), but is a consequence of the enormous numbers of electrons packed into a relatively small volume. Hence, cooling such material does not slow down the motions of the electrons as it would those in an ordinary gas, since all allowed lower-energy trajectories are already filled. Nor will cooling this degenerate matter cause it to contract. The electrons, forced into high-energy paths in which they move very rapidly, exert a large pressure. This pressure, P, depends only on the number of electrons, N, in a given volume, that is, on the density: $P \propto N^{5/3}$. Temperature does not play a role. If the electrons have speeds that are a large fraction of that of light, then $P \propto N^{4/3}$. Recall that the pressure of an ordinary gas depends on its density *and* temperature: $P \propto NT$.

In the foregoing we have described only the behavior of the electrons, but matter also contains atomic nuclei to which the electrons were formerly attached. How do they act? Helium comprises the largest number of nuclei in the core of a star evolving toward the giant region. Each nucleus contains two protons and two neutrons. Each of these has a mass about 1,835 times greater than the mass of the electron. When temperatures are high and the electrons are moving rapidly, the much more massive nuclei will be moving relatively slowly. In fact, helium nuclei would have velocities only about 2 percent of that of the electrons. Hence there are plenty of trajectories available for the nuclei, and they will exert only the pressure of an ordinary gas, which is much less than the degeneracy pressure exerted by the surrounding electrons. Consequently, they do not contribute significantly to the outward pressure.

When the electron gas is degenerate and its behavior does not depend on the temperature, then the safety-valve mechanism (see Chapter 13) cannot work. The gas will not contract when cooled nor will it expand when heated slightly. The safety valve will work again only if the temperature of the degenerate material becomes high enough to give a substantial fraction of electrons energies well above the lowest allowed by the Exclusion Principle. When that occurs we say that the degeneracy has been "lifted."

The behavior of degenerate material has important consequences for stellar evolution. For example, when the core temperature of a star of mass between about 0.5 and 4 solar masses reaches 10^8 K, the core is degenerate, and so does not expand when helium burning begins. Consequently the core becomes hotter, which increases the rate at which the triple-alpha process operates (recall that its rate is extremely temperature sensitive, about $E \propto T^{18}$). But this raises the core temperature even more, increasing the energy-generation rate, and so on. In general, this process is called a **thermal runaway** (in this particular case, it is called the **helium flash**), produced because the normal safety valve does not work in degenerate matter.

In a very short time the core is producing an enormous amount of energy, which, however, does not increase the star's luminosity, because the energy stays in the core. In fact, the energy radiated by the star actually decreases, and the star slides down from E to F (the 3-solar-mass track in Figure 15.10). This happens because, instead of being radiated to space, the energy produced in the helium flash goes into increasing the energy of the degenerate electrons until the degeneracy is lifted. At this point the perfect gas law is operative again, and the safety valve works, ending the helium flash. The star then settles down for its life as a giant, producing energy by the triple-alpha process. Stars with masses greater than 4 solar masses do not develop a helium flash, because they are sufficiently massive to produce helium-burning temperatures of 10^8 K *before* their cores become degenerate.

Also, note that the mass-luminosity relation, valid for all main-sequence stars, no longer holds for intermediate-mass stars off the main sequence, because of the marked increase in their luminosity as they evolve into the giant region.

Low-Mass Stars

Stars with masses of only a few tenths of a solar mass have less eventful lives. Convective as protostars, they remain convective throughout after they settle on the main sequence. Their masses are so small that their interior temperatures become hot enough to allow only the first two steps of the proton-proton chain to be carried out, that is, through the formation of He^3. Thus the hydrogen-burning energy available to them is only a few percent of that available to more massive stars. This is compensated for in part, at least, by the convective mixing, which brings fresh fuel into the hot core from the rest of the star. Thus, the entire mass of the star is available as nuclear fuel.

Finally, a star with a mass less than 0.08 $M_\oplus$ cannot develop a core sufficiently hot to enable the p-p reaction to occur. Consequently, it never has a main-sequence phase. For a time it derives a little energy from the reactions of protons with light nuclei such as deuterium and lithium (see Chapter 13). Eventually, however, such a star contracts until its interior density becomes large enough that electron degeneracy alone keeps it in equilibrium. Such an object is called a **brown dwarf**, an intrinsically very faint, very red star. Some astronomers speculate that brown dwarfs might make up a significant fraction of the mass of our galaxy. Despite intensive efforts so far, however, only one or two brown dwarfs have been positively identified.

If the mass is perhaps no more than a few thousandths of a solar mass, then the interior does not become degenerate, and a planet will form, its solid or liquid interior capable of supporting the small mass.

Interpreting Cluster HR Diagrams

With our present knowledge of near-main-sequence evolution, it is now possible to understand the various features of the HR diagrams of clusters of stars. Clusters serve as useful tests of stellar evolution theory, because all the stars in a given cluster formed at about the same time and so are about the same age, have about the same initial chemical composition, and are all at the same distance from us. Thus differences in apparent brightness of the member stars are also differences in their intrinsic brightness.

Stellar Populations

A useful generalization concerning the content of stellar systems was given in the late 1940s by the German-American astronomer Walter Baade. He called **Population I** systems those that contain young, hot, blue, intrinsically bright stars and massive clouds of gas and dust from which these stars are forming. A young open cluster such as NGC 2362 is a good example. **Population II** systems, globular clusters, for example, contain little gas and dust, and their intrinsically brightest stars are red. These are old systems that no longer have the raw materials from which their stars formed, and they are more advanced in their evolution that are those of Population I. The HR diagram of a Population I system like NGC 2362 differs markedly from that of M3 (see Figure 15.11).

Population characteristics are also attributed to individual stars: young, hot, intrinsically luminous stars belong to Population I; very old stars are Population II objects. It is the age of a star that determines its Population type, so Population characteristics are not an either-or matter, but form a continuous sequence. We speak of extreme (very young) Population I types, old Population I objects like the Sun, extreme (the oldest) Population II stars, etc. Population characteristics are also manifested in

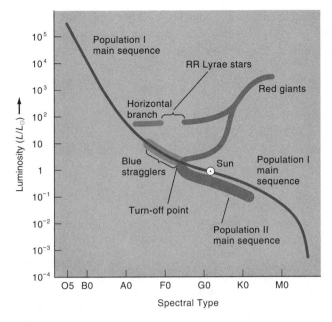

Figure 15.11. The HR diagram for a young Population I open cluster and the old Population II globular cluster M3. Notice how faint the M3 turn-off point is, indicating that the cluster must be at least several billion years old. Note also the RR Lyrae "gap"; any star in this region is an RR Lyrae variable. The so-called blue straggler region above the turn-off point in M3 would seem to violate the fundamental principle of stellar evolution, that the more massive the star, the more rapid its evolution. Instead, these are probably members of evolving close binaries in which mass has been transferred from one star to the other. The position of a straggler on the HR diagram is appropriate to its newly massive state.

other properties. Population I objects are located in or very close to the flat plane of the Milky Way, especially in the spiral arms, because that's where star formation is going on; Population II objects are located farther above or below the plane of our galaxy and have a spheroidal, rather than flat, distribution.

The Sun is located on the edge of a spiral arm; there are many open clusters (Population I) located nearby in our arm or in neighboring arms. By contrast, most globular clusters (Population II) are far from the Sun, surrounding the center of our galaxy or located far above or below the plane of the Milky Way.

Population I Clusters

Let us consider several stages in the evolution of a cluster, as shown schematically in the series of HR diagrams in Figure 15.12. We have already discussed diagram (A), in which the cluster is so young that the low-mass stars have not yet reached the ZAMS but are still contracting toward the main sequence. Some time later, the massive, rapidly evolving stars at the top of the main sequence will have begun to evolve away from it (B). These stars eventually evolve rapidly across the HR diagram (so that the probability of seeing a star in this region is small) and settle for a time as giants or supergiants (C). As time goes on (D), more and more of the upper main sequence disappears as the stars evolve to the giant region. This is never heavily populated, because

360

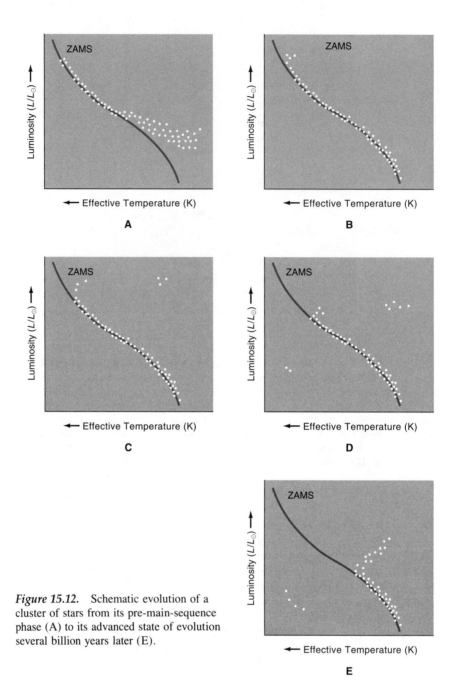

Figure 15.12. Schematic evolution of a cluster of stars from its pre-main-sequence phase (A) to its advanced state of evolution several billion years later (E).

stars don't live long as giants or supergiants, but soon move on toward the final, very low luminosity phases of their lives, perhaps as white dwarfs (see the next chapter). Finally, in (E), the cluster is so old that stars not even much more massive than the Sun have consumed so much hydrogen that their cores are helium-rich. Then they, too, evolve toward the giant region.

From model interior calculations we can find the length of time required by a star of any mass to begin to move off the main sequence. That is, starting from its first appearance on the ZAMS, we can calculate how old the star is when it has evolved to its turn-off point (see Figure 15.12C, for example). Hence this gives us a way to mea-

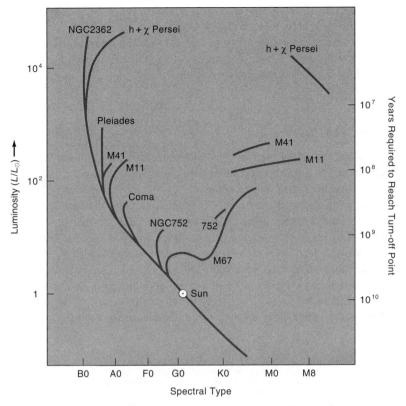

Figure 15.13. Several clusters plotted on the same HR diagram. They can be ordered easily by relative age, because the lower on the main sequence the turn-off point is, the older the cluster is. In addition, evolutionary theory allows absolute ages of the clusters to be estimated. The vertical scale on the right gives the age of the corresponding main-sequence turn-off points.

sure the age of a cluster. The "age" of a cluster is a little ambiguous, since this term implies that all the stars reached the zero-age main sequence at the same time. They do not, however, because, as we have seen, the rate at which protostars contract to the main sequence depends on their mass. As one considers older clusters, however, this spread in the formation time of stars in the cluster becomes a smaller and smaller fraction of the age of the cluster.

The absence of stars on the upper main sequence does not necessarily mean that the cluster is so old that these stars have evolved away. It could also mean that the cluster formed very recently but without any massive stars. It is clear, however, that when stars begin to move off the main sequence, they have been converting hydrogen to helium long enough that their structure is beginning to change. Thus, the turn-off point is a reliable indicator of the age of a cluster.

It is particularly instructive to plot several clusters on the same HR diagram, as in Figure 15.13. The clusters can be easily ordered in age; the higher the main-sequence turn-off point, the younger the cluster. NGC 2362 is so young that massive stars have not even begun to evolve off the main sequence. The double cluster h and χ Persei is only a little older, but the Coma cluster is much older, about a billion years old. Note the luminosities of the giant and supergiant sequences relative to the upper main sequences of the clusters to which they belong. Giants and upper-main-sequence stars in a younger Population I cluster have roughly the same luminosities. By contrast, the luminosity of the giant branch of the old Population I cluster M67 is much brighter than its upper-main-sequence stars, because it is so old that stars only a little more massive than the Sun are leaving the main sequence. As we have seen, such stars evolve in a rather different way than do more massive stars.

Population II Clusters

As mentioned above, model interior calculations enable us to calibrate the luminosity of the main-sequence turn-off point in terms of age. Since the model calculations are made for stars of different masses, to assign an age to a turn-off point requires knowing the masses of main-sequence stars. Furthermore, it assumes that the initial chemical compositions of the different open clusters were the same, so that they all have the same ZAMS. This may not be a bad assumption for the open clusters, but many globular clusters are deficient in the metals as compared with open clusters. This results in their main sequence being displaced to the blue, to the so-called **sub-dwarf main sequence**. Furthermore, the giant branch of a metal-poor cluster is also bodily shifted to the blue, the shift being greater the larger the metal deficiency, as shown in Figure 15.14. (The stars of M15 have a lower metal abundance than those of 47 Tucanae).

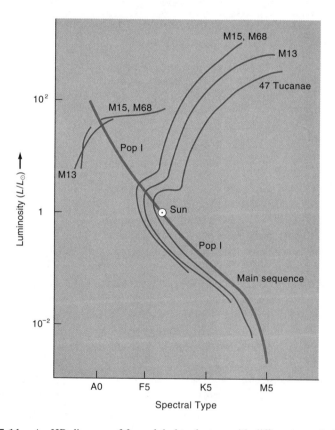

Figure 15.14. An HR diagram of four globular clusters with different metal abundances; M15 and M68 have only about 1 percent of the metal abundance of the Sun, M13 has about twice that, and 47 Tucanae has about 20 percent of the solar-metal abundance. Note that the more metal-poor the clusters are, the more their HR diagrams are shifted to the blue.

Figure 15.15.   The old open cluster, M67. It has survived for several billion years because it has a larger number of stars and therefore stronger self-gravitation than most open clusters.

Despite these uncertainties, the derived age of 13–14 billion years for 47 Tucanae is probably correct to within about 25 percent.

The globular cluster M3 shows a so-called **horizontal branch** blueward of its giant branch (see Figure 15.11). A horizontal branch is seen in many, but not all, globular clusters, and is made up of stars evolving from the red-giant phase toward their final end as white dwarfs. These stars derive their energy from the triple-alpha process. Within the horizontal branch is a region where no stable, nonvariable objects are found. The stars located there are the **RR Lyrae** variables, which are related to Cepheid variables in that they form the low-luminosity end of the period-luminosity relation. Thus RR Lyrae variability is a phase of stellar evolution just as is the Cepheid phase.

No young globular clusters are shown on diagrams like Figure 15.14, and most of the open clusters shown are much younger than the globulars. These differences can be generalized to all globular and open clusters: globular clusters are old, and open clusters are generally young. Open clusters surely must have formed several billion years ago, so why shouldn't we see them now as old clusters? The answer is simple. As the name implies, stars in open clusters are not nearly as densely packed as those in globular clusters. Consequently open clusters are much more easily pulled apart than are globular clusters by the gravitational forces of the galaxy's disk, by other clusters, etc. It can be shown, in fact, that most such clusters won't survive for more than a billion years or so; hence the paucity of old galactic clusters. An exception is M67 (Figure 15.15). It is old and has survived because it has more stars than most open clusters. By contrast, globular clusters, with tens of thousands of stars packed into a small volume, can survive this gravitational buffeting for billions of years. Hence the lack of young globulars means that they were formed a long time ago, when the galaxy was much younger and star formation was proceeding more actively. When we remember the large numbers of stars that each globular cluster contains, it is not surprising that they formed at an epoch when much more interstellar matter remained to be converted to stars than exists at present.

Post-Main-Sequence Evolution of the Sun

Let us conclude this part of our discussion of stellar evolution by considering the evolution of the Sun. Model calculations show that during its main-sequence life, the Sun's luminosity has been slowly increasing from its original ZAMS value, so far by about 30 percent. At the same time its diameter has increased by about 15 percent, so its temperature has remained essentially constant. The Sun is now at (3) on the HR diagram in Figure 15.16. Its mass, chemical composition, present temperature, radius, and luminosity are all well known. Model calculations indicate that to be in its present evolutionary state, the Sun must have arrived on the main sequence about 4.7 billion years ago (an age consistent with that of the oldest rocks we know of, about 4.1 billion years). As the Sun's luminosity slowly increases, so will the Earth's temperature. In about 1.5 billion years the polar ice caps will have melted and coastal areas will be reclaimed by the oceans. About 4 billion years from now the solar luminosity and radius will have increased over their present values by factors of 1.5 and 1.3 times, respectively; the Sun will be at (4) in Figure 15.16. (Total solar eclipses will have long since been things of the past.) Soon the core hydrogen will be exhausted, the core will contract, and hydrogen burning will take place in a shell surrounding the core. Much of the energy from the shell will go into expanding the outer layers of the Sun, so that the total luminosity will remain roughly constant and the Sun will move from (4) to (5) in about 3 billion years. At (5) the Sun will be about two times its present diameter. Because of the energy produced by the contracting core, the shell temperature will be-

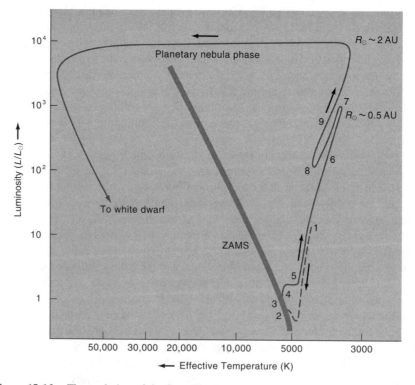

Figure 15.16. The evolution of the Sun. The Sun required about 10^7 years to contract from point 1 to the zero-age main sequence, 2, along the track indicated by dashes. In the 4.7 billion years since then the Sun has moved to its present position at 3. See the text for a description of subsequent evolution. The last phases are uncertain.

come very high. This will result in a big increase in the Sun's luminosity, so that in about another 0.5 billion years the Sun will climb to the red-giant region at (6). At the peak of the red-giant branch (7), its luminosity and radius will be about 1000 and 100 times, respectively, their present values. As seen from the Earth, the Sun will be a spectacular sight, an enormous red disk, which, when setting, will extend one-half of the way to the zenith! Mercury will be vaporized fairly quickly and become part of the red giant. Venus and the Earth will lose their atmospheres and the Earth its oceans. Venus and then the Earth will be engulfed.

The Sun's structure will differ drastically from its present main-sequence configuration. One-quarter of its mass will be contained within a radius only one-thousandth that of the red giant Sun. (Imagine a medium-size classroom about 10 meters long. If this represents the diameter of the star, then the core would have a diameter of only a centimeter or so!) The density of this tiny core is enormous, about fifty thousand times that of water, so that a cupful of core material would weigh about 10,000 kilograms.

When by contraction the core temperature reaches about 10^8 K, the triple-alpha process will begin, and the core will heat up. Since the core material is degenerate, however, its safety valve won't work, and the helium flash will occur. In a very short time, a large amount of energy will be produced in the core, all of which will go into heating it. Finally, the temperature becomes so high that the electron degeneracy is lifted, and the safety valve will operate again. Consequently, the core will expand and cool, decreasing the triple-alpha rate, and the luminosity of the Sun will decrease. It will move on the HR diagram from (7) to (8) in about 10,000 years. The temperature of the now nondegenerate core will stabilize at about 100 million degrees, and produce energy by conversion of helium to carbon. When the core becomes carbon-rich, the Sun's luminosity will again increase slowly (9), and the Sun will gradually leave the helium-burning main sequence, just as earlier it left the hydrogen-burning main sequence when its core hydrogen became depleted. The core is not sufficiently massive for temperatures to become hot enough for carbon burning, so it will have reached the end of its nuclear energy resources.

As we shall see in the next chapter, ultimately the Sun will lose mass in a planetary nebula phase and finally become a white dwarf having about half of its present mass, but only the diameter of the Earth. It will still maintain gravitational control over the solar system (diminished by three planets!), and the planets will continue in their endless orbits around the Sun. The white dwarf Sun's luminosity, however, perhaps 1 percent of its present value and slowly fading, will be too feeble to provide life-giving warmth to any of the planets.

Where Are the Sun's Neutrinos?

Though the structure and recent evolution of the Sun provide support for our ideas about evolution, the Sun is also the source of a long-standing challenge to stellar evolution theory or to related particle physics theory. First, a little background.

Because photons interact so readily with matter, energy generated in the Sun's core requires millions of years to reach the photosphere. During their long outward journey, the high-energy photons created by the fusion of hydrogen are degraded to those in the visual and infrared regions of the spectrum that we observe. Thus our knowledge of physical conditions in the Sun's center is both indirect and out of date. By contrast, neutrinos generated in the fusion process reach us only about eight minutes later, since they leave the core without interacting with solar material at all. So if we could detect these neutrinos, they would provide us with a window to the Sun's interior and tell us what is going on in the core *now*. The difficulty, of course, is that the same property that enables them to make a straight-line escape from the Sun also means that they are very difficult to detect, since detection requires interaction with matter. (To stop all the neutrinos in a given beam would require a slab of lead parsecs thick.) Furthermore, it is even more difficult to detect low-energy neutrinos than those of higher energy. Unfortunately, the enormous numbers of neutrinos produced by the Sun in the first step of the proton-proton chain have low energies. Only a tiny fraction of the total, produced in a side branch of the chain, are of the more easily detected high-energy variety. Nevertheless, a wonderful experiment was undertaken several years ago to detect just these neutrinos.

A tank containing about 100,000 gallons of cleaning fluid, C_2Cl_4, was set up deep in a gold mine in South Dakota to shield it from cosmic rays. If a chlorine nucleus captures a neutrino it is converted to radioactive argon by the following reaction:

$$^{37}Cl + \nu \rightarrow {}^{37}Ar + e^-.$$

The argon can then be detected because of its radioactivity. Despite the huge numbers of chlorine atoms in the tank, only about one argon atom per day was expected to be produced and measured! It would seem easier to look for a needle in a haystack. Nevertheless, the experiment was carried out with great care from 1968 to 1986, with the surprising result that only about one-third the expected number of neutrinos were found. There seems to be no question about the reliability of the result, so it has generated a great deal of discussion.

In addition, new solar neutrino experiments are now in operation in Japan and in the former Soviet Union. They are yielding the same general result—the expected number of neutrinos is not being detected.

Two types of explanations for this discrepancy have been proposed. One suggests that the standard solar model is in some way incorrect. For example, perhaps the temperature of the Sun's core varies, and it is now cooler than it has been in the past. The number of neutrinos and the energy generated would both decrease. The observed neutrino output would respond immediately to this change; however, the observed energy output would not show a corresponding decrease for millions of years, the time it takes for photons to travel out through the Sun. Or suppose that for some unknown reason the core abundance of elements heavier than helium is lower than expected. Then the interior temperature of the Sun would be lower, resulting in a decrease in the neutrino production. Neither these nor other similar possibilities are convincing, however.

The other type of explanation of this problem has to do with the properties of neutrinos. Three types of neutrinos exist, one of which is associated with electrons and is produced in the hydrogen-burning reactions in stars. It is this neutrino that is detected in the experiments we have described. The other two kinds of neutrinos are associated with particles we have not needed to discuss. If one or more of the three kinds of neutrinos has even a tiny mass, then it is possible that the ordinary neutrino produced in the hydrogen-burning reactions could be transformed to one of the other types. In that case it would not be detected, and there would appear to be a deficiency in the neutrino production rate. Such a result would have interesting consequences for theories that attempt to unify the electroweak and strong interactions that we will discuss in Chapter 22. It is a bit startling to realize that our familiar Sun might be the source of such exotic information.

Terms to Know

Contraction phase, accretion phase, accretion disk, protostar, T Tauri star, Herbig-Haro object, OB association. Hayashi track. Zero-age main sequence, turn-off point, thermal runaway, helium flash, brown dwarf. Sub-dwarf main sequence, horizontal-branch stars, Populations I and II.

Ideas to Understand

How clouds collapse; the formation and structure of protostars; the role played in star formation by magnetic fields; how stellar rotation is measured. The significance of the zero-age main sequence; the fundamental rule in stellar evolution; the evolution of stars of various masses. The cause and role of degenerate matter in evolution; why the safety-valve mechanism fails to work; the helium flash. The interpretation of HR diagrams of star clusters; the turn-off point; how ages of star clusters are found. The evolution of the Sun; the solar neutrino puzzle.

Questions

1. Describe the role played by dust first in the formation of a molecular cloud and later in the formation of a protostar.

2. A 10 $M_\odot$ and a 2.5 $M_\odot$ star are both on the main sequence. What can you say about their relative diameters, luminosities, temperatures, chronological ages, and evolutionary ages?

3. Why don't stars of earlier than O spectral type exist?

4. Explain why nuclear burning keeps the central temperature of a star lower than it would otherwise be.

5. Do all open clusters contain red giant or supergiant stars? Why or why not?

6. Describe briefly a situation in which you could be sure that a G2V star is very young, say, less than a few million years old.

7. Why are there only a relatively few galactic (open) clusters whose hottest main-sequence star is of spectral type F?

8. (a) Why do we believe that the Praesepe cluster is older than the Pleiades cluster? (Refer to the HR diagram for selected star clusters.)
   (b) It is generally possible to observe main-sequence stars of fainter intrinsic luminosity in open clusters than in globular clusters. Why do you suppose that is?

9. Consider three stars having the following masses: 30 $M_\odot$, 1.0 $M_\odot$, and 0.5 $M_\odot$. If all three formed in a cluster that is 5 billion years old, which would now be on the main sequence? Which would be beginning to evolve off the main sequence? Explain your answers briefly.

10. How would the overall (total light from all the stars) color of an open cluster change from the time its stars were first on the main sequence to the time the cluster was very old?

11. List the following stages in the life cycle of a solar-mass star from the youngest to the oldest stage: white dwarf, zero-age main sequence, T Tauri star, red giant after the helium flash. For each stage give the most abundant element in the center of the star and which element (if any) is undergoing nuclear fusion.

12. What is a neutrino? Why are astronomers interested in detecting them from the Sun?

Suggestions for Further Reading

Bahcall, J., "The Solar Neutrino Problem," *Scientific American*, **262**, p. 54, May 1990. What the problem is and its possible solution.

Bally, J., "Bipolar Gas Jets in Star-Forming Regions," *Sky & Telescope*, **66**, p. 94, 1987.

Boss, A., "Collapse and Formation of Stars," *Scientific American*, **252**, p. 40, January 1985. How star birth is studied with computers.

————"The Genesis of Binary Stars," *Astronomy*, **19**, p. 34, June 1991. How binary stars might form.

Fisher, D., "Closing In on the Solar Neutrino Problem," *Sky & Telescope*, **84**, p. 378, 1992.

Hartley, K., "A New Window on Star Birth," *Astronomy*, **17**, p. 32, March 1989. The power of submillimeter (between the IR and radio regions) astronomy for probing star-forming clouds.

Lada, C., "Deciphering the Mysteries of Stellar Origins," *Sky & Telescope*, **85**, p. 18, May 1993.

————"Star in the Making," *Sky & Telescope*, **72**, p. 334, 1986.

Stahler, S., "The Early Life of Stars," *Scientific American*, **265**, p. 48, July 1991.

CHAPTER 16

Stellar Evolution: The Late Phases

How do stars end their lives? What happens when all the nuclear fuel available to them is exhausted? Remember that it is the fuel supply in the stellar core that counts, because only there is it hot and dense enough for nuclear reactions to occur. It is significant that stars, which can display such a wide variety of characteristics during their energy generation phases, end their lives as one of only three kinds of objects: as white dwarfs, as neutron stars, or as black holes, collectively referred to as **compact objects**. The bizarre nature of these objects, however, more than compensates for their small number. During the last few decades an incredible variety of phenomena has been found associated with compact objects; there is no sign that we have come to the end of the discoveries. Let us consider in turn each of these three types of stellar corpses.

White Dwarfs

General Properties

In a general way, stellar evolution can be simply understood as the contest between inward-acting gravity and outward-acting pressure. For stellar masses, gravity can be countered by only two types of pressure: that produced by ordinary gas (augmented by radiation pressure in very luminous stars) or that exerted by degenerate matter. Gravity compresses and heats ordinary matter in the core of a star until nuclear reactions not only produce the energy by which the stars shine, but also maintain the gas pressure that supports the star. For a time gas pressure balances gravity, and the star is stable. When the nuclear fuel in the core is consumed, gravity—always acting—compresses the core even more, and its temperature and density increase. If the core temperature becomes hot enough to ignite the nuclear ash, then the star has another lease on life. Eventually, however, the star runs out of usable fuel and rapidly approaches the end of its active life.

In the previous chapter we described the primary energy-generation processes and the resulting changes in structure as stars evolved from main-sequence objects to red giants. Now we want to follow their post–red giant evolution to the stellar graveyard.

Suppose that the mass of the star is not great enough to compress (and consequently heat) the core to the point that the next set of energy-producing reactions can take place. The core of the star has then run out of its available nuclear fuel, and the gas pressure in the core will begin to drop. Without the energy provided by nuclear burning, ordinary gas pressure can't stop the compression, and the density increases as the core's radius decreases. Gravity will continue to squeeze the core until a coun-

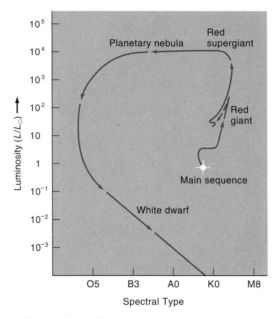

Figure 16.1. The schematic evolution of a low-mass star from the main sequence to its death as a white dwarf.

terbalancing pressure is produced sufficient for the star to be in mechanical equilibrium once again. For this to happen, however, the core density must be enormous, on the order of 10^6 gm/cm³. At this point, the pressure is provided not by an ordinary gas, but by degenerate electrons, and the star has become a white dwarf. Such a star is stable, and will not contract further if its mass is not greater than about 1.4 solar masses. The general track in the HR diagram that such a contracting star follows is shown in Figure 16.1.

Even though no energy is being generated, and the temperature of the star as a whole is dropping, the degenerate electrons cannot cool off and slow down because all the lower-energy states available to them are filled. Thus the pressure the electrons exert is undiminished. We saw in the last chapter that the pressure produced by degenerate electrons depends only on their density and not on their temperature. As a consequence white dwarfs display several curious properties. A relation between mass and luminosity does not exist for white dwarfs as it does for main-sequence stars. Their masses and radii are related, however, in the sense that the greater the mass, the *smaller* the radius of the star (see Table 16.1). This is just the opposite of the behavior of ordinary, non-degenerate matter, a chunk of which becomes *larger* as you add more material to it. (Think how frustrating it might be in a degenerate world trying to fill a bucket with sand!) Note also that the average density of a white dwarf increases rapidly as its mass increases. A 1.33 solar mass white dwarf is only six times more massive than one of 0.22 solar masses, but its average density has increased by more than 800 times! Although these are enormous densities, they are still many orders of magnitude less than that of an atomic nucleus (about 10^{14} gm/cm³). That is, even at the center of a white dwarf, the degenerate electrons are still widely separated from each other.

That the radius of a white dwarf shrinks as it becomes more massive results from another curious aspect of the behavior of degenerate matter. In increasingly massive white dwarfs, the electrons are packed closer and closer together, and so are forced to higher energy states. Consequently, they move more and more rapidly until their ve-

Table 16.1.   Relation between the mass and radius of a white dwarf (in solar units) and average density

Mass	Radius	Average Density (gm/cm^3)
0.22	0.020	3.8×10^4
0.40	0.0155	1.5×10^5
0.50	0.0138	2.6×10^5
0.74	0.0110	7.8×10^5
1.08	0.0071	4.2×10^6
1.33	0.00389	3.2×10^7
1.44	collapse	

locities approach that of light. When that happens, the pressure provided by degenerate electrons has decreased from $P \propto N^{5/3}$ to $P \propto N^{4/3}$, where N is the number of electrons per unit volume. A lower-density white dwarf can adjust its radius so that the gravitational and degenerate pressure forces are in balance. A white dwarf with a larger mass (and therefore higher density) produces a *smaller* degeneracy pressure relative to its mass, so that it is compressed further, and its radius decreases. With its greater mass and smaller radius, the star is subjected to an even greater gravitational force. Finally, a mass is reached such that its degeneracy pressure is unable to counter the greater gravity and the radius shrinks to zero! More precisely, degenerate electrons cannot support a mass greater than about 1.4 solar masses (see Table 16.1).[1] This is called the **Chandrasekhar limit**, after the Indian-American astrophysicist who developed the theory of white-dwarf structure (and for which he shared the Nobel prize in 1983). As the Chandrasekhar limit is approached, the central density of the white dwarf increases markedly, from about 10^6 gm/cm^3 for a 0.4-solar-mass dwarf to about 10^9 gm/cm^3 for a dwarf at its limiting mass.

What determines the mass of the white dwarf? Perhaps putting the question in another way might make the answer more apparent: what determines the mass of the degenerate core of a star in the last stages of its evolution? A little thought should convince you that the total mass of the star is the critical factor. The more massive the star, the more massive the white-dwarf core it will eventually produce, because in a massive star more matter will be brought to the high temperature required for its nuclear fusion. (Every appropriate fuel, of course, will ultimately be used up.) For example, the Sun will finally become a white dwarf having a mass of about 0.6 $M_\odot$, whereas a 6 $M_\odot$ star will produce a white dwarf of about 1.1 $M_\odot$. Stars of more than about 8 $M_\odot$ will produce dead cores too massive to be supported by degenerate electrons. Such cores collapse to much smaller and denser objects called neutron stars, producing a supernova outburst in the process. In the most extreme cases, the collapse results in a black hole. We will discuss neutron stars, black holes, and supernovae later in this chapter.

You might think that since their densities are so large, the interiors of white dwarfs would be extremely opaque, and there would be a marked temperature drop from the center outward. This is not the case, however; the temperature is the same throughout the interior of a white dwarf. Why is this? In ordinary stars, where energy is carried by radiation or convection, conduction is not important. (Recall that in conduction,

---

[1] Various small effects reduce this limiting mass to about 1.2 solar masses.

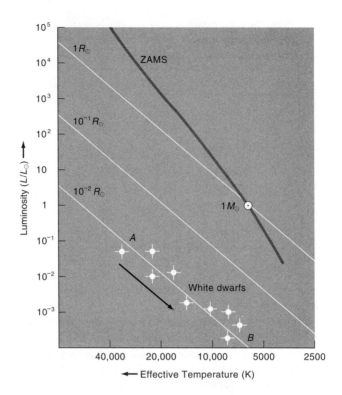

Figure 16.2. Lines of constant radius are shown on an HR diagram. Several white dwarfs with radii of about 0.01 solar radii are plotted. White dwarfs do not shrink as they cool, but maintain a constant radius and cool along the direction indicated by the arrow. White dwarfs of smaller radii will also cool along paths parallel to the arrow, but displaced to lower luminosities.

energy of motion—heat—is passed from one atom to its neighbor. The handle of a spoon in a hot bowl of soup becomes warm by conduction.) By contrast, conduction is very efficient in white-dwarf matter and is the primary mode of energy transport. In degenerate matter, electrons, carrying energy, can travel a long distance before interacting with their surroundings. Consequently, the largest portion of a white dwarf will have a uniform temperature of several million degrees throughout; the outer part of the star will not be much cooler than the core. It is only in the very thin (only a few kilometers) non-degenerate gaseous atmosphere surrounding the white dwarf that the temperature drops. Many white dwarfs show the Balmer lines of hydrogen, indicating an atmospheric temperature of only about 9,000 K.

The ultimate fate of a white dwarf is not at all spectacular; it simply fades away. Since energy can easily flow out of the star, the heat that it contains (which is a consequence of the motions of the atomic nuclei that are *not* degenerate) will steadily escape to space. The white dwarf will slide toward lower luminosities and temperatures, becoming fainter and redder (*A* to *B* on the HR diagram, Figure 16.2). As the ions cool, their energies of motion decrease, and eventually the electrical forces between them dominate. After about ten billion years the electrical forces cause the ions to form into a crystalline arrangement. Energy released in this process slows the cooling, but only temporarily, after which the white dwarf cools even more rapidly. Finally it cools to invisibility, a cosmic cinder.

Observational Evidence

So far, this discussion has been based on theoretical calculations. Observations can tell us a lot, however, because despite their intrinsic faintness, thousands of white dwarfs have been found. Approximately 10 percent of the stars in the vicinity of the Sun are white dwarfs. A few are members of nearby open clusters whose distances are known, and others are near enough to allow a reliable trigonometric parallax to be measured. From their distances we can find their luminosities, and their colors and spectra give the temperature of their radiating outermost layers. Their radii can then be found from $R = [L/(4\pi\sigma T^4)]^{1/2}$; these confirm the small values predicted by theory.

Careful searches have thus far turned up no white dwarfs fainter than about $10^{-4}L_\odot$. If this result holds up, it implies that not enough time has elapsed since the earliest epoch of star formation for white dwarfs to cool beyond this value. If the cooling times yielded by theory are accurate, then star formation in our galaxy out near the Sun (where we have found most of the white dwarfs) began about 10.5 billion years ago.

That in itself is an interesting result, but if you want to be adventurous you could take the argument one step further. If the formation of the Milky Way, including the disk where the Sun is located, began about a billion years after the universe itself was formed in the Big Bang (a reasonable guess, but only a guess nonetheless), then the age of the universe is only about 11.5 billion years. As we will see in Part III of this book this is a short time, in conflict with other evidence and not generally believed. It is not clear what the error(s) in this argument might be, but it is interesting that dead stars might give fundamental cosmological information.

Little direct information about the masses of white dwarfs is available, because only a few binary systems are known in which one component is a white dwarf. The best known of these is Sirius, a visual binary system for which the masses have been found, as we discussed in Chapter 12. The bright star that we see, Sirius A, has a mass of $2.2\,M_\odot$, but the mass of Sirius B, the white dwarf, is $1\,M_\odot$ (less than $1.2\,M_\odot$). White-dwarf masses have been determined in this manner for two other systems; both are under $1.2\,M_\odot$.

How Stars Become White Dwarfs

What kinds of stars become white dwarfs? The simplest answer would be, stars with masses less than 1.2 times that of the Sun. Although this is true, it is by no means the whole story, or even the most interesting part of the story. Consider Sirius again. Sirius B (the white dwarf) is certainly more advanced in its evolution than the bright A1V-type star, Sirius A. Consequently, its mass originally must have been *greater* than 2.2 $M_\odot$ in order for it to have evolved faster than its companion. Therefore Sirius B must have lost at least $(2.2 - 1.0) = 1.2\,M_\odot$ of material in becoming a white dwarf.

White dwarfs that are members of clusters also attest to the phenomenon of main-sequence or post-main-sequence mass loss. Several white dwarfs are found in the Hyades cluster. To have evolved into white dwarfs, these stars initially must have been at least as massive as the masses of the stars now at the main-sequence turn-off point. This is now at about spectral type F0, which implies a mass of about $1.7\,M_\odot$. Hence the Hyades white dwarfs also must have lost considerable mass. An even more extreme case is provided by the Pleiades cluster, where a white dwarf has been discovered. By the same arguments as used for the Hyades, the original mass of the white dwarf must have been greater than about 6 solar masses!

Stellar Winds. All this indicates that there must be efficient mechanisms whereby post-main-sequence stars lose a great deal of mass to become white dwarfs. Several different processes may be operative. As we have seen, many stars show stellar winds in their pre-main-sequence phases. Some stars, especially those of high luminosity, also show appreciable winds in their main-sequence and post-main-sequence phases. In extreme cases (the most luminous supergiants), mass streams out at 3,000 km/sec, with loss rates up to about 10^{-5} $M_\odot$/yr (about three Earth masses per year!). Mass loss of this magnitude can have profound effects on the structure and evolution of these stars. Toward the end of their lives, many intermediate mass stars will have lost enough material through a wind to be safely below the Chandrasekhar limit. Massive stars also lose large fractions of their mass by winds, though not enough to become white dwarfs. For example, stars with initial masses of 20 and 40 solar masses lose nearly half and three-quarters of their mass, respectively, by a wind. By contrast, the Sun produces only a stellar breeze of a few hundred km/sec and a mass-loss rate of only about 10^{-13} $M_\odot$/yr. Such a small rate has no effect whatsoever on the evolution of stars. In general, the more massive the star (and so the greater its luminosity and radiation pressure), and the later it is in its evolution, the larger its wind-driven mass loss.

Planetary Nebulae. One of the most effective (and easily visible) mass-loss processes in a pre-white-dwarf star produces a shell of gas we call a **planetary nebula**. (The mass-losing star eventually becomes a white dwarf.) Don't let the name fool you; such a nebula has nothing to do with planets. The term was given because these masses of gas reminded eighteenth-century astronomers more of disks similar to planets than of star-like points of light. Perhaps the best known of these is the Ring Nebula in Lyra (Figure 16.3 and Plate C4), an object easily visible in a small telescope.

Planetaries are emission nebulae with strong lines of hydrogen, helium, oxygen, etc. The energy to excite these atoms comes from the hot central star, which is also the physical source of the nebula. (That is, earlier the nebula was the outer part of the central star.) Since the effective temperature of the star is very high (up to 100,000 K), it radiates many ultraviolet photons, which ionize the gas of the surrounding nebula. As the atoms recombine, they emit the lines characteristic of H II regions.

From the roughly 2,000 planetary nebulae that have been discovered, it is estimated that there are perhaps 30,000 in our galaxy. This estimate is uncertain, because it depends on the distances to the planetaries, which are poorly known. From the sizes and densities of nebulae (the latter measured spectroscopically), the masses of some shells have been estimated. These are uncertain, because the size of a nebula also depends on its distance. In any case, it is clear that they range widely in mass. Values less than about 0.01 up to several tenths of a solar mass have been found. Perhaps 0.20 solar masses is a "typical" value. The diameter of our "typical" nebula is a few tenths of a parsec.

At first glance many planetary nebulae appear to be two-dimensional rings. A more careful analysis, however, especially by spectroscopy, shows that in reality they are fairly thin, very roughly spherical *shells* of gas, which are slowly expanding away from their central stars, as is indicated by the spectrum of such a "ring" sketched in Figure 16.4. Spectra taken with the slit across a diameter of the nebula give lines that are single at B and B'; toward the center, however, they are split into two lines. Gas at A' moving toward us produces blue-shifted lines, but gas at A causes the line to be red-shifted. The amount of the shift indicates a velocity of expansion of about 25 km/sec. In about 50,000 years the nebula will have expanded so much that it will become invisible, merging with the general interstellar medium. To lose an appreciable amount of mass a star must eject several shells. About half the planetaries do indeed seem to

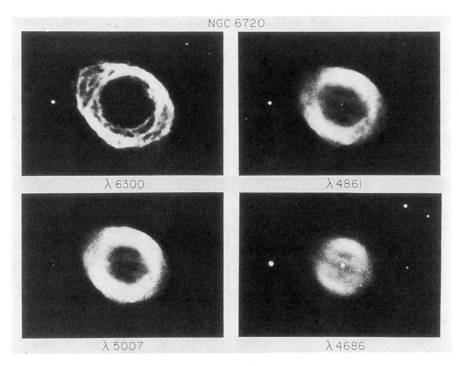

NGC 6720

λ 6300

λ 4861

λ 5007

λ 4686

Figure 16.3. The Ring Nebula photographed in the light of forbidden transitions of neutral and doubly ionized oxygen (6300 Å and 5007 Å, respectively), Hβ (4861 Å), and ionized helium (4686 Å). The helium image is the smallest because only the gas close to the central star will be hot enough for helium to be ionized.

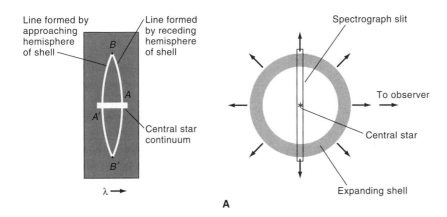

Line formed by approaching hemisphere of shell

Line formed by receding hemisphere of shell

Central star continuum

λ →

Spectrograph slit

To observer

Central star

Expanding shell

A

B

Figure 16.4. The spectral lines of an expanding shell of gas, as shown in (A), will be split by the Doppler motion of the shell toward and away from the observer. The central part of a line (A and A') will be split by twice the nebula's expansion speed, but the splitting will go to zero at B and B'. A real example of this splitting is shown in (B), along with the continuum of the central star.

consist of two or more shells ejected by the central star at different times. Recently, images of planetary nebula taken with very sensitive CCD detectors have revealed a faint "halo" of material surrounding the more easily seen shells, again indicating multiple mass loss episodes.

Several bits of evidence suggest that red giants are the progenitors, the immediate ancestors, of planetary nebulae. For example, the expansion velocity of planetaries, 20 or 30 km/sec, is also the escape velocity of matter out toward the edge of the envelope of a red-giant star. Dust grains exist in planetaries, probably formed in the cooler, more favorable environment of a red-giant atmosphere rather than in the hotter planetary nebula itself. The distribution of planetaries within the Milky Way—concentrated toward the center of our galaxy, but not strongly concentrated toward the plane—is the same as that of the older red-giant stars. Most significant, perhaps, is that the central stars of planetaries have properties similar to those of the cores of red giants: their composition is mostly carbon, oxygen, and heavier elements, rather than hydrogen and helium; their effective temperatures are high, up to 100,000 K; and their radii are about equal to the Earth's (note their location on the HR diagram in Figure 16.2). They evolve from the red-giant region with a constant luminosity about 6,000 times that of the Sun. After further cooling, these central stars become white dwarfs.[2]

It is not clear just how the shells are thrown off, but it should be emphasized that it is not a violent event like a nova outburst (see below). Instead, when the star is in its supergiant phase (Figure 16.1), instabilities occur in the helium-burning shell, resulting in so-called thermal pulses every few thousand years. These produce gusts of energy lasting several years, which, combined with a strong stellar wind, cause the star's bloated atmosphere to be blown away from the star. This shell of gas, when energized by radiation from the hot stellar core, is then seen as a planetary nebula. Theoretical calculations indicate that a star could evolve from the red supergiant stage to a planetary nebula in several thousand years.

An argument occasionally encountered for even more rapid red-giant evolution comes from an unlikely source. In the *Almagest*, Ptolemy lists the half-dozen brightest red stars in the sky. With one exception, these include stars like Antares, Betelgeuse, Aldebaran, etc., all of which are bright red giants or supergiants. The exception is Sirius, which today is distinctly bluish-white in color, consistent with its temperature of about 9,500 K. Many classical authors, including Horace and Seneca, for example, also make reference to a red Sirius. It seems unlikely that the star that is now the white dwarf in the Sirius system (the star that is more advanced in its evolution) could have been a red giant as recently as two thousand years ago. In addition, no shell or even wisps of gas have been found in the vicinity of Sirius. So what's going on? It turns out that Sirius (the so-called dog star) was associated by the Greeks and Romans with a hot and fiery wolf-like animal with a bronze (reddish) mouth. To prevent wheat rust, a serious agricultural disease associated with Sirius, the Romans sacrificed a red dog on about the day that Sirius sets with the Sun, April 25. Traditions and myths like these are most likely the origin of the red Sirius. However, a sixth-century manuscript giving instructions to monasteries on when to hold night services refers to a red Sirius. Because classical star names were not used in this document, it seems to be independent of the old traditions. Whether this really represents independent evidence for a red Sirius is hard to say, but the odds are against it (and the puzzle of a red Sirius is probably a red herring!). In any event, this is a good example of one of the many historical fascinations of an old science like astronomy. (The articles referred to above are given in *Suggestions for Further Reading* at the end of this chapter.)

---

[2] At the end of its red-giant phase, the Sun will lose nearly half its mass, leaving behind a white-dwarf core.

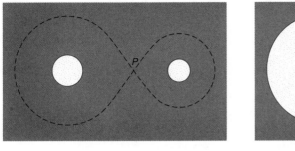
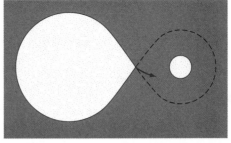

A   B

Figure 16.5. The Roche lobes for a binary star system are shown in (A). Matter at point *P* is equally attracted to each star and so could flow from one star to the other. In this system, however, little mass would ever reach the crossover region. If in its evolution to a giant, one of the stars filled its Roche lobe, as shown in (B), then substantial mass transfer could occur.

Incidentally, the expression *dog days*, referring to the hot days of mid- and late summer in the northern hemisphere, goes back to the sixteenth century when these days were counted starting with the heliacal rising of the dog-star Sirius.

Novae. Roughly half the stars in the sky are members of binary systems. The stars in most binaries are separated by a few AU or more, and so their periods of revolution about each other are measured in months or years. The gravitational field surrounding each star—the force felt by a particle at a distance *R* from each member of the binary system—is very nearly spherical, just as it is for a single star. It bulges only slightly toward the companion (see Figure 16.5). Now, at some point *P* between the two stars (halfway for equally massive components), the net gravitational force experienced by a particle is zero. This means that a particle, first gravitationally controlled by one star, could pass through this region and eventually be captured by the other star. (The two parts of this "figure eight" surface are called **Roche lobes**.) In most binary systems, however, the crossover region is so far away from either star that very little mass ever gets there. Thus in these systems the companion star has little effect on any mass lost by the other component. For example, only a very small fraction of the stellar wind streaming out from one star would be captured by the other.

More interesting things happen when the two stars are close together (a so-called close binary), so that their periods are measured in days, rather than months. In such systems a substantial part of the wind from one star could be gravitationally captured by the companion star, slowly increasing its mass. Another kind of mass transfer can take place in a close binary simply as a consequence of stellar evolution rather than a wind. As a star evolves away from the main sequence and becomes a giant or supergiant, it of course expands, so that its outer envelope gets closer and closer to the companion star. If it fills its Roche lobe, that is, the volume that is gravitationally connected to the other star, a substantial amount of matter can be transferred from the evolving star to its companion.

The material captured in this way would not immediately fall onto the companion star. Generally, the infalling matter would form an accretion disk around the companion for the same reason one is formed around a protostar (see Chapter 15); namely, the angular momentum of the captured material must be conserved (or remain constant). This means that the matter spins faster and faster as it approaches the companion, and so flattens into a disk. The disk would heat up as more material fell onto it. The disk

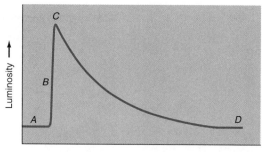

Figure 16.6. A representative nova light curve. *A* is the pre-nova phase. The outburst begins with a rapid increase in brightness, at *B*, taking from hours to a few days to reach maximum brightness at *C*. At this point, the system can be anywhere from about 1,500 to 1 million times brighter than it was in the pre-nova phase. Different novae fade at widely different rates; the fastest fade by a factor of 10 in as little as 10 days, whereas others may take 6 months. After years to decades, the nova has faded to its original brightness, at *D*.

is not stable, however. Blobs of matter in the disk would bump into each other, lose momentum, and eventually fall onto the companion, sometimes with dramatic results. Let us look at one of these results.

This is the **nova** phenomenon, a much more violent and spectacular event than the formation of a planetary nebula, but probably less effective as a mass-loss process. What we see as a nova is a previously faint (or even invisible) star that brightens rapidly over the course of a day or two, and reaches an intrinsic brightness as much as 10^5 times that of the Sun. It begins to fade almost immediately, taking several months to a year or two to settle back into its previous obscurity. This fading process is often irregular and may involve several brightenings or other variations (see Figure 16.6). The total energy output during the outburst may be as much as the Sun emits during a 10,000-year period. Spectroscopic evidence indicates that as a consequence of the outburst, a shell of gas expands away from the star with a velocity that might be as low as a few hundred kilometers per second or as high as 3,000 km/sec. The mass of the ejected material is relatively small, only about 10^{-3} to 10^{-4} $M_\odot$.

During its quiescent phase, the binary system in which the nova occurs typically shows only a late-type main-sequence (or near-main-sequence) spectrum with broad emission lines and a blue continuum superposed. The period of the binary system is short, often only hours long, indicating that the components are very close together. Apparently, the binary consists of a K- or M-type near-main-sequence star (producing the late-type spectrum) along with a hot object of low luminosity (producing no detectable visible radiation), that is, a white dwarf. (If the hot object's luminosity were even that of a main-sequence star, it would completely dominate the spectrum.) The broad emission lines and blue continuum are thought to arise in a hot, rapidly rotating accretion disk of gas surrounding the white dwarf (see Figure 16.7).

Such a binary system arises in an interesting way. The necessary conditions are a bit unusual but not extraordinary: a double star in which the two companions are close together and of unequal masses. Clearly, the white dwarf originally must have been more massive than its nearby companion in order to have evolved more rapidly. As it expanded into a red giant some of its material probably was captured by the companion star. After sufficient mass loss, the evolving star was left with its hot core, which cooled to become a white dwarf (see Figure 16.8). In its turn, the companion star eventually evolved off the main sequence and expanded. As it did, material from its outer

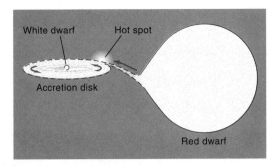

Figure 16.7. The type of binary system that becomes a nova. The late-type dwarf and white dwarf are close together, so that as the red star evolves and fills its Roche lobe, matter falls onto an accretion disk surrounding the white dwarf, producing a "hot spot" on the disk. Matter slowly dribbles from the disk to the white dwarf, where the fresh hydrogen eventually ignites explosively.

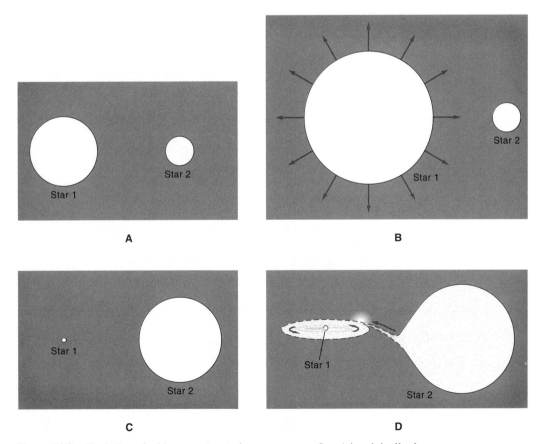

Figure 16.8. Evolution of a binary system to become a nova. Star 1 is originally the more massive of the pair (A) and so evolves more rapidly than its companion. As it evolves, it expands (B), develops a wind, and loses mass, some of which might be captured by Star 2. It eventually becomes a white dwarf. In its turn, Star 2 evolves (C), eventually transferring mass to the white dwarf, where it forms an accretion disk (D) from which matter falls onto the white dwarf and ignites explosively.

envelope eventually filled its Roche lobe and streamed toward the white dwarf, where it formed an accretion disk around it. Material from the inner part of the disk fell onto and slowly collected on the surface of the white dwarf. When enough of this hydrogen-rich matter sank to the high-temperature region of the dwarf it was ignited, that is, nuclear reactions occurred. (Remember that even the outer part of a white dwarf, just below its thin atmosphere, is very hot.) The freshly deposited hydrogen was explosively fused to helium, with the result that a shell was blown out of the system. The explosion and expanding shell is what we see as the nova phenomenon. That this is a superficial event for the star (in both senses of the word) is indicated by the observation that, only a few years after the outburst, the binary system appears to be as it was before the event.

The ejected shell is made up of the common elements—hydrogen, helium, carbon, nitrogen, oxygen, and iron. As the physical conditions in the nebula evolve during the course of the outburst, the appearance of the spectrum changes in a complicated fashion. It appears, however, that carbon, nitrogen, and oxygen are somewhat overabundant as a consequence of the nucleosynthesis that occurs during the explosive burning.

If this outburst happened only once during the history of such a double star, the event, though spectacular, would not be of much importance as a mass-loss mechanism. Two bits of evidence suggest that the process is repeated several times, however. Some binary systems called **dwarf novae** show repeated outbursts that last a few days and occur at intervals from weeks to months. These outbursts are similar to those of a nova, but much less energetic. They are a result of a sudden increase in mass transfer onto the accretion disk surrounding the white dwarf, rather than a nuclear explosion. In general, the more energetic such events are in a given system, the longer the time between outbursts. Extrapolating this relation to the violent novae explosions (which may not be justified) suggests that they should repeat at intervals of about 10,000 years. (A prediction safely beyond verification!)

More convincing evidence for repetition comes from the current estimate of how frequently novae occur in our galaxy, perhaps 50 each year.[3] If this rate were maintained for only the past few billion years, then nearly all the stars in our galaxy would have undergone a nova outburst. This is very unlikely. It is much more probable that only some stars become novae, but that these do so many times. The interval between explosions is simply the time required for the evolving red star to shed enough material onto its companion for it to sink to the high-temperature region of the white dwarf. To rid themselves of even a tenth of a solar mass, such systems must undergo several hundred explosions. This would require the nova phase of stellar evolution to last several million years, not a long time by stellar standards.

Let us conclude this section concerning white dwarfs as end products of evolution with a few comments. The maximum mass with which a star can begin its main-sequence life and still end as a white dwarf is uncertain; perhaps it is about 6 or 7 solar masses. The white dwarf limit of 1.2 $M_\odot$ applies to a non-rotating object. Rapid rotation could counter the force of gravity somewhat, enabling white dwarfs to exist with masses greater than this limit. No evidence for such stars exists at present, however. Finally, the fact that a star is initially formed under the white-dwarf limit of 1.2 $M_\odot$ does not necessarily preclude it from experiencing significant mass loss before becoming a white dwarf. The Sun, for example, will evolve into a white dwarf having a mass of about 0.6 $M_\odot$.

---

[3] The nova rate has been best measured in the Andromeda galaxy, a nearby system like our Milky Way. We can see Andromeda unhindered by the obscuring effects of interstellar dust, which makes it impossible to see novae in much of our galaxy.

Neutron Stars

General Properties

What happens if a star runs out of all nuclear fuel available to it, but has a mass greater than the Chandrasekhar limit? The pressure of degenerate electrons is not sufficient to counter the huge gravitational force, so the star cannot reach equilibrium as a white dwarf; it must contract further. As the density increases beyond 10^8 gm/cm^3, free electrons are driven into the atomic nuclei, where they combine with protons to form neutrons. As the density increases even more, the matter becomes increasingly neutron-rich, and more and more neutrons become degenerate. At densities of about 2×10^{14} gm/cm^3, the material consists nearly entirely of tightly packed degenerate neutrons with only a few charged particles remaining. At these enormous densities the strong nuclear force (see Chapter 9) no longer is attractive, but *repels* nuclear particles and stops the contraction. See how drastically the star has changed by the time this happens! Its density is nearly a billion times greater than the already huge density of a white dwarf, and its radius is only about a thousandth that of a white dwarf. This means that its diameter is about equal to the size of the District of Columbia! Not surprisingly, such an object is called a **neutron star**.

Composed mostly of matter squeezed to the density of atomic nuclei (one cupful would weigh over ten billion tons), a neutron star is an exotic beast. To further complicate matters, a neutron star has an enormous magnetic field, produced by the compression of the moderate field existing in its progenitor. For example, if the Sun, which has an extremely modest magnetic field, were to collapse to the size of a neutron star, its field strength would increase by about five billion times. Such intense magnetic fields strongly affect the behavior of the material they pervade.

A description in any detail of the structure of a neutron star is beyond the scope of this book, so we will simply point out a few general features. Model calculations show that the star is not uniform in density. The surface material is arranged in a sort of crystalline structure, so that it acts like a solid object, with consequences we shall describe shortly. A kilometer-thick layer at the surface, composed primarily of iron nuclei and degenerate electrons, has a density of 10^3 to 10^4 gm/cm^3 (see Figure 16.9).

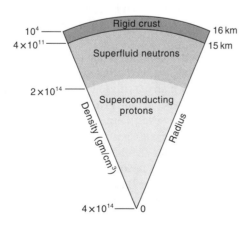

Figure 16.9. The crust of a neutron star consists of electrons and various heavy nuclei (mostly iron) that become extremely neutron-rich deeper in this very rigid layer. Beneath the crust there are no nuclei, just neutrons acting as a superfluid; that is, they flow without any friction. The relatively few protons remaining are in a corresponding superconducting state in which they have no electrical resistance. The enormous density in neutron stars gives rise to both phenomena.

Figure 16.10. A contemporary comment on the prediction that cosmic rays would be produced in a supernova that was the result of the collapse of an ordinary star to one only "14
miles thick," that is, a neutron star.

Beneath this is a transition zone a few kilometers thick that surrounds the main body
of the neutron star. This zone is composed of neutrons and a few protons and electrons,
at densities of about 10^{14} gm/cm³. The density of a single neutron is only a few times
greater than this, 4×10^{14} gm/cm³; so very little empty space exists in the interior of
neutron stars. The gravitational force exerted by such an object at its surface is huge.
Consequently, the speed required to escape from a neutron star is enormous, about half
the speed of light.

We are not sure what is the maximum mass that can be supported by neutron-rich
nuclear matter. Among other things, we need detailed knowledge of just how the strong
nuclear force behaves at extremely high densities. Today this is not known. Depending
on the assumptions made, model calculations give a range of up to about three solar
masses. Interestingly, rapid rotation can't help support the star, because the mass
equivalent ($m = E/c^2$) of the rotational energy is so great it simply adds to the problem.

That something like a neutron star should exist was first predicted in 1934, less
than two years after the discovery of the neutron itself. In proposing a connection between supernovae and cosmic rays, Walter Baade and Fritz Zwicky noted that if an ordinary star collapsed to a size such that its density equaled that of tightly packed neutrons, the gravitational energy thereby released would be about equal to the energy
output of a supernova explosion (see Figure 16.10). This remarkable hunch contains
the germ of the supernova process as we understand it today.[4] Four years later the prop-

---

[4] "With all reserve we advance the view that supernovae represent the transitions from ordinary
stars into *neutron stars*, which in their final stages consist of extremely closely packed neutrons"
(Baade and Zwicky).

erties of a star made up of neutrons were investigated theoretically, and it was found that there were no apparent reasons why such objects couldn't exist.

Some 30 years passed after their predicted existence before objects were found that could be identified as neutron stars. It is easy to understand why it took so long for them to be discovered. Neutron stars are simply too faint to be detected directly using conventional optical techniques, because their radiating areas, and consequently their luminosities, are so small. For example, a star only 10 parsecs distant with a 10-kilometer radius and the effective temperature of the Sun would be about 10 percent the brightness of the faintest star now detectable by the largest optical telescopes. For a brief time in the early 1960s astronomers thought that some of the newly discovered x-ray objects might be neutron stars. If one assumed that their temperatures were extremely high, say, 10^9 K, then, despite their small size, they would radiate copious amounts of x-ray radiation. Calculations soon showed, however, that such objects would radiate so much energy that they would cool very rapidly. Only a few months after their formation in a supernova event, their x-ray energy output would have become small. Since supernovae occur only rarely in a given galaxy, it was quite improbable that even one such x-ray-emitting neutron star should be caught immediately after its outburst, much less the many x-ray objects then discovered. Thus in the mid-1960s no evidence had been found for the existence of neutron stars. In fact, most astronomers believed then that all stars ended their lives as white dwarfs.

Pulsars: Their Discovery and Characteristics

It was only in 1967 that objects were discovered that were quickly shown to be neutron stars. These objects were named **pulsars**, after the most characteristic feature of the radiation they emit. They provide an excellent example of an accidental discovery in science, one that is worth describing. A group of Cambridge University radio astronomers, headed by Anthony Hewish, had undertaken a search for quasars by looking for radio sources that scintillated, or twinkled.[5] Remember that starlight, passing through the Earth's inhomogeneous atmosphere, twinkles, whereas planets, showing disks rather than points of light, don't. Similarly, point radio sources such as quasars scintillate when their radiation passes through the interplanetary medium near the Sun, whereas extended radio sources won't. Thus, even though a source might appear to be extended as observed by radio telescopes with their poor angular resolution, if it scintillates it would actually be a point source and hence likely be a quasar. This technique works best at long (meter) wavelengths, where, unfortunately, the sensitivity of radio receivers is low. Hence a telescope with a large collecting area had to be constructed. In fact, the Cambridge telescope was a fixed array of antenna wires tuned to a wavelength of 3.7 meters, and spread over 4.5 acres of the English countryside.[6] Furthermore, the radio receivers had to be capable of recording the rapidly changing (many times per second) intensity variations produced by scintillation. Such equipment had not before been necessary in radio astronomy, because source intensity fluctuations, if they occur at all, generally take place on time scales of days or longer, not fractions of a second. All in all, the Cambridge telescope was unusual, and just happened to be well suited for observing the then-unknown pulsars.

When the instrument was put into operation, Jocelyn Bell, then a graduate student in Hewish's research group, was given the responsibility of running the equipment and examining the data, recorded on about 100 feet of paper each day. Because the radio

---

[5] Quasars are the extremely luminous cores of some distant galaxies; see Chapter 20.

[6] Built by a half dozen graduate students, it required driving one thousand 9-foot-long wooden posts into the ground on which were suspended 120 miles of wire, cable, and cord.

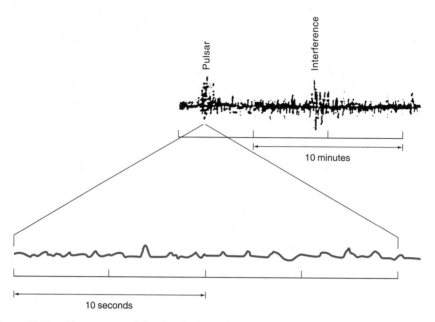

Figure 16.11. The top part of the sketch shows the "scruff" that caught the attention of Bell. Because the minute or so during which the signal was present extended for less than a half-inch on the paper record, nothing could be said about the time structure of the signal. However, a high-speed recording of the signal, as given on the lower portion of the figure, showed pulses separated by 1.33 seconds.

telescope was not pointable, it could detect objects only as their daily motion moved them across the one-half degree of sky (in the east-west direction) to which the antenna was most sensitive. A point source would be seen by the telescope for only a few minutes each day, and, given the speed with which the data recorders ran, corresponded to no more than one-half inch on the paper record. When a likely source was found, it was observed again but with the data recorders running about 100 times faster so that the signals, generally produced by scintillation, could be spread out on the paper. Bell noticed that occasionally, a ratty-looking signal she called "scruff" would appear that resembled neither the point sources she was looking for, nor locally produced electrical "noise" (see Figure 16.11). Though it would come and go, the source seemed to be fixed in the sky and so was not local. After many weeks of trying, a high-speed recording of the signals from this source was finally obtained. Instead of the randomly spaced brightness fluctuations of scintillation, there appeared a series of pulses that—though of varying amplitude—were regularly separated in time by 1.33 seconds! Such pulses were completely unexpected. Nothing remotely comparable had ever been observed in astronomy. The most rapidly varying astronomical objects then known were a few binary stars that had periods of about an hour. But this new object varied with great regularity a thousand times more rapidly. These signals seemed so bizarre that for a short time, the Cambridge group thought they might be artificially created, coming from distant civilizations! Soon, however, Bell noted three other sources with the same behavior as the first very fast regular pulses (one with a period of only a quarter of a second) and variable amplitudes. It became clear that these were a new class of astronomical objects.

With the announcement in 1968 of the discovery of pulsars, many astronomers immediately began searching for others and investigating these remarkable objects;

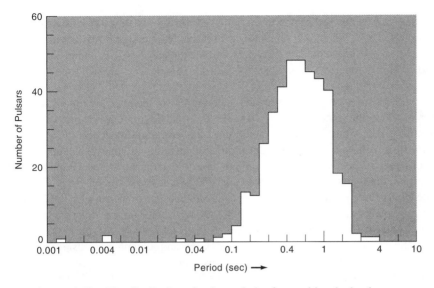

Figure 16.12. The distribution of pulse periods of several hundred pulsars.

some astronomers even found recordings of such signals that they had overlooked. Rather quickly the general features of pulsars were established. As of this writing more than 400 have been discovered; of these about 300 have been well observed. From their characteristics the following generalizations can be made.

(1) The pulse periods now known range from about 0.0016 sec to 4.3 sec, with the most commonly observed period being about 0.5 second (see Figure 16.12).

(2) The pulse amplitudes vary (because of scintillation), but their periods are remarkably constant—more so than all but the most sophisticated atomic clocks. This constancy, combined with their rapid repetition rate, enables their period to be measured with remarkable precision. For example, the first pulsar discovered has a period of 1.337301109 seconds!

(3) The pulse lasts for no more than 0.1 of the period, and more commonly for only a few percent of the period. For example, if the period is one second, the pulse width—the time during which the pulsar is "on"—is usually no more than a few hundredths of a second. This has an interesting implication. Suppose we are observing an object of radius R (see Figure 16.13). Light from A arrives before that from B and C,

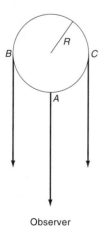

Figure 16.13. A luminous object of radius R cannot change its brightness within a time of less than R/c. See text for explanation.

since the latter is delayed by the time t required for light to travel the distance R, that is, a time $t = R/c$, where c is the velocity of light. Now imagine that the object were somehow turned off; would it disappear instantaneously? No, it would take a time R/c to go out. In other words an object of radius R cannot vary faster in time than about R/c.

So if a pulsar is on for 0.01 seconds, then the size of the emitting region is roughly $0.01c$ or a few thousand kilometers, smaller than the Earth. The volume of space in which the pulsar's radiation is generated must be very small indeed.

(4) In a given pulsar the pulse shape differs depending on the wavelength at which it is observed—the longer the wavelength, the wider the pulse (see Figure 16.14). Also, the radiation is highly polarized, up to 50 percent in some cases.

(5) Only three pulsars emit pulses outside the radio range, in the x-ray and gamma-ray regions, and only three have so far been found to radiate pulses in the visible region of the spectrum (the Crab in Taurus, one in Vela, and one in the Large Magellanic Cloud; the first two are also two of the three x-ray and gamma-ray pulsars).

(6) They are not uniformly distributed over the sky, but are concentrated toward the plane of the Milky Way, and hence must be within our galaxy. Most are within a few thousand parsecs of the Sun, and are on average within about 375 parsecs above or below the galactic plane. Their mean velocity away from the plane is about 120 km/sec. Note that both of these values are larger than the corresponding averages for young Population I objects.

What kind of object can display all these properties? We need to find some very rapid, repetitive process, taking place on a small object that can be associated somehow with the observed pulses of radiation. Let's first consider which processes might repeat themselves at a rate of about once per second. It is easy to eliminate all but one of the possibilities.

(1) Artificial signals (extraterrestrial civilizations) can be ruled out. Since we would expect the sources to be located on planets, the Doppler effect would produce a slow, systematic periodic change in the rate of arrival of the pulses as the planet orbited its star. No such change in the pulse period is observed. Though it is possible that the orbits of a few of these planets would be oriented face-on to our line of sight and, therefore, show no Doppler effect, it is completely unlikely that all 400 or so of the pulsars now known would be oriented in this way. It would be even more astonishing that there should be so many civilizations we could easily detect.

(2) Binary systems can be eliminated easily, because ordinary stars can't possibly revolve around each other so rapidly.

(3) Stellar pulsations, that is, the regular expansion and contraction of a star, can also be eliminated. You can see this in the following way. The rate of stellar oscillations depends on the average density of the star. Imagine you are holding one end of an ordinary door spring that has a weight suspended from the other. Pull the weight down and let it go. If the spring is easy to stretch, the weight will bounce up and down slowly; but if the spring is stiff, it will oscillate rapidly: the stiffer the spring the shorter its period of oscillation. The average density of a star is analogous to the stiffness of the spring, in that the greater the density, the "stiffer" the star and the shorter (faster) its pulsation period. The densities of ordinary stars are such that their natural pulsation periods are generally long, hours to days, like the periods of Cepheid variables. Even white-dwarf stars with their large densities would have periods of about ten seconds, too long for the pulsars. On the other hand, because their densities are so enormous, neutron-star pulsations would be too fast, on the order of 0.001 seconds. Pulsations of neither kind of compact object can produce the one-second pulsar periods.

(4) We are left only with stellar rotation. The Sun rotates once in about 28 days. In order to rotate once per second its equatorial velocity would have to be greater than the speed of light! Clearly, ordinary stars cannot rotate rapidly enough to produce a pulse each second. Even a white dwarf the size of the Earth would have to rotate with an equatorial velocity of about one-fifth the velocity of light, which is sufficient to disrupt the star.

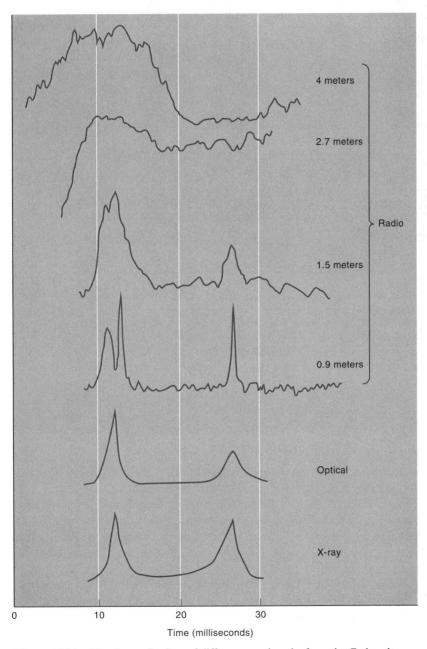

Figure 16.14. The shape of pulses of different wavelengths from the Crab pulsar.

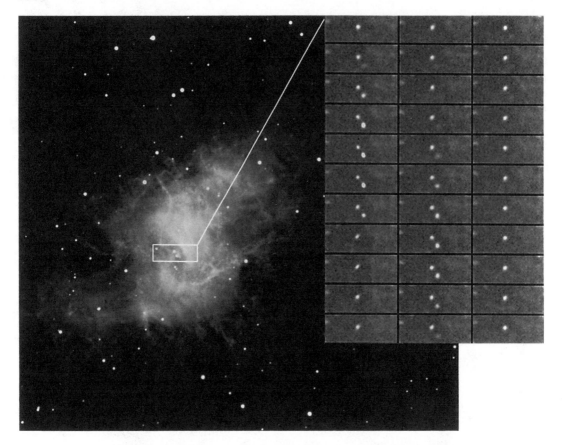

Figure 16.15. The photograph is of the Crab nebula and the pulsar that powers it. The inset consists of a sequence of 33 observations of the Crab pulsar and the nearby star, each about one millisecond long (so the sequence encompasses the period of the pulsar), running from top to bottom and from left to right. The first column shows the brighter, primary pulse from the pulsar; the sequence in the second column records the secondary pulse; the third column shows only the constant nearby star, the pulsar being "off." Another constant star is seen at the left edge of each frame.

Finally, could neutron stars safely rotate fast enough? Yes! Since they are so small, a rotation period of once per second translates to an equatorial velocity of only about 50 km/sec. This is an extremely modest velocity that causes no difficulties at all. So by a process of elimination we are left with this one plausible natural phenomenon that has a very regular periodicity of about one second, equal to the period of a typical pulsar. Pulsars are rapidly rotating neutron stars. Here was evidence, admittedly indirect, that neutron stars actually exist. Note, however, that we have not yet identified the mechanism producing the pulses, only the means by which a rapid repetition rate is achieved. For the discovery of pulsars, Hewish shared the Nobel prize in physics in 1974.

That these objects must somehow be associated with stellar death was strongly implied when a pulsar was discovered in the Crab nebula (see Figure 16.15 and Plate C4). The pulsar is the remnant object of the supernova outburst that produced the nebula. So it now appeared that there were two kinds of stellar corpses: white dwarfs and neutron stars.

Pulsars: Energy Generation

We still have to answer many questions, the most obvious of which is how the pulsed radiation is produced. Because the pulse width is quite narrow compared with the repetition rate (recall that the pulsar is "on" for only a small part of its period), the radiation must somehow be beamed, rather than emitted in all directions from the entire surface of the neutron star (see Figure 16.16). The flashes produced by a lighthouse

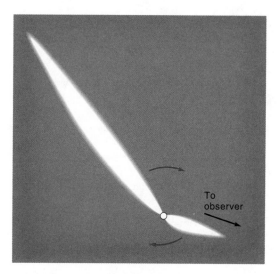

Figure 16.16. The angular size and relative intensity of the Crab pulsar's primary and secondary beams (seen from "above") as measured in radiation from about 1600 Å to 6000 Å. As the neutron star rotates, the two beams are swept across our line of sight, producing the pulses of light. Note that the pulsar is off most of the time.

or an airport beacon are familiar examples of this idea. The intense magnetic field of a neutron star (10^{12} or 10^{13} times stronger than the Earth's) probably plays a key role in beaming the radiation. Charged particles trapped in the magnetic field are strongly confined and somehow "focused" by the particularly powerful field at the two magnetic poles. Now suppose that the magnetic axis is not aligned with the rotational axis (see Figure 16.17). Then as the star rotates, each of its magnetic poles will sweep across a small strip of the sky once each rotation period. Radiation continuously generated by particles confined and accelerated by the polar magnetic fields will appear as the pulsar's flash if our line of sight to the pulsar happens to include one of its magnetic poles.

Since this general picture of the pulsar mechanism emerged, theoretical models of pulsars have been proposed at a rate (it sometimes seemed) only slightly slower than that of their discovery. Despite this great effort, the mechanism by which pulsars generate radio energy is still not clear. For example, the geometry of the situation is not certain. Is the energy produced near the surface of the neutron star, perhaps at the magnetic poles where the field is strongest? Or is it generated farther out from the star, where the field would force the particles to be swept around at very nearly the velocity of light? A model must also explain how the radiation can be strongly polarized.

The energy produced by a pulsar is very nearly incredible. Radio pulsars emit energy at a rate of 10^{-7} to 10^{-3} $L_\odot$, an enormous amount of energy to come from such

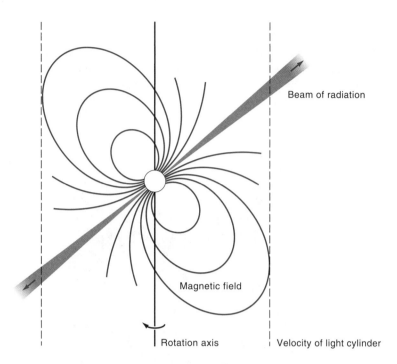

Beam of radiation

Magnetic field

Rotation axis | Velocity of light cylinder

Figure 16.17. A model for a pulsar. The narrow beam of radiation escapes along the direction of the magnetic axis, inclined to the rotation axis. At the dashed lines—which represent the speed-of-light cylinder—the magnetic field lines, rigidly tied to the neutron star, move at the speed of light.

a small object. (It may help bring the point home to mention that this radiated power is many hundreds of billions times the energy that all the radio and TV stations in the world could emit simultaneously!) The most intensively studied pulsar is located in the Crab nebula. Its radio radiation is intense, about 1 percent of the total luminosity of the Sun. In the visible spectral range, however, it radiates 100 times as much energy as in the radio region, and in the x-ray region 20,000 times as much as in the radio region. At the distance of the Sun, the Crab pulsar would quickly roast everything on the Earth.

Furthermore, all this energy is generated within a very small volume; the amount of energy per unit volume is enormous. The maximum volume in which energy is produced can be estimated because, whatever its exact nature, the radiation mechanism must be connected with the intense magnetic field associated with the neutron star. This field will be swept around by the star's rotation so that at increasing distances from the star the field moves with greater and greater velocities. Charged particles will be carried along by the magnetic field, so that their velocities will be larger and larger the farther they are from the star, until they move with very nearly the velocity of light. The surface defined by the distance from the neutron star at which particles travel with $V = c$ is called the **light cylinder**. The physical characteristics on this surface change abruptly, and set a limit within which the pulsar's energy must be generated. Typically this surface is about 10^4 km from the pulsar. For the Crab pulsar, the power generated within the volume of its light cylinder is roughly 100,000 times greater than that generated per unit volume by thermonuclear processes in the core of the Sun! If the pulsar's energy is generated within a much smaller volume near its surface, the energy

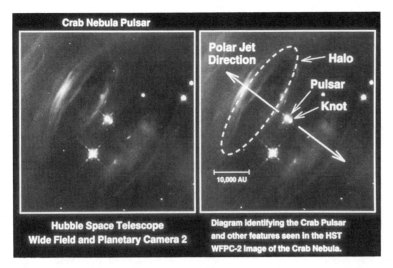

Figure 16.18. A high-resolution image of the pulsar and its immediate surroundings, obtained with the Wide Field and Planetary Camera on HST, reveals a previously undetected knot of emission just 1,500 AU from the pulsar. This knot and the pulsar are thought to lie along the rotation axis of the pulsar. In the opposite direction, wisps have been located about 10,000 AU from the pulsar that seem to form a ring of emission centered on its rotation axis. The interpretation of these features is still uncertain.

density is even greater. New data are becoming available from the repaired HST that may help answer some of these questions (see Figure 16.18).

Pulsar Spin Rates

Though we still don't know exactly how pulsars generate the energy they radiate, we can be fairly certain about their ultimate energy source, because pulsar periods are observed to increase very slowly over time; that is, the rate at which neutron stars are spinning is gradually decreasing (see Figure 16.19). The spin-down rate is exceedingly

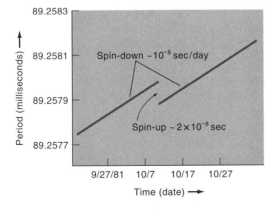

Figure 16.19. The rotation of the relatively young Vela pulsar is slowing (period increasing) by about 10^{-8} seconds each day. In October 1981, a glitch—a spin-up—occurred, and the period suddenly decreased by about 2×10^{-8} seconds. Over a period of weeks, the pulsar recovered its earlier spin-down rate.

slow, however, on the order of 10^{-10} sec/day, and so does not really contradict earlier statements concerning the constancy of pulsar periods. This spin-down, slow as it is, means that some of the neutron star's rotational energy is being lost. In fact, the energy lost as the Crab pulsar's spin slows down is just about enough to account for all the electromagnetic energy it radiates away, from the radio to the x-ray region of the spectrum. In other words, the pulsar's spin, a form of mechanical energy, is somehow being efficiently converted to radiation, a form of electromagnetic energy.

Though the rotation rates of pulsars are gradually slowing, a few have been observed to speed up suddenly in an event called a **glitch**. By several days after the spin-up, the pulsar usually has recovered and again steadily slows down, sometimes at a different rate than before, however. Very small spin-ups, like those observed in the Crab pulsar, can be accounted for if the radius of the neutron star suddenly decreases by a very small amount, less than one millimeter. Since the angular momentum mVr remains constant, if r decreases, V must increase. Apparently, the magnetically stiffened crust of the neutron star adjusts itself to strains produced by changing conditions—cooling perhaps—by suddenly shrinking in a sort of starquake.

Spin-ups about 100 times greater than those in the Crab have been observed in other pulsars and are more difficult to explain. The problem is that the crust cannot support the much greater strains required to produce the larger spin-up. Something else besides starquakes must be going on. In general, both kinds of glitches occur more often in the younger pulsars; older pulsars apparently become more stable.

The general slowing of the spin rates of pulsars not only suggests that the most rapidly rotating ones are the youngest, but it also gives a way of estimating their ages. The simplest assumption to make is that the decrease in the spin rate has been constant since the pulsars' formation. Then the time required to reach the presently observed period would be found simply by dividing the present period by how fast the period changes. That is, $t \propto P/\dot{P}$, where $\dot{P}$ is the rate of change of the period. However, one would expect that the spin rate decreased more rapidly soon after the birth of the neutron star than it is doing now. This leads to a time estimate given by

$$t = 0.5\frac{P}{\dot{P}}.$$

For the Crab pulsar (which until recently had the shortest known period), $P = 0.033$ seconds and $\dot{P} = 36.5 \times 10^{-9}$ sec/day $= 4.2 \times 10^{-13}$ sec/sec; that is, every second the period increases by 4.2×10^{-13} seconds. Thus,

$$t = \frac{(0.5)(0.033)}{4.2 \times 10^{13}} = 3.9 \times 10^{10} \text{ seconds} = 1{,}240 \text{ years.}$$

As we shall see, the Crab pulsar is thought to be about 930 years old; so the age we have just estimated is approximately correct. In this way, ages of pulsars are found to range from 10^3 to 10^9 years.

In the 1980s three pulsars were discovered that indicated that there can be exceptions to this relation between rotation rate and age. Their periods are 6.1, 5.4, and 1.6 milliseconds, all much faster than the Crab's 33-millisecond period. (Note that 1.6 milliseconds implies a spin rate of more than 600 times per second or 36,000 rpm! The surface of the neutron star at the equator must be moving with about 10 percent of the speed of light.) If these rates were a consequence of the pulsars' youth, they must have been born yesterday. However, no evidence of an outburst, like an expanding cloud of gas, exists around any of these objects. So why are they spinning so rapidly? An explanation for two of the three may be found in their membership in binary systems. It is speculated that in its evolution, the companion star has transferred mass and angular momentum to the neutron star, causing it to spin up to the presently observed rate.

The fastest of the three pulsars, however, does not appear to be in a binary system, at least at the present time. Perhaps it did have a companion at one time from which it acquired mass and spin, after which the newly active pulsar "evaporated" its companion. That is, radiation streaming out from the pulsar heated up the side of the companion facing it, so that matter was boiled off the star, eventually destroying it completely. Alternatively, if the companion evolved to a white dwarf after causing the neutron star to speed up, perhaps the two stars coalesced into one. In any case, it appears that old neutron stars can be rejuvenated, and so the age-pulse rate relation does not apply to these objects.

These so-called **millisecond pulsars** (of which many more have now been discovered) have much smaller magnetic fields than their slower cousins. As a result, they produce less radiation and do not interact as strongly with their surroundings. Thus their spin rates change much more slowly than do those of ordinary pulsars. By far the most accurate clock known in nature or in the laboratory is a millisecond pulsar with a spin-down rate of only 3×10^{-12} sec/yr, or about a billionth of a second every thousand years! This is to be compared with ordinary pulsars that spin down about 10^4 times more rapidly. Since they lose energy so much more slowly, millisecond pulsars last much longer as pulsars than do their "classical" counterparts. Fundamentally, it is the strength of the magnetic field that determines the properties of pulsars, including their lifetimes.

Incidentally, only a half dozen or so of the more than three hundred well-observed pulsars are in binary systems, a far smaller fraction than one would expect from the large number of such systems observed around us. It is not a question of being unable to detect possible binary pulsar systems. Since the period of a pulsar can be measured accurately, a periodic change in that period—a Doppler shift caused by orbital motion—can be detected easily. Because a pulsar rotates so many times in even a day, the sensitivity of the technique is extremely high; even if a companion were only of planetary mass, the regular period shift in the pulsar could be measured. Perhaps the process of formation of a neutron star—most probably a supernova explosion—somehow disrupts the binary system.

A millisecond pulsar (PSR 1257 + 12) has been found to have not one companion, but two.[7] Even more interesting, these two companions are of planetary, not stellar, mass! They have been inferred from tiny period changes as indicated above. Preliminary results give the mass of each object to be about three Earth masses, their orbits nearly circular, one located 0.36 AU, the other 0.47 AU from the pulsar, with periods of 66.6 and 98.2 days, respectively. As of this writing, this pulsar had been observed intensively for more than a year; observations over a longer time should confirm and possibly refine these results.

How these objects formed is unclear. Although they are planetary in mass, they are not likely to have originated like planets in our solar system. Perhaps they are the remnants of objects that were evaporated away as described above, or possibly they formed from the debris of the supernova explosion that produced the neutron star, or . . .

Only three pulsars—the Crab, Vela, and PSR 1509 − 58—are found in association with the remnant of a supernova, however, nor do the vast majority of supernovae remnants contain a pulsar. The first part of that statement can perhaps be accounted for by noting that the rapid periods and period change of these three pulsars give young ages

---

[7] This catchy name is derived from pulsar (PSR) with the following numbers giving its approximate coordinates in the sky.

that are consistent with the age of the expanding supernova shell; hence it is not surprising that they are still near the remnants of the exploded star. Supernovae shells expand rapidly, up to 10,000 km/sec just after the outburst. Consequently, they will fairly quickly move far away from the neutron star and eventually merge with the surrounding interstellar medium. For example, in a million years a supernova shell can have expanded to a diameter of 100 parsecs or so.

The second part of the statement above, that only a few supernovae remnants contain a pulsar, is a bit harder to understand. Several factors may be responsible, however. Since pulsar radiation is strongly beamed, we simply may not be in the beam direction and, therefore, see no pulses. Also, supernovae remnants are brighter radio sources than pulsars and hence easier to detect. Or perhaps some neutron stars don't turn on immediately after the explosion but only some tens of thousands of years later. In any case, the pulsar phenomenon must be a transient phase in the life of a neutron star, because by the time its rotation period has slowed to four seconds or so, the pulses no longer occur. With its pulsar signature gone, the faint neutron star, continuing to cool, is no longer observable, at least by present techniques. Finally, not all supernovae events may lead to the formation of a neutron star. Instead, a black hole might be formed, or the exploding star might be completely destroyed, leaving behind no stellar remnant whatsoever. Many questions, few answers.

Let us next turn our attention to the supernovae themselves.

How Stars Become Neutron Stars

Supernovae.   The **supernova** catastrophe is not only the most spectacular mass-loss mechanism and the most violent event a star can undergo, it is also the most violent occurrence anywhere in the universe. Its energy is exceeded only by the event that formed the universe itself. In just a few seconds a supernova releases about 100 times more energy than does the Sun over its entire 10-billion-year main-sequence life!

For many years supernovae were not recognized as such, but were thought to be the much less energetic ordinary novae. As we shall see in Part III, this contributed to the confusion about the true nature of the spiral nebulae (now known to be galaxies), since "novae" were observed within some of the nebulae. After it became clear that the spirals were separate galaxies far away from our own, however, it was obvious that to appear so bright at these great distances, some stars must undergo a much more powerful explosion than a mere nova outburst. This event was called, not too imaginatively, a supernova. S Andromedae, an 1885 supernova, was about one-sixth as bright as the entire Andromeda galaxy, which is a larger-than-average spiral; so for a time this one star was as bright as the combined light of billions of ordinary stars.

Supernovae are rare, and occur rather infrequently in any given galaxy, perhaps two or three each century. Only four have been recorded unambiguously as naked-eye objects in our own Milky Way galaxy. These were observed in 1006 (in the constellation of Lupus), 1054 (the Crab in Taurus), 1572 (Tycho's "nova" in Cassiopeia; see Figures 16.20 and 16.21), and 1604 (Kepler's in Ophiuchus). Presumably many more have occurred, but have been hidden from our view by interstellar dust.[8] This is the

---

[8] The brightest radio source in the sky, Cassiopeia A, is clearly a supernova remnant, and is located in a dusty region of the sky. After its discovery as a radio object, an optical shell was found from which an expansion velocity was measured, indicating that it exploded in about 1658. It is possible that it was observed then as a barely visible naked-eye object by the English Astronomer Royal, J. Flamsteed.

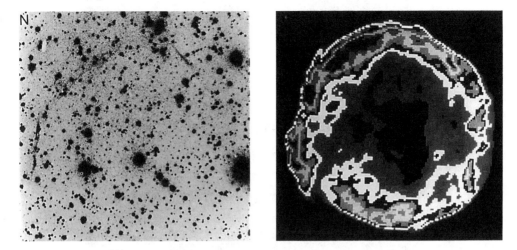

Figure 16.20. A long exposure barely shows a portion of the remnant of Tycho's supernova, but the radio image shows it in considerable detail. The diameters of the two images are both about 8 arc minutes.

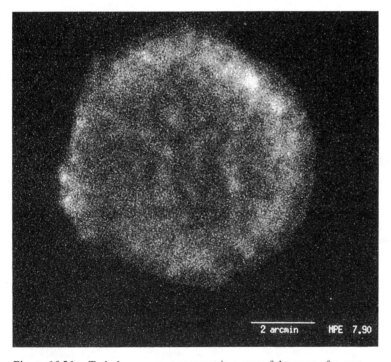

Figure 16.21. Tycho's supernova remnant is a powerful source of x-rays, as seen in this image from the German–U.S. x-ray satellite, ROSAT.

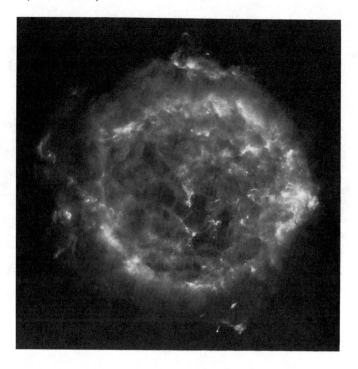

Figure 16.22. A radio image of the supernova remnant, Cassiopeia A, at a wavelength of 6 cm. In our own galaxy, however, the supernova was not bright because it was hidden from us by interstellar dust.

case, for example, with Cassiopeia A (see Figure 16.22). In any case, many galaxies must be searched in order for us to be able to observe a significant number of supernovae. Such searches, carried out since the 1930s, have resulted in the discovery of about 650 supernovae; as a result we now have a reasonably clear idea of their properties.

The basic classification system for supernovae is simple and is based on the non-appearance or the appearance of hydrogen lines in the visible spectrum; if the former, it is called a Type I; if the latter, it is a Type II. It would be a mistaken assumption to believe that the Balmer lines do not appear because the physical conditions in a Type I supernova are unfavorable for their production; these objects are really deficient in hydrogen.

Several sub-types are now distinguished, but for our purposes, it is sufficient to limit our discussion to the division of Type I into Ia and Ib, again according to their spectra. The characteristic feature of Type Ia is a prominent absorption line of Si II at about 6300 Å, which is seen at maximum light and for a few weeks afterwards. This line is absent in Type Ib supernovae; instead, they characteristically show lines of helium. Type Ia objects are found in all kinds of galaxies, spirals (like the Milky Way), ellipticals, and irregulars (like the Magellanic Clouds). In spirals, they are found in the halo and interarm regions, showing no particular preference for star-forming regions. By contrast, Types Ib and II are not found in ellipticals, but appear only in spirals and irregulars and are associated with star-forming regions like the spiral arms or H II regions. This suggests that Type Ia supernovae arise from a low-mass, old stellar population (Population II), while Types Ib and II come from young, massive stars (Population I).

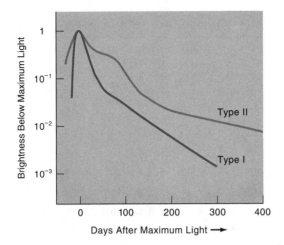

Figure 16.23. Typical light curves of supernovae of Type I (no hydrogen lines) and Type II (hydrogen lines). Note that the brightness of each light curve is set equal to 1 at maximum light.

The shapes of the light curves of Types Ia and Ib supernovae are similar, but those of Type II show considerable differences from one object to the next (see Figure 16.23). Type I objects take a few weeks to reach maximum light, decline quickly for perhaps a month, and then more slowly, with their brightness decreasing by half about every 75 days. Their energy during this phase probably comes from radioactive nuclei, made in the outburst, that decay to give the characteristic 75-day decline rate (that is, they spontaneously emit energetic particles; see Chapter 23). Type IIs rise to maximum much more quickly, taking only hours or a few days. Some remain near maximum brightness for a few months, while others begin declining immediately. After perhaps four months, they also show the characteristic 75-day decline rate.

Type Ia objects are generally three or four times brighter at maximum light than Type Ib. Type II outbursts vary considerably in maximum intrinsic brightness, and, in rare instances, will be as bright as Type Ia. In any case, all types liberate an incredible amount of energy at maximum, releasing many billions times as much optical radiation as the Sun.

Given that different stellar populations are the progenitors of Type Ia and of Types Ib and II supernovae, it is not surprising that the explosion mechanisms differ also. Let us first consider current ideas as to how a Type II supernova occurs.

Type II supernovae occur when a massive star has used up all possible nuclear fuels. Recall that low-mass stars cannot compress their cores sufficiently to enable anything heavier than hydrogen to burn. Massive stars, however, can produce such high core temperatures and densities that heavier and heavier elements, the ash of previous processes, can be used as fuel for the next. Stars of 10 to 12 $M_\odot$ and greater can produce the enormous core temperatures and densities large enough to enable carbon to burn successively into neon, oxygen, silicon, and ultimately to iron. Though very many different nuclear reactions take place, they do not produce much energy. The star rapidly approaches its end. Just how rapidly is shown in Table 16.2, where model calculations of the central temperatures, densities, and optical and neutrino luminosities are given at the onset of the fusion of each of the major nuclear fuels in a 15-solar-mass Population I star. Note how quickly the successive fuels are expended. The hy-

Table 16.2.   Lifetimes of various fuels for a 15-solar-mass star[a]

Fuel	Central TK	Central density	L_ν	$L_\odot$	Time
Hydrogen	3.4×10^7	5.9	0	2.1×10^4	1.2×10^7 yrs
Helium	1.6×10^8	1.3×10^3	1	6.0×10^4	1.3×10^6
Carbon	6.2×10^8	1.7×10^5	8.9×10^4	8.7×10^4	6.3×10^3
Neon	1.3×10^9	1.6×10^7	1.8×10^8	9.7×10^4	7.0
Oxygen	1.9×10^9	9.7×10^7	2.1×10^9	9.7×10^4	1.7
Silicon	3.1×10^9	2.3×10^8	8.9×10^{10}	9.7×10^4	6 days
Collapse	8.3×10^9	6.0×10^9	1.8×10^{15}	9.7×10^4	0.3 seconds

[a]L_ν and $L_\odot$ are the neutrino and ordinary (electromagnetic) luminosities, respectively, in units of the Sun's luminosity.

drogen (main-sequence) phase lasts for about 12 million years, and the helium phase for about a tenth of that. All subsequent fuels can maintain the star for only a few thousand more years. Silicon, the last major fuel before core collapse, lasts for only about a week!

See what is happening. Beginning with the onset of carbon burning, the star loses more energy by the emission of neutrinos than by optical radiation; very quickly the neutrinos drain off vast quantities of energy from the interior of the star. By the time silicon is ignited and fusing to iron, about a million times more energy is being lost by neutrino emission than by optical radiation! The neutrino energy loss causes the interior of the star to contract, raising the temperature, increasing the density, and quickly using up its nuclear fuel.

Recall (Chapter 13) that the nucleus of iron is the most tightly bound of all. Consequently, any nuclear reaction involving iron—either splitting it to lighter nuclei or fusing it to a heavier one—requires energy. Iron is the end of the line for energy production. The incredibly hot and dense stellar core grows as the surrounding matter is fused to iron. Though the core itself produces no energy, it is maintained for a short time by the pressure of degenerate electrons. Gravity, however, always acting, soon overcomes the pressure and causes the core to contract and become even hotter. When the core reaches a temperature of several billion degrees, high-energy photons—gamma-rays—interact with the iron nuclei, tearing them apart. At the same time, electrons are forced into nuclei, making them neutron-rich. These two processes use up enormous amounts of energy, and the iron core collapses catastrophically. A supernova is born. In a 20-solar-mass star, the iron core is about 1.4 solar masses, and just before collapse its radius is about half that of the Earth. In only a few tenths of a second, this core collapses to a radius of about 100 kilometers!

Huge numbers of neutrinos are produced by the reaction $p + e \rightarrow n + \nu$. In fact, they carry off 99 percent of the energy of the supernova. This enormous energy loss cools the core, so that it completes its collapse to its final state—a neutron star about 10 kilometers in radius.

When the inner part of the collapsing core reached a density several times that of nuclear matter, the strong nuclear force became repulsive, causing the core to bounce outward and collide with the rest of the still infalling core. Just what happens next is not clear. Somehow, perhaps aided by the enormous flux of neutrinos, a shock wave is produced that escapes the core. It travels rapidly outward through the star, heating up the layers outside the core, thereby triggering a wide variety of nuclear reactions. Most of the elements heavier than iron are produced very quickly in the envelope and spewed out into interstellar space. The energy released by the imploding core and the exploding envelope we see as a supernova; the neutron core we may see as a pulsar, and the rapidly expanding envelope (up to 10,000 km/sec) as a supernova remnant.

This whole supernova collapse process takes place in only seconds, so you can understand that it is difficult to model. Consequently many aspects are uncertain; for example, just how is the outer part of the core pushed outward to disrupt the surrounding star and produce the visible supernova? Is a neutron star always formed, or could a black hole be the end product in some instances? Or is it possible that sometimes no remnant whatsoever is left? In any case, it is remarkable that all the nuclear burning done by the star over its lifetime of millions of years is undone in just a few seconds; all the radiation emitted by the star is paid back with the gravitational energy released by the collapse of the iron core.

A Type Ib outburst is thought to arise by core collapse also, but in an originally very massive early-type star from which all of the outer layers, including the hydrogen-rich material, have been peeled off by a strong stellar wind. Only the central portion of the star is left, rapidly consuming its remaining fuel. Soon, the iron core collapses with the consequences just described. One of the observational differences is that, in contrast to the Type II outburst, the ejected material contains no hydrogen.

The ultimate energy source in Types Ib and II supernovae is gravitation: in a Type Ia outburst, however, it is nuclear energy. If a white dwarf with a mass near the Chandrasekhar limit (about $1.4\,M_\odot$) acquires additional material, for example, from an evolving companion, the temperature at which the degenerate carbon-rich core can ignite is lowered and it begins nuclear burning that quickly becomes a runaway. This is similar to the thermal runaway in the helium flash (the safety valve does not work), except that the end result is not to lift the degeneracy, but rather to blow up the star in a gigantic nuclear explosion.

Given these scenarios, it is understandable that Type II explosions should differ widely in their brightness at maximum light—they occur in stars of different masses—whereas Type Ia maxima are all very similar—they occur in white dwarfs having essentially identical masses. If, as may be the case, Type Ia supernovae are a truly homogeneous class of objects (that is, if they all reach the same brightness at maximum and if they can be unambiguously identified by their light curves and their spectra), then they could be powerful distance indicators once their intrinsic brightness was well determined. They could be used to establish the distance of galaxies several billion parsecs away!

The Crab was a Type II supernova, but Tycho's and Kepler's were probably Type I. The masses of the nebular remnants blown into space differ correspondingly. Those from Type I supernovae have masses of about 0.5 solar masses; those from Type II can have masses of up to 50 times the mass of the Sun. At first the nebular shells move through interstellar space with very high velocities, on the order of 10,000 km/sec; however, as they plow into more and more of the interstellar medium, they are gradually slowed. The energy lost by the expanding nebula goes into heating the low-density intercloud medium to temperatures of up to a million degrees, as we saw in Chapter 14.

The Crab Supernova Remnant. The Crab nebula (see Figure 16.15) is the best-known supernova remnant. It is the product of a supernova outburst that took place about 950 years ago in 1054 (by some accounts on July 4!). Interestingly, it was noted in contemporary Chinese, Korean, and Japanese records (and possibly by native North

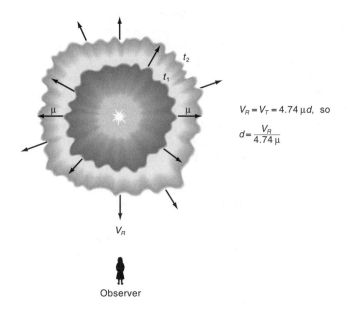

$$V_R = V_T = 4.74\,\mu d, \text{ so}$$

$$d = \frac{V_R}{4.74\,\mu}$$

Observer

Figure 16.24. Schematic representation of a uniformly expanding spherical shell at t_1 and then several years later, t_2. During this time, the shell has increased in angular size by an amount, μ, as a consequence of its radial velocity V_R km/sec (assumed to be constant). Hence the distance in parsecs to the shell is given by the formula $V_R/(4.74\mu)$, where μ is given in seconds of arc per year.

Americans), but no reference to it has been found in any records from Europe, despite the fact that it was visible in the daytime for about 23 days after its appearance, and for six months at night. Confirmation that the Crab nebula is the result of the 1054 event comes not only from its location in the sky, but also from the age of the expanding nebula. This is calculated by measuring the proper motion (in units of seconds of arcs per year, as you recall) of several filaments and knots in the nebula. Assuming that this rate of expansion has been maintained since the explosion, it is easy to estimate how much time has been required for the nebula to reach its present angular size. This turns out to be about 900 years, in reasonable agreement with the 1054 event. By associating the expansion velocity determined by the Doppler shift with the angular expansion measured from photographs taken decades apart, its distance from the Sun is found to be about 2,000 parsecs (see Figure 16.24).

The most prominent feature in the Crab nebula (and what gives it its name) is its filamentary structure. The physical conditions in the filaments are similar to those in H II regions. A photograph made in light that does not include the emission lines from the filaments gives a quite different impression, however (Figure 16.25). What one sees is an amorphous, bluish object that is about the same size overall as the filamentary structure.

Over the entire spectrum this nebula radiates about 100,000 times as much energy as does the Sun. If this energy were produced by ordinary thermal processes, an unreasonably large amount of nebular material would be required. Thus, the mechanism by which such radiation could be produced was a mystery until the 1950s, when it was realized by the Soviet astronomer I. S. Shklovsky that this was nonthermal radiation generated by high-energy electrons (moving at 99.999 percent or more the velocity of light) spiraling in a magnetic field. Such radiation can be strongly polarized. This was

Figure 16.25. An image of the Crab nebula in the light of continuous radiation (and excluding atomic line radiation) appears as an amorphous object, with no filamentary structure (compare with Figure 16.15). This bluish light is synchrotron radiation, produced by very high energy electrons spiraling in the magnetic field of the nebula.

an entirely new energy-generation process in astronomy; it has since become crucial to our understanding of a wide range of astronomical phenomena. Energy produced in this way is called **synchrotron** radiation, because this radiation process was studied by physicists trying to understand the energy losses in the high-energy particle accelerators (including synchrotrons) being designed in the late 1940s and 1950s. The pronounced polarization of the radiation from the Crab nebula supports this interpretation. The energy lost by the Crab pulsar as it slowly spins down is the ultimate energy source for the nebular radiation as well as for the pulsar's own energy output.

Supernova 1987A. On February 23, 1987, a supernova visible to the naked eye was discovered near the Tarantula nebula in the Large Magellanic Cloud, only about 50,000 parsecs away (Figure 16.26). This is the brightest supernova astronomers have seen since Kepler's 383 years earlier and the first naked-eye supernova since 1885, so they immediately began to observe it intensively. By the next night, southern-hemisphere radio and optical telescopes were observing it, and from its geosynchronous orbit the U.S.-European International Ultraviolet Observer was measuring the supernova's ultraviolet spectrum. By early March, the then recently launched Japanese x-ray satellite Ginga completed its checkout phase and began observing the supernova with its detectors sensitive to low-energy x-rays. The Soviet space station, Mir, obtained high-energy x-ray data, while the U.S. satellite Solar Maximum Mission measured γ-rays.

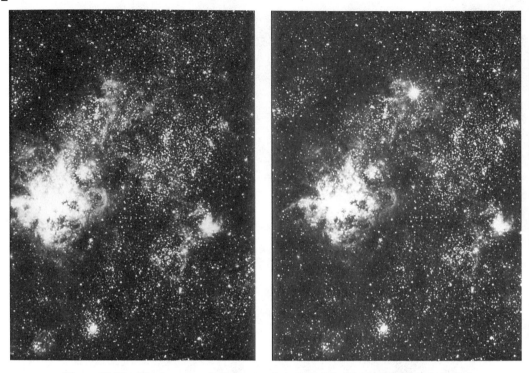

Figure 16.26. The image on the left was taken about 18 years before that on the right, which was made just a few days after the discovery of the supernova in the Large Magellanic Cloud.

Most exciting of all, two massive detectors, one in the U.S. and the other in Japan, measured a pulse of neutrinos from the supernova! As we will see, this observation confirmed the core-collapse theory of the supernova outburst. Finally, SN 1987A was discovered before reaching maximum brightness and is located in a well-studied portion of the Large Magellanic Cloud. As a consequence, not only was it observed over a broad wavelength range throughout its entire outburst period, but both photometric and spectroscopic data exist concerning the progenitor star as well. Little wonder that astronomers greeted the news of the discovery with such excitement and anticipation.

What has been learned so far? The progenitor star was a B3 supergiant, rather than the red supergiant usually predicted by stellar evolution theory. We are still not sure why such a star could become a supernova. The fact that the metal abundances in the Large Magellanic Cloud are lower than those in our galaxy might have affected the structure and evolution of the star in such a way that, about 40,000 years before exploding, the red-supergiant envelope began to contract, causing the star to evolve back to a blue supergiant.

In any case, as a blue supergiant for a few tens of thousands of years, it lost a lot of mass through a high-velocity wind. This matter eventually overtook and collided with the slower moving gas ejected earlier by the star when it was in its red-supergiant phase. The result is a ring of hot gas (the smallest of the three rings in Figure 16.27) measuring about 1.4 light-years in diameter and expanding at about 10 km/sec. It is not understood why this material is in the form of a ring and not a spherical shell. Even more puzzling is the nature of the two larger rings as seen in the figure, with one in front of and the other behind the supernova.

Models suggest that the supergiant had a mass of about 18 solar masses and a radius about 50 times that of the Sun. Such a star is much smaller in radius and has a

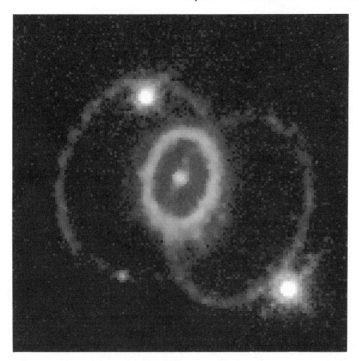

Figure 16.27.   This spectacular red-light image of the 1987A supernova was taken with the Wide Field Camera on the repaired HST. The smallest ring is centered on the supernova. It is thought to be material blown away from the progenitor of the 1987A supernova when it was in its blue-supergiant phase, which overtook the earlier, more slowly moving red-giant-phase wind. The origin of the larger, remarkably regular rings is unknown at present.

higher gravity than a late-type supergiant. Hence a greater portion of the outburst energy went into pushing the envelope out from the exploded star than into the radiation emitted by the supernova. In fact, early in its evolution it was only about a tenth as bright as the usual Type II supernova. In addition, this may explain why the radio energy emitted by the supernova was never very intense, and began declining in brightness immediately after its discovery, and why no x-rays from the supernova were detected for at least several months after the outburst.

After remaining constant in brightness for about a week, in early March the supernova began a slow ten-week-long brightening, reached a broad plateau, and then started to decline as shown in Figure 16.28. There are at least two possible energy sources that might have caused the supernova to become brighter. One is that the core of the B3I progenitor had become a rapidly rotating neutron star—a pulsar. As with the Crab pulsar, some of its rotational energy would be transformed to the electromagnetic radiation we see emitted from the envelope. A pulsar has not so far been detected, however.

Instead, the increased brightness was most likely a consequence of energy generated by the decay of radioactive nuclei produced in the outburst. If so, then after the outer envelope thinned further, the brightness should have declined at a rate characteristic of the half-life of the decaying radioactive elements that caused the brightening. (Half-life is simply the time required for half of a given quantity of a radioactive element to decay; see Chapter 23.) In particular, the following sequence would play the major role: ^{56}Ni, with a half-life of 6.1 days, decays to ^{56}Co, which has a half-life of 77 days before decaying to stable ^{56}Fe. What was observed? About 120 days after the

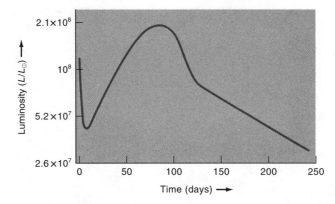

Figure 16.28.   The energy over the entire electromagnetic spectrum emitted by SN1987A during the first eight months of observations. See the text for a discussion of the light curve.

outburst, the brightness of the supernova began a steady decline with a half-life of 77 days. In August, two emission lines associated with the decay of cobalt to iron were observed in the x-ray region of the spectrum at about 0.015 Å and 0.010 Å by the Ginga and Mir spacecraft. The envelope was being powered by the radioactive decay of cobalt! The amount of nickel (the "parent" of the cobalt) that must have been produced in the explosion was calculated to be about 0.08 solar masses. Here was direct evidence of nucleosynthesis in a star.

More evidence appeared beginning in November when astronomers detected infrared lines produced by several elements, including oxygen, neon, silicon, sulfur, and magnesium, all in much greater quantity than would have been present in the B3 star. These elements were produced at the time of the outburst, as the shock wave from the collapsed core passed through and heated the outer part of the star, but were unobservable until the envelope had thinned considerably.

The neutrinos from the supernova were the first such particles observed from any celestial object other than the Sun. Interestingly, had the supernova been only a few parsecs closer to the Earth, so that the neutrinos (as well as all the electromagnetic radiation) would have arrived only a decade earlier, they would not have been measured, because no suitable detectors were in operation then. Both the U.S. and the Japanese neutrino detectors had been set up for something quite different, namely to see if the proton would decay as predicted by grand unification theories (see Chapter 21), and luckily, both were operating on February 23. The background neutrino count measured in these detectors is small; the U.S. detector measures only about one or two neutrinos per week that have the same energy as those from the supernova. During a 6-second period on February 23, however, it counted eight neutrinos! At the same time, the Japanese instrument detected eleven neutrinos during a slightly longer period. Even though these are small numbers, they are enormously greater than the background count. Given the difficulty in stopping neutrinos, this implies that a very large number were incident on the detectors. (Remember that neutrinos interact with matter only through the weak nuclear force, which is feeble indeed.)[9] The eight neutrinos counted by the U.S. detector indicate that something like 10^{16} neutrinos must have hit it!

Note that the neutrinos did not all appear at the same time, indicating that they did not travel through the collapsed core unimpeded. Instead, the core density was so

---

[9] Both of the detectors are in the northern hemisphere; so after traveling for 170,000 years, the neutrinos passed right through the Earth before hitting the U.S. and Japan.

enormous that the neutrinos required several seconds to work their way through a few kilometers of this material.

The total energy carried by the neutrinos can be estimated, with the result that they account for the bulk of the energy produced by the event: about two orders of magnitude more than the kinetic energy of the expanding envelope, and four orders of magnitude more than the electromagnetic radiation emitted by the supernova.

How was so much energy acquired by the neutrinos? We have seen that when a stellar core collapses and forms a neutron star, large numbers of neutrinos are produced as the core's protons are converted to neutrons. In fact, it has been estimated that 1987A emitted 10^{56} neutrinos in just a few seconds. Their energy—3×10^{53} ergs—just about equals the gravitational energy released when a 1.4-solar-mass core collapses to a neutron star. Here is convincing evidence that a core collapse did indeed take place! The mere fact that neutrinos escaped indicates that a black hole was not formed in the initial collapse,[10] otherwise they would have been trapped. Many astronomers feel that eventually we will see the neutron star as a pulsar. As of this writing, however, no pulsar has been detected. Perhaps the pulsar's beam does not sweep across the Earth, or perhaps the envelope has not yet expanded sufficiently to become transparent. If that is the case, a pulsar should become visible first in x-rays, then optical, and finally radio radiation. It will be interesting to see if this actually happens.

As if all that were not enough, by measuring the energy and relative time of arrival of each neutrino it is possible to make a rough estimate of its mass. The result is that the mass of the detected neutrinos is no larger than 2×10^{-5} the mass of the electron. Though the universe contains huge numbers of neutrinos, their total mass is too small to play any role in the evolution of the universe given the tiny mass of each. In particular, it is insufficient to slow the expansion of the universe appreciably (see Chapter 19).

There is little question that over the next several years this object will continue to yield new information concerning the nature of the supernova outburst.

Weird and Wonderful Binary Systems Containing Neutron Stars.

Let us close this section with brief descriptions of binary systems in which one component is a neutron star. (We will describe one more such system in Chapter 21.)

The salient feature of these systems is that they radiate in x-rays 10^3 to 10^5 times the total energy radiated by the Sun, and so are called **x-ray binaries**. The binary periods range from a few hours to many days, so they are close double stars. The visible companion star in these binary systems can be a massive early-type star or a low-mass, late-type object. In some of these systems the x-ray source is eclipsed by the companion, indicating that the orbital plane is nearly in our line of sight. When the companion star in an x-ray binary is losing mass (by a stellar wind or by evolution-driven expansion causing its Roche lobe to overflow), some of it is captured by the neutron star. The gas may form an accretion disk around the compact object, where it is compressed by the strong gravity and heated to high temperatures by infalling material, producing a copious flux of x-rays. The gravity of a white dwarf is insufficient to heat the accreted gas enough to produce x-rays, but that of a neutron star is. For example, a marshmallow at the end of its fall toward a neutron star would have an energy equal to about one million tons of TNT! (As we will see at the end of this chapter, the compact object in some x-ray binaries is likely to be a black hole rather than a neutron star.)

In other instances, the neutron star's strong magnetic field guides the inflowing ionized gas to the magnetic poles where it crashes into the surface of the star, pro-

---

[10] Conceivably, if after the initial collapse enough matter fell back onto the neutron star, it could collapse to a black hole; that is very uncertain, however.

ducing x-rays. As with pulsars, the axis of rotation is displaced from the magnetic axis so the x-rays appear and disappear with each rotation of the neutron star. That is, the neutron star is an x-ray pulsar.

Finally, some objects produce a steady x-ray output interrupted by intense bursts of x-rays, usually at intervals of hours to days. Individual bursts reach a maximum in about 1 second and typically last about 10 seconds, producing energy at roughly 50,000 times the Sun's rate during this short period. A reasonably successful model of these **x-ray bursters** postulates mass flow from a companion to a neutron star. Gravitational energy released by this infalling matter heats it so that it produces the steady stream of x-rays. When the temperature at the base of this accreted layer reaches about 10^9 K, catastrophic helium or carbon burning occurs, producing the powerful bursts. In some respects, the process can be thought of as the neutron star analog of the nova process. Note that in x-ray binaries, the bursts are the only phenomenon in which nuclear burning plays a role; in general, energy is produced by the more powerful gravitational mechanism.

In 1986 an unusual binary system was discovered in a globular cluster. It consists of a white dwarf and a neutron star separated by only 130,000 km and revolving about each other with a period of only 11.5 minutes (not hours or days!). Note that this binary system is so small that it could fit easily between the Earth and the Moon, as shown in Figure 16.29. This pair of stars, discovered by the European x-ray satellite

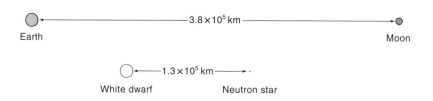

Figure 16.29. The orbital dimensions of the Earth–Moon system compared with those of the white-dwarf/neutron-star binary system.

Exosat, is a powerful x-ray source, radiating about 10^5 times more energy in x-rays than the Sun does over the whole spectrum! These x-rays are produced by matter being stripped from the white dwarf and falling into an accretion disk surrounding the neutron star, from which it crashes onto the star. Since matter falling onto a neutron star acquires a velocity that is a considerable fraction of that of light, it does not take much infalling material to produce the large x-ray output. Fundamentally, gravitational energy is being converted to radiation.

This system may have formed in an unusual manner. It is possible that the white dwarf and neutron star formed and evolved independently, but subsequently captured each other gravitationally. Such a capture is extremely improbable in the solar neighborhood, where the star density is low (ping-pong balls separated by 500 kilometers), but it could possibly happen in a globular cluster. There the number of stars per unit volume is about 10,000 times greater than it is in the solar neighborhood, so a capture is possible. Regardless of its origin, this system should tell us a lot about the behavior of interacting compact objects.

One of the strangest of all binary systems is SS 433. This object, a source of x-ray, optical, and radio radiation, lies near the center of a large radio object that is

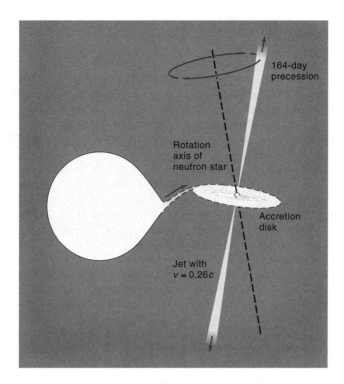

Figure 16.30. A model for SS 433. Matter from the giant star flows onto an accretion disk surrounding the neutron star. High-velocity jets of material stream out from the disk, their directions changing as the axis of the disk precesses with a 164-day period.

probably the remnant cloud of an old supernova. Optical spectra show three sets of emission lines. One set, due to hydrogen and helium, undergoes a modest variation in wavelength corresponding to a velocity change of about 70 km/sec with a period of 13.1 days. The other two sets of emission lines not only were at unfamiliar wavelengths, but also showed large and systematic changes over a 164-day period. These turned out to be hydrogen and helium lines, one set red-shifted, the other blue-shifted by amounts corresponding to up to 27 percent the velocity of light!

The model that emerges from these observations is of a 13.1-day-period binary system consisting of a fairly massive normal giant star and a compact object, probably a neutron star (Figure 16.30). Matter flows from the giant onto an accretion disk surrounding the compact object. The unusual feature of this system (in fact, so far it is unique) is that jets of matter are apparently being squirted out from the accretion disk at very high velocities in opposite directions. These jets produce the red- and blue-shifted emission lines. (The latter, by the way, hold the record blue shift in astronomy.) The axis of rotation of the neutron star precesses like a top with a 164-day period, causing the observed radial velocities of the jets to vary with the same periodicity. Polarized radiation from the jets is produced by the synchrotron mechanism, so a magnetic field must be present. The process by which blobs of hot gas are shot out along a well-defined direction is not understood. Nor is it known why this kind of object seems to be so rare, one of a kind at the moment, nor what its relation is (if any) to the apparent supernova event represented by the remnant in which SS 433 is embedded. There are many interesting questions to answer!

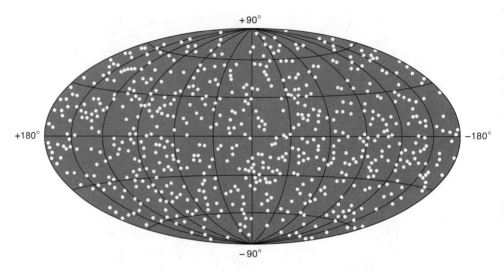

Figure 16.31. The distribution of bursters over the sky as located by detectors on the Compton Gamma Ray Observatory. The coordinates are based on the galaxy, so that the equator is the galactic plane and the galactic center is at the origin. Note how uniformly distributed the burst sources are—no concentration to the center or the plane of the Milky Way.

Gamma-Ray Bursters: An Astronomical Mystery. Back in the 1960s the U.S. launched the Vela series of satellites to monitor Soviet compliance with a nuclear test ban treaty. The satellites did indeed detect gamma-rays; however, it soon became clear they were coming not from nuclear bomb tests on Earth, but from space. Ever since the military disclosed these data in 1973, astronomers have been attempting to discover their origin, so far with little success. Though the directions to many bursters have been measured to within a few degrees, no candidate objects have been identified. It was speculated that some kind of an explosion on nearby neutron stars was the most likely prospect. When the Compton Gamma Ray Observatory was launched in 1991, astronomers expected that its sensitive instrumentation would enable it to identify at least some sources, but this has not yet happened.

What are the characteristics of these mysterious sources? The bursts occur suddenly, reach maximum in only milliseconds, and last typically for roughly 30 seconds. During outburst the energy output is not constant, but varies rapidly. As far as is known, no bursters have gone off more than once. The strongest burst observed is about 10,000 times stronger than the weakest. If all bursts are similar and produce about the same amount of energy, this suggests a range in distance for these objects of about a factor of 100. Finally, the more than 700 bursters so far detected by the Compton Observatory are uniformly distributed over the sky (see Figure 16.31). That is, they are not concentrated to the ecliptic, to the galactic plane, toward the center of our galaxy, etc.

This last result was unexpected, since Compton's sensitive detectors are able to see beyond the disk of our galaxy, in which neutron stars (the presumed sources) are located, and so were expected to reveal a nonuniform distribution of burst sources. Astronomers have suggested that the sources could be as near as the cloud of comets surrounding the solar system several thousand AU from the Sun, or located in extremely distant galaxies, or somewhere in between! Obviously, the intrinsic energy emitted differs enormously—by about 25 orders of magnitude, in fact—depending on the location of the sources. In other words, we have no idea where these sources are or what they are. They are discussed here simply because many speculations have associated them with neutron stars in one way or another.

Black Holes
General Properties

We have seen that stars that have initial masses of about 8 $M_\odot$ or less die as white dwarfs. Stars with main-sequence masses between 8 and perhaps 30 $M_\odot$ evolve an electron-degenerate core that collapses to a neutron star, at the same time producing a supernova. But how do very massive stars end their lives? Can a neutron star have an arbitrarily large mass, or is there a limit beyond which the star is unstable? As noted earlier, one of the uncertainties in the theory of compact objects is the maximum mass that neutron-rich nuclear matter can support. Whatever it may be, and theoretical estimates suggest it is only 2 or 3 $M_\odot$, suppose the dead core is more massive than this limit. What happens then? No state of matter is known that is capable of supporting itself under these circumstances; the mass must collapse indefinitely. Gravity triumphs—a **black hole** is formed. The notion of several solar masses of material collapsing to an arbitrarily small volume is so contrary to our experience and common sense that it seems bizarre indeed. This is just what happens, however, when no arrangement of matter is able to resist the force of its own gravity. There is no alternative.

Consider what happens as a massive burned-out core contracts. Its mass remains the same, but its radius decreases and so its surface gravity, $g = GM/R^2$, increases. Therefore the velocity required to overcome gravity and escape from the object increases. [Remember that the velocity of escape is given by $V = (2GM/R)^{1/2}$.] Light leaving the collapsing core becomes more and more red-shifted, since it loses energy in escaping from the rapidly strengthening gravitational field. Light leaving at a large angle, θ, to the perpendicular falls back into the object (see Figure 16.32). The cone through which light can escape becomes narrower and narrower until finally it closes;

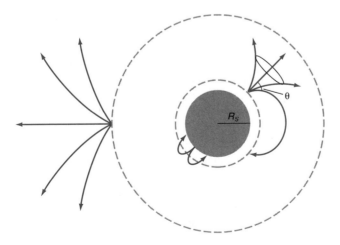

Figure 16.32. The shaded circle has the Schwarzschild radius, R_s, appropriate to the collapsed mass within. Photons cannot escape from this surface. Near R_s, photons can escape only through a narrow cone; this cone opens up as the distance from R_s increases until finally, all photons escape.

not even light can leave the object. The escape velocity from the collapsing object has exceeded the velocity of light; it has become a black hole![11]

The surface at which $V = c$ occurs is called the **event horizon**, because beyond that surface, that horizon, we can see nothing. The radius of the black hole at which the escape velocity equals the velocity of light is called the Schwarzschild radius, after the German astronomer who first solved Einstein's equations of general relativity appropriate to this situation. This radius is given by $R_s = (2GM)/c^2 = 3M$ km, where M is in solar masses. That is, a 10-solar-mass collapsing core would have a Schwarzschild radius of $3 \times 10 = 30$ kilometers. For the Earth (which could not collapse to a black hole under its own gravity, but would have to be compressed by some outside force), R_s is just 9 millimeters! Note that the formula for R_s derived from general relativity theory is just the Newtonian expression for the escape velocity when $V = c$. This accidental result is misleading, however, because Newtonian physics does not predict black holes. It is only in the context of relativity theory that there is anything special about $V = c$.

In fact, the very existence of black holes is explained as a consequence of Einstein's theory of gravity, some aspects of which we will describe in Chapter 21. When the radius of an object is within a factor of two or so of its Schwarzschild radius (the exact value is uncertain), gravity is so powerful that no state of matter can withstand it. For example, the Schwarzschild radius of a $3\,M_\odot$ object is only about 10 kilometers, about the same size as a neutron star of smaller mass. It would be strange indeed if nature did not take the final step to a black hole.

The star's density when it has collapsed to the Schwarzschild radius is given by $10^{18}/M^2$ gm/cm^3, where M is again in solar masses. Thus our 10-solar-mass black hole would have a density at the Schwarzschild radius of 10^{16} gm/cm^3, about 50 times the density of a neutron star.

As you might guess, unusual effects can occur around black holes, but only when you get close to them. Like the Earth, a burned-out Sun cannot become a black hole from its own gravitational force (nor will it ever become one). Let's imagine, however, that somehow or other the Sun experienced this ultimate collapse, leaving the planets unscathed. The gravitational behavior of the planets would not change. They would continue to orbit the Sun as they do now, their motions in no way altered despite the drastic difference in the Sun's condition. This might seem strange, but consider: neither the mass of the Sun nor any of the Sun-to-planet distances would have changed in our imaginary catastrophe. Thus, the gravitational force between the Sun and the planets would not have changed. Remember that R in Newton's law is measured from the center of the Earth to the center of the Sun (in the present example) because mass acts as if it were concentrated at the very center of the object. The Sun's "edge," the photosphere, is now 700,000 km from the center of the Sun. If the Sun were a black hole, however, we could get as close as three kilometers from its center and still be just outside the hole. The gravitational force there would be enormous. This is why effects beyond our experience occur near black holes.

What are some of these bizarre effects? Suppose we could watch (from a safe distance) a spaceship starting to fall into a stellar-mass black hole. At first it would accelerate, falling faster and faster. But as it approached the Schwarzschild radius, we would see it slow down and its color become increasingly red-shifted. Soon we would not see it by visible radiation, but would have to track it by the radio radiation it emit-

---

[11] Interestingly, at the end of the eighteenth century the French mathematician Laplace considered what would happen if the velocity of escape from an object equaled the velocity of light. He called such objects "dark bodies."

ted.[12] Even if we could follow it in this way, we would soon become bored, because it would appear to take an infinite amount of time to fall into the hole. As we will see later, these effects can be understood only with the help of general relativity theory. One of the consequences of that theory is that any periodic behavior—a clock, or a light wave, for example—slows down in a strong gravitational field as compared to its frequency in a weak field. Thus, we would never actually see an unfortunate astronaut fall into a black hole, because it would appear to us that her watch was ticking ever more slowly. The astronaut, however, would find herself falling faster and faster, and her watch continuing to run normally. Close to the black hole the gravitational force would be increasing so markedly with decreasing distance (or, in Einstein's picture, the curvature of spacetime would be so pronounced) that the force would change significantly within distances of only a meter or two. Gravity would be appreciably different on her feet than on her head. This difference would pull her apart in a sort of gravitational analog of a medieval rack. Were she able to survive this, however, she would notice nothing in particular as she crossed the Schwarzschild radius. She would, however, be forever cut off from the rest of the universe.

Matter falling into a black hole loses its identity; it is no longer an elephant or a kitchen sink.[13] The only properties that count are its mass, its rotation, and its electrical charge. Because most objects are electrically neutral, only the first two properties, mass and angular momentum, are astronomically important. As it swallows more mass, a black hole simply becomes larger, in the sense that its Schwarzschild radius increases in size.

Searching for Black Holes

Do black holes really exist? Perhaps the best answer at the moment is that no physical reason is known that would forbid their existence. Although a stellar-mass object that all astronomers would agree is a black hole has not yet been found, there is little reason to doubt that black holes have indeed been formed. (Remember that a neutron star is very nearly there.) In fact, it seems likely that many black holes exist, and with a wide range of masses. In the spring of 1994, rather convincing evidence was obtained by the Hubble Space Telescope indicating that a massive black hole exists at the center of the galaxy M87. Furthermore, most astronomers feel that massive black holes (about 10^8 solar masses) power quasistellar objects. In Chapter 20 we will describe how matter falling into such a black hole is thought to account for some of the properties of QSOs, in particular their enormous energy output.

How can we find black holes? Stellar black holes might be relatively nearby, and so perhaps offer a good opportunity for discovery. The problem is that by definition we can't see them directly, so we must detect them by their gravitational properties. We look for a double star with only one component visible spectroscopically. There are many of these, so next we find those with some marked peculiarity, like the strong x-ray emission seen in the x-ray binaries described previously. Analysis of the properties of these x-ray objects indicates that most of them have a neutron star as the unseen companion. In a few systems, however, a black hole seems to be required.

---

[12] Actually, the spaceship would quickly fade to blackness as it approached the event horizon, because light rays from the ship would be bent out of our line of sight. (This is an effect of general relativity that we will discuss later.)

[13] The phrase "black holes have no hair" is meant to signify this characteristic of black holes. This commonly used phrase has always seemed to me to be singularly unsuggestive, however.

Candidate Black Holes

Several binary systems are now known that are suspected of having a black hole as one component. Three of the more likely systems are Cygnus X-1, LMC X-3 (in the Large Magellanic Cloud), and A0620-00 (in Monoceros), a system detected by the British x-ray satellite Ariel 5. All three binary systems produce enormous amounts of rapidly varying x-ray energy, and all involve the transfer of matter from an evolving star onto an accretion disk surrounding a compact object. The nature of the evolving star is quite different in each of the three systems, however. In Cygnus X-1 the visible component is a B0 supergiant; in the LMC system it is a B3 main-sequence star; in the Monoceros object it is a K5 main-sequence star. (Figure 16.33 shows these three binaries roughly to scale.) Thus, most likely there are several evolutionary routes by which such systems form. In general, however, one component of the binary must be the kind of star that at the end of its evolutionary life not only becomes a supernova, but also leaves behind a black hole in the process. (At the moment we do not know just which stars fill this bill.) Also, since the two stars of the binary system are close together, it is likely that in its evolution the originally more massive star will transfer a substantial amount of matter to its companion before becoming a supernova. Later, as the companion star evolves, it returns mass to the black hole, forming an accretion disk in which x-rays are generated. One such scenario is shown in Figure 16.34.

Cygnus X-1 is thought by many astronomers to be one of the best black-hole binary candidates. What is known about this system? The spectrum of the visible star is B0, and its distance is thought to be about 2,500 parsecs. If this is so, the blue star is a supergiant, and has a mass about 30 times greater than the Sun, assuming it is a normal supergiant. Observations of the Doppler shift of the spectrum of the blue star show that its projected orbital velocity is 75 km/sec, and that it revolves around the companion in 5.6 days. The brightness of the star changes periodically by a few percent, probably as a result of the varying shape—and therefore varying radiating surface

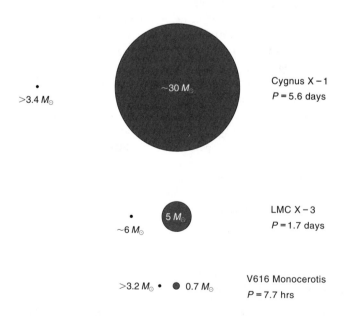

Figure 16.33. Approximate relative orbital dimensions of three black-hole candidate systems. Note the large variety of binary systems that may have a black-hole component.

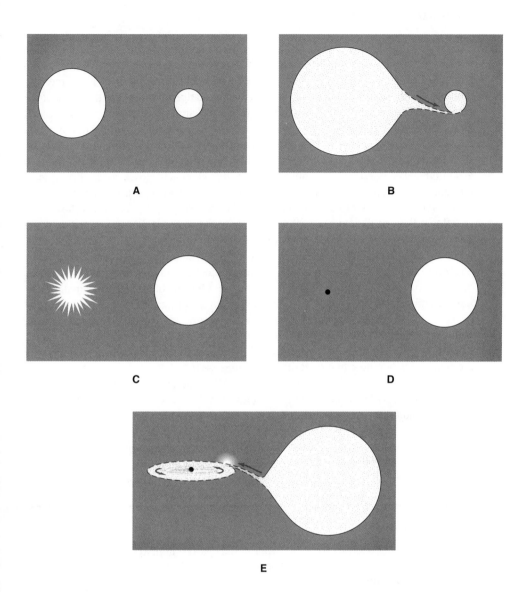

Figure 16.34. The evolution of a binary star that might result in one component becoming a black hole and the system becoming an x-ray binary. The more massive star in the binary (A) evolves and loses much of its mass to the companion (B). The originally more massive star, its fuel exhausted, becomes a supernova (C), leaving a black hole behind (D). Subsequent evolution of the companion could then cause an accretion disk to be formed around the black hole, which could be heated sufficiently by infalling matter to produce x-rays (E).

area—of the non-spherical star as it revolves around the nearly compact companion. X-rays are produced in irregular pulses as short as 0.001 seconds, indicating that they are produced in areas less than 300 km across. No eclipses of the x-ray-producing object are observed, so the orbital plane of the binary must be inclined to our line of sight. This means that the true orbital velocity of the visible star must be greater than 75 km/sec, but how much greater depends on the value of the orbital inclination.

These data do not allow us simply to calculate the masses of the two components. What we can do, however, is construct mathematical models of the system, assuming a series of values for the masses of the two objects and for the orbital inclination. Mass transfer is assumed to take place through the Roche surface joining the two objects, so a range of separations of the objects can be found. Those models that give agreement with the observed parameters then yield a plausible range of values for the mass of the compact object. For Cygnus X-1, this range is from about 9 to 15 solar masses, well above the mass limits of a neutron star. Nonetheless, it is not a completely convincing case. For example, can the mass of a standard supergiant be ascribed to the evolving star? Perhaps mass exchange with the companion earlier in its evolution has produced a non-typical supergiant? Or could the blue star be much closer to us than is now thought and therefore be neither as luminous nor as massive as a supergiant? So far uncertainties of this sort have prevented clear identification of it as a black hole.

Regardless of their previous lives, all stars end in the stellar graveyard as white dwarfs, neutron stars, or black holes (or possibly no remnant at all). In all likelihood, white dwarfs are the most common corpses and black holes the least, because the kinds of objects that can become white dwarfs are the most common. Once matter is tied up in a compact object, it will remain there forever. Only if the universe ultimately collapses, and all galaxies, stars, and interstellar clouds are crushed back to primordial energy, can the matter contained in compact objects be rejuvenated. Otherwise, there is no escape from the stellar graveyard, and it is only through their gravitational force that compact objects can play any role in subsequent events. For most purposes they are removed from the game of stellar evolution.

In the 50 years since the discovery of stellar energy-generation processes, remarkable progress has been made in our understanding of the structure and evolution of stars. In addition, new and exotic kinds of objects have been discovered, objects that were totally unexpected. With more sensitive telescopes and detectors becoming available in the near future, we are most likely in for many more surprises.

Next, we will return to the development of cosmology, picking up the story at about 1750 and quickly proceeding to modern evidence and ideas.

Terms to Know

Compact object, Chandrasekhar limit, neutron star, pulsar, millisecond pulsar, glitch, black hole; planetary nebula, Roche lobe, nova, supernova, synchrotron radiation. Event horizon, Schwarzschild radius.

Ideas to Understand

Properties of white dwarfs and neutron stars; why they form; what supports white dwarfs and neutron stars; why there is a limit to the masses of white dwarfs and neutron stars. How mass can be transferred between binary stars; characteristics of mass loss by planetary nebulae, stellar winds, novae, and supernovae. Nature of pulsars and their relation to neutron stars; the origin of millisecond pulsars. The causes of a supernova and the observed consequences of the outburst. The two fundamentally different energy-generation processes that occur in the explosion of supernovae. Properties of black holes; where black holes might be found and how they are searched for.

Questions

1. Which of the following would be the most interesting for an astronomer to observe, and why?

    (a) A pulsar with a period of 2 seconds.

    (b) An x-ray binary with component masses of 10 and 15 Suns.

    (c) A supernova with an expansion speed of 5,000 km/sec.

2. Explain what would be very puzzling about the discovery of a binary system consisting of a 20-solar-mass main-sequence star and a white dwarf.

3. Three stars, a main-sequence M-type, a giant M-type, and a supergiant M-type, are all found to have the same apparent brightness and the same proper motion. Explain your answers to the following questions.

    (a) How do the masses of these stars compare with that of the Sun?

    (b) Which star is least advanced in its evolution?

    (c) Which star is least likely to become a white dwarf?

4. Has the Sun ever been or will it ever be a main-sequence star? a neutron star? a red giant? a white dwarf? Explain your answers.

5. Do you think there are more white-dwarf stars than main-sequence stars with masses less than 1 solar mass in a globular cluster? Explain your answer.

6. Why are there only three kinds of dead stars?

7. Why isn't there a white dwarf in the center of a main-sequence star, an A-type star, for example?

8. Why can't the central temperature of a star increase indefinitely?

9. The line in Figure 16.35 is the main sequence. Of the stars marked, *B* is a supergiant, and *S* is the Sun. Answer the following with the appropriate star or "can't tell." Briefly explain your answers.

    (a) Which is the largest in diameter? Which is the smallest?

    (b) Which has or had the longest main-sequence lifetime?

    (c) Which is most likely to become the next supernova?

    (d) Which is similar to the final evolutionary stage of the Sun?

    (e) Which has the smallest average density? Which has the largest?

    (f) If it were a member of a binary system, which star might now be a nova?

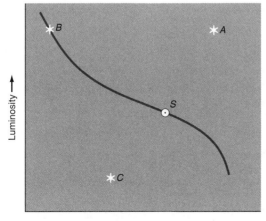

Figure 16.35. Sketch for Question 9.

10. Describe the three factors that could invalidate the argument given in the text that white dwarfs suggest that the age of the universe is only about 11.5 billion years.

11. Why does the estimated total number of planetary nebulae in our galaxy depend on the distances of those we have actually discovered?

12. Assume that the Sun's rotation period is 30 days, its radius is 7×10^5 km, and that it will not lose any mass as it becomes a white dwarf with a radius of 7×10^3 km.

(a) What would its period of revolution be as a white dwarf? Comment on the validity of the assumptions.

(b) By how many times would its average density increase?

(c) How many times greater would the acceleration of gravity be at the surface of the white dwarf than at that of the Sun? than at that of the Earth?

13. If an x-ray binary emits mainly x-rays with a wavelength of about 1 Å, what is the temperature of the radiating gas (assuming it radiates like a blackbody)? How is the gas heated to this temperature?

14. The shell of a particular planetary nebula is expanding at the rate of 20 km/sec, and its diameter is 0.1 parsec. How many years has this nebula been expanding?

15. Suppose neutron stars did not exist. What would happen to a dead stellar core that was more massive than the Chandrasekhar limit?

16. (a) Show that a 2-solar-mass sphere of neutrons, packed so that they are all in contact, has a radius of about 13 km. Take the radius and the mass of the neutron to be 10^{-13} cm and 1.7×10^{-24} gm, respectively.

(b) A neutron star like that in (a) is sometimes called a giant atomic nucleus. What would be the atomic number of such a nucleus? its atomic weight?

17. The period of PSR 1257 +12 (the object with the planet-mass objects) is 6.2 milliseconds. How many times does this pulsar rotate in one day?

18. (a) What is meant by the event horizon of a black hole? What is the Schwarzschild radius?

(b) What would be the radius of a black hole with the same mass as Jupiter? What would its density at the Schwarzschild radius be?

(c) Compare the results in (b) with the corresponding results for a 100-solar-mass black hole.

19. Review the material of Part II and list what you think are the three or four most significant or interesting unsolved problems (for example, the question of the Sun's neutrinos). Describe each problem and why you think it is important.

Suggestions for Further Reading

The following articles describe various aspects of white dwarfs, neutron stars, pulsars, and black holes.

Bell Burnell, J., "Little Green Men, White Dwarfs, or What?," *Sky & Telescope*, **55**, p. 218, 1978. A fascinating account of the discovery of pulsars.

Byrd, D., "Do Brown Dwarfs Really Exist?," *Astronomy*, **17**, p. 18, April 1989.

Croswell, K., "The Best Black Hole in the Galaxy," *Astronomy*, **20**, p. 30, March 1992. The case for A0620-00.

Graham-Smith, F., "Pulsars Today," *Sky & Telescope*, **80**, p. 240, 1990.

Hewish, A., "Pulsars After 20 Years," *Mercury*, **18**, p. 12, January/February 1989.

Kaler, J., "The Smallest Stars in the Universe," *Astronomy*, **19**, p. 50, November 1991. White dwarfs, neutron stars, and black holes.

————"Planetary Nebulae and Stellar Evolution," *Mercury*, **10**, p. 114, July/August 1981.

Kawaler, S. and Winget, D., "White Dwarfs: Fossil Stars," *Sky & Telescope*, **74**, p. 132, 1987.

Margon, B., "The Bizarre Spectrum of SS 433," *Scientific American*, **243**, p. 54, October 1980.

McClintock, J., "Do Black Holes Exist?," *Sky & Telescope*, **75**, p. 28, 1988. Some of the stellar possibilities.

Nather, E. and Winget, D., "Taking the Pulse of White Dwarfs," *Sky & Telescope*, **83**, p. 363, 1992. White dwarfs, the age of the universe, and stellar seismology.

Nicastro, A., "White Dwarfs: Big Things in Small Packages," *Astronomy*, **12**, p. 6, July 1984. A nice account of the history of the discovery of white dwarfs and their properties.

Parker, B., "In and Around Black Holes," *Astronomy*, **14**, p. 6, October 1986. Description of black holes having electrical charge and rotation, as well as the simplest variety having only mass.

Parker, B., "Those Amazing White Dwarfs," *Astronomy*, **12**, p. 15, July 1984.

Shaham, J., "The Oldest Pulsars in the Universe," *Scientific American*, **256**, p. 50, February 1987. Millisecond pulsars and their possible origins.

Soker, N., "Planetary Nebulae," *Scientific American*, **266**, p. 78, May 1992. Current ideas of origin and evolution of planetary nebulae.

Will, C., "The Binary Pulsar: Gravity Waves Exist," *Mercury*, **16**, p. 162, November/December 1987. An excellent account of the discovery of the binary pulsar and the information it has yielded.

A small sample of the articles and books on supernovae is listed below.

Filippenko, A., "A Supernova with an Identity Crisis," *Sky & Telescope*, **86**, p. 30, December 1993. The 1993 supernova in M81.

Marschall, L., *The Supernova Story*. New York: Plenum Press, 1988. A good introduction to the topic.

Murdin, P. and Murdin, L., *Supernovae*. Cambridge: Cambridge University Press, 1985. An excellent popular review, historical as well as technical, of the pre-1987A supernovae world.

Seward, F., Gorenstein, P., and Tucker, W., "Young Supernova Remnants," *Scientific American*, **253**, p. 88, August 1985.

Stephenson, F. and Clark, D., "Historical Supernovas," *Scientific American*, **234**, p. 100, June 1976. Early records of seven supernovae from 185 to 1604.

Straka, W., "The Cygnus Loop: An Older Supernova Remnant," *Mercury*, **16**, p. 150, September/October 1987.

Thorpe, A., "Giving Birth to Supernovae," *Astronomy*, **20**, p. 6, December 1992. Some well-known stars that might eventually become supernovae.

deVaucouleurs, G., "The Supernova of 1885 in Messier 31," *Sky & Telescope*, **253**, p. 115, 1985.

A few of the articles on supernova 1987A in the Large Magellanic Cloud are listed below.

Lattimore, J. and Burrows, A., "Neutrinos From Supernova 1987A," *Sky & Telescope*, **259**, p. 348, 1988.

Malin, D. and Allen, D., "Echoes of the Supernova," *Sky & Telescope*, **79**, p. 22, 1990. How light echoes arise and photographs of those surrounding the 1987A supernova.

Naeye, R., "Supernova 1987A Revisited," *Sky & Telescope*, **85**, p. 39, February 1993. An update on the best-observed supernova in history.

Woosley, S. and Weaver, T., "The Great Supernova of 1987," *Scientific American*, **261**, p. 32, August 1989.

An account of gamma-ray astronomy is given in the following:

Hurley, K., "Probing the Gamma-Ray Sky," *Sky & Telescope*, **84**, p. 631, 1992. An overview of results from the Gamma-Ray Observatory; includes a map of sources.

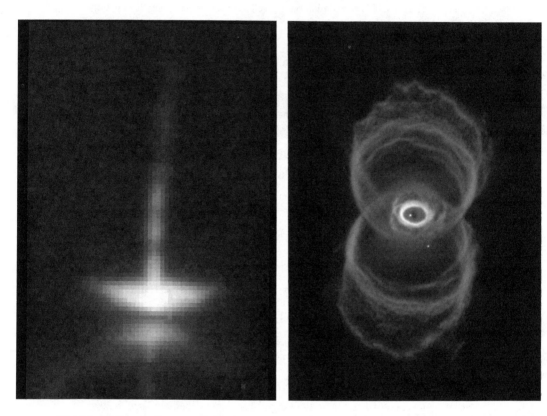

Hubble Gallery. Two remarkable images show star birth (left) and star death (right). Herbig-Haro 30 (left) clearly shows the main features of a star-forming system. The edge-on flattened disk of gas is cut in two by obscuring dust, and two narrow jets of material flow out at right angles to the plane of the disk, within which a star is forming. The Hourglass Nebula (right) gives clues as to how dying stars lose mass by means of stellar winds. Note that the mass-shedding star is slightly off center.

PART **III**

Cosmology: From Herschel to the Present

"(I feel) engulfed in the infinite immensity of spaces whereof I know nothing, and which know nothing of me, I am terrified. . . . The eternal silence of these infinite spaces alarms me."

—Blaise Pascal, 1657; French philosopher

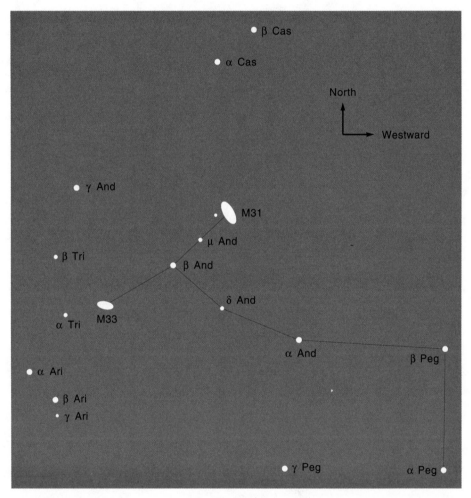

One of the impressive sights in the northern autumnal sky is M31, the Andromeda galaxy. This galaxy, about 2 million light-years away, is the most distant object that can be seen with the naked eye (if you have a dark sky; otherwise you may be able to see it with binoculars). Find the Great Square of Pegasus (note the angular scale) and work your way from α And to δ And to β And, and then northward to μ And and finally to M31. What you will see is a faint, hazy patch of light, the central bulge of the galaxy as it was before humans walked the Earth. With binoculars you can see another Local Group galaxy, M33, located about the same angular distance from β And, but in the opposite direction from M31. See the autumn star chart (4) for the location of Pegasus.

CHAPTER 17

The Discovery of Galaxies

In Part I of this book we considered various aspects of our planetary system and so never strayed very far from home. The light-travel time from Sun to Earth is only 8 minutes, and from the Sun to Pluto only about 5 hours. In Part II we left these cozy surroundings and traveled out to the incomparably larger world of stars. Here light-travel times must be measured in years, often thousands of years, not hours or minutes. Now, in the third part of this book, we will make the ultimate leap into the realm of galaxies and quasars, where light-travel times are measured in billions of years and distances are measured in millions and billions of parsecs.

The stars, clusters of stars, and interstellar clouds we have described are themselves organized into larger systems called galaxies. Our home galaxy, the Milky Way, is a highly flattened system like a thin platter, but with a bulge at its center (Figure 17.1). The visible part of our galaxy is about 30,000 parsecs across and contains

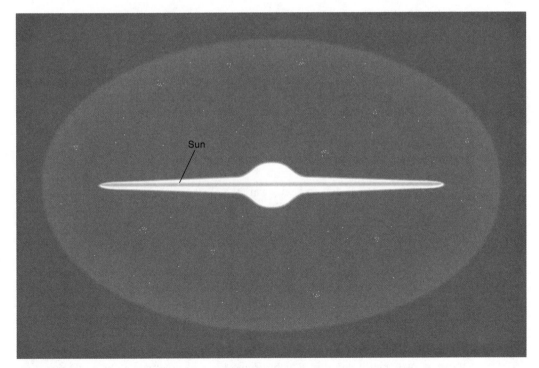

Figure 17.1. A sketch of our galaxy as it would appear seen edge-on. Shown are the central bulge surrounded by a flat disk of stars in which are embedded spiral arms, and a band of dust (shaded) in the plane of the galaxy. The position of the Sun, about 8 kpc from the center, is also shown. The entire galaxy is surrounded by a "halo" of stars, globular clusters, and very rarefied gas.

421

Figure 17.2. NGC 5364, a galaxy showing a bright central bulge and loosely wound spiral arms embedded in a faint disk. Note the dark lanes of interstellar dust.

roughly 200 billion stars. Most of these stars and interstellar clouds are confined to a thin region in the plane of the platter, only some hundreds of parsecs thick. The Sun is in this plane, about 8,000 parsecs from the center of the galaxy, about which it revolves approximately once every 250 million years.

Embedded in the galactic disk are "arms" defined by bright stars and gas clouds arranged in a rough spiral pattern, which gives its name to this type of galaxy (see Figure 17.2). Other galaxies, called ellipticals because of their apparent shape (see Figure 17.3), show no spiral structure, but may (or may not) have a plane of symmetry that suggests rotation about an axis perpendicular to the symmetry plane. The third commonly defined class of galaxies, the irregulars, have no plane of symmetry or apparent rotation axis (see Figure 17.4).

All these systems come in a wide variety of sizes and masses, and are often themselves arranged in groups and clusters, as shown in Figures 17.5 and 17.6. As with groups of stars, these range from double systems to giant clusters of a thousand or more galaxies. Our own Milky Way is itself a member of a small system of about 30 galaxies called the Local Group.

Galaxies in the Virgo cluster are about 20,000,000 parsecs away from us. Although this is an immense distance, it is just next door on the cosmic scale of things. The Coma cluster of galaxies is about 130,000,000 parsecs away, and 3C295 (see Figure 17.7), a fairly bright quasar (we think quasars are the incredibly bright nuclei of certain active galaxies), is about 1.4 billion parsecs distant. We know of many other such objects that are even farther away. In discussing galaxies and the structure of the universe, we enter a world in which distances are so great that light-travel times are enormous. This has important implications.

Remember that we know astronomical objects only by the radiation they emit. The light from 3C295 took 4.7×10^9 years to reach us, so we see it as it was 4.7×10^9 years ago. We have no direct knowledge of what it is like "now."

Figure 17.3.   An elliptical galaxy has no spiral arms and very little gas or dust. This elliptical, M49, is about 65 million light years away, has a diameter of about 100,000 light years, and is a member of the Virgo cluster of galaxies.

Figure 17.4.   The Small Magellanic Cloud is an irregular galaxy and is a satellite of our Milky Way. Located in the southern sky, it is only about 200,000 light years distant and is visible to the naked eye.

424

Figure 17.5. This small group of galaxies, Stephan's Quintet, is most likely a group of four, not five, gravitationally bound galaxies. The galaxy in the upper left is much farther away than the others.

Figure 17.6. The Centaurus cluster of galaxies is about 45,000,000 parsecs away and extends some 2° across the sky.

Figure 17.7.   Although nearly 5 billion light years away, the quasi-stellar object 3C295 is fairly bright. We see it as it was about the time the solar system was formed. As a group, quasars are the most distant objects we know.

Early cosmology dealt largely with the solar system; its observational basis was provided by naked-eye measurements. Galileo's telescopic observations were indeed influential in undermining the old ideas, but the new outlook provided by Copernicus and Kepler came about without the telescope. By contrast, exploration of the realm beyond the solar system would have been impossible without it. Larger and larger telescopes provided with the most sensitive and sophisticated instruments available continue to enlarge and deepen our view of the universe. In this part of the book we will see how our understanding of the universe of galaxies has developed. The nature of the questions we can ask has undergone dramatic, even spectacular, changes during the last two centuries of discovery. The universe as we understand it today is incomparably richer and more exciting than anyone could have dreamed even 50 years ago, and, as we shall see, our explorations have just begun. To provide a little background, we will begin with a brief description of ideas prevalent some 200 years ago.

Early Ideas on the Nature of Nebulae

By the middle of the eighteenth century, people were speculating about the nature of the objects beyond the solar system. There was great confusion about the structure and composition of the nebulae, those faint, fuzzy, amorphous patches of light scattered around the sky (see Figures 17.8a and 17.8b). Were they simply clouds of unresolved stars (as suggested by Galileo's resolution, by his small telescope, of portions of the Milky Way into stars); or were they composed of some sort of a special cloud-like substance, a "nebular fluid"; or were they perhaps a mixture of stars and fluid? Were these objects within the Milky Way or were they far beyond it? How large was the Milky Way? How large the universe? These questions, first seriously attacked about two centuries ago, were answered only in the second and third decades of our own century.

In 1750 Thomas Wright, a self-taught Englishman, proposed that the band of light we call the Milky Way (Figure 17.9) is a consequence of stars being located in the thin

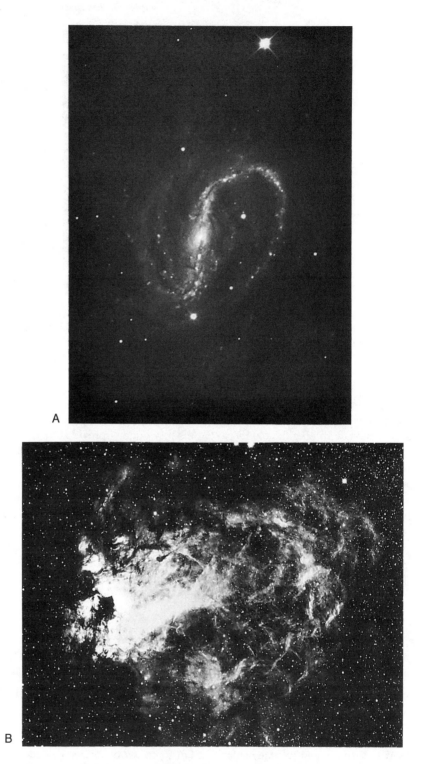

Figure 17.8. (A) NGC 7479 is a galaxy a little over 100 million light years away. It is a barred spiral, that is, the spiral arms arise from the ends of a central bar of stars. (B) M17, sometimes called the Omega nebula, is a gaseous nebula about 5,700 light years distant. It is perhaps not too difficult to imagine that to a visual observer these two objects might seem to be similar (though one would appear much brighter than the other).

Figure 17.9. The Sun is located within the thin disk of a spiral galaxy. When we look perpendicular to the plane of the disk, we see relatively few stars; when we look in the plane of the disk, we see many stars that form the band of light called the Milky Way.

shell contained between two spheres, as shown in Figure 17.10. If the solar system were located in the shell halfway between the two spheres, then many stars would be seen in the directions *AB*, few in the direction *CD*, and something like the Milky Way would be observed. That the Milky Way system is flattened is a reasonable inference from its appearance, which is that of a band cutting the sky into two equal parts. If the Sun were well out of the plane, we would see a faint disk in one part of the sky and very little in the other. The visual evidence, however, does not rule out the possibility that the Milky Way stars are arranged in a ring rather than in a slab.

Wright's model was motivated much more by theological considerations than by scientific ones. For example, Wright believed that the universe was spherically symmetric because the center of such a system was the appropriate location for God, in a way turning inside out the medieval universe in which God surrounded the central Earth. If the Milky Way was interpreted as one tangent slab in a shell, then the shell could accommodate many more "milky ways." He also speculated that every star was a Sun and the center of its own planetary system.

A few years later Wright had some wild second thoughts about this picture. He went back to an old notion that the celestial sphere was solid. Understandably impressed by the disastrous Lisbon earthquake of 1755, he invoked "geological" activity in the celestial sphere to account for various astronomical phenomena. For example, he asserted that novae were volcanic eruptions, and the Milky Way the result of hot flowing lava! Such speculations can hardly be considered as remarkable precursors of modern ideas. They do give a flavor of the background in which contemporary astronomers worked, however, and also are of interest because, incorrectly reported in a German newspaper, they stimulated the philosopher Kant's cosmological thinking.

Immanuel Kant (1724–1804) is best known as a philosopher, but in his younger days he dabbled in astronomy. In addition to his cosmological ideas sketched here, he

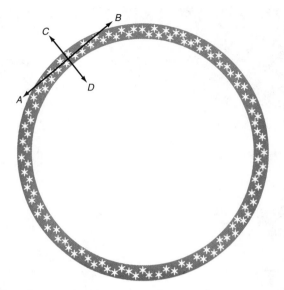

Figure 17.10. The universe as imagined by Wright consisted of a spherical shell filled with stars. The solar system was located at the intersection of *AB* and *CD*, from which an observer would see a band of stars (in the direction *AB*) across the sky, but few stars in the direction *CD*.

correctly proposed that tidal friction between the oceans and the Earth slowed the Earth's rotation. In 1755 he suggested that the Milky Way is a flattened disk and, like the flat solar system, rotated about its center. He also felt (though with no supporting observational evidence) that at least some of the nebulae represented galactic systems at enormous distances from us, so that the universe was populated by many Milky Way–like galaxies. Later this idea was popularly and somewhat misleadingly called the **island universe** hypothesis. Both kinds of systems—galaxies and solar systems—were formed from nebulae by the same process, he thought. In fact, Kant thought that all celestial objects somehow condensed out of nebulae, but could give no mechanism by which this took place. A third suggestion of Kant's was new—that the universe as a whole evolved, or changed with time. The contrary notion probably was a leftover from the old pre-Copernican attitude of the constant, unchanging starry skies. All these cosmological speculations are correct, though their correctness was shown only long after Kant.

Herschel's Work

Telescope Maker and Observer

The first large-scale observational attack on these problems was made by William Herschel (1738–1822), certainly one of the greatest of observational astronomers (see Figure 17.11). Herschel, a musician in a military band during his youth in Germany, emigrated to England when he was 19 years old, at just about the time of Wright's and Kant's publications. While supporting himself through a variety of musical activities—oboist, organist, conductor, teacher, composer—his interests turned to astronomy. Unable to afford a telescope, he taught himself how to grind mirrors. Soon he

Figure 17.11. William Herschel, the German-born English astronomer, was a remarkably energetic and talented telescope builder and observer.

Figure 17.12. Caroline Herschel was not only William's helpful sister, but was herself an accomplished observer.

was devoting more and more of his time to astronomy, both as an observer and as a telescope maker. Aided by a brother and particularly by his sister Caroline (see Figure 17.12), Herschel soon became prominent in both capacities. No mean observer herself, Caroline discovered eight comets in the little spare time she had when not helping her brother with his observational work. Their diligence and determination in observing are legendary. One night while running to help her brother, Caroline skewered her leg on a large iron hook. In her diary she recorded that "I had, however, the comfort to know that my Brother was no loser through this accident, for the remainder of the night was cloudy."

Herschel's reflecting telescopes were soon recognized as the best then available anywhere. His discovery of Uranus, which he first believed to be a comet, won him a small pension from King George III. (Incidentally, King George, the reigning British monarch at the time of the American Revolution, was an enthusiastic amateur astronomer and had his own observatory on the edge of London. It was well equipped with many instruments, including several telescopes built by Herschel.) With this pension and the income earned from selling his telescopes—he built more than 600 in all!—Herschel was able to give up music and devote all his time and enormous energies to astronomy. The largest telescope he built was a 40-inch reflector, but it was too clumsy to be used effectively. Much of his most important work was done with a reflector of only 18 inches in aperture (see Figure 17.13).

Herschel made four separate, systematic surveys of the whole sky available to him from England, noting and describing every object revealed by his excellent telescopes. This work resulted in the first important modern catalogs of nebulae; in all he recorded about 2,500 such objects. A more typical effort of the time was that by the French ob-

Figure 17.13.   Herschel did most of his important work with an 18-inch reflector, one of hundreds of telescopes that he built. The direction of the telescope was fixed in altitude by the block and tackle shown and the observer, perched several feet above the ground, observed objects as the Earth's rotation passed them through the field of view of the instrument.

server Charles Messier, who cataloged only 103 objects.[1] Herschel's achievement would have been remarkable under the best of circumstances; considering the poor astronomical quality of English skies, it was little short of incredible. His work, continued and extended to the southern hemisphere by his son John, forms the basis for the modern catalog of nebulous objects.[2]

Herschel's extensive cataloging work was not done just for its own sake, however. He felt that observational work should always be guided by specific questions. Throughout his career, Herschel attacked the fundamental problems of astronomy— what is the nature and extent of the universe, and how do celestial objects evolve? Astronomers are still working on these problems. In marked contrast to people like Wright, Herschel's attitudes are completely modern, and to read his scientific papers today requires little shifting of our intellectual gears.

The Size and Shape of the Milky Way

How did Herschel attempt to measure the extent in space of the Milky Way? He knew that the stars were too far away for a direct measurement of their distance to be made by their parallactic displacement caused by the Earth's revolution around the Sun. In any case, Herschel wanted the distances to a large number of stars, not just one or two, and so he needed a simple method. For this he invented the method of star counts, a

---

[1] Messier compiled his catalog to aid comet hunters, who often confused nebulous objects with comets. Since his catalog contains most of the brightest nebulae, we still use his designations; for example, the Orion nebula is M42.

[2] In 1864 John Herschel published his *General Catalog* of 5,000 nebulae; in 1888 John Dreyer had collected data on 13,000 nebulae, which he published in his *New General Catalog*. His NGC numbers are still in use today.

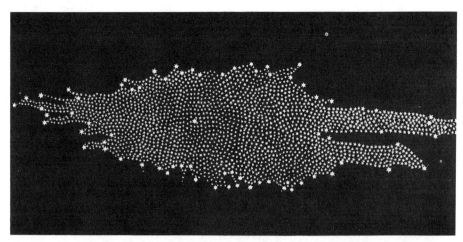

Figure 17.14. Herschel attempted to measure the extent of the Milky Way by counting all the stars he could see in 683 different small areas distributed around the sky. Assuming that the number of stars counted in each area was proportional to the extent of our star system in that direction, he arrived at the shape given in cross-section in the figure, with the Sun near the center.

technique widely used until the 1930s and occasionally employed even today. This marked the first use of statistical methods in astronomy.

He assumed, of course, that his telescopes could penetrate to the edge of the galaxy (otherwise there would be no point in undertaking the investigation), and that, on the whole, stars were distributed uniformly in space. The general idea of the method is simple: unless stars are clustered strongly, the number counted in a given direction should be proportional to the extent of the system in that direction. With his telescopes, Herschel counted the numbers of stars he could see in some 683 regions of the sky, each about one-quarter of the area of the full Moon. Using his two assumptions, the relative extent of the Milky Way as seen from the Sun in each of the 683 directions was simply proportional to the number of stars he counted in each of these areas. The distribution of stars given by this census is shown in cross-section in Figure 17.14. The Milky Way system is flattened, with two branches, and the Sun is near its center. In a general way this corresponds to a visual view of the Milky Way. The band of the Milky Way certainly suggests a flattened system, and the two branches are easily seen in the constellation of Cygnus (see Figure 17.9, upper right). This splitting of the Milky Way is a consequence of dust blocking the light from stars, rather than a lack of stars as Herschel thought; recall that the existence of interstellar dust was not firmly established until the 1930s.

Though this star census required an enormous amount of work, it actually covered less than 0.1 percent of the sky. This brings home the necessity for some sort of sampling or statistical techniques in investigations of this sort. If you think this kind of astronomical sampling is risky, just consider that today's opinion polls measuring the popularity of a politician, for example, are based typically on a sample of about 0.001 of 1 percent of the population.

This was the first attempt at a semiquantitative estimate of the size of the Milky Way and, not surprisingly, it was flawed. Herschel eventually concluded that his method was not reliable because, he found, stars are not distributed uniformly but do in fact cluster. Hence a large number of stars counted in a particular direction does not necessarily mean that the Milky Way extends a great distance in that direction.

Figure 17.15.   In 1790 Herschel discovered the planetary nebula shown here (now called NGC 1514). The star was so well centered within the nebula that he felt that the star and nebula must be physically associated and at the same distance from us. Because he could see one star in this system, he was forced to the view that the nebulosity could not be a mass of unresolved stars, but must be some sort of "nebular fluid." This convinced him that true nebulosity did indeed exist, rather than being simply a mass of stars too far away to be resolved.

Toward the end of his life, Herschel devised another method to estimate stellar distances. If all stars are equally bright intrinsically, then the fainter a given star appears to be, the farther away it really is. In particular, Herschel assumed all stars to have about the same intrinsic brightness as Sirius, the brightest-appearing star in the night sky. He thought that stars might differ in their intrinsic brightnesses only in the same small way that members of a given species might differ from each other. Given the state of knowledge then (not a single stellar parallax had yet been detected), this was all he could assume if he was to pursue this method at all. We have seen, however, that this assumption is wildly wrong, that stars can differ in their intrinsic brightnesses by factors of tens of millions or even more! In any case, applying this method with his largest telescope, he found stars that, on the basis of their apparent brightness, he thought were 2,300 times farther away than Sirius. Yet he could see stars that were still fainter (and presumably farther away); he had not reached the limits of the Milky Way and was forced to admit defeat.

The Nature of the Nebulae

What did Herschel think about the nature of those nebulae that were found in such abundance over the sky? On examining the objects in Messier's catalog, Herschel found with his telescope that 29 of the objects that Messier had said were nebulae and contained no stars were, in fact, collections of stars. This suggested to him that, given even larger high-quality telescopes, all nebulae might be shown to be clusters of stars at great distances from us. In turn, this raised the possibility, at least, that some of them might be outside our own galaxy. Later, however, he changed his mind when he found an object consisting of a faint amorphous ring with a star at its exact center, shown in Figure 17.15. Since there could be little doubt that the ring and the star were associated, here was an example of a nebula that apparently did not consist of many faint stars. Perhaps a "nebular fluid" really did exist after all. Herschel died with no clear opinion on the question. His uncertainty was shared by most astronomers throughout the nineteenth and early twentieth centuries.

Some of Herschel's work has been described here not only because of its pioneering nature, but also because it defined many of the problems astronomy faced in discovering the nature of the universe. Furthermore, Herschel established several of the techniques by which these problems were attacked for the next century.

More Confusion about the Nature of the Nebulae
Discovery of Spiral Structure

In 1845 the first telescope larger than any of Herschel's was put into operation. Built in Ireland by Lord Rosse (whom we encountered in Chapter 10), this 72-inch reflector was a monster for its time (see Figure 10.12). Perhaps the major discovery made with this instrument was that some of the nebulae showed a spiral structure (see Figure 17.16). Since M51 (the first nebula in which a spiral structure was discovered) was

Figure 17.16. A sketch of M51 made in 1845 with the 72-inch reflector constructed by Lord Rosse in Ireland, showing spiral structure for the first time. A modern photograph of M51 is also shown. Additional observations made with the 72-inch instrument revealed the spiral structure of many other nebulae.

thought to be similar to the Milky Way, the assumption was made that the spirals were galaxies, separate from our own. This inference was supported by the supposed resolution by the 72-inch of many nebulae into stars. Among the nebulae that were claimed to consist of large numbers of faint stars were the Ring nebula (in Lyra, Figure 16.3), the Crab (in Taurus, Figure 16.15), and the Orion nebula (Figure 14.3), all of which are in fact gaseous and not collections of many faint stars. Other observers soon claimed similar resolutions of nebulae into stars and some even wondered if any true nebulosity existed. These are good examples of how preconceived notions shape the interpretation of observations, especially those made at the limit of the telescope. By mid-century the notion that the nebulae were separate galaxies, "island universes," was commonly held. Toward the end of the nineteenth century, however, several developments caused the pendulum to swing back again.

The Distribution of Nebulae

For example, John Herschel extended the survey work of his father to the southern hemisphere and cataloged several thousand more nebulae. When the positions of all these nebulae were plotted on a map of the sky, it became obvious that most of them were not distributed randomly over the whole sky but were arranged symmetrically about the Milky Way. That is, most were found toward the north and south poles of our galaxy and not in the plane of the Milky Way. To some astronomers, this suggested that they must be part of our own system rather than independent objects outside the Milky Way. This was another red herring. The absence of nebulae in the band of the Milky Way, which resulted in their apparent concentration toward the galactic poles, did not reflect their true distribution. Instead, it was a consequence of interstellar dust particles (which are located primarily in the plane of the Milky Way) blocking out the light from distant objects.

Photographic Evidence

In the second half of the nineteenth century astronomical photography advanced to the point that good-quality pictures of nebulae could be made. Many objects thought to have been resolved into stars showed none in photographs. Furthermore, long exposures revealed how much larger some nebulae really are than the eye could detect. For example, in 1888 a long-exposure photograph taken in Britain showed that the limits of the Andromeda galaxy, M31, extended to its two small satellites (Figure 17.17). To some astronomers this was sensational and convincing evidence that this spiral nebula was really a solar system in formation, the galactic nucleus in the process of condensing to a star, and the two satellites becoming planets. Note the level of uncertainty: the same object could be interpreted as a huge galaxy at an enormous distance from the Milky Way, or as a small, nearby solar system in formation!

Spectroscopic Results

By the middle of the nineteenth century, the new technique of spectroscopy was being applied to astronomy, for example, by the English amateur Sir William Huggins. In the 1860s he found that some objects, such as the Orion nebula, showed a spectrum of emission lines. By Kirchhoff's laws (announced only a few years earlier), this indicated that the nebula emitting this radiation is a hot gas and not a collection of stars, which would have produced a continuous spectrum crossed by absorption lines. Many

Figure 17.17. In the late nineteenth century when photographs first showed its extensive outer portions, the nebula in Andromeda, M31, was interpreted by some astronomers as a nearby planetary system in formation: the bulge would become the central star, the two small elliptical objects would become planets. In fact, M31 is a large spiral galaxy, the nearest spiral to the Milky Way. Although it is 2,000,000 light-years away, its central bulge is just visible to the naked eye. The two "planets" accompanying it are small elliptically shaped satellite galaxies.

astronomers then generalized this result and believed that all nebulae were gaseous. This implied that the nebulae were not separate galaxies, but were objects within our own galaxy, so that the Milky Way encompassed all the universe.

Early in the twentieth century, V. M. Slipher at Lowell Observatory began the then very difficult task of obtaining photographic spectra of spiral nebulae. He did this at the direction of the observatory director, Percival Lowell (of life on Mars fame; see Part IV). Like many of his contemporaries, Lowell did not think of spirals as possible galaxies like our Milky Way, but rather as solar systems in formation. It was in this context that Lowell wanted spectra of spirals taken. By 1912 Slipher had managed to obtain the spectrum of M31, the nearby spiral nebula in Andromeda. On measuring the wavelengths of several absorption lines, he found that they indicated a velocity of approach for the nebula as a whole with respect to the Sun of 300 km/sec. This was by far the largest astronomical velocity measured up to that time. Since typical stellar velocities are in the range of 10 to 30 km/sec, this was a completely unexpected result, and was greeted with some skepticism at first. In 1915 Slipher presented radial velocity measurements for 15 spirals, one of which was moving at an astonishing 1,100 km/sec away from us. These unexpectedly large velocities, far greater than any astronomer had guessed possible, meant that the spirals probably could not be held gravitationally within our own galaxy. This suggested that spirals could be far away from us and, therefore, very large. Instead of investigating solar systems in formation, Slipher was onto something quite different, and some astronomers quickly took these large velocities as a strong argument for the island universe idea. And so the argument went on.

Further Attempts to Measure Our Galaxy
Star Counts: The Kapteyn Universe

Though after the 1830s trigonometric parallaxes of a few nearby stars could be measured, in order to determine the extent of our galaxy, distances to large numbers of stars had to be found. Consequently the star-counting techniques pioneered by Herschel were greatly refined and extended during the latter part of the nineteenth and early twentieth centuries. The application of photography to astronomy produced a vast increase in the amount of data available. Several hundred thousand stars over the entire sky were cataloged in the southern as well as northern hemisphere, and proper motions were becoming available for many of them.[3] Photographing and cataloging the whole sky is such an enormous undertaking that on various occasions astronomers from around the world have shared such burdens to make the task manageable. For reasons like these and also because, except for navigation and timekeeping, governments have never considered astronomy to be of much practical value, astronomers have a strong tradition of international cooperation and exchange that has been maintained through nearly all political and military crises.

Using statistical methods, astronomers attempted to improve Herschel's crude method by taking into account the large spread in the intrinsic brightness of stars, as well as the relative numbers of stars of different intrinsic brightness. Also, it was clearly realized that the presence of any interstellar matter that absorbed starlight would seriously distort the results by making stars appear to be farther away than they really are. Consequently, much effort was expended in searching for any indication of such matter, especially by the Dutch astronomer Cornelius Kapteyn (1851–1922). He found no convincing evidence for this absorption, however.

Several models of the Milky Way were produced by the star-counting techniques, culminating in 1922 in the so-called **Kapteyn universe**. According to Kapteyn's results, the Milky Way was about 15,000 parsecs in diameter and the Sun was located about 700 parsecs from the center, as sketched in Figure 17.18. Though this was a sizeable structure, the feeling among many astronomers at the time was that it was not sufficiently large to encompass all the nebulae observed; some of them, at least, must be separate systems outside the Milky Way. A bit of supporting evidence for this attitude was supplied by the novae, first discovered in our own galaxy. Nova-like objects appear in spiral nebulae. On the assumption that they are identical to the novae in our galaxy and, in particular, that their intrinsic brightnesses at maximum light are all the same, their distances were found to be about 100 times greater than galactic novae. Hence

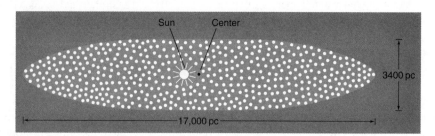

Figure 17.18. The cross-section of the distribution of stars in the galaxy derived by Kapteyn in 1922. It is fairly flat and the Sun is near, but not exactly at, the center of the system.

---

[3] Much of the southern-hemisphere work was done in South Africa by convict labor.

the spiral nebulae in which they were located must be outside our own system. This interpretation, though a step in the right direction, illustrates the fundamental danger in the method of similar objects (Chapter 12). These "novae" were in reality not ordinary novae, but were much more intrinsically luminous; they were a class of objects not then recognized, the supernovae.

In the early part of this century, trigonometric parallax was the only reliable direct technique for finding stellar distances. Unfortunately, this is useful for the nearest stars only, and so is completely inadequate for measuring the size of our galaxy. A powerful new technique, in fact, still one of the most important for cosmological investigations, was provided by the **Cepheid variables**.

Cepheids as Distance Indicators

Recall from Chapter 12 that Cepheids are intrinsically variable stars. They vary in brightness in a regular fashion, with periods of light variations from hours to about 50 days. A given Cepheid changes in brightness by a fixed amount, but the class as a whole shows a range in brightness change of anywhere from only a few percent to as much as five or six times. These stars are intrinsically very bright—the most luminous are supergiants about 10,000 times brighter than the Sun—and so are visible at great distances; furthermore, they are relatively easily found because of their light variations. Thus they are potentially useful distance indicators.

This potential was realized in 1908, when the Harvard astronomer Henrietta Leavitt investigated variable stars in the Magellanic Clouds, small satellite galaxies of our Milky Way. She noticed that the longer the time required for a Cepheid to go through its light cycle from maximum light to minimum and back to maximum again, the brighter the star appeared to be.[4] Since the stars in the Magellanic Clouds are all at approximately the same distance from us, any relation between their apparent brightness and period also holds for their intrinsic brightness and period. This is to say, if two stars, A and B, are equally distant from us, and if A appears to be twice as bright as B, then A is also two times brighter intrinsically than B. Leavitt soon established the form of this relationship, called the **period-luminosity relation**, shown in Figure 17.19. She realized that these stars could be useful distance indicators, because if the

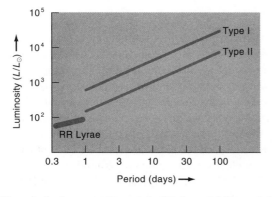

Figure 17.19. The relation between the period of light variability and the intrinsic luminosities of Cepheids. Note that there are two types of Cepheids that differ in intrinsic luminosity by a factor of about 4. The RR Lyrae variable stars are also shown.

---

[4] The brightness of a Cepheid is usually taken to be that at maximum light.

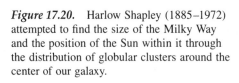

Figure 17.20. Harlow Shapley (1885–1972) attempted to find the size of the Milky Way and the position of the Sun within it through the distribution of globular clusters around the center of our galaxy.

distance to the Magellanic Clouds were known, then the curve given in Figure 17.19 would give the relation between the period and the *intrinsic* brightness of her Cepheids. With the intrinsic and apparent brightnesses of a Cepheid known (both measured at the same phase of its variation—maximum light, for example), its distance could be calculated easily.

It was quickly realized that Leavitt's variables were Magellanic counterparts to Cepheids already known in our galaxy, so, if all Cepheids behaved the same way, the period-luminosity (*P-L*) relation could be calibrated using either Magellanic or Milky Way Cepheids. The relation between period and luminosity of all Cepheids would then be known. To find the distance to any Cepheid, one would simply measure the period of its light variation; the *P-L* relation then would give its intrinsic brightness, thereby enabling its distance to be found. Here was a powerful new technique needing only (in principle) the distance to a single Cepheid to be known in order to calibrate the *P-L* relation. The trouble is that not one such star happens to be near enough to the Sun for its trigonometric parallax to be measured. This is not too surprising, since Cepheids are supergiants, which are rare objects. Hence indirect statistical techniques had to be used to find the distances to a few objects, but after a few years astronomers believed they had established the period-luminosity relation, so that Cepheids could in fact be used to give reliable distances. We will encounter several examples of its use.

Shapley's Method

An approach to the problem of the size of our Galaxy completely different from that of the star counters was developed during the second decade of this century by Harlow Shapley (Figure 17.20), working at Mount Wilson Observatory. It had been noticed earlier that globular clusters were distributed asymmetrically on the sky (see Figure 17.21). In studying these clusters, Shapley also was impressed by their strongly asymmetric distribution in space. Of the hundred or so then known, all but a few are located in the half of the sky toward the richest part of the Milky Way. Even more impressive is the fact that about two-thirds of the globular clusters are found within 30 degrees of the dense Milky Way star clouds, which might be assumed to be in the direction of the center of the galaxy. To find their distances, Shapley searched for RR Lyrae stars (variables that are "relatives" of Cepheids), many of which are located within some globular clusters. Knowing the intrinsic brightness of these stars, he found the distances to some of the clusters.

Unfortunately, RR Lyrae stars are the least luminous of Cepheid-like variables and so can be seen in only the nearer clusters. To find distances to the more distant globulars, Shapley had to use less direct means. For example, he calculated the average

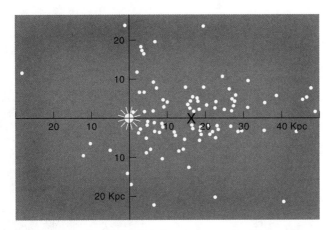

Figure 17.21. Shapley's distribution of globular clusters projected on a plane perpendicular to the disk of the Milky Way and through the Sun. Note the overwhelming preponderance of clusters in the right half of the diagram. The center of the distribution of the clusters is marked by the "x." The distances indicated are in kiloparsecs, so Shapley thought the Sun was about 16,000 parsecs from the center of the Milky Way, which is about twice the modern value.

diameter of the nearer clusters for which he had found distances using their RR Lyrae variables. By assuming that the more-distant clusters had the same diameter as the nearer ones, he found the distances to the former by calculating how far away they would have to be to appear as small as they did.

Shapley at first rejected an earlier suggestion that these clusters formed a subsystem surrounding the center of our galaxy, implying that the Sun was at a great distance from the center. The distances he was obtaining for the globulars placed them well outside the limits of our galaxy as determined by Kapteyn and others. In fact, by 1916 Shapley was fairly well convinced that some of the globular clusters were so far away that they were probably similar in size to our own galaxy. But if that were so, why would objects outside our galaxy be distributed over the sky in such an asymmetric fashion, so many in one direction, so few in the other?

A year later Shapley reversed himself. He proposed that the clusters did indeed form a subsystem around the center of our galaxy, and that their lopsided distribution was a consequence of the Sun being at a great distance from the center. This interpretation of the observations required a drastic change in orientation. First, the resulting diameter of our galaxy was enormous, about 100,000 parsecs by Shapley's estimate; second, the Sun was about 15,000 parsecs out toward the edge. Such an unheard-of diameter—about seven times that suggested by star counts—was not readily accepted by many astronomers. This huge size suggested to Shapley that the Milky Way made up the whole universe, but most astronomers were reluctant to claim that the Milky Way was anything other than an average spiral. Hence, if the spiral nebulae were equal in size to our galaxy, to appear as small as they do, the nebulae must be much farther away from us than even the distances given by the novae.

Shapley countered by arguing that one astronomer at least (Adriaan van Maanen, also at Mount Wilson) thought he had found small but measurable proper motions of objects within a few of the spiral nebulae. The appearance of spirals correctly suggests that they are rotating like pinwheels. If they were millions of parsecs away, van Maanen's proper motions would imply enormous speeds of rotation, in some cases greater than the speed of light. Hence, if these proper motions were real, they indicated that the spirals must be nearby, in fact, within Shapley's large galaxy. This seemed to be

Figure 17.22. Edwin Hubble (1889–1953) was the leading observer of galaxies in his day. In 1923, he found Cepheids in spiral nebulae and so showed that they were separate star systems outside the Milky Way. A few years later he established the redshift-distance relation for galaxies, indicating that the universe was expanding (see Chapter 19).

a strong argument for the idea that Shapley's galaxy encompassed all the objects of the universe.

Furthermore, suggestive evidence that the Sun is far from the center of our galaxy is given simply by the appearance of the Milky Way, especially as viewed from the southern hemisphere. It is much brighter toward Sagittarius than it is in directions well away from Sagittarius, for example, to the north (leftward in Figure 17.9). If the Sun were located near the center of the galaxy, however, the Milky Way should appear to be about equally bright all around the sky. If more astronomers had seen the southern Milky Way with their own eyes, perhaps they might have taken this argument more seriously.

Hubble Solves the Problem

Finally, the argument was unambiguously resolved by the Mount Wilson astronomer Edwin Hubble (Figure 17.22). Soon after undertaking a general survey of nearby nebulae to establish their properties, he discovered in 1923 the first Cepheids in the spiral nebulae M31 and M33 (Figure 17.23). Applying the period-luminosity relation to them (established by Shapley), he found that they were about 285,000 parsecs away, well outside even Shapley's huge Milky Way galaxy. With these distances he calculated that their diameters were about 10,000 parsecs. Though he communicated these results privately to a few astronomers, he was reluctant to publish them because they so completely contradicted the work of van Maanen, his Mount Wilson colleague, who had measured the apparent proper motions of stars in spiral nebulae. Finally, Hubble allowed these results to be announced at a meeting of the American Association for the Advancement of Science in December 1924; nearly immediately, most astronomers accepted them. This quick acceptance was probably a consequence of at least two factors. Hubble's distance measurements were made with Cepheid variables and the period-luminosity relation. This was felt to be a more direct and unambiguous technique than star counts or globular cluster distances, and certainly more reliable than barely detectable proper motions. Also, Hubble was using the most powerful telescope in the world at the time, the 100-inch Mount Wilson reflector.

Thus, with one stroke, another great problem in astronomy finally was solved. Kant's guess of space filled with "island universes" was correct. Suddenly the universe was realized to be of an immense size, possibly even infinite in extent, containing uncountable numbers of galaxies. Not only was the Earth not at the very center of a cozy Ptolemaic universe, nor the Sun located at the center of a single system that encom-

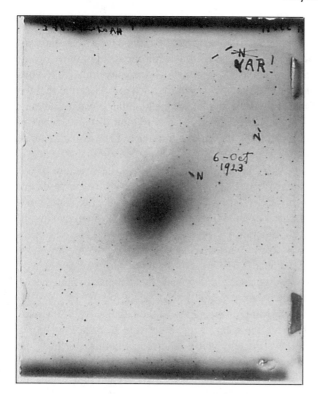

Figure 17.23. By comparing photographs of M31 taken at different times, Hubble could find which stars were variable. The stars marked "N" were novae, one of which at the top of the photograph was replaced by "VAR!," when Hubble realized it was a Cepheid variable instead; it was the first to be found in a spiral nebula.

passed the whole universe, but instead our planet was little more than a cosmic speck orbiting a commonplace star located out toward the edge of an ordinary galaxy, itself but one of countless numbers of such objects.[5]

Final Comments

It could be reasonably argued that this change in our view of the astronomical universe was the most radical readjustment yet required, but we do not speak of a "Hubbleian Revolution" as we do of a "Copernican Revolution." There are at least two reasons for this: the intellectual, theological, and social world of the 1920s was not the relatively simple structure that had prevailed 400 years earlier, and that had slowly disintegrated as various of its foundation blocks crumbled. Cosmology was no longer tied to an overall worldview, as it had tended to be in medieval times. Consequently, Hubble's work did not have the wide repercussions of the sort that Copernicus' did. Furthermore, since the eighteenth century, change has been generally regarded not only as inevitable, but also as desirable, indicating progress in all realms of human activity. Though we may now question the notion that all change is beneficial, there is little doubt that for most people Hubble's expansion of our views did indicate real and exciting progress in understanding the universe. These new ideas generally were applauded and accepted, not resisted.

It soon became clear not only that had both Shapley and Kapteyn been wrong, but also that their models shared an identical error. Contrary to the views of nearly all as-

---

[5] Keep in mind that though it was agreed that spiral nebulae were separate galaxies, some nebulae are indeed gas clouds within our own galaxy. This includes, for example, H II regions that we described in Chapter 14.

tronomers in 1925, interstellar matter that could dim and redden starlight really does exist, so when the star counters found that they could see no more stars toward the center of our galaxy, rather than having reached the boundary of our system, they had instead reached the limit at which stars could be seen through the interstellar fog. Thus their galaxy was too small. For his part, Shapley was correct when he asserted that the Sun was far from the center of the Milky Way, but he put the globular clusters and their Cepheids too far away. He attributed their faintness entirely to their distance, when in fact it was caused partially by absorption by the intervening dust. Thus his estimate of the size of the Milky Way was too large.

Interstellar dust makes it impossible to investigate the large-scale structure of our galaxy by star-counting techniques. We just can't see far enough in the plane of our galaxy. Since most globular clusters are well north or south of the galactic plane, however, they are not too greatly obscured by the interstellar fog. Hence, Shapley's methods could be made to work if the extinction by the dust was taken into account properly. This was begun in the 1930s, using photoelectrically measured colors of stars. Astronomers found the Sun to be about 10,000 parsecs from the center of our galaxy, and the Milky Way to be about 30,000 parsecs in diameter; the distance to the galactic center has recently been revised downward to about 8,000 parsecs.

The development of our understanding of the nature of the nebulae well illustrates how science often proceeds. There was no clear, direct path from Herschel to Hubble along which steady advances followed triumphantly, one after another. Instead, the available evidence was indirect and confusing. Some data suggested that the Milky Way was relatively small, some that it was large; other observations implied that the nebulae were distant systems, and yet other data that they were nearby objects. Some evidence turned out to be plain wrong, like van Maanen's still unexplained "measurements" of proper motions in spirals. Other evidence turned out to be misleading; for example, the objects observed in spirals and thought to be novae, by analogy with similar objects in our galaxy, actually were the incomparably brighter supernovae. Individual astronomers were often swayed one way, then the other, on these questions. Only when direct evidence was found by Hubble—Cepheids in a few spirals—was the matter clearly settled.

It is interesting to speculate why Shapley did not himself look for Cepheids in spiral nebulae. All the necessary tools were at his fingertips: the period-luminosity relation, the techniques for finding Cepheids in distant objects, and large telescopes. Perhaps his conviction that his enormous galaxy contained the whole universe prevented him from making the much more significant discovery of the nature of the spiral nebulae. If so, it would be a spectacular example of how what we discover is shaped and limited by our preconceptions.

Terms to Know

Cepheid variables, the period-luminosity relation, RR Lyrae stars; the Kapteyn universe, island universe.

Ideas to Understand

The gross structure of our galaxy as suggested by its naked-eye appearance. How Herschel attempted to find the size and shape of our galaxy; his work on the nature of the nebulae. How Kapteyn and Shapley attempted to find the size of our galaxy; difficulties in their methods. Hubble's determination of distances to the spiral nebulae.

Questions

1. List chronologically the observations that suggested that the universe was small (essentially everything within the Milky Way) and those indicating a large universe. Briefly give the reasons why they supported one or the other picture. Which observations were interpreted correctly, which incorrectly?

2. (a) Shapley used both the diameters and the integrated (total) brightness of globular clusters in finding their distances from the Sun. What basic assumption did he have to make in order to do so?

(b) Suppose globular cluster *A* has three times the diameter of cluster *B*. Which is farther away? By how much? Suppose cluster *C* has an apparent integrated brightness four times that of cluster *D*. Which is farther away? By how much?

3. Van Maanen thought he had measured proper motions of about 0.021 seconds of arc per year in the galaxy M101, which we see approximately face-on. This galaxy is about 7.6 Mpc away. What would be the tangential velocity corresponding to van Maanen's proper motion? Is this reasonable? How far away would the galaxy have to be for the tangential velocity to be only 200 km/sec?

4. The largest radial velocity measured by Slipher was +1,100 km/sec. What wavelength would he have measured for the Ca II line that has a rest wavelength of 3933 Å?

5. (a) Why are Cepheid variables such useful distance indicators?

(b) Suppose two Cepheids, *A* and *B*, both pulsate with a ten-day period, but Cepheid *A* has an apparent brightness four times greater than Cepheid *B*. Which star is farther away? By how much?

6. A certain G-type star (similar to the Sun) whose distance is 100 parsecs appears as bright as a certain Population I Cepheid variable with a period of 30 days. How far away is the Cepheid?

Suggestions for Further Reading

Berendzen, R., Hart, R., and Seeley, D., *Man Discovers the Galaxy*. New York: Science History Publications, 1976. An easy-to-read account of the discoveries and ideas that led to our present view of the universe of galaxies.

Gingerich, O., "Messier and His Catalogue," *Sky & Telescope*, **12**, pp. 255 and 288, 1953. A two-part article, well illustrated.

———"Observing the Messier Catalogue," *Sky & Telescope*, **13**, p. 157, 1954. Includes a map of the sky showing the positions of all the Messier objects.

———"The Missing Messier Objects," *Sky & Telescope*, **20**, p. 196, 1960. What happened to the few objects "missing" from the catalogue.

Gingerich, O. and Welther, B., "Harlow Shapley and the Cepheids," *Sky & Telescope*, **70**, p. 540, 1985. How Shapley's ideas about the nature of Cepheids and their use as distance indicators developed.

Hoskin, M., *Stellar Astronomy*. Chalfont St Giles: Science History Publications Ltd, 1982. A collection of Hoskin's historical writings, including a history of work on nebulae from Herschel to 1920 and an account of Wright's cosmological ideas.

Jones, K., *Messier's Nebulae and Star Clusters*. New York: American Elsevier Publishing Company, Inc., 1969. More background material is given than in the book by Mallas and Kreimer.

Mallas, J. and Kreimer, E., *The Messier Album*. Cambridge, MA: Sky Publishing Company, 1980. Includes photographs and finding charts for the Messier objects as well as condensations of the articles by Gingerich (listed above).

Smith, R., *The Expanding Universe: Astronomy's Great Debate 1900–1931*. Cambridge: Cambridge University Press, 1982. An authoritative account of the controversy that led to the recognition of the expanding universe.

Whitney, C., *The Discovery of Our Galaxy*. New York: Alfred A. Knopf, 1971. A well-written account from the beginnings to about 1970.

This barred spiral is a newly discovered member of the Local Group of galaxies. Though as little as about 10 million light-years away, it was discovered only in 1994 because the dust, gas, and stars in the plane of our Milky Way galaxy make it difficult to see. It is the first "standard" barred spiral found in the Local Group.

CHAPTER 18

The Universe of Galaxies

Types of Galaxies

Spirals

Hubble immediately set out to explore the new realm of galaxies. He soon established that most of them could be classified into one of three broad types: spirals, ellipticals, and irregulars. The **spirals** are visually the most spectacular of these because of their prominent arms. A careful examination of many different spirals suggests that they may be further characterized by the relative prominence of the arms and the **central bulge**, or **nucleus**, from which they emerge. Hubble denoted galaxies with a relatively large nucleus and tightly wound, indistinct arms as **Sa**; those with a small nucleus and well-defined, loosely wound arms as **Sc**; and those intermediate to those as **Sb** (see Figure 18.1, left).

Though the arms produce much of the light from spiral galaxies, they contribute only a small fraction of their mass. The light comes from bright, blue, young stars, often located in clusters, and from the clouds of gas and dust (H II regions such as the Orion nebula) from which they have recently formed, and that the young stars in turn heat and cause to glow. These objects form a relatively thin layer in the plane of the galaxy, as can be seen in photographs of spirals seen edge-on. A good example is NGC 4565 in Figure 18.2. The arms are embedded in a **disk** densely populated by billions of faint stars, each of fairly small mass, but that, together with the nucleus, make up most of the visible mass of the galaxy. The disk and nucleus are redder in color than the arms.

The middle row of Figure 18.2 shows three edge-on spirals with decreasing (left to right) bulge-to-disk prominence. The three galaxies in the bottom row show how the middle-row objects would probably appear if they were seen face-on. The disks in which spiral arms are embedded are indeed flat.

As suggested by their flattened form, spirals rotate about axes perpendicular to their planes of symmetry; they rotate in the sense that the arms trail. Since spirals have made perhaps 50 rotations since their formation, the arms would have long since wound up and been destroyed unless they were continuously being re-formed. We shall discuss this point later.

The **barred spiral** galaxies form a separate but parallel group to the ordinary spirals, and are denoted by **SBa**, **SBb**, and **SBc** (see Figure 18.1, right). As the name implies, a bar-like arrangement of stars across the nucleus is prominent in these objects. The arms emerge from the ends of the bar rather than from the circularly symmetric nucleus. In contrast to most of the nucleus and to most of the disk of the galaxy,

446

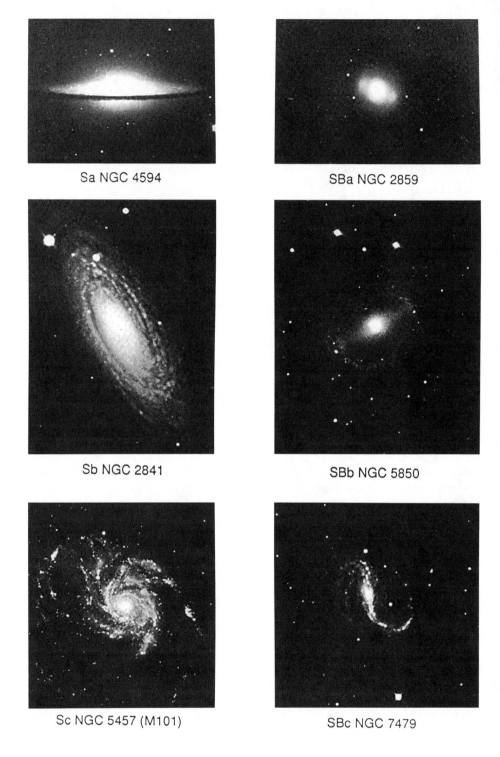

Sa NGC 4594

SBa NGC 2859

Sb NGC 2841

SBb NGC 5850

Sc NGC 5457 (M101)

SBc NGC 7479

Figure 18.1. The galaxies in the left column show the characteristics of the three types of ordinary spirals. Those in the right column are barred spirals in which the arms emerge from the ends of bars rather than from the central bulge. Their classification is analogous to that of the spirals.

Figure 18.2. The galaxies in the middle row are edge-on spirals Sa, Sb, and Sc (from left to right). The galaxies in the bottom row show how those directly above would probably appear if the latter were seen face-on. The top row are S0 galaxies; they have central bulges and disks but no prominent arms.

the bar, a collection of stars and gas, rotates as if it were a solid object (which, of course, it is not). There are about as many barred spirals as ordinary spirals.

In addition to the ordinary and barred spirals, a third kind of spiral exists, the S0 ("S-zero") systems. Although these galaxies have a central bulge and a disk, they do not have spiral arms or much (if any) dust, so their luminosity is smoothly distributed. Three examples are shown in the top row of Figure 18.2. As with the other galaxies in this figure, their bulges decrease in prominence with respect to their disks, from left to right.

Ellipticals

The second main group of galaxies consists of the **ellipticals**, so called because of their apparent shape (Figure 18.3, upper four). In contrast to the spirals, typically they show little internal structure, that is, no arms or lanes of particularly bright stars and gas, nor dark clouds of dust. In fact, they are nearly, though not entirely, free of interstellar matter. They are distinguished from spirals in another way: their brightest stars are red, not blue; that is, they are Population II or old Population I stars, not the young Population I objects of spiral arms.

Ellipticals are classified according to their apparent shape, which is given by $10(a - b)/a$, where a and b are the major and minor axes of the image, respectively. Thus an **E0** galaxy appears circular ($a = b$), and an **E7** (the most elongated elliptical) has a major axis about three times longer than its minor axis. No elliptical galaxy is as flat as an ordinary spiral, for which the major-to-minor axis ratio is closer to 10:1.

It is important to note that what is being classified is the two-dimensional *projected* shape on the sky of the three-dimensional galaxy. That is, an E0 may indeed be a truly spherical galaxy, or it could be an E7 seen face-on. A little thought will convince you that the true shape of the galaxy must be at least as elongated as its projected shape: an image classified as E5 could not in reality be an E3, but it might be an E6 or E7. Thus this classification gives only a limited physical description of these galaxies.

One might guess that the ellipticities result from the rotation of the galaxies: the faster the rotation, the greater the flattening. In this case, the polar axis would be at most as long as the diameter (for a sphere), but generally shorter, depending on the flattening. Some ellipticals may indeed be oblate spheroids (as these shapes are called), but recently it has been recognized that some ellipticals might be prolate spheroids in which the polar axis is always longer than the diameter, like a watermelon. Both of these shapes have axes of only two different lengths, the polar axis and the diameter. To complicate matters further, however, it is even likely that some elliptical galaxies are triaxial ellipsoids in which each axis is a different length! Evidently, rotation of the entire galaxy is not a major determinant of the shapes of ellipticals. We will return to this point later.

Irregulars

Irregular galaxies, the third general type, show no axis of symmetry, presumably indicating little or no systematic overall rotation (see Figure 18.3, lower two). They often contain many gas and dust clouds, and have large populations of bright, blue stars, but none of these seem to be organized into any large-scale structure. Their overall colors are the bluest of normal galaxies; in contrast to spirals, they are bluer toward their center than their edges. They are usually smaller, less massive, and fainter than the more commonly studied spirals and ellipticals. Irregulars are very numerous, however,

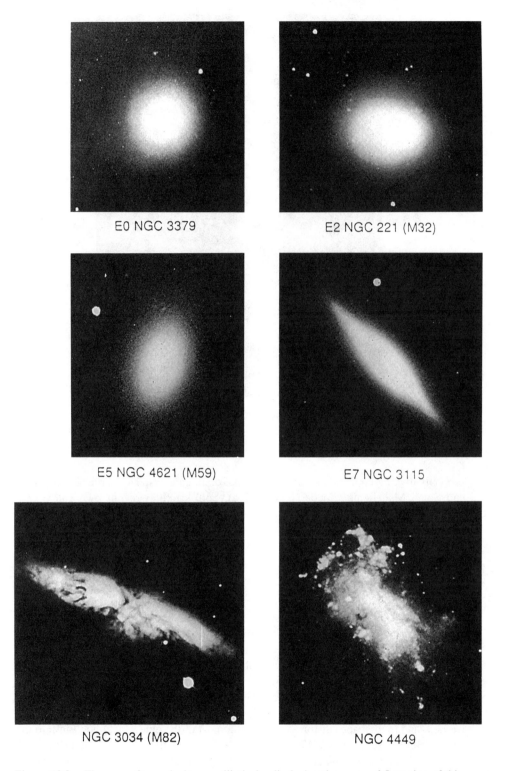

E0 NGC 3379

E2 NGC 221 (M32)

E5 NGC 4621 (M59)

E7 NGC 3115

NGC 3034 (M82)

NGC 4449

Figure 18.3. The upper four galaxies are ellipticals, displaying the range of flattening of this group. The two objects in the bottom row are irregular galaxies, showing no plane of symmetry.

Figure 18.4. The Large Magellanic Cloud in the southern sky is a satellite galaxy to the Milky Way. Only about 50 kpc distant, it is visible to the naked eye. Even though it is usually listed as an irregular galaxy, it is sometimes classified as a barred spiral.

probably making up from one-third to one-half of all galaxies. The Large (Figure 18.4) and Small Magellanic Clouds (Figure 17.4), both satellites of our own galaxy, are irregulars.[1]

Peculiar Galaxies

Most galaxies fit into the preceding simple classification scheme, but a small fraction show peculiarities that are consequences primarily of the scale of separation of galaxies. Recall that the ratio of the distances between stars to their own diameters is extremely large—the ratio of several hundred kilometers to the diameter of a ping-pong ball. Thus, even in star clusters collisions or even close encounters of two stars are rare. For galaxies, this ratio is strikingly different; if whole galaxies were now reduced to ping-pong balls, their neighbors would be only a half meter or so away! In contrast to stars, then, galaxies can come close enough to each other to be gravitationally affected, sometimes strongly so. In this way their nuclei can be disrupted, their planes warped, and their shapes distorted, producing most of the objects that don't fit into the simple Hubble scheme (see Figures 18.5–18.7). Recently evidence has been accumulating that these encounters may play an important role in the evolution of galaxies. We will discuss some of this evidence later.

---

[1] Much more elaborate galaxy classification systems have been invented, but for our purposes Hubble's simple system is adequate.

Figure 18.5. If the several dust lanes were not present, NGC 4753 would be an S0 galaxy instead of an S0 (pec) because no spiral arms are visible.

Figure 18.6. The spiral arms in NGC 2146, an Sab (pec) galaxy, have been badly distorted.

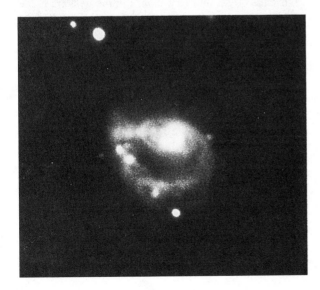

Figure 18.7. NGC 985 has just one arm, which is semi-circular in shape.

Classification by Content

Recall the idea of stellar populations, which you first encountered in Chapter 15 in connection with star clusters. OB stars are examples of the youngest Population I objects, whereas the Sun is an old Population I star; a red giant in a globular cluster is an example of Population II. The population concept can be applied not only to clusters of stars but also to galaxies. Spiral arms are defined by Population I objects—OB stars, H II regions, and dust clouds, for example. Population II systems, like elliptical galaxies, are characterized by bright-red supergiants and a relative absence of dust and gas. Interstellar gas may make up as much as 30 percent of the mass of an extreme Population I irregular galaxy, whereas an elliptical may have as little as 0.01 percent of interstellar matter. Most of the raw materials from which stars are formed are gone; so in that sense these systems must be more advanced in their evolution than those of Population I.

Now, taken as a whole, there is nothing to indicate that ellipticals are actually older than spirals. This does not, however, contradict the idea that ellipticals are more advanced than spirals in their evolution. A person's appearance may not be a good indicator of his or her age. Ellipticals may in some sense "live faster" than spirals and go through their evolutionary paces more rapidly than spiral galaxies. The brightest individual stars that characterize elliptical galaxies—bright-red giants—are more advanced in their evolution than those objects that are characteristic of spirals—gaseous nebulae and bright blue stars.

The distribution within a galaxy and the space motions of the two populations are quite different. Population I systems are flattened, but Population II systems are spheroidal. This points up the observation that many galaxies can be thought of as having one or both of two basic structures—a disk or a spheroid. A spiral has both (the spheroid being the galactic nucleus), but an elliptical has only the spheroidal component. We will better understand the origin of many of these properties when we discuss possible ways in which galaxies form and evolve.

Classification of the Milky Way

Because we can view our galaxy only from within it, it was until recently not easy to say what type it is. The narrow band of the Milky Way crossing the sky indicates that our system is flattened, and the existence of gas and dust clouds in the galactic plane means that it is some kind of a spiral. From our location near the galaxy's plane of symmetry, it is difficult to see the overall spiral structure. From the size of the galactic bulge compared to the galaxy as a whole, however, it was estimated that the Milky Way is an Sb, probably similar to M31, our neighboring galaxy in Andromeda. This classification was confirmed in 1990 by a spectacular image (shown in Figure 18.8) produced by the COBE (COsmic Background Explorer) Earth-orbiting satellite. This image, taken in near-infrared radiation (which penetrates the interstellar dust of the galactic plane), shows clearly the bulge and disk of an Sb galaxy.

Properties of Galaxies

The Mass of the Milky Way

It turns out to be easy to give a rough estimate of the mass of the galaxy if we consider only the mass that is contained within the Sun's orbit around the central bulge. Stars and gas clouds revolve around the center of our galaxy much as do the planets around

Figure 18.8. This COBE image projects most of the sky onto a flat surface (like the map of the world). It was made in three infrared wavelength bands (1.2 to 3.4 microns) that penetrate much of the dust in the galactic plane. The extent of the disk visible here extends nearly half-way across the sky.

our Sun, their motions described by Kepler's laws. That is, entire galaxies don't rotate like a rigid object—a wheel, for example—in which all parts go around in the same time. Instead, the stars and clouds move independently under the gravitational influence of the mass inside their orbits. Furthermore, just as the velocity of Pluto around the Sun is less than the orbital velocity of the Earth, stars much farther away from the center of the Milky Way are expected to move more slowly than stars nearer to the center.

Most stars in our vicinity travel around the center of the galaxy in very nearly circular orbits in the plane of the Milky Way (they are Population I objects). A few move along elliptical paths (intermediate Population), but the motions of all are determined by the gravitational force of all the mass closer to the center than they are. By applying Newton's generalization of Kepler's third law (Chapter 8), we can find how much mass there is in the galaxy inside the Sun's orbit around the center. Because the Sun is roughly halfway out from the nucleus (about 8,000 parsecs away), and the galaxy thins out markedly toward its edge, the Sun's orbit includes most of the visible mass of the galaxy.

To measure this mass, all we need to find is the period of revolution of the Sun around the galactic center. This can be done by measuring the radial velocity of the Sun with respect to the globular clusters. These travel around the galactic center along randomly oriented orbits, some clusters moving away from us, some toward us, and still others across our line of sight. If the Sun were not moving with respect to the clusters, the average of a large number of cluster radial velocities would be about zero, since as a group the globulars are neither moving toward the galactic center nor away from it. Thus a *non-zero* result must be a reflection of the Sun's velocity with respect to the whole group of clusters, and hence with respect to the galactic center. This turns out to be about 200 km/sec.

With this value, it is easy to estimate the galaxy's mass. The Sun's galactic orbit can be taken as circular, so the period of revolution is just $t = d/V$ = (circumference of Sun's orbit)/(Sun's velocity) $= 250 \times 10^6$ years. In the appropriate units (solar masses, years, and AU), Kepler's third law tells us that

$$(M_\odot + M_{MW})P^2 = a^3 \text{ and } M_{MW} = 0.7 \times 10^{11} M_\odot.$$

Much more sophisticated analyses (which also take into account the visible matter beyond the Sun's orbit) give about 1×10^{11} solar masses, not much different from this simple estimate. Thus, the visible Milky Way is about 30,000 parsecs in diameter, its mass is about 10^{11} solar masses, and the Sun is about 8,000 parsecs from the center.

Finding the Distances to Galaxies

The large spiral galaxy M31 and its two small satellite galaxies (both ellipticals, see Figure 17.17) suggest that there exists a large range of sizes, intrinsic brightnesses, and masses of galaxies. To derive these quantities we must know the distances to galaxies. Many different kinds of objects are used to measure their distances; we will mention only a few of the more prominent ones.

Assuming that the period-luminosity relation for Cepheid variables within our own galaxy is applicable to those in other galaxies, then Cepheids provide the most accurate method of finding distances to the nearer objects. By this method we can measure the distances to systems that are perhaps 10,000,000 parsecs (abbreviated as Mpc, megaparsecs) or 10 Mpc away. Within this distance are many galaxies that contain a variety of objects intrinsically brighter than Cepheids. The absolute brightnesses of these objects can now be calculated and used to extend the distance scale outward. For example, the brightest novae can be used out to about 25 Mpc. The brightest blue and red supergiants also can be seen that far away, but as a practical matter are not reliable at great distances, because they are generally found in star clusters; it is difficult to be sure that only a single star is being measured.

A new method using planetary nebulae shows promise out to perhaps 20 Mpc. It works like this. One of the prominent emission lines in a planetary is a forbidden transition at 5007 Å due to O III (see Figure 14.6). Measure the apparent brightness of this line in, say, 25 of the brightest planetaries in a nearby galaxy whose distance is known by Cepheids, for example. Form the average of these measurements and, knowing the distance to the galaxy, calculate the intrinsic brightness of the oxygen line. When these measurements are repeated for other nearby galaxies whose distances are also known, it turns out that the intrinsic brightness of the oxygen line is the same in all these galaxies. Hence, assuming that this intrinsic brightness holds for all galaxies (uniformity of nature again), it can be used to find the distances to other galaxies. A significant advantage of this technique is that planetary nebulae are easy to find in distant galaxies.

Another method receiving much attention involves Type Ia supernovae, which offers the prospect of finding distances to objects about 100 times farther away than ordinary novae. This requires, of course, that Type Ia supernovae all have the same intrinsic brightness at maximum light and that the latter be better known for this potentially powerful technique to be useful, however.

Equally large distances can be obtained by using galaxies themselves as the standard light source. We have to be very careful here, because even galaxies of the same type have a wide dispersion in their intrinsic brightness. However, Sb and Sc galaxies with well-defined arms seem to have a fairly narrow range of intrinsic brightness, and might be useful for distance determination.

Recently, the kinematics—motions—of stars and gas within elliptical and spiral galaxies, respectively, have been shown to be useful indicators of the intrinsic brightnesses of these galaxies. Very roughly speaking, this can be understood in the following way. Matter in a galaxy moves in response to the gravitational field it finds itself in; all else being equal, the greater the mass of the galaxy, the faster the matter will be revolving around the galactic center. Since the brightness of a galaxy should be related to the amount of matter contained within it, there should be a relation between the intrinsic brightness of the galaxy and the velocity of its stars or gas. Indeed, such a relation has been found empirically. Spectroscopic measurements of the 21-cm line of hydrogen give the appropriate velocities, which then indicate the intrinsic brightness of the galaxy. Once we know the galaxy's apparent and intrinsic brightness, its distance is easily found. This method has so far been applied out to distances of about 50 Mpc.

In all these methods we are of course relying on the basic assumption of the similar object method—the uniformity of nature. In addition, many of these methods depend fundamentally on the validity and accuracy of the Cepheid period-luminosity relation. Furthermore, though the error associated with each method may be relatively small, measuring large distances requires many different methods, each one making another step out in distance, but each one adding its own error. Despite the enormous effort that has been made in developing these techniques, as well as many others, there is still a spread of nearly a factor of two in the distances found to given galaxies by different observers.

Sizes and Masses of Galaxies

It is relatively easy—at least in principle—to measure a galaxy's angular diameter and its apparent brightness.[2] If its distance is known, it is a simple matter to convert those measures to the galaxy's linear diameter and intrinsic brightness, respectively. But finding the mass of a galaxy is more difficult. The method we described for the Milky Way can be applied to nearby galaxies, also. We find some bright object in the galaxy in question, like a very luminous star, and measure its orbital velocity. If the distance from the center of its galaxy is known (and for this we need to know how far away the galaxy is) then its period can be calculated. Kepler's third law can then be used as we used it to find the mass of the Milky Way. Again, this would give the mass of only the part of the galaxy located inside the orbit of the star.

In principle the method is simple enough, but there are difficulties. A relatively simple problem is that a correction must be made to the *observed* radial velocity of the object, which will, in general, not be its full orbital velocity, because we measure only that part of its orbital velocity that is along our line of sight. Imagine a spiral that is exactly face-on; the radial velocities of objects moving in the plane of the galaxy would be zero. Only if the galaxy were edge-on would we measure the full orbital velocity. Thus we must measure the inclination of the galaxy and correct the radial velocity appropriately.

A more basic problem is that reliable masses of galaxies cannot be calculated from the radial velocity of just one star. In practice velocities are measured at as many dif-

---

[2] Even here, however, a note of caution is appropriate. For example, ellipticals often appear to be smaller than they really are because the brightness of their outer regions decreases slowly with increasing distance from the center. For example, the angular diameter of the giant elliptical M87 is often given as about 3 arc minutes, but images made with the most sensitive detectors show it to be several times greater in extent.

ferent distances from the galactic center as possible. We then adjust a theoretical mass model of the galaxy, based on our best guesses as to the mass in the central bulge and in the disk (for a spiral), and adjust it so that it predicts the same run of rotation velocity with distance as the one we observed. The mass of the galaxy is then assumed to be the mass given by the model. A problem here is that it is often difficult to find bright objects (whose radial velocities we can measure) very far away from the center of the galaxy.

In any case, the mass of the visible matter in galaxies ranges from about 10^6 to 10^{13} solar masses; the former are dwarf galaxies, barely more massive than the richest globular clusters; the latter are typically giant ellipticals. Later in this chapter we will describe evidence that indicates that galaxies are much larger and more massive than is found from their visible matter.

Distribution of Galaxies in Space

Groups and Clusters of Galaxies

Probably more than half of all galaxies are now found within groups and clusters; perhaps in the past many more were so located. Thousands of clusters have been cataloged. As with stars, these groupings can range in number from double galaxies to giant clusters containing thousands of members. We are located in a small cluster, the so-called **Local Group** of galaxies, which includes about 30 currently known members (see Table 18.1). About half of these are ellipticals, only three are spirals, and the rest are irregulars. The local group is roughly 1 Mpc across; by contrast, giant clusters can be up to 10 Mpc in diameter.

The nearest example of a sizable cluster is located in the constellation of Virgo (see Figure 18.9); it consists of a few thousand galaxies within a volume about two Mpc across, and is about 20 Mpc away. The Virgo cluster contains all types of gal-

Figure 18.9. The central portion of the cluster of galaxies in Virgo. The four brightest galaxies are giant ellipticals, two of which are seen here. Although it is about 20 Mpc away, the cluster spreads over an area of the sky about 12 times the diameter of the Moon.

Table 18.1.   The Local Group

Name	Dimension	Type	Distance (kpc)	Diameter (kpc)
Andromeda galaxy M31	178′ × 63′	Sb	730	38
Milky Way		Sb	(8.5)	30
Triangulum galaxy	62 × 39	Sc	900	16
Large Magellanic Cloud	650 × 550	Ir	50	9.5
IC 10	5 × 4	Ir	1300	1.9
Small Magellanic Cloud	280 × 160	Ir	60	4.9
NGC 205	17 × 10	E6	730	3.6
NGC 221	8 × 6	E2	730	1.7
NGC 6822	10 × 10	Ir	520	1.5
NGC 185	12 × 10	dE0	730	2.5
NGC 147	13 × 8	dE4	730	2.8
IC 1613	12 × 11	Ir	740	2.6
WLM system	12 × 4	Ir	1600	5.6
Leo A	5 × 3	Ir	2300	3.3
Fornax dwarf galaxy	20: × 14:	dE3	130	2.2
IC 5152	5 × 3	Ir	1500	2.2
Pegasus dwarf galaxy	5 × 3	Ir	1300	1.9
Sculptor dwarf galaxy		dE3	85	1.5
Leo I	11 × 8	dE3	230	0.7
Andromeda I		dE0	730	
Andromeda II		dE0	730	
Andromeda III		dE2	730	
Aquarius dwarf galaxy		Ir	1500	
Sagittarius dwarf galaxy		Ir	1100	
Leo II	15 × 13	dE0	230	1.0
Ursa Minor dwarf galaxy	27 × 16	dE6	75	0.6
Draco dwarf galaxy	34 × 19	dE3	80	0.8
LGS 3	2	Ir	900	0.5
Carina dwarf galaxy		dE	170	

axies. Among the brightest galaxies spirals predominate, but the four brightest of all are ellipticals, including the powerful radio source M87. Like stars in a cluster, galaxies in a cluster are gravitationally bound to each other. Each galaxy has its own motion within a cluster, but superposed on this is a common radial velocity of the cluster as a whole. This common motion is used to establish membership in a cluster. Another nearby cluster is shown in Figure 17.6.

The space between galaxies in a cluster is usually not empty, but is pervaded by a tenuous hot gas, up to 10^8 K! It is detected because of the x-rays it emits. The hot gas in a rich cluster emits a large amount of energy in x-rays, as much as 10^{10} or 10^{11} solar luminosities. X-ray lines of some metals have been observed, indicating that the gas has been ejected from the galaxies, since the metals must be the product of stellar nucleosynthesis. The mass of intergalactic gas in a cluster can be large, up to as much as the mass of the galaxies themselves.

It might be thought that by taking a census of rich clusters of galaxies, we could establish the true relative numbers of galaxy types, masses, sizes, and intrinsic brightnesses. Unfortunately, this will not work because many galaxies, such as the dwarf ellipticals and irregulars, for example, are so intrinsically faint that they can barely be seen even within the Local Group. Some of these objects are so sparsely populated with stars that they are transparent; that is, more distant objects can be seen right through them! A good example is the Local Group dwarf elliptical, Leo II, shown in Figure 18.10. These objects have masses of only several hundred thousand to a few million solar masses, so they overlap the globular clusters at the low end of the mass range, but their stars are much more spread out than are those in globulars. Their HR diagrams are similar to those of globulars and ellipticals. The combined brightness of all the stars in the fainter dwarf galaxies is less than that of the brightest single stars known (!), so these systems are impossible to find beyond the Local Group. If the proportion of dwarf galaxies in the Local Group is typical (a big if!), then they are the most common galaxies in the universe. Unfortunately, we do not know the true percentages of galaxy types, masses, or brightnesses to any degree of accuracy. Our samples are always biased toward the brighter and more massive objects.

Superclusters and Voids

As more and more clusters of galaxies were identified, some astronomers began to suspect that clusters themselves were not randomly distributed over the sky. Instead it appeared that some clusters might be physically associated in even larger groupings. Because we would expect a certain amount of clumpiness of clusters to occur just by chance, it is not easy to establish the reality of a cluster of clusters. By the 1950s, however, it had become fairly clear that our own local group of galaxies, as well as many other nearby groupings and the large Virgo cluster, form an enormous **supercluster**.

Though the work is time consuming, many such superclusters have now been identified. The problem is to show that a group of galaxy clusters are really associated in space, and not just in the same direction but at widely different distances. Thus, radial velocity measurements must be made for large numbers of galaxies to see if they are in fact traveling together. Improvements in both optical and radio detectors have greatly increased the rate at which redshifts can be measured. For example, in 1975 redshifts for only about 1,000 galaxies had been measured; by the early 90s that number had increased to more than 40,000.

If a supercluster exists in a given direction, then the radial velocities of its members tend to fall within a relatively small range of values, corresponding to their relatively small spread in distance. On either side of this velocity spread there should be

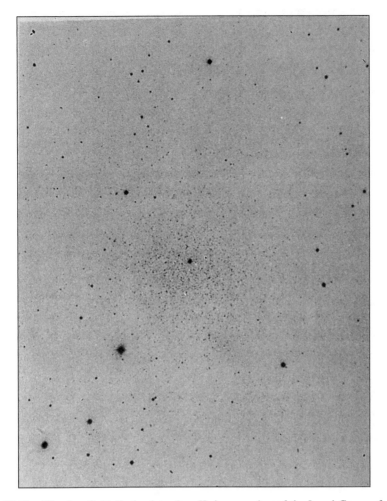

Figure 18.10. The dwarf elliptical galaxy, Leo II, is a member of the Local Group of galaxies. Dwarf ellipticals, though about as massive as rich globular clusters, are much larger than globulars, so that the space density of stars is very low. Consequently these systems are difficult to find.

a range of radial velocities in which few galaxies are found, indicating that the supercluster is indeed a discrete entity.

These superclusters are enormous: their diameters can be as large as 50 Mpc and their masses up to 10^{15} to 10^{16} solar masses. Most galaxies, those in clusters as well as those not, may well belong to superclusters. Rather than being roughly spherical in shape, superclusters seem to be arranged in elongated, even filamentary forms, as shown in Figure 18.11. They fill only a small volume of the universe, perhaps 10 percent. Huge volumes that are very nearly empty of matter (or at least of luminous matter, see below) exist between them. These are called **voids**, and just how empty they are remains uncertain.

In the late 1980s, as more and more redshifts became available, an enormous structure was discovered—the so-called **Great Wall**. It is a sheet of galaxies at least 170 Mpc long and 60 Mpc high, but only 5 Mpc thick (Figure 18.11). It is a spectacular example of the web-like structures in which galaxies seem to be organized.

The attraction of massive, nonuniformly distributed structures can cause the velocities of galaxies to depart from those of the smooth Hubble flow. That is, because

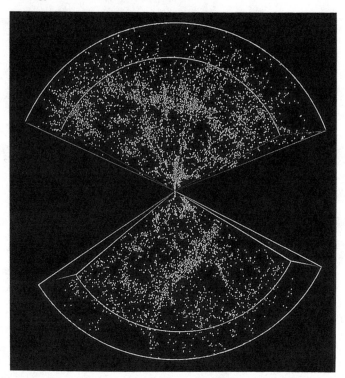

Figure 18.11.   The two pie-shaped slices together include the volume of space extending about 250° around the sky in wedges about 36° high. The Milky Way is at the apex of each slice. Each point represents a galaxy plotted according to its direction in the sky with its distance from the apex set by its redshift. About 9,500 galaxies that are brighter than a certain limit with redshifts of less than 15,000 km/sec are plotted. The Virgo supercluster is the collection of points from the apex heading toward "noon." The Great Wall is the bar of points extending from about 10:30 to 2:00. A similar feature, the Southern Wall, is seen in the lower slice. Conspicuous voids are seen adjacent to both walls.

of the gravitational force exerted by nearby superclusters, galaxy redshifts may differ from those predicted by their distances and uniform Hubble expansion. To establish this requires finding distances to galaxies by means other than their redshifts, such as the kinematic method described earlier in this chapter. In the next chapter we will see evidence that the Milky Way galaxy has an "excess" velocity of about 600 km/sec, that is, a velocity larger than that of the expected Hubble flow. This is a result of the supercluster of which we are a part moving—"falling"—toward the Hydra–Centaurus superclusters, which in turn also have a velocity excess. One interpretation of these velocities is that a huge, even more distant structure (nicknamed the **Great Attractor**) is the mass causing the larger-than-expected velocities of these superclusters. Clearly, these motions complicate the use of the Hubble relation as a distance indicator.

Since the 1970s or so we have come to realize that the smallest chunk of the universe that contains a representative sample of the whole must be enormously larger than had been previously supposed. Structure has been seen as large as the limiting size of the surveys so far made, which is now about 300 Mpc. Even this large scale may have to be revised upward when yet deeper surveys are made. The redshift surveys so far made extend out only about 150 Mpc, and cover a tiny fraction of the sky. Less than 0.1 percent of the observable universe has been mapped in three dimensions. What new structures will be found when deeper and more extensive surveys are made?

Nonluminous Matter

As mentioned above, it was always assumed that most of the mass of a spiral galaxy was located in and around the central bulge. According to Kepler's laws, stars located well beyond the bulge would have smaller orbital velocities the farther out from the galactic center they were (see Figure 18.12). (Recall that the orbital velocities of the planets decrease outward from the Sun.) Optical observations of galaxies at considerable distances from their centers are difficult because of their faintness. In the 1970s and 1980s, however, radio measurements made of interstellar gas, as well as some optical observations in many galaxies, including our own, showed that their velocities did not decrease, but remained constant or even increased with increasing distance from the galactic center, as shown in Figure 18.12. That is, the outer parts of galaxies are revolving too rapidly for the amount of visible mass they contain. How can this happen? Only if there is a *lot* of matter far beyond the limits of the visible galaxy, that is, a lot of dark, nonluminous matter, *matter we cannot see!*

So far the only way we know about it is through its gravitational effect. Just how it is distributed around galaxies and how much of it exists is uncertain. It can't be concentrated around the center of galaxies, because that would result in Keplerian rotation curves. To produce the observed rotation curves, it must extend far beyond the visible galaxy, but how far is not known. Mass estimates vary from one galaxy to the next, from three or four times as much dark matter as luminous material to perhaps 30 or 40 times as much. It has been estimated that were this material luminous, the thousand nearest galaxies would appear surrounded with halos as large as the full Moon.

So at this time we can see only a small fraction of the matter in the universe. Furthermore, we don't know what this dark matter is made of! Astronomers are in the frustrating (and slightly embarrassing) position of suddenly being able to account for only a small fraction of the universe. We do not yet understand all the implications of this very important new development.

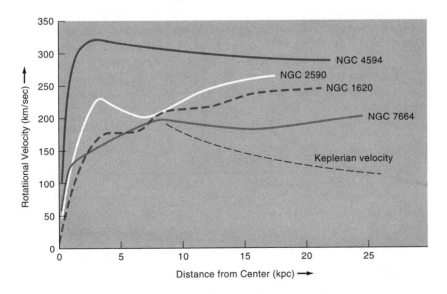

Figure 18.12. Curves of rotational velocity at increasing distances from the centers of four spiral galaxies. The velocities remain roughly flat instead of becoming smaller with greater distances from the center, as would be expected if most of the mass were located in and near the central bulge. The expected decrease, labeled "Keplerian velocity," is shown for NGC 7664. Contrast this with the observed velocity.

This "missing matter" presumably explains another problem with the masses of galaxies. If a cluster of galaxies is "glued" together by gravity (so that the galaxies will not escape, but remain in the cluster), then the average velocities of the member galaxies will be related to the total mass of the cluster. We can see this in the following way. If a galaxy is moving too rapidly, the gravitational pull of the rest of the cluster won't be sufficient to keep it from escaping. But if (as we believe) galaxies aren't escaping from a cluster, then there must be at least enough mass in the cluster to keep the member galaxies contained. Thus, there is a relation between the velocities of the galaxies within the cluster and the total cluster mass. (This is analogous to the kinematic determination of intrinsic brightnesses of individual galaxies mentioned earlier in this chapter.) When this idea is applied quantitatively to clusters of galaxies, the result is that the total cluster mass divided among the member galaxies gives as much as 50 times greater mass for each galaxy than is expected from the mass of its visible matter. To put it another way, the galaxies are moving too rapidly to be contained within the cluster by the mass we see. Again, this implies that there is a lot of dark, invisible matter in the cluster.

In contrast with isolated galaxies, however, this matter is not likely to be concentrated around galaxies in a cluster, since such halos would be disrupted by gravitational interactions with other halos. The dark matter is more likely to be distributed smoothly throughout the cluster of galaxies.

What this dark matter may consist of is unknown. That part made of ordinary matter like neutrons and protons simply may be very large numbers of "stars," perhaps brown dwarfs, with masses too small to ignite hydrogen, or perhaps Jupiter-size objects located in the halos of galaxies. These have been given the name **MACHO**s (Massive Compact Halo Objects).[3]

As we will see later, however, it is very unlikely that all or even most dark matter could be ordinary matter like neutrons and protons. Theoreticians have speculated that huge numbers of exotic particles may have been left over after the universe emerged from its incredibly hot, dense beginning. These would pervade all of space. Our ignorance is exemplified by the masses suggested for these particles: one is a group of light particles having masses in the range of a billionth the mass of an electron; another is a group of heavy particles with masses up to a thousand times that of a proton. These last are known collectively as **WIMP**s—Weakly Interacting Massive Particles. Since they do not interact strongly with matter, they are difficult to detect. Sensitive instruments are being developed to search for these particles. Obviously, the identification of this dark matter is one of the major problems of astrophysics.

In any case, Table 18.2 gives the present best estimates of the ranges of diameters,

Table 18.2. Properties of visible galaxies

	Spirals	*Ellipticals*	*Irregulars*
Mass ($M_\odot$)	10^9–10^{12}	10^6–10^{13}	10^8–3×10^{10}
Diameter (10^3 pc)	5–50	1–150	1–10
Luminosity ($L_\odot$)	10^8–2×10^{10}	3×10^5–10^{11}	10^7–2×10^9
Brightest stars	blue (Pop I)	red (old Pop I)	blue (Pop I)
Interstellar matter	gas and dust	very little	gas and dust
Ages of stars	young and old	old	young and old

---

[3] The unusual technique being used to search for these objects will be described in Chapter 21.

masses, etc., of the different galaxy types, based on the visible matter they contain. Their masses may well be greatly revised in the future. Perhaps their relative values among the three types of galaxies are approximately correct, but even this is not certain. With these uncertainties in mind, note that the largest ellipticals are larger than any of the spirals, and that both of these types are larger than the irregulars. Our own galaxy is seen to be a sizeable spiral, but much smaller in diameter and mass than the largest ellipticals.

To summarize, the universe contains billions of galaxies, at least half of which are in groups or clusters; the vast majority of galaxies can be classified into one of only three broad groups; galaxies exhibit a large range in their properties; and the visible matter is only a small fraction of the total mass of the universe. Are there any other properties of the universe we can measure? We will take up this question in the next chapter.

Terms to Know

Spiral, barred spiral, elliptical, dwarf elliptical, and irregular galaxy; central bulge, nucleus, disk. The Local Group, clusters, and superclusters; voids, the Great Wall, the Great Attractor. Nonluminous or dark matter; MACHO, WIMP.

Ideas to Understand

The classification scheme of galaxies; how distances, masses, diameters of galaxies are found; the range of properties of the various types of galaxies. Evidence for dark matter; implications. The idea of a representative sample of the universe.

Questions

1. Calculate the ratio of the major and minor axes (a/b) for an E7 galaxy. Next, look at the photograph of the spiral galaxy, NGC 4565 in Figure 18.2, and estimate the ratio of its major to minor axes. Compare results for the two galaxies.

2. Could an elliptical galaxy classified as an E3 really be spherical in shape? Could it really be more flattened than its E3 image? Explain your answers carefully.

3. Suppose a galaxy rotated like a solid body (like a wheel, for example). Sketch its rotation curve, that is, its velocity versus distance from the center of the galaxy. Compare this curve with that of a real galaxy (Figure 18.12).

4. Which type of galaxy in the Local Group is the most massive? Which has the largest diameter? Which is the most numerous? How do these types compare with the most massive, largest, and most numerous outside of the Local Group?

5. If the Large Magellanic Cloud is about 50,000 parsecs away, how large is its parallax in seconds of arc? Could this be measured by the trigonometric parallax method?

6. (a) Confirm the result given in this chapter that the period of revolution of the Sun about the center of our galaxy is about 250 million years. Take its velocity to be 200 km/sec, and its distance from the center of the galaxy to be 8,000 parsecs.

(b) If the age of the Sun is 5 billion years, how many circuits of the galaxy has it made?

7. A star at the edge of a small spiral galaxy travels around the center in a circular orbit with a velocity of about 120 km/sec. The radius of the galaxy is 5,000 parsecs.

(a) What is the period of revolution of the star in years?

(b) Calculate the approximate mass of the spiral galaxy.

Suggestions for Further Reading

Bertola, F., "What Shape Are Elliptical Galaxies," *Sky & Telescope*, **61**, p. 380, 1981. Observational evidence that ellipticals can have shapes other than oblate spheroids.

de Boer, K. and Savage, B., "The Coronas of Galaxies," *Scientific American*, **247**, p. 54, August 1982.

Burstein, D. and Manly, P., "Cosmic Tug of War," *Astronomy*, **21**, p. 40, July 1993.

Dyer, A., "A New Map of the Universe," *Astronomy*, **21**, p. 44, April 1993. A map plotting the positions of 14,000 galaxies within 500 million light years of our galaxy.

Ferris, T., *Galaxies*. New York: Stewart, Tabori and Chang, 1982. Lots of pictures.

Hirshfeld, A., "Inside Dwarf Galaxies," *Sky & Telescope*, **59**, p. 287, 1980. The properties and some of the curiosities of these strange systems.

Hodge, P., "The Local Group," *Mercury*, **16**, p. 2, 1987.

———"The Extragalactic Distance Scale: Agreement at Last?," *Sky & Telescope*, **86**, p. 16, October 1993.

Kiernan, V., "How Far to the Galaxies?," *Astronomy*, **17**, p. 48, June 1989. How distances to galaxies are found.

Lake, G., "Understanding the Hubble Sequence," *Sky & Telescope*, **83**, p. 515, 1992.

Marshall, L., "Superclusters: Giants of the Cosmos," *Astronomy*, **12**, p. 6, April 1984.

Rubin, V., "Dark Matter in Spiral Galaxies," *Scientific American*, **248**, p. 96, June 1983. Describes the measurements and their interpretation in detail.

Hubble Gallery. This image covers less than one percent of the area of the Moon and yet shows several hundred galaxies, many of which have never been seen before. Shown here is about a quarter of the total field imaged by HST over ten consecutive days, resulting in the deepest (faintest) view of the universe yet obtained.

"The world was created on 23rd October, 4004 B.C. at midday."
—Archbishop James Ussher (1581–1656); Irish prelate

One of several attempts to establish the age of the world using the best data then available.

These are HST images of increasingly distant (left to right) elliptical and spiral galaxies. The two galaxies in the left-most column (labeled "Today") are nearby; those in the columns labeled "9, 5, and 2 Billion Years" are about two-thirds, one-third, and nearly one-tenth, respectively, the age of the nearby galaxies. At an age of about 9 and even 5 billion years, ellipticals look pretty much as they do today, whereas the arms of spirals are increasingly poorly defined and ragged. When galaxies were only 2 billion years old, the distinction between spirals and ellipticals had about disappeared. This suggests that ellipticals formed surprisingly early in the universe, but spirals took longer.

CHAPTER 19

The Cosmological Clues

Generalizations about the Universe

Olbers' Paradox

We have been discussing some of the properties of the building blocks of the universe, the galaxies. Is there anything we can say about the universe as a whole? Can any characteristics be ascribed to the overall assemblage of matter and space we call the universe? If this seems to you to be a big order, you're right, it is. That it may not be hopeless, however, is suggested by a question called **Olbers' paradox**, after the nineteenth-century German scientist. Olbers' paradox dates from 1823, but it has a much longer history. It was discussed by the Swiss astronomer, J. P. L. de Cheseaux, in 1744. In 1720 Halley wrote two papers on the subject, and implied that he had heard others discussing it even earlier. He may have been referring to a discussion of Kepler's analysis of the problem, which he published in 1610.

The question discussed by these astronomers may be asked as "Why is the sky dark at night?" The immediate reply would most likely be "Because the Sun has set." The questioner persists, and asks again, "Do you really understand why the sky is dark at night?" Apparently there must be more to this than meets the eye; what could that be?

Let's begin by making some plausible assumptions and see where they take us. Assume (1) that the universe is infinite in extent, old, and static (that is, the galaxies that populate it are not moving with respect to each other), and that the galaxies don't change significantly with time (earlier astronomers would have spoken in terms of stars). Further assume (2) that on a large scale the galaxies are uniformly distributed throughout space; and (3) that at any particular time, the galaxies "way out there" are, on the average, like the galaxies "back here." (Remember that we see these galaxies today as they were when they emitted the light we now receive.) Finally, unless we want to make up our own rules of the game (in which case anything goes), we must assume (4) that the laws of physics are the same everywhere in the universe. In the absence of any evidence to the contrary, these are all perfectly reasonable assumptions. In fact, assumptions (2) and (3) make up what is called, somewhat grandly, the **cosmological principle**. It asserts that our part of the universe is on the average a good sample of the universe as a whole.

Now imagine two enormous but relatively thin shells centered on our galaxy, with radii R_1 and R_2 ($= 2R_1$) as shown in Figure 19.1; both have thickness t. The volume of the smaller shell is just the surface area of the shell times its thickness, or $V_1 = 4\pi R_1^2 t$; the larger shell has a volume, $V_2 = 4\pi R_2^2 t = 4\pi(4R_1^2)t$; that is, the volume of the larger shell is four times that of the smaller. Since we have assumed that galaxies

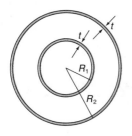

Figure 19.1. Imagine two enormous shells of radii R_1 and R_2 ($= 2R_1$), both of thickness t, centered on our own galaxy. The text describes how this construction is used to show that, under plausible assumptions, the sky should be bright both day and night.

are uniformly distributed in space, this means that there are four times as many galaxies in the larger shell than in the smaller. Now, the larger shell is twice as far away from us as the smaller one, and since the intensity of light falls off inversely as the square of the distance, the radiation reaching us from the larger shell is diminished to 1/4 of that from the smaller shell. The net effect of the larger, more distant shell, then, is that it contributes the *same* amount of radiation as the smaller, nearer one (four times as many galaxies, but each only 1/4 as bright as the galaxies in the nearer shell). Since we assumed the universe to be infinite in extent, we can divide it up into an infinite number of shells, each contributing the same amount of energy to the total. But this leads immediately to the conclusion that the sky should be infinitely bright all the time; there would be no night (which would be a matter of complete indifference to us, since we would have been vaporized!). Given our assumptions, this result is inescapable. The absurdity of the result can be slightly reduced when we realize that at some third distance R_3, our view of the sky would be completely covered by galaxies, so that the light of galaxies farther away from us never reaches us. Thus, instead of being infinitely bright, the sky everywhere would be only as bright as the average galaxy. Since galaxies shine by the stars they contain, the sky would be about as bright as the average star, for example, the Sun. Our conclusion, though, would be qualitatively the same as before.

Kepler used this result as an argument against an infinite universe, the possibility of which had been opened up by Copernicus' system. Olbers and de Cheseaux suggested that absorbing matter existed in space, blocking out the light and reducing its intensity to what we observe. Eventually, however, since the radiant energy is not destroyed, this matter would heat up and shine just as brightly as everything else, since, by our assumptions, it has been absorbing energy for an infinite amount of time. This way out of the "paradox" doesn't work. Instead we must look at the basic assumptions.

Resolution of Olbers' Paradox

Suppose the universe were not infinite but instead finite and small compared to R_3, but none of our other assumptions were changed. This would result in a finite amount of radiation and a dark sky here on Earth, which is what we are after. Such a universe (initially composed of unmoving galaxies) would be gravitationally unstable, however, and would eventually collapse on itself, because every galaxy in the universe is always attracting every other galaxy. Again, this is in contradiction to what we actually observe. The universe is not collapsing.

If distant galaxies were on the average not like those nearby, but were systematically fainter, the paradox could be avoided. Though our evidence concerning this possibility is rather meager, there is no reason to believe that distant galaxies are intrinsically fainter than nearby systems. In fact, it is likely that galaxies were more luminous in the distant past rather than less, because they probably contained more massive, bright young stars when they were younger systems.

We assumed that the galaxies were not moving, but suppose the universe expanded with a velocity increasing systematically with distance. Then photons from galaxies farther and farther away from us would be increasingly red-shifted, and would deliver less energy to us than if the universe were static. Thus the more distant shells would actually provide less energy than the nearer ones, and the paradox might be avoided. So does the obvious fact of the dark night sky indicate that the universe is expanding? What a wonderful result that would be from such a simple observation! Unfortunately, the dark sky indicates no such thing. The red shift is far too small an effect to account for the dark sky. Another factor is much more significant: galaxies, made up of stars, don't shine forever, but evolve or change with time and eventually die, as we have seen. Their luminous lifetimes would have to be 10^{23} years to enable them to fill the universe with their radiation. This is the key to the resolution of the paradox. Galaxies just don't shine long enough to produce all the necessary radiation.

Another way of looking at this is to imagine all the known matter in the universe to be completely converted into its energy equivalent according to Einstein's equation $E = mc^2$. Even this would produce enough radiation to raise the temperature everywhere to only a few tens of degrees K, far less than the 5,500 K or so (the temperature of the Sun, a typical star) implied by the paradox. The energy density of matter in the universe would have to be billions of times greater than it is for the sky to be as bright as the Sun.

The point of all this is that general properties of the universe do have consequences for us. The sky is dark at night because stars have relatively short lifetimes.

The Cosmological Principle

Another apparently simple notion, namely, the cosmological principle, also has surprisingly broad implications. As we mentioned above, it asserts the uniformity of nature on the largest scale; that is, the part of the universe that we can see is typical of the whole. If we were located on a distant galaxy, our view of the universe would be about what we see from the Milky Way. This sounds harmless enough, and is necessary if we are to play the cosmological game at all. Let's look at it in a little more detail, however. Observations show that on a very large scale the universe looks pretty much the same in all directions. If our line of sight extends to large enough distances, it encounters about as many galaxies, clusters of galaxies, and superclusters in one direction as in any other. We say that the universe is **isotropic**, the same in all directions.

But would the universe look the same regardless of *where* we were? Maybe not. Isotropy does not guarantee that the universe is **homogeneous**, the same everywhere. Suppose, for example, that the numbers of galaxies decreased with distance from us in the same way in all directions (see Figure 19.2). Such a universe would be isotropic, but it would not be homogeneous. In a homogeneous universe there can be no "special" place—no center, no edge—since such a place would differ from other places. So adopting the apparently reasonable cosmological principle that the universe is isotropic and homogeneous forces us to give up the familiar notions of center and edge!

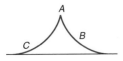

Figure 19.2. Imagine a perfectly symmetric mountain. The view from the top, *A*, down along the mountain, would be the same in all directions. We would say that such a view is isotropic. The mountain's surface is not homogeneous, however, because the view of the mountain from some point *B* would not be the same as from *C*.

This in turn helps to define the ways in which we can characterize the universe. We will return to these seemingly strange ideas later.

Are Physical Laws Universally Applicable?

We have always assumed that the laws of physics are the same everywhere; lacking evidence to the contrary, that is all we can do. But why do the basic physical constants in these laws, such as the velocity of light or Planck's constant or the strength of the gravitational force, have the values they do? (Shades of Pythagoras!) Are the values of these fundamental constants, which determine the nature of the universe, somehow tied to or related to that universe? But the universe is evolving, so maybe the physical constants are also changing with time. If so, we cannot logically venture back into cosmic time riding on the present values of these constants. So what can we say about the invariability of physical constants or of the laws of nature elsewhere in the universe?

Obviously, we can't go to a galaxy a billion parsecs away and perform experiments directly, but we can see if the ratios of certain constants change with time or not. Let's look at an example: the ratio of the size of the universe to the size of the electron and the ratio of the electromagnetic to the gravitational force both have the value of about 10^{40}. We will soon see that the universe has been increasing in size since its beginning, so that if, as has been suggested, the equality of these ratios is basic and not just a coincidence, then the fundamental constants of gravitation and of atomic physics are somehow related to each other and to the age of the universe. If so, the relative rates of a gravitational clock (one based on some aspect of gravity) and an atomic clock (based on the frequency of light emitted by a particular kind of atom) would change over time. So far experiments using both kinds of clocks have shown that the gravitational constant G has changed at a rate of no more than about 10 parts per trillion per year. Even over the age of the universe this would amount to a change of only about 1 percent, which is probably too small to be significant. Studies of possible change in other physical constants have also failed to find any significant results. The assumption of the universality of physical laws seems reasonable.

Perhaps this discussion suggests to you that there may be some hope of finding phenomena that will enable us to specify, at least broadly, the properties of the universe as a whole. In fact, we now know of three such phenomena, three cosmological clues, that tell us about the nature of the universe. These are the redshifts of galaxies, the 3 K background radiation, and the abundance of helium. We will look at all three of these clues in this chapter.

The First Cosmological Clue
Redshifts of Galaxies

In Chapter 17 we saw that beginning in 1912 V. M. Slipher obtained photographic spectra of several spiral nebulae from which he measured their radial velocities. By 1925 radial velocities had been measured for 45 spirals, all but five by Slipher. Of these the vast majority showed velocities away from us; that is, their spectra were shifted to the red. Four years later, Hubble had found the distances of 18 of these. His plot of their velocities against their distances (see Figure 19.3) showed fairly convincingly that these two quantities were related. (Actually this had been indicated a year earlier by the American mathematician H. P. Robertson, who had used Hubble's galaxy dis-

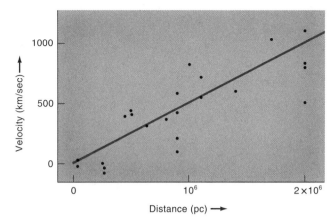

Figure 19.3. Hubble's 1929 redshift-distance plot for nearby galaxies that showed the expansion of the universe. The four most distant points represent galaxies in the Virgo cluster. All galaxies on this plot are now believed to be about ten times farther away than Hubble's distances.

tances and Slipher's velocities. His work, published in a British physics journal, was overlooked by astronomers, however. Work by several astronomers even prior to Robertson's suggested such a relationship. As almost always happens with important discoveries, the notion of a relation between velocities and distances of galaxies did not come out of nowhere; it had been "in the air" for several years.) Two years later Hubble and his colleague M. Humason, who did most of the radial-velocity work, had velocities and distances for many more distant galaxies, and these clinched the reality of what has come to be known as the **velocity-distance relation**. Here was the first observational phenomenon that had to do with a property of the universe as a whole.

Velocity-Distance Relation

The situation is as follows: all but the very nearest galaxies have spectra shifted to the red; furthermore, the greater the distance of the galaxy, the larger its redshift (see Figure 19.4). (That a few of the galaxies nearest to us have blue-shifted spectra is easily explained. Like stars, galaxies have their own "peculiar" motions that can be larger than the systematic redshift when the latter is small.) It seemed natural to interpret the redshifts as Doppler shifts caused by the physical velocity of the galaxies away from us, although Hubble and other astronomers were cautious on this point. Thus, one could write $V \propto d$ or, equivalently,

$$V = Hd,$$

where H is called the **Hubble constant**. Since $H = V/d$, the units of H must be a velocity per distance; these are usually given as (kilometers/second) per megaparsec. The Hubble constant describes just how the velocities of galaxies increase as we consider systems at greater and greater distances from us.

Though this interpretation of the redshifts as a Doppler shift is a natural one, there is a better way to look at it, which is to consider the galaxies to be at rest in their local surroundings; it is *space itself* that is expanding, carrying the galaxies along with it. The observed redshifts are a result of this expansion of space; we will call the velocities corresponding to these redshifts **expansion velocities**. By contrast, Doppler shifts

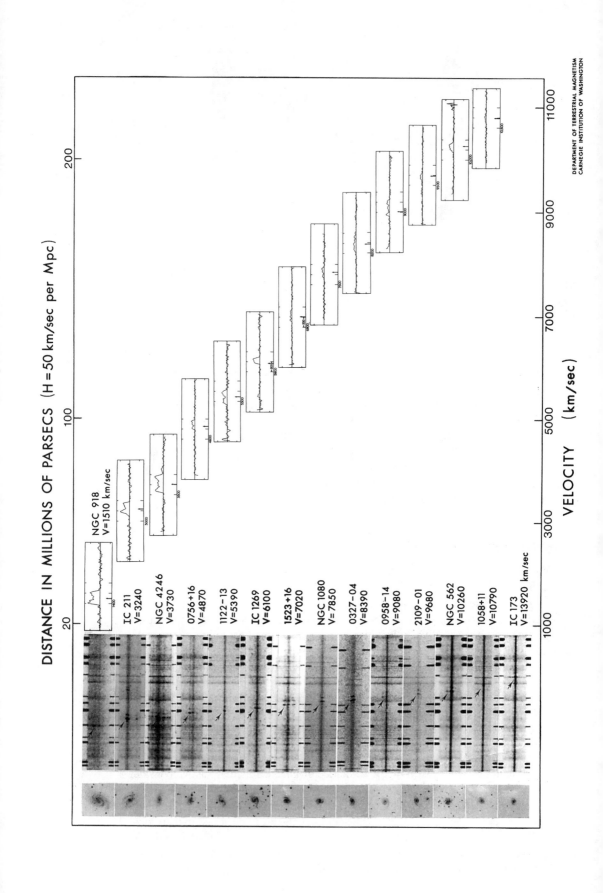

DISTANCE IN MILLIONS OF PARSECS (H = 50 km/sec per Mpc)

NGC 918
V=1510 km/sec

IC 211
V=3240

NGC 4246
V=3730

0756+16
V=4870

1122−13
V=5390

IC 1269
V=6100

1523+16
V=7020

NGC 1080
V=7850

0327−04
V=8390

0958−14
V=9080

2109−01
V=9680

NGC 562
V=10260

1058+11
V=10790

IC 173
V=13920 km/sec

VELOCITY (km/sec)

DEPARTMENT OF TERRESTRIAL MAGNETISM
CARNEGIE INSTITUTION OF WASHINGTON

Figure 19.4.   The image, optical spectrum, and 21-cm spectrum for 14 spiral galaxies at increasing distances from the Milky Way. The prominent emission lines above and below a galaxy spectrum are comparison lines that establish the wavelength scale. The strongest emission line (and the one to which the arrow points) in the galaxy spectrum is H-alpha. Clearly visible on the spectrum of IC 211 are two emission lines on either side of H-alpha due to forbidden nitrogen and two forbidden sulfur lines further to the red. (Redward of the sulfur lines is a fainter pair of lines extending across the spectrum that originate in the Earth's atmosphere.) Note how spectral features in each galaxy spectrum are increasingly shifted to the red as one progresses down the sequence. Velocities are indicated on the radio spectra; the tic marks are 200 km/sec apart.

(which we call **Doppler velocities**) are produced by the relative motions *through* space of light source and observer. The difference between the two kinds of velocities can be visualized easily by imagining a rubber band (the "space" of the universe) on which two beans (galaxies) have been glued. As we stretch the rubber band, the beans move apart with an expansion velocity corresponding to the rate at which the band is stretched. The Doppler velocity of the beans with respect to each other is zero; to be otherwise the beans would have to move on the rubber band.

Let us look at the observations in a bit more detail. Hubble and Humason found that the redshifts of galaxies located over the sky are systematically larger the greater the distances of the galaxies from us. This indicates that galaxies are all being carried away from us, the more distant ones faster than the nearer ones. We interpret this as an expansion of the space of the whole universe.

To us it looks as if we were at the center of the expansion, but we are not; the galaxies are *not* being carried away from us alone. The same expansion would be seen from *any* other galaxy, as can be seen by studying Figure 19.5. This gives schematically the redshift-distance effect in two directions, first with the observer at galaxy *A*, then with the observer located one galaxy to the right, at *D* (shown on the second line in the figure). The lengths of the arrows are proportional to the redshift of the galaxies measured first from *A*, then from *D*. As seen from *D*, *A* is moving away with the redshift *D* had when measured from *A*, but now, of course, in the opposite direction; the redshifts of *B* and *C* as measured from *D* are greater than they were with respect to *A* by the relative redshift between *A* and *D*; those of *E* and *F* with respect to *D* are less than they were with respect to *A* by the relative shift between *A* and *D*. It should be clear that observers on *D* will measure the same redshift-distance relation as did their counterparts on *A*, so that the same expansion will be seen from any galaxy. There is no center to the expansion; it is expanding in the same way everywhere.

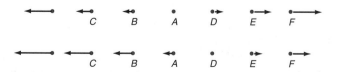

Figure 19.5.   The points *A*, *B*, *C* . . . represent galaxies and the lengths of their arrows represent their redshifts as measured first from galaxy *A*, then (second line) from galaxy *D*. The same redshift–distance relation is obtained in both cases; we are not at the center of the expansion.

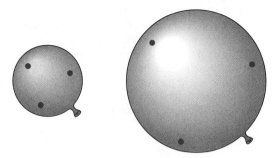

Figure 19.6.   A balloon on which three beans (galaxies) have been glued. As the balloon is blown up, the beans move away from each other.

Let us return to our simple model of a rubber band (the "space") on which now several beans (galaxies) are stuck. The beans don't move on the band but are carried along by it as it is stretched. As the band becomes longer, the distance between any two beans increases. Imagine that in, say, 10 seconds, the length of the rubber band is doubled. This means, of course, that all lengths on the band double. In 10 seconds the separation between two beans initially 1 centimeter apart doubles; similarly, the distance between two beans that originally were 3 centimeters apart also doubles. The expansion velocity between the two latter beans obviously must be greater than that between the first pair, in fact, three times greater. The relation between distance and redshift (expansion velocity) is said to be linear; that is, double one and you double the other, triple one and you triple the other.

Another, somewhat more instructive, example is that of a balloon being steadily blown up, with several bean-galaxies stuck on its surface. As the balloon expands, the distance between *any* two beans, measured along the *surface* of the balloon, increases. If the circumference of the balloon doubles, all distances between beans double (Figure 19.6). We will see later that this balloon model can be extended to represent some of the properties of an expanding universe.

What Redshifts Tell Us

Since the redshifts are consequences of the expansion of the universe, they must tell us something about that expansion. If all distances in the universe double during the time that light from a distant galaxy makes its long journey to us, then the wavelength of the light from that galaxy doubles also. Expanding space stretches the wavelength of the light by the same factor as that by which the space itself expands. In other words, the redshift tells us how the *scale* of the universe has increased in the time it takes light to travel from the distant galaxy to us. To see this, let us again return to our rubber band with two beans glued to its surface (Figure 19.7). Each bean has a particular position on the band, and the two are separated by a given distance at time t_1. At time t_2 later their positions on the band will not have changed—they are glued to the same points—but their separation will have increased by a certain amount, say, 10 percent. That is, in the time $\Delta t = t_2 - t_1$, all lengths on the band have increased by 10 percent. We can abbreviate this statement by saying that the rubber band's scale factor has increased by 10 percent. If the light traveling from one bean to the other took the same time $\Delta t = t_2 - t_1$ to make the trip, the change in the scale factor would produce an increase in the wavelength of the light by 10 percent also.

Figure 19.7. Two bean-galaxies are glued to a rubber band. At time t_2, the rubber band has been stretched so that the separation between the two beans has been increased by 10 percent more than it was at time t_1. That is, the band's scale factor has increased by a factor of 1.10. See text.

It is easy to put this into quantitative terms. The redshift (customarily denoted by z) is the ratio of the increase in wavelength to the original wavelength:

$$\text{redshift} = z = \frac{\lambda_{\text{rec}} - \lambda_{\text{em}}}{\lambda_{\text{em}}} = \frac{R_{\text{rec}} - R_{\text{em}}}{R_{\text{em}}} = \frac{R_{\text{rec}}}{R_{\text{em}}} - 1; \text{ so } 1 + z = \frac{R_{\text{rec}}}{R_{\text{em}}},$$

where λ_{em} and λ_{rec} are the wavelengths, and R_{em} and R_{rec} are the scale factors at the time the light was emitted and at the time it was received, respectively. If a galaxy has a redshift of 0.10, this means that all distances in the universe are 1.10 times larger now than when the light we measure was emitted by the galaxy.

Knowing the Hubble constant H implies that we know V, the so-called speed of recession of a galaxy at a given distance, d, from us. Strictly speaking, however, the redshift tells us only how much the universe has expanded since a particular galaxy emitted the radiation we measure. To calculate V from the observed redshift expansion velocity, we must work within a particular mathematical model of the universe. Different models, all based on Einstein's theory of general relativity (a few aspects of which we will describe briefly in Chapter 21), give different answers for the relation between expansion velocity and Doppler velocity, V. Happily, for relatively small cosmic distances (up to a billion parsecs or so), all models predict that the Doppler velocities are just about the same as the values of the expansion velocities. So as a practical matter, within this limit we can use Hubble's equation, $V = Hd$, directly. Keep in mind, however, that V should be interpreted as an expansion velocity of space itself rather than as a velocity of galaxies through space.

The Hubble Constant: How Fast?

As we have noted, the Hubble constant, H, is equal to V/d and is usually given in km/sec per million parsecs. When in 1929 Hubble first calculated H, he found a value of about 520 km/sec per Mpc. This meant that galaxies with recessional speeds of about 520 km/sec were 1,000,000 parsecs distant; those with speeds of about 1,040 km/sec were 2,000,000 parsecs away; etc. Today, however, we think that the correct value of H is between 50 and 100 (km/sec)/Mpc. For a given redshift a galaxy is now put at a much greater distance than before. Why this big change? It is certainly not because the expansion of the universe is slowing down that rapidly! Instead, it has to do with observational difficulties. To determine H requires finding two numbers for every galaxy: its radial velocity, measured from the redshift of its spectrum, and its distance, determined from a long chain of assumptions and bootstrap techniques. Radial velocities are relatively easy to measure and are quite accurate. You should realize by now, however, that distance measurements are extremely difficult and so subject to large errors.

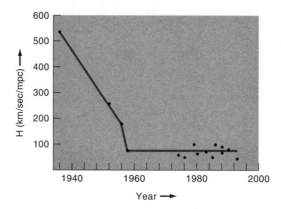

Figure 19.8.   Derived values of the Hubble constant from 1936 to the present. Since the large decrease ending in the late fifties, most values have ranged between 50 and 100 (km/sec)/Mpc. We have adopted 75 (km/sec)/Mpc in the text.

The decrease in *H* to a value between 1/5 and 1/10 of that originally found by Hubble and shown in Figure 19.8 is a consequence of several revisions of the distance scale. These revisions have resulted in the more recently determined distances to galaxies being systematically larger, by five to ten times, than the earlier values. Several factors were responsible for these large corrections. One of the most significant was the realization in the late 1940s that there are two types of Cepheid variables, each with a different but parallel period-luminosity relation, displaced from each other by about a factor of four in intrinsic brightness. It so happened that Hubble had used the period-luminosity relation appropriate for the intrinsically fainter kind of Cepheids to find distances to galaxies, whereas in actuality he was observing the intrinsically brighter variables. This one factor led him to underestimate the distances to the galaxies he observed by a factor of two. Another source of error was the confusion by Hubble of distant H II regions (so distant that they appeared star-like) with very luminous blue stars. The latter are intrinsically fainter than the former, and so again distances to galaxies were underestimated. The moral of this story is obvious: one must be careful with "similar" objects! It would be foolhardy to believe that we have seen the last of these changes in the distance scale.

In what follows we will adopt *H* = 75 (km/sec)/Mpc, with the understanding that this value is uncertain by as much as 30 percent. The Hubble constant could be as large as 100 or as small as 50 (km/sec)/Mpc, reflecting a corresponding uncertainty in our knowledge of the distances to galaxies.

The Hubble Constant: For How Long?

If the universe is expanding it must have been smaller in the past than it is now, and smaller now than it will be in the future. If we assume that the rate of expansion has remained unchanged, we can easily calculate how long it has taken for the universe to expand from some arbitrarily small size to its present size, a time called the **expansion age** of the universe.[1] Consider two galaxies separated by a distance *d* and moving apart

---

[1] This is also called the "Hubble time."

with a velocity V ever since the expansion began. Then the time during which they have been moving apart (the expansion age) is just

$$t = \frac{d}{V} = \frac{1}{H} = \frac{1}{75 \frac{\text{km/sec}}{\text{Mpc}}}$$

$$= \frac{1}{75} \frac{\text{Mpc}}{\text{km/sec}} \times \left(3 \times 10^{19} \frac{\text{km}}{\text{Mpc}} \right)$$

$$= 4 \times 10^{17} \text{ sec} = \frac{4 \times 10^{17} \text{ sec}}{3 \times 10^7 \text{ sec/yr}}$$

$$= 1.3 \times 10^{10} \text{ yrs, or 13 billion years.}$$

Hubble's original value for the expansion constant was about seven times greater than the modern value. Thus the expansion age was only 1/7 of what we now think it is, or only about 2×10^9 years—but Earth rocks were known to be older than that. This awkward situation was a consequence of the incorrect distance scale for galaxies.

The expansion age is sometimes called the "age of the universe," but it should be remembered that it really is the *maximum* time that has elapsed since the expansion we now observe began. It is a maximum time because the rate of expansion might have been appreciably faster in the past, and been slowed to the present rate by the gravitational attraction of the galaxies for each other. Thus, the actual expansion age of the universe could be appreciably smaller than 13×10^9 years. We shall see that this possibility has profound implications for the future evolution of the universe.

Finally, if we think we know the value of H, we can use the Hubble law to find distances to galaxies. All that is necessary is to measure the Doppler velocity of a galaxy; its distance comes from $d = V/H$. For example, if the redshift gives a velocity of 10,000 km/sec, then

$$d = \frac{V}{H}$$

and

$$d = \frac{10,000 \text{ km/sec}}{(75 \text{ km/sec})/\text{Mpc}} = 133 \text{ Mpc}.$$

Modern telescopes and detectors have enabled us to discover galaxies so distant that their spectra are enormously red-shifted. By the simple Doppler formula, a redshift of $z = 0.5$ implies a speed half that of light, and $z = 1$ a velocity equal to the speed of light. Yet z's of 2, 3, and even 4 have been measured! So what's going on? Are we really measuring a speed that is several times greater than that of light? It is time that we touched on some of the ideas of the theory of special relativity.

The Special Theory of Relativity

Albert Einstein published his special theory of relativity in 1905, when he was a 26-year-old patent examiner in the Swiss patent office at Bern (Figure 19.9). This was only one of four papers he published in that year. Among the others was the paper in which Einstein showed the equivalence of mass and energy; another in which he gave a theory of Brownian motion (the erratic motion of very small particles suspended in a fluid) that showed that their movements were a result of being bombarded by mol-

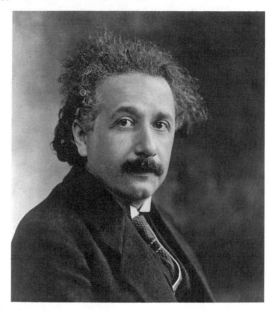

Figure 19.9. Albert Einstein (1879–1955), a German-born physicist and certainly the best-known scientist of the twentieth century. He received the Nobel prize in 1921 for his work on the quantum nature of radiation.

ecules of the fluid and hence proved the reality of molecules; and a paper in which he showed that light itself was discontinuous and had to be considered to be made up of packets of energy, now called photons. This was one of the foundations of quantum physics, and Einstein received the 1921 Nobel prize for this work. Such a burst of creativity had not been seen since Newton's enforced vacation from Cambridge. Indeed, Einstein was a rare scientist, one of the few who can rank with Newton. In contrast to the latter, however, Einstein was a much warmer person, enjoyed the company of close friends, and was concerned with what went on in the world as well as in science.

The Assumptions of Special Relativity

The special theory of relativity is based on two postulates: (1) the speed of light in a vacuum is measured to be the same by *all* observers, regardless of the relative motions of the light source and the observer; and (2) only *relative* uniform motions are measurable. These postulates imply that no matter or energy can travel faster than the speed of light, and that there is no "absolute" frame of reference (or coordinate system) in which "absolute" motion can be measured. Such a coordinate system had been assumed to exist since the time of Newton, however. Einstein claimed that nature's laws are just as valid in one uniformly moving reference frame as in another. All this sounds harmless enough, but special relativity forced fundamental changes in our concepts of space and time. Along with the quantum theory, it produced a true revolution in our concept of the physical world.

The first postulate has been confirmed experimentally any number of times; there is no getting around it. We will come back to it in a moment. You have all validated the second postulate, though perhaps you were not aware of doing so. Imagine that while having a very smooth ride in a bus (use your imagination!), you perform all sorts of simple experiments like simultaneously dropping a marble and a baseball to the floor, tossing a ball straight up in the air, and timing the swing of a pendulum. The results of these experiments will be exactly the same in the bus as on the ground: the marble and baseball will hit the floor together, the tossed ball will fall right back into your hand, and the period of the pendulum will be the same in the bus as on the ground. The laws of physics are the same in the uniformly moving bus and on the ground. Without looking out the window, there is no way you can tell that you are moving. Nothing strange here, since, after all, by doing the same sorts of experiments we can't even tell that the Earth is moving around the Sun at 108,000 km/hr!

Now, let's get a bit more elaborate and ask a friend to stand on the side of the road and tell you what she thinks the speed of your baseball is as you toss it up the aisle of the bus. Your friend would measure the speed of the ball as the sum of the ball's speed within the bus plus the speed of the bus itself, say, 10 km/hr plus 70 km/hr, or 80 km/hr. When you tossed it in the opposite direction, she would measure the ball's speed as 60 km/hr. Again, there is nothing strange here. We would all agree that the speed of the ball measured from the ground must include the speed of the bus from which it is tossed.

Now, however, suppose that you ask your friend to measure the speed of a light beam you flash in the direction of the bus's motion. Her result would not be 70 km/hr plus 300,000 km/sec; it would be just 300,000 km/sec! Fire the beam in the opposite direction, and your friend would again measure 300,000 km/sec for the light's speed. Regardless of how fast the bus traveled, the speed of light would always be found to be 300,000 km/sec. Now that is strange! There must be something special about the speed of light, because the ordinary rule for combining speeds does not work when speeds become a large fraction of that of light. Evidently, observers moving with respect to each other do not perceive space and time in the same way; otherwise the speed of light would depend on the motion of the bus.

Contrary to what Newton and his successors up to the time of Einstein had assumed, there is no such thing as *absolute* space or time existing independently, indifferent to the velocity of the observer. Instead, the motion of the observer plays an important role in determining distances and times. For example, an observer watching a rapidly moving spaceship whiz by would find that the spaceship's mass had increased, its length in the direction of motion had decreased, and a clock on the ship had slowed down relative to the values of these quantities measured when the spaceship was at rest. Furthermore, all these changes would be by the same numerical factor, $1/\sqrt{(1 - V^2/c^2)}$, which depends only on the velocity, V, of the spaceship with respect to the observer. (Recall that c denotes the speed of light.) Notice that when velocities are small (as in our everyday world), V/c is nearly 0, so that this factor very nearly equals 1, and the effects of special relativity are undetectable. As the velocity approaches that of light, however, V/c increases toward 1, the quantity in the square root becomes smaller and smaller, the factor becomes larger, and the consequences of relativity become significant.

Strange as these results may seem, there is no question of their validity. They seem bizarre to us only because we live in a world where speeds are very small compared to that of light; we are not aware of the consequences of special relativity in our daily lives.

What We Can All Agree On

Suppose that two observers, moving nearly at the speed of light with respect to each other, see two flashes of lightning. Using our everyday notions of space and time, would they agree on how far apart in space and in time the two flashes occurred? No,

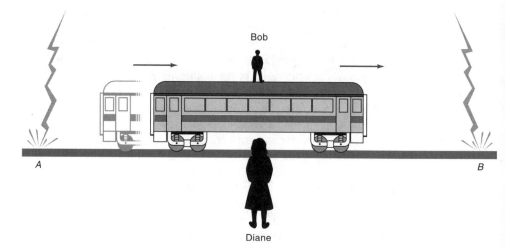

Figure 19.10. Bob is on a train, moving at a very high speed to the right, while Diane is standing by the track. Two strokes of lightning hit the tracks; to Diane they appear to be simultaneous. In the brief time it takes the light from the flashes to reach Bob, however, the train has carried him a short distance to the right; to Bob they will not appear to be simultaneous. Instead he will see flash *B* before flash *A*.

they would not (see Figure 19.10). Now, a high-speed world would be a chaotic place if its inhabitants could never agree about any of the events taking place in it. So is there any way in which they could agree? Yes, there is. Einstein showed that to specify an event, one must say when it occurred as well as where. Furthermore, events do not unfold in space and time separately, but take place in **spacetime**, where space and time are on equal footing and must be considered in combination. Time becomes a fourth number required to specify events. Spacetime is a four-dimensional quantity; that is, it takes four numbers to specify the position of a point in spacetime.

At a certain level, there is nothing strange about spacetime—we use it all the time. For example, you and a friend decide to meet for dinner at the restaurant on 5th and Main Streets at 6:00 P.M. You have used the position in both space and time to specify an event. Now, suppose that after dinner the two of you want to be at the coliseum for the hockey game at 8:00 P.M. You would think of the difference in distance and the difference in time between the two events quite separately. In our everyday world, time and distance can be separated. In the high-speed world ruled by relativity, however, time and distance must be taken together to specify the so-called spacetime "interval" between the two events of dinner and the hockey game.

In the everyday world we specify a distance between two points by the Pythagorean relation: $s = \sqrt{(x^2 + y^2 + z^2)}$. The separation in time between two events is just their difference in time. In the high-speed world, however, we combine the space and time coordinates; but since we must combine apples and apples, we can't just add distance and time. Instead, we convert the time to a distance by multiplying it by the velocity of light. That is, a time difference is given by the distance light travels during the time in question. Given the special role the speed of light plays in relativity, this seems natural enough. The spacetime interval, then, is $\sqrt{(x^2 + y^2 + z^2 - c^2t^2)}$. Note the minus sign; that means a spacetime interval between two events could be 0.

If observers moving with different velocities with respect to each other describe the separation of events in spacetime in terms of this combination of space and time, their descriptions will all agree. The commonly heard summation of relativity theory, that "everything is relative," is particularly misleading here. Instead, Einstein showed how all uniformly moving observers, measuring events in spacetime, can agree in their description, how they can calculate the same spacetime interval between two events. In spacetime all observers, regardless of how fast they are moving, will measure the speed of light in a vacuum to be 300,000 km/sec. Unless we think of events in the universe as unfolding in spacetime, we will not agree about what is happening.

Some Consequences of Special Relativity

Special relativity introduces another wrinkle relevant to our discussion. For redshifts that are not too large (say, no more than 10 percent), we can convert a wavelength shift to a speed by using the simple Doppler formula, $V = cz$, where c is the speed of light. Recall that according to the familiar formula,

$$\frac{\lambda_{obs} - \lambda_{lab}}{\lambda_{lab}} = z = V/c \ .$$

For larger speeds, however, Einstein's theory requires this approximate formula to be replaced by the exact form, namely,

$$1 + z = \sqrt{\frac{1 + \frac{V}{c}}{1 - \frac{V}{c}}}$$

or, equivalently,

$$\frac{V}{c} = \frac{(1 + z)^2 - 1}{(1 + z)^2 + 1}.$$

This last form of the Doppler formula shows that V is always less than c (the denominator is always larger than the numerator), as required by special relativity. This formula also reduces to the familiar one when velocities are small compared to those of light, that is, when V/c is very small. Conversely, when V is nearly as large as c, that is, when V/c is nearly 1, the value of z is correctly predicted by Einstein's formula to become very large. In fact, z becomes infinite when $V = c$. By contrast, the simple Doppler formula predicts $z = 1$ when $V = c$, which is incorrect (see Figure 19.11). There are many instances in cosmology when the exact Doppler formula must be used.

Finally, one consequence of special relativity that we have already referred to many times in this book (particularly in connection with the production of nuclear energy in stars) is that matter and energy are two aspects of the same thing. Fundamentally, they are equivalent; matter is "congealed" energy. The energy equivalence of matter is given by the famous equation $E = mc^2$, where E is the energy of a given mass m. It represents the amount of energy obtained when a piece of matter is completely converted to energy. The velocity of light, c, is a large number, so the energy equivalence of even 1 gram of matter is enormous, about equal to the chemical energy content of 500,000 gallons of gasoline.

Incidentally, note that we use the term "theory" of special relativity to refer to a set of postulates and their consequences that completely agree with experiments. Clearly, the frequently heard phrase "It's just a theory," meaning that it's just speculation and need not be taken seriously, does not apply to the theories of special relativity, general relativity, electromagnetism, or atomic spectra, to name but a few. These are all well established and form parts of the structure by which we describe physical

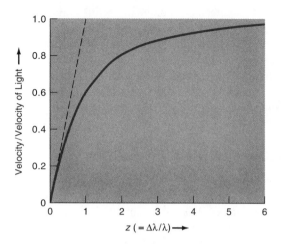

Figure 19.11.   A comparison of the Doppler formula correct for all values of z and the formula correct for small z only (dashed line). Differences become appreciable at values of z of about 0.3, corresponding to velocities about 25 percent that of light. Note that regardless of the value of z, V is always less than c.

phenomena. Unfortunately, "theory" is commonly used in both the senses referred to here; be sure not to confuse the two. After this brief detour, let's return to the cosmological clues.

The expansion of the universe, the first of the cosmological clues, is surely one of the major discoveries in the history of astronomy. It seems to force the astonishing implication that at some time in the distant past, *all* the space in the universe, with *all* its matter and energy, comprised only an unimaginably tiny volume! According to this picture, the present universe had its origin in the expansion of an incredibly dense, blindingly hot point of energy, an expansion that we see continuing today. For obvious reasons this kind of scenario for the beginning of the universe is called the **Big Bang** model. To say the least, it seems a bit risky to base such an astounding conclusion on just one piece of evidence. Fortunately, there is additional observational support for this model. Let us consider the second cosmological clue.

The Second Cosmological Clue

The 3 K Background Radiation

We will see in Chapter 21 that for a time one could interpret the expansion of the universe in other than Big Bang terms. In 1965, however, an accidental discovery provided crucial evidence for the Big Bang picture. Two radio astronomers, Arno Penzias and Robert Wilson, working at Bell Telephone Laboratories, were preparing a radio telescope for astronomical use. The instrument, located in New Jersey, was originally constructed for communication satellite work. Penzias and Wilson found that this telescope, tuned to a wavelength of about 7 centimeters, constantly measured a slightly greater intensity than expected. Significantly, the weak signal was not associated with the Sun, bright stars, or any other celestial objects, but appeared with the same intensity in all directions in the sky and at all times of the day or night. The temperature of this radiation was very low, only a few degrees above absolute zero. Try as they might, they could not eliminate this faint radio whisper. (Their efforts included cleaning out the droppings left by pigeons nesting in the antenna. Astronomy sometimes makes strange demands!)

Quite by accident they heard of research on Big Bang theory that had recently been done by a group of physicists only a few miles away, at Princeton University. This work predicted that there should now exist background radio radiation, uniform in intensity over the whole sky, which had its origin early in the hot Big Bang. Calculations indicated that because of the expansion of the universe, this radiation has become enormously diluted and has cooled to about 10 K. Acting on this prediction, the Princeton group was building a detector with which to search for the radiation. When Penzias and Wilson contacted the Princeton scientists, it quickly became clear that the mysterious radiation their telescope was detecting must have had its origin in the Big Bang. The radiation was an ancient relic of the original fireball from the birth of the universe! Its measured temperature of about 3.5 K was fairly near that predicted, and its uniform distribution over the sky was just what was to be expected from radiation originating in the early universe. Penzias and Wilson and the Princeton workers published the observational discovery and its theoretical interpretation, respectively, in two companion papers in 1965. Thirteen years later, Penzias and Wilson shared the Nobel prize for their crucial discovery.

Its Spectrum

Although most astronomers were convinced that this radiation had originated in the Big Bang, there remained the possibility that it had a quite different origin, perhaps in very distant galaxies, in a hot intergalactic gas, or in some other, unknown source. To clinch the Big Bang interpretation beyond dispute, the spectrum of the radiation had to be established; that is, it was necessary to measure how much energy is radiated at different wavelengths. According to the theory of its origin, this background radiation should have the spectrum of a blackbody. Recall that the important feature of blackbody radiation is that its characteristics depend only on its temperature and on nothing else. At a given temperature any two blackbodies will emit the same spectrum. (Review the relevant section in Chapter 9 if you need to refresh your memory about blackbody radiation.)

Why is it expected that the spectrum of the cosmic background radiation should be that of a blackbody? The argument goes as follows. Early in its history a Big Bang universe would be extremely hot, dense, and thoroughly mixed. A much greater portion of its mass-energy was in the form of radiation than in that of matter. That is, the faint cosmic background radiation we now detect once dominated the universe. Furthermore, because of its high temperature, matter in the early universe (mostly hydrogen) would be separated into its constituent protons and electrons. In this form matter absorbs radiation extremely efficiently. A photon emitted by the hot gas would be almost immediately reabsorbed; photons could not travel very far before disappearing. In other words, the hot cosmic cloud was extremely foggy, the matter and radiation constantly interacting. Under these conditions it is easy to show that this radiation would have a spectrum appropriate to that of a blackbody at the very high temperature then prevailing.

As the universe expanded, it cooled, but it was still hot enough that the matter and blackbody radiation interacted so energetically that none of the particles could attach together to form atoms. The matter and radiation remained thoroughly mixed and cooled together until, about one-half million years after the Big Bang, the temperature had fallen to 3,000 K. At this temperature, the protons of the gas were able to capture and retain an electron, forming neutral hydrogen. This had an important consequence: except at wavelengths corresponding to line radiation, neutral hydrogen does not interact with radiation and so becomes transparent to it. As a consequence, most of the radiation does not "know" that there is any matter there. Therefore, in this phase of the evolution of the universe, photons traveled indefinitely without being absorbed. The matter and radiation, previously completely mixed, no longer interacted, evolved along different paths, and had quite different fates. The smoothly distributed matter somehow became nonuniform and broke into clumps that eventually formed galaxies. The radiation, however, remained uniformly distributed, and simply cooled down in inverse proportion to the expansion of the universe. Note that the background radiation comes to us from a time before there were galaxies; it is the oldest radiation we have detected and this gives us, essentially, a snapshot of the universe at a very early time in its history.

Big Bang model calculations show that when matter and radiation no longer interacted and went their separate ways, the value of the redshift, z, was about 1,000. That is, the scale factor of the universe was only 0.001 what it is today; equivalently, the universe has expanded by a factor of 1,000 since then. Now, it turns out that a redshifted blackbody spectrum is also a blackbody spectrum, but one with a lower temperature. The temperature of the background radiation should now be about

$$\frac{3,000 \text{ K}}{1,000} = 3 \text{ K} ,$$

and its spectrum should have the characteristic blackbody shape. Consequently, to establish unequivocally the origin of the background radiation, its spectrum had to be established. Observations to do so are difficult, because the radiation is weak, and the Earth's atmosphere absorbs it at many wavelengths. Many measurements at a variety of wavelengths strongly suggested that the spectrum has the predicted blackbody shape with a temperature of about 2.7 K. In 1990 the COBE satellite (Figure 19.12) produced the beautiful data given in Figure 19.13, showing conclusively the blackbody nature of the spectrum.

Though for a few years after its discovery, attempts were made to find a noncosmological origin for the 3 K radiation, it soon became clear that this "primordial fireball radiation," as it is sometimes colorfully called, could arise only from the Big Bang class of models. It provides direct evidence that a Big Bang did indeed occur and that the universe must be evolving. This radiation provides us, then, with our second cosmological clue.

Given the origin of this radiation, we would expect it to fill the universe and to have pretty much the same intensity in all directions; that is, it should be very nearly isotropic. Penzias and Wilson found this to be true to within about 10 percent. It is important to establish the uniformity of the radiation with greater precision, because small-scale patchiness would threaten the Big Bang interpretation of its origin. Careful analysis of the COBE measurements show that the radiation is isotropic to about one part in 10^4. This remarkable result convincingly supports the cosmological interpretation of the 3 K radiation. The matter and energy of the early universe were indeed uniformly distributed.

There are two exceptions to this uniformity. The first is that toward one direction in the sky, the radiation is slightly warmer (by one part in 10^3) than the average, and in the opposite direction it is slightly cooler. This is just what we would expect if we were moving with respect to the background radiation. Our motion produces a small Doppler shift in the spectrum of the radiation, making it appear like that of a slightly warmer (blue-shifted) blackbody in the direction of approach, and a slightly cooler one (red-shifted) in the opposite direction. These measurements indicate that our galaxy is

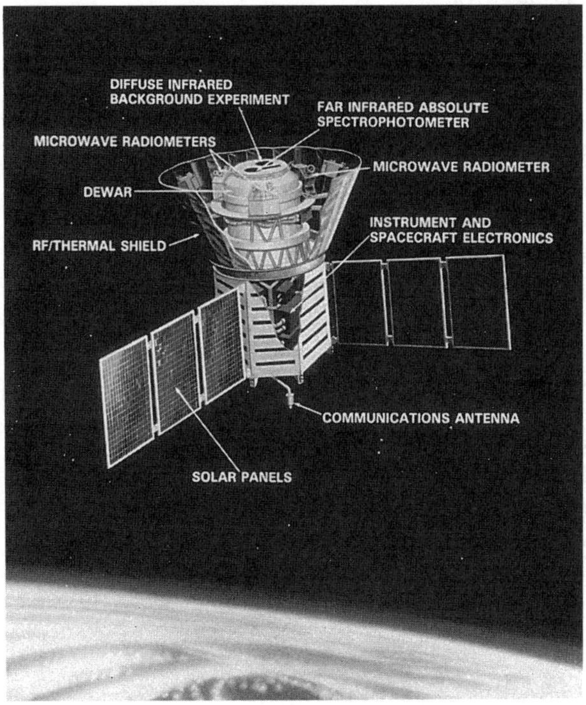
DIFFUSE INFRARED BACKGROUND EXPERIMENT
FAR INFRARED ABSOLUTE SPECTROPHOTOMETER
MICROWAVE RADIOMETERS
MICROWAVE RADIOMETER
DEWAR
INSTRUMENT AND SPACECRAFT ELECTRONICS
RF/THERMAL SHIELD
COMMUNICATIONS ANTENNA
SOLAR PANELS

Figure 19.12. The U.S. Cosmic Background Explorer (COBE) was launched in November, 1989. One instrument, operating at 3.3, 5.7, and 9.6 millimeters, was designed to determine whether the cosmic background radiation was the same in all directions. Two instruments measured infrared and microwave radiation. One of these, the Diffuse Infrared Background Experiment, was sensitive to the infrared from 1 to 300 microns. With it, the image of our galaxy was made (Figure 18.8). The Far Infrared Absolute Spectrophotometer (FIRAS) measured the background radiation from 0.1 to 10 millimeters, in 100 bands. It produced the spectacular data shown in Figure 19.13.

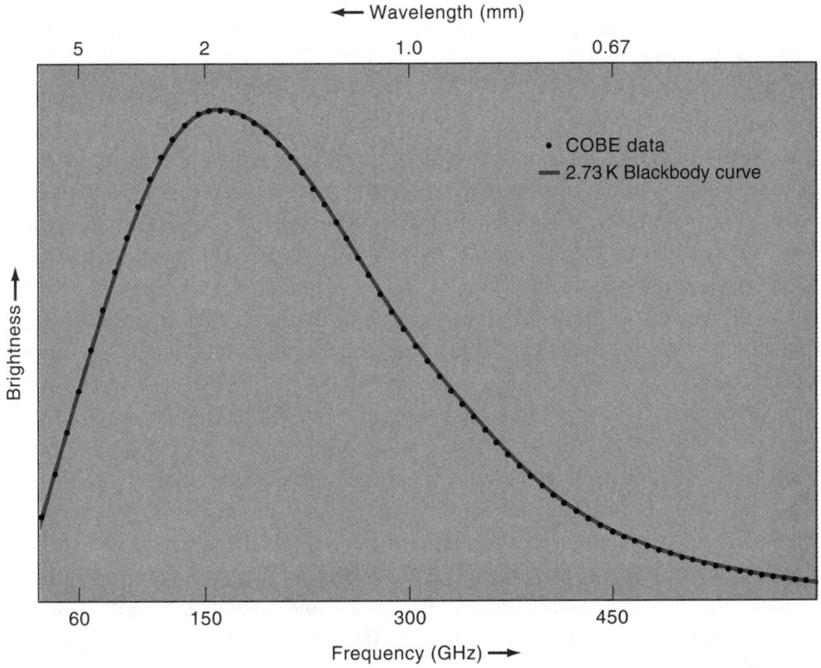
← Wavelength (mm)
5
2
1.0
0.67
• COBE data
— 2.73 K Blackbody curve
Brightness →
60
150
300
450
Frequency (GHz) →

Figure 19.13. The points represent COBE measurements of the brightness of a patch of the sky taken at wavelengths from 0.1 to 10 millimeters. The line is a blackbody curve with a temperature of 2.73 K. The data points deviate from the line by no more than 0.25 percent. In effect, these data validate Planck's law to a much greater accuracy than has been possible in the laboratory. (The relation you learned earlier in this course, between the wavelength at which a blackbody curve is a maximum and the temperature, does not hold for this curve because it is given in frequency rather than wavelength units.)

moving with a velocity of about 600 km/sec in the direction of the constellation of Centaurus.

The other kind of departure from uniformity detected in COBE data is a large-scale patchiness amounting to only one part in 100,000 in the temperature. This is the first indication of structure in the early universe. Unfortunately, the best angular resolution of COBE's instruments was only 7 degrees, corresponding to much larger structures than any we see today, including the Great Wall. Many groups are attempting to detect patchiness on much smaller scales, so far with no definitive results. We will return to the question of structure in Chapter 22.

Not only does the 3 K radiation confirm the Big Bang model for the origin of the universe and support the cosmological principle, it also enables us to deduce the thermal history of the universe. Since the 3 K radiation cools as the scale factor of the universe increases, given the present temperature of the background radiation we can extrapolate back in time and calculate its temperature at any epoch we wish. In particular, it is easy to find the physical conditions in the universe as early as 0.01 seconds after the Big Bang. The discovery of the background radiation has made the early universe an object of scientific investigation, rather than just speculation. In addition, this understanding of the physical conditions soon after the Big Bang provided by the 3 K radiation leads us to the third cosmological clue, described below.

The First Prediction of the Background Radiation

Interestingly, the existence of the background radiation had been predicted by the Russian-American physicist George Gamow in 1946 in a discussion of his Big Bang model (Figure 19.14). Later, his co-workers R. Alpher and R. Herman calculated a temperature of about 25 K for this radiation; in 1953 Gamow estimated its temperature to be about 7 K. These predictions were ignored at the time for at least two reasons: the technology of radio astronomy was too primitive to enable the required measurements to be made, and the Big Bang model itself was not taken very seriously, because at first the expansion age of the universe was too short to be believed (recall that the Hubble constant was in serious error). Also, Gamow's Big Bang model had failed in one of its basic aims—to account for the origin of the chemical elements. As a consequence, all aspects of the theory tended to be discounted. That the predictions were ignored at the time is understandable, but that they were completely forgotten is deplorable.

The discovery of the 3 K radiation exemplifies an interesting characteristic of science and of its sociology. Namely, nature is there for anyone to investigate, describe, or measure. If scientist *A* does not make a particular discovery, then *B* very likely will. The vast majority of contributions made by individual scientists are not unique in the same sense that a symphony or a sonnet perhaps is. Had Penzias and Wilson not accidentally detected faint radiation of unknown origin, the Princeton group certainly would have done so (and indeed did so!) only a few months later. Scientists may differ in their styles or in the elegance of their work, but their discoveries, taken from nature, are available to all.

This accounts for a sociological aspect of science: the value given to priority in discovery. A novelist does not rush into print for fear that he may be scooped; scientists often do. Though scientists arguing about priority is not a very attractive spectacle, it is at least understandable. Priority of discovery is the coin of the scientific realm, and though its value is often inflated and generally decreases with time, it is the only coin there is.

Figure 19.14.   The Russian-American physicist George Gamow (1904–1968). While attempting to account for the production of all the chemical elements in the early universe, he realized that one of the consequences of the Big Bang would be what we now know as cosmic background radiation. Gamow was also a talented popularizer of science, writing many books for laypeople.

The Third Cosmological Clue
The Origin of Helium

As we saw in Chapter 11, measurements of the chemical composition of the Sun and stars show the following: the simplest element, hydrogen, is by far the most abundant, amounting to roughly 75 percent by mass; helium makes up about 23 percent of the total; all the other elements combined, the "metals," make up only about 2 percent. Furthermore, analyses of stars of various ages show that the oldest stars in the galaxy have a very low abundance of metals, but that this abundance is larger among stars that were formed more recently. The metal abundance has not changed much during the last few billion years, since it is about the same in the youngest stars as in the Sun. In stars much older than the Sun, however, the metal abundance may be somewhat less than the solar value all the way down to 0.01 or even 0.0001 of that in the Sun.

What about the abundances of hydrogen and helium; are they much different in old stars than in young stars? No, the oldest stars have just about the same hydrogen and helium abundances as the most recently formed objects. Old, young, and middle-aged stars all show an abundance of roughly 75 percent hydrogen and 25 percent helium.[2] Evidently, hydrogen and most of the helium have existed since early in the his-

---

[2] Actually, the helium abundance has increased slightly since the Big Bang as a result of hydrogen fusion, but the increase is small, and we will neglect it for now. It will be discussed further in Chapter 22.

tory of the universe, before the oldest stars formed. (Recall that the measured abundances of the elements in stars refer to those in the outer layers, the atmosphere of the star. Though the core of the star is becoming increasingly helium-rich, this usually has little effect on the composition of the atmosphere, because core-atmosphere mixing is of only small importance in the stars used for this kind of abundance determination. The atmospheric abundances of these stars are a pretty accurate measure of their original composition.)

The age-independent abundances of hydrogen and helium contrast sharply with the age-dependent metal abundance and indicate that they must have different origins.

The nucleus of hydrogen consists of a single proton, and as such can be easily accounted for in any Big Bang model. The real problem is helium, the nucleus of which consists of two protons and two neutrons. We have just seen that it must have been made early in the history of the universe, since its abundance is the same now as it was billions of years ago. Perhaps it was formed in the Big Bang itself. After all, at the beginning the temperature and density of the universe were very high indeed, and matter existed in the form of simple particles such as electrons, protons, and neutrons. Could helium have been formed from this primordial brew? Though our answer is given by calculations and not by direct observation, it is reliable because, thanks to the 3 K radiation, the physical conditions in this phase of the Big Bang are well understood. Such calculations show that 0.01 seconds after the beginning, the universe was very hot, about 10^{11} K; very dense, several billion times denser than water; and very simple, consisting primarily of radiation, with only relatively few particles such as electrons and even fewer protons and neutrons.

As was said earlier, the behavior of the universe at that epoch was completely dominated by radiation, which means that protons and neutrons could not combine to form heavier nuclei because the radiation would blast them apart immediately. When the universe was about 100 seconds old, however, its temperature had cooled enough for deuterium nuclei (one proton and one neutron) to remain fused. As soon as this happened, reactions very quickly went on to form small amounts of tritium (deuterium capturing a second neutron), ^{3}He (deuterium capturing a second proton), and even some lithium, but mostly ^{4}He. This continued only as long as the temperature and density remained high enough for the large repulsive forces between the nuclei to be overcome. About four minutes after the beginning, the universe had cooled and the density decreased to the point that helium could no longer be produced. During these few minutes, however, calculations show that about 25 percent of the matter was converted to helium nuclei, with the rest remaining as protons, along with tiny amounts of deuterium, tritium, ^{3}He and lithium. The agreement with the observed abundance of helium is so good that it is impossible to avoid the conclusion that it was made in the hot furnace of the early Big Bang. (The helium that fills advertising blimps or our party balloons was not manufactured in the Big Bang, however. Instead, it comes from the radioactive decay of uranium and thorium in rocks.)

Helium and the Density of the Universe

In addition, Big Bang nucleosynthesis tells us that there were about a billion photons for every proton and neutron. These photons cooled and eventually became the cosmic background radiation of which there are now about 550,000 in each liter of space. With this number, and the ratio of the abundances of photons relative to protons and neutrons, we have a way to calculate the average density of the universe (about 5×10^{-31} grams per cubic centimeter).

It is easy to understand qualitatively how the abundance of deuterium, for example, depends on the density. The rate at which deuterium was transformed to heavier nuclei was established by how frequently it collided with other particles. In turn, this depended on (among other things) the number per unit volume of such particles, that is, on the density. The greater the density, the more likely it was that collisions (and thus nuclear transformations) occurred, so the smaller the final abundance of deuterium would be. Thus, a deuterium abundance can be translated into a measurement of the density of the universe.

Could deuterium also have been made in stars, complicating the argument? No; deuterium is so easily transformed into heavier nuclei that it can only be destroyed, not manufactured, in stars. Therefore the measured abundance of deuterium (about 2×10^{-5} that of ordinary hydrogen) means that *at least* this amount of deuterium was made early in the Big Bang. In turn, this means that the density of the universe is *no more* than the 5×10^{-31} grams per cubic centimeter given above. The observed abundances of the three other kinds of nuclei made during the first few minutes of the Big Bang lead to the same value for the density. Because these abundances span some nine orders of magnitude, from very common helium to rare lithium, this is indeed an impressive success of the Big Bang model.

So now we have three cosmological clues, all supporting the idea that the universe began in a hot Big Bang: the systematic redshift-distance relation for galaxies, the 3 K background radiation, and the abundance of helium. More than that, the exact amount of helium or deuterium produced depends on the density of matter in the universe.

Before considering in greater detail how to fit these observational clues into a model of the universe, we should first consider a class of recently discovered objects, those with so-called active galactic nuclei. These objects not only may yield additional clues about the nature and history of the universe, but also are of great interest in themselves.

Terms to Know

Olbers' paradox, the cosmological principle, isotropic, homogeneous; redshifts of the galaxies, velocity-distance relation, Hubble constant, expansion and Doppler velocities, spacetime, expansion age; 3 K cosmic background.

Ideas to Understand

The significance of Olbers' paradox and its resolution, the implication of the cosmological principle; interpretation of the redshifts of galaxies; relation between the Hubble constant and the expansion age. Properties, origin, and implications of the 3 K background radiation and the helium abundance.

Questions

1. Suppose an absorption line with a rest (stationary) wavelength of 5000 Å is observed at a wavelength of 5500 Å in the spectrum of a distant galaxy.

(a) Is the galaxy approaching us or receding from us? How can you tell?

(b) What is its velocity?

2. A certain spectral line is measured in the laboratory to be at 5000 Å. At what wavelength would you expect it to be in the spectrum of a galaxy that is 100 megaparsecs away from us? [Take $H = 75$ (km/sec)/Mpc.]

3. A galaxy is observed to have a redshift corresponding to a velocity of 7,500 km/sec.

(a) How far away is it in pc? [Take $H = 75$ (km/sec)/Mpc.]

(b) How many years ago was the radiation we now receive emitted by this galaxy? (This is called the look-back time.)

(c) How big is the universe now compared with its size when the light was emitted?

4. If the universe is expanding, why does M31 show a blue-shifted radial velocity of about 300 km/sec?

5. (a) In 1940 the Hubble constant was thought to be about 500 (km/sec)/Mpc. Today we think it is about 75 (km/sec)/Mpc. Was the universe then thought to be larger or smaller than we now think it to be? by about how many times? younger or older? by about how many times? Explain your answers.

(b) Suppose you discovered a very distant galaxy in which a particular spectral line that has a rest wavelength of 5000 Å is observed to have a wavelength of 4000 Å. Would you call the *New York Times*? If yes, why?

6. If the Hubble constant $H = 75$ (km/sec)/Mpc, then the expansion age of the universe is about 13 billion years. Why is this a maximum value for the expansion age?

7. (a) Use Hubble's first published data (Figure 19.3) to derive the 1929 value for the Hubble constant.

(b) In the optical spectra shown in Figure 19.4, several emission lines remain fixed in wavelength regardless of the redshift of the galaxy. Explain why this is so. Why do these lines extend across the spectrum but the galaxy lines do not?

(c) What is the value of z for NGC 1080 (Figure 19.4)? At what wavelength is H-alpha observed in the spectrum of this galaxy?

8. Suppose it were shown that the Hubble constant for very distant galaxies (say, about 3 billion parsecs away) was 150 (km/sec)/Mpc, whereas for galaxies within a billion parsecs it has the value we have been assuming, 75 (km/sec)/Mpc. What would this suggest about the cosmic expansion—would it be speeding up, slowing down, or not changing as the universe evolves? What would this suggest about the future of the universe?

9. Suppose the spectrum of a distant galaxy shows a redshift of $z = 1.0$.

(a) By what factor has the scale of the universe increased during the time it took light to travel from the galaxy to us?

(b) According to the ordinary (nonrelativistic) Doppler formula, what is the velocity of this galaxy? What is its velocity according to the relativistic formula?

10. Suppose that galaxies evolve so that they are brightest soon after they are formed and steadily decrease in brightness afterward. Suppose further that we don't know this, and instead assume that their luminosity has remained constant with time. As a result, would the universe really be larger or smaller than we think it is? Explain your answer.

11. Why is it difficult to study the cosmic background radiation from the ground?

12. Describe the observed characteristics of the cosmic background radiation. What evidence indicates that it originated in the Big Bang?

13. (a) What observations enable us to say that helium was produced in the Big Bang?

(b) How does the background radiation enable us to calculate when helium was produced?

Suggestions for Further Reading

The spectacular developments in cosmology have produced a small explosion of books on the topic. A few are listed below; in general they apply to Chapters 19–22.

Cohen, N., *Gravity's Lens: Views of the New Cosmology*. New York: John Wiley and Sons, Inc., 1988.

Cornell, J. (ed.), *Bubbles, Voids and Bumps in Time: The New Cosmology*. Cambridge: Cambridge University Press, 1991. Six articles on various cosmological topics.

Davies, P., *The Runaway Universe*. New York: Penguin Books, 1980.

———*The Forces of Nature*. Cambridge: Cambridge University Press, 1986.

Gribbin, J., *In Search of the Big Bang: Quantum Physics and Cosmology*. Toronto: Bantam Books, 1986.

Overbye, D., *Lonely Hearts of the Cosmos*. New York: Harper Perennial, 1992.

Pagels, H., *Perfect Symmetry*. Toronto: Bantam Books, 1985.

Riordan, M. and Schramm, D., *The Shadows of Creation*. New York: W. H. Freeman and Company, 1991.

Trefil, J., *The Moment of Creation*. New York: Charles Scribner's Sons, 1983.

———*The Dark Side of the Universe*. New York: Charles Scribner's Sons, 1988.

Kippenhahn, R., "Light from the Depths of Time," *Sky & Telescope*, **73**, p. 140, 1987. The cosmic background radiation.

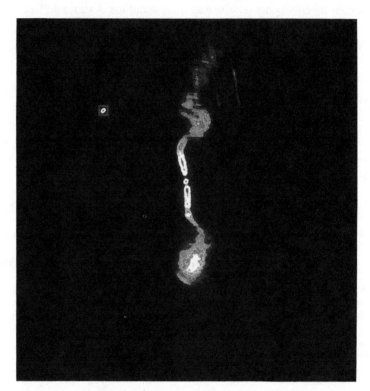

At the center of this radio galaxy (3C 449) is a large elliptical galaxy from which two radio jets emerge on opposite sides of the central source and spread into prominent lobes. 3C 449 is about 235 million light-years away and its radio emission extends more than 100,000 light-years out from each side of the central galaxy.

CHAPTER 20

Violent Events in Galaxies

Radio Galaxies

In the 1960s, a new class of objects was discovered that are extremely luminous, up to several hundred times the intrinsic brightness of even the brightest normal galaxy. Furthermore, their brightness varied significantly over timescales from years to hours, short by astronomical standards. These discoveries have led astronomers to recognize that incredibly energetic events are occurring at the centers of many different kinds of galaxies. In this chapter we will describe this galactic violence, which was first discovered by radio astronomy. In addition, we will consider some of the cosmological implications of these objects.

In 1909, long before the nature of the spiral nebulae had been established, the first spiral (NGC 1068 in Figure 20.1) was discovered that had emission lines in its spectrum as well as the usual absorption lines. The emission lines were the same as those that are prominent in planetary nebulae spectra (Figure 20.2). Hubble described three of these spirals in 1926, and in 1943 Carl Seyfert studied a small number of these objects, pointing out their characteristic features: a luminous, star-like nucleus and emission lines of the abundant elements: hydrogen, neutral and ionized helium, oxygen, neon, etc. In some objects the emission lines are very broad, indicating random motions of up to several thousand km/sec. Little further attention was given to these spirals (now called **Seyfert galaxies**) until radio sources began to be identified with galaxies in the 1950s.

Figure 20.1. NGC 1068 was the first spiral galaxy discovered to have emission lines in its spectrum. It is now recognized as a so-called Seyfert galaxy, and it has a powerful energy source at its center. Its spectrum is shown in Figure 20.2.

493

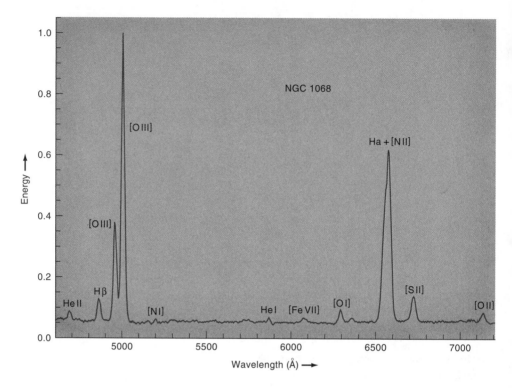

Figure 20.2. Usually the integrated spectrum of an entire galaxy consists of absorption lines produced by the combined light of the brightest stars. The spectrum of NGC 1068 was the first discovered showing prominent emission lines. These are formed in the hot gas at the center of the galaxy. Note the forbidden transitions (those in brackets). What do they tell you about the density of the gas producing them?

Thermal and Nonthermal Radio Sources

Surveys of the sky in the late 1940s and 1950s with the new telescopes sensitive to radio waves revealed many strong sources, most of which could not at first be associated with any visible object, because radio telescopes of the day had very poor angular resolution. As a consequence, the identification of many radio sources with visible objects proceeded slowly. Astronomers were quickly able to show, however, that sources have two kinds of spectra. One type, called a thermal source, is just the radio spectrum that a gas cloud emits by virtue of its temperature, typically about 8,000–10,000 K. The other kind of radio source has a markedly different spectrum, one that increases in intensity toward longer wavelengths (see Figure 20.3). In addition, its radiation is often polarized, sometimes strongly so. From these characteristics it was clear that this radiation is not produced as a consequence of the temperature of a gas; instead, it is synchrotron radiation, generated by a nonthermal process. We have already seen that the radiation producing the amorphous part of the supernova remnant, the Crab nebula, is generated by the synchrotron mechanism. Because the spectrum of synchrotron radiation becomes weaker toward shorter wavelengths, that is, toward optical wavelengths, a nonthermal radio source might well be invisible in ordinary light.

Characteristics of Radio Galaxies. The nature of the radio sources outside our galaxy was soon established. Some objects such as the Andromeda galaxy and our own

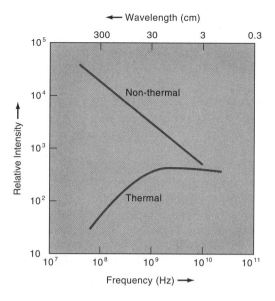

Figure 20.3. Typical spectra in the radio region of a thermal source—a gaseous nebula, for example—and a nonthermal source such as a radio galaxy.

galaxy were found to be weak radio objects, emitting only one-millionth or so of their total energy in the radio range, primarily by thermal processes. Other objects were found to be much more powerful radio sources, however, emitting larger fractions of their total radiation in the radio region. These are called **radio galaxies**, a good example of which is Centaurus A, shown in Figure 20.4 and Plate C5; it emits about 10,000 times as much radio radiation as does M31. The main properties of radio gal-

Figure 20.4. Centaurus A (NGC 5128) is the nearest powerful source of radio energy. About 4 million parsecs away, it is a giant elliptical (E0) galaxy crossed by a band of dust. The dust band rotates about an axis perpendicular to its plane, but the E0 rotates about an axis in the plane of the band. Apparently, this system was formed by a merger of a spiral and an elliptical galaxy. See also Plate C5.

axies are:

(1) by definition, they radiate an appreciable amount of their total energy in the radio region of the spectrum;

(2) the radiation is nonthermal, polarized, and generated by the synchrotron process;

(3) the radio sources associated with the galaxy are often double, sometimes single, and sometimes have a jet-like structure; and

(4) many, but by no means all, of these radio galaxies have optical spectra similar to those of Seyfert galaxies.

We have already touched on the first two points; let us next turn to the third. Many sources are double. That is to say, their radio emission comes from two enormous regions called **lobes** sometimes extending for a million light-years or so on either side of the nearly centrally located visible galaxy. Examples are given in Figures 20.5, 20.6, and 20.7. The central galaxy is often an elliptical. No visible matter related to the galaxy is to be found in these lobes. Recall that synchrotron radiation is produced by very high-energy electrons (invisible) spiraling around a magnetic field (also invisible). This structure strongly suggests that the high-energy particles and magnetic field were somehow ejected from the visible central galaxy. Often the lobes differ from each other in structure, and are not completely symmetric with respect to the central galaxy. This is a consequence of the motion of the radio galaxy through the thin intergalactic medium. As tenuous as it is, the medium resists this motion and distorts the shapes and positions of the lobes (see Figure 20.8).

Other radio galaxies don't show the huge double-lobe pattern; instead, much of their radio energy is emitted from a jet-like structure that apparently originates at the very center of the visible galaxy and extends some distance out from it but still within the galaxy. M87, a giant elliptical in the Virgo cluster of galaxies, is a good example

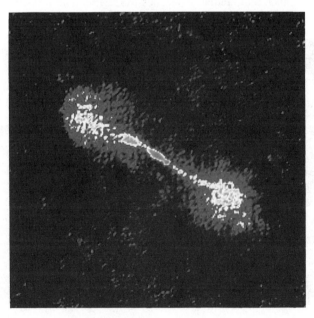

Figure 20.5. A typical double-lobed radio source (2354 + 471) with jets carrying energetic electrons and the magnetic field extending outward from the central source and ending in large clouds. The latter form, perhaps, as the jets emerge from the higher-density regions within the source galaxy and expand into the lower-density intergalactic medium. This radio image was taken at a wavelength of 20 centimeters.

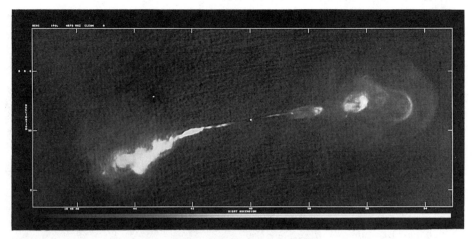

Figure 20.6. This is an image of Hercules A taken with the Very Large Array in radiation of 6-centimeter wavelength. It is one of the brightest radio sources in the sky and is associated with a faint galaxy in a distant cluster. The two structures each extend about 500,000 light years on either side of the central core. Note the difference in the two: the western (right) structure consists of several loops—each of which is larger than our own galaxy—while the eastern jet is continuous but irregularly deformed.

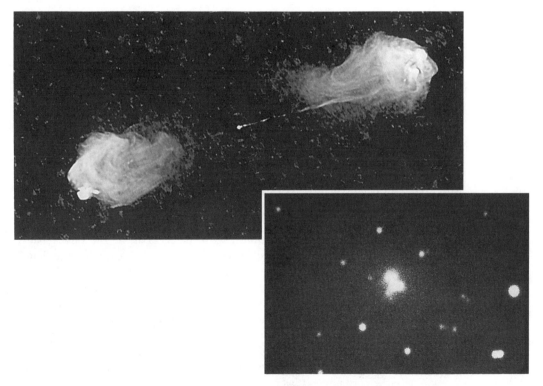

Figure 20.7. Cygnus A, another very bright radio source, has as its parent galaxy a disturbed elliptical galaxy, shown in the inset. The galaxy, centered on the point in the radio image, would approximately fill the space between the two lobes. Note that the radio emission (at a wavelength of 6 centimeters) is particularly intense at the leading edges of the lobes. This suggests that the particles and magnetic field there are being compressed as they run into intergalactic gas, causing them to become brighter.

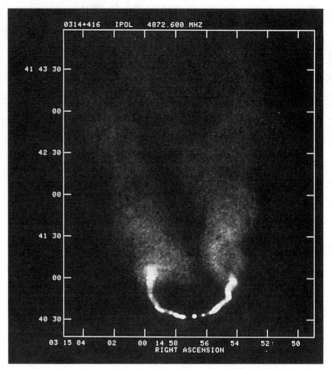

Figure 20.8. The parent galaxy of this radio source (NGC 1265) is located in the Perseus cluster of galaxies. As the galaxy moves at about 2,000 km/sec through the intergalactic medium of the cluster, its radio jets are bent backward by the force of the collision with the cluster gas. This radio image was taken at a wavelength of 6 centimeters.

of this kind of source (see Figure 20.9). Several knots of matter can be seen very near the center of the galaxy and falling along a line extending from it. They suggest that several separate events in the nucleus of the galaxy were responsible for the ejection of the knots. Perhaps millions of years from now there will be enormous lobes of radio emission on either side of M87. Still other radio galaxies show neither of these structures, but are simply single sources. In any case, these radio galaxies all seem to indicate that very energetic events, ultimately responsible for the radio emission, take place at or near the center of the visible galaxy.

The most powerful radio sources emit in radio wavelengths about 10^{12} times the total solar luminosity, or five to ten times more than the visible energy emitted by the Milky Way. In one second each of these radio galaxies emits as much energy as the U.S. would consume in about 10^{18} years! Given the spectrum and distance of the source, the theory of synchrotron radiation enables us to calculate the minimum total energy carried by the high-speed electrons and the magnetic field. The values that result are incredible. The energy that is carried by the electrons and magnetic field in such a source is equivalent to the energy the Sun would radiate in about 10^{19} years. This is the energy equivalent of the mass of a million Suns completely converted to energy according to the formula $E = mc^2$. (This is not to say that mass-to-energy conversion is the mechanism by which the energy is generated. It is simply a graphic illustration of the enormous energy contained in these sources.) Clearly, there must be an *extremely* powerful source of energy at the center of the visible galaxy associated with the radio sources. We are not sure how this huge reservoir of energy is produced. We will describe current ideas later in this chapter.

Keep in mind the two aspects of the source we have referred to: the ultimate energy source (the reservoir), and the mechanism by which energy from that source is converted to radio radiation (the nonthermal synchrotron process). These are roughly analogous to a power station (the energy reservoir) providing electricity to a light bulb (which radiates by heating a tungsten filament, a thermal process).

A

B

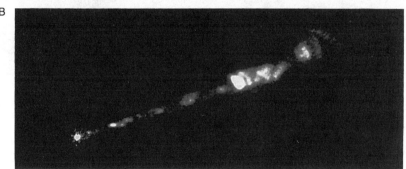

C

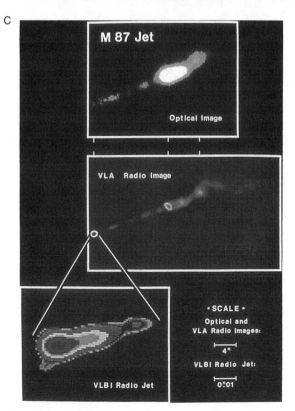

Figure 20.9. (A) The giant elliptical galaxy in the Virgo cluster, M87. It is about 3 arc-minutes in diameter. At its center is a powerful energy source producing a jet shown in (B). The length of the jet is only about 20 arcseconds, so it is buried deep within the galaxy. (C) shows a radio image of the jet and an extremely high resolution image of the central source. Note how small the latter is.

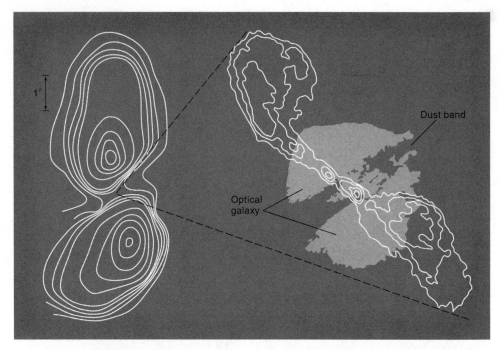

Figure 20.10.   Centaurus A is an enormous radio object. Two lobes of radio emission emerge from a radio jet at the center of the optical galaxy, as shown in the 6-centimeter radio image on the right. These two lobes are buried at the center of the 31-centimeter radio image on the left, which extends for several degrees across the sky, or for a distance of about 700,000 parsecs!

Let us look at one of the best known of these radio galaxies, Centaurus A (Figure 20.10). As is apparent in Figure 20.4, the central visible galaxy, about 1/6 of a degree in diameter, is a pathological object resembling a giant elliptical galaxy crossed by a prominent, distorted band of dust. The galaxy (cataloged as NGC 5128) is rotating about an axis in the plane of the dust band. At opposite sides of the visible galaxy are two relatively small, intense radio sources. Extending about 5 degrees on either side of the galaxy are two huge lobes of radio radiation. Thus, this is a double-double source. If the distance to Centaurus A is about 4 million parsecs, then the two inner components are separated by about 8,000 parsecs. The two outer components extend over a distance of about 700,000 parsecs! The distorted appearance of these huge lobes suggests that the system is encountering intergalactic matter.

Quasi-Stellar Objects

Not all radio sources are associated with visible galaxies. In 1960, a star-like object was identified with a radio source (3C48—the 48th object in the third Cambridge catalog of radio sources). Astronomers made the natural conclusion that the first radio star had been discovered. Since the Sun and other stars radiate very little energy in the radio region of the spectrum, the apparent discovery of a radio-bright stellar object was very interesting. This "star" (and others that were soon found, for example, 3C273) had an enigmatic spectrum, however; its visible-light spectrum had no absorption lines, but consisted of emission lines corresponding to transitions in no known chemical elements! Astronomers puzzled over this until 1963, when it was realized that the spectrum was just that of ordinary elements like hydrogen, but red-shifted by an enormous amount. For example, in 3C273 a spectral line due to Hβ, normally at a wavelength of 4861 Å, is red-shifted by 812 Å! 3C273 and 3C48 are relatively bright and easily seen, even with a telescope of modest aperture; yet these objects have redshifts of about 16 percent and 37 percent of the speed of light, respectively. If the redshifts are cosmological, that is, if they result from the expansion of the universe, then these objects are at enormous distances. For example, 3C273 must be about 600 Mpc away! To be at such enormous distances and still be easily visible with small telescopes, these objects must be incredibly bright—hundreds of times brighter intrinsically than an ordinary bright galaxy. This discovery caused great excitement among astronomers. Maybe these were the distant objects that would enable us to discover the properties of the early universe. A QSO with a much greater redshift is shown in Figure 20.11.

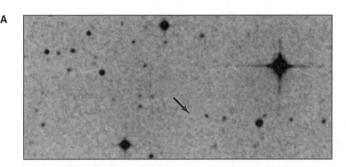

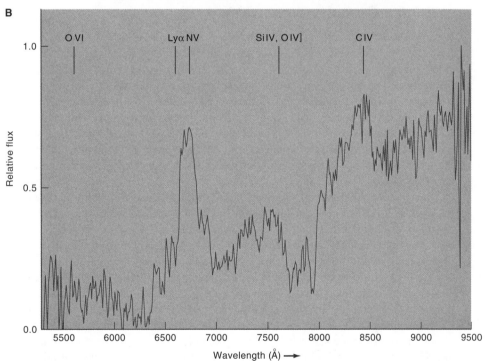

Figure 20.11. The spectrum of the very distant quasar Q0051-279 is shown at (B); (A) shows the object itself. Notice the enormous redshift; for example, the rest wavelength of the Lyman line of hydrogen is 1216 Å, but it has been red-shifted nearly 5400 Å to about 6600 Å!

Properties of QSOs

Because of their appearance, these objects were soon called quasi-stellar objects (QSOs or quasars). They have been found to have the following characteristics:

(1) their appearance is usually star-like, but sometimes they show a very faint structure;

(2) their spectra show broad emission lines;

(3) their spectra are all red-shifted, some enormously so; apparently none are nearby, because none have been found with small redshifts;

(4) they vary in brightness, the variations occurring sometimes as rapidly as in days;

(5) some are strong radio emitters, but most are not; and

(6) they are often extremely bright in the infrared region of the spectrum.

Let us consider their light variability first. Since they are so bright intrinsically, any significant increase in their brightness implies that a huge additional amount of energy must have been generated. Furthermore, by the same argument we used in discussing pulsar variability in Chapter 16, we infer that the relatively rapid light variability of QSOs has an astonishing implication: a quasi-stellar object that varies on a timescale of one day must be no more than about one light-day across; but one light-day is only about 175 AU, or roughly the size of the solar system! Evidently, some QSOs produce within a volume about the size of the solar system several hundred times as much radiation as a whole galaxy. This mind-boggling result led some astronomers to wonder if QSO redshifts were reliable indicators of their distances. Perhaps they were produced by some process other than the expansion of the universe.

Redshifts: Cosmological or Not?

If their redshifts had nothing to do with their distances and if the QSOs were not very far away, then their intrinsic luminosities would be much smaller and more "reasonable." So perhaps they are objects ejected at very high speeds from nearby galaxies. But if this were the case, why do we not see blue-shifted quasars, quasars coming toward us? Searches for such quasars have been fruitless. Well, the argument went, maybe the QSOs were ejected from the center of our galaxy some time ago, have passed us, and now are all moving away from us. But why would these thousands of objects have been ejected from the center of our galaxy and not from any others? Furthermore, there is no evidence for all this violence that supposedly occurred in the center of the Milky Way.

Let's take a different tack; suppose the redshift is the result of a physical phenomenon so far unknown to us. A few astronomers claim to have observational evidence for this idea. They have found instances in which a QSO with one redshift seems to be physically associated with a galaxy having quite a different redshift. Clearly, if both objects really are at the same distance and not in the same line of sight just by chance, one (or both) of the redshifts must be noncosmological. But to establish that the QSO and the galaxy are really next to each other in space, really physically associated, is difficult. The statistical analysis is tricky, but a certain number of chance alignments of QSOs and galaxies are to be expected. (The same effect is found with stars; occasionally two stars are separated on the sky by only a second or two of arc, but in fact are separated in space by large distances and not physically associated at all.) Most astronomers believe that the galaxy-QSO associations are of this type. Even if the evidence for a noncosmological redshift were compelling, there would still remain an obvious question. Just what could this new physics be, this process by which such large redshifts would be produced, but which has nothing to do with the gravitational redshift, the Doppler shift, or the expansion of the universe? No convincing answer has been given to this question.

Virtually all astronomers now believe that QSO redshifts are cosmological and that they really are as far away as their enormous redshifts indicate. As we have seen, this means that they are remarkably luminous and produce incredible amounts of energy from a volume approximately the size of the solar system. The energy generated is so enormous that, again, some scientists speculated that a new physics is needed, this time to account for the enormous energy production instead of the large redshifts. (They are also extremely rare, as indicated in Table 20.1.)

Table 20.1. Relative numbers of ordinary and active galaxies

Kind of Galaxy	Relative number
Field galaxies	1
Bright spirals	10^{-1}
Seyfert galaxies	10^{-3}
Radio galaxies	10^{-5}
QSOs	10^{-6}
Radio-quiet QSOs	10^{-8}

As long as QSOs were thought to be an anomalous class of objects, unrelated to any other objects and hundreds of times brighter than anything else astronomers knew about, such desperate speculations might seem justified. Soon, however, it became apparent that QSOs were the brightest objects on a sequence of increasingly luminous objects extending from normal galaxies to quasi-stellar objects. Collectively these objects are now referred to as **active galactic nuclei** (or AGNs), because their distinguishing features are associated with the nuclei of these objects. Among the "intermediate" objects are Seyfert galaxies, radio galaxies (often giant ellipticals), and **BL Lacertae** objects (Figure 20.12), quasar-like systems that often vary rapidly in brightness (even in a day) and have strongly polarized, but nearly featureless, spectra.

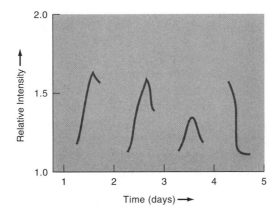

Figure 20.12.   The variability of a BL Lacertae object in the visible region of the spectrum. The gaps in the data occur during daytime hours.

Incidentally, many years ago BL Lacertae caught the notice of astronomers who thought it was a variable star within our own galaxy. The designation they gave it indicates two of the salient features of these active galaxies: star-like appearance and variability. (Roman letters as a prefix to a constellation are an arcane way of denoting variable stars in order of their discovery.)

A short-exposure photograph of a Seyfert galaxy shows only the tiny, star-like nucleus and not the surrounding disk and spiral arms, indicating that the nucleus is much brighter than the rest of the galaxy. The redshifts of Seyferts are relatively small, so they are nearby. The total amount of energy radiated by the nucleus, though much larger than that from an ordinary galactic center, is smaller than that from a QSO. That is, the contrast in brightness of the nucleus and surrounding galaxy is much smaller in a Seyfert galaxy than it is in a QSO.

Unlike normal galaxies, objects with active nuclei generally have bright, broad emission lines, indicating that a hot gas is in rapid motion with velocities of thousands of kilometers per second. These large, turbulent gas velocities, toward us and away from us, broaden the emission lines by the Doppler effect. Some sort of powerful "engine" must be located at the centers of these objects, stirring up the gas. In addition, some active nuclei emit huge amounts of x-ray and infrared radiation, again quite uncharacteristic of ordinary galaxies.

Before describing current ideas about the nature of quasars, it is worth emphasizing what we don't know:

- What is the central "engine"?
- What is the fuel supply of the central engine, and how long does it last?
- Is the quasar phenomenon a phase through which most galaxies pass soon after their formation, or do only a small fraction of galaxies become quasars? Can a QSO be "rejuvenated"?
- What is the relation—if any—between quasars and less energetic AGN objects such as Seyfert galaxies?

The Nature of QSOs

The picture that has emerged is that in the nuclei of some galaxies a very energetic phenomenon occurs that produces an enormous amount of radiation, tens or hundreds of times more than is emitted by the rest of the galaxy. All that we can see is the bright nucleus (which we call a QSO), because the rest of the galaxy is just too faint to be seen at the great distances of the QSOs. There is observational support for this notion; for example, very faint galaxy-like glows have been detected around relatively nearby objects such as 3C48, 3C273, BL Lac, and others, but not around more distant QSOs where the surrounding galaxy would be too faint.

The compactness of a QSO, the occasional luminosity outbursts that exceed the conversion of a solar mass into energy, and the rapid increase in brightness and velocity of material toward the center of Seyfert galaxies all point toward a powerful gravitational-energy source for an AGN. Nuclear-energy sources don't work; they can't produce the large and variable amounts of energy observed. Other models have invoked large numbers of supernovae, collisions between stars, and other such extreme circumstances. Today, however, most astronomers believe that a massive black hole (10^8 solar masses or greater) at the center of a galaxy is the engine that powers the QSO. (Recall that black holes are objects that combine a large mass and small radius so that the velocity required to escape from them is greater than the speed of light.)

A 10^8-solar-mass black hole would have a Schwarzschild radius of only two AUs and a density at R_s of 100 gm/cm^3 (see Chapter 16). Just as the whole galaxy rotates

Figure 20.13. Hubble Space Telescope observations of the bright Virgo cluster galaxy, NGC 4261, provide our best look yet at the central "engine" of an active galaxy. The illustration on the left is a combined optical and radio image showing the central galaxy as the origin of a typical double-lobe radio source. The optical image on the right shows a 100-parsec-diameter disk extending to within a few AU of the central source, which is fed by infalling matter from the disk. Apparently, the central engine ejects matter in streams that are perpendicular to the plane of the disk, which in turn feed the radio jets.

around its center, matter rapidly rotates around the black hole. The orbital velocity of matter near the black hole will increase as it spirals in, because its angular momentum (mVr) must remain constant: r decreases, so V increases. Soon the velocity of the gas is so great that its infall toward the hole is slowed and finally nearly stops altogether. As new matter spirals in toward the black hole, it runs into the material that is already there and an accretion disk of gas is built up. This acts as a sort of reservoir from which matter eventually falls into the hole. (The picture is similar to that for a nova, in which an evolving star dribbles matter onto an accretion disk surrounding its white-dwarf companion.)

Images of the bright Virgo cluster galaxy, NGC 4261, support this model. The left part of Figure 20.13 shows the elliptical galaxy in visible light combined with its double-lobed radio image. An HST image of the central region of the galaxy (right) reveals a disk of gas and dust about 300 light-years in diameter. This disk extends into the center of the galaxy to the presumed hot accretion disk/black hole structure. The radio structures, perpendicular to the disk, are apparently squirted out through the poles of the disk.

Collisions between blobs of gas heat the disk and cause the gas to lose angular momentum. As it does so it moves toward the inner radius of the disk and finally disappears into the black hole. The hot disk produces the light we see and its variability results from nonuniform mass flow onto it. It has been estimated that a few solar masses must be lost to the black hole each year to produce the radiation we observe

Figure 20.14. An HST image of the central region of M87. Remember that the jet extends only 20 arcseconds along the 100-arcsecond radius of the optical galaxy, so that the increase in the brightness of the galaxy toward its center is very pronounced. The bright spot at the center is not starlight; rather, it is visible light from the strong radio source. The bright dots scattered over the image are globular clusters.

from a typical QSO. According to this model, then, it is the conversion into radiation of the very large gravitational energy of material falling into a massive black hole via an accretion disk that is the ultimate power source of a QSO.

Despite considerable effort by many astronomers, strong evidence of a black hole powering an AGN was found only in 1994. M87, the giant elliptical in the Virgo cluster, a powerful source of radiation with a jet originating from its core (Figure 20.9) where the star density is at least 300 times greater than in a normal giant elliptical (Figure 20.14), has long been a prime candidate for a black hole. Images made with one of the HST cameras (which has been fitted with optics to correct the primary mirror's aberration) show a thin spiral-shaped disk of gas about 500 light years in diameter at the core of M87. HST spectra of gas 60 light years from the center of the disk show that it is revolving with a speed of about 550 kilometers per second; closer to the center, the velocity reaches 800 kilometers per second. It is easy to calculate that this requires that there be two or three billion solar masses within the disk. This enormous amount of matter cannot be stuffed into a solar system–sized volume in the form of stars; a massive black hole is most likely at the core of the galaxy.

That there are few nearby QSOs indicates that this process occurs earlier in the evolution of a galaxy rather than later. In addition, only a handful of QSOs have been found with redshifts greater than about $z = 3.5$. (As of this writing, the largest redshift is $z = 4.897$.) The number of QSOs falls off at apparently great distances, indicating that few were formed earlier than the time corresponding to that distance, or earlier than roughly 10 billion years ago (see Figure 20.15). This suggests that there is a well-defined QSO era, that the QSO phenomenon begins rather soon in the life of a galaxy, and that it is a transient phase of galactic youth lasting several hundred million years or so. Perhaps by that time little matter is left near the black hole to power it. If the

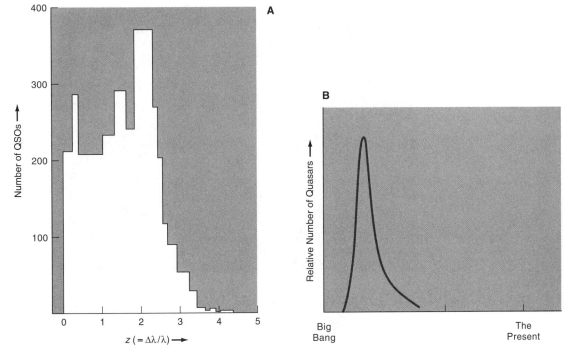

Figure 20.15. (A) The distribution of redshifts of the 7,300 quasars known in 1992. The largest number of quasars are at a z of about 2, corresponding to a distance, and hence a time, of about 1/5 of the present. Note the rapid decrease of QSOs with redshifts greater than about $z = 2.3$, indicating that relatively few were formed very early in the universe. (B) A schematic representation of the quasar era. The number of quasars is plotted against look-back time, which is not given explicitly because it depends on the value of the Hubble constant and the model of the universe assumed. The important point to note is that the quasar era lasted only a few billion years.

redshift surveys are correct, the number of QSOs peaked one or two billion years after the QSO era began. On this picture Seyfert galaxies (which, you recall, are relatively nearby and closer to us in time than QSOs) are perhaps dying QSOs in the process of becoming ordinary galaxies.

Interestingly, though there are no QSOs relatively nearby, there might be massive black holes next door. The Andromeda galaxy, M31, one of its companions, M32, and even the nucleus of our own Milky Way are all suspected of harboring a massive object. Among the pieces of supporting observational evidence is that stars very close to the center of M31 and M32, as well as a ring of molecular hydrogen around the center of our galaxy, are all moving very rapidly. The masses implied by these motions at the observed distance from the centers of these galaxies can be calculated, and are about 50×10^6, 5×10^6, and 10^6 solar masses for M31, M32, and the Milky Way, respectively. Supporting evidence is seen in M32 where the star density increases dramatically toward the center of the galaxy (Figure 20.16). The central mass in our galaxy could perhaps be a dense compact cluster of stars, rather than a black hole. Though none of these objects could have ever powered a luminous QSO, they might have produced some low-energy AGN-like phenomena. It may be that nearly all galaxies experience something of this sort.

Figure 20.16. An HST optical image of the central portion of M32 (one of the two elliptical galaxy companions of M31) showing the marked increase in the number of stars toward the center of the galaxy. It has been estimated that the central star density is more than one hundred million times that in the vicinity of the Sun. This perhaps supports the notion that a massive object lies at the center of the galaxy. The area pictured is 175 light-years on a side.

Unfortunately, we know of no direct method by which we can get the distances of QSOs without using their redshifts. They have a wide range of intrinsic brightnesses, so that their apparent brightnesses are not correlated with their distance. And since they are little more than points in the sky, we cannot get distance information from their apparent size. Thus we are still unable to use them as cosmological probes of the distant universe.

Black holes are very much in fashion in astronomy today and have been invoked to account for a wide variety of energetic phenomena. Though we now have good evidence for one at the center of M87, and many other plausible candidates for black holes most likely do in fact contain them, we should be cautious in claiming a black hole to be the culprit in *every* astronomical mystery.

Matter Between Us and Quasars

QSOs can serve as probes of another kind, however: they are powerful light bulbs producing a spectrum against which absorption lines formed in matter between us and the QSO can be seen. That is, matter along the line of sight to a distant QSO will be revealed by its absorption of the QSO radiation. Now, the mere fact that we see high-z objects means that the intergalactic medium can't contain much dust or neutral hydrogen; otherwise Lyman-alpha emission lines would be absorbed, for example.

Nevertheless, two kinds of absorption-line systems are observed, always with redshifts less than that of the background QSO. One kind of system shows lines of metals such as oxygen and magnesium (in emission as well as absorption: see Figure 20.17). They probably arise from the interstellar gas in galaxies that happen to be along the line of sight to the quasar. By studying the abundances of these lines, we should be able to calculate how much heavy-element production took place at various times in the past.

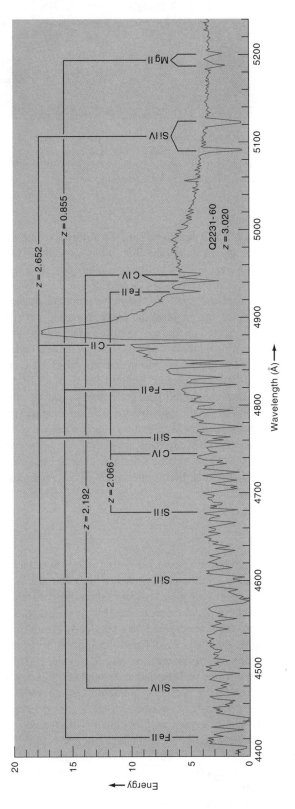

Figure 20.17. A portion of the spectrum of a distant quasar with a redshift z = 3.020. The strong emission line at about 4880 Å is Lyman-alpha, red-shifted from its rest wavelength of 1216 Å. The quasar spectrum shows many absorption lines produced by discrete clouds of intergalactic gas or by halos of galaxies that happen to fall along the line of sight to the quasar. The majority of the metal lines are marked. Note that they arise from galaxies with four different redshifts. Most of the features shortward of the QSOs Lyman-alpha line are themselves hydrogen Lyman series absorption lines (the Lyman-alpha forest) with redshifts smaller than 3.020.

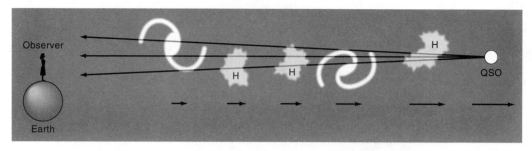

Figure 20.18. The line of sight from a distant quasar passes through several clouds of hydrogen-rich material (labeled "H") as well as some galaxies, all given with arrows schematically indicating their redshifts. Lyman-alpha lines are probably produced in the former and metal lines in the latter. (The sketch is not to scale.)

The other and much more numerous kind of absorption-line system is produced by hydrogen, primarily the Lyman-alpha line, which, recall, arises from the ground state to the first excited state ($n = 1$–2) of neutral hydrogen. This hydrogen must be in clouds of gas between us and the QSO. That these are discrete clouds rather than a universe-filling medium is indicated by the large numbers of narrow Lyman-alpha lines, separated in velocity, seen against the QSO spectrum, rather than a continuous absorption dip (see Figure 20.17). In fact, the lines are so numerous that they are referred to as the **Lyman-alpha forest**. Perhaps some of these clouds are in the process of becoming galaxies, or perhaps they will eventually diffuse into space and lose their identity (see Figure 20.18). In any case, studies of both kinds of absorption-line systems, still in their infancy, should tell us a great deal about the early universe.

The Hubble Space Telescope has provided a result of great interest. Ultraviolet spectra of nearby quasars such as 3C273 show many more lines of the Lyman-alpha forest than had been expected, simply by extrapolating from observations of more distant quasars (Figure 20.19). This indicates that whatever the absorbing clouds may be, there are many more nearby than had been thought. These clouds should provide observers with a rich lode to mine.

Final Comments

Observational cosmology has reached a frustrating point. Extremely luminous objects exist that we can detect at enormous distances. If only the distances to these objects could be found independently of their redshifts, then those distances combined with their redshifts would enable us to measure the Hubble constant. Since QSOs are far away, the values of H so found would refer to a very early epoch of the universe, when the expansion was faster than it is now. This would enable us to make a much better estimate of the expansion age of the universe and to calculate just how rapidly the expansion is slowing down; as we will see in the next chapter, this would tell us the future of the universe. This tantalizing prospect cannot yet be achieved.

Perhaps new large ground- and space-based telescopes under construction, along with increasingly sensitive detectors, will provide us with the necessary data. Perhaps new insights will show the way, like "grand unified theories" of the fundamental forces and "inflationary" universes coming from the realm of high-energy physics (see Chapter 21). In any case, the next several years should provide us with plenty of excitement.

In the next chapter, we will look at the kinds of models of the universe we can build, given our present state of knowledge.

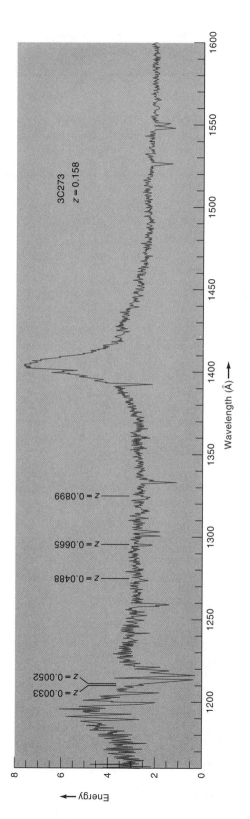

Figure 20.19. A portion of the spectrum of the nearby quasar, 3C273. The strong emission line is Lyman-alpha red-shifted to z = 0.158. The five faint absorption lines marked with their redshifts are Lyman-alpha lines formed in gas clouds along the line of sight to the quasar. Astronomers were surprised that so many clouds were found along such a relatively short sightline. The other absorption lines are formed by metals in the halo of our own galaxy. Note the Lyman-alpha line at 1216; a cloud of hydrogen surrounding the Earth forms the sharp emission line within the absorption line formed by Milky Way gas.

511

Terms to Know

Radio galaxy, lobe, thermal and nonthermal radiation. Active galactic nucleus, QSO, Seyfert galaxy, BL Lac object; Lyman-alpha forest.

Ideas to Understand

Characteristics of radio galaxies, QSOs, and Seyfert galaxies; arguments for and against QSOs being at cosmological distances; implication of rapid light variability; how active galaxies may generate their energy; evidence for a black hole in M87.

Questions

1. Describe the basic forms shown by radio galaxies. How are these forms produced?

2. Suppose you discover an object that looks like a star, but emits many times more radio energy than any true star of the same apparent magnitude. What should you measure or look for to decide whether this is a quasar?

3. What observational evidence suggests that radio galaxies undergo more than one violent event or "explosion"?

4. Suppose a quasar is observed to increase its brightness by about 50 percent in just 24 hours. What is the approximate diameter of the quasar's emitting region in AU?

5. 3C273 is the brightest-appearing quasar, about 10^{-6} times as bright as Sirius, the brightest star in the sky. The rest wavelength of Hβ = 4861 Å, but in 3C273 it is redshifted by 812 Å.
   (a) What is the expansion velocity of 3C273?
   (b) With H = 75 km/sec/mpc, how far away is it? What is the look-back time to 3C273?
   (c) If the quasar were only as far away as Sirius (which is about 3 parsecs distant), how many times brighter than Sirius would it appear to be?

6. It is impossible to observe the Lyman lines of hydrogen in the spectrum of even the nearest stars using a telescope on the ground. With that same telescope, however, we can observe Lyman lines in the spectra of quasars. Explain why.

7. Quasars are not uniformly distributed through space, but are much more numerous at large than at small redshifts. Why don't they violate the cosmological principle?

8. Refer to the quasar spectrum in Figure 20.11.
   (a) Verify that the redshift of this QSO,

$$z = \frac{\Delta\lambda}{\lambda},$$

is about 4.4. The rest wavelengths of the Ly-α, N V, and C IV lines (identified in the upper row in the figure) are 1216 Å, 1240 Å, and 1550 Å, respectively. Use all three lines in your calculation.
   (b) What is the expansion velocity of this QSO?

9. Show that the 100-million-solar-mass black hole mentioned in this chapter would indeed have a Schwarzschild radius of 2 AU and a density at that radius of 100 gm/cm^3.

10. Refer to Figure 20.17. Why must more than one metal-absorption line be identified at a given redshift to make a convincing case that the absorption is really occurring in a galaxy with that redshift?

11. Why, in Figure 20.19, are there no absorption lines of Ly-α with redshifts greater than z = 0.158?

Suggestions for Further Reading

Balik, B., "Quasars with Fuzz," *Mercury*, **12**, p. 81, May/June 1983. What the spectrum of "fuzz" surrounding a QSO tells us.

Burns, J., "Chasing the Monster's Tail: New Views of Cosmic Jets," *Astronomy*, **18**, p. 28, August 1990. Structure and origin of jets in galaxies.

Burns, J. and Price, M., "Centaurus A: The Nearest Active Galaxy," *Scientific American*, **249**, p. 56, November 1983.

Croswell, K., "Have Astronomers Solved the Quasar Enigma?" *Astronomy*, **21**, p. 28, February 1993.

Finkbeiner, A., "Active Galactic Nuclei: Sorting Out the Mess," *Sky & Telescope*, **83**, p. 138, August 1992. Describes a model that accounts for many of the different types of AGNs.

Geballe, T., "The Central Parsec of the Galaxy," *Scientific American*, **241**, p. 60, July 1979.

Kanipe, J., "M87: Describing the Indescribable," *Astronomy*, **15**, p. 6, May 1987.

———"Quest for the Most Distant Objects in the Universe," *Astronomy*, **16**, p. 20, June 1988.

Osmer, P., "Quasars as Probes of the Distant and Early Universe," *Scientific American*, **246**, p. 126, February 1982.

Preston, R., "Beacons in Time: Maarten Schmidt and the Discovery of Quasars," *Astronomy*, **17**, p. 2, January/February, 1988. An account of the discoveries and the astronomers involved.

Rees, M., "Black Holes in Galactic Centers," *Scientific American*, **263**, p. 56, November 1990.

Schorn, R., "The Extragalactic Zoo," *Sky & Telescope*, **75**, p. 23, p. 376, **76**, p. 36, 1988. Three articles on active galaxies essentially in the form of an extended glossary. Useful.

Weedman, D., "Quasars: A Progress Report," *Mercury*, **17**, p. 12, January/February 1988. An overview by an active researcher in the field. Follows the article by Preston, above.

Wilkes, B., "The Emerging Picture of Quasars," *Astronomy*, **19**, p. 34, December 1991.

Wyckoff, S. and Wehinger, P., "Are Quasars Luminous Nuclei of Galaxies?" *Sky & Telescope*, **61**, p. 200, 1981. Describes some of the early observations indicating that they are.

Radio techniques first made a big impact in astronomy when applied to the study of galaxies. Listed below are a few books and articles on the general subject.

Downes, A., "Radio Galaxies," *Mercury*, **15**, p. 66, March/April 1986. A good review of the topic.

Hey, J., *The Evolution of Radio Astronomy*. New York: Neal Watson Publications, 1973. A history of the field by one of its pioneers.

Lovell, B., *The Story of Jodrell Bank*. New York: Harper and Row, 1984. The building of one of the largest radio telescopes in the world, told by its builder.

Verschuur, G., *The Invisible Universe Revealed*. New York: Springer-Verlag, 1987.

"Henceforth space by itself, and time by itself, are doomed to fade away into mere shadows and only a kind of union of the two will preserve an independent reality."

—H. Minkowski, 1908; German mathematician

"Geometry tells matter how to move, and matter tells geometry how to curve."

—C. Misner, K. Thorne, and J. Wheeler, 1972; U.S. physicists

CHAPTER 21

Models of the Universe

The purpose of this chapter is to see what kind of a model of the universe we can construct that is consistent with the current observational evidence. Even though there are big gaps in our understanding (there is still plenty for you to discover!), we will be able to see fairly clearly where our problems are.

Just a little reflection should convince you that in trying to understand the structure and evolution of the universe—the largest entity we can possibly consider—some of the everyday ideas we take for granted may not be adequate. In fact, what is a bit surprising is that so many of the concepts we have derived in our Earth-bound laboratories seem to work well when applied to the universe as a whole.

We know something about galaxies, the building blocks of the universe; they come in a variety of sizes, shapes, and colors, and most of them are organized into groups, clusters and superclusters. We know that the universe is expanding, that all space is filled with the cosmic background radiation, and that the hydrogen and helium that make up the stars were produced a few minutes after the beginning. All this supports the idea that the universe originated in a hot Big Bang, a simple theory that nonetheless has considerable predictive power.

But now we want to consider the overall structure, the "shape" of the universe, and how this structure changes with time. We might expect that some fundamental, basic idea might be involved in the organization of the physical universe—and indeed one is. Unlikely as it may appear, the geometry that characterizes the universe as a whole reveals its key structural element. To understand why this is so, we will first consider the properties of different kinds of geometries, and then see which is appropriate for describing the universe. These ideas were developed in the context of Einstein's theory of general relativity, which is a theory of gravity with broader applicability than Newton's. We will describe some of the tests that support this newer concept of gravity. Then we will discuss another aspect of the universe: its origin and evolution, how it is changing with time. Though it may seem that these two aspects—geometry and evolution—have little to do with each other, we will see that, in fact, they are intimately related.

Finally, the conventional Big Bang picture has difficulty in accounting for several features of the universe and so is very likely incomplete. We will look at these problems and see how they might be resolved by means of a modification—called inflation—to the Big Bang very early in the history of the universe.

Different Kinds of Geometries

Flat Space

Our immediate, everyday experience takes place in what we call Euclidean space (because it is described by Euclid's geometry, the plane geometry many of us learned in high school). Euclidean space is flat and has three **dimensions**. What do we mean by dimension? Simply, the dimension of a space is the number of numbers required to specify uniquely a point in that space. For example, on a blackboard draw two intersecting lines perpendicular to each other; call one the x-axis and the other the y-axis (a Cartesian coordinate system; see Figure 21.1). The position of any point on the

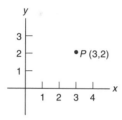

Figure 21.1.   The position of the point P is given by the two numbers, $x = 3$, $y = 2$. The surface on which the $x - y$ coordinate system is drawn is a two-dimensional surface.

blackboard can be specified by its signed (that is, plus or minus) distances from the two axes. Since two numbers suffice to specify the point's position, we say the blackboard is a two-dimensional space. In a similar manner, it requires three numbers to specify the position of any point in a room. Thus, a room encompasses a three-dimensional space.

Let's consider some of the characteristics of flat, two-dimensional space. First of all, the shortest distance between two points on a blackboard, for example, is a straight line. In general we call the path in a given space on which the distance between two points is shortest a **geodesic** of that space; the geodesic in flat space is a **straight line**. Note that the polygons of flat space (triangles, squares, etc.) are made up of geodesics.

Next, consider a well-known property of flat space. Given a straight line (a geodesic) and a point P outside that straight line (Figure 21.2), we can draw only one line

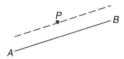

Figure 21.2.   In flat (Euclidean) geometry, only one line can be drawn through the point P parallel to the line AB.

through P that is parallel to the straight line. Any other line would eventually intersect the first line and, therefore, could not be parallel to it. Other familiar properties of flat space are that the interior angles of a triangle add up to 180°, and that the circumference of a circle and its area increase as its radius and as its radius squared, respectively.

All this probably seems pretty obvious—it's just the way things are. However, there are circumstances even in our own experience when this isn't the way things are. Suppose you want to measure the distance between two points in your hometown from a city map that had an appropriately large scale. If there were no changes in elevation between the two points, the result of this measurement would agree well with the answer you would get if you actually took a steel tape and carefully measured the same distance directly. Now suppose you wanted to do the same thing for two points within your home state. If you were careful the two results would again agree fairly well. But now suppose you wanted to measure the shortest distance between your hometown and Paris. Then the straight-line distance on a conventional map would differ substantially from the shortest distance you would measure in reality. This is so because the surface of the Earth is curved, not flat, and we can't use flat geometry to measure distances accurately over a spherical surface.

Geometry of Curved Space

Let's see what some of the geometric characteristics are of a curved *surface*, like that of the Earth. To specify the position of any point on this surface, we need only two numbers, the latitude and longitude of the point. Thus the *surface* of a sphere is a curved, two-dimensional space. In the curved space, the line giving the shortest distance between two points, the geodesic, is a **great circle**. Remember that a great circle on the surface of a sphere is formed by the intersection of a plane through the center of the sphere with the surface of the sphere, as shown in Figure 21.3. (Trans-Atlantic

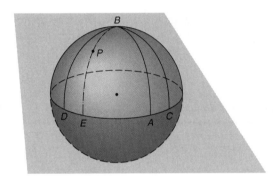

Figure 21.3. The shortest distance between two points on the surface of a sphere is defined by the cut on the sphere's surface (*D-E-A-C*) made by a plane that passes through the center of the sphere. *AB*, *CB*, and *DB* are also geodesics. On Earth, the former is the equator, and the latter three are lines of longitude. Note that there is no geodesic through *P*, which will not intersect other geodesics, for example, *AB*.

or trans-Pacific flights used to be advertised as following "great circle routes," indicating that they are the shortest routes between the two cities in question.) Any polygons that we draw on the surface of the sphere must be made up of geodesics—great circles—just as triangles on a blackboard must be made up of straight lines. With that in mind, a glance at Figure 21.3 will make it clear that on the surface of a sphere there is no geodesic through a point *P* that will not intersect line *AB*; there are no parallel geodesics on the Earth's surface. On the surface of the Earth, lines of longitude intersect at the poles; the equator, another great circle, intersects all the lines of longitude. Circles of latitude can be made parallel to each other, but they are not geodesics (arcs of great circles), because the planes forming them do not pass through the Earth's center.

Next, consider the angles in a spherical triangle, a triangle whose sides are made up of arcs of three great circles. In Figure 21.4 the triangle is made of two lines of longitude and the equator. The longitude lines intersect the equator at right angles, so

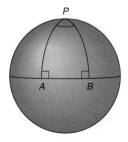

Figure 21.4. The angles at *A* and *B* are always 90°; so regardless of how narrow the spherical triangle *ABP* is, that is, how small the angle at *P*, the sum of the interior angles will always be greater than 180°.

the sum of the interior angles of this triangle must always exceed 180°, regardless of how skinny we make it. To see how the circumference or area of a circle drawn on a sphere differs from that drawn on a plane, imagine the figure resulting from a flat piece of paper folded over a hemisphere with no gaps or overlaps. It would look like that shown in Figure 21.5. Clearly, the area and circumference of a spherical circle of ra-

Figure 21.5.   You can make a spherical cap out of a flat piece of paper by cutting it as shown and gluing the edges together. The circumference and area of the spherical circle are less than those of a flat circle.

dius R are less than those of a flat circle of the same radius. So the geometry appropriate to a curved surface—in this case, a sphere—has properties quite different from those of the geometry of a flat surface, such as a blackboard.

Table 21.1 lists the results we have so far obtained. Flat (Euclidean) geometry has

Table 21.1.   *Different geometries in two dimensions*

| Space | Geometries in two dimensions | | |
	Flat	Spherical	Hyperbolic
Curvature	zero	positive	negative
Example	surface of blackboard	sphere's surface	saddle's surface
Geodesic	straight line	great circle	no common name
Number parallel lines	one	none	many
Sum of angles in triangle	180°	$>180°$	$<180°$
Area proportional to	r^2	$<r^2$	$>r^2$

zero curvature, whereas spherical geometry in which the surface curves back on itself (like the Earth's) is said to have constant positive curvature. It is not difficult to guess that there must be a geometry of negative curvature as well. Such a geometry is usually exemplified in two dimensions by the surface of a saddle. This curves outward in such a way that it never closes back on itself, but continues opening up to infinity. Such a space is said to be hyperbolic, and some of its properties are suggested in Figure 21.6.

You may have noticed that we have discussed the properties of two-dimensional curved spaces and not those of three dimensions. We can visualize two-dimensional curved space from a third dimension; we could not imagine three-dimensional curved space except from a fourth spatial dimension, of which we have no direct experience. However, the properties of two-dimensional surfaces can be extended to those of three dimensions intellectually, if not visually; this is the best we can do. For example, it is easy to see that on the surface of a sphere (a two-dimensional space; so the "interior" of the sphere has no meaning), there is no center and no boundary, yet the area of the surface is finite; that is, it has a perfectly definite value. (Recall the balloon model of the expanding universe we described in Chapter 19.)

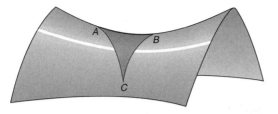

Figure 21.6. The surface at the center of the saddle-shaped figure is an approximation to two-dimensional hyperbolic space. The interior angles of the triangle at *A, B,* and *C* always add to less than 180°.

Though we cannot visualize them, the same properties apply to a three-dimensional, positively curved space—it has no center, is unbounded, yet has a finite volume. Similarly, a three-dimensional negatively curved space has no center and is unbounded, but has an infinite volume. (Remember that the saddle surface does not close back on itself. Such a surface does have a "center," however, in the sense that there is a unique point, a point different from all other points in the space. In this sense, it is an imperfect model for the hyperbolic geometry.) Asking what lies "outside" these curved three-dimensional spaces is not very useful, because that implies the existence of a fourth spatial dimension to reach the "outside."

As long as we are describing only small areas on the surface of the Earth, flat geometry gives accurate results. As the area becomes larger, flat geometry becomes a poorer and poorer approximation to the real state of affairs; finally, we must use spherical geometry. It's true that there are mountains and valleys, plateaus and canyons on the Earth's surface, but these are just minor ripples—small perturbations—superposed on the overall, nearly constant positive curvature of the Earth's surface. With this brief outline of the properties of various geometries in mind, we can return to the problem of the large-scale structure of the universe.

Geometry of the Universe

The obvious question is, what is the geometry appropriate for describing the universe as a whole, neglecting all the small perturbations in spacetime caused by galaxies and clusters of galaxies? In dealing with the universe, our ordinary notions of space and time are no longer appropriate, and we must think in terms of space and time merged together into spacetime (see Chapter 19). We can't continue to work with geometries of three-dimensional space independent of the fourth dimension, time. In the context of the universe, the geometries we have been describing might be thought of as three-dimensional projections of the four-dimensional structure of spacetime (just as a shadow is a two-dimensional projection of a three-dimensional object), or perhaps as relatively simple analogies to four-dimensional curved spacetime. Much of our discussion will of necessity be an approximation to the real situation. In any case, what we want to know is the curvature of spacetime appropriate to the universe as a whole.

Within relatively small distances (like several hundred million parsecs), the structure of spacetime in the universe might be very nearly flat (just as a small area of the Earth's surface is flat enough to be characterized by plane geometry). But what about the universe as a whole? If its spacetime structure has positive curvature, that implies that the observable universe is finite, that it has a definite volume; we would say that it is **closed**. But if the curvature is negative, the universe would be infinitely large, or

open. Knowing the spacetime curvature of the universe is equivalent to knowing its overall, global structure. Here is the connection between the geometry and the overall structure of the universe mentioned at the beginning of this chapter.

Geometry and Gravity

Next we want to know why one geometry is appropriate and not another. What determines whether the spacetime of the universe is positively or negatively curved or exactly flat? Since these geometries imply radically different structures (a finite or infinite universe), the curvature must have to do with a fundamental property of the universe. The answers to these questions are found in Einstein's general theory of relativity. This theory, completed by 1916, has to do with the nature of gravity. In this work, Einstein removed the restriction of special relativity, that is, of uniformly moving observers, and considered coordinate systems in arbitrary accelerated motion.

Recall Newton's first law of motion (Chapter 8) and apply it to a situation in which gravity is the acting force. According to Newton, a body at rest remains at rest and a body in motion remains in uniform motion in a *straight line* unless acted upon by a force, in this case, gravity. Einstein took a radically different point of view, namely, that gravity was not to be understood as a force. In effect, he would shorten Newton's statement to "a body at rest remains at rest and a body in motion remains in uniform motion in a straight line," with no mention of gravitational forces. So how does Einstein account for the motion of, say, the Earth around the Sun, which we have ascribed to the action of gravity? The key is in what Einstein meant by "uniform motion in a straight line." We have seen that geometries other than Euclid's are possible, and that they have different kinds of geodesics. What we call a "straight line" applies only to flat, zero-curvature geometry. If, for example, a region of spacetime were positively curved, then an object left to itself would travel in what Einstein understood by a "straight line"; that is, it would follow a path given by a geodesic appropriate to the curvature of spacetime in that region of space. The Earth's orbit around the Sun is such a geodesic, the particular geodesic appropriate to the circumstances of the Earth's formation—its velocity and location when it and the rest of the solar system formed.

Now what is it that determines the kind of geometry that prevails at any point in spacetime? Einstein's answer is that it is the *presence of matter and how it is distributed* that is responsible for the geometry at any point. Since mass plays the fundamental role in Newton's gravity (and in all but extreme circumstances his theory gives an accurate description of what is going on), it should come as no surprise that mass is fundamental to Einstein's view, also. In Newton's system, mass gives rise to the force of gravity, whereas in Einstein's picture, mass determines the geometry of spacetime. Gravity is replaced by curved spacetime. Spacetime empty of matter should be flat, but matter can deform spacetime to produce positive or negative curvature, depending on how that matter is distributed. Spacetime near a star is positively curved: the greater the mass concentrated in a particular volume, the greater the curvature of the surrounding spacetime.

It has been said that Einstein geometrized gravity; that is, he replaced gravity by mass-determined geometry. (Recall the Greek notion that geometry might be fundamental to understanding the nature of the universe!) These are wonderful ideas. If they seem strange, remember that the notion of Newtonian gravity acting over immense empty distances with enormous force is also strange. It has become "reasonable" after 300 years of successful use.

Earlier in this century the kind of mathematics Einstein found convenient to use in his theory was unfamiliar to most scientists (and to most mathematicians, too). It was primarily for this reason that general relativity acquired its reputation as a very difficult theory, understood—it was said in the 1920s—by only three people! Today, college students study it. Nonetheless, Einstein's formulation is much more complicated mathematically than Newton's, and we know that Newton's theory accurately describes an enormous range of physical situations involving interacting masses. When what we call gravity is weak, Einstein's theory reduces to that of Newton's and gives the same answers. Now, if general relativity theory merely gave the same results as Newton's, there would certainly be no reason to use it or perhaps even to ascribe to it a closer correspondence to physical "reality" than Newton's much simpler theory. The crucial requirement for any new theory is that it not only accounts for all known phenomena it attempts to represent, but also predicts new results that can be checked experimentally. Does general relativity predict any measurable phenomena that Newton's theory cannot? The answer to this is yes. Einstein suggested three tests of his theory, one of which explained a well-known discrepancy in gravitational theory, the other two of which involved previously unobserved phenomena. Let us consider these tests now.

Test of General Relativity

The Precession of Mercury's Orbit. The first of these tests has to do with the precession of the orbit of Mercury around the Sun. The orientation of Mercury's elliptical orbit does not remain fixed in space, but slowly rotates (or precesses) about 574 arcsecs per century (see Figure 21.7). Thus, in about 226,000 years, Mercury's orbit

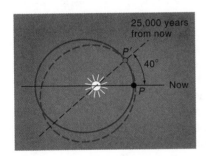

Figure 21.7. In 25,000 years Mercury's orbit rotates about 40°; that is, Mercury's closest approach to the Sun (its perihelion point) moves from *P* to *P'*.

will swing completely around once. Most of this motion—about 532 arcsecs per century—can be accounted for by Newtonian gravitational perturbations on Mercury by the other planets; the remainder—about 43 arc seconds per century—cannot. (Note how small this effect is: only 0.43 arc seconds in one year!) This led nineteenth-century astronomers to the not unreasonable conclusion that another planet must revolve around the Sun, disturbing Mercury's motion. Because such a planet, presumably far inside Mercury's orbit, would never appear more than a few degrees from the Sun, it would be hard to find. Its reality was such in the minds of some astronomers that it was even given a name, "Vulcan," appropriate to its proximity to the Sun. Vulcan was "discovered" on several different occasions by many astronomers, some of whom even saw more than one Vulcan! By the turn of this century, however, it had become apparent that the "discovery" observations were inconsistent with each other, and that in all likelihood the sightings were consequences of the difficulties of the observations combined with a certain amount of wishful thinking. Thus the discrepancy remained.

Einstein, however, showed that the missing 43 arcsecs per century was accounted for by general relativity, his theory of gravity. It is a result of the positive curvature of spacetime near the Sun and not a consequence of the gravitational disturbance of an unknown planet.[1] The orbits of Venus and Earth also show this effect but to a lesser degree, since they are farther from the Sun and so the curvature is less pronounced. Table 21.2 gives the observed and predicted values of the precession for Mercury, Venus, and Earth. Thus general relativity solved an old problem in planetary motion.

Table 21.2. Precession of the orbits of Mercury, Venus, and Earth

Planet	Predicted	Observed
Mercury	43.03 sec/century	43.1 ± 0.45 sec/century
Venus	8.6	8.4 ± 0.48
Earth	3.8	5.0 ± 1.2

PSR1913+16.   A recently discovered—and much more spectacular—example of the same effect is worth mentioning here. In 1974 a pulsar was found with the very short pulse period of 0.05903 seconds. It was soon discovered that its period varied in a systematic way, becoming shorter, then longer, then shorter again over about an eight-hour time span. The most natural interpretation of this behavior is that the pulsar is a member of a binary system, and that its variable period is caused by its orbital motion. As it approaches us, the pulsar period is shortened, and as it moves away, the period is lengthened. The orbit is quite elliptical, having an eccentricity of 0.617131. The size of the orbit can be calculated from changes in the pulse period, and has been found to be about the diameter of the Sun. Thus the companion cannot be a normal main sequence star, or else it would eclipse the pulsar. Hence the companion is either a white dwarf or another neutron star. Since the general relativistic effects depend on the masses of the interacting stars, these have been derived and are 1.441 and 1.387 solar masses for the pulsar and its companion, respectively. These are probably the most accurate stellar masses so far measured, apart from the Sun.

Since the objects in this system are moving in an intense gravitational field (note the small size of the orbit), we should expect to see effects produced by the strongly curved spacetime, and so we do. We have just seen that the perihelion of Mercury's orbit advances about 43 seconds of arc per century because of the weak curvature of spacetime near the Sun. The corresponding precession of the pulsar's orbit is 4.22662° per year (!), not accounted for in Newton's gravity, but in good agreement with that predicted by general relativity.

An even more interesting phenomenon is taking place in this system. The binary orbital period is slowly shortening, about 10^{-7} seconds per revolution. Since there are about three orbital revolutions each day and, as with a clock, the effect is cumulative, this small period change is easily detected and measured accurately after several years of observations. What causes the period to shorten? Just as accelerated electric charges give off electromagnetic waves, accelerated masses radiate **gravitational waves**.

---

[1] Since the geometry of space near the Sun is positive, the circumference of Mercury's orbit is actually slightly smaller than its length would be in flat space (see Table 21.1). Now, Kepler's third law is appropriate for flat space; hence in one sidereal period, the planet travels a distance that carries it a little more than once around the real orbit. Consequently, the perihelion point advances.

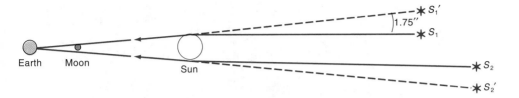

Figure 21.8. S_1 and S_2 are the true directions to the two stars. Light rays from the stars (which are very closely aligned with the Sun) will be deflected as they pass through the curved spacetime near the Sun, making them appear to be in the directions S_1' and S_2'. The maximum amount of the deflection, only 1.75 arcsec, is much exaggerated in the sketch.

These can be thought of as disturbances or ripples in spacetime. Since gravity is such a weak force, gravitational radiation is extremely feeble and difficult to detect; it remained an unverified prediction of Einstein's theory. This pulsar binary system has changed that, however. The rapid motions of the compact objects in this binary produce gravitational radiation at the expense of the orbital motion of the system. As a consequence of this energy loss, the period of revolution decreases. So far, this period decrease is taking place at just the rate predicted by Einstein's theory. We have yet to detect gravitational radiation directly, so this is the best indication we have so far that gravitational waves really do exist. Astronomy is a fruitful laboratory for the testing of general relativity.[2]

The Bending of Starlight.   Einstein's second test of his theory predicted that rays of starlight that pass close to the Sun would seem bent because of the curvature of the spacetime surrounding the Sun. (This bending is also predicted by Newtonian theory, but the amount is different.) Since light rays are often taken to define "straight lines," their apparent bending near the Sun, by the right amount, would constitute a particularly graphic confirmation of general relativity. Until recently, however, this was a difficult measurement to make, possible only during solar eclipses. If during a total solar eclipse there happened to be several stars located in nearly the same line of sight as the Sun, they could be photographed in the few minutes of darkness during totality. Some months later, after the Sun had moved along the ecliptic, these stars could again be photographed, this time at night. As shown in Figure 21.8, the stars closest to the direction of the Sun would appear to be shifted outward from the Sun in the eclipse photograph as compared to the night photograph. The predicted amount of this shift is extremely small, amounting at most to 1.75 arcsec for starlight just grazing the Sun. Starlight that passed the Sun at a greater angular distance would be bent less, because the spacetime through which it traveled would be less curved.

The first opportunity to make this test after the end of World War I came in 1919. It just so happened that the Sun was to be eclipsed when it was in the same direction as the Hyades star cluster. This was a happy circumstance, since the many cluster members greatly increased the likelihood of having suitable test stars. A British expedition to the eclipse path in West Africa observed the effect with just about the magnitude calculated by Einstein, 1.63 compared with the predicted 1.75 arcsec. Since to-

---

[2] Russell Hulse and Joseph Taylor were awarded the 1993 Nobel prize in physics for their discovery and analysis of this system.

Figure 21.9. The right-hand figure shows the cluster of galaxies (AC114), four billion light-years away, which is the gravitational lens for an even more distant galaxy. The left side of the figure shows the two lensed images, each surrounded by similar structure. The symmetry and identical colors of the two objects indicate that they are images of the same object. (The pair of objects close together in the center of the figure are most likely galaxies in the lensing cluster.) The wide separation of the lensed images indicates that the cluster has a massive core. Studies of such objects should give information concerning the dark matter contained within clusters of galaxies.

tality lasts only a few minutes and eclipse expeditions must generally operate with makeshift facilities, they provide wonderful opportunities to demonstrate Murphy's law (anything that can go wrong will go wrong). It was remarkable that on the first attempt to make these difficult measurements, things went as well as they did. Though the errors were rather large, the results were taken as confirming Einstein's theory.

The announcement of the eclipse results created a worldwide sensation. Einstein became a public figure, wined and dined by heads of state, easily the best-known scientist of the twentieth century. General relativity became a fashionable topic for discussion everywhere, in newspapers and magazines, at cocktail parties and academic seminars.[3]

Observations at several more eclipses did not succeed in much reducing the errors. Recent measurements of the more easily measured "bending" of radio waves that pass very near the Sun on their way to us from distant quasars, however, give the deflection predicted by general relativity to within about 1 percent. Not only has general relativity passed its second test, but an even more exotic example of the bending of light has been discovered.

In 1979 two QSOs were found in Ursa Major separated by only about six seconds of arc. Since QSOs are relatively rare objects, it was surprising that two should be found in so nearly the same direction. It was even more astonishing that the spectra and redshifts of the two objects turned out to be essentially identical. These circumstances led to the suggestion that what was being observed was not two QSOs, but a single object made to appear double by the action of a **gravitational lens**. An example is given in Figure 21.9.

---

[3] Interestingly, few experiments testing general relativity were performed during the next several decades. This led to the comment that it was one of the best known and least verified theories in physics. Its acceptance was due in large measure to its conceptual elegance.

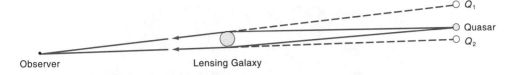

Observer   Lensing Galaxy

Figure 21.10. A massive galaxy will distort spacetime in its vicinity and cause rays of light from a distant quasar to be bent as shown. As a result, two images of the single quasar can be formed.

The notion of a gravitational lens comes from general relativity theory. Einstein felt it would never be possible to detect such an effect, one of the rare instances when his intuition failed him. The lens action occurs for the same reason that images of stars in the direction of the Sun are slightly displaced. In its simplest form it is easy to visualize, as you can see in Figure 21.10. Imagine a distant QSO, an intervening massive galaxy, and you, the observer, all in a line. Light rays from the QSO grazing the intervening galaxy are slightly bent by the curved space near the galaxy. Because of this bending, you see the two images slightly separated from each other. Several other such QSO pairs have now been found since the first one was discovered in 1979; gravitational lenses do indeed exist.

If the distant QSO and galaxy were exactly in our line of sight, and if the galaxy's mass were uniformly distributed so that it could be represented by a point mass, then we would observe a ring instead of two (or more) star-like images. Usually the gravitational "optics" are not exactly aligned, and in addition have "aberrations." Occasionally, however, nature cooperates. In 1987, the discovery of two luminous arcs found in a visible light survey of galaxy clusters was announced (Figure 21.11). Both

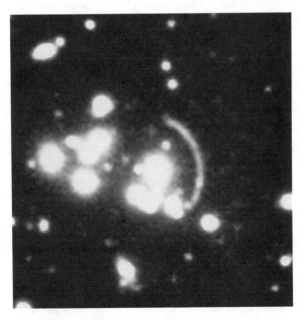

Figure 21.11. This visible light arc is the gravitational image of a distant galaxy formed by the cluster 2242-02. It is thought that dark matter plays a role in the lensing because just the visible matter in the cluster does not seem sufficient to form the image.

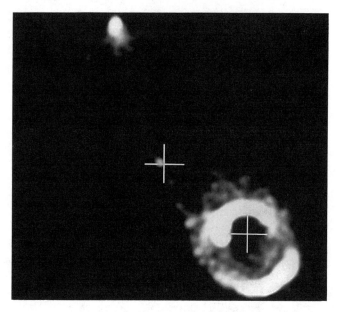

Figure 21.12.   The central cross marks the optical location of a distant ($z = 1.75$) radio quasar. It is a double-lobed QSO, the lower lobe of which has been gravitationally imaged (and brightened) into a ring by a foreground galaxy ($z = 0.254$) at the position indicated by the cross within the ring. The upper radio lobe is not affected by the galaxy. The two arcs form a nearly complete ellipse, indicating that the quasar lobe and the foreground galaxy are very nearly perfectly aligned. This 3.6-cm radio image was made by the Very Large Array.

are produced by gravitational lensing. Two years later a radio source was discovered that is in the form of an elliptical ring (Figure 21.12), a most unusual shape for a radio object. It seems likely that this too is an image formed by a gravitational lens.

This phenomenon is often observed with QSOs and very distant galaxies because, unless an object is extremely far away, there is little chance that another galaxy will be very nearly in the same direction and act as the gravitational lens.

Because the two light paths will not be exactly equal, changes in the QSO will not be seen at exactly the same time in its images. This time difference, which may be months or years, gives the difference in length between the two light paths to the observer. This path-length difference, along with the angular separation of the two images, can give the distance to the deflecting mass (the intervening galaxy) independent of the Hubble constant. Obviously, it would be extremely important to be able to calculate H in this way, independently of the redshift. As you might well guess, however, the imperfect geometry and nonuniform mass distributions of the lens galaxy complicate the interpretation. Nonetheless, there is some hope that eventually it may be possible to find distances to the deflecting galaxy in a few gravitational lens systems.

Recall that in Chapter 18, we mentioned MACHOs as one of the dark matter possibilities. Searches for these objects began in 1990 and make use of gravitational lensing. The idea is that on very rare occasions, a small object will move precisely across the observer's line of sight to a more distant star, causing the latter to become brighter by the gravitational lens effect. This brightening should be the same in all wavelengths, should produce a symmetrically shaped light curve, and should last for days to weeks for low-mass lensing objects. The increase in brightening can be large, by as much as two or three times, but the probability of occurrence is calculated to be extremely small because of the exact alignment required. Many millions of stars must be monitored every year if even a handful of events are to be detected. Obviously, a program of this sort requires fast computers and efficient data reduction programs. So far, a half-dozen events have been found that have the characteristics listed above. The lensed objects must be monitored to make certain they are not simply a new type of variable star, and many more events must be found involving a variety of types of lensed stars before positive identification of a MACHO can be claimed. It is remarkable that a recently verified exotic prediction of general relativity is already being exploited as an observational tool.

Gravitational Redshift.   Finally, Einstein's theory predicts the existence of an effect called the **gravitational redshift**. Radiation emitted by atoms located in a strong gravitational field (where the curvature is significant) will be shifted to the red as measured by a distant observer located in a weaker gravitational field. The greater the difference in the field strength between atoms and observer, the greater the redshift. The result looks like that produced by the Doppler effect when source and observer are receding from each other, but the cause is completely different. You can think of the gravitational redshift in the context of Newtonian physics by imagining that the stronger the gravitational field a particle finds itself in, the more energy it will have to expend to escape. If a photon is the escaping particle, its loss of energy is indicated by its radiation shifting to lower frequencies, that is, toward the red. This effect is small in the Sun (only about 0.01 Å) and difficult to observe in stars, whose absorption lines are intrinsically broad compared to the size of the effect. Furthermore, motions of the gases in stellar atmospheres produce their own Doppler shifts and confuse the result. Fortunately, a nonastronomical test for the gravitational redshift is possible. Radioactive nuclei of iron and cobalt emit extremely sharp, narrow lines of high-energy photons, gamma-rays, in fact. In 1959, the frequency of such radiation from cobalt was measured on the ground floor of a laboratory building and again on a floor only 70 feet above, where the curvature of space by the Earth's mass is very slightly less than that on the ground. A tiny difference in frequency was detected; it was within 1 percent of what was predicted by Einstein's theory. So the third of Einstein's predictions was validated.

The gravitational redshift is just a particular case of a more general result from general relativity, namely, that *any* periodic function (that is, regularly time-varying)—the ticking of a clock, for example—is slowed more in a strong gravitational field than in a weaker field. We have already referred to this effect when we described how a spaceship falling into the strong gravity of a black hole would appear increasingly redshifted to an observer a safe distance away; she would also see a clock on the ship slow down.

At present there exists no evidence suggesting that Einstein's theory is incorrect. It is more general than Newton's, able to account for phenomena that the latter theory cannot. Its replacement of gravitational force by spacetime curvature must be a deeper, more profound viewpoint. In circumstances in which gravity is stronger and stronger, Newton's theory becomes less and less accurate, and only Einstein's gives results that agree with measurements. Since these circumstances are not met in our ordinary experience, and since Newton's equation for gravity is so much simpler mathematically than Einstein's, we continue to talk about the "force of gravity" and use Newton's law for most purposes. But it is Einstein's theory that provides the basic framework for cosmological investigations, and it is to these that we now turn.

Observational Tests for the Geometry of the Universe

What tests can we perform to discover the geometry of the universe? Let us consider two of the several tests that have been proposed. One involves counts of galaxies, and the other the velocity-distance relation; both are simple enough in principle.

First, galaxy counts. Despite their clustering tendencies, let us assume that galaxies can be considered to be uniformly distributed if a sufficiently large volume of space is taken. Suppose we count N galaxies within a suitably large sphere of radius R. Next, count all the galaxies within a sphere of radius $2R$. Now, in ordinary flat space the volume increases as R^3, so we would have $8N$ galaxies within the larger sphere. If, however, fewer than $8N$ galaxies were found, that would mean that the volume increased less rapidly than R^3, which is a characteristic of closed, positively curved space. Conversely, if more than $8N$ galaxies were counted, the curvature would be negative, and the space would be open. Analogous situations occur in four-dimensional spacetime. Unfortunately, however, this apparently straightforward test can't be carried out, because so many galaxies are intrinsically faint and cannot be seen at great distances. As a consequence we cannot be sure that we have counted all the galaxies within even a few million parsecs, much less all those within the several hundred million parsecs that the test would require.

The velocity-distance relation itself should show the effects of the deceleration of the expansion if it could be extended to sufficiently large distances (see Figure 21.13).

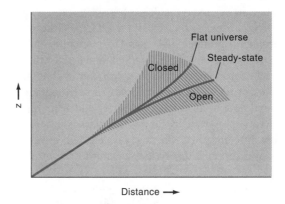

Figure 21.13. The velocity-distance relation for an open, closed, and flat universe. The differences become large only at great distances.

At present, measurements of velocities and distances of galaxies just barely reach such distances, but the possible errors in the distances of galaxies are so large that no firm conclusion can be reached. We are not now able to decide by observation which geometry represents the universe. Such a decision may become possible when larger space and ground-based telescopes become available.

Origin and Evolution of the Universe

The Big Bang Model

Now let us turn from cosmology, the structure of the universe, to cosmogony, its origin and evolution. For its evolution, there are only two possibilities: either the universe, in one or more of its basic attributes, is changing with time, or it isn't. That it is evolving comes from a simple interpretation of the redshift, namely, that the universe is increasing in size, so that its average density steadily decreases as time goes on. Furthermore, recall that the value of the Hubble constant gives the maximum time (about 13 billion years for $H = 75$ [km/sec]/Mpc) that has elapsed since the universe began to expand from an unimaginably dense state. In such a Big Bang model, the expansion has only one of three possible outcomes: either it will continue forever, although at an increasingly slower rate; or it will slow down and, after an infinite time, stop; or it will not only slow down, but eventually stop, reverse itself, and collapse. If either of the first two possibilities is correct, and the scale of the universe continues to increase, stars, having used all their energy stores, will eventually burn out and fade away, and the luminosity of galaxies will steadily decrease. The visible universe will eventually disappear, leaving behind only cold matter and very low energy radiation. We can refer to such a model as a **one-shot** universe.

On the other hand, if the last possibility is correct and the expansion will eventually reverse itself, the presently observed redshifted universe will be followed by a blue-shifted phase. Ultimately the universe will return to an ultrahigh-density state in which all substructures such as galaxies, stars, and planets will be destroyed, and all matter and energy will exist only in its basic, primordial forms. Enthusiasts for this scenario then speculate that, since the universe is now apparently expanding from such a state, it would do so again. They suggest that the universe is a gigantic oscillating or **pulsating** system going through an unending series of Big Bangs, expansions, collapses, and Big Bangs. There is no physical basis for this notion, however.

The connection between structure and evolution becomes apparent now. If the universe expands forever, it is open, and its spacetime curvature is negative. If, after an infinite time, the expansion finally stops, then its curvature is zero, and spacetime is flat. A pulsating Big Bang universe expands to some finite size, only to collapse again after a very long time. It is closed, and its curvature is positive.

What determines whether the universe is pulsating or is one of the one-shot systems? The only force capable of stopping the expansion is gravity, the strength of which depends on the amount of mass present in a given volume, that is, on the average density of the universe. If the density is large enough, the universe will attract itself enough to stop the expansion eventually (or more accurately, to make the curvature of spacetime positive). If not, the expansion, though slowing, will continue forever or will, after an infinity of time, come to a halt.

A crude analogy may be helpful here. Throw a ball up into the air; as it rises, it will slow down, stop, then fall back to Earth (Figure 21.14). Now imagine that the ball is thrown with greater than the escape velocity; even though it is always slowing down, it will never stop rising. If it is thrown with exactly the escape velocity it will ultimately (after an infinitely long time) come to a halt. The first case (the ball falling back) is analogous to a closed universe, the second to an open universe, and the third to a flat universe. The connection of these analogies with the density of the universe

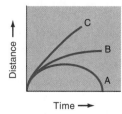

Distance →

Time →

Figure 21.14. The paths of a ball thrown upward with speed less than, equal to, and greater than the escape speed are shown in curves (A), (B), and (C), respectively. In (A), the ball falls back to Earth, in (B) it just barely escapes from Earth, and in (C) it escapes with plenty of energy to spare. Case (A) corresponds to a closed universe—it will ultimately collapse; (B) to an expanding universe that will slow and, after an infinite time, stop; (C) to an ever-expanding universe.

is straightforward. If the density of the Earth were larger (and so its mass greater), in the first instance the ball would fall back to the Earth sooner, and in the second and third, a larger velocity of escape would be needed.

How much mass is just enough to ultimately stop the expansion; that is, what is the so-called **critical density** of the universe? Imagine all matter—stars, clouds, and galaxies—spread uniformly throughout the universe. What the average density of this matter must be so that the universe is just marginally closed is uncertain; it depends on the value of the Hubble constant. The current, tentative answer is about 10^{-29} gm/cm^3, or about one hydrogen atom in each cubic meter of space. This is an incredibly small density; a volume equal to that of the Earth would contain about 0.001 grams of matter! Nevertheless, if the average density is greater than this, the universe is closed; if less, it is open. The observations are extremely difficult, and the present evidence is skimpy. Estimates suggest that the critical density is from 10 to perhaps 30 times greater than the density we measure (excluding the dark matter), indicating that we live in a one-shot universe. Recall, however, that we don't know the masses of galaxies very well, because we don't know how much dark matter they might contain. Thus, we don't really know the average density of the universe and hence, its future.

The Steady-State Model

Before 1965, that is, before the discovery of the cosmic background radiation, an approach quite different from the one just described could be taken. Though this alternative theory has been discredited, it is instructive to follow it for a moment. The idea was this: so far we have assumed that the universe is evolving, but suppose that on the average the universe does *not* change with time. What then? Where will that idea take us? Given the expansion of the universe, this may sound like a crazy notion, but perhaps it is not as crazy as it seems. Recall that in discussing the nature of the universe, we have assumed the cosmological principle, namely, that the part of the universe we are able to investigate is typical of the universe as a whole. In the 1950s some cosmologists suggested that we extend this principle. They postulated that not only is the average appearance of the universe independent of the particular place from which we examine it, but that it also is independent of *when* we look. That is, they suggested that we are located neither in any special place nor at any special time in the history of the universe. This assumption they called (with medieval accent) the **perfect cosmological principle**.

The consequences of such an assumption are interesting. Obviously, it implies that the universe does not, on the average, change with time. But everyone agrees that the universe is expanding, that the scale of the universe is constantly increasing. Hence, the average density of matter in a large volume must be steadily decreasing with time, apparently contradicting the perfect cosmological principle. Therefore, to accept the principle, one must postulate that matter is **continuously created** at such a rate that the average density remains constant in time. By this constant creation of new matter, the universe is continuously replenished and maintained in a steady state, one that is on average unchanging with time. Appropriately, this was called a **steady-state** model, in basic contrast to the evolving Big Bang model.

When the steady-state model was proposed, the cosmic background radiation had not been discovered, and so the model could not be dismissed out of hand. Instead one had to ask what it required of nature. How fertile must space be to satisfy the perfect cosmological principle? The rate at which new matter must pop into existence to maintain a constant average density for the universe turned out to be very small by terrestrial standards. Only about one new atom every few thousand years must appear in a volume equal to that of a medium-sized classroom. This is completely undetectable. Thus the principle that matter (or, more accurately, mass-energy) is neither created nor destroyed would still be valid on the local scale, the only circumstance under which we can test it.

Adopting the perfect cosmological principle meant that the universe had no beginning and would have no end. Since the steady-state advocates assumed that the universe did not change with time, the problem of origin was simply postulated away. It had always existed in the past, and would always exist in the future with, on average, about the appearance it has now.[4] But suppose we protest: this picture of a universe expanding yet remaining essentially unchanged because of the continuous creation of matter out of nothing is just too bizarre to contemplate! The supporters of the steady-state model asked, with considerable justification, if that is any stranger than the idea that once upon a time the entire mass and energy of the universe appeared in a blinding instant.

In discussing the nature and history of the universe and other such modest topics, "common sense" is not only suspect, but is often likely to be completely misleading as any sort of guide. Instead, any theory must stand or fall by its predictions; is it or is it not verified by the observational evidence? The predictive power of a theory is crucial. If it cannot be tested (at least in principle), it is not a scientific theory.

Observational Tests

Could any observational tests have been made that would have enabled us to distinguish between the Big Bang and steady-state models, between an evolving and a non-evolving universe? In principle, yes. What must be done is to compare some property of nearby galaxies with the same property in a number of increasingly distant galaxies. If this characteristic changes *systematically* with distance (and therefore with time), then this would be evidence that the universe evolves. For example, suppose we knew how to tell the age of a galaxy. On the Big Bang picture, galaxies farther and farther away from us would appear to be younger and younger. According to the steady-state model, however, any large volume of space would contain a mixture of old, young, and middle-aged galaxies, because galaxies would be constantly forming out of the newly created matter. Thus, there would be no systematic change of galaxy age with distance. Unfortunately, we did not know how to measure such a useful property as the age of a galaxy (the detection of age-related changes in normal galaxies is only now possible), and so this test could not be carried out.[5]

Though the steady-state model predicted a different form of the redshift-distance relation, observations cannot be made accurately enough to distinguish among models. Even a measurement of the average density of the universe would not have distinguished between the open, one-shot, Big Bang model and the open, steady-state model. Thus, before the discovery of the cosmic background radiation, the available evidence did not make it possible to decide even between an evolving and a non-evolving universe. It was hard to take cosmological speculations seriously.

This situation changed dramatically after the discovery of the background radiation. The steady-state theory does not predict it. Many attempts were made to account for it in the context of the steady-state model, but none of these were convincing. It soon became clear that the 3 K radiation could arise only from a hot Big Bang and that

---

[4] In Big Bang cosmology, the concept of the observer's horizon is meaningful. That is, we can see back only as far as light has traveled since the expansion began. In a steady-state universe that has always existed, however, the horizon has no meaning.

[5] Today, the fact that most QSOs are far away and few nearby indicates evolution.

the universe must be evolving. In addition, though the abundances of heavy elements could have been accounted for in the steady-state model, the theory would have had difficulty accounting for the absence of large changes over time in the hydrogen and helium abundances.

The Current Picture

So where do we stand now? It is clear that we live in an evolving universe. The hot Big Bang model seems inescapable. The 3 K radiation not only arises naturally in the model, but also enables us to calculate the physical conditions in the early universe back to within a fraction of a second of its beginning. Knowing these conditions allows us to predict that helium was produced with its observed abundance within a few minutes of the beginning. In addition, it enables us to calculate the density of ordinary matter in the universe. We must remember, however, that dark matter of as-yet-unknown composition makes up most of the mass of the universe.

Note also that our three fundamental cosmological clues give us direct information about the universe at three quite different epochs. Big Bang nucleosynthesis takes place from the first seconds to the first minutes; the 3 K radiation tells us what the universe was like a few hundred thousand years after the beginning; Hubble's law describes the universe from one or two billion years later to the present. All of this provides convincing support for the Big Bang model from the first seconds to the present. It is not yet a complete theory, however. Many aspects are unsatisfactory and it is certain that various features of the model will change. Let us look now at the shortcomings of the Big Bang model.

Problems with Big Bang Cosmology
The Flatness Problem

The fact that the observed average density of the universe is already fairly close to the critical density and that nonluminous matter would bring it even closer to critical density is rather curious. After all, the critical density is just one of the infinite number of densities the universe could, in principle, have. Physically, one might think that it is the most improbable value, because it is the fine dividing line between continuing expansion and ultimate collapse, between a finite and infinite life, the balance point of the universe, so to speak.

The situation is even more striking, however. Earlier in this chapter we used, as analogies for a closed, flat, and open universe, a ball thrown vertically into the air with a speed less than, equal to, or greater than the velocity of escape, respectively. Now imagine a ball thrown upward with only 0.1 the escape speed. You can guess that after a given time it will be moving with only a small fraction of its initial speed; it will quickly decelerate. Next let our strong-arm pitcher throw three balls up with 0.99, 0.999, and 0.9999 the escape speed. After the same length of time the first ball will not have slowed much at all, but will still be moving with a very large fraction of the escape speed; the second ball will have slowed even less, and the third even less than that.

What does this have to do with the universe? Well, after a very long time (15 billion years or so), the universe is still expanding at a good fraction of its escape speed (since the average density is near the critical density). It has not slowed down much at all, which effectively rules out a strongly closed universe. Instead, the universe is acting like one of the last three balls, so that it must have started with very nearly the escape speed. In fact, if you extrapolate back into time closer and closer to the beginning, you find that the speed gets closer and closer to that appropriate for a flat universe. So unlikely as it may seem (and assuming that there isn't something wrong with these arguments), it is possible that the initial expansion speed exactly equaled the escape speed. (Remember that just as with the earlier ball analogies, there is a close connection between the escape speed and the density of the Earth—or universe; to a critical density there corresponds a critical speed.) A corresponding conclusion is that the density of the universe is nearly exactly equal to the critical density; according to this argument, the universe is flat!

So can we understand why the early universe apparently emerged from the Big Bang with its geometry so nearly Euclidean, closely balanced between expansion and collapse? Or was this just the way it happened to be, what is called an "initial condition"? Conventional Big Bang cosmology cannot provide a fundamental answer to this **flatness problem**, and must assume it as one of the initial conditions under which the universe was born.

The Expansion Age of the Universe

The conclusion that the universe is flat or nearly so has a troubling implication, however. As you know, the reciprocal of the Hubble constant gives the maximum time since the expansion began; maximum because the fact that the rate of expansion must have been slowing since the beginning is ignored in this simple calculation. To calculate a more accurate age, we can no longer ignore this and must consider the age question in the context of a particular model of the universe.

The amount by which the universe has been decelerating depends on the amount of mass it contains, which is expressed as its average density. Remember that the critical density is such that if the average density of the universe is greater than the critical density, the universe is closed; less and it is open; equal and it is flat. A useful parameter is the ratio of the average density to the critical density, usually given by the Greek letter Ω (omega).

Figure 21.15 shows the look-back times calculated for three values of the Hubble constant and two values of Ω. When $\Omega = 1$, the density of the universe equals the critical density. Values greater than 1 correspond to a closed universe, and Ω values less than 1 correspond to an open universe. Note that the curves all flatten out at large redshifts so that the times given for $z = 40$, for example, are essentially the expansion ages. If, as we have assumed, the Hubble constant really is about 75 kilometers per second per megaparsec, then the maximum age of the universe is about 13 billion years ($\Omega = 0$). But since it seems likely that Ω is close to 1 (that is, the universe is near the critical density), then the expansion age of the universe is well below 10 billion years. Recall, however, that stars in globular clusters are much older than this, perhaps 15–18 billion years old. How can this discrepancy be resolved?

Globular cluster stars are difficult to model accurately and some uncertainties in the theory of their structure and evolution still exist. For example, their metal abundances, critical to their opacities, are rather uncertain. Nevertheless, it seems unlikely that their ages could be in error by 5 to 8 billion years. A smaller Hubble constant of, say, 50, with $\Omega = 1$, would imply an age of about 13 billion years. This just might be compatible with stellar ages if the latter are too large by only a few billion years. Other possibilities are that the Hubble constant is even less than 50, or that the cosmological models for which the look-back times in Figure 21.15 were calculated are too simple. It will be interesting to see how these apparently contradictory ages are finally resolved.

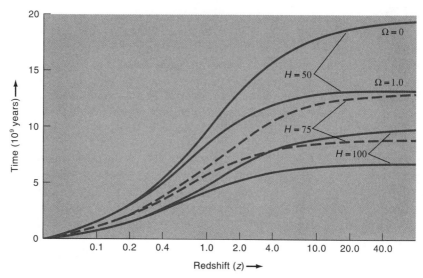

Figure 21.15. Look-back times calculated for $H = 100$, 75, and 50 km/sec/Mpc, and for $\Omega = 0$ (an extremely low density universe) and $\Omega = 1$ (the universe at critical density). For each value of H, the upper curve corresponds to $\Omega = 0$, and the lower to $\Omega = 1$. The $\Omega = 0$ curves for large z (where they flatten out) give the expansion ages obtained simply by taking the reciprocal of the Hubble constant.

The Horizon Problem

Another difficulty with the standard model is that of the isotropy of the background radiation. Its remarkable uniformity over the whole sky presents a real problem. Matter and energy can travel no faster than the vacuum speed of light. This means that if two regions of space are to interact, to be thoroughly mixed with each other (so that, for example, they have the same temperature), they must be separated by no more than the distance that light can travel during the age of the universe at the epoch in question. They must be within each other's reach, within each other's "horizon." If they aren't, they will not interact and smooth out their temperature differences. At the present epoch, our horizon is about 15 billion light-years, so two regions in opposite directions in the sky are separated by 30 billion light years and could not have been in contact with each other when matter and radiation went their separate ways. In fact, when we look now at two patches of the background radiation more than one or two degrees apart, we are looking at two regions that never "knew" about each other. It is difficult to understand how they have the same temperature to a precision of a few parts in ten thousand, or, even more striking, how radiation from the whole sky has the same temperature. To assume as yet another initial condition that the temperature was uniform from the beginning seems unsatisfactory and even artificial. But the ordinary Big Bang theory provides no explanation for this **horizon problem**. Could a solution be found in conditions that existed before Big Bang theory is applicable?

Galaxy Formation Problem

Although it is not clear that this is a Big Bang problem per se, we want to know why galaxies exist at all, which is the **galaxy formation problem**. This is the other side of the horizon problem, in that we must account for today's obviously lumpy distribution of matter in the universe—stars, galaxies, and clusters of galaxies. Recall that for the first half million to million years after the beginning, matter and radiation interacted so strongly that matter could not form permanent clumps. Consequently matter was as smoothly distributed as the radiation. After matter and radiation no longer interacted, a very uniform distribution of matter had somehow to break into clumps that could eventually become galaxy size. It is possible to understand how this might have occurred. Over the course of time, atoms would randomly come close enough together to capture each other gravitationally. A few of these small clumps would continue to grow and, as they became larger, they became more efficient in capturing new matter. Finally, a mass large enough to evolve into a galaxy would be formed. The trouble is that this mechanism takes too much time. In fact, calculations show that the universe has not existed long enough to form any galaxies by this process. Without some initial clumpiness, matter won't form into galaxies in the time since the Big Bang, but distant QSOs tell us that galaxies formed quite soon after the beginning. So either some unknown mechanism must be operative, or the universe must have formed with a certain lumpiness built in, another initial condition for the Big Bang. But again, perhaps something happened before the Big Bang era that could provide the necessary initial clumping.

The Matter-Antimatter Problem

The last problem with the conventional Big Bang model is that it cannot account for the existence of matter itself.[6] The world around us is made up of ordinary matter, for example, negatively charged electrons and positively charged protons. But we have seen that another kind of matter, called **antimatter**, can exist. In this form electrons are positively charged, and protons are negatively charged. Antimatter is just as legitimate as matter, and though it has been found in cosmic rays (positive electrons or positrons) and has been produced in high-energy particle accelerators, it is extremely uncommon. Particles of matter and antimatter cannot exist in close proximity, because they would completely annihilate each other, transforming all their mass into two high-energy photons, two gamma-rays, with their energy given by the formula $E = mc^2$. Consequently, large amounts of antimatter could exist elsewhere only if strictly segregated from ordinary matter, otherwise we would detect a large number of gamma-rays produced by the destructive interaction of the two kinds of matter. Such a strict segregation seems most unlikely, especially eons ago, when the universe was much smaller than it is now. In all likelihood, antimatter does not exist with appreciable abundance anywhere.

But why should the universe consist of matter to the exclusion of antimatter? It would seem much more natural that the universe would have been formed with equal quantities of the two kinds of matter. Had that happened, however, we would not be here, since all the matter and antimatter would have annihilated, leaving only high-energy photons behind. Therefore, there had to be an imbalance in the two kinds of matter from the very beginning of the universe. Now, that imbalance need not have been large. Calculations show that if for every 1,000,000,000 particles of antimatter, there were 1,000,000,001 particles of matter, that would have been sufficient to produce the matter world we live in! Again, Big Bang cosmology cannot account for even this tiny imbalance, but must assume it. Putting it another way, the imbalance must have been produced before standard Big Bang theory provides a good description of the universe, that is, sometime before about 0.001 seconds after the beginning.

---

[6] Another way of putting this is to say it can't account for the billion-to-one ratio of photons to protons and neutrons, found from the nucleosynthesis epoch.

Thus several challenging deficiencies of the Big Bang model remain. The only "solution" so far has been to assume them away, that is, to include them as part of the prevailing initial conditions. This is equivalent to throwing up your hands and saying, "Just because, that's why!" when insistently asked a question by an inquisitive child; it is not the most satisfactory answer. Various ways out have been proposed. It should not come as a surprise that basic physical laws are involved, but it is striking that the answer might involve an intimate relation between the properties of the universe—the largest physical entity—and those of the fundamental particles—the smallest physical units. In other words, the macrocosm and the microcosm may be closely related.

Can the Big Bang's Problems Be Solved?
The Fundamental Forces

To address this question, we must first take what might seem to be an unlikely detour, and briefly discuss the nature of the basic forces of nature. We have seen that there are only four fundamental forces: the so-called strong and weak forces having to do with the atomic nucleus, and the electromagnetic and gravitational forces. Their relative strength and the ranges over which they act differ enormously, as is shown in Table 21.3.

Table 21.3.   Relative strengths and ranges of the fundamental forces

Force	Relative Strength	Range
Strong nuclear	100	about 10^{-13} cm
Weak nuclear	10^{-12}	about 10^{-15} cm
Electromagnetic	1	infinite ($1/R^2$)
Gravitation	10^{-37}	infinite ($1/R^2$)

The strong force overcomes the huge electrical repulsion between the protons (the electromagnetic force) in a nucleus, and binds the protons and neutrons together. The weak nuclear force governs the decay of the neutron, interactions involving neutrinos, and the spontaneous emission of particles such as electrons from nuclei, that is, radioactivity. Both of these forces have very short ranges, the strong force acting only over a distance equal in size to an atomic nucleus; the weak force has an even shorter range. Beyond their ranges, these forces fall off to zero very quickly.

In marked contrast, the electromagnetic and gravitational forces act over infinite distances, since their strength decreases inversely as the square of the distance from the charge or the mass. The electromagnetic force binds electrons to positively charged nuclei to form atoms, which can then combine into molecules and people. It determines chemistry, biology, the electromagnetic spectrum, electronics, etc. Gravitation, though by far the weakest of the fundamental forces, keeps us securely tied to the Earth, determines the motions of planets, stars, and galaxies, and the structure of the universe as a whole. It, rather than electromagnetism, rules the universe, because cosmic objects overall are electrically neutral and so experience no net electromagnetic force. Putting it in another way, there is just one kind of mass, so gravitation is an attractive force only, and always makes its presence felt. By contrast, there are two kinds of charges, so the force of electromagnetism can be attractive or repulsive, and the charges often cancel each other.

These four forces ultimately account for all the interactions that produce the incredible variety of objects and phenomena in the universe. Realize that, by "all," we really mean *everything*, from the flight of a spaceship to the plodding of a turtle, from the explosion of a hydrogen bomb to that of a firecracker, from the radiation produced by a star to that produced by a light bulb. This is a remarkable fact, and one that you should think about. A very minor sidelight of this result is that one of the many reasons that scientists dismiss astrology or the claims of mental-energy spoon benders, for example, is that such phenomena fall outside the action of the four basic forces; another force would be required and none is known. (Admittedly, this argument is akin to using dynamite to kill a fly; astrology and other such aberrations fall apart under much lighter artillery.)

Scientists have, of course, speculated about the possibility of other forces, for example, one that is repulsive and operates only over very large distances. Such a force might be responsible for some fraction of the galaxy redshifts we observe, so that the idea of a Big Bang would have to be modified. The origin of the 3 K background radiation and of the time-independent helium abundance would then have to be accounted for in some way other than in the hot Big Bang, however. No convincing evidence for a fifth force has so far been found.

Unifying the Forces

One of the basic goals of physics is to invent a *single* theoretical law from which all four forces can be shown to arise. That is, physicists hope to "unify" the forces, to find a theoretical approach or structure that encompasses all the forces, sometimes called a "theory of everything." Given the widely different characteristics of the four forces, this might seem an unlikely prospect. In the late 1960s, however, two of these forces, the weak and the electromagnetic, quite different in their characteristics, were shown to be two manifestations of a more general "electroweak" force. Actually, this was not the first time that such a basic unification had taken place. The first was by Newton, who showed that Aristotle's two kinds of physics—celestial and terrestrial—could be combined into one, that of universal gravitation and Newtonian mechanics. In the nineteenth century another basic unification was made by the Scottish physicist Clerk Maxwell, who showed that electrical and magnetic phenomena and light could all be understood as various aspects of the electromagnetic force.

Recent work in theoretical physics indicates that under conditions of very high energy (or equivalently high temperature), the three nongravitational forces would become unified, that is, arise from a single interaction. First, the electromagnetic and the weak force are unified, then at higher temperatures, the electroweak and the strong nuclear force are unified (see Figure 21.16). The three forces we see today are the low-energy manifestations of the single force, according to these ideas. The most powerful accelerators in the world are able to produce particles with energies great enough to demonstrate the first unification, that is, the electromagnetic and weak forces into the electroweak force. Predictions of the theory have been experimentally verified, so that this unification is reasonably well understood. The next step, combining all three nongravitational forces, requires a billion times as much energy, and so is completely beyond the possibility of laboratory accelerators. Thus, the relevant formalisms, called **grand unified theories** (inelegantly abbreviated to **GUTs**), have not yet been shown to be correct. (Note the plural; GUTs refers to a class of theories of which one might be correct.)

Grand unified theories make a startling prediction: protons aren't forever. That is, after about 10^{31} years (according to the simplest version of the theory), protons will

538

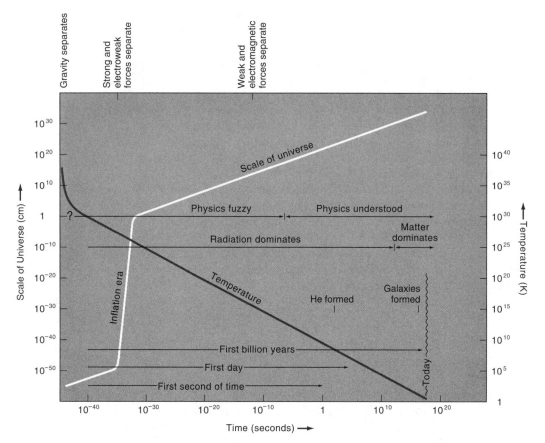

Figure 21.16. A "history" of the universe from about 10^{-40} seconds after the Big Bang to the present. Both the temperature (right scale) and the increase in the size of the universe (left scale) are given as they change with time. Note how much happened within the first second after the beginning. "Physics fuzzy" means that our understanding of what is happening becomes increasingly uncertain as one goes back in time.

spontaneously decay into other particles. The material world will no longer exist. Can this apocalyptic prediction be tested? Obviously, we have to do better than simply wait for the predicted length of time. If 10^{31} protons are monitored, however, then on the average, one should decay every year or so. A thousand tons of water contain enough protons to increase the expected number of decays to a dozen or so per year, a detectable number. This is equivalent to saying that sometime in a whale's lifetime, one of its protons might decay.

Experiments have been underway for several years to test GUTs in this way, but so far no evidence of proton decays has been reported, indicating that the proton lifetime must be longer than 10^{33} years. This eliminates the simplest versions of GUTs, but not the general approach. (Incidentally, it was to search for proton decay that the U.S. and Japanese detectors that recorded neutrinos from the 1987a supernova were set up.)

What about gravity, the force that "governs" the universe? Was it ever combined with the other three forces so that the "dreams of a final theory" of the forces might be realized? Current speculations predict that it could have been, but only at energies thousands of times greater than those required for the GUTs unification of the electroweak and strong forces. At the moment, this last unification is based more on hope than anything else and requires that theorists go beyond GUTs. That is, the three nongravitational GUTs forces are described in terms of quantum theory in which energy, for example, is not a continuous quantity but is made up of discrete packets. So far, however, no one has found a way to describe gravitation as a quantum force; it remains fundamentally different from the other three. Consequently, a theory (much less its experimental verification) in which all four forces arise from a single basic force has not yet been invented.

How might these theories be tested eventually? Though the high energies necessary to test GUTs (much less the final unification with gravity!) cannot be produced in the laboratory, they can be found in nature. For example, extremely high energies (corresponding to a temperature of about 10^{28} K!) did exist in the very early universe, about 10^{-35} seconds after the beginning. At this temperature, the GUTs unification would have taken place. Even earlier, 10^{-43} seconds after the beginning, the temperature of the universe was 10^{32} K, high enough for the final unificaton with gravity to have occurred.

Thus, the earliest instants of the Big Bang may provide us with a sort of laboratory in which the properties of the basic forces of the universe can be investigated. This in itself is a remarkable development. The nature of the fundamental forces, which involves the properties of the basic particles and their interactions (including those that produced the dark matter particles), apparently plays a crucial role in determining the properties of the universe.

The Inflationary Universe

The circumstances by which the strong nuclear force decoupled (separated) from the electroweak force as the very early universe cooled led to an interesting new idea in cosmological theory, that of a brief but critically important **inflationary** phase in the evolution of the *very early* universe. If this phase occurred, it did so before the time when the standard hot Big Bang model became an adequate description of the universe.

The idea is this. After the very young universe expanded and cooled to the point that the gravitational force manifested itself as a force separate from the other three, the universe entered the GUTs era. During the next decoupling—the strong from the electroweak force—a sort of phase transition is supposed to have occurred. Familiar examples of phase transitions are steam condensing to water (gas phase to liquid phase) or water freezing (liquid to solid phase). In order for water to cool to ice, it must get rid of a considerable amount of energy, its so-called latent heat. Under certain conditions, water can become supercooled, that is, it can cool below its freezing point without turning to ice. Then very suddenly it freezes, quickly releasing its latent heat.

Suppose that the phase transition associated with the strong and electroweak decoupling did not take place as soon as was possible (when the temperature was about 10^{27} K), but was delayed. The temperature continued to drop and the universe entered a supercooled state, called the false vacuum. This is the lowest energy state possible, and even as the universe expanded the energy of the false vacuum did not decrease, but remained constant. Several very strange things happened as a consequence. Gravity, instead of being an attractive force, acted like a strongly repulsive one, causing a huge and very rapidly accelerating expansion of the universe to occur. About every 10^{-34} seconds, the scale of the universe doubled. Very quickly it inflated by the enormous factor of about 10^{50}, growing from 10^{-36} the size of a proton to the size of a softball! This is an incomparably greater increase than could have resulted from the Big Bang. To get an idea of just how huge this expansion is, a hydrogen atom expanded by a factor of 10^{38} would be the size of the presently observed universe, and yet the proposed inflation was 10^{12} times greater than that! (Note that the inflation is of space itself, rather than of matter through space, and so is not limited to the speed of light.)

At about 10^{-32} seconds after the beginning, the universe left the supercooled state and the delayed phase transition occurred. As a result, the latent heat of the transition was released and triggered the formation of a vast flood of hot particles that filled the universe. All the matter and radiation we see today originated then. The attractive gravitational force was reestablished. The expansion of the universe returned to the much more leisurely Big Bang rate, a consequence of all the momentum produced during the short inflationary period.

If the inflation lasted for just the right amount of time at the right rate, several problems with the conventional Big Bang model could be solved. We would not have to invoke special initial conditions or some unknown processes; instead, the circumstances would have developed naturally from GUTs and the inflationary phase. For example, GUTs predict that, just after the strong and electroweak forces become distinct, 10^{-35} seconds after the beginning, certain particles will decay in a way that produces the very slight matter-antimatter imbalance to which we and the rest of the material universe owe our existence. This particle decay has at its root the same cause as the decay of the proton mentioned earlier.

Inflation solves the horizon problem presented by the smoothness of the 3 K radiation. Since the present universe would have evolved from a *much, much* smaller region than it would have in conventional Big Bang cosmology, all locations could have been in communication with all other locations at the very early time before the inflation took place. Hence temperature differences would have been smoothed out then and would have remained so afterward. The flatness problem would also be solved, because the huge expansion would have made the observable universe appear flat, just as a small area on Earth appears flat. That is, after spacetime expands by a factor of 10^{50}, it doesn't matter what curvature it started out with, because it will appear flat. In other words, inflation predicts that the average density of the universe equals the critical density. (Remember the nagging question of the age of the universe, however.)

Finally, it appears that during the inflationary period, extremely small fluctuations could be produced that would eventually lead to galaxy formation. Interestingly, inflation apparently produces both the overall uniformity seen in the cosmic background radiation and the departures from that uniformity that produce the clumpy universe we see around us!

Though inflationary ideas seem promising, we must remember that so far a completely satisfactory model has yet to be produced, much less verified. The theory seems to require a lot of "fine tuning" to give satisfactory results. Maybe inflation will work out, or maybe a flaw will be discovered that will force us to discard these ideas and turn to completely different approaches. Right now we just don't know.

But look at what has happened in this century. Around 1900 most astronomers felt that the universe was a relatively small, static structure. The state of astronomical observation and physical theory were such that scientific speculation about the origin and evolution of the universe was essentially impossible. Within just a few decades, however, we were beginning to realize that we live an enormous, dynamic universe that, by mid-century, a few astronomers and physicists were attempting to model. These models were of both the Big Bang and the steady-state varieties, because we could not tell whether, on the large scale, the universe was evolving or not. The discovery of the cosmic background radiation, however, resolved that question, and allowed us to understand how helium was produced during the first few minutes after the beginning.

An understanding of the history of the cosmos earlier than this seemed beyond our reach, however.

Only a couple of decades ago, few would have dreamed that it would be possible to ask questions about the possible nature and evolution of the universe only 10^{-36} seconds after its beginning. Even more remarkable, today it does not appear to be crazy to ask how the universe came into existence in the first place. In fact, scientists are attempting to construct mathematical models in which the universe is created out of literally nothing—no matter, no space, no time! (This has been called the ultimate free lunch.) Where all this will lead is unclear. Perhaps a deeper understanding of theory or a completely unexpected observation will force us to give up many of these new ideas and take some completely different tack. Nevertheless, it does appear that we are developing at least some of the tools and basic data, so that we can ask, if not yet resolve, the most basic questions possible about the universe.

An obvious lesson from all this is that it is extremely risky to say that we have come hard against a wall that bars us from further understanding of the world in which we live. In particular, if we base a philosophical or religious position on today's ignorance, we must be prepared to abandon that position tomorrow.

Terms to Know

Dimension; geodesic; flat, positive, and negative curvatures; open and closed universe; critical density. Orbital precession, bending of starlight, gravitational redshift, waves, and lens. Big Bang and steady-state models of the universe; perfect cosmological principle: continuous creation. The age, flatness, horizon, galaxy formation, and matter-antimatter problems; inflationary phase; GUTs.

Ideas to Understand

The properties of flat and curved geometries; relation between geometry and fate of the universe. Relation between mass and space curvature; between Newton's theory of gravity and Einstein's; tests of general relativity. Tests that distinguish between Big Bang and steady-state cosmologies. Problems with Big Bang cosmology; their possible solutions.

Questions

1. Explain in what sense Einstein "geometrized" gravity.

2. The Earth moves around the Sun in a nearly circular orbit. Contrast Einstein's interpretation of this fact with Newton's.

3. Describe two astronomical tests that confirm Einstein's theory of gravity.

4. (a) State the cosmological principle. Why does it require that the universe be isotropic and homogeneous?

(b) Compare it with the perfect cosmological principle. Why does the 3 K background radiation rule it out?

5. Describe the steady-state cosmology. How does it differ from the Big Bang picture with respect to an evolving universe? with respect to the age of the universe?

6. What is meant by the term "geodesic"? What is it called in flat space? on the surface of a sphere?

7. Suppose the universe is open. Describe what will happen to the background radiation. Suppose the universe is closed; what will happen to it then?

8. In what sense is general relativity a generalization of both Newtonian mechanics and special relativity?

9. Describe two problems with Big Bang cosmology.

10. Describe how inflation solves two problems of Big Bang cosmology.

11. The electromagnetic force is incomparably stronger than gravity, yet it is gravity that governs the universe. Explain.

12. When all the Newtonian effects of planetary perturbations that cause Mercury's orbit to precess are removed, Mercury's orbit shows more precession than Earth's. Why?

13. Describe carefully how you could find the size of the orbit of the pulsar PSR1913 + 16, which has a pulse period of 0.05903 seconds. (Hint: make a sketch showing how the frequency of the pulsar's pulses' changes as it orbits its companion. Assume the orbit is circular.)

Suppose now the orbit is quite elliptical (as it in fact is). What would your sketch look like now?

Suggestions for Further Reading

See also the books listed at the end of Chapter 19.

A nice series of four articles on aspects of modern cosmology titled "Our Cosmic Horizons" is given in successive issues of *Astronomy*, **16**, 1988, beginning with the February issue:

Scherrer, R., "Part One: From the Cradle of Creation," February, p. 40.
Trimble, V., "Part Two: The Search for Dark Matter," March, p. 18.
Gregory, S., "Part Three: The Structure of the Visible Universe," April, p. 42.
Melott, A., "Part Four: Recreating the Universe," May, p. 42. What computer simulations tell us.
Dressler, A., "Galaxies Far Away and Long Ago," *Sky & Telescope*, **85**, p. 22, 1993. Implications of some of the early galaxy images from the Hubble Space Telescope.
Powell, C., "The Golden Age of Cosmology," *Scientific American*, **267**, p. 17, July 1992. Some of the implications of COBE's discovery of nonuniformities in the background radiation.
Silk, J., *The Big Bang*. San Francisco: W. H. Freeman and Company, 1980. A readable account by a well-known astronomer.
————"Probing the Primeval Fireball," *Sky & Telescope*, **79**, p. 600, 1990. Pre-COBE and early COBE background radiation results.

An explanation of the cosmological meaning of various geometries is given in the following:

Albers, D., "The Meaning of Curved Space," *Mercury*, **4**, p. 16, July/August 1975.
Callahan, J., "The Curvature of Space in a Finite Universe," *Scientific American*, **235**, p. 90, August 1976. Geometry of the universe and general relativity.
LoPresto, J., "The Geometry of Space and Time," *Astronomy*, **15**, p. 6, October 1987.
Shu, F., "The Expanding Universe and the Large-Scale Geometry of Space-Time," *Mercury*, **12**, p. 162, November/December 1983. A good overview, excerpted from Shu's text, *The Physical Universe*.

High-energy physics or elementary-particle physics has become an integral part of cosmology. The following books and articles give relevant background and descriptions of this new field on the border of physics and astronomy:

Carrigan, R. and Trower, W., *Particle Physics in the Cosmos*. New York: W. H. Freeman and Company, 1989. A collection of a dozen articles that appeared originally in *Scientific American*. The articles deal with dark matter, large-scale structure in the universe, the fundamental forces and their unification, and matter and antimatter.

Davies, P., "Particle Physics for Everybody," *Sky & Telescope*, **74**, p. 582, 1987.

Wagoner, R. and Goldsmith, D., "Quarks, Leptons, and Bosons: A Particle Physics Primer," *Mercury*, **12**, p. 98, July/August 1983. Current view on the constituents of matter, adapted from their book, *Cosmic Horizons*.

Some of the new cosmological models and ideas building on the relation between particle physics and astronomy are described in the following:

Bartusiak, M., "The Cosmic Burp: Genesis of the Inflationary Universe Hypothesis," *Mercury*, **16**, p. 34, March/April 1987. A nice account, adapted from her book, *Thursday's Universe*.

Davies, P., "Relics of Creation," *Sky & Telescope*, **69**, p. 112, 1985. Seeds of structure arising from the early universe.

——— "The New Physics and the Big Bang," *Sky & Telescope*, **70**, p. 406, 1985. Inflation, supersymmetry, and superstrings.

Dicus, D., et al., "The Future of the Universe," *Scientific American*, **248**, p. 90, March 1983. What might happen to the universe in 10^{100} years (!). (Also in Carrigan and Trower, above.)

Guth, A. and Steinhardt, P., "The Inflationary Universe," *Scientific American*, **250**, p. 116, May 1984. Detailed account of the inflationary theory by its original inventor and by one of its modifiers. (Also in Carrigan and Trower, above.)

Halliwell, J., "Quantum Cosmology and the Creation of the Universe," *Scientific American*, **265**, p. 76, December 1991. Quantum theory applied to the universe as a whole.

MacRobert, A., "Beyond the Big Bang," *Sky & Telescope*, **65**, p. 211, 1983. Where did the universe come from?

Odenwald, S., "The Planck Era," *Astronomy*, **12**, p. 66, March 1984. The very earliest epoch of the universe.

Weinberg, S., *Dreams of a Final Theory*. New York: Vintage Books, 1993. Describes the search for a unifying theory of nature.

Some aspects of relativity theory are described in the following:

Einstein, A., *Relativity: The Special and General Theory*. New York: Crown Publishers, 1961. This book has gone through nearly 20 printings for a good reason. In only 150 pages it gives the clearest popular exposition of the special and general theories of relativity that I know of.

Odenwald, S., "Einstein's Fudge Factor," *Sky & Telescope*, **81**, p. 362, 1991. The cosmological constant, another possible parameter to characterize the universe.

Trimble, V., "Gravity Waves: A Progress Report," *Sky & Telescope*, **74**, p. 364, 1987. What gravity waves are, how they are produced, and how they might be detected.

Will, C., *Was Einstein Right? Putting General Relativity to the Test*. New York: Basic Books, 1986.

——— "The Binary Pulsar: Gravity Waves Exist," *Mercury*, **16**, p. 162, November/December 1987. Interesting account of pulsars in general and the discovery and analysis of the binary pulsar in particular.

Two unconventional cosmological views:

Arp, H. and Block, D., "The Myth of Overgrown Spirals," *Sky & Telescope*, **81**, p. 373, 1991. Arp has been the most persistent critic of the cosmological interpretation of galaxy redshifts. Here he gives another argument for his point of view.

Peratt, A., "Plasma Cosmology," *Sky & Telescope*, **83**, p. 136, February 1992. An alternative to the Big Bang model.

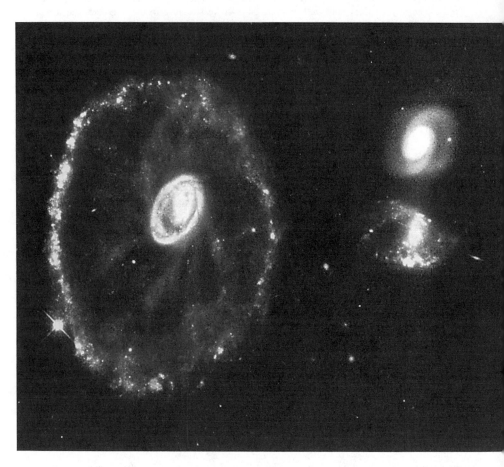

A spectacular Hubble Space Telescope image of a ring galaxy about 500 million light-years away in the constellation of Sculptor. It has undergone a collision (possibly with one of the two galaxies to the right) that disrupted its presumed original spiral structure and formed the prominent ring in which billions of stars are now forming. At the same time, spiral arms—still faint—seem to be developing.

CHAPTER 22

The Evolution of Matter

The world around us is constantly changing, not only from day to day and with the seasons, but also over much longer periods of time. We are familiar with the ideas of biological and geological evolution. Both processes act very slowly and usually very gradually, so that from one year to the next nothing much seems to happen. Over long periods of time, however, continents drift apart, mountain chains are uplifted, and plant and animal species appear and disappear; the world is completely transformed.

So far in this book we have discussed the evolution of the astronomical universe as a whole as well as that of individual stars. We don't know if the direction of the evolution of the universe will forever remain the same, or if sometime in the distant future that direction will reverse itself, so that the universe will collapse and perhaps be reborn, though that seems unlikely. We do know, however, that even the "constant" stars must eventually die. A few of them will make a dazzling display of celestial fireworks on their way to the stellar graveyard, and many others will glow quietly with a feeble light for many billions of years, but die they must.

There is yet another kind of astronomical evolution going on, the evolution of matter itself. As we shall see in this chapter, the *composition* of matter has changed over time; that is, the relative proportions of the chemical elements are different now than they were ten billion years ago. Remarkably, it is this change that has enabled biological evolution to occur; one consequence of this is that we are here, contemplating our ultimate origins. In addition, over the eons, matter has evolved in another way: the dominant *form* in which matter is organized has changed, from an incredibly hot and uniform soup of particles and radiation soon after the Big Bang, to the cold, lumpy universe of stars and galaxies we see around us (see Figure 22.1). How did the universe

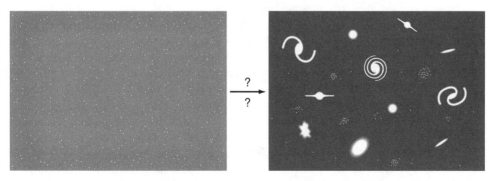

Figure 22.1. How was the universe transformed from an extremely smooth state (as revealed by the cosmic background radiation) to one of extreme clumpiness?

transform itself from near-perfect smoothness to a vast collection of mass concentrations—stars, galaxies, clusters, and superclusters of galaxies, bound together by gravity? The aim of this chapter is to describe some of the current ideas about how matter has changed over time; how matter itself has evolved in its composition and organization. Let us begin by considering changes in the composition of matter.

Changes in the Composition of Matter
Cosmic Abundances

On several occasions we have referred to cosmic abundances, which are based primarily on fairly accurate measurements of the abundances of the chemical elements in meteorites, the Sun, and a relatively small number of nearby stars. These have been given (by mass) as 72 to 75 percent hydrogen, 26 to 23 percent helium, and about 2 percent everything else, which we call the metals (though, as we will see, some stars show departures from this standard recipe). The relative abundances of the elements within the metals group are also fairly well determined (see Table 11.2). With our understanding of stellar evolution, we can determine the ages of stars and ask if the abundances of the elements have changed with time. We find that this cosmic recipe holds pretty well for stars such as the Sun and younger, that is, for stars formed within the last five billion years or so in our part of the Milky Way.

But what about stars older than the Sun, for example, those in globular clusters? Since the clusters are far away and consequently the stars are faint, it is more difficult to make a detailed chemical analysis of their composition. It is possible, however, to find the abundances of several key elements such as iron that serve as indicators of the metal abundances as a whole. In this way it has been found that stars in many globular clusters—in particular, those well above or below the galactic plane—are markedly deficient in the metals compared with the Sun. In some of these stars, the metals amount to only 0.1, 0.01, or even 0.001 percent of the total, not the "standard" 2 percent. The same situation prevails among single stars high above the galactic disk, the so-called **halo stars**. These stars are also known to be old objects. Thus the metal abundances 12 or 15 billion years ago were considerably smaller than they were about 5 billion years ago, when the Sun formed.

As we saw in Chapter 19, the abundance of helium in stars formed today is just about the same as in the oldest stars we know of, indicating that all the helium and all the metals could not have been formed by the same processes. We saw that the Big Bang itself provided conditions of temperature and density such that only a few minutes after the beginning, the cosmic helium abundance had been manufactured in the rapidly cooling and expanding fireball.

In addition, we saw that this Big Bang nucleosynthesis leads to an estimate of the average density of the universe of about 5×10^{-31} grams per cubic centimeter. This means that the density of ordinary matter such as neutrons and protons (collectively called baryons) is no more than about 10 percent of the critical density, and probably even less. (The value of the critical density depends on that of the Hubble constant, which is itself uncertain, hence the waffling.) Estimates of the density of the matter that we can see amount to only a percent or so of the critical density and lead to the notion that at least a few percent of the dark matter are made of baryons (MACHOs, Chapter 18). In addition, if the inflationary picture is correct and the density of the universe exactly equals the critical density, then at least 90 percent of the mass of the universe must consist of some exotic, nonbaryonic and so far unknown kinds of particles. Thus, the physical conditions early in the Big Bang not only tell us where helium comes from, but they are also the best evidence that most of the mass of the universe could be nonbaryonic.

We are still left with the question of the origin of the metals. They could not have been formed in the Big Bang, because the universe was expanding so rapidly that the high temperatures and densities necessary for the formation of heavier elements from lighter ones did not last long enough. Furthermore, the creation of metals in the Big Bang would not have accounted for their increased abundance among more recently formed stars.

Where else could the elements have been formed? Where in nature do high temperatures and densities last for long periods of time? Well, what about the stars themselves? Perhaps they are the element factories. We have seen that advanced energy-generation processes produce heavy elements, and that mechanisms (supernovae explosions, for example) exist that expel matter from stars back into the interstellar medium. Stars forming subsequently from the interstellar medium would, therefore, contain a pinch or two of this processed material that includes the heavier elements. As time went on, the products of stellar evolution slowly increased the metal content of the interstellar medium, and hence the metal abundances in stars forming from it. Stars would seem to hold great promise, then, as the source of the metals.

But isn't there a problem here? The major source of stellar energy is the conversion of hydrogen to helium. Helium, as well as the metals, is cooked inside stars. So why isn't this the major source of helium in the universe? There are several reasons why it isn't. Since hydrogen burning can take place only in the relatively small, hot stellar core, most of the star's hydrogen remains hydrogen; it is not processed. Stars with masses of about half a solar mass or less go directly from the main sequence to the white-dwarf region, with none of the core being expelled to the surrounding interstellar medium. Stars up to about four solar masses burn helium to carbon. If the core material is ejected (and usually it won't be), it is primarily in the form of carbon, not helium. Finally, the most massive stars may become supernovae and explosively blow much of their matter back to the interstellar medium. That matter, however, would have been transformed far beyond helium to much heavier elements. Model calculations show that in order to produce 25 percent or so of the helium in the universe, stars would have produced a metal abundance of much more than 2 percent. Stars cannot produce all that helium without at the same time making a lot more carbon, oxygen, and other heavy elements than are observed. Thus, although a little star-produced helium does find its way back to the interstellar medium, it is no more than a "contaminant," less than 10 percent of the total. The vast majority of helium in the universe was not produced in stars, but in the Big Bang.

The Evolution of the Elements

So what is our current picture of element formation? A few minutes after the Big Bang, the universe consisted of hydrogen, helium, very small quantities of light isotopes such as heavy hydrogen, and radiation. No significant quantity of heavier elements was made, because the temperature and density dropped too rapidly as a consequence of the expansion. We don't know when star formation began, but it must have been at least a million years later, when the background radiation and matter no longer interacted, and probably much later than that. Whenever it formed, the first generation of stars (sometimes called Population III) consisted of hydrogen and helium only, with helium amounting to about 25 percent by mass of the total. Since dust is made of the metals, none existed by which the contracting protostars could have been cooled, thereby speeding the contraction. Consequently, these first stars should have been ex-

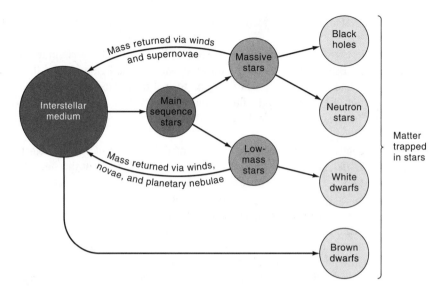

Figure 22.2. Schematic representation of the interaction between the interstellar medium that stars form from, and evolving stars that return matter—some of which is enriched with metals—back to the interstellar medium.

tremely massive (to overcome the internal pressure) in order to form at all. Regardless of their mass, all these stars generated energy by the proton-proton chain, since no carbon existed to enable the carbon cycle to operate. Attempts to find these ancient Population III objects or their relics have so far been unsuccessful, however, probably because they are rare and do not have any easily identifiable characteristics.

The more massive first-generation stars produced elements heavier than helium. Those that became red supergiants or supernovae produced a wide variety of nuclei, some of which were blown back into and mixed with the interstellar medium. Eventually new stars formed from this metal-enriched material. Those that were more massive than the Sun could now generate energy by the carbon cycle. Toward the end of their evolutionary lives, these stars in turn enriched the interstellar medium with metals. New stars formed from this material; the more massive stars cooked more metals, and eventually returned them to the interstellar medium. It is this interaction between the medium and stars through star formation, element production, and material ejection that has resulted in the buildup of the metal content in the universe (see Figure 22.2). This buildup should not be thought of as being uniformly distributed over a given galaxy, resulting in a direct and unique relation between the ages and metallicities of stars. Instead, the metal enhancement was rather spotty with some regions of a galaxy becoming enriched faster than others. As a result, stars of the same age might differ significantly in their metal abundances.

The metal enrichment must have been fairly rapid once star formation began, because no stars have been found with extremely small metal abundances of, say, less than 10^{-4} of the solar value. Furthermore, the spectra of quasars so far away that we see them as they were only a few billion years after the Big Bang show little depletion of the metals. For example, a quasar with a redshift of 4.73 has a normal spectrum. This early and relatively rapid increase in metallicity was due to at least two factors. It is quite likely that the star formation rate (and therefore the rate at which metals were produced) was greater then than now simply because more star-forming raw material was available. (The Milky Way provides a local example in that the heavy element abundances in the disk generally decrease from center to edge, in part, at least, because the density of interstellar gas in the primordial cloud fell off in the same way.) Also, when the metallicity was very small, it did not take much stellar processing to double it. Even with a constant production rate, the *fractional* increase of the metal abundances will, of course, decrease with time. To double it now would require as much evolution as all that has already taken place. Thus, there is little systematic difference in the metal abundance of the Sun and of stars several billion years younger. Careful studies of the nuclear processes that produce the heavier elements show that this scenario accounts for the observed elemental abundances in a reasonably satisfactory way.

What of the future? The interstellar medium in our galaxy is being replenished at the rate of a few solar masses per year by planetary nebulae, stellar winds, etc. The already low star-formation rate will decrease only slowly, but for the reasons given above, the fractional rate of increase in the overall metal abundance will be small. Most galactic stars that form a few billion years from now will be made from much the same cosmic mix as the Sun's.

This general trend toward diminishing star-formation rates and small metal enhancements can be slowed or temporarily interrupted in a given galaxy, however. This could happen by a galaxy accreting surrounding gas or by interacting with another galaxy. For example, recall that x-ray observations have shown that some clusters of galaxies contain huge amounts of hot intergalactic matter, sometimes amounting to as much (or even more) mass as the galaxies themselves. This gas falling into a galaxy could provide the raw material from which new stars can form and might keep some cluster galaxies "young" for a long time.

Examples of star formation stimulated by galaxy interactions are given by infrared observations made by IRAS at wavelengths between 12 and 100 microns. These show that some galaxies radiate as much as 99 percent of their energy in the infrared, whereas most galaxies radiate less than half of their energy in this part of the spectrum. In extreme cases, these galaxies can emit 100 times more energy in the infrared than a normal galaxy does over the entire spectrum, making them comparable to quasars in their total luminosities. Such systems are called **starburst galaxies** because a sudden burst of star formation is responsible for their extraordinary IR brightness (see Figure 22.3). Apparently the starburst phenomenon is triggered in a galaxy when it experi-

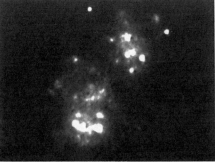

Figure 22.3.   The left panel shows an optical image of the starburst galaxy NGC 1741, taken from the ground. This galaxy is about 150 million light-years away and its structure suggests that it has experienced an encounter with another galaxy recently. The right panel shows an ultraviolet image of the galaxy's center taken with HST. Separated by about 3,000 light years are two starburst centers, each consisting of several compact regions of star formation just a few hundred light-years across. Each region contains several hundred stars that formed only a few million years ago.

ences a strong gravitational encounter with another galaxy. This causes interstellar gas to be compressed and driven to the central region of the galaxy, where stars are formed in great numbers. The huge output of infrared energy comes from dust warmed by the newly forming stars. The very large increase in star formation soon results in an increased production of metals that enrich the interstellar medium. Stars formed from this material may show a larger abundance of metals than stars formed at the same time in other parts of the galaxy. Despite these starburst events, however, when most of the interstellar matter becomes permanently bound up in stars, the star-formation and metal-production rates must fall to near zero.

Consider what this picture of the origin of the metals implies. With the exception of the hydrogen in water, nearly everything that we see around us consists of the "metals." Dirt consists largely of silicon and oxygen, along with trace elements such as iron; the cellulose in trees and plants contains hydrogen, carbon, and oxygen; our automobiles are manufactured from iron, nickel, aluminum, carbon, etc. The DNA in living things is made of hydrogen, carbon, oxygen, nitrogen, and phosphorus. Except for hydrogen, all these elements were produced inside stars. Most of the atoms of our bodies and of nearly everything around us were processed perhaps two or three times in the cores of red-giant stars or supernovae. We would not be here were it not for stellar evolution! In a very real sense, we are the offspring of the stars. Our environment is much broader than we usually suppose, and our connection with the universe is profound and crucial to our very existence.

Changes in the Organization of Matter

One of the striking characteristics of the astronomical universe today is how "clumpy" it is. Most normal matter appears to be organized into stars and clusters of stars; these are arranged in galaxies, and most galaxies are parts of groups or clusters of galaxies, many of which are organized into huge superclusters. The average, smoothed-out density of clusters of galaxies is about ten times the average density of the universe; that of galaxies about 10^6, and that of stars about 10^{30} times that of the universe. The hierarchic arrangement of matter in the universe is pronounced.

Now, recall that until about a million years after the Big Bang, matter and radiation were thoroughly mixed and very smoothly distributed. When the temperature of the expanding universe fell to about 3,000 K, however, protons and electrons combined to form hydrogen, which is transparent, so matter (hydrogen) and radiation no longer interacted. The radiation remained uniformly distributed, but the gas eventually became clumped into the galaxies we see now. How did the universe change from one in which the density of matter was very nearly the same everywhere to one in which there are enormous density contrasts? When was the universe transformed from a state of uniformity to one of pronounced clumpiness?

Formation of Galaxies

Just as stars are the fundamental building blocks of galaxies, galaxies can be taken as the basic units of the universe. We can observe stars in all phases of their lives, from birth to death. Our knowledge of stellar structure is reasonably good and we understand the basic characteristics of stars, such as why they cannot have arbitrarily small or large masses and diameters. In marked contrast, there is no equivalent of a Hertzsprung-Russell diagram for galaxies. We cannot identify with certainty galaxies that are forming, although we know of many that are being transformed. To make matters worse, we don't know what of importance for galaxy formation might have happened during an early, unobservable phase in the Big Bang, nor do we understand the (probably crucial) role played by dark matter. We don't know how galaxies form or even *why* they form. Is there only one scenario by which galaxies form, or are there several? Did they form quickly in less than, say, a billion years, or did the process extend over a much longer time? If the latter is true, then, in contrast with stars again, the distinction between the formation of a galaxy and its evolution would be fuzzy.

We might reasonably have expected that galaxies, or at least clusters of galaxies, would be fairly uniformly distributed in space. Instead, extensive redshift surveys of thousands of galaxies that are up to many hundreds of millions of light-years away have revealed that they are arranged in frothy, sponge-like distributions, somewhat as if they were stuck on the surfaces of gigantic cosmic bubbles millions of light-years across (see Figure 18.11). How did this come about? And what processes can create voids—enormous volumes of space nearly devoid of any galaxies—such as that in the constellation of Bootes, or the huge structures of galaxies such as the Great Wall, or the structure inferred from the apparent common motion of the Local Group and the Virgo and Hydra-Centaurus clusters toward some as-yet-unidentified concentration of galaxies?

We have only the most general ideas of what determines the masses and sizes into which a huge cloud of gas may fragment. Did large structures form first and break up into smaller ones: for example, did a galaxy cluster–size mass separate out and break up into galaxy-size units that in turn fragment to stellar cluster masses in a "top-down" process? Or did the organization of matter go in the other direction, with larger units being built up from smaller ones and galaxies forming from star clusters in a "bottom-up" process?

We mentioned in Chapter 21 that in response to its own gravitation, an infinite medium of uniform density will break into huge clouds whose sizes depend on the initial density and temperature of the medium. This process favors the top-down scenario: the initial formation of large-scale masses become clusters of galaxies, break up into protogalaxies, and ultimately become protoclusters and protostars. Though this sounds reasonable, it is not understood how this process could get started. One or both of two basic conditions are necessary: there must be a lot of time for the protoclusters to form from an initially quite smooth medium, or there must be large density fluctuations in the original medium so that the process goes rapidly, or some combination of the two. The smoothness of the cosmic radiation background (Chapter 19) rules out large initial density fluctuations; and with galaxies appearing perhaps as soon as one or two billion years after matter and radiation decoupled, there is far too little time for the process to work in a smooth medium.

The realization that most of the mass of the universe is dark matter opened up new possibilities for understanding the development of structure. Remember that most of the mass of the universe may not be the ordinary visible matter that constitutes the people, planets, stars, and galaxies we see around us. Instead, it is composed of some unknown particles, probably relics (such as WIMPs) of a very early phase of the universe. So far, we know about these particles only through their gravitational interaction with ordinary matter; no other interactions have as yet been detected. This means that, in contrast with ordinary matter, a clumpy distribution of dark matter might not show up as corresponding clumps in the background radiation.

The general idea is, therefore, that early in the Big Bang, long before the electrons and protons of ordinary matter had recombined, dark matter—whatever it might be—somehow developed a clumpiness. The gravitational force exerted by these density fluctuations would cause corresponding fluctuations in the ordinary matter. Before recombination, however, ordinary matter and radiation interacted so violently that no substantial density fluctuations could be maintained. After recombination, when ordinary matter and radiation went their separate ways, the denser clumps formed by the dark matter "seeds" could be maintained and grow into huge masses.

Two general kinds of models have been proposed. In one, the dark matter is assumed to consist of particles having a very small mass and traveling with speeds close to that of light, at least in the early universe. Because of its speed, it is called **hot dark matter**. (Neutrinos are a popular candidate, but evidence for even a tiny neutrino mass is weak.) Such rapidly moving matter would wash out any small-scale irregularities such as those that would give rise to galaxies, but it might account for structure on the very largest scale, for example, giant clusters of galaxies. The other type of model is called **cold dark matter** because its relatively massive particles move slowly. It would have acted as a catalyst for the formation of structure by means of the bottom-up process, with smaller masses forming first and collecting into larger units. Though theories based on dark matter have had some successes, they also appear to have difficulties, for example, in accounting for huge structures such as the Great Wall. Cosmologists are turning to models consisting of both hot and cold dark matter and perhaps one of these may work. As of this writing, however, we do not really understand how matter in the universe came to have its structure, and to be organized as it is.

Will any of this structure be hinted at by very slight nonuniformities in the temperature of the 3 K background radiation when precise measurements become available over angular scales of minutes of arc instead of COBE's scale of several degrees? Perhaps so, because the various kinds of dark matter collecting in space under the influence of different "seeds" should give rise to structures having different characteristics.

In any case, galaxies did form, and many of them formed at an early time in the history of the universe. What are some of the general characteristics of galaxies that might be clues to their formation? Probably their most fundamental property is their mass. In marked contrast to stars, whose masses range over a factor of about 500, galaxies show a spread of a factor of about one million in their visible masses. What accounts for this huge range? Another basic property must be the ratio of baryonic to dark matter in galaxies. Is this constant or is there a significant spread in this ratio? Perhaps less fundamental but more easily seen is the fact that regular galaxies come in two forms, ellipticals and spirals, or more generally, spheroids (ellipticals and the central bulges of spirals) and disks (of spirals).

Even if we don't know how the protogalactic blobs formed in the first place, can we at least understand how these two kinds of structures could have developed? For example, could a galaxy of one form evolve into one of a different form—an elliptical, for example, into a spiral or vice versa? Such evolution of galaxy form is most unlikely, however, because an isolated system of revolving stars in a galaxy is generally stable. Barring strong disruptive forces from the outside, the motions and so the shape of stellar systems will not change much. Furthermore, how could a gas-poor elliptical transform itself to a gas-rich spiral? Irregular galaxies—though containing a lot of interstellar matter—can't be young galaxies about to be transformed to spirals because they have old as well as young stars.

Perhaps the process of galaxy formation itself accounts for the different types (Figure 22.4). The contraction of a huge, very slowly rotating protogalactic cloud involves a contest between the inward-pulling gravitational force and the outward-acting pressure forces, just as in a protostellar cloud. The outward force in a protogalaxy is provided by its internal motions, which are of two kinds: random (for example, the motions of atoms) and organized (orbiting clouds of gas with ranges of eccentricities and inclinations). Much of the energy of these motions is eventually lost through collisions. Atoms collide and radiate their energy away; gas clouds collide and merge or

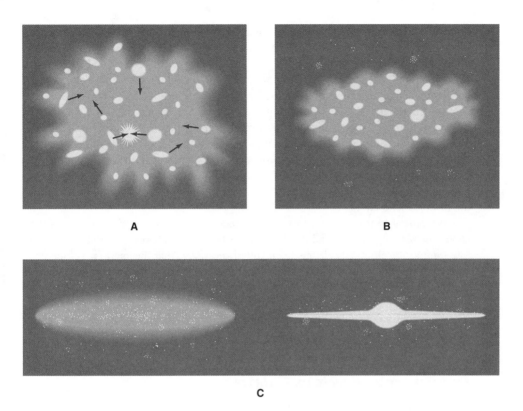

Figure 22.4. A schematic representation of some aspects of galaxy formation. (A) A huge cloud of material may be made up of many subclouds moving in a wide variety of orbits. When two such subclouds collide, they lose energy and either merge or disperse. (B) A few of the subclouds in the outer part of the halo form stars and globular clusters, some of which move in retrograde orbits reflecting their original motions. Eventually the matter in the denser central region of the cloud moves in noncolliding orbits and it contracts, rotates more rapidly, and forms a central bulge. (C) Over the next few billion years, the rest of the cloud slowly falls towards the plane of the galaxy, forming more globular clusters and stars. If all of the star-forming material is used up, the galaxy becomes a low-luminosity elliptical; if not, the rest of the matter falls to the plane of the galaxy, forming a disk and spiral arms.

disperse until they are all following noncolliding orbits. As the gas loses its energy, it sinks to a disk perpendicular to the axis of rotation, and at the same time the rotation of the protogalaxy speeds up, because its total angular momentum, mVr, must remain constant (r decreases, so V increases).

But how does this process lead to a spiral or to an elliptical galaxy? We have seen that a system of stars will not collapse, but one of gas and dust clouds will. If a lot of star formation occurred before the protogalaxy completely collapsed, an elliptical galaxy would result. If, however, the opposite was the case and the protogalaxy collapsed to a disk before all of the star-forming material was used up, a spiral would be formed.

Now, the Hubble sequence, ellipticals-Sa-Sb-Sc-irregulars, forms a progression of galaxy types in which an increasing fraction of the total visible mass is in the form of interstellar matter. Stars in the bulges of galaxies are quite old and it is generally thought that galaxy bulges formed first. This is reasonable because the dense bulge would contract most rapidly in the protogalactic cloud. Ellipticals consist only of a bulge and contain very few young stars and very little interstellar matter for further star formation. Evidently, star formation in ellipticals was efficient, occurred early, and declined rapidly. By contrast, irregular galaxies have no central bulges and so no initial epoch of rapid star formation. Their stars formed relatively slowly and at a roughly constant rate over their entire lifetimes. Even now, they still have a fair amount of star-forming material left. The spiral sequence, Sa-Sb-Sc, is one of decreasing central bulge prominence, suggesting that the overall star-formation rate in Sc galaxies was less rapid and continued for longer times than it did in Sa galaxies. In this way, the different galaxy forms would be consequences of differences in the rates at which gas clouds turned into stars. Why such marked differences in rate should occur is not understood, however. Perhaps it has to do with the initial density or the turbulence of the protogalactic cloud.

According to this picture, the more flattened a spheroidal system is, the faster it should be rotating. This is indeed the case for the central bulges of spirals and for the low-luminosity ellipticals. It has been found, however, that the random velocities of stars in giant ellipticals are larger than their systematic motions around the center of their galaxy. This indicates that the flattening of giant ellipticals is not caused by rotation alone; the rotation rates of bright ellipticals are not proportional to their flattening. Thus it is an oversimplification to lump all spheroids together as a single family. There must be more to galaxy formation than we have so far considered.

Environmental Effects on Galaxies

Evidence that collisions and mergers of galaxies are important factors in their evolution was presented as long ago as 1940, but it was not until the 1970s that the idea was taken up and developed. Recall that galaxies are separated by distances only 10 to 100 times greater than their diameters, so that strong gravitational interactions and even collisions between galaxies must be relatively common. Numerical computations show that during a close encounter of two galaxies, tidal distortions are produced in both systems that cause them to lose energy (Figures 22.5 and 22.6). This energy loss, which can be likened to a sort of friction, will cause their relative orbits to shrink. If the two galaxies are moving slowly enough, they may even merge, just as friction between the upper atmosphere and an artificial satellite causes the satellite to lose altitude and eventually plunge to Earth. Under the changing gravitational field, the stars, now becoming a single system, will be quickly redistributed into a new arrangement. It is interesting to note that the resulting brightness distribution with distance from the center resembles that in an elliptical galaxy regardless of the types of galaxies merging. This suggests that at least some ellipticals might have been formed by mergers.

Furthermore, a larger proportion of ellipticals compared with spirals seems to be found in dense clusters of galaxies than in sparsely populated ones (early in their history mergers were more likely in dense clusters). This supports the notion that they result from mergers of spirals. More direct evidence is provided by some ellipticals that contain a disk of gas and dust rotating more rapidly than, and at a large angle with, the surrounding stars. Such a disk could be the remnant of a recent merger with a spiral galaxy. Formation of the brighter ellipticals by merger would also account for their slow rotation, since they would have been made by stars revolving in a variety of inclinations and senses.

Computer simulations of such encounters give results that closely mimic the appearance of real galaxies. For example, depending on the relative masses, velocities, separations, and orientations of the two galaxies, objects with rings oriented in the

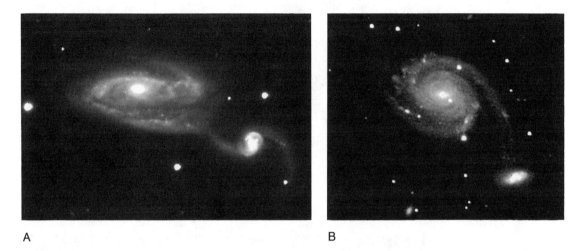

A   B

Figure 22.5. Both of these pairs of interacting galaxies resemble M51 (Figure 17.16).
(A) NGC 5394/5395 is about 110 million light-years away. The two galaxies are 100 thousand and 50 thousand light-years in diameter. Note the bright nucleus with surrounding
arcs in the smaller galaxy, presumably a consequence of the encounter of the two galaxies.
(B) NGC 7752/7753 is a system about 150 million light-years distant. Note the nuclear bar
in the larger galaxy and the brightness clumps in the smaller.

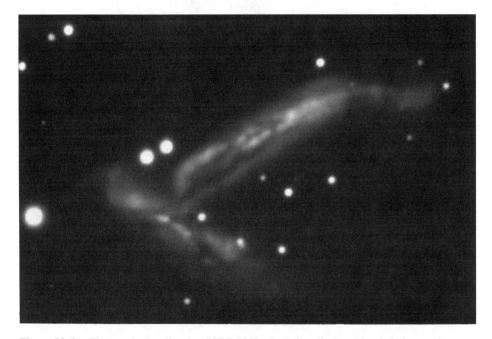

Figure 22.6. These galaxies, listed as NGC 7253, are being disrupted by their interaction.
It is not clear whether the smaller component is itself an interacting chain of three galaxies.

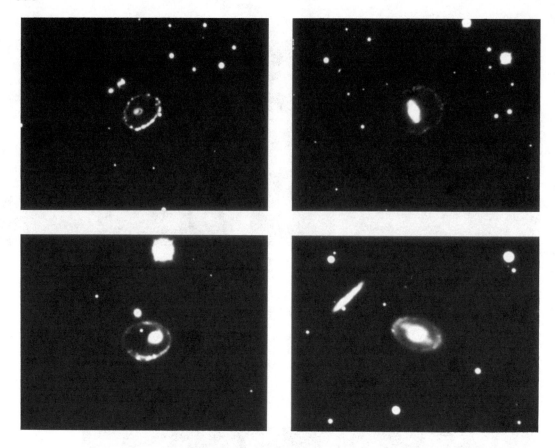

Figure 22.7. Four examples of ring galaxies, a rare class formed by interactions of two galaxies.

plane of the enclosed galaxies (Figure 22.7), systems with long tadpole-like tails, or galaxies surrounded by faint, outer arcs or ripples (Figure 22.8), and completely merged elliptical galaxies can all be produced in these simulations. The tadpole tails facilitate the merger by carrying a large amount of angular momentum away from the main body of the system, and the ripples are apparently remnants of what was originally disk material in the merged spiral.

The peculiar galaxy NGC 7252 exhibits many of these characteristics (Figure 22.9): it has two tails but only one body, where the distribution of brightness across it resembles that of an elliptical galaxy; the main body of the galaxy is surrounded by ripples; and at the core of the galaxy is a small spiral, rotating in the sense opposite to that of the main body, apparently one of the consequences of an earlier merger.

Mergers or the consequences of mergers could also account for some of the other observed properties of galaxies. For example, the cores of ellipticals contain hundreds or even thousands of times as many stars per unit volume as do the disks of spirals (see Figure 22.10). Where do all these stars come from if mergers with spirals are important? The starburst effect (the rapid formation of large numbers of stars as a result of an encounter with another galaxy) may well be responsible for the high densities of stars in the centers of ellipticals. As long as the newly formed stars are surrounded by dust, they are not seen directly. When the smog of dust dissipates, we see the high star density that is characteristic of the cores of ellipticals.

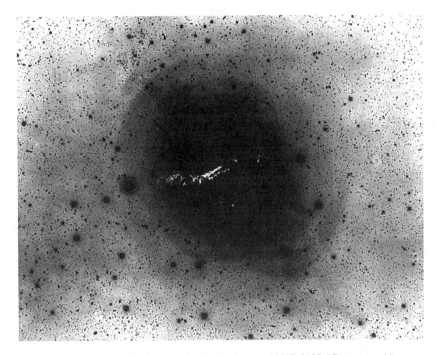

Figure 22.8. This is a negative image of a long exposure NGC 5128 (Centaurus A). Note the faint outer shells or ripples, leftovers of the merger between the giant elliptical and a spiral galaxy that also produced the dark band of material across the elliptical. Figure 20.4 is a positive image of this galaxy.

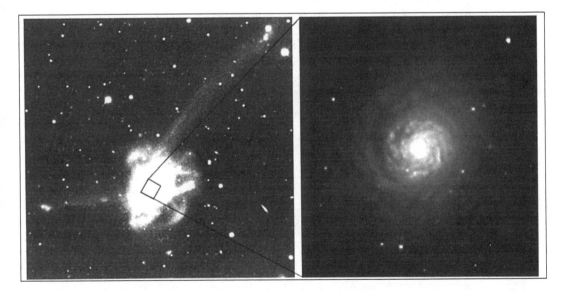

Figure 22.9. The image on the left, taken from the ground, is of NGC 7252, where two galaxies are merging. The long tails, which are characteristic of many close encounters between galaxies, are clearly shown. The right HST image is of the central region of the merged system. It shows a small spiral and about 40 bright globular clusters. The bright part of the spiral is about 10,000 light-years across.

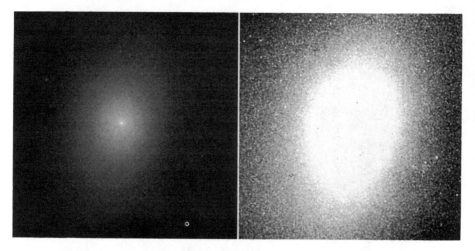

Figure 22.10. NGC 7457 is an elliptical galaxy that is only about 40 million light-years away. The image on the right, taken with a camera on the HST, shows the increasing star density toward the center of the galaxy, as is characteristic of ellipticals. The contrast of the image on the left has been adjusted to show the very bright, small core at the center of the galaxy, where it is estimated the star density is enormous—at least 30,000 times greater than that in the neighborhood of the Sun. Unresolved at 0.1 arcsecond, the core's diameter must be less than 20 light-years. It is not certain how such a dense core was formed, but because this is an ordinary elliptical galaxy, such cores are likely to be common.

Some composition characteristics of spheroidal systems also can be explained by the same mechanism. In particular, the metal abundances of stars in spheroidal systems are often observed to decrease steadily from center to edge, and the overall average metal abundance of stars is larger the more massive the galaxy. A bright elliptical with these properties is not the simple sum of smaller galaxies, since the metal abundances would be too low and the center-to-edge abundance effect would be destroyed by the violence of the merger itself. Both properties, however, could be the result of the starburst process just described.

Yet another property of ellipticals may be accounted for by this process. Elliptical galaxies have many more globular clusters than do comparable spirals. Where do all those globular clusters come from if ellipticals formed from relatively globular-poor spirals? Perhaps they are formed in the violence of the merger process itself. NGC 7252 has about 40 globulars that are bluish in color, not yellow or red as seen in the usual old globulars, suggesting that they are recently formed. Also, HST observations of NGC 1275 reveal many blue, newly formed globulars (Figure 22.11). The appearance of this galaxy suggests that it had a strong interaction with another galaxy. Perhaps this precipitated a burst of star formation that could account for the observed density of globular clusters.

Interactions of a galaxy with gas clouds rather than with another galaxy may also be important. S0 galaxies have the disk and central bulge of a spiral, but have very little if any interstellar gas or dust. They are much more often found in rich clusters than not (see Figure 22.12). Many clusters contain hot x-ray-emitting gas that may have been ejected from cluster members or may have fallen into the cluster from the outside. In any case, there is a lot of it, and as a spiral moves within the cluster, this intergalactic matter strips away the spiral's gas, transforming it into an essentially gas-free spiral: an S0 galaxy.

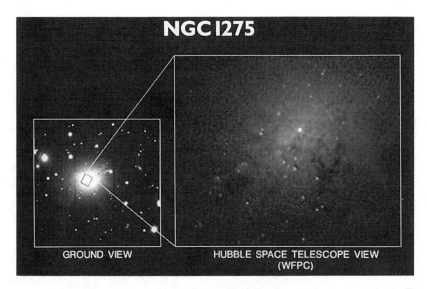

NGC 1275

GROUND VIEW

HUBBLE SPACE TELESCOPE VIEW
(WFPC)

Figure 22.11. The large peculiar galaxy, NGC 1275, is located at the center of the Perseus cluster of galaxies, about 200 million light-years away. Its strange shape, as shown in the ground-based image on the left, suggests that it is the result of the merger of a giant elliptical with a spiral galaxy. The HST image on the right shows about 40 blue (and presumably young) globular clusters, which are seen here as dots. This suggests that such clusters form either by galaxy–galaxy encounters or by the interaction of the galaxy with gas within the cluster.

Figure 22.12. The giant cluster of galaxies in Coma Berenices contains a large number of S0 spirals and ellipticals, as well as a few giant ellipticals.

Figure 22.13.   Abell 2199 is a cluster of galaxies more than 300 million light-years distant. The cD galaxy, NGC 6166, is prominent at the center right. The bright spots seen against the center of this galaxy are probably foreground objects and not multiple nuclei of the cD galaxy, as had once been thought.

Supergiant ellipticals that are found only at the center of some rich clusters of galaxies may provide another example of galaxy interactions with gas (Figure 22.13). For historical reasons these objects are denoted as **cD galaxies**, which can be thought of as indicating "central dominant" (although that was not the original meaning). These extremely luminous objects have enormous faint envelopes extending throughout much of the entire cluster. Some of them seem to have several nuclei that have been interpreted as the cores of galaxies captured by the supergiant. However, these may well be small, independent galaxies seen against the supergiant rather than merged objects. Their enormous envelopes might consist of captured debris stripped off other cluster members by gravitational interactions.

So, do we understand how galaxies formed? Obviously we don't, but progress has been made. Probably what is most important is the realization that no single process can account for the formation and structure of all galaxies; several mechanisms play a role. Faint ellipticals and the bulges of spirals are small and rotate about as rapidly as would be expected from their shapes, so presumably mergers did not play a role in their formation. Thus, at the very least, the smaller spheroidal systems must have formed by "condensation" out of a primordial cloud, as discussed earlier in this chapter. However, we see other galaxies evolving by merging with smaller "mature" galaxies. Could the larger galaxies have been formed originally by a condensation process and then built up by one or two mergers, or were they built up by many mergers? In other words, how much mass of an elliptical comes from mergers? What role, if any, do mergers play in the formation of spiral galaxies? The disks of spirals seem to be fragile and easily disrupted. Could a disk survive such a violent encounter? Many questions remain to be answered; no doubt many surprises await us.

The Formation of the Milky Way

Let us conclude this section with a brief description of the formation and evolution of our own galaxy. Though much of this is speculative, it is clear that not just one but many of the processes we have discussed played a role in forming the Milky Way. We begin with a huge protogalactic cloud, itself perhaps formed from several other fragments, embedded in a significant mass of dark matter. Motions within the cloud were nearly random; there was only a slight net rotation, perhaps induced by torques from neighboring clouds. As the cloud contracted, the central denser portion collapsed more quickly than the rest of the cloud and formed the rotating bulge of the galaxy where star formation began. Massive stars composed of hydrogen and helium only must have formed early, evolved, and produced heavy elements, which were blown into the interstellar medium, slightly enriching the surrounding primordial gas.

Just how the halo and disk were formed is uncertain, but observations suggest that the halo did not form as part of the bulge. For example, some Sc spirals have practically no central bulge at all, indicating that the halo is not a simple extension of the bulge. In our galaxy, some of the oldest stars and clusters in the halo rotate in the sense opposite from that of the bulge. In contrast with the latter, the halo has little net rotation, which suggests that it was formed from large masses of randomly moving material captured by the protogalaxy and did not participate in the cloud contraction that produced the bulge. This material might have already undergone some star formation and therefore some chemical evolution, thereby accounting for some spottiness in the metal abundance in the galaxy.

Ages of globular clusters, found by fitting their color-magnitude diagrams with stellar evolution models (Chapter 15), tell an interesting story. The very oldest halo clusters (those with the smallest heavy element content) formed quickly, over a time span of only one-half billion years. However, younger clusters (with greater metallicity) have a spread in age of about four billion years, indicating that the halo formed over a long period of time. Roughly two-thirds of the globular clusters have an average metal abundance of about 0.1 percent instead of the Sun's 2 percent. The rest, which generally form a flatter system closer to the disk than the others, have metal abundances about one-third of solar, more nearly like the abundances in the disk. Again, this supports the idea that the halo and central bulge evolved independently and that the halo formed over a long period of time.

What about the disk? Presumably gas settled into a rotating disk soon after the bulge formed. When did star formation begin in it? Two lines of evidence suggest that the stellar disk is significantly younger than the halo. Recall that if our understanding of how they cool is correct (Chapter 16), it is possible to estimate the ages of the faintest white dwarfs (and so the age of the oldest stars) in the galactic disk. This gives an age of about ten billion years, significantly less than the age of the halo. In addition, the oldest known clusters in the disk are three or four billion years younger than the youngest clusters in the halo. Perhaps supernovae outbursts, strong stellar winds, etc., prevented significant star formation from taking place in the disk for a time. If these age differences are real (and that's a big "if"), they suggest that it took the galaxy a significant fraction of the age of its oldest stars (15–17 billion years) to complete its formation (10 billion years ago). In any case, star formation did take place in the disk, with more occurring near the center than in the outer regions, because the heavy element abundance generally decreases from the center outward.

What will happen in the future? More and more of the galaxy's matter will be tied up in stars. Right now, only a few solar masses of material are being returned to the interstellar medium each year, by novae, supernovae, planetary nebulae, and winds from massive stars. As more of the interstellar matter is tied up in low-mass stars, less will be returned to the medium, and the star-formation rate will slowly decline. With the decrease in density of the remaining gas, it becomes less likely that massive stars will form. The upper main sequence of our galactic HR diagram will be nibbled away as the massive stars evolve into giants and supergiants. With a steadily decreasing number of massive stars, there will be a corresponding decrease in the importance of the mechanisms that help in star formation—supernovae, stellar winds, and radiation pressure. Ultimately, massive stars will add to the population of white dwarfs and neutron stars, and spiral arms will be less and less prominent. The familiar constellations, quickly becoming unrecognizable because of the stars' proper motions, will finally disappear entirely, since they are delineated predominantly by the more-massive, shorter-lived objects.

The galaxy will be reinvigorated, temporarily at least, when it acquires the so-called Magellanic stream. This is a group of a half-dozen gas clouds extending over a 100° arc in the sky from the Magellanic Clouds to our galaxy, presumably a result of tidal interactions between the Clouds and the Milky Way. Calculations indicate that eventually the stream and these two satellite galaxies will merge with the Milky Way, probably triggering a burst of star formation. The Milky Way has already been affected by an interaction with the Magellanic Clouds, which produced a warp or tilt in the outer part of the disk.

The gravitational behavior of the galaxy will be unchanged, however, the dead and dying stars remaining in their orbits around the galactic center. With no hot stars to energize them, no spectacular H II regions will exist. Only a few wisps of gas will be seen here and there. The galaxy will literally fade away, even if the universe is closed, because the eventual collapse and possible rebirth of the universe in another Big Bang would happen (if at all) tens of billions of years after our galaxy dies. All in all, a rather drab ending for our lovely Milky Way.

We've come a long way from the tiny, Earth-centered world of our ancestors, a world that was thought to have been created just for our benefit. Now, however, our claim to fame, our uniqueness, no longer depends on any special position we occupy—we aren't the center of anything in the universe. Nor can we boast of any special relation to it. Instead, we can perhaps take satisfaction that we have been able to get some inkling of the nature of the universe, some ideas—which tomorrow may seem primitive—of its structure and evolution. These ideas are based on concepts that we have invented in response to phenomena we experience; we must not forget that our understanding may always be limited, that we may be fundamentally incapable of creating the concepts required for a real understanding of the physical universe. J. B. S. Haldane said it nicely: "The universe is not only queerer than we suppose, it is queerer than we *can* suppose." Nonetheless, the attempt to understand is one of the hallmarks and pleasures of our humanity.

Terms to Know

Halo stars, galactic disks, and spheroids; cD galaxies; starburst galaxies; baryons; hot and cold dark matter.

Ideas to Understand

The origin of the chemical elements and how their abundances change with time; the role played by the Big Bang. Why most of the helium we observe was not produced in stars. Our physical relation to the stars. The hierarchy of astronomical masses and the clumpiness problem; possible schemes for its solution; the role played by the environment in galaxy formation; galactic mergers. The formation and fate of the Milky Way.

Questions

1. Could the very first massive stars formed in our galaxy have generated energy by the carbon cycle? By the proton-proton chain? Explain.

2. (a) What was the chemical composition of the universe five minutes after the Big Bang?
(b) How does it differ from the composition today?
(c) How did this difference come about?

3. (a) What will our galaxy look like in the next 10 billion years?
(b) How will M31 (our spiral-galaxy neighbor in Andromeda) look in 10 billion years?

4. (a) Compare how matter was distributed in the very early universe with what we see today.
(b) What role might dark matter have played in this transformation?

5. Compare the likelihood of a collision between two stars in the solar neighborhood with that of two galaxies in a cluster. What does this imply for the formation of peculiar galaxies and giant galaxies?

6. (a) After reviewing Part III, list the major unsolved problems having to do with galaxy formation and evolution, AGNs, and cosmology.

(b) Consider three or four of these in detail. Describe why they are important and how they might be solved.

Suggestions for Further Reading

Bennett, G., "The Cosmic Origin of the Elements," *Astronomy*, **16**, p. 18, August 1988. The manufacture of the elements in the Big Bang and in stars.

Croswell, K., "Galactic Archaeology," *Astronomy*, **20**, p. 28, July 1992. Formation of the elements and the evolution of our galaxy.

Twarog, B., "Chemical Evolution of the Galaxy," *Mercury*, **XIV**, p. 107, July/August 1985. A review of the observations and a model that attempts to account for them.

Many of the books listed at the end of Chapter 19 discuss the formation and interactions of galaxies. Another useful book is by the following:

Parker, B., *Colliding Galaxies: The Universe in Turmoil*. New York: Plenum Press, 1990. Though the emphasis is on galaxy collisions, this book gives an account of the development of much of modern cosmology.

In addition, the following articles discuss various aspects of galaxy formation and interactions:

Allen, D., "Star Formation and IRAS Galaxies," *Sky & Telescope*, **73**, p. 372, 1987. A chatty and personal account of observing some of the bright infrared galaxies discovered by IRAS.

Barnes, J., Hernquist, L., and Schweizer, F., "Colliding Galaxies," *Scientific American*, **265**, p. 430, August 1991.

Dressler, A., "Observing Galaxies Through Time," *Sky & Telescope*, **82**, p. 126, 1991. How galaxies change with time.

Hartley, K., "Elliptical Galaxies Forged by Collision," *Astronomy*, **17**, p. 42, May 1989.

Keel, W., "Crashing Galaxies, Cosmic Fireworks," *Sky & Telescope*, **77**, p. 18, 1989. What happens when galaxies interact.

Silk, J., "Formation of the Galaxies," *Sky & Telescope*, **72**, p. 582, 1986. A good account of modern ideas.

Smith, D., "Secrets of Galaxy Clusters," *Sky & Telescope*, **73**, p. 377, 1987. Some of the characteristics of clusters.

Sulentic, J., "Odd Couples," *Astronomy*, **20**, p. 36, November 1992. The formation and evolution of binary galaxies.

Some aspects of large-scale structure and its formation are described in the following articles. The first two articles describe the streaming of nearby clusters of galaxies toward some enormous mass concentration:

Burstein, D. and Manly, P., "Cosmic Tug of War," *Astronomy*, **21**, p. 40, July 1993.

Dressler, A., "The Large-Scale Streaming of Galaxies," *Scientific American*, **257**, p. 46, September 1987.

Geller, M. and Huchra, J., "Mapping the Universe," *Sky & Telescope*, **82**, p. 134, 1991. How maps of galaxies are made and what they tell us, by two of the leaders in the field.

Gregory, S. and Morrison, N., "The Largest Supercluster Filament," *Mercury*, **XV**, p. 54, March/April 1986.

Schramm, D., "The Origin of Cosmic Structure," *Sky & Telescope*, **82**, p. 140, 1991. Formation of structure on various size scales, role of dark matter, GUTs.

van den Bergh, S. and Hesser, J., "How the Milky Way Formed," *Scientific American*, **268**, p. 72, January 1993. A good account of current ideas.

PART IV

Worlds Beyond the Earth

"As for the Yankees, they have no other ambition than to take possession of this new continent of the sky (the Moon), and to plant upon the summit of its highest elevation the star-spangled banner of the United States."

—Jules Verne, 1865; French science-fiction author

One of the over 60 launches of captured World War II German V-2 rockets from the White Sands (New Mexico) Proving Grounds in the late 1940s and early 1950s. These launches began the era of space astronomy.

CHAPTER 23

Planetary Preliminaries

In this portion of the book we return home to our solar system, though in the last chapter we will look outward again. Primarily we will be concerned with what we know about the physical properties of the major constituents of the solar system. When we first discussed the planets, we considered them only as points of light moving somewhat erratically across the sky. With the Copernican view and Kepler's laws it became possible to predict their motions accurately, and to calculate their distances from the Sun in terms of the Earth-Sun distance. Telescopic observations allowed us to measure the planets' angular diameters and, when their absolute distances were known, their linear diameters. Such observations also often enabled their shapes and rotation periods to be determined. Then, with Newton's laws of motion and gravity we learned how to find the masses of the planets. Thus we already know many of their gross properties.

Now, however, we want to describe the physical properties—the temperature, pressure, composition, structure, etc.—of the atmospheres, surfaces, and as far as is possible, the interiors of the major objects of the solar system. We will do this always keeping in mind the question of their suitability for life or their possible connection with life's origin. The objects of our solar system present a wide variety of environments, as well as give us clues about how such a system of star and planets formed in the first place. Surface and atmospheric properties bear directly on a planet's life-supporting ability, and its interior gives us clues about its formation and evolution. In Chapter 24 we will be taking a fresh and much more detailed look at the planets than we have so far done.

In the last chapter of the book we will consider not only the old and intriguing question of the possibility of life elsewhere in our solar system, but, even more ambitiously, the possibilities of its existence beyond our solar system. We will bring our current understanding of conditions within the solar system together with various ideas already encountered in this book, and try to assess the possibility that environments favorable for the support of life exist outside our own solar system. Most of this is speculative; for example, our understanding of how life might form is hazy at best. Even though we cannot yet give even a tentative answer to the question of life elsewhere, it is instructive—and fascinating—to see where our present understanding takes us.

We will begin Part IV with a general introduction to the modern study of the solar system. The latter part of this century will be remembered as the time when we took

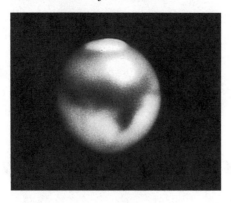
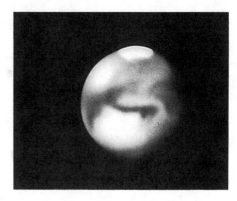

Figure 23.1. Two photographs of Mars taken at the 1971 opposition with the 36-inch refractor at Lick Observatory.

the first tentative steps beyond the cozy confines of our birth planet into the fringes of space. Probably of greater scientific importance, however, are the remarkable un-manned voyages of exploration that have been undertaken. The technological devel-opments that enable us to put scientific instruments above our atmosphere and to send instrumented spacecraft to the planets have produced an explosive growth in our knowledge of the planets and their satellites. Consequently, we will begin with a brief account of some of the principles of spaceflight, which are essentially an application of Kepler's and Newton's laws of planetary motions. This is followed by a brief dis-cussion of the kinds of instrumentation carried by spacecraft. The rest of this chapter is devoted to an overview of the major properties of the solar system, along with some of the general processes operative on the planets and their satellites, and finally to a consideration of their origin.

The Space Age and the Solar System

The full impact of the space age on astronomy was first felt in solar-system studies. Ever since the beginning of astrophysics early in the twentieth century, most astron-omers had concentrated their attention on the stars, using the powerful new tools of accurate photometry and spectroscopy, along with the new theories of matter and ra-diation being developed by physicists. Solar-system objects were not studied inten-sively for several reasons.

First, the quality of much of the data was poor. Detailed photographs are frus-tratingly difficult to take because of the blurring produced by the Earth's atmosphere. Only for the nearby Moon can the best photographs reveal features as small as about one kilometer. Photographs of Mars taken when the planet is nearest the Earth, and with one arc second of seeing, show features no smaller than about 275 kilometers across (see Figure 23.1). As seen from the Earth, Jupiter's angular diameter is at best only about 45 arc seconds, so that it is difficult to record details much smaller than about one-quarter the size of the Earth (Figure 23.2). Visual observations made at mo-ments of good seeing are very tricky, and led to outlandish claims by some observers, for example, that Mars was inhabited by a dying civilization. Such fantasies tended to discredit solar-system studies in general.

Second, the physical structure of planets is more difficult to understand than that of stars, because planets are not gaseous throughout, but are solid and liquid as well.

Figure 23.2. An excellent ground-based photograph of Jupiter, taken with the 2.1-meter reflector at Kitt Peak National Observatory.

As we have pointed out before, the equations of state of solids and liquids are much more complicated and uncertain than that of a gas. The theoretical framework, as well as much of the necessary data, on which to build an understanding of planetary structure was weak. Finally, astronomy, like most other human activities, has its fads and fashions, and solar-system studies were simply out of fashion, not considered "serious" work by most astronomers.

Much of this changed dramatically when we were able to overcome the limitations imposed by our atmosphere, that is, when it became possible to send astronomical instruments above the Earth's atmosphere. Spacecraft have landed on the Moon, Venus, and Mars; they have also been put into orbit around these objects, carrying out detailed reconnaissances of their surfaces. Other spacecraft have passed close by Mercury, Jupiter, Saturn, Uranus, and Neptune and many of their satellites, as well as by comets. We now have volumes of wonderful high-resolution images of the surfaces of these objects (see Figures 23.3 and 23.4). (Interestingly, the resolution so far achieved for most

Figure 23.3. An image of Mars taken by one of the Viking spacecraft as it approached the planet. Compare the detail here with that in Figure 23.1!

Figure 23.4. Jupiter as recorded by Voyager 1 from a distance of 32.7 million kilometers on February 1, 1979. The two cameras on Voyager had apertures of only 6.5 and 17.5 centimeters. The former camera produced this image.

of the planets and their satellites is roughly equivalent to that possible for the Moon from the Earth. Think of the surprises if we could land instruments on some of these objects.) In addition, we have made measurements of temperatures, densities, and compositions of the surfaces and cloud layers of the planets and their major satellites. We can study these bodies in ways that were undreamt of only a few decades ago. In fact, the planets are becoming as much the province of geologists (inappropriate term!) and meteorologists as of astronomers. This specialization is a direct consequence of the vast increase in our knowledge of the solar system brought about by space exploration. Since interplanetary voyages have been fundamental to our new understanding of the solar system, it is appropriate to consider some of the orbital aspects of spaceflight.

Spaceflight: General Considerations

What are some of the factors that must be taken into account in sending a spacecraft to a planet? What kinds of orbits are followed? Though the details are complex, the principles of the process are simple. Recall Newton's generalization of Kepler's third law of planetary motion given in Chapter 8:

$$(M_1 + M_2)P^2 = \left(\frac{4\pi^2}{G}\right)a^3.$$

Let us apply it to a satellite in orbit around the Earth. Let M_1 be the mass of the Earth, which we know; and M_2 that of the spacecraft, which is completely negligible by comparison with the Earth; $(4\pi^2/G)$ is just a number, the value of which depends on the units in which G, the constant of gravitation, is given. With this formula, we can calculate what the period would be of an Earth-circling satellite in an orbit with

*Table 23.1. The speed and period
of a satellite at various altitudes*

Altitude (km)	Period (min)	Speed (km/hr)
0	84.4	28,400
200	88.4	28,000
400	92.4	27,600
600	96.5	27,200
1,000	105.	26,400
35,900	24 hrs	9,390

semimajor axis a. Table 23.1 gives the satellite's period and speed in a circular orbit for a few altitudes above the Earth's surface. The semimajor axes of these orbits are, of course, measured from the center of the Earth.

Zero altitude is given for illustrative purposes; to say the least, it would be a bit tricky to fly a spacecraft around the Earth at sea level. Notice that the period of a near-Earth satellite is about an hour and a half.[1] The Earth rotates inside the satellite's orbit so that spacecraft with periods less than 24 hours rise in the west and set in the east (assuming the launch is to the east, as it is from Cape Kennedy in Florida).

Geosynchronous Orbit.   Note that as a satellite orbit becomes larger and larger, its period increases and its speed decreases. When its semimajor axis is about 42,200 km (so that the satellite is about 35,900 km above the Earth), its period is the same as the rotation period of the Earth, so that the satellite stays above the same point on the Earth. Such an object is said to be in a **geosynchronous orbit**. Communication satellites are put into such orbits, because they can then be in continuous contact with the same ground stations (see Figure 23.5). It is also advantageous for a space observatory to be in a geosynchronous orbit. Not only is the satellite always in direct contact with a ground station and so easier to operate, but much less of the sky is blocked by the Earth, since it is so far away from the spacecraft. In a near-Earth orbit, half the sky is occulted by the Earth, whereas in a geosynchronous orbit the Earth subtends an angle of less than 20 degrees. As a consequence, the amount of observing time available is about three times greater than in a near-Earth orbit. The difficulty, of course, is that it requires much more energy to boost a spacecraft to such a high orbit. The remarkably successful International Ultraviolet Explorer (as of this writing it has been operating for 15 years!) with its 46-cm telescope feeding ultraviolet-sensitive spectrographs, is in a geosynchronous orbit (Figure 10.31).

Near-Earth orbiting satellites won't stay up forever because, though the atmospheric density is very low, it is not zero. Consequently there is some drag, some friction with the residual air, which cause satellites to lose energy and eventually spiral into the Earth. Objects in orbit only a few hundred kilometers above the Earth will remain there for just a few years; those at altitudes of about 800 km or higher will stay in orbit for centuries. Near space is becoming increasingly populated with thousands of pieces of orbiting junk, ranging in size from nuts and bolts to whole satellites.

---

[1] Although the orbit of such a satellite precesses slowly (its orbital plane makes one rotation in about two months), compared with the much more rapid rotation of the Earth it is nearly fixed in space.

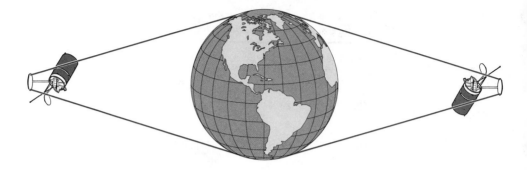

Figure 23.5. Two communication satellites in geosynchronous (24-hour) orbits can transmit and receive messages to and from very nearly the entire Earth.

Minimum Energy Orbit.   Now, suppose that a satellite in a circular orbit at an altitude of 200 km above the Earth is given a speed greater than 28,000 km/hr. Its orbit would then become elliptical, and the greater the speed it was given, the larger would be the size of its orbit. This suggests that we can think of the launching of an Earth-orbiting satellite to consist of two steps: first, get it to the desired altitude; second, give it a velocity increase in the right direction that will result in the orbit of the desired eccentricity and size.

Next, let's suppose that we want to leave Earth orbit and journey to some distant planet in the solar system. When we are in a circular orbit around the Earth, our speed is about 8 km/sec with respect to the Earth. To escape from Earth orbit and go into orbit around the Sun, but still at a distance of 1 AU from it, we would increase this speed by only about 3 km/sec. As you can see, the first step is the hardest. It takes a speed of 8 km/sec to get into an Earth orbit, but only 3 km/sec more to leave the Earth behind and go into an orbit around the Sun.

If we want to go to another planet rather than just staying in orbit around the Sun, however, we would have to change our speed yet again. For a trip to Mercury or Venus we would have to lose energy (by slowing down), but to travel to Mars or Jupiter we would have to gain energy. Either way, we would have to fire rockets and use fuel; that is, slowing down as well as speeding up requires energy. Once the appropriate velocity (the right speed in the proper direction, and so the proper orbit) is reached, however, we could just sit back and coast to the vicinity of our destination. Such a trip generally follows a so-called **minimum-energy orbit** because it requires the least velocity change and so the least energy; it takes the longest time, however. Many interplanetary space voyages have followed approximate minimum-energy orbits. Finally, to land on the planet or go into orbit around it would require additional velocity changes and, therefore, more energy yet.

A Trip to Mars.   Let's apply these ideas to a trip to Mars, step by step. First of all, the spacecraft is launched into an Earth orbit. To take maximum advantage of the velocity of the rotating Earth, 0.46 km/sec at the equator, we launch in a generally eastward direction. (For safety reasons, the launch site should be on a coast with ocean to the east; hence the choice of the Florida Atlantic coast for U.S. launch facilities.) Now we are in a near-Earth circular orbit. To fly to Mars with the least expenditure of energy, we must go into an elliptical orbit around the Sun such that, at its greatest distance from the Sun, the spacecraft just reaches the orbit of Mars for which $a = 1.52$

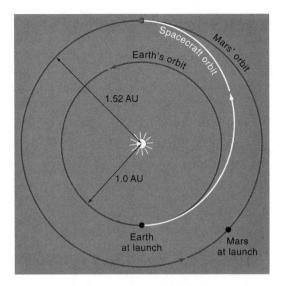

Figure 23.6.   The spacecraft orbit shown requires the least fuel to go from Earth to Mars. Such voyages can be made only when the Earth and Mars are oriented as shown, that is, at intervals of the synodic period of Mars or about every 2.1 years. See text for description.

AU. To make as much use as possible of the Earth's orbital velocity, we fire our rockets in a direction tangential to the Earth's orbit (see Figure 23.6). The closest approach to the Sun (perihelion) is at 1.0 AU, whereas the greatest distance (aphelion) is 1.52 AU. Hence, the semimajor axis of the orbit is given by

$$a = 1/2 \ (1.0 + 1.52) = 1.26 \ \text{AU}.$$

Using the formula given in Chapter 8,

$$V = \sqrt{GM\left(\frac{2}{r} - \frac{1}{a}\right)}$$

with $GM = 1.33 \times 10^{26}$ in units of centimeters, grams, and seconds, and also $r = 1.0 \ \text{AU} = 1.5 \times 10^{13}$ cm, $a = 1.88 \times 10^{13}$ cm, we can calculate that this orbit requires a speed at perihelion of 32.6 km/sec. The speed of an object in a 1-AU circular orbit, however, is only 29.6 km/sec. Thus we need to give our spacecraft an additional speed of 32.6 − 29.6 = 3.0 km/sec, tangential to the Earth's orbit. With this successfully accomplished, we would then be in a solar orbit with a semimajor axis of 1.26 AU. With no further velocity changes (and barring a collision with Mars), we would simply continue to travel around the Sun in this orbit.

   How long will the trip to Mars take? By Kepler's third law, $P^2 = a^3 = (1.26)^3 = 2.0$; so $P = \sqrt{2} = 1.4$ years. This is the time required for a complete trip around the orbit, but to get to Mars we need to go only halfway around. Hence the trip will take 0.7 years, or about 8.4 months, roughly the time taken by Mariner 4, which was launched on November 28, 1964, and arrived at Mars on July 15, 1965. At the time of launch, then, Mars must be about where it is indicated in Figure 23.7; otherwise when we arrive at Mars' orbit, Mars won't be there. This is one of the serious limitations of a minimum-energy orbit: it restricts the period of time during which a spacecraft can be launched if it is to rendezvous with a particular object. Typically these

launch windows are open for a few weeks to a month or two. Similarly, to leave Mars so that we arrive at a point in the Earth's orbit when the Earth is also there means that the departure from Mars must again be carefully timed. Such launches must take place only when the Sun, Earth, and Mars are in, or close to, a particular configuration. If a launch opportunity is missed, we must wait until that same configuration occurs again; that is, we must wait one synodic period. For Mars, minimum-energy orbit launch opportunities occur every 2.1 years.

We left our spacecraft coasting toward the point where Mars will be when we arrive at its orbit. At the intersection of the two orbits, the spacecraft's velocity will be 21.6 km/sec, whereas Mars' orbital speed is about 24 km/sec. The speed difference is a consequence of the two different orbits: our spacecraft is in an elliptical orbit with $a = 1.26$ AU, whereas Mars is in a more nearly circular orbit, with $a = 1.52$ AU. To transfer from our orbit to Mars', we must increase our velocity by about 2.4 km/sec. Finally, if we want to land safely, we must counter the escape velocity of Mars (which is also the velocity we would acquire by falling into the planet from a great distance) by another velocity adjustment, this one of about 5.1 km/sec. In this way we could make a soft landing on Mars.

To summarize, four orbital changes are required for our journey: from Earth's surface to Earth orbit; from Earth orbit to Sun orbit with $a = 1.26$ AU; from Sun orbit to Mars orbit with $a = 1.52$ AU; finally, from Mars orbit to Mars' surface.

The positions and velocities of spacecraft can be carefully measured from the Earth throughout a flight, enabling course corrections to be made with great precision. Solar-system navigation is remarkably accurate, and for most purposes does not require the presence of astronauts in the spacecraft.

Gravity Assist.   Exploring the inner solar system by traveling along a minimum-energy orbit does not take too long, since the distances are relatively small. Getting to the outer planets beyond Jupiter is another matter, however; distances are much greater, and flight times become many years rather than months. Under special circumstances, however, there is a quicker way to travel, called the **gravity-assist** technique. The idea is simple. Send a spacecraft out toward a planet, Jupiter, say. As it approaches the planet it speeds up, its trajectory is changed, and it gains energy at Jupiter's expense. That is, the energy gained by the spacecraft is lost by Jupiter. The spacecraft then heads toward its destination along a new trajectory. Obviously, the spacecraft's ultimate destination and the planet or planets from which energy is acquired must be oriented in an appropriate way for this technique to be useful.

A spectacular example of this method is the flight of Voyager 2, which took advantage of the unusual circumstance that all the giant planets were on the same side of the Sun, to visit each one, and to use each for a gravity assist to the next (see Figure 23.7). Voyager reached Neptune 12 years after launch, abut 160 years sooner than it would have done by following a minimum-energy orbit. Because a suitable second-stage rocket was not available, the Galileo probe to Jupiter, launched by the space shuttle in 1989, has used gravity assist by the Earth and Venus to reach its destination in 1995.

Spacecraft Instrumentation

There is little point in just sending a spacecraft to a planet; we must have instruments on board to make the kinds of measurements we want. Generally the payload, be it scientific instruments or people, is only a tiny fraction of the mass of the spacecraft. Most of the latter's mass is made up of structure and particularly of fuel.

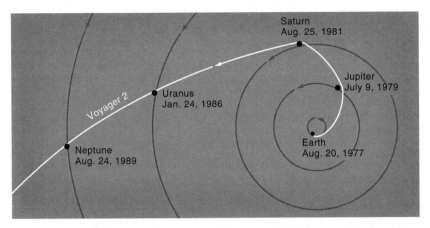

Figure 23.7. The flight path taken by Voyager 2 on its remarkable trip to the giant planets. It received gravity assists (it acquired energy) from the planets, especially from Jupiter and Saturn.

Subject to limitations of weight, volume, and electrical power, just about any kind of instrument can be, and has been, flown in space. Cameras, spectrographs, photometers, all sensitive to radiation from the radio to the x-ray region of the spectrum, instruments that directly measure the magnetic fields and the composition and energies of particles in interplanetary space or in planetary atmospheres, and devices that perform biochemical experiments on planetary soil have all been sent into space. These instruments are operated by radio commands from the ground, and the data obtained must be transmitted back to the ground. This means, of course, that the on-board data must be transformed into a series of electrical signals. Cameras, for example, don't use film; instead, they are usually some kind of a television or CCD camera.

The techniques for building reliable equipment for spaceflights are fairly well understood, and instruments nearly as sophisticated as any on the ground can now be flown, and usually they work. We can make the full range of observations from space just as we can from the ground. Thousands of detailed images of planetary surfaces and atmospheric cloud layers have been taken and entire planets mapped; these now allow geologists and meteorologists to investigate worlds other than the Earth; the compositions, temperatures, and densities of planetary atmospheres have been measured; the energies and constituents of the solar wind have been determined by direct sampling; the list is nearly endless.

Some of these observations have a serious limitation, however. The highest-resolution photographs of Jupiter and Saturn, for example, were made by the Voyager fly-bys; that is, the spacecraft did not go into orbit around the planets, but flew by them. Consequently, the period of time during which the most detailed images were obtained was only a few days long. Ideally, spacecraft would be placed into orbit around a planet for a long period of time. In this way it would be possible to monitor the Jovian weather, for example, in great detail.

Exploration by Radar. A relatively new ground-based technique—**radar**—has become increasingly important for planetary work since World War II and has already been used extensively in space. A radar telescope consists of a parabolic dish at the focus of which is a high-powered source of pulsed radio waves. Radio waves are used because they can penetrate the atmospheres of planets. Just as incoming parallel rays of light from a distant object are brought to a focus at the focal point of a parabolic

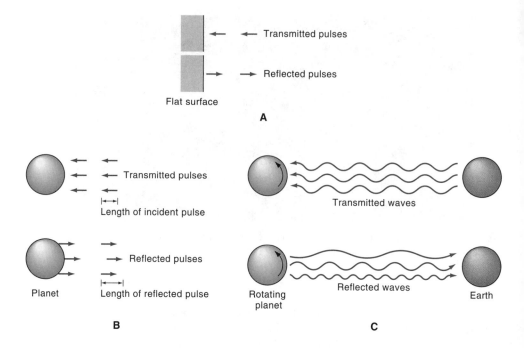

Figure 23.8. By timing how long a short pulse of radio waves takes to reach and return from its target, the latter's distance can be determined (this is essentially how a radar altimeter works). (A) When the target is flat, the reflected pulses are of the same duration as the transmitted pulses. (B) When the target is a planet, however, the part of the pulse reflected from the equator arrives sooner than that reflected by the rest of the planet. Hence, the reflected pulses last longer than those transmitted. The amount the pulse is spread is related to the radius of the planet. (C) Because the planet is rotating, the wavelength of the reflected pulses will be lengthened (according to the Doppler effect) when reflected from the receding limb of the planet, and shortened when reflected by the approaching limb. This effect enables the rotation rate of the planet to be measured.

dish, radiation from a transmitter at the focus of a paraboloid will be reflected out in a parallel stream from the dish. The dish is pointed toward a planet, Mercury, for example, and the return signal that bounces off Mercury is detected by the same dish. Great sensitivity is required, since only a tiny fraction of the transmitted signal returns to the antenna. By an analysis of the returned signal, the distance, rotation rate, orientation of the spin axis, and radius of Mercury can be measured. For a nearby object such as Venus, it is even possible to produce a topographic map of the planet's cloud-shrouded surface.

It is easy to see how in principle, at least, this kind of information can be obtained. Radio waves reflected by a flat wall perpendicular to the line of sight will come back to the dish (after twice the light-travel time from the antenna to the wall) as a pulse of the same duration as that which was transmitted (see Figure 23.8). Now, however, suppose that a spherical planet reflects the pulse. The return pulse will be lengthened by the time required for it to bounce back from different parts of its surface. In other words, as compared with the incident radiation, the reflected pulse will be stretched out in space (and so in time) by a distance equal to the radius of the planet. In this way the radius of the perpetually cloud-covered body of Venus was measured.

If the planet is rotating, then the limb moving toward us reflects radiation back at a higher frequency than the transmitted pulse, whereas the limb moving away from us

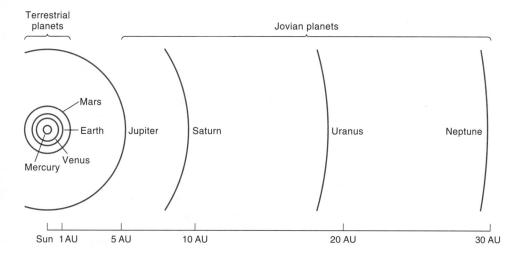

Mars

Earth

Jupiter

Saturn

Uranus

Neptune

Venus

Mercury

Sun 1 AU   5 AU   10 AU   20 AU   30 AU

Figure 23.9. The small, low-mass terrestrial planets are located very near to the Sun, whereas the massive Jovian planets are far from the Sun.

Doppler shifts the radiation to lower frequencies. The velocity deduced by the Doppler shift can be combined with the size of the planet to yield its rotation period. Beginning in the 1960s this technique was used to measure the rotation period of Mercury and Venus.

Now that we have some idea of how voyages in the solar system are made and the kinds of observations that have been carried out with spacecraft, let us turn our attention next to the general characteristics of the only planetary system we know about, how it might have formed, and what processes are active in shaping its members.

Solar-System Properties and Processes

A Quick Look at the Solar System

The Planets.   Since one of the aims of this chapter is to consider the origin of the solar system, let's very quickly describe its members and some of its important features. Recall that the planets revolve around the Sun in the same sense (counterclockwise as seen from the north) and in about the same plane (the ecliptic), which is also very nearly the plane of the Sun's equator. Except for Venus, which slowly rotates in the clockwise sense, the planets' and Sun's axial rotations are also counterclockwise (although both Uranus and Pluto are tipped over on their axes so that their "north" poles are south of the ecliptic and their "south" poles rotate in the clockwise sense).

The Sun makes up about 99.9 percent of the mass of the solar system. In that sense, the planets are little more than bits of cosmic debris remaining from the star-formation process. One of the most striking features of this tiny family is that eight of the nine planets fall naturally into two groups: the relatively nearby terrestrial (Earth-like) planets—Mercury, Venus, Earth, and Mars, all within about 1.5 AU of the Sun; and the more-distant gaseous, giant, Jovian (Jupiter-like) planets—Jupiter, Saturn, Uranus, and Neptune (Figure 23.9). By contrast with the terrestrial planets, the Jovian planets are much more widely spaced, their orbits extending from about 5 AU to 30 AU from the Sun (see Appendix F).

All the terrestrial objects are small (Earth is the largest of the group), and not very massive, but have rather high average densities, three to five times that of water. They all have solid surfaces, and Venus, Earth, and Mars have atmospheres. Venus' atmosphere is so deep and dense that it completely hides the surface; all we see is a nearly featureless cloud cover. Though Mars' atmosphere is thin, it is dense enough to stir up dust storms, which occasionally make it impossible to see its surface, sometimes for weeks at a time.

All together, the terrestrial planets have only three satellites: our Moon and the two tiny satellites of Mars. These last two are irregularly shaped objects with average radii of only about 11 km (Phobos) and 6 km (Deimos); they are most likely asteroids captured by Mars. In some ways it may be more appropriate to consider the Earth-Moon system to be a double planet rather than a planet and its satellite. In this sense, then, the terrestrial planets could be considered to have no satellites at all.

Though two of them break the clean separation by solar distance, three more objects should be added to the group of terrestrial planets: our own Moon and two satellites of Jupiter, Io and Europa. Not only do they have solid surfaces and densities similar to those of the terrestrial planets, but their radii are about 1,000 km or larger. As we shall see, this means that they are sufficiently massive to have undergone significant internal evolution, as have the Earth-like planets themselves.

In marked contrast, the Jovian planets are much larger and more massive than the terrestrial planets, and have low densities; Saturn would float in a sufficiently large ocean. Jupiter, the largest, has a diameter eleven times the Earth's; the diameter of the smallest Jovian planet, Neptune, is still nearly four times the Earth's. They rotate rapidly, Jupiter, for example, in only about ten hours. As a consequence it has a pronounced equatorial bulge. In addition, it rotates somewhat more rapidly at the equator than it does at the poles. All this exemplifies the fact that we are seeing a thick gaseous envelope rather than a solid surface. In fact, the giant planets do not have solid surfaces.

The Jovian planets have large families of satellites: so far 16 have been discovered orbiting Jupiter, 19 around Saturn, 15 around Uranus, and 8 belonging to Neptune. It is quite likely that there are still undiscovered small satellites in orbit around these planets. Titan and Ganymede, moons of Saturn and Jupiter, respectively, are both larger than Mercury; Callisto, another satellite of Jupiter, is only a few kilometers smaller in diameter than Mercury. The four giant planets have ring systems as well, though only Saturn's is prominent. Its rings are made up of myriads of chunks of ice, confined to an extremely thin plane, all in orbit around their parent planet.

Pluto, the ninth and most distant planet, is something of a misfit. Its small size excludes it from the group of giant planets, but its low density is not characteristic of terrestrial objects. Its properties resemble those of a Jovian satellite, which led to the suggestion that it is an escaped satellite of Neptune. The recent discovery that Pluto has a satellite makes this idea less plausible, however.

Small Objects.   About 200 years ago it was pointed out that the spacing of the planets then known could be given by a simple arithmetic rule (see Table 23.2). This rule is often referred to as Bode's "law," which is somewhat misleading, because it apparently does not arise from any physical effect; furthermore, as the table shows, it does not work well for Neptune and Pluto. Nonetheless, it was sufficiently suggestive to cause astronomers to organize a search for the planet that the "law" predicted to exist between Mars and Jupiter. Before the search began, however, a small object, Ceres, was accidentally discovered in 1801. With a semimajor axis of 2.8 AU, it was at the distance given by the rule and was identified as the "missing" planet. (Interestingly, at first it was announced to be a comet, just as happened with Uranus.) Soon after, however, many more small objects, now collectively called **asteroids**, were

Table 23.2. Bode's "law"

Double after 0	Add 4	Divide by 10	Actual a (AU)	Planet
0	4	0.4	0.39	Mercury
3	7	0.7	0.72	Venus
6	10	1.0	1.00	Earth
12	16	1.6	1.52	Mars
24	28	2.8	2.77	Ceres
48	52	5.2	5.20	Jupiter
96	100	10.0	9.56	Saturn
192	196	19.6	19.22	Uranus
384	388	38.8	30.11	Neptune
768	772	77.2	39.44	Pluto

found in the Mars–Jupiter gap; the orbits of about 5,000 of the larger objects have been cataloged (see Figure 23.10). With a diameter of about 1,000 km, Ceres is the largest and amounts to nearly half the total mass; the smaller members of the group are much more numerous than the larger ones, there being perhaps 100,000 with diameters of a kilometer or more. In sum, however, their mass is no more than about 5×10^{-4} that of the Earth's. Clearly, there never was a full-size planet in the asteroid belt.

Like the planets, the asteroids travel around the Sun in direct orbits, but their orbital eccentricities (largely between 0.1 and 0.3) and inclinations (up to 30°) are larger than those of most of the planets.

Comets are often characterized as "dirty snowballs," that is, a mixture of dust particles and ices of various kinds. Most comets exist in the outermost regions of the solar system where they are impossible to see because their nuclei (the snowball) are only a few kilometers in diameter. Only when one approaches within a few AU of the Sun does solar radiation cause its ices to vaporize, releasing dust particles at the same time

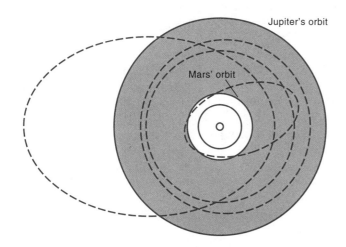

Figure 23.10. Most, but not all, asteroids (with orbits shown here as dashed lines) are located between the orbits of Mars and Jupiter (shaded area).

Figure 23.11. Halley's comet as it appeared on June 6 (left) and June 7 (right), in 1910. Note the marked changes in the tail. Because it was near the Sun when it was close to the Earth, the comet was a spectacular sight in 1910, in contrast to its appearance in 1985–86.

and producing the often spectacular tails we associate with comets (Figure 23.11). Comets are not only of interest in themselves, but also because they are very likely part of the original matter from which the solar system formed, unprocessed in any significant way.

Objects similar to asteroids and comets bombarded all the planets early in the history of the solar system. Even today, the Earth's mass is increased by about 10,000 tons of extraterrestrial debris each year. Small orbiting particles called **meteoroids** fall toward the Earth usually at speeds of 10 to 30 kilometers per second and glow briefly as they are heated by friction with the gas of the upper atmosphere and destroyed. We call these tracks **meteors** or, more commonly perhaps, "shooting stars" or, for the brightest, "fireballs." Those that survive their plunges through the atmosphere and strike the Earth's surface we call **meteorites**. These are of particular interest to us because, apart from the Earth and Moon, they give us our only direct samples of solar-system matter.

Most of the meteors we see are caused by tiny fragments of cometary material, too fragile to survive their atmospheric entries. As we will see in the next chapter, meteorites are samples of the asteroids, rocks that are able to reach the Earth's surface.

So the Sun has a small retinue of planets and their satellites, along with remnants of the cloud from which the solar system formed. Before considering how that might have happened, let's see how old the system is.

The Age of the Solar System

An important property of the solar system is its age. Clearly, the time available plays a significant role in any attempt to understand the formation and evolution of our solar system. In Chapter 15 we saw that according to stellar evolution theory the Sun is no more than five billion years old. The planets (as planets) must be younger than that, of course. The oldest rocks so far found on Earth have ages of about 4.1 billion years, the oldest Moon rocks are about 4.4 billion years old, and the sample of asteroids recovered on Earth are about 4.6 billion years old. The latter may be the sort of matter from which the planets formed. Thus solar-system objects condensed soon after the formation of the Sun.

How do we measure the ages of rocks? Relative ages (this feature is younger than that one) of many geological features can be found just by seeing which layers of sedimentary rock, for example, lie above others and so are more recently deposited. Small numbers of impact craters can often be arranged in an age sequence according to their overlap, the younger craters falling across the older.

Absolute ages of rocks (this rock is 4.2 billion years old) can be found from the **radioactive decay** of heavy atomic nuclei contained in the rocks. Some nuclei, such as uranium, thorium, and potassium, are not stable, but are said to decay, that is, spontaneously to emit alpha, beta, or gamma radiation (recall that these are helium nuclei, electrons, or high-energy photons, respectively). The rate at which these processes occur is independent of external circumstances such as temperature or pressure. These nuclear decays continue until a stable end-product is formed. For U^{238} the stable final product is Pb^{206}, which has 10 fewer protons and 22 fewer neutrons than uranium. It is impossible to predict which particular nucleus will decay at any given time. One can measure only the **half-life** of a particular element, that is, the time required for half the nuclei in a sample to be transformed to their final product (see Figure 23.12). The half-lives and decay products for some radioactive nuclei are given in Table 23.3.

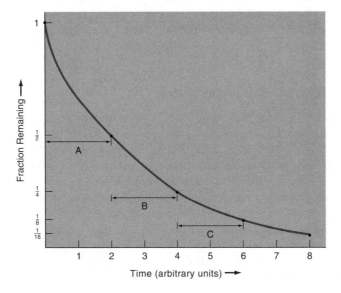

Figure 23.12. Suppose the time for half of the nuclei in a sample to emit one or more particles and be transformed to the nonradioactive, stable product is two units of time (two billion years, for example). After two time units (span A) one-half of the sample is transformed; after another half-life (span B), half of the remaining sample is transformed (the fraction of the sample is decreased from one-half to one-quarter), etc. Note that in each half-life, it is the same fraction of the nuclei in a sample that is transformed, but not the same number.

*Table 23.3.   Half-lives and decay products for
long-lived radioactive nuclei*

Nucleus	Half-life (10^9 yr)	Final decay product
Uranium 238	4.47	Lead 206
Uranium 235	0.70	Lead 207
Thorium 232	14.0	Lead 208
Rubidium 87	48.8	Strontium 87
Potassium 40	1.25	Argon 40
Iodine 129	0.016	Xenon 129

Imagine that a rock has just solidified, so that no more atoms can be combined with it. Suppose a particular sample of the rock contains radioactive K^{40}, but does not contain any Ar^{40}, which is produced by the decay of the potassium. As time passes, nuclei of potassium will disappear and those of argon will appear, until after about 1.3 billion years half the potassium will have decayed into argon; after 2.6 billion years, only one-quarter of the nuclei will be potassium, and three-quarters will be argon; and so on. Thus, from the ratio of the numbers of potassium and argon nuclei, we could calculate the age of the rock since its solidification.

This is straightforward enough, but what about the obvious possibility that a little argon already existed in the rock when it formed? Unless this were taken into account, we would infer too great an age for the sample. Hence we need to measure the amount of Ar^{40} that exists in nonradioactive rocks to establish the relative abundance of this isotope of argon that is not produced by radioactivity. This turns out to be easy to do, and a reliable correction can be made for the nonradioactively produced argon. Thus, ages of rocks can be fairly accurately established and a minimum age of the solar system determined.

The Formation of Planetary Systems

What Is a Planet?   Since we know of only one planetary system, our discussion of origins will necessarily center on it. First, however, let's ask what we mean by a planet. This is not as trivial a question as it might appear, because we must be careful not to let the Earth (or even the solar system) completely determine our ideas. For example, if we were to define (carelessly!) a planet as having a solid surface, then the Jovian planets are disqualified. If a planet is an object producing absolutely no energy of its own, then not only are the Jovian planets eliminated, but the Earth is as well. The former produce much infrared energy of their own, and radioactivity generates a very small outflow of energy from the Earth. Generally speaking, when a planet is said to be an object that produces no luminosity of its own, we mean that it produces no visible radiation, no radiation in the wavelength region where the object is brightest in reflected sunlight.

Size is not a good criterion. Many satellites are larger than Pluto and even if the latter's status is ambiguous, everyone agrees that Mercury is a planet, but two satellites, Ganymede (Jupiter) and Titan (Saturn), are larger than it is.

The recently discovered objects in orbit around a pulsar (Chapter 16) raise interesting questions. If a body of planetary mass orbiting a star defines a planet, then these objects qualify as planets. But is that too general? For example, should the formation process be part of the definition? Our planetary system formed from the same nebula that produced the Sun. Since the pulsar presumably was the remnant of a supernova explosion, it is unlikely that the pulsar planets formed as ours did and existed before the parent star became a supernova. Just how they did form is unknown, but one possibility has them as the tiny cores of companion stars whose envelopes have been boiled away by the pulsar radiation. In that case their densities are likely to be very large by the standards of ordinary planets and their matter degenerate, possibly. If so, are they planets?[2]

We saw in Chapter 13 that a self-gravitating sphere of gas with a mass greater than about 0.08 solar masses will burn hydrogen and be a star, not a planet. Objects with masses less than 0.08 but greater than about 0.01 solar masses can't burn hydrogen, but could burn light elements such as lithium. Their average density would be hundreds of times greater than that of any planet in our solar system; they would be exotic planetary cousins, indeed. Could such an object, usually called a brown dwarf, be thought of as a planet or should it be disqualified also? If so, then any self-gravitating object with a mass greater than about 0.01 solar masses would not be a planet. But this might be unnecessarily restrictive. Must a planet be defined as an object that relies on a star for its primary energy source? Perhaps the atmosphere of a slowly cooling brown dwarf might be at a life-supporting temperature for a long time. Should a single, isolated object of this sort (or one orbiting a real star) be called a planet?

Everyone will admit that there is an upper limit to the mass of a planet, though perhaps not agreeing as to what that mass is. Is there a mass *below* which an object shouldn't be considered a planet? Obviously a rock isn't a planet, but is a 50-kilometer-diameter asteroid a planet? Or a 100-kilometer asteroid? Perhaps an object should be thought of as a planet when its self-gravity is large enough to overcome the strength of its material (rock, for example) and it crushes itself into a spherical shape. Depending on the relative amounts of iron and rocky material, this requires a minimum mass of about 10^{23} grams of planetary material, or very roughly 0.01 percent the mass of the Earth and a diameter of about 500 kilometers. This would eliminate most, but not all, asteroids (Ceres' diameter is 1,000 km), as well as many small satellites such as those of Mars, but not our own Moon.

Even though in this chapter we will take our solar system and its planets as defining the norm, the examples above should remind us that perhaps rather different definitions are possible.

Establishing a theoretical understanding of the origin of our solar system is an extremely difficult problem; by no means do we have all the answers, although considerable progress has been made in the last two or three decades. One of the questions we want to keep in mind is whether it is reasonable to think that planetary systems are easily and commonly formed. Or, from a somewhat different point of view, do we know any physical reasons that make the formation of planets unlikely? Unfortunately, we know of only one planetary system, and conclusions based on a sample of one cannot be too firm. Nonetheless, let's see what can be said.

General Difficulties.   Although we know the present positions and velocities of the major components of the solar system quite accurately, it is not possible to use Newton's laws to trace their motions all the way back to the epoch of formation. For one

---

[2] The possibility of discovering planets outside our solar system holds great fascination for astronomers as well as for the public. Given the history of this subject, the latter would be wise to show greater restraint in believing claims of discovery than the former sometimes display in making them!

thing, we do not know their present motions with sufficient precision to allow us to extrapolate back in time more than a small fraction of the age of the system. More importantly, much of the mass and momentum of the material from which the planets formed have long since been spread to great distances from the Sun or left the solar system entirely, and we have no way of knowing the exact role they played. Also, the formation of the system involves not only the gaseous state, but the solid and liquid forms of matter as well, which considerably complicates the problem. Finally, we have only one example of a planetary system to guide and constrain our speculations. If we found even one other system it might enable us to eliminate aspects of competing theories, assuming, of course, that all planetary systems formed in much the same way.

Nonetheless, we do know many features that must be accounted for by any theory if it is to claim our attention. Most obviously, the orbital regularities and compositional differences of the planets must be explained. Chance capture by the Sun of planets that formed elsewhere could not result in the system's striking orbital regularities. Also, the ages of the Sun and various samples of solar-system material are the same. The Sun and the planets must have formed from the same solar nebula. That the planets fall into the terrestrial and Jovian groups with each having markedly different densities (indicating corresponding differences in composition), and each occupying entirely separate regions of the solar system, also make it clear that only a single sequence of events could have been responsible for the origin of the planets. The composition of the low-density giant planets roughly approximates that of the Sun, with hydrogen and helium making up most of their mass, whereas the terrestrial planets have lost those light gases and are made up of the heavier elements only. These differences must fall naturally from any theory worthy of the name.

From the eighteenth century until well into the twentieth century, speculation on the origin of the solar system centered around two different approaches: the planetary system formed after and independently of the Sun, or the Sun and planets formed as part of the same process. In the first instance, a rare interaction of the Sun and a passing object was usually postulated to have produced the system; in the second instance, the solar system was held to be just a natural by-product of the process by which the Sun (and presumably other stars) was formed.

Collision Theories.   An example of the first type of theory was the idea that a near collision of the Sun with a passing comet pulled matter out of the Sun (see Figure 23.13). This material went into orbit around the Sun, cooled, and the planets formed from it. When it was realized how little mass a comet contained, and consequently how it could not pull out matter from the Sun, a variation on this notion was suggested. A star, not a comet, made a close approach to the Sun and was responsible for extracting matter from it, perhaps aided by huge solar eruptions that were then thought to be possible. Is such a near collision likely? Recall that the distribution of stars in our part of the galaxy can be represented by ping-pong balls separated by several hundred kilometers. Hence, any theory invoking a near collision of the Sun and a star implies that the planetary system must be exceedingly rare in our galaxy, possibly even unique. Depending on one's taste, this may or may not be a difficulty with this kind of theory. A definite problem, however, is why the star should have passed by the Sun in the latter's equatorial plane. This is required in order that the ecliptic coincide with the Sun's equator. And why shouldn't the hot gas drawn from the Sun have simply expanded into the vacuum of space rather than have condensed into planets? These are major difficulties, and models of this type are not in favor today.

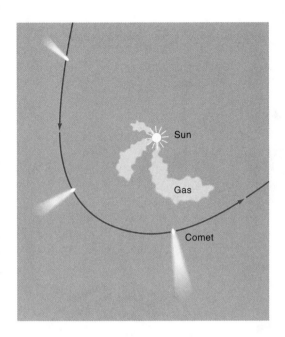

Figure 23.13. This illustrates a type of collision theory for the origin of planets that was popular around the turn of the century. As it passed near the Sun, a comet was postulated to pull huge clouds of gas from the Sun, and the gas then condensed to form the planets.

Nebular Theories. In the late eighteenth century, the German philosopher Kant (who suggested the "island universe" idea) and especially the French mathematician Laplace suggested a mechanism by which planetary systems might form about as often as stars do. This process has become known as the **nebular hypothesis** (Figure 23.14). In its original form it went something like this. A mass of very slowly spinning gas contracted. As more and more mass became concentrated toward the center and to the plane of the nebula, its angular velocity about the center increased in order that the total angular momentum remain constant. As the nebula continued to contract, rings of material were left behind as more material fell into the central regions. The gas in these rings coalesced and formed the planets, while that in the center became the Sun.

On this picture, the Sun would end up rotating very rapidly. However, the present Sun, containing 99.9 percent of the mass of the solar system, is actually rotating very slowly and contains only 2 percent of the angular momentum of the solar system. Most of the angular momentum resides in the planets; mostly in Jupiter, as a matter of fact. So how did the Sun lose so much of its angular momentum? This was a major problem with nebular theories of solar-system origin (in fact with the origin of stars, as well) until the roles of magnetic fields and stellar winds in the early star-formation process became appreciated. As we saw in Chapter 15, early in its history the Sun developed a strong solar wind consisting primarily of electrons and protons. These charged particles were swept around the Sun by its magnetic field and, as they moved outward, carried angular momentum with them. Calculations indicate that this mechanism was sufficient to slow the Sun's rotation to its present rate, and so removed a major problem with the nebular approach.[3] Let us return to our fragment of a molecular cloud, slowly pulling itself together.

---

[3] A further stimulus to solar-system-formation theory was provided by work done by a group of Soviet astronomers that came to the attention of western astronomers only in the early 1970s, even though this work had been ongoing for more than 20 years.

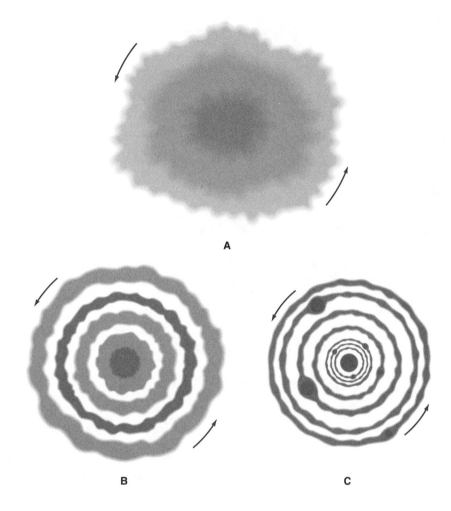

Figure 23.14. About two centuries ago, the first nebular theory for the origin of the solar system was worked out. According to this picture, both the Sun and the planets were formed from the same cloud of material. As more matter became concentrated around the center and it rotated more rapidly, the nebula left behind rings of matter from which the planets formed.

The contracting core of the nebula (which became the proto-Sun) was heated by the release of gravitational energy until the surrounding gas formed a warm (about 2,000 K) disk rotating around the proto-Sun. The contraction slowed, less gravitational energy was released and more infrared energy was radiated away, and the nebula cooled. As soon as the disk temperature allowed, material that could condense from the gaseous state to a liquid or solid (becoming tiny grains) did so. At first, the particles grew as atoms condensed onto the grains that acted as seeds for particle growth. As the particles became larger, any that collided with relative velocities of even one or two kilometers per second would break up. Thus, in order to grow, the relative velocities must have been very slow, on the order of meters per second, so that the particles, perhaps held together by electrical forces, formed loose spongy aggregates rather than hard, tiny marbles.

As the accretion continued and the particles grew, they reached a point when they were no longer buffeted about by the nebular gas, but were massive enough to settle into a thin central plane of the nebula (the ecliptic). Thus concentrated, these plane-tesimals grew more rapidly. When its density became sufficiently large, the disk broke up into cold, gravitationally stable blobs of planetesimals up to perhaps a few kilo-meters in diameter. Computer simulations indicate that to make objects this size took only about ten thousand years. These blobs coalesced and grew into protoplanets. As before, fast head-on collisions would have shattered them. As time went on, however, more and more of the protoplanets were revolving around the protosun in nearly non-colliding orbits so that their relative velocities were small, and they slowly merged and grew, aided now by their increasing gravity. The largest objects grew the fastest, re-sulting in just one planet in each "ring" of the nebula. Lunar-sized objects took per-haps a million years to form and the final planets ten to one hundred million years. Thus the entire process was rather rapid.

According to computer simulations, planetary growth by accretion of planetesi-mals can account for various features of the planetary system. For example, collisions by millions of small planetesimals would force all the planets to be rotating in the counterclockwise sense with their axes more or less perpendicular to the ecliptic, as is the case for all the planets except Venus, Uranus, and Pluto. However, a collision with just one large planetesimal could have knocked Uranus over so that its rotational axis is nearly in the ecliptic; such a collision could also have caused Venus' slow clockwise rotation. Except when tidally affected (like the Earth–Moon system), the rotations of the planets and satellites range from a few hours to about 20 hours. This is half or less of the rotational speed at which loose equatorial material would be thrown off. Again, simulations show that this slow rotational speed would have resulted from the myriad impacts by small objects.

Finally, growth by accretion accounts for the large numbers of impact craters seen on nearly every solid surface in the solar system. All of the larger craters are 4 billion years or older, indicating that planetary growth came to an end about then. Indeed, capture of large objects by the gravitational pull of the just-formed planets cleared the solar system of these large objects. Small particles were driven out when the Sun be-came luminous and radiation pressure slowly pushed them. Also, the solar wind from the early Sun helped drive out gas and dust, especially from the inner part of the solar system, because by expanding into a larger and larger volume, the wind would have been far less effective on the more distant protoplanets.

What materials made up the protoplanets? Imagine a mass of gas of solar com-position slowly cooling from an initial temperature of about 2,000 K. What would hap-pen to it? Elements such as tungsten and osmium with the highest condensation tem-peratures (from about 1,500 to 2,000 K) would form (condense into) solid particles first, but are quite rare and so of little importance in later stages. Oxides of the more abundant metals such as calcium and aluminum form at about 1,500 K, and at tem-peratures only a few hundred degrees cooler, metallic iron and nickel and magnesium silicates (for example, $MgSiO_3$ or enstatite) form. Because these are abundant, solid grains would be composed primarily of these elements. At about 1,000 K the feld-spars (common in the Earth's crust) would condense. These are aluminum silicate compounds ($AlSi_3O_8$) combined with sodium, potassium or calcium, for example, $CaAlSi_3O_8$. Between 500 and 1,000 K, the gas and solid particles would react and form compounds such as FeS (troilite), FeO, Fe_2SiO_4, etc. At even lower tempera-tures, carbon would condense and combine with silicates to form various organic mol-ecules. At 200 K and below, first water ice would form, which would then combine with ammonia and methane ice.

That the solar nebula underwent a **condensation sequence** of this sort is supported by various pieces of evidence. For example, instead of considering a cloud of gas with the same temperature throughout but cooling slowly, now imagine, at a given time, the

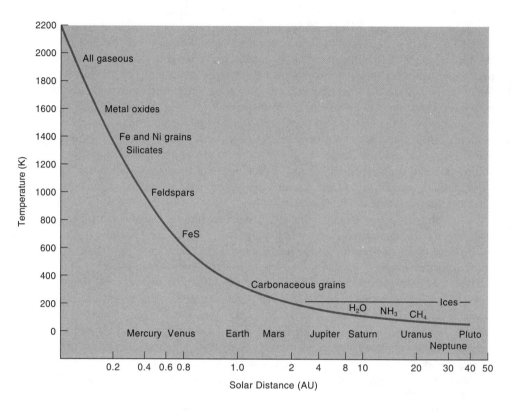

Figure 23.15. Model calculations give the temperature of the solar nebula with distance from the Sun. The temperature (and so the solar distance) at which grains of various material will begin to condense out of the nebula are shown, along with the positions of the planets. Note the approximate correspondence of planetary composition and position in the nebula at which grain formation begins.

cloud in the shape of a flat disk extending out from the Sun. Since the temperature of the cloud would decrease with increasing solar distance, the entire condensation sequence would be happening at the same time with different materials condensing at different solar distances. Near the Sun only high-temperature condensates would form, but far from the Sun where the temperature was low, all of the minerals would have condensed as well as the ices. The latter, formed from the abundant hydrogen, oxygen, carbon, and nitrogen, would be by far the primary constituents. By combining the composition of the condensates with their solar abundances, one can calculate an approximate density of the material at any distance from the Sun. It turns out that this matches the observed densities of the planets very well. The way the planets actually formed must have been more complicated, however; for example, material from different solar distances must have become mixed to some extent. Nevertheless, the condensation sequence fits well the pattern of the average planetary density decreasing with increasing distance from the Sun (Figure 23.15).

Meteorites provide additional evidence that the condensation process took place. For example, aluminum-rich inclusions having condensation temperatures as high as 1,600 K have been found in some meteorites; others have large amounts of entstatite that condensed when the temperature was a few hundred degrees cooler. Their structure indicates that these minerals have not undergone subsequent heating, implying that they condensed in the cooling solar nebula.

Most of the nebular matter could not condense in the high-temperature environment near the Sun, so the terrestrial planets had only a tiny fraction of the mass of the nebula from which to form. Because of their small gravities and the higher temperature near the Sun, they were unable to hold onto the hydrogen and helium that made up most of the nebula and so their final masses are small. The more distant giant planets formed where the nebula was cool enough for ices of the abundant elements to condense. They aggregated from large quantities of condensates and ices and were able to capture gravitationally even more of the surrounding gas. Jupiter is the largest planet in the system, perhaps because it formed in the region where ices were first able to condense and the nebula had not thinned out as it did farther from the Sun.

Because the planets have compositions that depart markedly from that of the nebula from which they formed, the mass of the latter must have been much greater than simply the sum total of the present planetary masses. To end up with their present composition (mostly metals), the terrestrial planets needed a nebular mass about 400 times greater than their present masses. Even the giant planets that more closely approximate solar composition needed nebular masses that are 10 to 60 times their present value. Adding these up gives a minimum nebular mass of about 0.03 solar masses. How much more massive than this the nebula might have been is uncertain.

How did the satellites form? Those that revolve around their primary in the "standard" counterclockwise sense in circular, low-inclination orbits were probably formed from nebular material left around their parent planet in a manner similar to the formation of the planets themselves. Examples are the four Galilean satellites of Jupiter. In fact, these show a trend in their sizes and densities similar to that shown by the planets; namely, that the inner two—Io and Europa—are smaller and denser than the more distant Ganymede and Callisto. The young Jupiter must have been made quite hot by the energy released as matter fell into it. Perhaps Io and Europa were too warm to be able to keep hydrogen and helium or form large quantities of ice so they remained small, resembling terrestrial objects more than their Jovian cousins.

The generally smaller satellites in retrograde orbits or with high inclinations are most likely captured asteroids; examples are the eight outermost of Jupiter's moons as well as the two tiny satellites of Mars.

Asteroids and comets are very likely some of the debris of the formation process, with asteroids originating primarily in the inner, higher density region of the nebula, and comets originating farther out where ices predominated. As we will see in the next chapter, some asteroids are apparently primordial, but others show the effects of differentiation. As far as we know, comets, conglomerates of ice and dust, have not been significantly processed since formation.

Nothing in this account of the formation of the solar system seems to require special conditions or circumstances. Recall also that many protostars are apparently surrounded by disks of gas and dust, the raw material from which planets form. In addition, the IRAS satellite has detected stronger than expected infrared radiation from many ordinary main-sequence stars, including Vega (A0V), Fomalhaut (A3V), ϵ Eridani (K2V), and β Pictoris (A3V). This radiation is not from the star, but from a disk-like distribution of dust warmed by radiation from the embedded star. For example, Vega is surrounded by some 300 Earth masses of dust having a temperature of about 100 K and a radius of roughly 80 AU. Note that these are "mature" stars, with ages probably from 100 million to perhaps a billion years. Thus, planets should have formed by now, if indeed they did form. The dust shell surrounding β Pictoris apparently is sparse within the first 30 AU from the star, suggesting (but not proving!) that the dust that was there has accreted into planets. On this picture, the outlying dust could have been ejected from a planetary system as it formed, or continually released from cometary material.

Thus, the present view is that a planetary system is just one of the possible results of the process by which stars form. Admittedly, it is not clear yet how the conditions that give rise to a star plus planets differ from those that result in one or more stars and no planets. It does not appear, however, that extremely unusual circumstances are required to form planets.

Yet a nagging doubt remains. On one hand, we have stars of about 0.08 solar masses, the least massive objects that can burn hydrogen; on the other hand, we have a giant planet such as Jupiter that would have to be about 80 times more massive to be a star. But what about the objects in between, the more massive of which are the so-called brown dwarfs? Admittedly they are faint and are therefore hard to find, but astronomers have looked very hard for them with no certain success so far. Does this suggest that objects less massive than ordinary stars are not formed like ordinary stars, but in some significantly different way? Are we really on firm ground when we claim that planet formation is just one of the paths taken in star formation? The discovery of several brown dwarfs would be reassuring.[4]

Origin of the Moon.   Before we think we understand more than we really do, however, it is worth pointing out that the origin of our nearest neighbor, the Moon, is still something of a mystery. Four possibilities have been put forward over the last century: the Moon was spun off of a rapidly rotating Earth; it was formed elsewhere in the solar system and then captured by the Earth; the Earth and the Moon formed together as a double planet; or, finally, it coagulated from the debris produced by a giant impact of a Mars-sized object on the proto-Earth (see Figure 23.16).

All of these proposals have difficulties. For example, it is not easy to understand how the Earth could have acquired a rotation period of two hours, which would be necessary if it was to spin off the Moon. Also, the angular momentum required is about four times as much as is contained in the present Earth–Moon system. So what happened to it? And how are the significant differences in the compositions of Earth and Moon rocks to be understood, for example, the lack of water in the latter? The capture idea suffers from the difficulty of getting rid of enough kinetic energy in the encounter to enable the Moon to be captured, rather than simply having its orbit changed and continuing on its independent way. The kinetic energy difficulty can be avoided by having the Moon form very near the Earth with low relative velocity, in which case it would more likely collide with the Earth than be captured by it. And if the Earth and Moon formed as a double planet, why are their densities (see Table 24.1) and hence their compositions so different?

These and other problems (rather than any new data) revived interest in the mid-1980s in the possibility that a dramatic event, the collision of a Mars-sized planetesimal with the forming Earth, made the debris that collected into the Moon. In order that the fragments of the impactor and the proto-Earth have the angular momentum required by the system, the impact must have been a glancing one at a relative velocity of about 10 kilometers per second. If the impactor was differentiated into an iron core and a silicate mantle (not unlikely, given its assumed size), its core would have fallen into the Earth. Some of its mantle, along with fragments from the Earth, would have gone into orbit around the Earth. The heat produced by the impact would have driven water and volatile elements from the impactor fragments, giving them lunar compositions. Furthermore, if the Earth had already been differentiated so that its fragments

---

[4] Late in 1995, astronomers announced the discovery of a faint companion to the star, Gliese 229, which is about 6 parsecs away in the constellation of Lepus. The temperature of the companion is about 1000 K, its luminosity is no more than about 4×10^{-6} that of the Sun, and its mass is no more than 50 Jupiter masses. This object is most likely a brown dwarf.

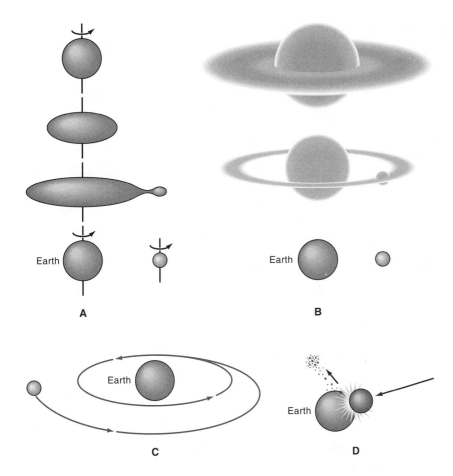

Figure 23.16. Four suggestions about the origin of the Moon: (A) the Moon was split off from a rapidly rotating Earth; (B) the Moon and Earth were formed together as a double planet; (C) the Moon was formed apart from the Earth, someplace else in the solar system, and it was subsequently captured by the Earth; (D) the Moon was formed from the debris produced after a Mars-sized object collided with the early Earth.

would also have been silicates, then its contribution to the proto-Moon would have been the appropriately low-density silicates. Because the mass of the assumed impactor is about 8 to 10 times that of the Moon, the capture process need not have been very efficient and most of the debris would have either fallen into the Earth or would have been lost to the system. The remainder orbiting the Earth would in a matter of only a day or so gravitationally clump together into a lunar mass object. Tidal interactions would then have caused the Moon to move to its present distance from the Earth.

Since late in the formation of the planets it is possible that most of the nebular material would have been collected into relatively few large objects, impacts of the sort required here may well have occurred. Computer simulations suggest that the impact circumstances, the assumed mass and composition of the impactor, and the relative contributions of the Earth and impactor can be chosen to support this picture of lunar formation, but it would be reassuring to have direct evidence of the event. For example, did the impact leave any distinctive geological or geochemical traces that could be looked for? A mass colliding at about 10 kilometers per second has a kinetic energy equal to its mass in TNT, so that the assumed collision would likely have melted the early Earth. It is not clear if such melting is consistent with the geochemical evidence, however. Because of the nature of the event, it will be difficult to show that the Moon actually did form in this way, but perhaps this theory will be more successful than the others.

It has also been suggested that impacts by massive objects late in the formation of the solar system could have not only knocked Uranus and Pluto on their sides, but could also have resulted in the Earth's axial inclination.

Despite our incomplete understanding of all the processes, most astronomers feel that planets are indeed a common by-product of stellar formation.

Next, let us look at some of the processes that enable us to account for the physical properties of the planets and their satellites. By understanding these, you will have a good idea of how many of the salient features of the solar system came about.

General Properties of Planetary Interiors

How We Learn About Interiors.   We know much less about the internal structure of the planets than we do about that of the stars, even though the former are incomparably nearer to us than the latter. As we noted in Chapter 13, this is because planetary interiors are solid and liquid rather than gaseous, for which the equation of state is very simple. Nonetheless, several clues concerning planetary interiors are available. The observational data relevant to the interior properties of a planet are its size and mass, which give its average density. By itself, this does not tell us much unless this happens to be very small or very large. If the former, say an average density about equal to that of water, then we would know that the object must consist mostly of ices; an appreciable amount of heavier material would increase the density too much. Similarly, if the average density were five or six times that of water (the density of iron-rich minerals), then we would know that not much of the object could be made of ice. An intermediate average density could be made up of various (and unknown) proportions of ices, rocks (2.8 to 3.9 times that of water), or iron-rich material (about 8 times water).[5] If, however, we somehow knew what its surface material was, that would help us to estimate the interior density and, therefore, its composition.

A planet's deformation from sphericity caused by axial rotation or by tidal forces exerted by its own satellites tells us about the distribution of mass in its interior and its plasticity (ability to flow) under mechanical stress. A planet with mass strongly concentrated toward its center will deform less under rotation than one in which the mass is distributed uniformly. The departure from sphericity of its gravitational field deduced from the motions of its satellites or rings also gives information about the mass distribution.

The strength and shape of a planet's magnetic field reveals something about the nature of its interior material, such as its electrical conductivity and fluidity. The outflow of heat (if any) compared to that received from the Sun gives information about possible energy-producing processes (such as radioactivity) as well as the thermal conductivity of the material.

Analysis of the vibrations caused by earthquakes and recorded by seismometers distributed over the Earth's surface has given us most of our information about the in-

---

[5] You should realize that in saying an object is made of rocks or ices, we don't mean ordinary stones or ice cubes. Instead, rock implies compounds of elements such as iron, silicon and oxygen and ice implies those of carbon, nitrogen and oxygen combined with hydrogen. Except at the surface of a planet, these "rocks" and "ices" would not be recognizable as such because they are transformed to gases, liquids or high-density solids by the high pressures and temperatures in planetary interiors.

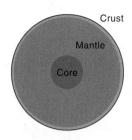

Figure 23.17. General considerations suggest that many planets should have a three-part structure: an outer crust, an interior mantle, and a central core. See text.

terior structure of the Earth. To date, only a few seismometers have been placed on two other objects, the Moon and Mars. Consequently our knowledge of the interiors of the other terrestrial planets is based to a considerable degree on analogy with the Earth's structure.

Internal Structure. Like the stars, the planets are very nearly in mechanical equilibrium—neither shrinking nor expanding—under inward-acting gravity and outward-acting internal forces. Unlike the stars, however, the force that counters gravity is not the pressure exerted by a high-temperature gas, but rather the pressure of the solids, liquids, and relatively cold gases that make up planetary interiors.

The outer part of a planet is likely to be composed of material somewhat different in nature from the more compressed inner region. The surface of the Earth, for example, is covered with familiar rocky material. When rocks are under huge pressures, however, they crumble and become a different kind of material, even though made of the same chemical elements. Very near the surface of a solid planet the gravitational compression force (the weight of overlying matter) is not large, and a thin layer a few tens of kilometers thick is able to sustain itself simply by its own strength (it takes a large force to crush a rock). This material is called the **crust**.

Obviously, the density of the interior will increase with depth, because the material is compressed by the overlying layers. In addition, if the interior is at least somewhat plastic, able to flow, material of higher density will settle slowly toward the center, displacing less dense matter. In this way, the planet's composition becomes **differentiated**, that is, it varies with depth, the denser material being deeper in the interior. This differentiation will result in a high-density core. On these general grounds, then, one might expect planets to have three distinct regions: a crust, a central **core**, and a region in between called the **mantle**. This simple picture does indeed hold in many planets (Figure 23.17).

The interiors of the terrestrial planets are primarily solid (though some have a liquid core), whereas those of the Jovian planets are largely liquid or gaseous surrounding solid cores, a consequence of their solar abundances of hydrogen and helium. The centers of the planets are warmer than their surfaces, but nowhere near hot enough to fuse hydrogen, which, of course, is why they are planets and not stars. As time goes on, planetary interiors are becoming more stable, because internal energy sources are diminishing, and differentiation has nearly been completed.

Planetary Magnetism. Since the temperature of a planet's interior may be a few tens of thousands of degrees, any large-scale magnetic field it has cannot be due to anything resembling a permanent magnet (like an ordinary bar magnet) deep in its interior; the high temperature would destroy the magnetism. It appears that a planet generates a magnetic field only if it has a liquid interior and rotates fairly rapidly. The details are not well understood, but the idea is that rotational energy of the planet combined with thermally driven convection within the core sets up fluid currents in the molten interior. The hot, slowly flowing material is ionized, and so is in effect an elec-

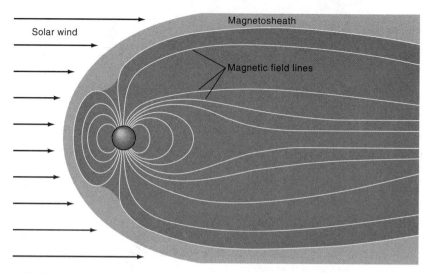

Figure 23.18. A generic magnetosphere. The magnetic field surrounding a planet is distorted by interactions with the charged particles of the solar wind. The upwind field is compressed and a long magnetic tail is formed on the downwind side.

tric current; an electric current produces a magnetic field. The currents in the liquid interior are not likely to be well-defined, regular systems, but are probably twisted and complex loops. As long as there is a net flow of ionized material, however, a magnetic field will result. This kind of model is called a **dynamo theory**, since a dynamo converts energy of motion into an electric current.

On Earth, a compass needle easily detects this field extending from the north to the south magnetic poles. These do not coincide with the geographic poles, but are inclined at an angle of about 11° with them. Apparently, the molten core material does not flow symmetrically about the rotation axis. That these currents are irregular is indicated by the wandering of the position of the magnetic poles, changes in the average strength of the field, and reversals in the direction of the Earth's field. All these events are preserved in the magnetized material in rocks going back roughly 100 million years.

With an external magnetic field, a planet is able to trap charged particles, primarily electrons and protons, flowing out from the Sun in its weak wind. The region of space in which the motions of charged particles are influenced by the planet's magnetic field is called the **magnetosphere**. The magnetosphere can deflect charged particles away or constrain them to remain in belts or in fairly well defined regions surrounding the planet. Charged particles in the solar wind that collide head-on with the magnetic field cannot easily penetrate it, but do compress the magnetosphere on the sunlit side of the planet. On the night side, a long tail is formed (see Figure 23.18).

Forming Planetary Surfaces

With the condensation sequence in mind, we can understand why the terrestrial objects have solid surfaces whereas the Jovian planets don't. At ordinary pressures, heavy chemical elements become solid at much higher temperatures than do the hydrogen or helium that largely compose the giant planets. Since the terrestrial planets are composed mostly of heavy elements, we live on a planet with a solid surface.

Figure 23.19. A crater on the far side of the Moon, 75 kilometers in diameter and 4 kilometers deep. Note the central peak, terraced walls, flat floor, and ejecta surrounding the crater, all of which are typical of impact craters.

Processes that shape the surfaces of planets and satellites once these have been formed can be divided into two groups: those that come from outside the planet, and those that originate on or within the planet itself.

Collisions. In the first group is impact cratering by meteorites and other solar-system debris. Impact craters are by far the most common surface feature on planets and satellites (Figure 23.19). These collisions can be extremely powerful. The impact speed of a meteorite and a planet is made up of two parts: the relative orbital velocity of the two objects plus the velocity that the meteorite acquires as it falls into the planet. This last is equal to the planet's escape velocity. For the Earth, this is 11.2 km/sec; if the relative orbital velocity with an asteroid is, say, 5 km/sec, then the asteroid would strike the Earth moving at about 16 km/sec. With this speed it has an energy 30 times that of the same mass of TNT! This would produce a crater having a volume about a thousand times that of the impacting asteroid. Most of the in-falling object would be vaporized as it struck, resulting in an explosively expanding gas bubble, a shock wave. The crater excavated by this gas bubble is nearly always circular, because its very rapid expansion is independent of the direction of motion of the much more slowly moving asteroid. Only if the planet were struck at a near-grazing angle might the crater be somewhat elongated.

Now, the maximum possible relative orbital velocity is not 5 km/sec, but about ten times that value for an object in a highly elliptical retrograde orbit (clockwise as seen from the north). The energy yield in such a collision would be about 500 times that of the same mass of TNT. Thus it is clear that impacts by massive objects can have an enormous effect on the landscape of a planet. For example, such explosive collisions are responsible for the most obvious features on the surface of the Moon—the dark maria. In fact, there is abundant evidence for such impact cratering throughout the solar system. Large and more complex multi-ring impact craters are found on Mercury, Venus, Mars, several satellites including the Moon, and possibly the Earth.

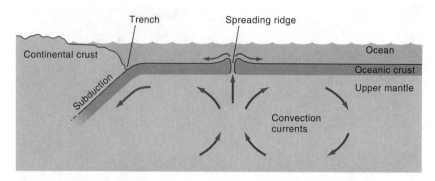

Figure 23.20. Slow convection currents in the upper part of Earth's mantle can force material between two continental plates, pushing them apart. When one plate slides below another (in a subduction zone), a trench is formed.

Internal Processes. Surface-shaping processes arising on or within the planets themselves are those of traditional geology: the motions of planetary mantles (the layer just below the crust), volcanic action, and erosion by water and atmosphere. All of these are familiar processes here on Earth. Mantle motions cause the half-dozen huge and several small plates into which the Earth's surface is divided to drift, slowly but inexorably to smash into each other and form mountains; volcanism produced much of the Earth's atmosphere; and winds and running water constantly erode away mountains and other features in the landscape. The consequences of one or more of these processes are evident on most of the planets and satellites with solid surfaces.

The mechanism by which **continental plates**[6] are driven apart is convection in the mantle. Recall that in Chapter 13 we encountered convection in stars. It occurs in a star when radiation is unable to carry all the energy outward, toward the surface. Since planetary interiors are solid or liquid rather than gaseous, conduction rather than radiation is the major heat-transport mechanism. But when conduction can't carry out all the energy, convection currents set in. Obviously, currents in a thick, gooey, nearly solid material are much slower than those in a gas. Nonetheless, warmer matter slowly wells up from the mantle between the plates and pushes them apart, while colder material sinks toward the center (see Figure 23.20).

Internal Heat. The energy that establishes convection currents comes from the heat produced by several sources. When denser materials lie above less-dense matter and the interior of the planet is sufficiently yielding to allow the denser material to sink slowly toward the center, then **heat of differentiation** can be produced. That is, energy is released as the slowly moving (but denser) material falls toward the center of the planet, converting its kinetic energy into heat. If this process is to be significant, at least some of the interior of the planet must be somewhat plastic, which means that it must already be heated to some extent. This can be accomplished by the accretion process (mentioned above), when large, rapidly moving masses bombard a planet, thereby converting their gravitational kinetic energy to heat, primarily near the surface of the planet. In addition, energetic particles emitted by the radioactive decay of heavy elements in the interiors of planets contribute to the heating.

Both the accretion process and the radioactive decay of elements such as iodine were most effective early in the history of the solar system rather than later. The radioactive decay of long-lived elements such as uranium, thorium, and potassium, how-

---

[6] Huge thin slabs of crustal material that cover the Earth (see the next chapter).

ever, continues to produce energy right up to the present. This heat has kept the Earth's mantle plastic, so that even though it is solid it can flow, albeit slowly. As we have noted, adjacent continental plates can be driven apart by mantle material rising up between plate boundaries. This, of course, causes them to push against other plates. Volcanoes are found along these boundaries or occasionally at "hot spots" within the plate. These hot spots apparently are plumes of mantle material rising through a vent or fissure in a plate. Thus geological activity will occur if the interior of a planet has been warmed sufficiently.

Two Generalizations. We can extend this to the following generalization: if two planets have the same proportion of radioactive material in their interiors, the larger planet is more likely to show geologic activity than the smaller. Earth is geologically active, but the Moon is not. To understand why, consider how a planet loses its heat, how it cools off. The heat must of course escape through the surface. Now the volume of a sphere (and hence the amount of radioactive matter and so its heat content) depends on R^3 ($V = 4/3\pi R^3$), but its ability to cool, that is, its surface area, goes as R^2 ($A = 4\pi R^2$). The ratio of heat content to cooling ability ($R^3/R^2 = R$) is greater for a larger sphere than for a smaller one. A familiar example of this is that a pebble, heated in the summer Sun, cools off quickly at night, whereas a large rock remains warm much longer. Similarly, it is harder for the heat to escape from a larger planet, and so its interior will become warmer and stay warmer for a longer time, and so be more likely to produce geological activity than will the interior of the smaller object.

This leads directly to another generalization, namely, that the larger planets have younger surface features than smaller ones. This is so because geological activity slowly destroys the surface of a planet; in fact, the first billion or so years of Earth's history have disappeared as a consequence of mantle motions.

Planetary Atmospheres

General Considerations: Observational Data. From the mass and radius of a planet we can calculate its surface gravity, that is, the force of gravity at the surface of the planet tending to compress the atmosphere. Just as for stellar atmospheres, planetary atmospheres are thin compared with the diameter of planets, and so the surface gravity is essentially the same through the atmosphere. Knowing the distance of the planet from the Sun, we can calculate how much solar energy strikes the planet. By measuring the brightness of the planet we can calculate how much energy it reflects and hence its **albedo**, the fraction of solar radiation reflected. The rest is absorbed by the atmosphere and surface. The albedo of Venus is high, nearly 60 percent, whereas that of the Moon is low, about 11 percent.

By comparing the solar spectrum with that of the light reflected by the planet, we can see what absorption features are produced by the planetary atmosphere and so identify the absorbing gases. Just as with stellar spectra, a detailed analysis of the planetary absorption features can often reveal the temperature and pressure of the atmospheric levels where the features are formed. Direct measurements by balloons, rockets and satellites have given us the temperatures, densities, and compositions of Earth's atmosphere, and space probes have given such data for the atmospheres of Mars and Venus. Finally, images show the large-scale circulation patterns and atmospheric features such as Jupiter's Great Red Spot, and how they change with time.

General Considerations: Physical Principles. To learn more about planetary atmospheres we must apply some of the same physical principles and ideas that we have used to study other astronomical objects (see Chapter 13). For example, a planet's atmosphere is in hydrostatic (or mechanical) equilibrium; on the average it is neither expanding nor contracting, with the gravitational pull inward being balanced by the pressure forces outward. In addition, some regions of a planetary atmosphere may be convective, with blobs of warmer gas rising, cooler gas falling. Note the extremely wide range of applicability of the concepts of hydrostatic equilibrium and convection: hot, low-density gases of a star's atmosphere; the much hotter, denser, but still gaseous stellar interiors; relatively cool and solid or liquid interior of planets; and now the cold gaseous, liquid, and solid planetary atmospheres.

Also, whether encountered in a hot stellar interior or a cold planetary atmosphere, gases obey the simple gas equation of state. The concept of the opacity of a gas (its resistance to the flow of radiation) that is important in stars is also applicable to a planetary atmosphere. Only if an atmosphere has some opacity will it absorb incoming solar energy and be heated by it. As in stars, the opacity of an atmospheric constituent depends on the wavelength as well as on its abundance. In the Earth's atmosphere, for example, water vapor is transparent to visible light, but absorbs infrared radiation, producing the greenhouse effect we will describe later in this section. In addition, the same sort of condensation process that is important in forming the planets is responsible for the clouds in planetary atmospheres.

Like planetary interiors, planetary atmospheres are more complicated than their stellar counterparts. Planetary atmospheres contain liquids (cloud droplets) and solids (ices) as well as gases; these can condense or evaporate, releasing or absorbing energy. Furthermore, clouds can provide a sharp barrier to the flow of radiation. The main planetary heat source is from the outside (the Sun) rather than from the inside, and varies with the day/night cycle. Three of the giant planets produce a significant amount of energy in the infrared, which has a considerable effect on the heat balance of these planets. Global circulation currents (such as the band structure of Jupiter) play a more important role in planetary than stellar atmospheres.

Despite these complications, mathematical models of planetary atmospheres are constructed in much the same way as they are of stellar atmospheres. That is, given various starting conditions such as the surface gravity, heat input, and chemical composition, the run of temperature, density, pressure, etc. with depth are calculated, subject to various requirements such as hydrostatic equilibrium, the equation of state, and the relevant opacities.

A couple of simple examples should clarify some of the processes that affect the temperature structure of a planetary atmosphere. First, assume a planet surrounded by an atmosphere transparent to all wavelengths of radiation. Solar radiation passes through the atmosphere unimpeded and strikes the planet's surface, which absorbs some of it. The surface temperature increases until it becomes warm enough to reradiate in the infrared as much energy as it absorbs. All of this energy escapes because the opacity of the atmosphere is zero. The atmospheric temperature near the ground would be raised by conduction from the warm surface and convection currents might become established, but the transparent upper atmosphere would be as cold as space. The very highest part of the atmosphere might be heated by collisions with particles of the solar wind (see Figure 23.21A).

Next, consider a more realistic atmosphere that contains a fairly well defined absorbing layer, for example one of ozone that absorbs ultraviolet energy. Visible radiation passes through the atmosphere heating the surface as before. Incident ultraviolet radiation is absorbed, however, so in that region of the atmosphere the temperature gradient will be reversed for some kilometers (that is, the temperature increases) before continuing to drop. Suppose that in addition, atoms in the uppermost part of the atmosphere can absorb some incident sunlight. The temperature there would be raised.

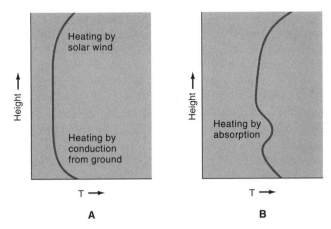

Figure 23.21. Diagram (A) shows schematically how the temperature of an atmosphere transparent to all wavelengths might vary with height. All of the incident energy gets to the surface and warms it, which in turn might warm the bottom of the atmosphere by conduction. The rest of the atmosphere would be as cold as space except for the topmost, which might be heated by collisions with the solar wind. (B) Here, the effect of an absorbing layer (for example, ozone) is shown. The temperature of the energy-absorbing layer increases.

The greater the number of absorbing regions or of heat sources, the more complex the structure of the atmosphere (Figure 23.21B).

Water and carbon dioxide in a planet's atmosphere are transparent to visible radiation but absorb the infrared. Consequently, the planet's surface temperature is raised, increasing the total energy radiated ($E = \sigma T^4$) and decreasing slightly the fraction radiated in the infrared (recall, $\lambda_{max}T$ = constant; as T increases, λ_{max} decreases). The heating continues until equilibrium is achieved, that is, until as much energy escapes from the surface as it receives (Figure 23.22). The equilibrium temperature can

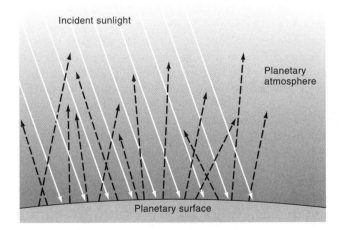

Figure 23.22. If the constituents of an atmosphere are transparent to visible sunlight, but absorb infrared radiation, a greenhouse effect will result. Incoming sunlight heats the surface of the planet, which then radiates at infrared wavelengths appropriate to its temperature. Most of this radiation is absorbed by the atmosphere (dashed lines), heating it and the planetary surface and causing more infrared radiation to be emitted to repeat the cycle. Quickly, a balance is established and the planet, at a higher temperature than it would be without the greenhouse effect, radiates away as much energy as it receives.

be much higher than it would otherwise be because of this **greenhouse effect** (so-called since a greenhouse is warmed because its glass transmits visible radiation but is opaque to the infrared). We will see that Venus provides a spectacular example of this phenomenon and that there would not have been life on Earth without it.

When both hydrogen and oxygen are present in an atmosphere, water is most likely to form since both elements are reactive. If most of the hydrogen is tied up in water and substantial oxygen still remains, oxygen-rich compounds are formed, such as CO (carbon monoxide) and CO_2 (carbon dioxide). This kind of an atmosphere is called **oxidizing** and is the situation in the terrestrial planets. If, however, all of the oxygen is used up and hydrogen still remains, hydrogen-rich compounds are formed with relatively common elements such as carbon and nitrogen, for example, CH_4 (methane), and NH_3 (ammonia). Such an atmosphere is said to be **reducing** and occurs when the relative numbers of elements approximate cosmic abundances, as in the giant planets.

When the temperature and pressure in an atmosphere are appropriate, a particular gas may condense into its liquid or icy state, forming a cloud layer. For example, in Jupiter clouds of ammonia, ammonium hydrosulfide, and water form where the atmospheric temperature is about 150 K, 200 K, and 275 K, respectively. Saturn's cloud structure is similar, but is spread over a wider range of altitudes than is Jupiter's because of the latter's greater surface gravity, which compresses the atmosphere.

Finally, what determines whether a planet keeps an atmosphere or not? The larger a planet's surface gravity and the colder its atmospheric temperature, the more likely it is to retain its atmosphere once it has formed. A large gravity means that the escape velocity is high; the colder the atmosphere, the slower the average velocity of its constituent particles. Deep down in a planetary atmosphere an atom or molecule will most likely collide with another particle before it moves very far. These collisions are as likely to send the atom back deeper into the atmosphere as to propel it outward. It is only at the fringes of the atmosphere, the **exosphere**, where the densities are small, that the atom has a good chance to escape the planet (Figure 23.23). Even in the exosphere, however, only those atoms that have outward velocities exceeding the escape speed will actually leave the planet. The escape speed [$v_{esc} = \sqrt{(2GM/R)}$] is determined by the gravity of the planet, and the particle speeds are set by the temperature of the atmosphere. At any given temperature [which, recall, is just a measure of the kinetic energy, $(1/2)mV^2$, of the gas atoms or molecules], light particles will have greater speeds than heavier ones and so will be more likely to escape. The Earth's atmosphere, for example, has very little hydrogen, but a lot of nitrogen. Lighter gases such as hydrogen and helium tend to be concentrated at higher altitudes in an atmosphere.

With this general background of solar-system properties and processes well in mind, let us next consider in more detail the physical characteristics of the major components of our own system.

Terms to Know

Geosynchronous orbit, minimum-energy orbit, launch window, gravity assist, radar. Terrestrial and Jovian planets, Bode's law. Meteoroids, meteors, meteor showers, meteorites, asteroids, comets; exosphere, albedo greenhouse effect, oxidizing and reducing atmospheres. Crust, core, mantle, continental plates, heat of differentiation, accretion; radioactive decay, half-life; dynamo theory, magnetosphere; planetesimals. Condensation sequence.

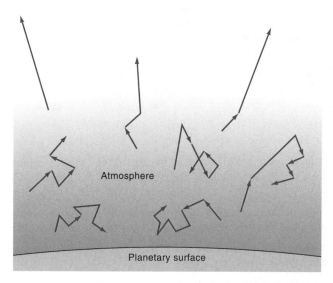

Figure 23.23. Atoms and molecules deep in the atmosphere where the particle density is large undergo many collisions and do not move very far between impacts. On average, they remain within the same atmospheric layer. Higher in the atmosphere, however, the density is lower, collisions are less frequent, and particles can travel longer distances between impacts. The part of the atmosphere from which particles can travel far enough to escape is called the exosphere.

Ideas to Understand

How a spacecraft is flown to a planet on a minimum-energy orbit; by gravity assist. How radar is used in solar-system work. The general properties of the terrestrial and Jovian planets; processes that shape planetary surfaces and atmospheres; characteristics of the magnetosphere. How ages of rocks are measured. How a planet might be defined; why large objects are spherical; why large objects cool off more slowly than small ones; collisional and nebular theories of the origin of the solar system; origin of the Moon.

Questions

1. (a) About how many times more massive is Jupiter than the combined mass of the rest of the planets?

(b) If the mass of the Sun is 332,000 Earth masses, about how many times more massive is the Sun than all the planets?

(c) The radius of our Moon is 1,738 km. Which moons in the solar system are larger and which planets do they orbit?

2. Compare the major properties—mass, diameter, average density, and numbers of satellites—of the terrestrial and Jovian planets. In a general way, how can these differences be accounted for?

3. Suppose you want to travel to Venus along a minimum-energy orbit.

(a) What is the semimajor axis of such an orbit?

(b) How long would the trip take? About how often could you begin such a journey? Draw a diagram showing the relative positions of Venus and the Earth at the time of launch.

(c) You have been launched into a 1.0-AU circular orbit around the Sun. What would you have to do to transfer to the orbit that will take you to Venus?

4. Calculate the sidereal period (in minutes) of a spacecraft that orbits the Earth at a height of 500 km above the equator. The radius of the Earth at the equator is 6,378 km, its mass is 6×10^{27} gm, and the constant of gravitation, G, is 6.67×10^{-8} if you express distance in centimeters, time in seconds, and mass in grams.

5. Suppose we want to build an observatory in a satellite orbiting the Earth just above the Earth's atmosphere. How large would a telescope have to be to resolve features on the Moon 20 meters across? (One second of arc at the Earth–Moon distance corresponds to 2 kilometers.)

6. Show that a 24-hour orbit is achieved with the altitude and speed given in Table 23.1.

7. It has been suggested that a planet orbits the Sun beyond Pluto. Assuming that Bode's law applies, what would the semimajor axis of the orbit of this object be? What would its period be?

8. In the interiors of stars, energy is carried outward by photons, by convection, and by conduction. Are all of these important in the interiors of planets? If not, why not?

9. The velocities of escape from Mars and Mercury are about the same; yet Mars has an atmosphere, and Mercury doesn't. Explain.

10. The mass of Saturn is about 95 times that of the Earth's, but its surface gravity is about the same as the Earth's. Calculate Saturn's radius.

11. If the CO_2 content of the atmosphere continues to increase, how will the Earth's climate be affected? What might happen to this country's grain belt?

12. Suppose someone claimed that a rock sample has a ratio of U^{238} to Pb^{206} of 1 to 7. How long ago did it solidify from a molten state? What assumptions must you make to work this problem with just the information given? Would you believe your answer in light of the age of the Earth?

13. How would you define a planet? Justify your answer.

14. Why do we know that the original mass of the Earth must have been much greater than it is now?

15. Explain why it is unlikely that a planet could have a massive high-density layer on top of a low-density layer.

16. Suppose that on planet A the surface rocks have a density of about 2.8 gm/cm³ and the average density of the planet as a whole is 3.0 gm/cm³. What would you surmise about the interior of this planet? Now suppose that the surface and average densities of planet B are 2.8 and 6.0 gm/cm³, respectively. What would you say about the interior of planet B?

Suggestions for Further Reading

Hartmann, W., *Moons and Planets*, Third Edition. Belmont, CA: Wadsworth Publishing Company, 1993. An excellent introduction to the fundamental ideas and principles of planetary science.

Kivelson, M. (ed.), *The Solar System: Observations and Interpretations*. Englewood Cliffs, NJ: Prentice-Hall, 1986. Although rather advanced in parts for the introductory student, this book presents much useful comparative planetology.

Morrison, D., *Exploring Planetary Worlds*. New York: Scientific American Library, 1993. A concise, nicely illustrated, and up-to-date description of the solar system.

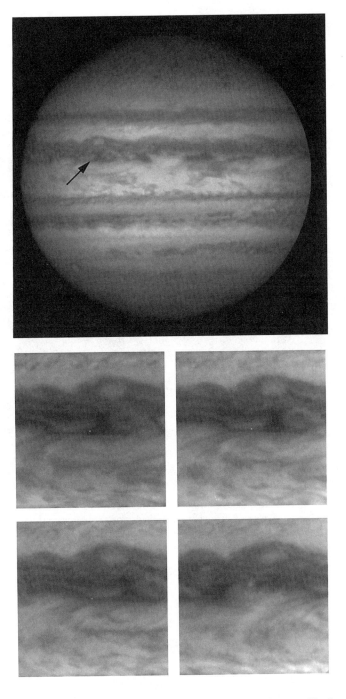

Hubble Gallery. The arrow on the left side of the upper image of Jupiter shows the predicted entry point of the Galileo probe. The entry point is at the exact center of the four images on the lower half of the page, each one of which is about three Earth diameters across. This four-part series was taken 10, 20, and 60 hours after the first image on the upper left; it shows winds sweeping clouds about 24,000 kilometers eastward.

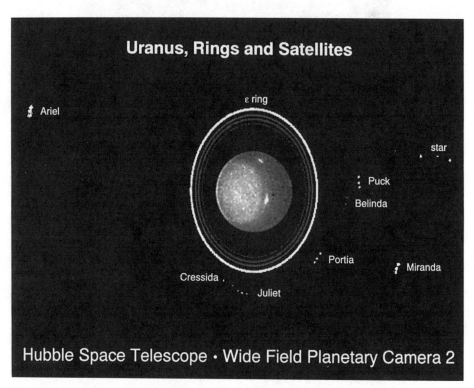

Uranus, Rings and Satellites

Ariel • ε ring • star • Puck • Belinda • Portia • Miranda • Cressida • Juliet

Hubble Space Telescope · Wide Field Planetary Camera 2

A Hubble Space Telescope image of Uranus, its rings, and some of its satellites. This is a composite of three exposures taken about 6 minutes apart in August, 1994. Even in that short time, the satellites move an appreciable distance, as indicated by the three spots for each satellite.

CHAPTER 24

Solar-System Planetology

In this chapter we will describe the physical properties of the larger objects of the solar system. We want to understand as best we can how they came to have their present characteristics and, in particular, how differences among these objects came about. Though many of the answers are incomplete, they are helpful in establishing the sensitivity of planetary environments to various physical factors such as the size of the planet or its distance from the Sun. This will be important when we attempt to assess the likelihood that life could develop on planets elsewhere.

The Terrestrial Planets

Let us consider the terrestrial planets first. As mentioned in the previous chapter, we will include in this group not only the four best-known members, Mercury, Venus, Earth, and Mars, but also our own Moon and two of the four Galilean satellites of Jupiter—Io and Europa. These satellites have densities greater than about 3 gm/cm³ and radii greater than 1,000 km (see Figure 24.1 and Plate C6). Why do we extend the definition of terrestrial objects in this way? Because if a planet or satellite has an average density greater than about three times that of water, then to a crude approximation its composition can be said to be largely rock (density about 2.5 to 3) and iron (iron and similar metals have densities about 8 times that of water). If its radius is greater than about 1,000 km, then it is large enough that its interior has undergone significant evolution. (Recall the discussion of internal heating in the previous chapter.)

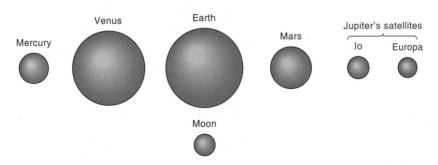

Figure 24.1. The relative diameters of the four planets and the three satellites that can be considered to make up the family of terrestrial objects. See Plate C6.

The other Galilean satellites of Jupiter—Ganymede and Callisto—and Saturn's largest moon, Titan, meet the size requirement, but have low densities, typically 1.5 to 2 gm/cm³. This density indicates that the composition of these objects differs fundamentally from that of the terrestrial bodies, with a corresponding difference in their structure. Very roughly speaking, these satellites are composed of rock and ice. Hence they are not included among the terrestrial objects.

Because we know most about the Earth, its interior and atmosphere will serve as the prototypes for this group. Its surface, however, is largely covered by water, and the area that isn't has been greatly modified by geological activity as well as by wind and water erosion; it does not give us a good record of the intense bombardment that was important in the early history of the solar system. The Moon's surface, having no water and very little erosion, presents us with a long, continuous record; we shall use it as the prototype for the surfaces of terrestrial objects. After describing these "standards," we will compare them with the other members of the group, pointing out the differences and, when possible, the factors that gave rise to these differences.

Table 24.1 gives some of the physical properties of the terrestrial objects. The headings are self-explanatory except for the **reduced density**. This is the estimated

*Table 24.1. **Masses and densities of the terrestrial planets***

Body	Mass ($\oplus$ = 1)	Mean radius (km)	Mean density (gm/cm³)	Reduced density (gm/cm³)
Earth	1.00	6371	5.51	4.03
Moon	0.012	1737	3.34	3.34
Mercury	0.053	2439	5.43	5.31
Venus	0.815	6051	5.24	3.95
Mars	0.107	3390	3.93	3.71
Io	0.015	1815	3.55	3.55
Europa	0.008	1569	3.96	3.96

density of the planet when its material is not compressed by the very high pressures prevailing in the interiors of massive objects, but instead is subjected to only enough pressure to eliminate voids. The reduced density is a more representative value of the average density of the material making up the object. Note that Venus and Earth, the most massive of the terrestrial objects (and therefore those that exert the greatest interior pressures), show the largest decrease in their reduced density as compared to their mean density. Though the Earth has the largest mean density, Mercury has the largest average reduced density of any solar-system object. Note also that the reduced densities of the low-mass objects are the same as their mean densities, because they are not sufficiently massive to compress their interiors appreciably.

The Interior of the Earth

Information from Density.   Our direct knowledge of the Earth's interior is extremely limited. The deepest holes bored into the Earth go down only about 15 km; geological activity such as volcanoes brings material up from depths of no more than 100 km. Nonetheless, we can infer a fairly detailed picture of our planet's interior.

First of all, the average density of surface rocks is only about 2.7 gm/cm^3, much less than the average density of the Earth as a whole (see Table 24.1). Hence its interior density must be greater than the average. That is, the Earth's interior is not uniform, is not composed of the same material at all depths, but must be differentiated. This implies that the interior must have been molten at one time to enable denser material to sink toward the center and lighter material to rise toward the surface.

We could make a mathematical model of the Earth's interior by assuming that a certain fraction of its volume was occupied by materials having a density X, another shell by a compound with density Y, etc., and adjusting the volumes until the overall density agreed with that measured. (Table 24.2 gives the densities of common con-

Table 24.2.
Densities of common
planetary material

Material	Density
Water	1.0 gm/cm^3
Basalt	2.9
FeO	5.7
Nickel	8.9
Granite	2.6
FeS	4.8
Iron	7.9

stituents of the Earth.) For example, the outermost shell could be a mixture of granite and basalt having an average density of 2.7, a central shell consisting of iron oxide, and a core of nickel and iron. Unfortunately, the data we have so far would be satisfied by many different models. More data are needed in order to derive a unique model, or at least to reduce appreciably the number of possibilities.

Information from Earthquakes.   Most of these additional data are derived from analyses of earthquake waves, which, in fact, are the major source of our knowledge of the Earth's interior. The Earth is covered by a half dozen major and several minor crustal slabs or plates, which are slowly driven with respect to each other (see Figure 24.2). Often the motion of one plate against another is not smooth and continuous. Instead, portions of the two plates "stick" against each other, not moving appreciably for decades or even centuries, until the stresses become so large that they finally slip and produce a sudden motion, an earthquake. These violent readjustments produce several types of waves: **surface waves**, which, as the term indicates, travel along the Earth's surface and carry most of the energy of a shallow earthquake; and **P-waves** and **S-waves**, which travel in the body of the Earth and so give us information about its interior. P-waves are an oscillation, a back-and-forth disturbance, in the direction of travel of the wave, whereas the disturbance in an S-wave is perpendicular to the direction of motion of the wave. In contrast to P-waves, S-waves cannot travel through a liquid; when one layer of liquid shears against another, there is no restoring force tending to move the layers in the opposite sense and so no continuing wave motion. The three kinds of waves travel with different velocities, surface waves being the slowest and P-waves the fastest. Furthermore, P- and S-waves are refracted by the varying

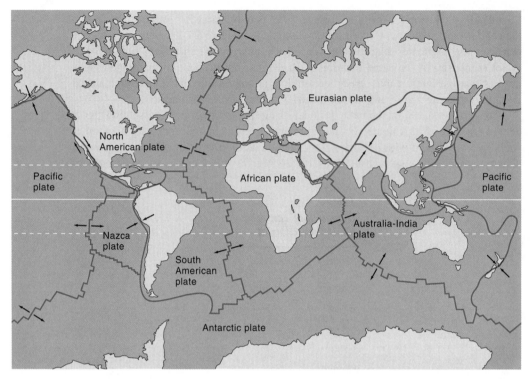

Figure 24.2. The largest tectonic plates into which the lithosphere (the crust and upper mantle) of the Earth is broken. The arrows indicate the directions of motion of the plates, which typically move a few centimeters per year.

density and composition of the Earth's material (see Figure 24.3). This is analogous to the bending of light rays as they pass through materials having different indices of refraction.

By studying the times of arrival and amplitudes of the waves as measured by seismographs all around the world, geologists can not only find the positions and strengths of earthquakes, but can also map out the structure of the Earth's interior. For example, the shadow zone shown on Figure 24.3 is a region where, for the most part, neither P-waves nor S-waves are detected. This enables the size of the Earth's core to be determined. The detection of extremely weak P-waves in the shadow zone indicates that the core is not entirely liquid, but has a small solid volume at its center.

Were it not for earthquakes, our knowledge of the Earth's interior would be meager indeed. The Moon and Mars are the only other objects that have been studied seismically, but the seismographs left on the Moon and Mars are in no way comparable to the extensive network of seismographs over the Earth's surface.

The picture that has emerged from earthquake and other studies is that the body of the Earth is strikingly layered, both in its chemical composition and in its mechanical structure. By composition, it is divided into three main sections, an outer crust, a mantle, and a central core (see Figure 24.4). The crust is just that, a very thin layer on average only about 7 kilometers deep over the oceans, and 30 kilometers or so thick over the continents. Both types of crust are made of igneous rocks,[1] but those forming the continents are granites, whereas over the oceans they are basalts, similar to those

---

[1] Rocks formed by cooling from hot, molten rock.

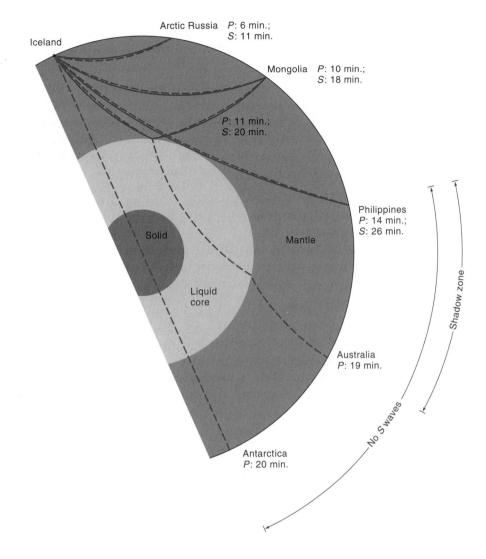

Figure 24.3. Earthquakes produce two types of waves that penetrate the interior of the Earth: P-waves (dashed lines), which can travel through both solids and liquids, and S-waves (solid lines), which cannot propagate through a liquid. Typical travel times for some P- and S-waves are given. Data from a worldwide network of instruments (seismometers) that detect these waves have enabled geologists to infer the existence of a liquid core, for example, from the absence of S-waves on the side of the Earth opposite a quake.

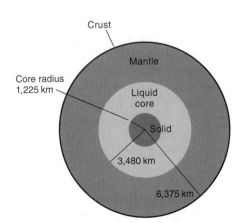

Figure 24.4. The main regions of the Earth's interior, which are shown approximately to scale.

on the Moon. The **continental crust**, which provides us with the oldest Earth rocks, is a mixture of CaO, Al_2O_3, and SiO_2 (called calcaluminous silicates). The **oceanic crust** consists of both calcaluminous and ferromagnesian (FeO, MgO, SiO_2) silicates. These rocks form from hot mantle material forced upward by convection. The molten rock reaches the surface through the rifts between the huge plates, constantly driving them apart by a few centimeters each year. This part of the crust is young, having been solidified for less than 200 million years. The mantle, which makes up about two-thirds of the mass of the Earth, is a mixture of the ferromagnesian silicates. The **inner** and **outer cores** together make up one-third of the Earth's mass. They are composed of about 80 percent iron, with nickel and other elements making up the balance, probably in different proportions in the two regions. The Earth has been chemically separated or differentiated.

How do we know the chemical composition of the interior? As mentioned earlier, geological activity gives us samples of material from the upper mantle; these are largely silicates. Why do we think the core is iron and not, for example, zinc, which has the same density as iron? Because we have no reason to believe that the nebula from which the solar system formed was composed of anything but the elements in their cosmic abundances. Since iron is about a thousand times more abundant than zinc, the core is much more likely to consist of iron. Indeed, the elements that make up most of the Earth—oxygen, silicon, magnesium, and iron—are all among the most abundant of the metals (see Table 11.2). A model can be made guided by cosmic abundances and by the composition of the crust, and constructed according to the data given by seismographic analyses of earthquakes, that reproduces the average density of the Earth.

Structural Regions of the Interior.   Because of the wide variety of temperatures, pressures, and densities found in the interior, the Earth is also structurally or mechanically differentiated (see Figure 24.5). The inner part of the iron core is solid, but its outer portion does not transmit S-waves and so must be liquid. The material of the mantle has a melting temperature higher than that which prevails in the outer part of the Earth's interior, so the mantle is solid. The melting point of the iron in the core, however, is much lower than the temperature at the bottom of the mantle, and so the outer core is liquid. So how is it that there is a solid inner core? The melting point of matter depends on the pressure to which it is subjected: the greater the pressure, the higher the melting point. The pressure increases with depth throughout the Earth, and in the region of the inner core it has reached more than three million atmospheres (that is, three million times the atmospheric pressure at the surface of the Earth). This high pressure causes the melting point of iron to be higher than the (approximately) 6,000-K temperature of the inner core. Consequently, a solid inner core has formed.

In addition to the solid-liquid division of the Earth's structure, it is also useful to distinguish two regions within the crust and mantle: the **lithosphere** (the rock sphere) and the **asthenosphere** (the weak sphere). The former includes the crust and the upper portion of the mantle; the asthenosphere includes more of the mantle. Why this distinction? Though the crust and mantle are solid, the degree of their solidity varies. The Earth's lithosphere is relatively rigid and has low plasticity, that is, little tendency to flow, whereas the boundary of the asthenosphere is marked by a sudden increase in plasticity, in the ability of the material to flow under pressure. The slowly flowing asthenosphere caused the brittle lithosphere to crack into the several plates as mentioned earlier, and continues to push these plates apart or slowly crush them against each other. They float and slide on the hot, plastic upper portion of the asthenosphere.

Convection currents in the asthenosphere slowly carry hot material up and cooler material down, creating new crust, mostly in the oceans. Volcanoes occur when mol-

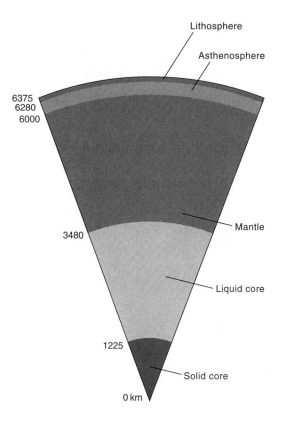

Lithosphere

Asthenosphere

6375
6280
6000

Mantle

3480

Liquid core

1225

Solid core

0 km

Figure 24.5. Some of the regions of the Earth distinguished by their mechanical properties. The relatively rigid lithosphere slides on the plastic asthenosphere, the region of the upper mantle where convection takes place. The depth of the lower boundary of the asthenosphere is uncertain.

ten rock escapes to the surface in the rifts between plates. Ground water can be heated by this rock to produce hot springs, such as those in Yellowstone Park or in Iceland. Just how deep into the mantle the slow convection that drives this activity extends is one of the unanswered questions about the Earth's structure. Perhaps it extends downward about 400 kilometers, since at greater depths the high pressure in the mantle probably increases the density of the material enough that convection can no longer occur.

Mountain Building and Continental Drift. When cool, sinking asthenospheric material occurs at plate boundaries, one plate may itself sink beneath its neighbor in a **subduction zone**. In this way deep trenches, such as those off Japan or Chile, can be formed. Mountains are the "crinkles" produced when one plate slowly collides with or sinks below another. The Himalayas and the Andes mountains, for example, are only a few tens of millions years old. They are being formed by the collisions of the Indian and Eurasian and the Nazca and American plates, respectively. About 200 million years ago most of the land mass of the Earth happened to be collected into one giant continent, named **Pangaea** (*pan* means "all," *gaia* "the Earth"). Since then, tectonic activity has produced the arrangement we now see (see Figure 24.6). Because more recent plate motions destroy evidence of older plate configurations, it is impossible to reconstruct continental arrangements that occurred more than about 0.5 billion years ago.

In the last chapter we listed the major energy sources that drive the convective currents in the mantle responsible for tectonic activity. Both external and internal heating made the young Earth much more active 4 billion years ago than it is now. This is sup-

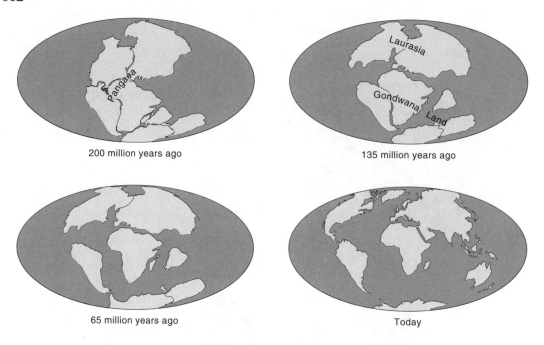

200 million years ago

135 million years ago

65 million years ago

Today

Figure 24.6. About 200 million years ago the Earth's land mass formed one large continent, Pangaea. Continental drift, still continuing, slowly rearranged the continents to create what we see today. Note that the land mass that became India moved northward about 9,000 km in 200 million years!

ported by the fact that the oldest known Earth rock solidified about 4.1 billion years ago, about half a billion years after the Earth formed. This early heating raised temperatures to the melting point, and allowed elements such as iron and nickel to trickle to the center and form the liquid core.

After this period of intense activity, the Earth's mantle solidified (though not to the point of rigidity) from the bottom outward. Large atoms (those of larger atomic weight) cannot exist in solid minerals under high pressure; they are "squeezed" outward toward the crust, which, as a consequence, has a higher abundance of heavy elements such as sodium as well as the radioactive elements, such as uranium and thorium. Since the radioactive elements are closer to the surface, the heat they produce escapes the Earth more easily than it would if the elements were still uniformly distributed within the Earth. They are still producing an appreciable amount of heat, however, and consequently a significant amount of geological activity. For now the Earth, though slowly cooling off, is still geologically active, as we might expect for the largest of the terrestrial planets.

What will happen over the next several billion years as the Earth's interior continues to cool? The temperature drop will be most marked near the surface, so the top of the asthenosphere will lose its plasticity and effectively become part of the lithosphere. As the lithosphere thickens, geological activity will decline. The constantly eroding continents will steadily decrease in size and eventually disappear, since they will not be replaced by new crust brought up from the mantle. Ultimately, the core will solidify and the Earth's magnetic field will weaken and finally disappear.

The Earth's Atmosphere

Its Origin and Early Evolution. It is clear that the Earth's present atmosphere (see Table 24.3) could not have been the original one, since that would have all been lost during the Earth's early high-temperature phase. That is, lighter atoms had sufficiently

Table 24.3. *Atmospheric composition of the terrestrial planets*[a]

Body	Pressure (atm.)	Atmospheric composition by percent				
		N_2	O_2	CO_2	H_2O	Trace
Venus	96	3.4	—	96.4	0.14	Ar, CO, SO_2
Earth	1.0	78.1	20.9	0.03	0.8	Ar, Ne
Mars	0.01	2.7	0.13	95.3	0.03	Ar
Io	—	—	—	—	—	Na, SO_2

[a]Mercury, the Moon, and Europa have essentially no atmosphere.

large velocities to escape the Earth's gravitational pull. Support for this notion is given by the low abundance in our atmosphere of the noble gases—elements such as helium, argon, and neon. In the Sun, neon is nearly as abundant as nitrogen (see Chapter 11). In the Earth's atmosphere, however, neon is just a trace element, whereas nitrogen is the most abundant constituent, more than 10,000 times more abundant than neon.

The low atmospheric abundance of the noble elements can be understood in the following way. The noble gases do not readily form molecules, so they are not tied up in atmospheric compounds or in surface rocks. Furthermore, with the exception of helium, these are relatively heavy atoms, and a high temperature is required to give them the requisite escape velocity. Therefore they must have been lost long ago, when the Earth was very hot. If neon was lost, then nitrogen, which is lighter than neon, must also have been lost. Thus the present nitrogen-rich atmosphere was formed after the Earth cooled.

The nature of the Earth's atmosphere during its first two billion years or so is speculative, and we will not describe the possibilities in any detail. Suffice it to say that the active crust must have outgassed great quantities of gas, just as volcanoes do on a much smaller scale today. Most prominent among these gases were carbon dioxide (CO_2), water vapor (H_2O), with smaller amounts of molecular nitrogen (N_2) and various oxides of sulfur. There is some uncertainty about how much hydrogen-rich compounds such as methane (CH_4) and ammonia (NH_3) were present in the early atmosphere. Because of the overwhelming cosmic abundance of hydrogen, some scientists have argued that hydrogen compounds must have been dominant. Even if they were, this phase probably could not have lasted more than a few hundred million years or so, for the following reason. No ozone (O_3) existed in the upper atmosphere as it does today, shielding the Earth's surface from solar ultraviolet light. Consequently, energetic photons from the Sun penetrated the atmosphere and broke up the ammonia, methane, and water vapor, allowing the free hydrogen released in this way to escape into space.

Two additional comments are appropriate here. Recall that the zero-age main-sequence Sun was about 70 percent as luminous as the present-day, slightly evolved Sun. This means that sometime after the end of Earth's accretion phase (which caused its surface to be heated), the Earth's temperature was much lower than it is today. In fact, the oceans would have been frozen and would probably have remained so until perhaps two billion years ago, long after geologic evidence indicates that liquid water did in fact exist. Consequently, there must have been a strong greenhouse effect heating the early Earth. This could have been produced either by an atmospheric abundance of carbon dioxide ten to a hundred times greater than exists now, or by a small amount of ammonia, which absorbs infrared radiation very effectively.

The second comment is that laboratory experiments (to be described in the next chapter) that attempt to replicate conditions on the early Earth produce interesting organic substances associated with life when their initial ingredients include hydrogen-rich compounds such as methane and ammonia. Such experiments starting with oxygen-rich compounds are not quite as successful. As we have seen, however, it appears likely that hydrogen-rich compounds were largely destroyed fairly soon after the Earth's formation. Could enough methane and ammonia have survived to provide the most favorable circumstances for the emergence of life, as well as the ammonia that could have kept the oceans from freezing through its greenhouse effect? We don't know.

Biological Component. In any case, biological and geological evidence suggest that about 3.5 billion years ago simple single-cell organisms appeared that derived energy from the Sun. They probably used some non-oxygen-producing process (rather than converting carbon dioxide and water into carbohydrates and oxygen), because analysis of rocks formed more than 2.5 billion years ago indicate that they formed in the absence of oxygen. By about 2 billion years ago, however, blue-green algae had produced so much oxygen by photosynthesis that it began to be a significant component of the atmosphere. Also, marine organisms used carbon dioxide dissolved in the oceans to build their shells. These ultimately formed carbonate rocks. Probably by 600 million years ago or so the atmospheric composition was pretty much the same composition we enjoy today. Thus both geological and biological processes were (and continue to be) responsible for our atmosphere. Oxygen is very reactive, combining readily with other elements and compounds. It has been estimated that in the absence of plant life (which ingests carbon dioxide and emits oxygen), the free oxygen in our atmosphere would disappear in just millions of years. Without plant life, warm-blooded creatures like us would not have been able to exist in the first place, nor could we continue to exist today.

Greenhouse Effect. Since we know how much solar energy reaches the Earth, it is easy to calculate that without an atmosphere, the Earth's average temperature would be about $-20°C$. The actual temperature is considerably warmer than that, however, about $+15°C$. This is a consequence of the greenhouse effect. Except for water in the form of small droplets that reflect light (that is, clouds), our atmosphere is transparent to incoming visible solar radiation. Carbon dioxide and water vapor are not transparent to the infrared energy radiated by the Earth, however, and water vapor in particular absorbs much of it. This causes the temperature of the Earth's surface to be increased by 35 degrees to its present value. Our biosphere could not survive without the higher temperature produced by the greenhouse effect; it is of great significance to us.

As you know, there is concern that even though it is a minor source of infrared absorption compared to water, the carbon dioxide added to our atmosphere by industrial activity may eventually increase the greenhouse effect significantly. The consequences this may have on our climate are not yet clear, because the Earth's atmosphere, oceans, land masses, and polar caps form a very complex system, with the various parts acting and reacting on all the rest. Nonetheless, we should be concerned that any kind of human activity might have long-term global consequences.

The Moon's Surface

General Features. We have described our "standard" terrestrial interior and atmosphere, those of the Earth. Let us now turn our attention to our "standard" surface for these objects, that of the Moon. That as children we talked about the "man in the Moon" means that even to the unaided eye, the Moon shows contrasting dark and bright features (Figure 24.7). Before the telescope, the former were thought to be seas or oceans (and denoted by the Latin word for seas, *maria*); the bright regions were called *terrae*, meaning lands. They are now generally referred to as "highlands." Even a small telescope reveals that though the dark regions are relatively flat, they also show hills, ridges, and an occasional crater. Therefore, the maria can't be bodies of liquid. The bright regions are seen to be quite rugged, often mountainous and heavily cratered. They reflect about 15 percent of the incident sunlight, twice as much as do the darker maria. Thus these areas differ not only in their terrain, but also in their chemical composition.

Some of the lunar mountains (the heights of which were first measured by Galileo) rise by as much as 8,000 meters above their surroundings. When compared to the diameter of the parent body, these are several times higher than Mt. Everest. We will see that they were formed by quite a different process than were terrestrial mountains.

Figure 24.7. The full Moon as photographed from Apollo 17. Note the dark maria and the bright terrae (now called "highlands").

Figure 24.8. An Apollo image of the Moon. The prominent crater at the bottom is about 6 kilometers in diameter. Note the large numbers of craters only a few hundred meters in diameter.

Craters.   One of the most striking features of the Moon is its heavily cratered surface. With telescopes on Earth we can see about 30,000 craters, but from space many more are visible (Figure 24.8). They range in size from centimeters to about 250 kilometers in diameter. The overwhelming majority of them are the result of meteoritic impacts rather than internal activity. It is only in this century that lunar craters were recognized to be caused primarily by impacts. Astronomers and geologists were misled by the lack of both terrestrial impact craters and large objects in the inner solar system capable of causing lunar craters. By contrast, there are many volcanoes on Earth, and so volcanic activity was incorrectly thought to have played the major role in shaping the Moon's surface.

All told, the Moon's surface records about 200,000 craters with diameters of a kilometer or greater. That the number of craters per square kilometer is about 20 times greater on the highlands than on the maria indicates that the latter formed more recently than the former. Just how much younger the mare craters are could be established only with data from the Apollo program. Over a three-and-one-half-year period beginning in July, 1969, six of the Apollo missions placed men on the Moon for a few days of exploration. Ultimately, 392 kilograms of lunar rock were returned to Earth for analysis. These data from the Moon rocks have enabled us to establish a fairly accurate chronology of the early history of the Moon, which is described below.

Most of the Moon is covered with layers of debris (called the **regolith**) from the many impacts experienced early in its history. Because there are relatively few craters less than 15 meters deep, it has been inferred that about 15 meters of lunar surface have been worn away by impacts or covered by debris since its formation. Because the Moon has no atmosphere and therefore no winds and no running water, little erosion

Figure 24.9. The Apollo 17 astronauts in the Taurus-Littrow valley. The surface of the Moon is covered with fine sand and the remains of small rocks, which were pulverized by the continuing bombardment of tiny meteorites, along with a few larger rocks and a boulder. The latter may be debris from the impact that produced the Serenitatis basin.

has occurred except for bombardment by tiny meteorites, which continue to strike the surface in great numbers. These have pulverized the upper layers of the regolith, turning all but some rocks and boulders into a layer of fine sand a few meters thick (Figure 24.9). Old craters show the effects of this erosion in having shallow basins that sometimes contain newer craters, or rims partly obliterated by subsequent impacts and debris. Relatively new craters that have not undergone as much of this wearing away process have sharper, more clearly defined rims surrounding deep craters.

Occasionally, several craters are found in a straight line. These **crater chains** are among the few craters formed by volcanic action and are generally found along narrow trenches where the Moon's crust has fractured. These features, called **rilles** (Figure 24.10), can be quite straight (those resulting from surface faulting) or rather winding, somewhat reminiscent—misleadingly—of riverbeds. Rilles are usually no more than a few hundred meters deep, but can be up to five kilometers wide and extend for several hundred kilometers. Since the uphill end usually starts in a crater, these rilles were most likely formed by flowing lava produced by impact heating.

Radiating out from some craters are bright streaks of material called **rays**. These can be seen to best advantage at full Moon, when the Sun is directly above them and we see most of the light they reflect.[2] One of the most prominent ray systems is associated with the relatively young crater Copernicus (Figure 24.11). We know it is young because the rays are deposited on top of older features. The rays are produced by many small secondary impact craters and by reflection from glass-like material. This was molten when ejected by the impact, and solidified to a glassy state. In contrast to primary craters, secondary craters can be somewhat elongated rather than circular, because they are formed by relatively small masses of ejected material that travel much more slowly and hence with less energy than the primary impacting objects.

---

[2] Craters and mountains are best seen at the quarter Moons, when they cast deep shadows.

618

Figure 24.10. Hadley Rille, about 1.5 kilometers wide and 300 meters deep, as imaged from Apollo 15. Hadley Rille is probably an old lava channel. The Apollo 17 landing site was next to the lowest part of the rille in this image.

Figure 24.11. The crater Copernicus is in the upper right corner of this photograph taken with the 100-inch telescope. Note the white streaks (or rays) radiating out from the crater, resulting from ejecta from the impact that formed Copernicus. The dark area below Copernicus is the Imbrium basin. Figure 24.12 shows where on the Moon the latter is located.

Origin of the Maria and Highlands. A significant Apollo finding is that much of the surface of the Moon is covered by basalts (formed by cooling of molten lava). The Moon's surface could have been made molten by meteoritic impacts early after its formation, or by radioactive heating somewhat later. The ages of lunar rocks showed that both processes must have been important. That the maria are younger than the highlands is clear from the few craters seen in the maria. Indeed, rocks from the bright highlands were found to be 4 billion or more years old (the oldest is about 4.2 billion years), whereas those from the darker maria formed about 3.5 billion years ago (3.1 to 3.9 billion years). Remember that radioactive dating gives the age of a rock from the time of its solidification. Calculations have shown that about one billion years were required for radioactivity to have sufficiently heated the Moon for it to be molten. Thus, the highlands (originally the lunar crust) were completely pulverized and heated by impacts associated with the formation and very early history of the Moon, whereas the younger basalts of the maria were heated by internal radioactivity.

During the impact era, large basins were formed by gigantic collisions early in the Moon's history. These events (and not plate motions) also produced the major lunar mountains bordering the basins. From the statistics of lunar craters and the ages of

Figure 24.12. The Imbrium basin is the large, roughly circular dark area at the bottom left of this 100-inch photograph. Is is the result of an impact with a 100-kilometer object followed by extensive lava flooding. Compare lightly cratered Imbrium with the lunar highlands in the upper left of the photograph. The rayed crater on the upper edge of Imbrium is Copernicus.

Figure 24.13. Mare Orientale is the youngest of the large impact basins. It was produced by a collision with an object of about the same size as the one that produced the Imbrium basin, but lava flooding was less extensive here. The outer ring of mountains is about 900 kilometers in diameter. Mare Orientale is located on the limb of the Moon, and though it was discovered from the Earth, it is seen best from space, as is shown in this image by Lunar Orbiter IV.

Apollo rocks, it is apparent that the period of intense bombardment ended about 3.8 billion years ago. Since then the rate has been essentially what we observe today.

Soon after the end of the era of rapid impacts, heating by radioactivity melted the interior and huge amounts of lava flowed through fractures in the basins, eventually nearly filling them. This kind of volcanic activity, in which cones or mountains are not formed, occurs on Earth as well. The range in ages of maria rocks indicates that a high level of volcanism lasted about 800 million years. The Moon has been essentially dead for the last three billion years, with no geological (or selenological) activity to speak of.[3] (From our point of view, the most interesting thing to happen on the Moon in the last three billion years was the arrival of the astronauts.)

The Imbrium and Orientale basins are two of the largest and youngest of the giant craters (see Figures 24.12 and 24.13). The Imbrium event took place about 3.9 billion years ago, and the Orientale impact about one hundred million years later. The impacting object must have been about 100 kilometers in diameter in both events. The

---

[3] Despite extensive searches, Earth-based observers have been able to collect only two well-documented instances of lunar activity, both in 1963, when a small amount of gas escaped from a vent near the crater Aristarchus.

Figure 24.14. A small portion of the heavily cratered far side of the Moon.

Imbrium basin was flooded with lava, but Orientale was not. We can still see the latter's "bull's-eye" appearance, caused by the two concentric rings of mountains formed by the impact.

Since the Moon keeps the same face toward us as it revolves around the Earth, we knew nothing about the far side of the Moon until Soviet and U.S. spacecraft photographed it. Interestingly, there are significant differences between the two hemispheres. There are few large, lava-filled basins on the far side; instead, it resembles the rugged, heavily cratered lunar highlands (Figure 24.14). Now, both hemispheres should have experienced the same bombardment by comet and asteroidal debris soon after the Moon's formation. However, for reasons not yet clear, the crust is thicker on the far side than on the Earth-facing hemisphere, so lava was less likely to penetrate the far-side crust and fewer dark maria were formed. Lava-free Orientale is just on the "edge" of the near and far sides.

No trace of water in any form has been found in rocks, nor has any organic matter been detected. Lunar rocks are also depleted in iron and volatile compounds compared with the Earth's crust. The day-night variation of the Moon's surface temperature is very large, from 130°C to −170°C, because it has no moderating oceans or atmosphere, and because day and night each last two weeks.

Intercomparison of Terrestrial Objects

Let us next see how the various terrestrial objects compare with each other.

Rotation Periods. The rotation periods of the Moon and of Mars have long been known. The Moon's rotation is locked with its revolution about the Earth, so that it

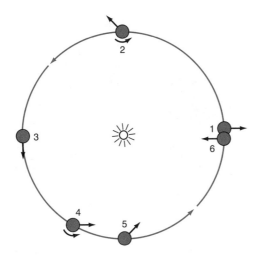

Figure 24.15. Starting from position 1, positions 2, 3, 5, and 6 show Mercury's orbital position at intervals of 22 days; that is, at intervals of about one-quarter of its period of revolution. In 22 days, Mercury undergoes 22/58.7 = 0.37 rotations, or 0.37 × 360° = 135°, as indicated by the arrows. At position 4, Mercury has rotated once with respect to the stars and at position 6 (after one sidereal year) it has rotated 1.5 times. In one more sidereal year, the arrow will again point in the same direction as it did at position 1. Thus in two of its years, Mercury makes three complete rotations with respect to the stars, but only one (midnight to midnight, as shown by the arrow) with respect to the Sun; its solar day is 176 Earth days.

always presents the same face toward us. It requires 27.3 days to rotate once (its sidereal period). The length of the Martian day is just slightly over 24 hours. Both rotations are in the direct sense, that is, counterclockwise as seen from the north. By contrast, the rotation period of neither Venus nor Mercury can be measured from the ground in visible light. Thick clouds on Venus cover its surface; Mercury has no clouds and its surface markings are not large and distinct enough to be followed as the planet rotates. Astronomers estimated the rotation periods of both planets to be anywhere from several days to their year, 88 and 225 days for Mercury and Venus, respectively. These long periods implied, of course, that the two planets always kept the same face toward the Sun, just as the Moon does toward the Earth. In the 1960s, however, radar measurements showed that Mercury's period is 58.6 days, exactly 2/3 of its orbital period. By the same technique, Venus' rotation period was found to be 243 days, longer than its year; perhaps even more surprising, it rotates in the retrograde sense, that is, clockwise as seen from the north.

With benefit of hindsight, the rotation period of Mercury is to be expected, and demonstrates an interesting kind of mechanical resonance fairly common in the solar system. The most familiar example is that of the Moon's rotation. As you know, the gravitational pull of the Earth on the Moon's tidal bulge keeps the Moon's axial rotation period locked with its orbital period. Now, Mercury differs from the Moon, in that it rotates on its axis three times for every two trips it makes around the Sun. Consequently, the same face points toward the Sun only at every other perihelion passage, not at every passage. Its tidal bulge, however, is aligned along the Mercury-Sun line at every perihelion passage (see Figure 24.15). Since Mercury's orbit is moderately eccentric, the gravitational pull of the Sun on Mercury's tidal bulge is most effective at perihelion and over time produced the observed resonance.

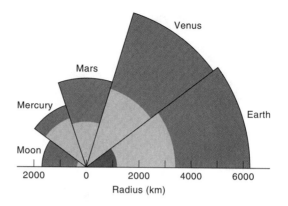

Figure 24.16. The interiors of the Moon, Mercury, Earth, Venus, and Mars on the same scale. It is not certain if Venus and the Moon have cores, or if the Moon and Mercury have crusts differing in composition from their mantles. Only the Earth has an inner core. Note the relatively thin mantle on Mercury.

We have to be careful when we talk about the length of the day on Mercury. We could define a "day" as the length of time required for it to make one complete rotation with respect to the stars, that is, its rotation period of 58.6 days. A more familiar definition uses the Sun as the marker; the solar day is the time from one noon (or sunrise, etc.) to the next. As we just saw, this requires 3 × 58.65 = 176 Earth days. Mercury's solar day is two Mercurian years long!

The explanation for Venus' slow retrograde rotation is not certain. Perhaps soon after its formation a large object collided with it in such a way as to reverse its spin. The length of the solar day on Venus is about 117 days, or just a bit over half its year. Now, the synodic period of Venus (say, from one closest approach of Venus and Earth to the next) is 585 days. During this time, Venus rotates nearly exactly five times with respect to the Sun and four times as seen from the Earth. It very nearly—but not quite—presents the same face to the Earth at every perihelion passage. To be exactly in synchronism, its rotation period would have to be 243.16 days, just a little less than three hours longer than it actually is. Apparently, this near coincidence is just that— a coincidence. For many years at a time, then, the same hemisphere of Venus faces the Earth at each closest approach. Since the best radar maps made from Earth are made during those times, we had until recently a much better knowledge of one side of Venus than the other.

The Interiors of the Moon, Mercury, Venus, and Mars. Recall that Mercury has the largest reduced density of any of the terrestrial planets. Its relative content of heavy elements, probably largely iron, is greater than the Earth's, probably because Mercury, being so near the Sun, was too hot for certain kinds of silicates to condense. Since Mercury is most likely chemically differentiated, its iron core must be very large, about three-fourths of its radius, or about the size of the Moon (see Figure 24.16). It has a weak magnetic field, only about 1 percent of the Earth's, but this is still stronger than that found around the Moon, Venus, or Mars. The field suggests that at least part of the core is still molten, but this is thought unlikely since Mercury, with its small diameter, cooled fairly quickly. Perhaps the field is just residual magnetism in its rocks that formed when the core was still molten, but the field was weak because of Mercury's slow rotation rate.

Seismometers placed at five different locations on the Moon by the Apollo astronauts show that it is very quiet geologically; moonquakes release energy at a rate less than a billionth of that produced by quakes on Earth. Furthermore, moonquakes originate at depths of 800 to 1,000 km below the surface (about ten times deeper than do earthquakes), indicating that the Moon's lithosphere is some 1,000 km thick. Overlying the lithosphere is a crust about 60 km thick (Figure 24.16). There is a little evidence for a liquid core on the Moon about 600 km in radius, but the magnetic field generated there is only about 10^{-4} that of the Earth's field.

Analysis of the motion of spacecraft orbiting the Moon shows that the Moon's gravitational field is not uniform; there are areas of increased strength at specific points below the surface. These so-called gravity anomalies are caused by mass concentrations of somewhat higher density (probably basalts) than the surrounding material. If they were on the Earth, the gooey mantle could not support them. That they exist on the Moon is additional evidence for its thick, rigid lithosphere. The origin of these concentrations is not clear.

As we will see below, the surface structures of the Moon, Mercury, and Mars differ fundamentally from the Earth's in that each is completely covered by only a single plate. Internal heat does not get to the surface by convection as it does on Earth; it must escape by conduction or volcanism. Since these objects are all smaller than the Earth, they lost their internal heat more rapidly than did our planet; hence their lithospheres are sufficiently thick that we should not expect them to show the kind of plate-tectonic activity so important on the Earth.

Because of Venus' thick cloud cover, we have had to rely on radar measurements (especially those from the Magellan mission, described later in this chapter) for most of our information concerning its surface and, by inference, its structure. However, since Venus and Earth have very nearly the same mass, radius, and mean density, we would expect their internal structure and evolution to be nearly the same. Therefore Venus presumably has a sizable liquid core and its very weak magnetic field (only 0.01 percent of Earth's) is most likely a consequence of its slow axial rotation. Although radar images give evidence of much tectonic activity, the lengthy ridges, rifts, or long, folded mountain ranges that would delineate huge Earth-like global plates are not seen. Surprisingly, Venus appears to be a one-plate planet. Why there should be such a fundamental difference in the crustal structure of two such similar planets as Earth and Venus is an interesting and as-yet-unanswered question. If plate activity on Earth has anything to do with water, possibly acting as a lubricant, then perhaps Venus' dryness accounts for the difference. This is highly speculative, however.

There is plenty of evidence of local tectonic activity in areas of cracks and ridges, for example. Other features caused by tectonic activity are the so-called **coronae** (Figure 24.17), first found by Soviet spacecraft. About 400 have been identified so far ranging in diameter from 60 to 2,600 kilometers, although few are more than 400 kilometers across. Circular or oval in outline, they generally rise no more than a kilometer above the surrounding plain and often are surrounded by a depression or moat. They are probably formed by upwelling mantle plumes or hot spots pushing on the crust. Material under the moat may even be sinking back toward the mantle, forming a tiny subduction zone. Coronae are thought to be small examples of the process that forms the much larger volcanic rises. Even Venus' mountains are probably formed by yet larger mantle plumes.

The picture that suggests itself is of a large number of tectonic centers, differing widely in their activities, dotting the surface, rather than being along the cracks between plates as is generally the case on Earth. Most of Earth's internal heat results in tectonic activity, but neither mantle upwellings nor volcanic activity on Venus appears sufficient to account for the loss of its internal heat. At the present time, anyhow, most

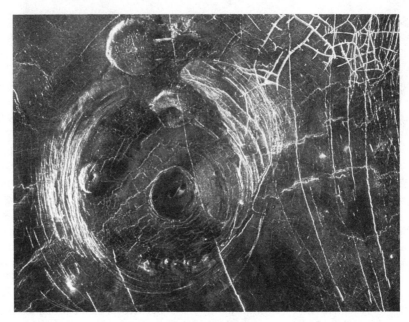

Figure 24.17. The dark circular feature near the center of this radar image of Venus is a corona, about 200 kilometers in diameter. The slightly uplifted coronae are probably the result of underlying mantle plumes pushing upward on the crust.

of the heat escapes by conduction to the surface (as is the case for smaller terrestrial bodies).

Surface Features of Mars and Mercury.   It is significant that all the planetary surfaces in the solar system that are capable of recording meteor impacts (that is, which have solid surfaces) show that they have undergone such bombardment. This includes satellites associated with the giant planets, as well as the terrestrial objects. This is not too surprising, since the very process by which planets and satellites formed involved accretion of matter by impacts. A billion years or so after the solar system formed, most of the colliding objects had been "cleaned out," and only occasional collisions have occurred since then.

The enormous number of craters that dominate the Moon's appearance is in marked contrast to the hundred or so that have been found on Earth. The Earth must have been as heavily bombarded as the Moon, but erosion and the Earth's active crust have long since destroyed any evidence of the early events. More recent impacts most likely struck the oceans, which cover six-sevenths of the Earth's surface. The best-known meteor crater on Earth, though by no means the largest, is the Barringer crater in Arizona (Figure 24.18). It was formed perhaps 25,000 years ago by the impact of an iron meteorite only about 25 meters in diameter. Nearly all of the recently discussed terrestrial craters have been found from the air since, even when severely eroded, a crater's outline can be fairly obvious from above.

The surfaces of Mercury and Mars show ample evidence of the intense early bombardment that all planets experienced. Some Martian craters are huge, up to 1,800 km wide and 4 km deep. Craters dominate the southern hemisphere and equatorial regions; the northern hemisphere is much less cratered, more because of modifications produced by greater volcanic activity in the northern half of the planet than because of a real decrease in cratering density. In fact, all the Martian craters show the effects of volcanic action and erosion (Figure 24.19).

Figure 24.18. The best-preserved impact crater on Earth is the Barringer crater in the Arizona desert. The impacting object was a house-sized iron meteorite that produced an explosion equivalent to that of a 3-megaton bomb, resulting in a crater 1.2 kilometers in diameter and about 200 meters deep.

Figure 24.19. The Argyre basin, with a diameter of 700 kilometers, is the second-largest impact crater on Mars. Notice the many craters in the basin that were made after the Argyre impact, as well as the mountainous terrain, which is debris resulting from the impact. Two cloud layers, probably ice crystals of carbon dioxide, can be seen in the background.

Figure 24.20. The extensively cratered surface of Mercury. Note its resemblance to the lunar highlands. This image is a mosaic made by Mariner 10 in 1974.

Most of Mercury closely resembles the lunar highlands, but with some differences in detail (Figure 24.20). Because the surface gravity on Mercury is twice that on the Moon, debris ejected by an impact (for example, the bright rays associated with some craters) covers only about one-sixth the area on Mercury that it would on the Moon. For the same reason, secondary craters formed by boulders ejected from the primary crater are much more closely clustered around the primary one on Mercury than they are on the Moon. Interestingly, only one large multi-ring crater, Caloris (similar to the Moon's Orientale), is known that at all resembles a lunar mare (Figure 24.21). Its floor, however, is neither as smooth nor as dark as lava-flooded basins on the Moon (in fact, lava flows are no darker than the rest of Mercury's surface); its composition is uncertain. The large impact craters that must have been made on Mercury were most likely erased by lava flows.

Intermediate in diameter between Earth and the Moon, Mercury cooled less rapidly than the latter so its lithosphere should not be as thick as the Moon's. Therefore, it should show some evidence of greater tectonic activity than does the Moon. This is seen in Mercury's many **lobate scarps** (Figure 24.22), that is, scalloped cliffs, some hundreds of kilometers long. These indicate that Mercury's lithosphere has contracted by a few kilometers and cracked. This shrinkage was caused by surface cooling of a few hundred degrees Celsius, which in turn indicates that differentiation—core formation and the upward forcing of radioactive elements—occurred early in Mercury's history.

Mars, however, has long canyons, the most spectacular of which is called Valles Marineris after the spacecraft that discovered it (Figure 24.23); it is about 5,000 km long, 75 km wide, and 6 km deep! These canyons indicate that Mars' surface has been split by tensional forces. That is, the surface has been expanding slightly, because it was warmed in the fairly recent past. This implies that differentiation and core formation occurred later on Mars than on Mercury, perhaps two rather than four billion

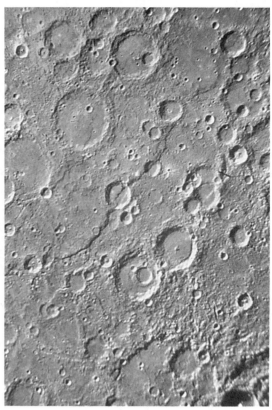

Figure 24.21. The Caloris basin is the only large multi-ring impact crater on Mercury. The center of the crater is off the image to the left. With a diameter of about 1,000 kilometers, it is similar in size to the Moon's Imbrium basin. This image was made by Mariner 10.

Figure 24.22. One of Mercury's lobate scarps is seen at the center of this image. Cliffs such as these were formed by surface shrinkage, a process apparently unique to Mercury. On the Earth, Moon, and Mars, such features are the result of tensional (that is, expansion) forces, splitting the crust.

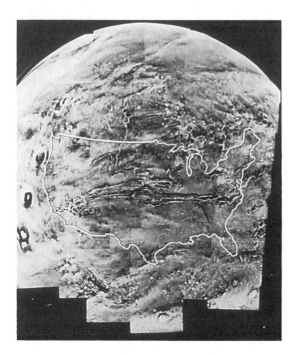

Figure 24.23. Mars' Valles Marineris, a spectacular canyon that is so long it would extend across the entire United States, was formed by tensional forces splitting the crust.

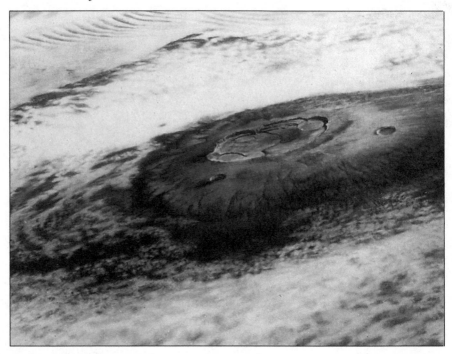

Figure 24.24. Olympus Mons on Mars is probably the largest volcano in the solar system. It is 25 kilometers high, its base is more than 500 kilometers wide, and the caldera measures about 70 kilometers in diameter.

years ago. Support for the idea of recent surface heating comes from the relative lack of craters in the large Martian lava flows, indicating that they are no more than about one billion years old.

Photographs by the Mariner and Viking spacecraft sent to Mars revealed not only large canyons, but also the existence of about a dozen inactive volcanoes. The Tharsis region, in the northern hemisphere, contains four huge volcanoes. The largest of these, Olympus Mons (Figure 24.24), is as large at its base as the state of Missouri; it is 25 km high, and about 70 km across the top of its cone! This is much larger than the largest volcanic cone on Earth, the island of Hawaii. The huge size of the Martian volcanoes is due largely to the fact that the crust stays put and lava flows up through the same vents. An enormous volume of solidified lava surrounding these volcanoes testifies to the immobility of the Martian crust. By contrast, motions of Earth plates prevent the build-up of huge amounts of lava at any one location. The Hawaiian islands are a consequence of plate motion over a fixed "hot spot" from which lava is vented. In addition, the smaller surface gravity on Mars as compared with the Earth makes it possible for a very large pile of lava to support itself instead of collapsing under its own weight.

Mars shows large-scale surface features not seen clearly on any other planet except the Earth, namely, polar caps. In contrast to those on Earth, the Martian caps are made of frozen carbon dioxide (dry ice), but with water ice underneath. The northern cap extends southward to the fiftieth parallel in the depths of winter, but the dry ice sublimates to the atmosphere every summer, exposing the water ice, which remains frozen. The southern dry-ice cap may well be permanent. Other large-scale features are bright areas that are desert-like but much drier, colder, and more boulder-strewn than any terrestrial deserts. Martian soil contains iron oxides that give it its reddish color.

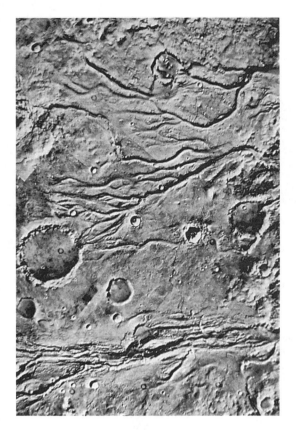

Figure 24.25. One of the most interesting discoveries on Mars is a large number of channels with tributaries that look like terrestrial river valleys. This image shows examples in the Lunae Planum region of Mars. Apparently there was running water on Mars sometime in the past.

Perhaps the most surprising recent discovery is of several Martian valleys with many smaller intersecting gulleys that appear to be quite similar to terrestrial river valleys and their tributaries (Figure 24.25). They are not thought to be caused by lava flows, because they are found in nonvolcanic regions, but there is certainly no liquid water on Mars now. In fact, if any did exist, during a summer day it would evaporate at a rate of about an inch per hour. Perhaps Mars' climate has changed considerably and it was much warmer in the past. Later in this chapter we will return to this problem of water on Mars.

The Surface of Venus. What can we say about the landscape of Venus? Before the development of radar mapping techniques, the answer would have been "nearly nothing"; thick clouds prevent us from seeing the surface. Radar operating at centimeter wavelengths can penetrate the thick cloud cover, however, and be reflected back from the terrain beneath to a receiver on an orbiting spacecraft. Our early spacecraft views of Venus' surface came to us from the Pioneer Venus radar maps, as well as those made by the Soviet spacecraft Venera 15 and 16. These have been superseded by the many higher-resolution radar images returned by the Magellan spacecraft. We now have a good idea of the surface features of Venus. In fact, Venus' surface is more completely mapped than is Earth's.

Launched in May 1989, the Magellan spacecraft arrived at Venus in August 1990, and was placed in a 3.3-hour near-polar elliptical orbit around the planet. Every orbit during the 37 minutes of its closest approach to Venus, Magellan's radar antenna recorded an image 20 kilometers wide by 17,000 kilometers long, from Venus' north pole to about 70 degrees south latitude. After eight months (one Venusian day), the

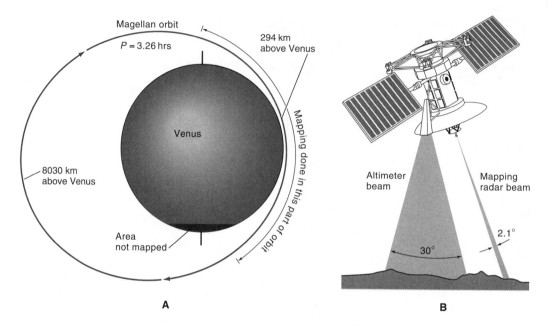

Figure 24.26. (A) The elliptical orbit of the Magellan spacecraft around Venus. Mapping is done when the spacecraft is closest to Venus. The data are then transmitted back to Earth during the part of the orbit when the spacecraft is farther away from Venus. (B) During each orbit, the radar beam maps a swath 20 kilometers wide and about 17,000 kilometers long. Nearly the entire planet is mapped as Venus rotates under Magellan's orbit. Altimeter data are later combined with the mapping data.

planet had rotated once under Magellan's orbit, enabling a map of more than 70 percent of the surface to be produced. By 1992, 98 percent of the surface had been mapped. The best resolution was about 120 meters. At the same time, an altimeter measured the relative heights of features to better than 50 meters along a different but parallel swath of the surface (see Figure 24.26).

Although its extremes of elevation about equal those of the Earth, much of Venus' surface is relatively flat. On Earth huge areas—the ocean basins—are 15 to 20 kilometers below high plateaus and mountains. On Venus, however, about 80 percent of the surface does not vary more than a kilometer from Venus' average radius of 6,052 kilometers. It has been suggested that Venus' lithosphere, warmed substantially by the high atmospheric temperature, is somewhat elastic and tends to smooth out elevation differences. The surface showing greater relief is divided between lowlands as much as 2 kilometers below the flat plains, and highlands that rise appreciably above the plains. Most prominent among the highland regions are Aphrodite, in area about equal to half of Africa, and the Australia-sized Ishtar, which with the Maxwell Mountains[4] rising about 12 kilometers above the average radius, contain the highest points on Venus (see Figure 24.27). The peak of Maxwell Montes is somewhat taller than Mount Everest, the highest mountain on Earth.

Volcanism and tectonic activity have been the dominant agents forming Venus' surface, with the former covering more than 80 percent of the planet. As we have seen,

---

[4] The physicist J. C. Maxwell is the exception to the women-only rule in naming Venusian surface features. He provided the theoretical description of electromagnetic radiation that made possible—among many other things—the radar technique by which Venus has been mapped.

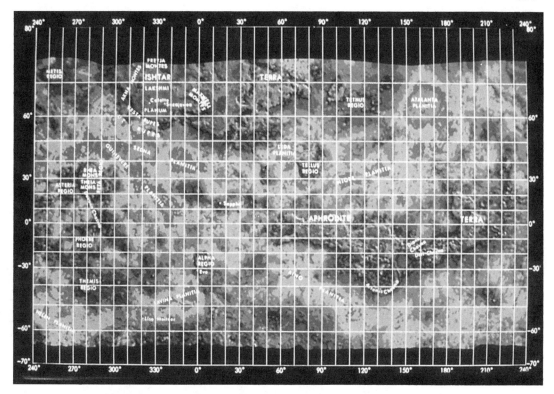

Figure 24.27. A radar map of the surface of Venus with 50-kilometer resolution, made by the Pioneer Venus spacecraft. Most of Venus' surface is relatively flat. Only two continent-sized areas rise appreciably (about 6 kilometers) above the surrounding plains. The highest point on Venus, Maxwell Montes on Ishtar, is about 12 kilometers high.

tectonic activity has not produced any Earth-like continental plates, but rather deformations resulting from strains in the crust produced by mantle motions. These appear as belts of mountains or ridges extending over large areas, formed by crustal compression (see Figures 24.28 and 24.29).

Volcanism has had a profound effect on most of the surface of Venus, most recently in the last few hundred million years. Essentially all its rocks are volcanic in origin, producing a landscape of unrelieved bleakness. Much of the lava was thin and fluid, able to run long distances down slopes as shallow as a 1-centimeter drop in about 6 meters. Hot lava has carved long channels that are somewhat reminiscent of riverbeds. In fact, one of these, running for 6,800 kilometers, is the longest channel known in the solar system (Figure 24.30). It is not clear how these are formed; they imply a very fast rate of lava flow, which seems inconsistent with the narrow, uniform width of the channels.

A few dozen pancake domes about 30 to 60 kilometers in diameter and only a few hundred meters high (Figure 24.31) have been found on Venus so far. They are considerably larger than similar structures on Earth, and were formed by much thicker lava than that making the channels. On Earth, thick lava has an abnormally high silica content.

Hundreds of thousands of small shield volcanoes dot the surface of Venus. These are only a few kilometers in diameter and roughly 100 meters high. Apparently they are associated with the formation of the flat plains that cover most of Venus. Despite

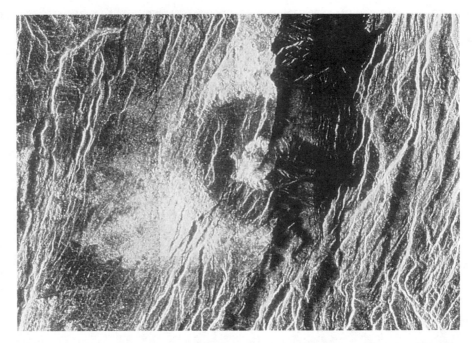

Figure 24.28. A 37-kilometer diameter impact crater in Beta Regio on Venus has been cut by many faults since it was formed. The largest fault valley is about 20 kilometers wide and several kilometers deep. The faults and fractures shown here and in Figure 24.29 are the result of tectonic activity.

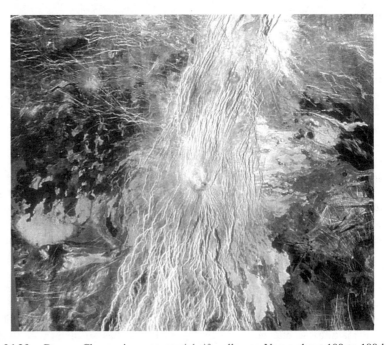

Figure 24.29. Devana Chasma is an equatorial rift valley on Venus about 100 to 180 kilometers wide with many faults and fractures.

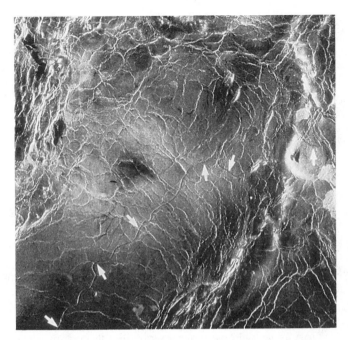

Figure 24.30. The arrows on this image identify a 600-kilometer-long segment of a channel that runs for 6,800 kilometers, the longest in the solar system. Its width is quite uniform and averages about 1.8 kilometers. Its origin is uncertain, though other similar channels on Venus end in lava flows, suggesting that they were formed by rapidly moving lava.

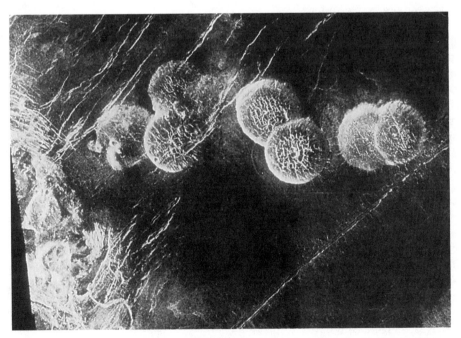

Figure 24.31. Several volcanic domes ("pancake" domes) are shown in this image. Even though they can be up to 60 kilometers across, they are only a few hundred meters high. Their walls are rather steep, indicating that the upwelling lava that formed them was thicker than that which presumably formed the channels on Venus. Notice how younger domes overlap older ones, and how some have been cracked by faults.

Figure 24.32. Views of the landscape of Venus surrounding the two Soviet landers, Venera 13 and 14. Note the flat rocks and their sharp edges. The surface illumination is similar to that on a heavily overcast day on Earth. That the equipment on these two landers survived the high surface temperature and pressure long enough to take and transmit these images is remarkable.

all the evidence for much volcanic activity in the past, no currently active volcanoes have been discovered so far.[5]

A striking aspect of the surface is the sharpness of its features (Figure 24.32). Although there is no water on Venus to erode the surface (nor is there any evidence of ancient oceans or running water), it was thought that winds in its thick atmosphere might blow enough dust to cause at least some erosion. Because the atmosphere is so thick, however, the surface winds are very slow and cause little erosion.

No features older than a billion years have been found on Venus, and since it is much less cratered than even the lunar maria (Figure 24.33), the average age of the surface is estimated to be about 400 million years. This is considerably older than the 100-million-year age of Earth's ocean floors or even the average age of its surface, only 200 million years. The surfaces of the Moon, Mars, and Mercury are a few billion years old. It is likely that volcanic action produced huge amounts of lava, which obliterated most of the old features on Venus.

Apparently there are few impact craters on Venus smaller than about three kilometers and no more than 1,000 larger than that. The small craters are probably absent because meteroids less than about 100 meters in diameter never make it to the surface. The thick atmosphere produces such a large pressure difference between the entering and trailing sides that the object is crushed before striking the surface. This produces a shock wave, which, though not making a crater, can pulverize the surface into gravel. These areas do not reflect radar well, and so appear as dark blotches on the surface of Venus (see Figure 24.34).

---

[5] Shield volcanoes are common on Earth. Mauna Loa and Mauna Kea on the island of Hawaii are among the largest, about 200 kilometers in diameter and reaching 9 kilometers above the ocean floor. Mauna Loa is still active, but Mauna Kea (on which several large telescopes are located, including the two Keck 10-meter instruments) is inactive.

Figure 24.33. A radar view of one hemisphere of Venus. Note how few impact craters there are (compare this for example with Mercury, Figure 24.20), but how extensive the crustal fracturing is.

Figure 24.34. Note the two dark splotches at the right of this radar image of Venus. The radar-bright spot at the center of the lower spot is an impact crater, surrounded by dark debris. The upper dark splotch has no bright center, indicating that the impacting object was broken up by the dense atmosphere before it struck the surface. The interaction between the atmosphere and incoming object produced such a powerful shock wave that, on striking the surface, it pulverized the material, making it dark in radar images.

Figure 24.35.   The uppermost crater in this image, called Danilova, shows the flower petal structure unique to Venus. It is about 50 kilometers in diameter.

Another kind of crater, described as shaped like a butterfly or a flower petal, is unique to Venus. Both of these patterns are consequences of the thick atmosphere. On the Moon, ejecta can travel hundreds of kilometers, forming a smooth, generally symmetric apron around the crater. Venus' dense atmosphere prevents the impact debris from traveling very far, however, and causes it to break up into petal-like lobes (Figure 24.35). When craters are formed by a meteoroid striking the surface obliquely (see Figure 24.36), the shock wave accompanying the meteoroid prevents debris from being thrown back in the direction of the incoming wave, producing an asymmetric pattern.

The largest crater on Venus, called Mead, is about 275 kilometers in diameter and is one of the rare multi-ringed craters on the planet (see Figure 24.37). It is likely that the inner ring was formed by the impact and the outer ring by material that subsequently collapsed. Such large basins are found in considerable number on planets with older surfaces that still show the effects of bombardment by the large asteroids required to form these structures, but few are seen on Venus because they were destroyed by lava flows.

Terrestrial Atmospheres: The Peculiar Case of Venus.   As we have seen, until recently Venus was thought to be a twin of the Earth. Not only are its gross physical properties the same, but it should have had the same early history as the Earth, both in terms of external impacts and internal heating. With similar tectonic activity, the same primordial atmosphere should have been produced. Because Venus is a little nearer to the Sun, it receives about 1.9 times as much solar radiation as does the Earth. Venus' perpetual cloud cover reflects about 75 percent of the radiation it receives, however, so it was thought that the surface temperature might well be comfortable. It certainly seemed possible that conditions on Venus' surface could be similar to those on Earth.

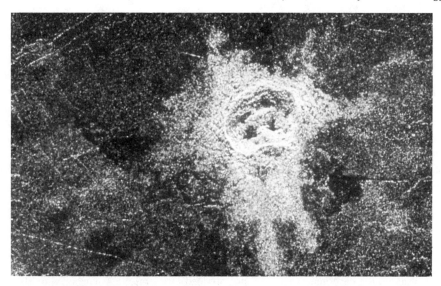

Figure 24.36. This impact crater, 12 kilometers across its longest dimension, is quite asymmetric. It may have been formed by an object striking the surface obliquely or by one that broke apart before impact.

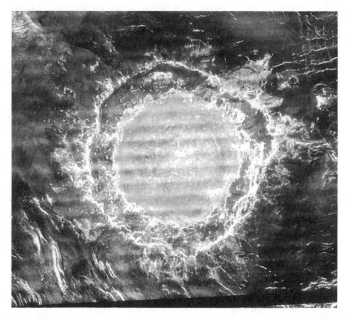

Figure 24.37. Mead is the largest crater on Venus. This double-ring feature is 275 kilometers in diameter. Only a small number of large craters are seen on Venus, presumably because its surface is relatively young and few impacts by large objects have occurred within the last few hundred million years.

The reality turns out to be wildly different. Venus' surface temperature, day and night, is about 500° C! As on Mars, carbon dioxide makes up 95 percent of the atmosphere, but the atmospheric pressure is 96 times that of the Earth's or about equal to that under 1,000 meters of water. The clouds we see from the Earth are made of sulfuric acid, hydrochloric acid, hydrofluoric acid, and carbon monoxide, all most unpleasant compounds.

How can we account for such a strikingly different environment? You should be able to make a good guess now. Because of the huge amount of carbon dioxide in its atmosphere (along with small but crucial amounts of sulfur dioxide and water vapor), Venus has a very strong greenhouse effect, which raises its surface temperature to the melting point of lead. All of this carbon dioxide comes from surface outgassing and from volcanoes, just as on Earth, where probably as much carbon dioxide was produced as on Venus. So the question really is, why is carbon dioxide such a small part of the Earth's atmosphere? The answer is simple: here on Earth, the carbon dioxide combined with silicates to form carbonates, such as limestone and dolomite. For example, in the presence of water, $CO_2 + CaSiO_3 \rightarrow SiO_2 + CaCO_3$. In fact, it has been estimated that the equivalent of about 70 atmospheres of carbon dioxide is tied up in Earth rocks.

Why didn't this happen on Venus? Venus is closer to the Sun than the Earth is, and so was just enough warmer that there was not much surface water. Hence the reactions that produce the carbonate rocks did not proceed efficiently; less carbon dioxide was removed from the atmosphere; the greenhouse effect became more pronounced; the temperature became warmer, depleting the surface water even more, which decreased the rate at which carbon dioxide was removed, etc. A runaway greenhouse effect was produced that finally came to a halt when all the surface water was evaporated. See how a small difference in the distance of a planet from the Sun produced a huge difference in the ultimate environment of that object!

We still have one more circumstance to explain. The result of the runaway greenhouse effect on Venus was that huge quantities—literally oceans—of surface water were evaporated into its atmosphere. But we now see even less water vapor on Venus than we do in the Earth's atmosphere. What probably happened was that solar ultraviolet radiation dissociated the water vapor, enabling the hydrogen to escape from Venus. The oxygen combined with other atoms or molecules that, when near the top of the atmosphere, were dissociated or underwent reactions with other constituents, ultimately escaping also.

The clouds on Venus float from about 30 to 60 km above its surface (Figure 24.38). At the equator, they rotate around the planet in only four days, producing a "jet stream" moving at about 350 km/hr; at the surface, however, the winds blow at only a few km/hr. The circulation of the thick atmosphere very nearly evens out any equator-to-pole and seasonal variations in the temperature of Venus' surface. Thus the climate is pretty much the same from one location to another. Although the clouds are very nearly featureless in visible light, they show considerable and fairly permanent markings when imaged in ultraviolet radiation. These dark patterns, composed of unknown constituents, must be a consequence of atmospheric circulation patterns.

Below 30 kilometers the atmosphere is clear and is about as bright as the Earth on a very heavily overcast day. Because the atmosphere absorbs blue light more effectively than yellow, from the surface the sky would appear orange in color. The strong refraction of light produced by the thick atmosphere makes it possible to see well beyond the horizon. In fact, since the refraction is greater than the curvature of Venus' horizon, someone on the surface of Venus would always have the impression of being in a bowl. By Earth standards, a strange place indeed.

The Atmosphere of Mars.   Since the early histories of the terrestrial planets were similar in that they all experienced intense heating by impacts, the original outgassed atmospheres were also likely quite similar. The primitive atmosphere on Mars was probably composed primarily of carbon dioxide, nitrogen, and water vapor. It has been

Figure 24.38. Thick clouds cover Venus, making it impossible to see the surface. Although the clouds are nearly featureless in visible light, they show the circulation pattern of the upper atmosphere at wavelengths around 3000 Å, as seen in this Mariner 10 image.

estimated that enough water was outgassed to have covered Mars to a depth of perhaps a few hundred meters. There is certainly plenty of water on the Earth now, and the river-like channels discovered on Mars suggest there was flowing water on that planet sometime in its past. Furthermore, Mars' present atmosphere contains about as much water vapor as it can, given its low surface temperature. This suggests that perhaps much water exists below the surface, in a solid form. Why isn't there any liquid water on Mars now? Today no place on Mars stays above freezing for 24 hours, the length of the Martian day. The maximum and minimum temperatures are about 21°C and -73°C, respectively, and the global average temperature is about -50°C. It is simply too cold (and the atmospheric pressure is too small) for liquid water to exist.

The present Martian atmosphere is composed of about 96 percent carbon dioxide, 2 or 3 percent nitrogen, and the rest argon and a few trace constituents. But the mass of the atmosphere over each square centimeter of the Martian surface is only about 1/60 of the corresponding mass of the Earth's atmosphere. (It is thick enough, however, to convert a small amount of solar energy to winds, which slowly erode the surface.) Today's thin Martian atmosphere provides practically no warming by the greenhouse effect, only about 5°C.

If some of the Martian valleys were cut by flowing water, then at some time in the past the average temperature must have been above freezing. When might that have been? From the numbers of impact craters that are superposed on the river valleys, it can be estimated that the valleys were cut during the first half billion years after Mars formed. Calculations indicate that if the atmospheric content of carbon dioxide had been at least 150 times greater than it is now, the greenhouse effect would have been sufficient to allow liquid water on the surface of Mars.

We still have a problem, however. Although the early atmosphere of the Earth most likely contained a great deal of carbon dioxide, there is little left today, because it was dissolved in rain and in the oceans and tied up in carbonate rocks such as limestone. (The small amount of atmospheric carbon dioxide still present is a result of tectonic activity, which continually replenishes that which disappears into rocks.) The same process taking place in an early, warmer Mars would have removed the carbon dioxide in only 100 million years or so, not long enough to allow running water to carve the river channels we observe today. Furthermore, Mars is a one-plate planet; its tectonic activity was insufficient to produce enough carbon dioxide to replenish that which combined into carbonate rocks or which may be tied up in the fine-grained Martian soil.

So how could there have been flowing water on Mars? Perhaps the early bombardment by solar-system debris provided the necessary energy to outgas enough carbon dioxide to maintain the greenhouse effect for the half billion years or so needed. Or maybe cyclical changes in the eccentricity of Mars' orbit, in the angle of tilt and direction of its rotation axis (precession), cause corresponding periodic changes in the climate of Mars. Just how large these effects may be is not known, however. Such cyclical changes in the Earth's orbital parameters are now fairly well established as the cause of our ice ages. Since Mars' orbital parameters show wider variations than do the Earth's, perhaps their effects might be significant, though probably not sufficient to account for a climate that could produce flowing water. In any case, the relatively short time during which water existed may account for the negative results obtained by the Viking lander experiments, which were designed to detect signs of life on the surface of Mars.

In its geology and its evolution, Mars is intermediate between dead worlds such as the Moon and (to a somewhat lesser extent) Mercury, and a geologically active planet such as the Earth. In fact, Mars is more nearly like the Earth than are any of the other terrestrial planets. The appearance of dead planets, unchanged over the last few billion years, is dominated by impact craters with little evidence of internal activity. By contrast, the Earth continues to evolve actively through geological and atmospheric forces. The origin of mountains on the terrestrial planets is instructive in this regard. Debris from large impacts (an external process) produced the mountains on the Moon and Mercury, whereas on Mars the highest features are volcanoes (a result of modest internal activity). Although some Earth mountains are produced by volcanic action, most are the result of plate motion (extensive internal activity).

The Atmospheres of the Moon and Mercury.   The atmospheres of the Moon and Mercury are markedly different from those of the Earth and Mars, and can be simply described. Apart from a few atoms captured from the solar wind, these objects have no atmospheres. A simple observation shows this to be true for the Moon: stars occulted by the Moon show no refraction effects and disappear nearly instantaneously rather than gradually, as they would if the Moon had an atmosphere. Like the Earth, the Moon and Mercury were heated by impacts early in their histories; the gases released escaped because the surface gravities of these objects are too low to retain even heavy elements. Mercury, being the planet closest to the Sun, is warmed the most by it, which makes it even more difficult for the planet to hold an atmosphere. Mercury has the largest temperature range of any planet in the solar system; at noon it is about 430°C, and at local midnight it is less than −170°C. As we have seen, temperatures on Mars are not nearly as extreme. Since it is farther from the Sun, it receives less solar energy, and so doesn't become as hot in the daytime; and since it has an atmosphere, it does not cool off so much at night.

The Future of the Terrestrial Planets.   We have seen that Venus, Earth, and Mars all produced large amounts of carbon dioxide that today makes up a large fraction of

Figure 24.39. A Voyager 1 image of Io, the most volcanically active object in the solar system. Its overall orange color is due to flows of sulfur. Dark areas are probably lava lakes of liquid sulfur and bright areas are deposits of sulfur dioxide frost. The doughnut-shaped feature in the center of the image is an active volcano seen from directly overhead. Material ejected out of the dark vent at the center forms the bright ring on the surface. See also Plate C8.

the atmosphere of Venus and Mars, but only a small part of Earth's. Venus was slightly warmer than Earth to begin with, its water mostly in the form of vapor, so it experienced a powerful greenhouse effect. The small greenhouse effect on Mars will slowly increase its temperature over the next billion years, perhaps making the planet more suitable for habitation then. The important point to note is that in our solar system there is only a very narrow distance range from the Sun where environments friendly to life might exist. Let us turn now to the most distant members of the terrestrial family.

Io and Europa.   Two of the Galilean satellites of Jupiter, Io and Europa, fit our definition of terrestrial objects. They are about the same size as our Moon, and their densities bracket the lunar density. Io probably has more iron than the Moon, but more significantly, it has an energy source that drives tectonic activity. Europa has no such activity.

Io (the pizza satellite) is a remarkable object (Figure 24.39 and Plate C8). It was expected to be heavily cratered like our own Moon. Instead, no craters were seen down to the one-kilometer resolution of Voyager 1, indicating that the present surface is very young. If the cratering rate at Jupiter was about the same as at the terrestrial planets, then Io's surface can be no more than about one million years old. What could be forming and reforming Io's surface? Volcanoes! About a hundred features more than 25 km in diameter have been interpreted as calderas, collapsed volcanic craters. Even more to the point, Voyager photographed eight active volcanoes, all spewing clouds of material up to 300 km above the surrounding surface (see Figures 24.40 and 24.41). These eruptions apparently continue for much longer periods than they do on Earth. Six of the eight active volcanoes discovered by Voyager 1 in March 1979 were still erupting when Voyager 2 passed by in July. Io is the most volcanically active object

Figure 24.40. One of the largest of Io's volcanoes is seen on the limb (computer-enhanced to make it more visible). Its plume rises more than 100 kilometers above the surface.

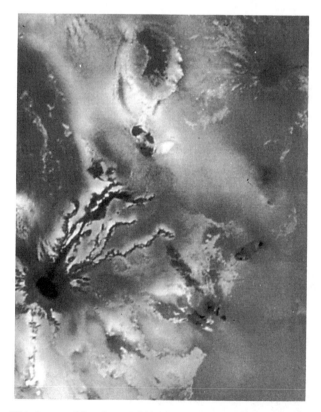

Figure 24.41. This image of Io, about 1,000 kilometers across, shows extensive sulfur lava flows from the volcanic vent at the lower left. No impact craters are visible, indicating that the surface is very young, probably no more than a million years old.

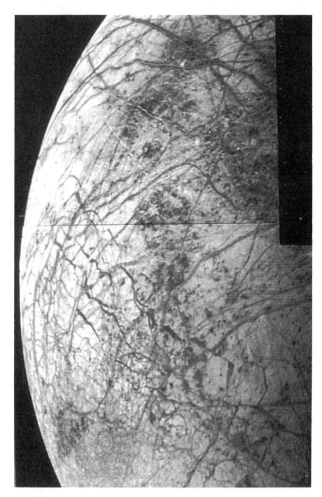

Figure 24.42. Note the remarkable smoothness of Europa's surface of ice, unbroken by impact craters or mountains. Instead, a complex network of cracks that are 10 to 50 kilometers wide and filled with dark material cover the satellite. Some of the cracks can be followed nearly halfway around the satellite. This Voyager 2 mosaic of Europa has a resolution of about 2 kilometers.

in the solar system. Though its size and density are comparable to the Moon's, Io's heat flow is at least 50 times greater than the Moon's.

Now, volcanoes imply a molten interior, but since Io is small, the internal melting cannot be by radioactive heating. Instead, a sort of tidal pumping is taking place. If Io had no companions, it would keep the same face toward Jupiter all the time, just as the Moon does with respect to the Earth. As a consequence, a constant bulge on Io toward Jupiter would be produced. But Europa and Ganymede continually perturb Io's orbit, increasing and decreasing the gravitational tide produced by Jupiter and so increasing and decreasing Io's bulge by about 100 meters every orbit. Consequently, Io's interior is continually stirred up and heated by tidally driven frictional forces, with volcanoes as one result. (Interestingly, this bizarre situation was predicted just before the Voyager discovery of volcanoes.) Rather than the silicate lava, steam, and carbon dioxide ejected by terrestrial volcanoes, sulfur and compounds such as sulfur dioxide are produced by this volcanic action and are found on the surface. Lava flows, probably of sulfur, extend for hundreds of kilometers. An extremely tenuous atmosphere of sulfur dioxide has been detected.

Europa is not subjected to the kind of gravitational buffeting that leads to a molten interior and volcanic action like that on Io. Instead, its surface is quite smooth, in fact, the smoothest known in the solar system (Figure 24.42). It has been suggested that the surface floats on an ocean of water covering the entire satellite. It is criss-crossed by many long, narrow, dark stripes or cracks, which may be filled-in fractures in Europa's

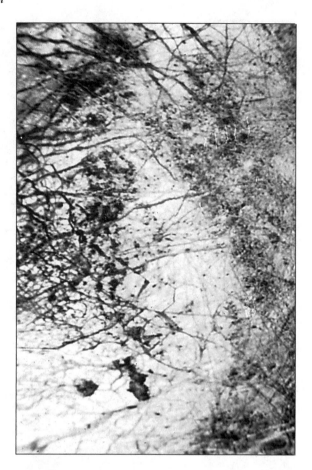

Figure 24.43. A close-up view of Europa's surface. The surface layer of ice may be about 100 kilometers thick, covering a mantle of water that surrounds a rocky core. The lack of craters implies that Europa is active. Perhaps a weak tidal interaction with Jupiter melts the ice, allowing water from below to break through and recoat the surface. The contrast of this image has been increased to bring out details.

icy crust (Figure 24.43). The stripes can be hundreds of kilometers long and tens of kilometers wide, but they are only about a hundred meters deep. Very few impact craters were detected by Voyager 2, so its surface is young. It must have frozen after the era of heavy bombardment was over. Though Europa has some ice, the bulk of the satellite is composed primarily of silicate rock.

Meteorites. Meteorites are of great interest because they give us samples of some of the asteroid-like planetesimals that formed in the early solar system. There are three major groups: the **stones**, the **irons**, and the **stony-irons**. The stony meteorites are by far the most common, making up about 95 percent of the total. Of these, roughly 80 percent are the **chondrites**, so named because they contain very old **chondrules**. These are millimeter-sized spheres made of minerals such as olivine or enstatite. It is not clear how they formed; one possibility is that they condensed from the solar nebula, and another is that they are the debris of impacts between planetesimals. In any case, they were melted and then rapidly solidified before becoming embedded in a meteorite. Since chondrules would have melted at temperatures above 1,300 K, chondritic meteorites cannot ever have been hot and are probably primitive objects. High temperature or impacts did destroy the chondrules in **achondrite** meteorites, however, indicating that they formed in an object that melted and differentiated. Achondrites are lava-like in appearance.

Making up only 5 percent of the chondrites, the **carbonaceous chondrites** are in some ways the most interesting. Their carbon-based minerals make them black; their reflectivity is very low. Not only do they contain chondrules, they also contain volatile compounds (those easily vaporized) such as water, indicating that they have never been even warm. Their low density, about 2.4 gm/cm^3 compared with about 3.6 gm/cm^3 for the other chondrites, indicates that neither have they been compressed. Thus, carbonaceous chondrites have been least altered of all the meteorites.

Although many iron meteorites are found in museums (because they are more unusual in appearance and so more likely to be found than the other types of meteorites), they comprise no more than 5 percent of the total. They are primarily an alloy of iron and nickel with small additions of rare elements such as gallium and germanium. Their average density is about 7.6 gm/cm^3. The crystalline structure of the alloy shows that it cooled very slowly, a few tens of degrees per million years. This means that iron meteorites must have been deep inside a larger object, otherwise they would have cooled more rapidly. Because the irons show no high-pressure effects, the parent body could not have been very large, suggesting that they originated inside objects only a few tens of kilometers in diameter. In addition, because they can be classified into only a few groups according to their composition, it has been suggested that the irons so far recovered on Earth originated in only a half dozen or so planetesimals.

One or two percent of meteorites are the stony-irons, a mixture of iron and stone as the name implies. The combination could have formed at the boundary of a mantle and iron core or perhaps as a result of collisions.

Where do meteorites come from? We have several clues. Their ages from radioactive dating indicate they are very old, and some of them, such as the carbonaceous chondrites, have not been changed much in the intervening time. Also, the irons and achondrites must have formed in objects that were hot enough to have differentiated, but yet were not very large. In the last chapter we saw that, other things being equal, small objects cool off more rapidly than large ones, so how could a body of, say, only 30 kilometers in diameter become hot enough to melt? Radioactive elements with long half-lives such as uranium and thorium release energy so slowly that it would be conducted away before the small object could become hot. But if a small parent object contained enough Al^{26}, its interior could have been melted because the aluminum, decaying to Mg^{26} by positron emission with a half-life of only 7 seconds, releases its energy quickly. The interior of the parent object would melt before it could get rid of the heat.

Chondrites and carbonaceous chondrites presumably formed farther out in the solar nebula where it was cool enough that the former could retain chondrules and the latter volatiles as well.

It has been known for some time by reflection spectroscopy that Vesta, the second-largest asteroid, has a basaltic surface.[6] This indicates that its surface consists of relatively low-density material that was once molten and subsequently differentiated. In fact, Vesta is the only large main-belt asteroid with this kind of surface material. Now, the spectrum of Vesta exactly matches that of a class of basaltic achondrite meteorites called the **eucrites**, suggesting a direct relation between the two. The problem was that Vesta's location in the asteroid belt is not near the zone from which objects can be perturbed into Earth-crossing orbits.

In the early 1990s, however, three small basaltic achondrite asteroids in near-Earth orbits were discovered. Objects with orbits that cross the Earth's will collide with it after only 10 to 100 million years. Therefore, they would have all disappeared by now unless the supply is constantly replenished. That this must be happening is shown by the relatively short time a meteoroid is exposed to cosmic rays. The cosmic ray exposure

---

[6] Minerals differ in their colors, that is, in the amount of light they reflect at different wavelengths. Thus the spectrum of light reflected by a solid object gives a clue to its composition (but is in no way comparable to the detailed information given by atomic spectroscopy).

time can be determined from the production of isotopes such as He^3 and Ne^{20} that are produced when high-energy particles strike a meteoroid, or by the microscopic tracks cosmic rays leave on penetrating an object. These give exposure times of only some tens of millions of years, indicating that before that time they were shielded from cosmic rays by at least a meter of material. So where did the three asteroid fragments of Vesta come from?

Vesta's family of fragments has recently been found to extend far beyond that previously established, all the way in to the instability zone where the three near-Earth asteroids were located. Chunks knocked off these small asteroids can be injected into an Earth-crossing orbit. Evidently, the eucrites are fragments of Vesta. For the first time, a convincing relation between a particular asteroid and class of meteorites has been established.

How did Vesta become hot enough to form a basaltic surface? Perhaps by the energy injected by radioactive Al^{26}, as described above, or perhaps by electromagnetic heating. This occurred when the early Sun's strong wind dragged its magnetic field across the planets, inducing electrical currents within them. It is possible that these currents could have melted the interiors of Vesta, producing its basaltic surface.

Thus, the evidence indicates that meteorites are fragments from very old objects—planetesimals—up to about 50 kilometers in diameter that were broken up by collisions occurring no more than about 100 million years ago (and continuing today). These planetesimals could not be the low-density dust and ice comets, but (as was stated without justification in the last chapter), must be the asteroids. So the next time you look at a meteorite, ponder its ancient age and the fact that it came from some planetesimal, somewhere within the solar system, to make up part of the original material that formed the planets.

Many meteorites have been chemically analyzed for evidence of life or at least organic molecules. Early results were invariably contaminated by terrestrial compounds, but techniques have improved so that reliable results can now be obtained. For example, a meteorite was recently recovered in Australia soon after it was seen to fall. It contained five amino acids found in organic proteins, as well as eleven other amino acids not common in living things. Other meteorites have been found to contain 17 or more amino acids, half of which may be biologically significant, as well as guanine and adenine (found in RNA and DNA). The available evidence indicates that none of these is likely to be of organic origin, but it does suggest that it is not hard to form these simple molecules.

Asteroids. Although much of our information about asteroids comes from meteorites, there are other sources as well. In October 1991, the Jupiter-bound spacecraft Galileo passed within 1,600 kilometers of the small object Gaspra (see Figure 24.44), and gave us our first detailed look at an asteroid (as opposed to asteroid-like satellites such as Mars' Phobos and Deimos, Jupiter's Amalthea, and Saturn's Phoebe). It is irregularly shaped, about 12 by 20 kilometers (intermediate in size between the two satellites of Mars), and dotted with craters ranging in size from 2 kilometers down to about 160 meters, the resolution limit of the image. Gaspra seems to be covered with a layer of crushed rock about a meter deep. Its rather small number of craters indicates that its surface is young. Perhaps a major impact destroyed most of its earlier craters about 200 million years ago.

In August 1993, Galileo took several images of another asteroid, Ida (Figure 24.45). It is elongated, irregularly shaped, and rotates in 4.6 hours. With a length of 56 kilometers, it is more than twice the size of Gaspra; it is an S-type asteroid (see below). The density of craters 1 kilometer in diameter or greater is about five times larger on Ida than on Gaspra. Since it is likely that they were both subjected to the same

Figure 24.44. The first close-up view of the asteroid Gaspra. This image was taken by the spacecraft Galileo on its way to Jupiter. This 20-kilometer rock rotates once every 7 hours; it may be a fragment broken off from a larger body.

Figure 24.45. Ida, the second asteroid imaged by the Galileo spacecraft. Considerably larger than Gaspra, it is about 56 by 24 kilometers in size. Note the much greater crater density on Ida than on Gaspra, indicating that the latter is younger than Ida.

impact rate, this suggests that Ida's surface is five times older than Gaspra's, or about a billion years old. Interestingly, Ida was found to have a tiny companion only 1.4 kilometers in diameter and about 100 kilometers away. It is likely that a collision destroyed the asteroidal "parent" of both Ida and its companion. Several other Ida-sized asteroids travel in similar orbits and presumably resulted from the same collision.

Radar observations have been made of more than 30 asteroids. The reflected signal is strongly affected by the composition (metallic or rock) and nature of the surface (for example, smooth or not), and so can be used to classify a particular asteroid as well as to measure its diameter. Optical and infrared spectroscopy by reflected light have enabled the compositions of more than 500 asteroids to be estimated. By these means, several classes of asteroids have been identified, of which the three containing most members are designated as S-, M-, or C-types. As suggested by the letters, S-types are stony and perhaps chondrites, M-types are irons or stony-irons and probably the most abundant asteroid type, and C-types are carbonaceous chondrites.

That these types are concentrated in the inner, middle, and outer parts of the asteroid belt, respectively, must be telling us something, but exactly what is not clear. The simplest interpretation is that their distribution in space is a consequence of their formation history, namely, the temperature distribution in the solar nebula when planetesimals formed. Another interpretation suggests that perturbations by Jupiter and fragmentations must have so jumbled the asteroids that their positional relationship with the nebula is long gone, and that the distribution of types is a consequence of their subsequent geological evolution (heating and differentiation, for example). Probably a combination of both approaches will be required to provide a full understanding of this effect.

Although most of the asteroids are located between Mars and Jupiter, several thousand are estimated to have orbits that bring them inside Mars' orbit. Some of these must collide with Mars, although there is no evidence of a major recent impact. Similarly, perhaps some hundreds of asteroids with diameters greater than 1 kilometer have been perturbed by Mars or Jupiter into Earth-crossing orbits. About 100 have been cataloged so far. These are called **Apollo asteroids** after a prominent member. It is estimated that a 100-meter object hits the Earth about every 3,000 years, most likely landing in the oceans that cover 70 percent of the surface.

A few asteroids have very large orbits. For example, Hildalgo, with a semimajor axis of 5.9 AU and an eccentricity of 0.66, has its aphelion beyond Saturn; Chiron's semimajor axis and eccentricity are 13.9 AU and 0.38, respectively, and its aphelion is nearly at Uranus' orbit. Chiron is particularly interesting: in 1988, 11 years after its discovery, it was found to double in brightness and produce a small atmosphere. The asteroid had become a comet! This points up the somewhat fuzzy distinction between an asteroid and a comet. Asteroids originating in the cooler regions of the solar nebula may have comet-like characteristics.

Comets.   Most of the planetesimal debris still left in the solar system is found in one of three regions: rocky material is found in the asteroid belt; icy matter—comets—is found pretty much where it formed, in the region extending about 100 AU out from Neptune; icy matter is also found in a huge cloud surrounding the solar system, several tens of thousands of AU from the Sun. The first of these comet reservoirs is called the Kuiper belt (after a Dutch-American pioneer of modern solar-system studies); the second is called the Oort cloud (after the Dutch astronomer who first proposed it). The Oort cloud was populated by plantesimals gravitationally ejected by encounters with the giant planets.

Since they are small objects only a few kilometers in diameter, the Oort cloud comets are completely invisible to us. Usually they must come within 3 or 4 AU of the Sun to brighten appreciably.[7] This happens when the gravitational effect of a passing star perturbs the motion of a comet very slightly, slowly sending it in toward the inner solar system. After its trip through the planetary part of the system, the comet effectively disappears, since it requires thousands or even millions of years to complete one trip around its huge orbit. In addition to these long-period comets, we see short-period comets (periods less than about 200 years) coming from the Kuiper belt, often forced into smaller orbits by a gravitational encounter with Jupiter. Halley's comet is in such a periodic, 75-year orbit, as noted in Chapter 8.

Long-period comets have orbits with nearly random inclinations, with many moving in retrograde orbits, and eccentricities near one (that is, their elongated elliptical

---

[7] A few comets, however, exhibit activity even when they are beyond Jupiter. Radio observations of comet Schwassmann-Wachmann showed that its activity was caused by carbon monoxide (which sublimates at a lower temperature than water) streaming out of its nucleus.

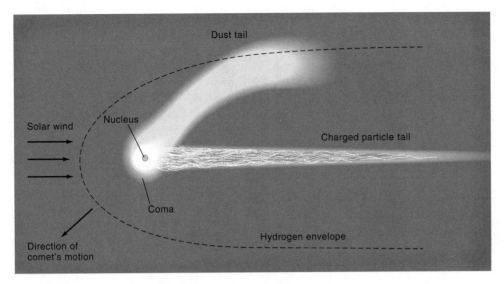

Figure 24.46. Most of the mass of a comet is in its nucleus, only a few kilometers wide. This material, heated by the Sun, evaporates and forms a coma with a diameter of tens of thousands of kilometers. The ion tail, swept directly away from the Sun by the solar wind, and the dust tail can be millions of kilometers long.

orbits are nearly indistinguishable from parabolic orbits), reflecting their spherical distribution in the comet cloud. The orbits of the approximately 100 short-period comets known are somewhat more regular, falling between those of the long-period comets and the planets. They reflect their location in the outer reaches of the original solar nebula.

Only when these objects approach within a few AU of the Sun do they take on what we ordinarily think of as a comet-like appearance. Solar energy causes the comet's ices in its **nucleus** to sublimate (the transition from a solid to a gas), forming a gaseous atmosphere from 10^5 to 10^6 kilometers in diameter called the **coma** (Figure 24.46). A long, often spectacular **tail** can develop, generally consisting of gas and tiny solid particles. Finally, the coma can be surrounded by a huge **hydrogen envelope** perhaps ten times the diameter of the coma.

Analysis of the gases in the coma and tail show that the nucleus consists of ices of water, carbon monoxide, and carbon dioxide. These ices form a matrix in which are embedded grains of silicates and carbonaceous material. The latter make the nucleus very dark, as shown by images of Halley's comet taken by the European Giotto spacecraft. Halley's nucleus was about 8 by 13 kilometers in size with a mass of about 6×10^{17} grams (or 10^{-10} that of the Earth), assuming its density to be about that of water. The hydrogen envelope comes from water vapor that has been dissociated into hydrogen and oxygen.

Solar UV radiation ionizes the gas in the tail of a comet (so that the H_2O in the coma becomes H_2O^+ in the tail), while the magnetic field in the solar wind interacts with it and sweeps it straight back from the Sun.[8] Thus, the tail leads the comet as the latter recedes from the Sun. The magnetic field can cause the structure of the gas tail to become twisted and knotty and even break off from the comet. The dust tail, however, is smooth in appearance. Here the small particles (typically 10^{-3} millimeters in

---

[8] Kepler was apparently the first to suggest that something from the Sun was blowing comet tails directly away from it.

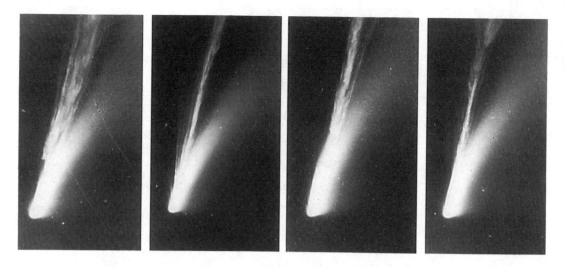

Figure 24.47. Four photographs of Comet Mrkos taken over a six-day period. Compare the relatively unchanging, broad, curving, and diffuse dust tail with the rapidly changing, straight gas tail. Interactions of the ionized tail gas with the particles and magnetic fields in the solar wind cause these changes. The four photographs in this figure were taken on August 22, August 24, August 26, and August 27, 1957, shown here from left to right.

diameter) are pushed slowly away from the Sun by the force of radiation pressure produced by solar photons. The dust tail is often curved because the dust and the comet have comparable velocities. Although it can be several tenths of an AU long, a comet's tail is "a whole lot of nothing" since its density is no more than a few thousand particles per cubic centimeter, far better than the best laboratory vacuum. Its light is partially sunlight reflected by its dust and partially sunlight reradiated by the gas (see Figure 24.47).

Because a comet loses mass every time it comes close to the Sun, it cannot last forever. Slowly losing its ices and solids, it will, after some hundreds of orbits, disappear entirely or perhaps become a burned-out comet. Chiron may be an example of the latter, with little ice now left. Some astronomers believe that the Apollo asteroids are also burned-out comets. Perhaps one of the main differences between comets and asteroids is in their ice content.

As a comet disintegrates, it leaves a myriad of dust particles along its orbit (Figure 24.48). When this part of the orbit intersects the Earth, the dust particles, heated in their high-velocity flight in the atmosphere, produce a meteor shower. The better known ones are denoted by the constellation from which they appear to come; for example, the Perseids come from the direction of the constellation Perseus (see Table 24.4).

A spectacular example of a fragmented comet is that of Comet Shoemaker-Levy 9, discovered in March of 1993 as a string of comet-like objects (Figure 24.49). Analysis of its orbit showed that in July 1992 the comet, captured by Jupiter, passed within 1.4 Jovian radii of the planet's center, at which time tidal forces broke the comet into many pieces. What is even more interesting is that the fragments were predicted to crash into Jupiter in July 1994, which is the first time such an event would have been seen. Indeed, over a period of six days beginning on 16 July 1994, 21 fragments

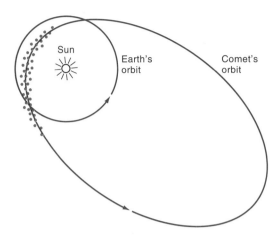

Figure 24.48. A comet slowly fragments, distributing dust particles around its orbit. When the Earth passes through this swarm of particles, a meteor shower results.

Region near Brightest Nucleus
January 1994
After Servicing Mission

Region near Brightest Nucleus
July 1993
Before Servicing Mission

Figure 24.49. The upper panel is an HST mosaic image of the fragments of Comet Shoemaker-Levy 9 after it broke up in the summer of 1992 when it passed close to Jupiter. In July 1994, these fragments crashed into Jupiter. The lower panels show some of the same fragments imaged six months apart. Note the marked change in their separation and orientation.

Table 24.4. Major meteor showers[a]

Shower	When appears	Associated comet
Quadrantids	January	—
Lyrids	April	1861 I
Eta Aquarids	May	Halley
Delta Aquarids	July	—
Perseids	August	1862 III
Draconids	October	Giacobini-Zinner
Orionids	October	Halley
Taurids	November	Encke
Leonids	November	1861 I
Geminds	December	—

[a]Of these, the Perseids and Geminds usually produce the most impressive displays.

smashed into Jupiter with velocities of about 60 kilometers per second. Even though the fragments were no more than a few kilometers in diameter, their high velocity gave them enormous kinetic energy, which was equivalent to some 10^8 megatons of TNT! Nevertheless, as we will see later in this chapter, they had only a rather small effect on Jupiter.

The Jovian Planets
General Characteristics

Our knowledge of the giant planets increased manyfold as a result of data acquired by the flybys, beginning with Pioneer 10 in 1973 and continuing through Voyagers 1 and 2 in 1989, each of which carried 11 scientific instruments, including spectrographs and cameras. Voyager 2 in particular, which visited Jupiter, Saturn, Uranus, and Neptune, surely made one of the most successful voyages of discovery in history.

Together the giant planets constitute about 99 percent of all the solar-system mass outside the Sun (see Table 24.5 and Figure 24.50); Jupiter alone makes up more than two-thirds of the total. Had Jupiter been about 80 times more massive, its interior would have been able to sustain hydrogen burning. The solar system would then have

Table 24.5. Some properties of the Jovian planets

Body	Mass ($\oplus$ = 1)	Mean radius (km)	Mean density (gm/cm³)	Temperature[a] (K)
Jupiter	318.0	69,800	1.33	170 ± 20
Saturn	95.1	58,300	0.69	135 ± 15
Uranus	14.6	25,500	1.26	75–80?
Neptune	17.2	24,500	1.67	70–75?

[a]This is the temperature in the cloud layer where the pressure is 1 Earth atmosphere.

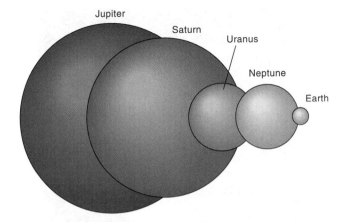

Figure 24.50. The diameters of the four giant planets compared with that of Earth. Uranus and Neptune have very nearly the same diameter, but are only about one-third the diameter of Jupiter and Saturn. See also Plate C7.

consisted of a binary star, comprising a G2 and a very cool M-type star, along with eight planets having properties rather different from those they have today. Let's briefly review the characteristics the giants have in common.

Densities. The average densities of all four planets are smaller than those of the Earth-like planets, because, like the Sun, the giants are composed mostly of hydrogen and helium. With an orbital semimajor axis of about 5 AU, Jupiter receives only 1/25 the amount of solar radiation that strikes the Earth. Thus the material from which it formed was not as warm and was easier to retain gravitationally than on inner planets such as the Earth. Even the light gases such as hydrogen and helium did not escape. As a consequence, Jupiter has an abundance of molecules containing hydrogen, such as CH_4, NH_3, and H_2. The other giants, of course, receive less solar radiation and so are even colder.

Their low average densities mean that they can't have the same structure as the Earth. None of the giants have solid surfaces, nor do they have large iron cores, for example; instead, they have cores of rock-like composition, though these are compressed to high densities.

Rotation. All four of the giants rotate on their axes more rapidly than the terrestrial planets; Jupiter, the fastest, takes only about 10 hours for one rotation, and shows the most pronounced equatorial bulge as a consequence (Figure 24.51). Since Jupiter's diameter is about ten times the Earth's and its rotation period is only about 1/2.5 of the Earth's, its velocity at any latitude is about 25 times that of the Earth at the corresponding latitude. For example, at the equator the Earth's velocity is about 1,650 km/hr, whereas on Jupiter the equatorial velocity is about 42,000 km/hr! Even relatively slowly rotating Uranus has an equatorial velocity nearly six times faster than Earth's. This rapid day/night cycle, along with their thick atmospheres, means there is little difference between the temperatures of the day and night hemispheres on each of the giant planets.

Figure 24.51. An image of Jupiter taken in 1991 by the Wide Field Camera on HST. Note the pronounced equatorial bulge caused by Jupiter's rapid rotation.

Magnetospheres and Rings. All the giants produce magnetic fields that trap charged particles (similar to the Van Allen belts surrounding the Earth). Since the charged particles come primarily from the solar wind, their density in the magnetospheres falls off with the planets' distance from the Sun. Jupiter's surface magnetic field is by far the strongest in the solar system, more than ten times that of the Earth, and about twenty times that of Saturn and Uranus.

All four have ring systems, though only Saturn's is prominent. All the Jovian planets radiate more energy than they receive from the Sun, though Uranus just barely does so.

Differences. There are significant differences, however, between Jupiter and Saturn, on the one hand, and Uranus and Neptune, on the other. Both of the nearer pair are much more massive and larger in diameter than the more distant two (Figure 24.50). In fact, Uranus and Neptune are roughly midway in their masses between the terrestrial planets and Jupiter. Though hydrogen and helium make up most of the mass of all four planets, helium is depleted with respect to hydrogen in Saturn and to a lesser degree in Jupiter; Uranus and Neptune, however, contain a somewhat higher proportion of heavy elements than do Jupiter or Saturn. The latter two rotate more rapidly than the outer pair—about 10 hours compared with about 16 hours. Uranus is peculiar in that its equator is tilted with respect to its orbital plane by nearly 98 degrees. Its poles are nearly in the ecliptic.

The interior structure of the outer pair is different, apparently, from that of Jupiter and Saturn. The magnetic field axes of Uranus and Neptune are unique in the solar system, in that they are roughly a third of a radius away from the center of the planet and are grossly tilted with respect to the planet's rotation axis. Finally, although the spectacularly successful Voyager 2 flyby has redressed the balance somewhat, we know less about Uranus and Neptune because they are so much farther away than the nearer pair. Let's look at these exotic objects in a little more detail.

Figure 24.52. An image of Saturn showing the atmospheric bands produced by its rapid rotation. Saturn's bands are somewhat less marked than those of Jupiter (compare this image with Figure 24.55).

Atmospheres of the Giants

Jupiter and Saturn. Differences in the appearances of the atmospheres among the giants are caused primarily by differences in their temperatures. As we saw in Chapter 23, reducing compounds condense out to form cloud layers at heights determined by atmospheric pressures and temperatures. These clouds of ice crystals are stretched into bands parallel to the equator and wrapped around the planet because of the planet's rapid rotation (see Figure 24.52). (If the Earth rotated much more rapidly, its high- and low-pressure systems would be stretched into bands similar to those on Jupiter.) On Jupiter and Saturn the clouds are, in order of decreasing altitude, those of ammonia, ammonium hydrosulfide, and water. Above all the clouds on the giant planets is a haze of hydrocarbons—for example acetylene—formed when sunlight breaks up methane carried to high altitudes by convection (Figure 24.53).

There is little difference between Jupiter's night and day cloud temperature because its thick atmosphere can store a lot of heat, and the rapid rotation does not allow the planet much time to cool off during the short night. The top of the cloud layer has a temperature of about $-150°C$, but below the obvious cloud bands the temperature increases because of the greenhouse effect and internal heating (see below). Models indicate that the lower atmosphere (below the cloud decks) consists of a mixture of liquid ammonia and water-ice crystals. This region may be clear enough to allow sufficient sunlight to penetrate that the probe from the spacecraft Galileo could photograph it when it arrives at Jupiter in 1995. Some astronomers have speculated that life could form in these warmer clouds.

Jupiter does not rotate as a solid body. Near the equator the rotation period of the clouds is about 9 hours, 50.5 minutes, and it increases at higher latitudes to about 9 hours, 55 minutes. The most pronounced features seen on any of the giants are Jupiter's alternating light and dark cloud bands. These bands change in detail, but persist as major features on timescales of decades. White bands are rising clouds of ammonia crystals; they are high-pressure systems in comparison with the dark bands. These are

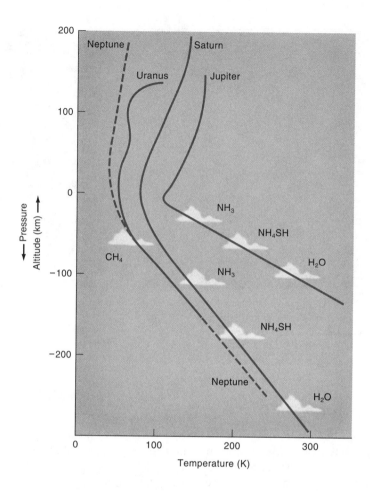

Figure 24.53. The run of temperature with altitude in the atmospheres of the giant planets. The pressures at the top and bottom of the diagram are about 0.001 and 10 times the pressure at the surface of the Earth, respectively; the zero-point of altitude is at 0.1 Earth atmospheres. Note that the atmospheric temperatures decrease with increasing planetary distance from the Sun, except for Neptune, which has about the same atmospheric temperature as Uranus because of its significant internal heat. The vertical extent of the cloud layers on Jupiter is smaller than that on Saturn, a consequence of Jupiter's greater gravity. Because Saturn's ammonia clouds are formed deeper in the atmosphere, they are not as prominent as they are on Jupiter. A high-altitude haze layer absorbs sunlight, causing atmospheric temperatures to increase in that region.

probably clouds of ammonium hydrosulfide or ammonium sulfide crystals falling back toward Jupiter. What chemicals give the clouds their orange, red, or brown colors are still unknown.

A striking feature in the atmosphere of Jupiter is the **Great Red Spot**, which has been visible since its discovery in 1664 (Figure 24.54). The Spot varies in size and has been as large as 40,000 km long and 13,000 km wide (three Earths would fit across it!); in 1979, when the Voyagers passed by Jupiter, the Spot was considerably smaller, only about the size of the Earth. It remains some 10 km above the cloud layer and at about the same latitude, but wanders in longitude, so it can't be tied to any "surface." It is a meteorological phenomenon, a super-high-pressure region, most likely. On this

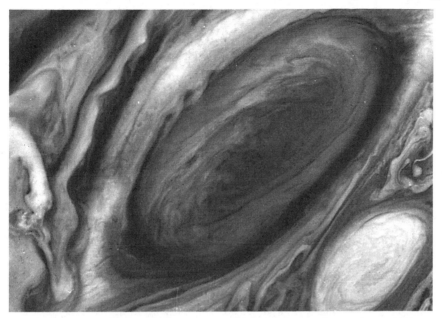

Figure 24.54. Jupiter's Great Red Spot is a meteorological feature that has been visible since its discovery more than 300 years ago. Both the Red Spot and the white oval below it are high-pressure regions rotating counterclockwise. What causes the red colors is unknown.

interpretation, what we see is the top of a rotating gas column, the center of a complex circulation pattern. Because it lasts such a long time, this flow must be stable and must be energized by some mechanism still not agreed upon by meteorologists.

Many more dark and white spots, all smaller than the Red Spot with shorter lifetimes (still decades, in some cases), were found by Pioneers 10 and 11 (see Figure 24.55). These are probably less dramatic examples of the same kind of storm that is producing the Red Spot. The longevity of Jupiter's atmospheric features—the colored

Figure 24.55. Several of Jupiter's dark and bright spots can be seen. Compare the sizes of the spots with those of the Earth, shown on the right.

bands, Great Red Spot, smaller dark and white spots—is one of the remarkable features of Jovian meteorology. That, unlike Earth, Jupiter has no land masses, no mountains, no solid surface at all, is probably a contributing factor to the long lifetimes of these features.

As of this writing, analysis of the images and spectroscopic data acquired when Comet Shoemaker-Levy 9's 21 fragments smashed into Jupiter has only begun. The impacts produced spectacular fireballs rising as much as several thousand kilometers above Jupiter and left large scars in its atmosphere (see Figures 24.56 and 24.57). These were easily seen with small telescopes from the ground, persisted for many weeks, and became somewhat elongated by Jupiter's rapid rotation. The plumes and the fallout pattern of the debris indicated that the largest cometary fragments had diameters of 2 or 3 kilometers, about what was expected. Despite the very large kinetic energy of such chunks, they apparently failed to stir up Jupiter's atmosphere to any great extent, a striking indication of the enormous energy content of the atmosphere.

Astronomers had hoped that the impacts would have dredged up material from deep in the atmosphere, giving them their first direct measurements of its composition. However, what they observed were infrared emission features from hydrocarbons such as methane and acetylene, which increased in intensity by as much as ten times after the impact, and ammonia in emission. Now, ammonia ice forms the uppermost visible cloud layer and the hydrocarbons are located in a haze layer above that, so only the topmost portion of the atmosphere was affected by most of the impacts. Spectroscopic data suggested that the effect of the largest impact might have reached the second cloud layer down (ammonium hydrosulfide), but no water was seen, either from the deepest cloud deck or from the comet itself. Why the impacts did not reach deeper into the atmosphere is a puzzle. Perhaps the fragments were not solid chunks of ice, but a loose collection of smaller pieces with less penetrating power. Definitive results from this unique event should be interesting.[9]

Saturn's atmospheric features are not as markedly colored as Jupiter's, nor are there as many bands or spots or any feature corresponding to the Great Red Spot (see Figure 24.52). Because Saturn's gravity is smaller, its atmosphere is not as compressed, and since it is farther from the Sun, the atmospheric temperatures are colder than Jupiter's.

The abundance of helium in the atmosphere of Saturn is about half that of Jupiter's. It has been suggested that although Saturn was formed with the usual abundance mix of hydrogen and helium, the helium has been sinking toward the center rather than remaining mixed throughout the planet, as it does on Jupiter. We will see the reason for this difference is the colder interior temperature of Saturn, a consequence of its smaller size and greater distance from the Sun.

Uranus and Neptune.   The blue color of Uranus and Neptune differs strikingly from the colors of Jupiter and Saturn. Basically, their low temperatures are responsible. Presumably, Uranus and Neptune have cloud layers of hydrogen-rich compounds such as ammonia, similar to those in Jupiter and Saturn, but because they receive less energy

---

[9] More information was obtained from a probe released by the Galileo spacecraft on December 7, 1995, when it survived for 57 minutes as it descended through 600 kilometers of Jupiter's atmosphere. Six instruments radioed data back to the Galileo orbiter suggesting that, among other results, the atmospheric density and temperatures were higher and winds faster than expected; the abundance of several elements including helium, neon and carbon was lower than anticipated; and the expected three-layer atmospheric structure was not present (at least at the probe's position). Whether these preliminary findings will hold up when a complete analysis of the data is finished remains to be seen.

Figure 24.56. Violet and ultraviolet images of three impacts of Comet Shoemaker-Levy 9 with Jupiter, taken about 20 minutes apart. Violet light, reflected off the tops of Jupiter's clouds, gives the familiar image of Jupiter, but ultraviolet light comes from hundreds of kilometers above the cloud bands where Jupiter is dark. In the ultraviolet, the impact sites are darker and larger, perhaps because of increased absorbers at high altitudes and strong winds in those regions that spread the debris. Note the Great Red Spot at the lower right on the violet image as well as the satellite Io, the dark spot on the upper left. The ultraviolet image shows the Jovian aurora clearly, especially around the north magnetic pole.

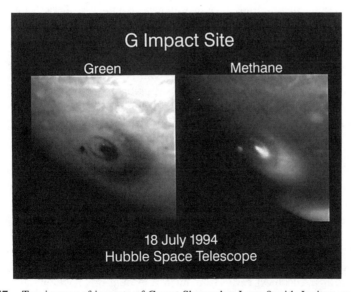

Figure 24.57. Two images of impacts of Comet Shoemaker-Levy 9 with Jupiter; one is in green light around 5550 Å (left) and the other is at 8890 Å, sensitive to a transition in the methane molecule. The smaller impact (to the left of the larger) occurred the day before the latter. The central dark spot and the inner edge of the outermost broad ring surrounding the impact in the green image are approximately the diameters of the Moon and Earth, respectively. The dark areas in the methane image indicate absorption by that molecule, whereas the bright impact area is due to reflection of sunlight by material blown above the methane.

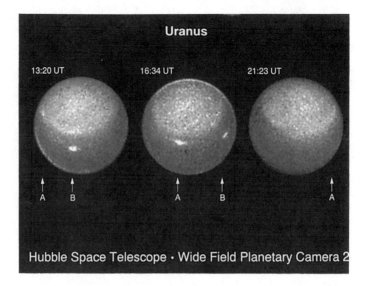

Figure 24.58.   This image of Uranus was made by the Hubble Space Telescope in August, 1994, when two bright clouds and a large polar cap were visible. The clouds and the haze forming the cap are high in the atmosphere. The clouds are primarily carried along by Uranus' rotation, although some motion may also be due to winds. The longest dimension of the larger cloud is about 4,300 kilometers, somewhat larger than our Moon.

from the Sun, these cloud layers are formed deeper in their atmospheres than in the nearer pair. Consequently, to be reflected by these clouds, sunlight must travel through more atmosphere that, in the outer two planets, contains an appreciable quantity of methane. This compound absorbs red light strongly so that mostly blue light gets back to us. Also, like Earth's atmosphere, those of Uranus and Neptune scatter blue sunlight more efficiently than red.

From Earth, Uranus is so featureless—with no Jupiter-like bands or spots to be seen—that we could not reliably measure its rotation period. Even the flyby of Uranus by Voyager 2 in 1986 did not reveal much detail in the planet's clouds. Only five distinguishable wisps of clouds broke the uniformity of its appearance (Figure 24.58). The bulk of the atmosphere is composed of hydrogen and helium, with various hydrocarbons produced in the upper atmosphere by the action of sunlight. Condensed methane forms the uppermost cloud layer with perhaps hydrogen sulfide or ammonia beneath that. Unlike the other three giant planets, Uranus produces relatively little internal heat. Why this is so is not clear, but with little energy coming from within the planet, the atmosphere is not convective but stable. As a consequence, there is very little structure to be glimpsed through the hydrocarbon haze.

Uranus' day–night cycle is strange. Recall that, unlike the other major planets, Uranus' equator is tipped by 98° to its orbital plane, presumably the result of a collision with an Earth-sized object early in its history. (Could this have anything to do with Uranus' small internal heat output?) In any case, each pole is in sunlight for about 42 years at a time (Figure 24.59). Even with these long periods of day and night, the heat content of Uranus' atmosphere is so large that there is essentially no difference in the temperature of the sunlit and dark hemispheres.

With its layer of hydrocarbon haze above clouds of methane, Neptune's atmo-

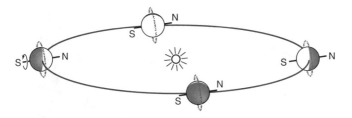

Figure 24.59. Uranus' axis of rotation is very nearly in its orbital plane, resulting in long periods of day and night in the northern and southern hemispheres.

sphere is similar to that of Uranus (see Figure 24.53). Despite its greater distance from the Sun, Neptune shows more structural details than does Uranus (Figure 24.60). The reason for this is that, like Saturn and Jupiter, Neptune radiates more energy than it receives from the Sun, about 1.8 times as much. This internal heat drives convection currents of warm gas above the methane cloud deck where they form dark and bright clouds.

Among the features discovered by Voyager 2 is an elliptical dark spot, roughly 10,000 kilometers across but variable in size and shape, spinning counterclockwise, and located about 22° south of Neptune's equator (Figures 24.60 and 24.61). Like Jupiter's Great Red Spot, this seems to be a high-pressure system. It was named, predictably, the Great Dark Spot.

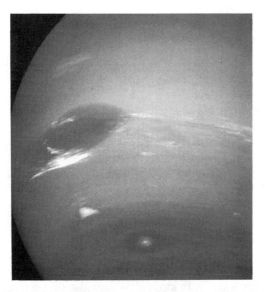

Figure 24.60. This image of Neptune was made by Voyager 2. In its longest dimension, the Great Dark Spot is about the size of Earth. About 50 kilometers above it are bright, wispy clouds of methane ice crystals. A smaller dark spot appears in the prominent cloud bands around the south polar region. Like the other giant planets, Neptune's atmosphere rotates differentially, but in contrast with the other giants, its atmosphere rotates more slowly than does its interior.

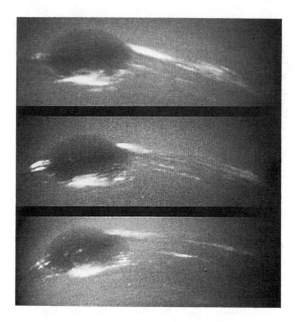

Figure 24.61. These three images of the Great Dark Spot were taken about 18 hours apart. The Dark Spot is at the same latitude as Jupiter's Red Spot (22° south), and, relative to Neptune's diameter, it is the same size as the Red Spot. Like the latter, it is also a high-pressure system with counterclockwise circulation. Although the Dark Spot remains constant over the 36 hours spanned by these images, the bright wisps change rapidly, indicating that the weather in this region of Neptune is quite variable.

The Interiors of the Giants

Jupiter and Saturn.   For obvious reasons, our knowledge of the interiors of the giants must depend heavily on model calculations. A model with hydrogen and helium as its main constituents is necessary if it is to agree with the measured mass, density, and polar flattening of the planet. Such a model probably gives us a fairly good idea of Jupiter's structure.

By our standards, the interior of Jupiter is very strange. It has no well-defined hard surface. Only a few thousand kilometers below the clouds, the pressure is so large that gases liquify. Its body consists of an outer layer of liquid molecular hydrogen and a deeper, electrically conducting layer of liquid metallic hydrogen (see Figure 24.62; note that the atmosphere is only a tenth of a percent or so of the planet's radius and so is not shown on the figure). What is **metallic hydrogen**? When hydrogen molecules, H_2, are squeezed close together by high pressure, they tend to lose their identity, the hydrogen atoms being attracted to neighboring atoms as much as to their original partners. Similarly, the electrons are no longer bound to a given molecule and instead become a sort of electron gas, with properties like those of a metal, such as the ability to carry an electrical current; hence the name metallic hydrogen.

A relatively small rocky core of 15–20 Earth masses exists at the center of Jupiter. Its central temperature is about 25,000 K, which is warm, but not nearly hot enough for any nuclear reactions to take place. Nonetheless, Jupiter radiates about 1.7 to 1.8 times as much energy as it receives from the Sun. The internal-energy source is probably residual heat; Jupiter is just cooling off.

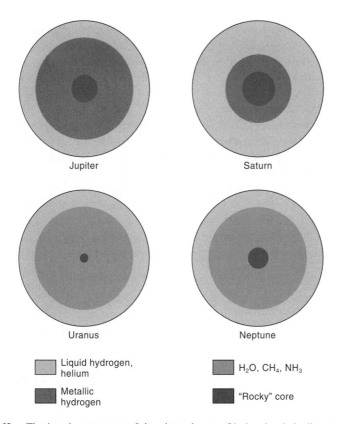

Jupiter Saturn

Uranus Neptune

| | Liquid hydrogen, helium | | H_2O, CH_4, NH_3 |
| | Metallic hydrogen | | "Rocky" core |

Figure 24.62. The interior structure of the giant planets. Obviously, their diameters are not drawn to scale, but the relative sizes of the various regions with respect to each planet can be directly compared. The interiors of Uranus and Neptune are rather uncertain. Uranus' rocky core is small and contains little mass. Most of the mass of Uranus and Neptune is contained within the intermediate layer of water, methane, and ammonia. Despite their names, all of the various layers act like liquids.

Saturn's interior is thought to be similar to Jupiter's, but its region of metallic hydrogen is smaller because of its lower gravity and internal pressure. Like Jupiter, Saturn also radiates more energy than it receives; unlike Jupiter, however, in addition to simple cooling, Saturn seems to have another heat source. Recall that helium is deficient in Saturn's atmosphere, and that Saturn's interior is colder than Jupiter's. Because of the lower temperature, helium condenses from the gaseous to the liquid state and "rains" onto the top of the metallic hydrogen layer. In this picture, the extra energy source in Saturn's interior is just the gravitational energy lost by the helium rain. This process has probably not yet begun in Jupiter, because its internal temperature is still high enough that metallic hydrogen and helium remain mixed. When its temperature cools sufficiently, the helium will begin to separate out and release gravitational energy as it seeps down.

Uranus and Neptune. The internal structures of Uranus and Neptune are still rather uncertain; we cannot yet derive unique models of their interiors. Some qualitative arguments can be made, however. Because they are considerably less massive, their interiors should not be as compressed as those of Jupiter and Saturn. Consequently, if their compositions were similar to Jupiter's, their average densities should be less than Jupiter's 1.33 gm/cm³. Instead, their densities are comparable to or greater than that, indicating that their compositions differ, that Uranus and Neptune must contain a larger proportion of heavy elements. But note also that even though Uranus and Neptune are nearly identical in diameter, Neptune is significantly more dense, and so must have a larger proportion of metals than Uranus.

In addition to these density data, some information concerning the distribution of matter inside the two planets can be inferred from measuring how spherical their surrounding gravitational fields are. This has been done by measuring the precessional motions of Uranus' ring system and of Neptune's satellite Triton. Model calculations suggest that, like Jupiter and Saturn, Uranus and Neptune have a rocky core, but probably of only a few Earth masses (see Figure 24.62). Surrounding this core is a mixture of the elements and compounds one would expect—water, ammonia, methane, hydrogen, and helium. In addition, because of Uranus' and Neptune's smaller masses, pressures in their interiors are not high enough for liquid metallic hydrogen to form.

Their Magnetospheres.   With its interior of electrically conducting liquid metallic hydrogen and its rapid rotation, Jupiter would be expected to have a strong magnetic field, and, indeed, it has the strongest field of any of the planets. Its surface strength is roughly 15 times the Earth's surface field, and because its extent is so great, its total energy is about 4×10^8 times that of the Earth's. Jupiter's magnetosphere is the largest structure in the solar system, extending well beyond the Galilean satellites. If somehow we could see it from the Earth, it would appear to be as large as the full Moon! Tied to Jupiter's interior, the magnetosphere rotates in only $9^h 55.5^m$, which makes it more flattened than Earth's. This strong magnetic field has formed large and intense trapped radiation belts filled with charged particles not only from the solar wind, but also from Io's volcanoes. These high-energy particles make exploration of this region of the solar system lethal for humans and hazardous for equipment. The trapped electrons spiral around the strong magnetic field and produce radio radiation by the synchroton mechanism. The particle-filled region is also responsible for strong meter-wavelength radio bursts having a million times greater energy than a terrestrial lightning bolt. (The ancients chose better than they knew when they named this planet after the god who hurled lightning bolts from Mt. Olympus.)

Saturn's magnetosphere is only about one-third the size of Jupiter's, and its magnetic field is much weaker than Jupiter's; in fact, it is weaker than the Earth's. The axes of the Earth's and Jupiter's fields are tipped by about 10 degrees with respect to their rotation axes (Figure 24.63). Models describing the generation of fields predict tilts of roughly this amount. Surprisingly, the axis of symmetry of Saturn's magnetic field is aligned with its axis of rotation. This may have something to do with its helium-enriched core, mentioned above. Because electrons in its magnetosphere are lost through collisions with Saturn's rings and inner satellites, the radiation belts do not generate much synchrotron radiation.

Uranus' magnetic field is about equal in strength to Saturn's, but Neptune's is five times stronger, second only to Jupiter's. The tilts of the field axes of Uranus and Neptune—59 and 47 degrees, respectively—are strikingly different from those of the other planets. In addition, the axis of symmetry of Uranus' field is displaced from the center of the planet by about a third of its radius, and that of Neptune by more than half its radius (see Figure 24.63). When these features of Uranus' field were discovered, they were thought to be connected with the event that knocked Uranus on its side. After Voyager found similar circumstances at Neptune, however, this argument could no longer be maintained. The origin of these magnetic field features is not understood.

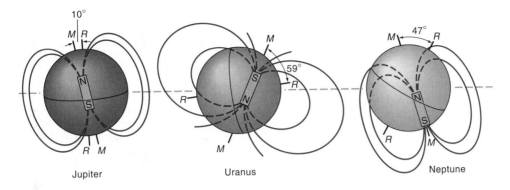

10°

Jupiter

Uranus

59°

47°

Neptune

Figure 24.63. Jupiter's magnetic field axis (*M*), like that of Earth, is only slightly inclined with respect to its rotation axis (*R*) and is centered within the planet. By contrast, the magnetic fields of Uranus and Neptune are not centered within the planets (being displaced by 0.3 and 0.55 planetary radii, respectively), and both are inclined by large angles with respect to their rotation axes. Note than the planets' magnetic fields are structured only *as if* a bar magnet were located as indicated. In actuality, the fields are produced by currents flowing in the interior of the planet, not by bar magnets.

Perhaps poor electrical conductivity of the deep interior forces the dynamo currents producing the field to be much closer to the surface of these planets. The odd geometry of Uranus' field combined with its rotation axis being in its orbital plane produces a strange magnetosphere that turns and twists as the planet rotates.

Satellites of the Giant Planets

The innermost group of eight Jovian satellites (which includes the four Galilean objects), all moving in direct, low-inclination orbits, are very likely the only moons formed with Jupiter. The high orbital inclinations of the four immediately beyond the Galilean satellite and the retrograde motions of the four outermost satellites strongly suggest that they are asteroids captured after Jupiter formed (Figure 24.64). In fact,

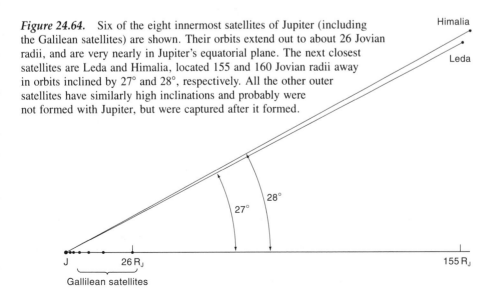

Figure 24.64. Six of the eight innermost satellites of Jupiter (including the Galilean satellites) are shown. Their orbits extend out to about 26 Jovian radii, and are very nearly in Jupiter's equatorial plane. The next closest satellites are Leda and Himalia, located 155 and 160 Jovian radii away in orbits inclined by 27° and 28°, respectively. All the other outer satellites have similarly high inclinations and probably were not formed with Jupiter, but were captured after it formed.

Himalia

Leda

28°

27°

J   26 R$_J$ 155 R$_J$

Gallilean satellites

Figure 24.65. Callisto's surface is the most heavily cratered of the Galilean satellites. Processes that would destroy craters are nearly absent on Callisto. Bright areas are ejecta from impacts. Note in particular the spectacular multi-ring crater, Valhalla, 1,500 kilometers in diameter; the diameter of the bright central region is about 600 kilometers.

the eight outer satellites have reflectivities and colors similar to the carbonaceous chondrites. With only one exception, Saturn's satellite system apparently formed with the planet. The captured object is Phoebe, which is the outermost satellite and orbits Saturn in the retrograde sense.

Two of the Galilean satellites of Jupiter, Ganymede and Callisto, as well as Titan, the largest satellite of Saturn, all have densities about twice that of water. This indicates that their compositions must differ from that of Earth-like Io and Europa, containing more water and less iron. We saw in the previous chapter that the early Jupiter was most likely fairly hot, making it impossible for Io and Europa to retain their hydrogen and helium. The more distant Ganymede and Callisto were cooler, but must have been warm enough that their silicates and ices differentiated, the former becoming the core while the latter became the outer region of these objects.

As expected, all of the satellites have craters as evidence of early bombardment. They show other features, however, some of them quite enigmatic. Callisto's surface must be very old because it is more heavily cratered than any of the other Galilean satellites (Figure 24.65). Its craters are shallow, a consequence of the way ice slumps and tends to fill a depression. Roughly half of Ganymede's surface is also covered with old craters, but the other half shows the effects of more recent internal activity (Figure 24.66). This apparently caused some of its ice to melt and fill in older craters. Ganymede also has an area covered with several shallow grooves that might have resulted from faulting (Figure 24.67).

Among the satellites of the solar system, Saturn's moon Titan is exceeded in size only by Ganymede (see Figure 24.68). Titan is of particular interest, since it is the only satellite with a thick atmosphere. It is composed primarily of nitrogen (about 90% by number), argon (about 10%), and methane (about 1%). The pressure at the surface is about 1.6 times that of the Earth's atmosphere. Since the surface gravity on Titan is approximately one-seventh that of the Earth, however, the mass of the atmosphere over each square kilometer of Titan is some ten times greater than on Earth. In fact, the atmosphere, consisting of low-level methane clouds below a high-altitude smog of organic molecules, completely shrouds Titan's surface from view. Despite the thick at-

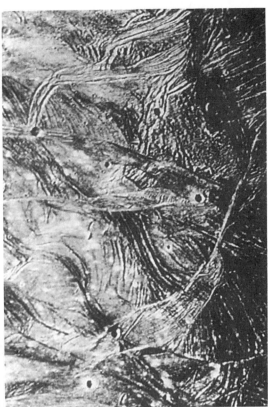

Figure 24.66. This Voyager 2 image of Ganymede shows the hemisphere that faces away from Jupiter. Note the prominent circular dark region.

Figure 24.67. Sizable areas of Ganymede are covered by a grooved terrain. In this Voyager 1 image, the valleys are about 10 to 15 kilometers apart and separated by ridges about 1 kilometer high.

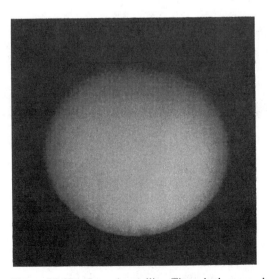

Figure 24.68. Saturn's satellite, Titan, is the second largest in the solar system. It is covered with a thick nitrogen-rich atmosphere that completely hides the surface.

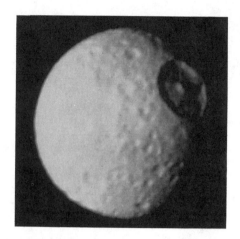

Figure 24.69. Rhea.

Figure 24.70. Mimas.

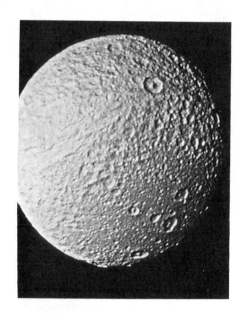

Figure 24.71. Dione.

Figure 24.72. Tethys.

Figure 24.69–72. Voyager images of four of Saturn's icy satellites are shown here. They are all heavily cratered and show no recent geologic activity. Note the large impact crater on Mimas; it is about 120 kilometers in diameter (or about a quarter of the diameter of Mimas itself). A slightly larger impact would have probably destroyed the satellite. Note also the narrow valley on Dione and the large complex valley (upper left) on Tethys; it extends about three-quarters of the way around the satellite. All of Saturn's satellites except Phoebe are tidally locked to Saturn.

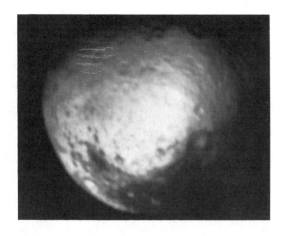

Figure 24.73. Part of Iapetus' surface is cratered like that of Saturn's other icy satellites, but its leading hemisphere is covered with a dark substance of unknown origin or composition.

mosphere, there is only a small greenhouse effect, so that its surface temperature is very cold, about − 180°C. Although this makes the prospects for life there now extremely dim, that may not be the case several billion years from now. As the Sun evolves and its luminosity increases, Titan's temperature will increase enough that it might come to resemble the early Earth.

Saturn has six other satellites with diameters between 400 and 1,600 kilometers. Like their parent, these objects have small densities; in fact, the densities of the smallest of the major satellites are 1.2 to 1.4 times that of water. This indicates that half or more of their volume must be ice. Four of them—Rhea, Mimas, Dione, and Tethys (Figures 24.69–24.72)—have bright, heavily cratered surfaces with no indication of any internal modifications for the last few billion years. Their impact craters are deep and not smoothed out by ice since they are so far from the Sun that ice is very hard and rock-like. Note the crater on Mimas covering a significant fraction of its surface.

Iapetus (Figure 24.73) and Enceladus (Figure 24.74) show exceptions to this uniform cratering, however. Tidal forces keep the same face of Iapetus locked toward Sat-

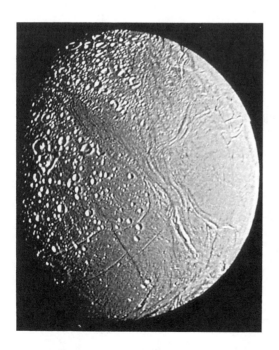

Figure 24.74. In striking contrast to the other icy satellites of Saturn, large areas of Enceladus are not cratered, but show evidence of recent geologic activity, probably volcanic. Why Enceladus should differ in this way is not understood.

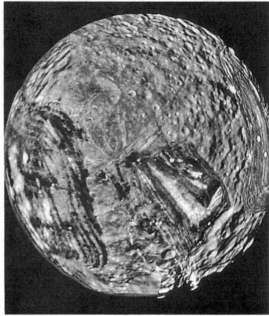

Figure 24.75. Voyager 2 obtained this image of Umbriel when it was a little over a half million kilometers away. Note the large numbers of impact craters and the lack of bright features on its dark surface. The brightest features are an 80-kilometer-diameter crater rim at the top and the central peak of a crater below and to the right. No geologic features are evident.

Figure 24.76. Miranda, which is nearest to Uranus, is the smallest of its five major satellites and has the strangest surface. Cratered areas are broken by ridges, troughs, and long faults of uncertain origin. Nor is it clear why such a small object should have undergone so much geologic activity. With its low density, Miranda must have a greater proportion of ice than the other large satellites of Uranus. The four other satellites also have larger proportions of rocky material than do the icy satellites of Saturn.

urn, so the hemisphere 90° away always leads Iapetus in its orbit. That face is largely covered with very dark material with a reflectivity of only about 4 percent. This is possibly original material of the satellite that has been revealed by the impacts that cleaned off the surface ice. Part of the surface of Enceladus is cratered in the usual manner, but part has only a few craters, suggesting that internal activity (perhaps resulting in the release of water) has resurfaced this part of the satellite within the last few billion years. This is surprising, since an object having a diameter of only 500 kilometers should have cooled long ago and become stable.

In contrast to Jupiter and Saturn, neither Uranus nor Neptune has a satellite as large as Earth's. Voyager 2 discovered ten small, dark moons orbiting Uranus to add to the five larger ones already known. All the latter show many impact craters, and all but Umbriel (Figure 24.75) show evidence of tectonic activity like deep groves and valleys. Miranda (Figure 24.76) perhaps presents the strangest appearance. Its heavily cratered surface is interrupted by large areas of banded and grooved terrain, deep fault canyons, and strange, nearly rectangular features. The origin of these features is not yet understood. If they resulted from geological activity, why should the smallest of the five major satellites of Uranus have been more active than the larger moons? Perhaps they resulted from successive large impacts, followed by the satellite's readjustment, which is not yet complete.

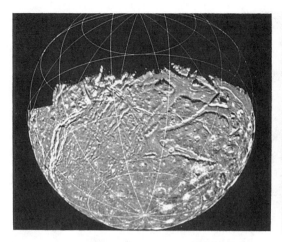

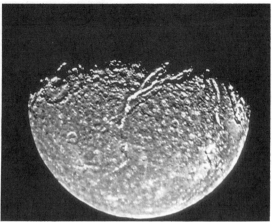

Figure 24.77. A latitude–longitude grid has been superimposed on this Voyager 2 image of Ariel. Ariel's southern hemisphere is in sunlight. Its impact-cratered surface has been broken by more recent geologic fault valleys and fractures. Note the partially filled-in valleys in the upper right.

Figure 24.78. A mosaic of Titania, Uranus' largest satellite. The images were made by Voyager 2 when it was 369,000 kilometers from the satellite. The southern hemisphere of Titania is in sunlight. The heavily cratered surface has been cut by long scarps 2 to 5 kilometers high that are of more recent origin than most of the craters.

Ariel's surface is the brightest and geologically the youngest of Uranus' satellites (Figure 24.77). Like Titania (Figure 24.78), it has few craters larger than 50 km in diameter, suggesting that they were formed by a later bombardment than that which produced the craters seen on Oberon (Figure 24.79) and Umbriel and throughout the solar system. Both of these moons have extensive faults, some of which are several kilometers deep.

Figure 24.79. Oberon is the second largest of Uranus' satellites. It shows many impact craters, some with bright ejecta. Particularly striking is the peak at about 5 o'clock on the limb. It is more than 20 kilometers high and may be the central peak of an enormous crater.

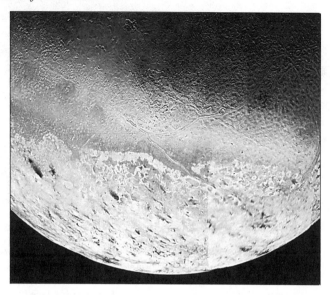

Figure 24.80.   Triton, Neptune's largest satellite, as recorded by Voyager 2. Not very many impact craters are visible, indicating that the surface is young. The lower part of the image is the large south polar cap (in sunlight for 30 years), perhaps covered by a layer of evaporating nitrogen frost deposited during the previous winter. The northernmost darker area is separated from the polar cap by a brighter region of unknown composition. Note the narrow fractures in the crustal ice in this region. Dark splotches on the polar cap may be methane turned dark by exposure to sunlight after it was released from below the surface in a "geyser" consisting largely of nitrogen.

In its flyby of Neptune, Voyager added six small moons to the two—Nereid and Triton—already known. The new satellites range in diameter from 50 to 400 kilometers, and, like Uranus' small moons, are dark, having reflectivities of only a few percent. They are too small to have undergone appreciable geological evolution. It is Triton, Neptune's largest satellite, that is of most interest. As noted earlier, the large satellites of the giants formed with their parents, but Triton, with its retrograde orbit, was most likely captured by Neptune. In addition, ground-based observations have revealed the presence of an atmosphere of methane and nitrogen. Triton may be similar to Pluto, but that will remain speculative until Pluto is explored by spacecraft.

Voyager images of Triton, with its large and bright south polar cap, are indeed striking (see Figure 24.80). About 80 percent of the solar radiation incident on Triton is reflected out to space, making the satellite bright but cold, about −230°C, only 40° above absolute zero. It is, in fact, the coldest object known in the solar system, and as a consequence most of its atmospheric gas is frozen out in the polar caps. Only a very thin, transparent, mostly nitrogen atmosphere, having a pressure of about 10^{-5} that of Earth, is left.

Triton's surface has relatively few small craters and no large ones. This suggests that geological activity destroyed evidence of the early bombardment (see Figure 24.81). Triton is too small to be able to generate much heat by radioactive decay, nor is its interior being stirred up like Io's. Instead, it has been suggested that Triton was captured in an elliptical orbit, and that its periodic swings by Neptune caused its interior to heat up and melt the satellite. According to this idea, by the time Triton's orbit was circularized and the internal heating ceased, the early bombardment by massive objects was over.

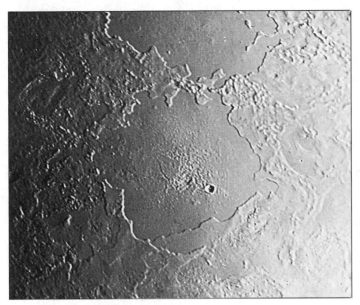

Figure 24.81. Two basins on Triton, which were possibly caused by impacts. These have apparently been flooded several times by now-frozen liquid from the interior. The rough area in the center of the lower basin may be the site of the most recent eruption. This Voyager 2 image shows an area about 500 kilometers in diameter.

Dark streaks on the bright polar cap have been attributed to a kind of volcanic activity in which nitrogen, warmed by the Sun, escapes from below the crust in a geyser rising as much as 10 kilometers above the surface. Exposed to sunlight, some of this material would form dark compounds which, falling back to the satellite, we see as dark smears on the south polar cap.

Ring Systems

The Rings of the Giants. All the Jovian planets are now known to have ring systems, but Saturn's is by far the most spectacular (Figures 24.82 and 24.83). None of these rings is solid; for example, Saturn's are made up of huge numbers of ice- or snowballs up to a few meters in size, all in Keplerian orbits (periods from 2 to 15 hours). Saturn's rings are about 280,000 kilometers in diameter, more than twice the diameter of the planet itself, but the band of snowballs is incredibly thin, no more than a few tens of meters. Stars can be seen through the rings. In striking contrast, Jupiter's very faint ring system, discovered by Voyager I, is a several-kilometer thick cloud of micron-sized dust particles only about 9,000 kilometers wide (Figure 24.84). Its inner portion extends into Jupiter's upper atmosphere and so must be falling into it. Consequently, something must be replenishing the ring. That "something" may be the erosion of a few boulder-sized objects within the rings, or perhaps erosion from Jupiter's two innermost satellites, whose orbits are just outside the ring.

In 1977, Uranus' ring system was discovered by accident when astronomers were preparing to observe the occultation of a bright star by the planet (Figure 24.85). With the two discovered by Voyager, 11 rings are now known. The ring particles of Saturn, covered with water ice, reflect as much as 60 percent of the light falling on them. The particles of Uranus' rings are also relatively large (centimeters to a few meters) but

Figure 24.82.   An image of a crescent Saturn and its spectacular ring system made by Voyager 1 on its way out of the solar system. Saturn can be seen through the rings at the bottom of the image. The prominent dark band and the thin dark band near the edge of the ring system are Cassini's and Encke's divisions, respectively; these are regions of relatively few ring particles. The width of Cassini's division is about 3,500 kilometers. Because a particle in Cassini's division would orbit the planet twice in the time Mimas orbits Saturn once, the particle would be subjected to a small perturbation every two orbits. This would ultimately change its orbit and remove it from the division. Resonances of this sort between particles and satellites produce the other divisions.

Figure 24.83.   As Saturn revolves around the Sun every 30 years, it maintains a constant orientation in space. As seen from the Earth every 15 years, the rings will be edge-on and therefore very thin; at other times, the northern or southern faces of the rings will be visible. The photograph shows the rings edge-on, nearly edge-on, and at their maximum "opening."

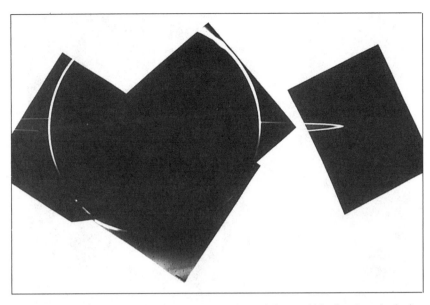

Figure 24.84. Jupiter's thin ring as imaged by Voyager 2 from within the planet's shadow (that is, looking back at the planet with the Sun behind it). Jupiter is outlined by sunlight scattered in its upper atmosphere. That the ring is visible means that its particles must be scattering sunlight forward from the Sun to Voyager. Large objects such as basketballs would simply block out the light. Only small particles about as large as the wavelength of light are efficient forward scatterers. Thus Jupiter's ring-dust particles are quite different from Saturn's snowballs.

Figure 24.85. This Voyager 2 mosaic shows 9 of Uranus' 11 rings. The two discovered by Voyager are too faint to be seen in this image. The rings are widely spaced and narrow. The outermost ring is at most about 100 kilometers wide, but the rest are only a few kilometers across. The ring particles are coal black, reflecting only a few percent of the light striking them. Images taken so that the sunlight is forward scattered show that the whole ring area is filled with very small particle ringlets, which, however, make up only a small fraction of the total ring mass.

Figure 24.86. Two images of the ring system of Neptune, taken a little more than an hour apart, show the two narrow rings, arcs of which had been detected from the ground, and two broad rings discovered by Voyager. One of the new rings is between the planet and the first ring, and the other is between the two bright rings. It is interesting to note that because its orbital motion is in the retrograde sense, the tidal interaction of Triton with Neptune will cause Triton to slowly approach the planet and eventually shatter in its gravitational field. Perhaps then (in about 100 million years), Neptune may develop a ring system to rival Saturn's.

quite dark, reflecting less than 5 percent of the incident light, and in this respect are similar to Uranus' newly discovered satellites. Both systems probably consist of carbonaceous material. Most of Uranus' rings are narrow, only a few kilometers wide; the outermost ring is also the broadest, varying in width from about 25 to nearly 100 kilometers. All are as thin as Saturn's.

Before the Voyager flyby, stellar occultation observations of Neptune had indicated that it had fragments of a ring system, incomplete arcs of material orbiting the planet. Voyager discovered two narrow and faint but complete rings, one of which included the three arc segments discovered earlier, which turned out to be concentrations of particles along the ring (Figure 24.86). In addition, Voyager found three broad dust bands, one of which extends down toward Neptune's clouds. The total amount of material in Neptune's rings is small, no more than would fill a 1-kilometer sphere; Uranus' ring matter might make up a ball five times that diameter, but Saturn's would amount to several 100-kilometer objects.

The detailed images obtained by Voyager showed that Saturn's rings have much more structure and are more dynamic than previously realized. For example, Saturn's broad, apparently featureless rings consist of thousands of individual ringlets, with some having more particles than others and others with twists and kinks (Figures 24.87 and 24.88). In addition, sometimes the rings are not completely flat, but show corrugations (like cardboard). Most of this structure can be explained as waves gravitationally induced in the surface of the rings by Saturn's satellites. (Interestingly, they are analogous to waves generated in spiral galaxies!) The overall wave pattern is constant, although changes do occur in its small-scale structure.

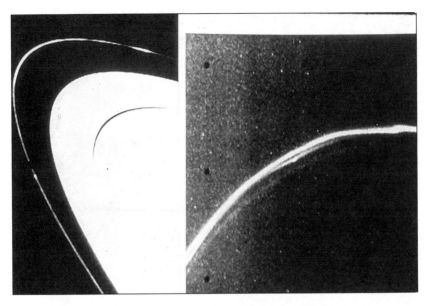

Figure 24.87. One of Saturn's narrow outer rings, the F ring. The right-hand image shows some of the structure in the ring, which is stabilized by two shepherd satellites, one of which can be seen at the lower part of the image on the left.

Figure 24.88. Rather than being uniform sheets, Saturn's rings were resolved into tens of thousands of separate rings, a sample of which is seen in this image of part of the B ring.

Figure 24.89.   Other newly discovered features of Saturn's rings are "spokes" that extend radially outward in the B ring. Because particles orbit with Keplerian speeds appropriate to their distance from Saturn, an individual spoke cannot last more than a few hours unless it is reformed. The image on the right was made with the Sun behind the spacecraft and the spokes are dark, whereas, when illuminated from behind, they are bright; that is, they forward scatter sunlight efficiently. This means that they are very small particles of ice, lifted out of the ring plane.

Another kind of structure in Saturn consists of **spokes**, extending radially outward in the rings (Figure 24.89). Clearly, differential rotation of the rings should quickly destroy these, so something must be maintaining the spokes. What that may be is not clear, but perhaps it is some sort of electromagnetic interaction with Saturn's magnetic field.

A major mystery is why the very narrow rings (only kilometers in width) discovered at Saturn, Uranus, and Neptune exist at all. Something must be keeping ring particles from slowly spreading apart and disappearing. That something may be small satellites that confine ring particles to their narrow orbits. Such objects, called **shepherd satellites**, apparently do play such a role for a few of the narrow rings, but similar objects have not been found associated with most of the narrow rings.

Origin and Evolution of Rings.   Although much remains to be learned, the origin of ring systems has been clarified somewhat by the recent discoveries. When Saturn's rings were the only ones known, it was thought that the rings were made of material that was never able to form into a larger body, implying that the rings were very old. Voyager results, in particular the detection of rings or bands of small dust particles, suggest other possibilities as well. Ring particles are constantly being ground down to smaller and smaller sizes by collisions with micrometeorites and other small objects. As time goes by, a larger and larger proportion of the ring material is transformed to dust, which eventually falls into the planet, and the rings become less and less prominent. This process can take place relatively rapidly, in just a few hundred million years. In this picture, Neptune's rings are in an advanced state of decay, as indicated by the small amount of material left in the rings, the large fraction of which is dust.

Saturn's much more massive rings, of which only a small fraction is dust, are still in their prime. Rings are dynamic entities, continuously evolving.

If the rings did not form at the same time as their planet, and since it is unlikely that we just happen to be witnessing a unique epoch of rings, they must form anew. This might happen when a small satellite of one of the giant planets is battered by a collision with another object, perhaps a passing comet. Since the giants have a number of small satellites, it is possible that eventually one of them will be shattered and a new Saturn-like ring system will be formed from its debris, along with the remnants of the smashed comet.

Pluto

The Discovery of Pluto.   Finally, how does Pluto, the most distant planet, fit with the rest of the planetary family? It is peculiar in many respects, not least in the way it was found. The great discovery of Neptune, made from an analysis of the motion of Uranus, inspired astronomers to wonder if there were any other planets even farther out than Neptune. Although Neptune accounted for most of the difference between the observed and predicted motion of Uranus, it appeared to some that a tiny part of Uranus' motion might still be unaccounted for. (Neptune had not been observed long enough to see whether its observed and predicted motions differed as a result of perturbations by the supposed new planet.) This led a few astronomers to try to calculate where the disturbing object might be. The most persistent of these workers was Percival Lowell, who, you will recall, founded an observatory in Arizona in 1894 primarily for the study of Mars and the search for the supposed ninth planet. Lowell died in 1916, after a decade of effort had yielded no new planet. The search at Lowell Observatory continued, however, and was finally successful in 1930, when Pluto was found within six degrees of its predicted position.

This seemed to vindicate those few astronomers who believed that the discrepancies in Uranus' motion were real. Doubts soon arose, however, as various indirect methods indicated a small mass for Pluto. When Pluto's moon was discovered and a reliable mass was calculated that was even smaller than earlier estimates, it was clear that Pluto could not have disturbed Uranus' motion. Thus its discovery near its predicted position based on those discrepancies could be nothing more than a strange coincidence!

The Discovery of Charon.   Pluto is so small and so far away that its apparent angular diameter is only about 0.1 seconds of arc. As a result, until recently we knew practically nothing about its physical characteristics; even its mass and diameter could only be estimated (and, as it turns out, quite incorrectly so). About all that had been found out was that its orbit is the most eccentric and the most inclined of any of the planets; its rotation period is 6.4 days; it had methane ice on its surface and a thin methane atmosphere; and it was probably even smaller than Earth's Moon. This situation changed dramatically in 1978, when a satellite of Pluto was discovered as a slight bulge on an image of its parent (Figure 24.90). Because Pluto and Charon are never more than 0.9 seconds of arc apart, it is difficult to separate them clearly with ground-based telescopes. Compare the ground-based image with one taken in 1991 from the Hubble Space Telescope.

Charon's motion around Pluto takes it north and south—nearly perpendicular—to Pluto's orbital plane, rather than east and west like most moons. But because of the Pluto-Charon double synchronism (described below), we know that Charon's orbit

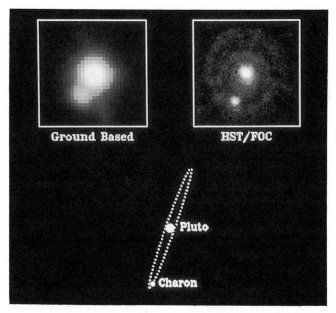

Figure 24.90. A good ground-based image of Pluto and Charon and one taken from the Hubble Space Telescope (before the servicing mission that improved the optics). The angular separation of Pluto and its satellite is about 0.9 arcseconds.

must be in Pluto's equatorial plane. Like Uranus, therefore, Pluto is tipped over on its side; its north pole is 32° south of its orbital plane (see Figure 24.91).

Twice every Plutonian year or about every 125 Earth years, Charon's orbit around Pluto is aligned toward the Earth, and occultations occur every 6.4 days. Luckily for us, not only did this rare alignment begin in 1985, soon after Charon's discovery, and continue for about five years, but during this time Pluto was near its closest approach to the Sun and so most favorably located for observation. During the next series of eclipses, beginning in about the year 2109, Pluto will be nearly at its greatest distance from the Sun and so more difficult to observe.

What Charon Has Taught Us. What have we learned so far? Charon orbits Pluto at a distance of only about 20,000 km, and the two objects are in a unique sort of double synchronism. First, Charon revolves once around Pluto in the same time that it rotates once on its axis; so, like our Moon and many other satellites in the solar system, it always presents the same face toward Pluto. Second, Pluto's 6.4-day rotation period is the same as Charon's period of revolution around Pluto; hence the same hemisphere always faces Charon.

Pluto is even smaller than our Moon, only about 2,400 km in diameter; Charon's diameter is about one-half that. The masses of Pluto and Charon are roughly 1/400 and 1/4,000 the Earth's, respectively, giving densities of about 1.8 and 1.2 times that of water, respectively. Charon's composition—which is mostly ice—differs significantly from Pluto's. Also, tiny as it is, Charon displaces our Moon as the satellite having the largest mass relative to its primary, about 0.1 compared to 1/81 for the Earth–Moon system. Because of its large relative mass and its proximity to Pluto, Charon is able to raise a tidal bulge on Pluto just as Pluto does on Charon, producing the "double synchronism" mentioned above.

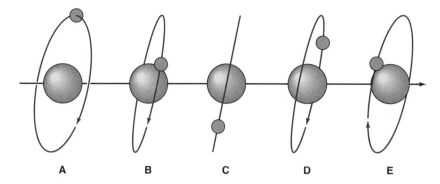

A	B	C	D	E

Figure 24.91. The changing orientation of Charon's orbit around Pluto as seen from the Earth. (A) The orientation in 1978 when Charon was discovered; (B) in 1985 when eclipses began; (C) in 1988 when eclipses were central; (D) in 1990 when eclipses ended; (E) the orientation in 1994. Eclipses will begin again in about 2109.

Analysis of the way light is dimmed as Charon and Pluto eclipsed each other indicates that Pluto has bright polar caps, most likely of methane ice. When Pluto is close to the Sun (as it was in the 1980s) some of this ice is vaporized and enters the atmosphere, only to freeze out again as Pluto recedes from the Sun. Eclipse results also suggest the existence of a broad, dark equatorial band that is at least partially free of the nitrogen ice that covers much of the surface.

The Origin of Pluto.   Ever since its discovery, Pluto's origin has been a puzzle. It does not fit with the giant planet family in size, composition, or spacing from the Sun. (In fact, given its small size and density it doesn't fit the terrestrial family, either.) Since its eccentric orbit periodically takes Pluto closer to the Sun than Neptune is, it was suggested long ago that Pluto may have once been a satellite of that planet. Dynamically, it appears possible that Neptune's satellite system could have been disrupted by a passing object in such a way as to liberate Pluto. The notion of a planetary satellite having its own satellite seemed a bit bizarre, however.

Now that we know that Charon's average density is substantially smaller than Pluto's, it appears that Charon might have been formed in the same way some astronomers think our own Moon was formed: as the result of a giant impact on Pluto that blasted off a considerable portion of its ice mantle. (The small companion of the asteroid Ida may have resulted from a similar process.) Furthermore, Chiron and Triton, along with a few other recently discovered objects beyond Neptune, suggest that there may be several sizable objects in this region, plantesimals formed in the solar nebula more than 4 billion years ago. By now, most of these would have been cleaned out from the planetary portion of the solar system. Perhaps Pluto and the others are among the few that are left.

There still remains the question of the possibly discrepant motion of Uranus. If it is real, what is causing it? Could there be another real planet (as opposed to a planetesimal) in our solar system? Or could the Sun have a faint, distant stellar companion[10]

---

[10] To account for possibly periodic (every 25 million years or so) extinction events on Earth, some scientists have suggested that the Sun has a faint companion that stirs up comets in the Oort cloud, causing one or more to crash into the Earth with consequences similar to those that did in the dinosaurs.

that barely affects the motions of the outermost planets? Or perhaps the discrepant motions are not real, but simply small errors? All of these positions have been advanced, but no definite answer has emerged. The last is probably most likely.

The clear message that this brief description of the solar system gives us is the incredible diversity of its members. Hardly any two planets or satellites look alike. It is remarkable how, beginning with the same primordial material differing only in its solar distance, the system could form and evolve into such a diverse collection of objects. This realization is in complete contrast to the view that prevailed before space exploration began. For example, no one expected that the magnetospheres of the giants would differ as they do, or that Venus, called Earth's twin, would be so unlike Earth in its geology and its atmosphere, or that Mars would show evidence of running water. To find the solutions to many of these puzzles will likely require understanding events that took place several billion years ago, when the planets formed or soon after. Needless to say, that will not be easy, but it should be interesting.

This diversity makes it difficult to generalize about the life-sustaining possibilities of planets that might exist in other systems. Nevertheless, in the next chapter, we will briefly discuss this and related matters relevant to life elsewhere.

Terms to Know

Reduced density, surface waves, P-waves, and S-waves. Earth's crust and inner and outer core; lithosphere, asthenosphere, subduction zone, Pangaea. Regolith, rilles, rays, lobate scarps, corona. Iron, stony irons, chondrites, chondrules, achondrites, and carbonaceous chrondrite meteorites. Comet nucleus, coma, tail, hydrogen envelope; the Great Red Spot, metallic hydrogen, the Great Dark Spot; shepherd satellite; Apollo asteroids.

Ideas to Understand

General properties of the terrestrial and Jovian planets. Earthquake waves and the Earth's interior structure, sources of internal heat. Possible origin and evolution of the Earth's atmosphere, the greenhouse effect, how atoms escape from an atmosphere. The nature of the Moon's surface. How and why Mercury, Mars, and Venus differ from the Earth and Moon; how they are similar; evidence for a warmer Mars in the past. Characteristics of Jupiter's surface, interior, and magnetosphere; of the satellites of Jupiter, Saturn, and Uranus; origin of Io's activity. Relation of asteroids, comets, and planetesimals.

Questions

1. Suppose that the Earth's atmosphere reflected visible light much more efficiently than it now does, and that it were more transparent to infrared radiation than it now is. What would be the effect of these changes on the Earth's surface temperature? Explain your reasoning briefly.

2. What evidence do we have that the entire planetary system was subjected to heavy bombardment by asteroid-sized objects? Why do we think that this occurred early in the history of the solar system?

3. Explain how we can detect CO_2 in the atmosphere of Mars when we make these ground-based observations through our atmosphere, which also contains CO_2. Hint: does Mars move with respect to the Earth?

4. Why can we use the spectrum of Venus to establish that its atmosphere contains carbon dioxide, but we can't find out the composition of Moon rocks from their spectrum (in reflected light)?

5. Briefly describe the planetary features below and name the planet on which each is located.
  (a) Olympus Mons
  (b) Great Red Spot
  (c) Valles Marineris
  (d) Mauna Loa

6. (a) Why is the dark side of the Moon so much colder than the Earth?
  (b) Why is the sunlit side of the Moon so much hotter than the Earth?

7. Suppose Io were the only large satellite of Jupiter. Explain how it would differ from its present state.

8. Explain the likely origin of each of the following:
  (a) The moons of Mars
  (b) Earth's atmospheric oxygen
  (c) Jupiter's strong magnetic field

9. (a) Asteroids are often not spherical in shape, but planets are always so (apart from tidal or rotational bulges). Why is this so? (Hint: compare the masses of planets and asteroids).
  (b) How many times larger are the diameter and volume of Earth than those of Ceres?

10. Why do we know Saturn's rings to be made up of many small, independently orbiting objects?

11. How long (in hours) does it take the particles at the outer edge of Saturn's A-ring (speed 16 km/sec) to make one trip around the planet? How long does it take for particles at the inner edge of the B-ring (speed 20 km/sec)?

12. Show that Pluto sometimes comes closer to the Sun than Uranus (use the information in Appendix F).

13. If the mass of Jupiter's ring is about 10^{10} kilograms and the density of its dust particles is about 2.5 times that of water, how large a sphere would contain all this dust?

14. Answer the following with "greater than," "the same as," or "less than," and briefly explain each answer.

If the albedo of Jupiter were only half its present value:

(a) The wavelength of the maximum intensity of reflected sunlight would be _____ it is now.

(b) The intensity of the reflected sunlight would be _____ it is now.

(c) The wavelength of the maximum of the emitted radiation would be _____ now.

(d) The intensity of emitted radiation would be _____ now.

15. The following questions are concerned with Jupiter and Saturn.

(a) What gas is most common in the atmosphere of Jupiter? of Saturn?

(b) Which has the larger percentage of helium in its atmosphere? the larger fraction of ammonia?

(c) Which planet is more massive? Which is more flattened in shape because of rotation?

(d) Does Saturn have any satellites larger than ours? Does Jupiter? If so, which one(s)?

(e) Are the moons of Saturn made mostly of rock or ice? How do we know?

(f) List the Galilean satellites of Jupiter in order of increasing size. Which of these have water or ice mantles inside?

16. (a) What is the angular diameter of the Sun as seen from Mercury? How does that compare with the diameter of the Sun we measure?

(b) How many times more energy does Earth receive from the Sun than does Pluto? What is the angular diameter of the Sun as seen from Pluto?

Suggestions for Further Reading

The planetary flybys generated great popular interest in the planets. A sampling of books is given below. The first group is on the solar system as a whole.

Beatty, J., Chaikin, A. (ed.), *The New Solar System*, *Third Edition*. Cambridge: Cambridge University Press, 1990. A well-illustrated collection of popular articles.

Frazier, K., *Solar System*. New York: Time-Life Books, 1985.

Greeley, R., *Planetary Landscapes*. Boston: Allen and Unwin, 1987. A well-illustrated description of planetary surfaces.

Morrison, D. and Owen, T., *The Planetary System*. Reading, MA: Addison-Wesley Publishing Company, 1988.

Sheehan, W., *Worlds in the Sky*. Tucson: The University of Arizona Press, 1992. The book is well described by its subtitle, *Planetary Discovery From Earliest Times Through Voyager and Magellan*. Contains much about the people involved.

Wagner, J., *Introduction to the Solar System*. Philadelphia; Saunders College Publishing, 1991.

Some books on the terrestrial planets, starting with a good overview, are listed below organized by planet:

Chapman, C., *Planets of Fire and Ice*. New York: Charles Scribner's Sons, 1982. A nice introduction to the terrestrial planets.

Dunne, J. and Burgess, E., *The Voyage of Mariner 10*. Washington: NASA SP-424, 1978. A description of the mission to Mercury (three visits) and to Venus (one visit).

Burgess, E., *Venus, An Errant Twin*. New York: Columbia University Press, 1985. Data obtained by U.S. and Soviet flights with descriptions of instruments and spacecraft through the Pioneer Venus missions.

Cattermole, P., *Venus: The Geological Story*. Baltimore: The Johns Hopkins University Press, 1994. Includes the Magellan images.

Fimmel, R., Colin, L. and Burgess, E., *Pioneer Venus*. Washington: NASA SP-461, 1983. Description of the spacecraft, the instruments, the mission, and some results.

Masursky, H., Colton, G. and El-Baz, F. (eds.), *Apollo Over the Moon: A View from Orbit*. Washington: NASA SP-362, 1978. A description of the Apollo missions 15–17, the instruments carried, and a sample of the images obtained.

Baker, V., *The Channels of Mars*. Austin: The University of Texas Press, 1982. Gives some historical background on the "canals" and interpretative descriptions of modern images.

Carr, M., *The Surface of Mars*. New Haven: Yale University Press, 1981. A well-illustrated description.

Cattermole, P., *Mars: The Story of the Red Planet*. London: Chapman and Hall, 1992. An up-to-date description of all aspects of Mars.

Viking Lander Imaging Team, *The Martian Landscape*. Washington: NASA SP-425, 1978. Instruments and results from Vikings 1 and 2.

Viking Orbiter Imaging Team, *Viking Orbiter Views of Mars*. Washington: NASA SP-441, 1980. Images of Mars from the Viking orbiters.

Chapman, C. and Morrison, D., *Cosmic Catastrophes*. New York: Plenum, 1989. Solar system violence.

Davies, J., *Cosmic Impact*. New York: St. Martin's Press, 1986. A popular account of meteors and asteroids.

Hutchison, R., *The Search for Our Beginnings*. London: British Museum and Oxford University Press, 1983. A thorough, popular account of meteorites, their properties and origin.

Books on the outer planets are listed below.

Elliot, J. and Kerr, R., *Rings. Discoveries from Galileo to Voyager*. Cambridge, MA: The MIT Press, 1984. Our current understanding of planetary rings.

Fimmel, R., Van Allen, J. and Burgess, E., *Pioneer. First to Jupiter, Saturn and Beyond*. Washington: NASA SP-445, 1980. The flights of Pioneers 10 and 11.

Hunt, G. and Moore, P., *Jupiter*. New York: Rand McNally and Company, 1981. Good overview with emphasis on Voyager results.

———*Atlas of Neptune*. Cambridge: Cambridge University Press, 1994. Historical introduction to the planet and a description of many Voyager 2 images.

Littmann, M., *Planets Beyond: The Outer Solar System*. New York: John Wiley and Sons, Inc., 1990.

Morrison, D., *Voyages to Saturn*. Washington: NASA SP-451, 1982. The missions of Voyagers 1 and 2.

"Space-travel is utter bilge."
—Sir Richard Wooley, 1956; UK Astronomer Royal

"Do there exist many worlds, or is there but a single world? This is one of the most noble and exalted questions in the study of Nature."
—Albertus Magnus (about 1200–1280);
German philosopher and theologian

THE PIERRE GUZMAN PRIZE

In 1899, Mme. Guzman, a Frenchwoman of means, gave to the French Academy of Sciences a substantial sum for a prize to be awarded to the person who found a method of communicating with another world. The prize was in memory of her son, Pierre Guzman. He was an enthusiastic amateur astronomer, who, like many others at that time, was so convinced that the existence of life on Mars had been established that his mother excluded communication with that planet from the competition! Unfortunately, the competition was never successfully concluded.

CHAPTER 25

Life Elsewhere

Does life exist elsewhere in our solar system? At a nearby star? Somewhere far away in the galaxy? Are we the only beings in the galaxy who wonder about the existence of intelligence elsewhere? These are very old questions, which have stimulated speculation, fantasies, and dreams for 2,500 years. Is it absurd to think that we are the only inhabitants of this enormous universe? Or is it even more foolish to think that we are not? We do not yet have answers to these questions, but just as we can for the biological aspects of the origin of life, we can probably define the astronomical context more tightly than we could only a few decades ago. One aim of this chapter is to apply what we now know about the stars and planets to the question of life elsewhere.

We will begin this chapter by describing some of the thoughts people have had over the past two millennia about the possibilities of life elsewhere than the Earth. Not only are these of interest in themselves, but they also give us a little perspective on our modern ideas, some of which resemble the old notions. After this introduction, we will consider in a very general way the biological requirements for life as we now understand them, and how these define and limit the appropriate, supportive astronomical circumstances. We will then see what kinds of estimates we can make about the probability of life elsewhere. As we proceed along this path, we will be moving into increasingly speculative territory. We will end with an account of the search so far, and a brief discussion of what success or failure in this search might imply. With what you have learned in this course, you will be as much of an "expert" on many of these matters as anyone else.

Life in the Solar System
Early Historical Background

The possibility that there are other worlds beyond our own, some of which might be inhabited, is a very old idea, going back to the Greeks. For example, in the fourth century B.C.E., Metrodorus of Chios wrote, "To consider the Earth as the only populated world in infinite space is as absurd as to assert that on a vast plain only one stalk of grain will grow." This kind of argument makes two assertions: that the universe is enormous, and that what happened here is not unique and will happen again someplace else. Today, proponents of life elsewhere make the same argument.

Not surprisingly, the Moon figured prominently in the speculations of early writers concerning other worlds. For example, Plutarch (about 46–120 C.E.), best known as a biographer and essayist, wrote of the Moon as being so similar to the Earth that it might be inhabited.

By contrast, the early Christian Church followed the Greek philosophers in believing that the universe was finite and consisted of only one Earth. (Recall that Aristotle's physics required a finite universe; he taught that the natural place for all "Earth" was at the center of the system, and so Earth was unique.) Also, Christianity taught that the universe was created for man, implying that there was only one inhabited world. Interestingly, in the Middle Ages this notion came into question because it seemed to compromise the power of God. In fact, in 1277 the Bishop of Paris condemned as heretical 219 propositions, often taught in the universities of the day, which seemed to limit God's power. Among these was the doctrine that God could not create other worlds. As a consequence, some scholars began to consider not only the possibility of other worlds, but also the possibility that some of these might even be inhabited. Note that this interest in life elsewhere came not as a result of any new knowledge, but rather as a change in attitude. This is a common phenomenon, even in science and even today.

Speculations After Copernicus

Speculations of this sort became easier when Copernicus removed the Earth from its central location, so that it became just another planet. In addition, Galileo's telescopic discoveries of mountains on the Moon and of moons orbiting Jupiter suggested to many (but not to Galileo, who argued against the idea) that the Earth was not unique and that other planets might be inhabited. Kepler himself wrote about the Moon and its possible life forms in a book called *Somnium*, recently translated into English as *Kepler's Dream*. He also made one of the first translations into Latin of Plutarch's book on the Moon, mentioned above. Only a few years after Kepler's volume, Bishop Wilkins, an Englishman, published *The Discovery of a World in the Moone*. He wrote that the Moon was Earth-like and probably inhabited, "but of what kinde they are is uncertaine."

At the end of the seventeenth century, the Frenchman Bernard de Fontenelle published a book of conversations in which, among related matters, he discussed the characteristics of the possible inhabitants on the planets of the solar system. This exemplified the idea that, since the Earth was just another planet and it was inhabited, then the other planets must also be populated. His book was widely read for well over a century, and had a considerable influence on popular attitudes.

By the end of the eighteenth century, the idea that there were other inhabited worlds was accepted by nearly all scientists, even though there was not a shred of supporting evidence. (As we shall see, this is very nearly the situation today!)

Some thought was given to the possibility of communicating with intelligence "out there." For example, in the nineteenth century the great German mathematician, C. F. Gauss (1777–1855), suggested that an attempt be made to send a signal to possible extraterrestrial beings. He proposed to do this by creating a right-angle triangle several hundred miles on a side. The triangle was to be outlined by forests grown in some large, flat area, such as Siberia. Gauss argued that such a figure could not be mistaken for a natural formation.

With the spread of Darwin's ideas on biological evolution during the second half of the nineteenth century, another aspect of this question came into focus. If the life around us was the product of evolution, then life must have originated somewhere a long time ago. Darwin himself felt that inquiring into the origin of life was as difficult and fruitless as asking about the origin of matter. (As you know, we have made considerable progress on the latter topic!) To scientists of a century ago, the beginning of life seemed to be such an insurmountable problem that they tended to push it as far away in time and space as possible. It was suggested, for example, that life began on another planet in our system, perhaps on the supposed planet that shattered into the thousands of asteroids in orbit between Mars and Jupiter. This provided an incentive to examine meteorites for any evidence of life. Positive results were often claimed, but were ultimately found to be the result of contamination from the Earth, or, in some instances, just the product of wishful thinking.

Svante Arrhenius (1859–1927), the Swedish chemist and Nobel laureate, further extended the idea that life formed elsewhere. Early in this century he suggested that life had been seeded on Earth by spores that floated throughout the galaxy. These spores, gently propelled about by the force of starlight, could survive for very long times and eventually carry life throughout the galaxy. This idea Arrhenius called **panspermia**. Though it had a considerable appeal, panspermia actually solves nothing. It effectively makes the problem of the origin of life insoluble, by removing it to a place at some great distance about which we can never know anything. Nonetheless, this idea has been revived recently.

The notion that the Moon might support life died hard. In the 1830s careful observations of the occultation of stars by the Moon showed none of the refraction effects that would be associated with an atmosphere. By contrast, such effects are pronounced in our atmosphere. It was estimated that the Moon's atmospheric density could be no more than about 1/2,000 that of the Earth, hardly sufficient for oxygen-breathing animals. Still, the question of the reality of changes on the Moon's surface and their possible significance provided considerable impetus for its study.

The introduction of photography in the second half of the nineteenth century was an enormous benefit for lunar studies as well as for other fields of astronomy. A single photograph was a more reliable document than sketches produced during many long nights at the telescope. As has been mentioned, however, more detail could be seen at moments of good seeing than could be recorded during the long exposures required by the grainy and insensitive photographic plates of the day. As late as the 1920s, a Harvard astronomer interpreted barely detectable changes in the lunar surface as being caused by colonies of primitive organisms carrying out their life cycle in the 14 days of sunshine: "We find here next door to us a living world, with life entirely different from anything on our planet."

The Mars Fad

Generally, astronomers who were inclined toward the possibility of life in the solar system found Mars a more attractive prospect than the Moon. Because of its greater distance, however, the observational situation is much more difficult (skeptics said impossible) and treacherous. Observers could discern whitish polar caps and large dark features against a brighter, yellowish background. They agreed on the approximate sizes and shapes of the larger and many of the smaller features. The polar caps were assumed to be water ice, the brighter areas desert, and the darker regions possibly vegetation. Apparent seasonal effects in which, during the Martian summer, a polar cap decreased in size and the dark areas became more prominent, supported the view of the proponents of life. They argued that the melting polar cap provided moisture for growing vegetation (the dark areas). Mars' rotation period of about 24.6 hours makes its day-night cycle similar to the Earth's. The 24-degree inclination of its equator to its orbital plane is nearly identical to Earth's, so that seasonal heating effects should be relatively the same. It became popular to think of Mars as a near twin of the Earth— to be sure, a little cooler and drier with a thinner atmosphere, but nonetheless a close kin. The notion of life there did not seem impossible. As a consequence, much public attention was paid to the close approach of Mars to Earth in 1877.

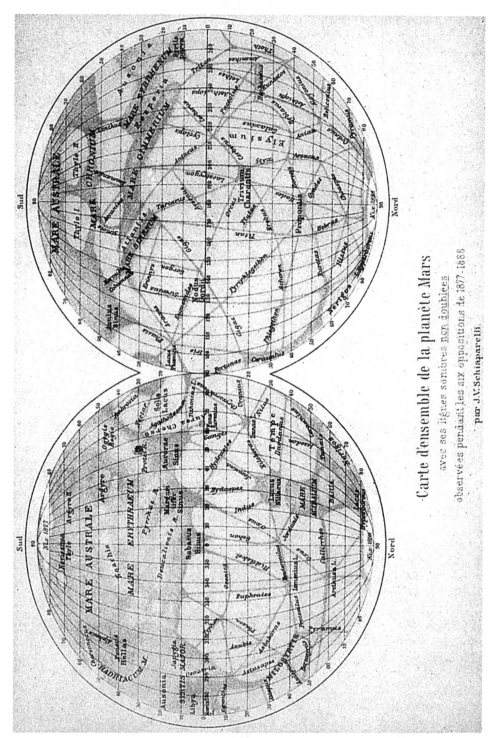

Figure 25.1. Maps of Mars by Schiaparelli, based on his observations made during the opposition of 1877–78. Note the linear features, the "canali."

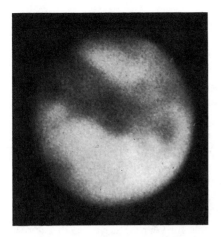 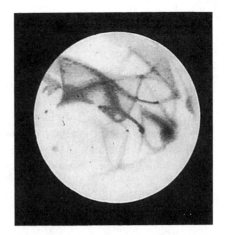

Figure 25.2. Comparison of a drawing and photograph of the same hemisphere of Mars made in 1926. At moments of best seeing, a telescopic image will be sharp for a small fraction of a second, after which the astronomer tries to sketch what he or she saw. The brain tends to link small indistinct features into lines.

This event did indeed bring a new result. Giovanni Schiaparelli, an Italian astronomer and a good visual observer, reported seeing about 40 long, straight, narrow markings for which he used the word *canali*[1] (Figure 25.1). In Italian this has the fairly neutral meaning of "channel," but it obviously sounds like the much more suggestive English word "canal." Even though very few other astronomers had seen these markings, Schiaparelli popularized them and many of the general public not only took them to be real objects, but also assumed they were of artificial origin. By the next favorable approach in 1892, other observers had seen the canals, others had seen none, and yet others saw double canals!

Most astronomers, however, remained unconvinced, partly because of experiments such as one reported in 1903. British schoolboys were asked to sketch carefully what they saw in various representations of Mars they were shown, placed over a range of distances. They, of course, were not told what they were drawing. The results strongly supported the notion that the eye tends to connect small features that are too small to be seen distinctly. Thus the Mars canal observers reported what they saw; unfortunately, what they saw were, for the most part, illusions (see Figure 25.2).

Perhaps the person who did most to popularize the notion of canals and their implications was the amateur astronomer Percival Lowell, whom we have already encountered several times. Lowell claimed that an elaborate network of canals had been built by intelligent Martians to bring irrigation water from the polar caps to the arid regions. In many articles and books he spun an elaborate tale of a dying planet whose inhabitants were desperately trying to survive in the face of environmental disaster. These sensational ideas captured the public's imagination (and continued to do so until the planet was explored by spacecraft). The British author H. G. Wells capitalized on this public interest in Mars when, in 1897, he published his science-fiction novel *The War of the Worlds*. This book, still in print, describes an invasion of Earth by a dying Martian civilization. Another example of popular enthusiasm for the idea of life on Mars was demonstrated in the summer of 1924 through the then-new marvel of radio. After a claim that coded signals had been received from Mars, radio stations in New

---

[1] This term, applied to dark streaks, had been introduced a few years earlier by another Italian astronomer, possibly to be consistent with the practice of naming dark areas for bodies of water.

Jersey were turned off a few minutes each day while people listened—unsuccessfully—for these interplanetary messages. In 1938 on Halloween night, Orson Welles managed to frighten a sizable portion of the U.S. population when he broadcast a supposed invasion by Martians.

It is easy to laugh at the enthusiasts for life on the Moon or on Mars. We should remember, however, that the few astronomers who took the idea seriously were neither stupid nor incompetent. They were working at (and unfortunately, often beyond) the technical limits of their instruments; their craft became more of an art than a science. Furthermore, the possibility of life elsewhere has fascinated us for 2,500 years. Is it so surprising that some astronomers, working with the tools they felt could give answers to this old question, became victims of their preconceived notions on the subject? Are our attempts to search for life similarly compromised? Are we sure we aren't fooling ourselves in some more subtle way? Can we stand back from our work and ideas, and examine them objectively? Try to decide for yourself as you study the rest of this chapter.

Compounds Basic to Life

A rigorous definition of life involves difficulties and subtleties that would lead us too far afield. For our purposes we will simply say that matter is alive if it can reproduce itself and can evolve, that is, change over time. This definition implies that life requires complex molecules that can store and transmit information, and also provide the large number of responses necessary for survival and adaptation to a changing environment.

On Earth these large molecules are composed of two basic types of compounds that are important in all life: 20 different **amino acids** (containing from 10 to only 27 atoms), and four different **nucleotides** (Figure 25.3). It is remarkable that of the enormous number of these compounds that are possible, only a tiny fraction are significant for life. The amino acids can form very large molecules called **proteins**, and the nucleotides are linked together into long chains called **nucleic acids**. The most famous of these is the double-helix deoxyribonucleic acid, or **DNA**, which carries the blueprint of the organism by determining which proteins are to be formed. A protein may contain from 100 to 1,000 amino acids. Their arrangement in forming a protein determines its function, either to set structure or to manufacture an enzyme that controls the rate at which chemical reactions occur. The differences between humans and hamsters depend on the proteins that are assembled by the nucleic acids. These same amino acids and nucleotides are found in all organisms here; this remarkable fact suggests a common origin for all life on Earth, a point we will return to later.

The relative number of atoms that make up our bodies are given in Table 25.1. Other living things have similar compositions. It is striking that only a few elements make up the bulk of the recipe for life, though tiny concentrations of many other elements are also required. Note that the most abundant element is hydrogen, which is barely present in the Earth as a whole. However, hydrogen and oxygen exist primarily as water in living things, and water is plentiful on the Earth's surface. With the exception of nitrogen, the other elements are among the most abundant on Earth as well as being the most common of the metals in the Sun and stars, as shown in Table 11.1. Interestingly, the trace elements such as zinc, copper, and manganese exist in living things in the same relative abundances as in sea water. This argues against the idea mentioned earlier in this chapter that life arose elsewhere and was somehow transported to the Earth.[2]

---

[2] Perhaps a counterargument could be made that alien life, evolving in response to local conditions, would eventually have taken on local chemical abundances.

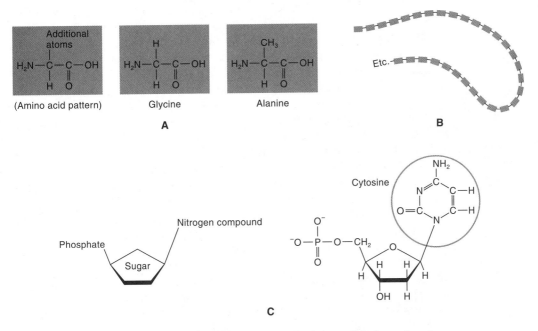

Figure 25.3. (A) The 20 different amino acids found in proteins are all built on the same pattern and differ only in their "additional atoms." For example, glycine is produced if the additional atom is H, alanine is produced if the addition is CH_3, etc. (B) The various amino acids (each represented by a rectangle) are linked together to form large protein molecules. (C) A nucleotide is made up of a sugar, a phosphate, and a nitrogen compound. The first two are always the same in the four nucleotides that make up DNA. It is only the nitrogen compounds that differ (shown enclosed in a circle in the structure of cytosine).

Table 25.1. Relative abundances of chemical elements in our bodies

Element	Relative abundance
Hydrogen	1,000
Oxygen	425
Carbon	170
Nitrogen	39
Calcium	3.8
Phosphorous	2.1
Sulfur	2.1

Modern Work

It was not until the 1950s that much work was being done on the origin of life that could be characterized as scientific rather than speculative in nature. The little research that was carried out earlier seemed to indicate that life elsewhere was not very probable. The argument usually emphasized the vanishingly small probability that large molecules could form from random encounters of their constituent atoms in some pre-biological broth. When the problem is put in this form, the prospects seem to be extremely poor, indeed. That this may not be the way to look at the question is suggested by a simple example. The students in your astronomy class probably come from all over the state and possibly from many other states and even other countries as well. Now, what is the probability that just by chance, just by their random motions, all these people would appear in a particular place at a particular time? I think you can safely guess that on that basis the mere existence of your astronomy class would look like a miracle! The point is, of course, that the process is not completely random, that there are many "forces" that bring you all together at a certain place and time, including your desire for an education, interest in astronomy, the location of the classroom, etc.

Similarly, several factors can combine to provide an environment favorable for the formation of life. For example, shallow tidal pools provide basins into which essential components may be concentrated by wave action or by flowing rain water; water is a good medium for mixing molecules; a thin surface layer of water (it need be only a centimeter or so thick) protects molecules from disruption by solar ultraviolet photons; at the same time, the pools are shallow and warmed by the Sun, thus facilitating chemical reactions. Finally, characteristics of atoms determine that molecules of certain shapes and kinds are more likely to form than others. For example, the structure of oxygen and nitrogen enable them to share more than one electron with carbon, making such compounds relatively stable.

In 1953 (the year the structure of DNA was established) an experiment performed at the University of Chicago showed that laboratory conditions, at least, could be made quite favorable to the formation of some of the basic molecules (Figure 25.4). A mixture of water, ammonia, methane, and molecular hydrogen—all simple molecules that might have been present on the early Earth—was zapped by an electric spark (to provide energy). After only a week or so, chemical analysis revealed that various amino acids had formed in this soup. In subsequent experiments, nucleotides as well as other amino acids were formed. These experimental results were surprising—the first steps leading to the formation of complex molecules turned out to be quite easy. The probability argument against the formation of such molecules was somewhat weakened. Unfortunately, however, it has so far proven impossible to produce very large molecules by experiments of this sort, much less a living organism. Remember, too, that DNA in a human cell has about eight billion nucleotide pairs, quite a different structure from compounds with ten or twenty atoms. Nonetheless, these and similar experiments showed that it was rather easy to assemble simple molecules, at least, in environments that probably resemble those on the early Earth.

Another factor that contributed to the change in attitude toward the possibility of life elsewhere was the discovery of a large variety of molecules existing in the unfriendly environment of interstellar space. That these molecules consist of more than two or three atoms was another big surprise. The largest molecule so far found has 13 atoms (three more than glycine, the simplest amino acid). Amino acids have also been found in meteorites. Nature seems to have little trouble in taking the first step down the *very* long road toward the formation of large molecules.

Though none of these laboratory experiments and astronomical discoveries actually show how life formed or that it exists elsewhere, they suggest that the possibil-

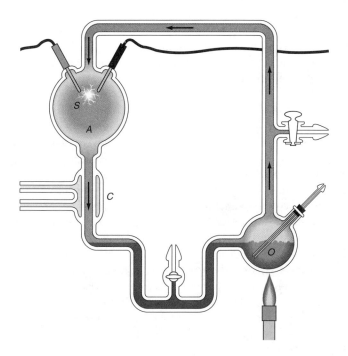

Figure 25.4. The Urey-Miller experiment. A mixture of water, methane, and ammonia (the "ocean," *O*) is heated and in gaseous form passes to a flask (the "atmosphere," *A*) where it is subjected to an electric spark, *S*. The material is then condensed (*C*) back to a liquid and collected after a time. Simple amino acids and nucleotides can be formed in this way.

ities are not completely crazy, that it is not only worth thinking about, but that it even may be appropriate to carry out searches. This view was formally recognized by astronomers in 1982, when they formed a new research section on bioastronomy within the International Astronomical Union (the international organization of professional astronomers).

General Biological Requirements for Life

Let's consider some of the biological aspects first. Life requires complex molecules. In order for them to form and continue to exist, several conditions must be met.

1. Chemical elements that can form large molecules must be available. On Earth life is carbon-based; the structure of carbon enables it to form huge molecules, because carbon-carbon bonds are strong, not easily broken, and so carbon can form long molecular chains. Also, carbon readily combines with hydrogen to form an enormous number of complex compounds. On Earth, carbon, nitrogen, oxygen, phosphorus, sulfur, and hydrogen form DNA; iron, potassium, and magnesium are also important. All these elements are among the most abundant of the metals. Thus the elements needed for life are those that are common in the galaxy generally, and not those that are rare or somehow peculiar to the Earth. Note that helium, the second most abundant element, is chemically inert and does not play a part in the chemistry of life.

Some have speculated that in a different environment, life based on silicon might be another possibility. Indeed, like carbon, silicon can form long chains, but they are much more easily disrupted than those of carbon. More stable chains can be formed by silicon and oxygen, however. In fact, silicones are such compounds, much in use today. Even though such chains as well as hydrogen-rich silicon compounds such as SiH_4 (analogous to CH_4) can be formed in the laboratory, none have been found elsewhere in the solar system, in the interstellar medium, or in stars. Apparently only simple compounds such as silicon dioxide, SiO_2, and various silicates form in astronomical environments.

2. It is not enough simply to have an environment with an abundance of the right kinds of atoms. Nothing will happen unless they can be brought together and mixed for long periods of time. This function is provided by a solvent, which we usually think of as a liquid, because on Earth water (the "universal solvent") serves this purpose. Whatever it might be, the solvent should be abundant and capable of dissolving and carrying large quantities of atoms and molecules. Like the Sun, most parent stars would emit quantities of ultraviolet photons, which can break up molecules; hence the medium should also provide a shield against this radiation. The solvent should remain in the same physical state (for example, as a liquid) over as wide a temperature range as possible; that is, its freezing and boiling temperatures should be as low and as high as possible, respectively. Also, the larger the amount of heat required to change its temperature or to change its state from liquid to gaseous, the better. Such a solvent would better protect an organism against changes in the temperature of its environment. It turns out that, overall, water satisfies these requirements better than any other. Nonetheless, other media might work, for example, a thick gaseous atmosphere such as Jupiter's.

3. The temperature of the environment should be "moderate," and temperature variations must be relatively small. Just what that moderate temperature should be is a bit fuzzy, but the general idea is fairly clear: if the temperature is too low, chemical reactions would go very slowly (or not at all); if too high, molecules would be broken apart. Temperature variations must be such that the solvent should not freeze or boil, as discussed in the previous paragraph. This restriction may be bypassed under some circumstances, because we do know of organisms that can survive temperatures greater than 100°C and less than 0°C.

4. These environmental conditions must last for an appreciable length of time, another slightly fuzzy notion. Although the exact time is uncertain, the oldest life forms on Earth appeared about 1 billion years after its formation. We will take this period of time as the minimum necessary for the simplest forms of life to develop, and another 3.5 billion years as the minimum time for intelligent life to evolve.

From the description of the planets in the previous chapter, it is clear that the prospects for life in our solar system other than on Earth are not at all good. Mercury and the Moon have no solvent and undergo large temperature variations; Venus' surface is very hot; at the present time Mars has no solvent, though there is some evidence that in the past water existed on its surface. At the present time, its sandy surface, stirred up by winds, is constantly subjected to lethal doses of solar UV radiation. The giant planets have thick gaseous atmospheres, which might serve as a solvent and within a relatively narrow range of atmospheric depth might provide an appropriate temperature. Perhaps primitive life might have developed there, but that possibility seems remote.

Life Beyond the Solar System

If life in the solar system elsewhere than on Earth is unlikely, is it possible that life might arise far away, on a planet orbiting some other star? The direct answer is that we have no knowledge whatsoever of any life "out there," nor do we even have any good idea what the chances of that possibility are. We are now better able to define the questions that bear on this, however, and establish at least a rudimentary foundation from which to examine the problem. Let us now turn to some of the relevant astronomical considerations.

Astronomical Conditions on Parent Stars

Although other kinds of objects may be capable of supporting life, it is probably reasonable to assume that planets provide the most promising environments. Let's see what requirements stars must meet in order for their planets (we have to assume they have some) to have environments favorable for life, even intelligent life. All stars are not equally good parents in this respect.

1. It should not be a first- or probably even a second-generation star. In order to make complex molecules, elements other than hydrogen and helium are necessary. This means appreciable element synthesis must have first taken place in stars to produce the carbon, oxygen, nitrogen, etc., needed for life, and these elements must have been dispersed into the interstellar clouds from which new stars form. Metal-rich stars should be much better candidates than metal-poor objects.

2. Stars with long main-sequence lifetimes are necessary. If we would like intelligent life to develop (which on Earth required more than three billion years of evolution after the simplest organisms formed), the parent star must be of spectral type later than about F5. Earlier-type stars would evolve off the main sequence in less time, and thus would not have maintained the stable environment required long enough for intelligence to evolve. Strictly speaking, we should not include F5 or later stars as possible parents if they reached the main sequence only recently, because such a long time after planet formation is required for life to begin in the first place. Given the size of the other uncertainties, however, we won't worry about this point in making our estimates.

3. There must be a "reasonably" wide zone of moderate temperatures around the parent star, so that there is a decent probability that a planet will be located within it. (Circular orbits would, of course, be advantageous since they provide a constant radiation environment.) The hottest stars would have the widest zones, but their short main-sequence lifetimes eliminate them as possible parent stars. M-type stars live on the main sequence for extremely long times, but they are so cool that their life-supporting zone is very narrow, only a few hundredths of an AU wide. The probability of a planet forming within this tiny distance range is quite small. Rather arbitrarily, we will take K5 stars as the coolest ones whose favorable temperature zone is wide enough that a planet would probably have formed within it. In doing so, we are eliminating a large fraction of all stars, since low-mass stars are the most common (see Figure 25.5).

4. Single stars are probably more favorable than double systems, because they provide a stable radiation environment. We can imagine situations where this is not a significant factor, for example, planets orbiting one member of a binary system in which the two stars are widely separated. The second star could be so far away that its radiation was of little consequence to its companion's planets. We will be conservative, however, and assume that all binaries—about half of all stars—are unfit parents.

5. Finally, the star-formation process must produce planets at least some of the time. If this were not so, if planets were formed only by rare, one-of-a-kind events, then the question of life elsewhere would be effectively settled. Discovery of even one other planet or planetary system would put to rest this unlikely possibility. Also, finding another system could enable us to begin to determine which are the important, general properties of such systems. As with so many aspects of this topic, our ignorance is profound.

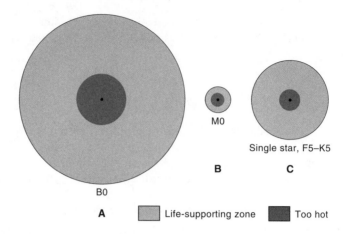

Figure 25.5. (A) The life-supporting zone around an early-type star is very large, increasing the probability that a planet might be formed within it. However, the main-sequence lifetime of such a star is too short for life to have formed. (B) An M0 star has a very long main-sequence life, but a very small life-supporting zone. (C) Favorable conditions are found among F5 to K5 stars, roughly. These have main-sequence lifetimes of several billion years as well as moderately wide life-supporting zones.

What we can say now is that we don't know of any reason why planetary systems *won't* form in considerable numbers. It may be relevant here that the separation of two stars in a binary system is quite commonly the same as the distance of the giant planets from the Sun. That is, the size scale of our solar system and of binary stars are roughly the same.

Life Elsewhere: Playing with Numbers

We can try to put all of this together with estimates for each of the relevant factors, and see what they tell us about the number of life-bearing systems. You should understand that the estimates we make in this chapter are uncertain in the extreme; they are useless as predictors. They do, however, point up the important elements in a discussion of this sort. In any case, let's bravely (or foolishly) plow ahead and make the guesses shown in Table 25.2. The first three factors listed are the only ones with some basis in fact. For the rest, we will give two guesses for each; an optimistic one and a pessimistic one, just to point out the huge uncertainties involved.

We don't know how likely it is that planets will appear as part of the star-formation process. As mentioned in Chapter 23, cold, flattened nebulae have been discovered around several nearby stars. If, as some astronomers feel, these nebulae are the sort that have formed planets, then additional infrared observations may make it possible to estimate how common the planet-forming process is. At the moment, however, we have certain knowledge of only one planetary system; the optimistic guess we made above concerning the probability that planetary systems form must remain just that—a guess.

A similar guess for the probability that, given a planetary system, the environment of one such object will be suitable for life is based on our one sample, the Earth, in a system of nine objects. Obviously, we have no way of knowing how good an estimate this is. Not only should a planet's orbit fall within the moderate temperature zone of the parent star, but it must have an appropriate surface gravity. That is, it is probably disadvantageous if its surface gravity is so large that the planet has retained all its pri-

Table 25.2. *Factors relevant to probability of life elsewhere*

	Reasonably well known	Guesses	
Factors		*Optimistic*	*Pessimistic*
Fraction of high-metal stars	0.5		
Fraction of F5–K5 stars	0.2		
Fraction of single stars	0.5		
Probability that these stars form planets		0.1	0.01
Probability that environment is OK for life		0.1	0.01
Probability that life *does* develop		0.1	0.0001
Probability that intelligent life develops		0.1	10^{-6}
Number of planets with intelligent life in the galaxy		10^6	10^{-4}

mordial hydrogen and helium; however, its gravity must be large enough that some at-mosphere is retained.

Mars and Venus are just on the outer and inner fringes, respectively, of the Sun's life-supporting region, so perhaps we should up the odds to three in nine. Mars in par-ticular might have been warmer in the past because of a stronger greenhouse effect than is now operative. Its weak surface gravity makes it difficult to retain an atmosphere, however. If Earth were in Mars' orbit, its thicker atmosphere might warm the surface sufficiently to compensate for the smaller amount of solar energy it would receive. Just remember, however, how similar the solar distances, masses, and radii are of Venus and the Earth, and how different their surface conditions are. Or recall the wide range of environments displayed by the four Galilean satellites of Jupiter. We should also re-member that, even though our Moon is within the range of acceptable temperature, it is not life-supporting, because its surface gravity is too small to retain an atmosphere. We will be conservative and adopt the one-in-ten estimate. Quite arbitrarily, we will take 0.01 for the pessimistic guess for each of these two factors.

The next factor, the probability that life does form given a suitable environment, will remain even more of a guess for some time to come, barring the actual discovery of life elsewhere. Our understanding of the origin of life is still far too primitive to do any better. Many biologists believe that life originated on Earth only once. In support of this assertion, they cite the fact that, as far as is known, the genetic code is uni-versal. That is, all organisms use the same DNA. Perhaps an even more significant fact has to do with the left or right "handedness" of a molecule. An asymmetric molecule can be of two forms, with each having exactly the same chemical composition, but the arrangement of the atoms in one is the mirror image of the arrangement in the other, like a pair of gloves. It is remarkable that all of the amino acids found in life are of the left-handed variety, rather than some organisms having right-handed amino acids and others having the left-handed molecules. This also strongly implies that life on Earth arose only once.

Now, when we have only one sample taken from a random collection of objects, it is impossible to make a meaningful guess as to how often that same sample will oc-cur. (Drawing one penny from a bag containing a large and unknown assortment of many different kinds of coins tells you nothing about the probability of drawing another penny.) Hence, the argument goes, it is impossible to make *any* meaningful guess as to the probability that life will develop on some other planet with a favorable environment. That is why finding even one more planet with life is so important, and why, for example, it would be extremely interesting to make a thorough search of Mars for evidence of any life forms. The often-taken guess of 0.1 for this probability (and the value we will take) could be orders of magnitude too large; we just don't know. For the pessimistic guess we will take 0.0001; some scientists feel that this is still too large.

If we multiply the optimistic guesses together, we get 5×10^{-5} as the fraction of stars that might have planets with life. If there are 2×10^{11} stars in our galaxy, then there are 10^7 stars with planets on which life has developed. Although this number must be taken with a healthy dose of salt, it does illustrate the common argument that, since our galaxy contains a huge number of stars, there may be a significant possibility that life exists somewhere out there. If, however, the pessimistic guesses are more nearly correct, there are only 100 life-bearing planets in the galaxy, a number so small that it could just as well be zero.

Improving the Estimates

Is there any hope of improving the accuracy of these estimates? In particular, what are the chances of actually discovering another planet or planetary system? Direct observation is extremely difficult, because a planet, shining by reflected light, is much fainter than its parent star. This, in combination with its proximity to its star, makes it extremely difficult to detect. For example, if Jupiter were in orbit around the nearest star (α Centauri) at the same distance it is from the Sun, it would appear to be about one billionth as bright as its parent and be less than 4 arcsecs from it. Detecting Jupiter from the Earth under such circumstances is out of the question with present techniques. A large telescope with excellent optics and stable pointing system in orbit above the Earth's atmosphere might have a chance, however. Observing at wavelengths that allow the planet to stand out more prominently against the star would be advantageous. For example, Jupiter's magnetosphere produces powerful bursts of radio waves that rival the Sun's output at the same frequencies. Unfortunately, none of these or similar search techniques have given positive results.

Another search method would be to look for the small radial velocity variations in the spectrum of the parent star as it orbits about the center of mass of the star–planet system. The velocity variation would have a period equal to that of the orbiting planet, so that observations would have to be made for years or even decades. Since the star would be much more massive than the planet, the velocity variations of the star would be extremely small and difficult to detect. For example, a planet with Jupiter's mass orbiting a star such as the Sun would produce a radial velocity variation of only about 0.01 km/sec, just barely within present observational capabilities. Such searches are underway with no convincing results so far.[3]

Astrometric techniques may be somewhat more promising. Just as a planet orbiting its parent star produces radial-velocity variations in the latter, it also produces de-

---

[3] Over a period of a few months beginning in late 1995, announcements were made of the discovery of three possible planets orbiting the stars 51 Peg, 47 UMa, and 70 Vir, all of which are similar to the Sun. The existence of planets is inferred from periodic radial velocity variations of the parent stars having amplitudes on the order of only 50 meters/sec! Preliminary results give these objects about 1, 3, and 8 Jupiter masses, and semimajor axes of 0.05, 2, and 0.6 AU, respectively. The masses are derived on the assumption that the planets' orbital planes are in our line of sight so that the observed velocity curves measure the full velocity.

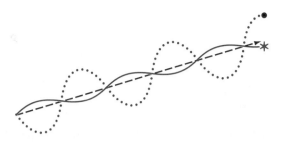

Figure 25.6. The proper motion of a star having a planetary companion is not a straight line, but a slightly wavy line, shown here by the solid line. The path of the unseen companion is shown by the dotted line, and the motion of the center of mass of the system is given by the dashed line.

partures from a straight-line path in the star's proper motion. If measurements of the proper motion of a star showed that it was moving not in a straight line but rather along a periodically wavy path, this fact would indicate the presence of an unseen companion in orbit around the star. This technique has been used to discover faint, low-mass stellar companions to visible stars or to determine the masses of visual binary systems (recall how the masses of Sirius A and B were measured in Chapter 12). Planetary objects have much smaller masses than stars, and so produce correspondingly small wiggles in the motion of the star (Figure 25.6). Furthermore, observations over many years would be required to establish the period and amplitude of the star's wavy path. With such measurements, however, it would be possible to estimate the mass of the companion and find out if it was star- or planet-sized. A few examples of this sort have been claimed, the best known of which is Barnard's star, but their interpretation as implying planetary systems is not generally credited. Again, a space telescope with excellent optics would enable much more precise measurements of this sort to be made, and, with a lot of luck, convincing evidence for a planetary system might be found.

It is true that pulsar "clocks" provide the kind of observational accuracy necessary for detecting objects of small mass. We saw in Chapter 16 that it is likely that two planetary-mass objects have been found in orbit around a pulsar. As indicated there, however, their likely origin, structure, and environment are most un–planet-like as far as their being possible abodes of life. Until more is known about these objects, it does not seem reasonable to count them as planets in this sense.

In summary, then, despite increasingly sensitive tools and techniques, we have yet to find clear evidence for the existence of objects with planet-like characteristics beyond our solar system.

More Playing with Numbers: Intelligent Life Elsewhere

Now let us require intelligence to be out there, rather than just any kind of life. This greatly reduces the number of possibilities and so makes success even more difficult to achieve.

Nonetheless, let's play this more demanding numbers game, and ask how many life-bearing planets there might be with which we could communicate. For this, we need not only life, and not only intelligent life, but technologically capable life. In this context, intelligence is only slightly less difficult to define than life itself, and we will not dwell on this question. For the sake of concreteness, however, let's simply take intelligence as the ability to store and manipulate information. With the attitude that what happened here will happen again, the optimistic might assume that if life develops, then about, say, 10 percent of the time it will go the whole way to intelligent life (see Table 25.2). That would mean about one million planets in our galaxy would be supporting intelligent life, a sizable number.

The pessimist, convinced that too much biology is being glossed over, might make the following comments. First, because it is so important to us, we often assume that intelligence is an advantage in the evolutionary struggle. Clearly, however, intelligence is not necessary for survival or even to thrive. Insects have been remarkably successful here, and though some species, ants for example, display much social organization, they hardly fit our definition of intelligence. Indeed, very unintelligent creatures will almost certainly outlive us. Furthermore, there is nothing in evolution by natural selection that forces life to become increasingly intelligent. Natural selection simply tests accidental genetic changes (mutations) against the environment. Those mutations that enable the organism to adapt better to its environment may be retained; others probably won't be.

Second, the number of possible random mutations is so incredibly large that only a tiny fraction of them could have been tested in the development of life on Earth. Consequently, if the whole evolutionary process were to begin again here on an early Earth, after several billion years, life (even if it developed) would look little like it does now except possibly in its most general characteristics. There would be no organisms recognizable as "people," for example.

Putting this in a different way, the process of biological evolution cannot be thought of as being deterministic in quite the same way we regard the laws of classical physics. In contrast to an experiment in physics, it is too simplistic to say that in an "evolutionary experiment" what happened once will certainly happen again. It is true that at each step of the way, the laws of physics and chemistry do determine what happens. The number of possible random variations is so huge, however, that there is no guarantee that, *of necessity*, the outcome of a long sequence of events will be repeated the next time around.

It is true, however, that many examples can be found of **convergent evolution**, the independent development of similar characteristics by organisms living in different environments. For example, porpoises, descendants from mammals, have come to resemble large fish-like sharks; the toucans of South America and the hornbills of Africa have both developed prominent bills used for crushing nuts; insects, fish, reptiles, birds, and mammals have independently developed means of flight. But counterexamples can also be given; in particular, for the last 200 million years Australia has been isolated from other land masses, as has South America for about 130 million years. No primates developed in Australia, and although monkeys did arise in South America, no great apes did.

In any case, life and intelligence required a large number of crucial biological milestones, realized in a specific sequence, in order to have developed here. To illustrate the point, let us briefly describe a few of these. For example, one was the development of aerobic photosynthesis (photosynthesis producing oxygen as a byproduct), the same process used by plants today. Cells that were able to use this (initially poisonous) oxygen in their metabolic processes, as ours do, produced much more energy than those that couldn't. This must have been an important factor in their survival and development. Furthermore, the oxygen released in photosynthesis eventually began building up in the atmosphere, forming ozone (O_3) that protected the Earth's surface from solar ultraviolet radiation.

Only about 1.5 billion years ago, a cell having a well-defined nucleus appeared. This was critical for the development of sexual reproduction, which in turn made possible a myriad of genetic combinations and so greatly speeded up evolution. Up to this point, organisms were still single-celled, but finally, only a little more than half a billion years ago, multicell organisms appeared, but not before DNA acquired two additional amino acids necessary for multicellularity. This resulted in the so-called Cambrian explosion of life forms.

A hundred million years ago or so, life moved to land. To be able to do so successfully, plants and animals both had to develop means of support to replace their lost buoyancy along with a way of absorbing oxygen. The animals did so by developing backbones and lungs.

A much more recent example is that, of the four related primates with prehensile hands (gorillas, orangutans, chimps, and protohumans), only the latter learned to walk on two legs, thereby freeing the arms and enabling the hands to be used for a wide variety of functions.

Even from this incomplete account, you can see that a large number of critical developments, none of which was inevitable, had to occur in order for intelligent life to appear. Just how many may be debatable, but let's say there were only ten. If we guess that each of the ten has a one in four chance of successfully occurring, that would mean that if life develops, then only one time in a million—$(1/4)^{10}$—would it go the whole way to intelligent life. With the other pessimistic guesses made above, that would mean about *ten thousand* galaxies would have to be searched to find even one example of an intelligence-supporting planet. Thus our optimistic and pessimistic guesses differ by ten orders of magnitude!

We are not done yet, however. Another big problem is not biological, but societal. That is, what is the *overlap* time during which we have had the technology to detect such signals and the other beings have been capable of sending them? Life has existed here for about 3.5 billion years, but we have gotten to a level of technological sophistication adequate for interstellar communication in only the last 50 years or so. How much longer will we continue to exist, at least at this technological level? More generally, how long do civilizations capable of interstellar communication, and so probably capable of destroying themselves by one means or another, actually last? We all hope we don't find out at first hand, but, clearly, it is an important factor in this discussion.

An example will make this more concrete. For the sake of argument, let's assume rather pessimistically that such technologically advanced civilizations, including the Earth's, last only about 1,000 years. So until about the year 3000 we would be capable of receiving messages from other civilizations. But 1,000 years is about 10^{-6} to 10^{-7} of our parent star's lifetime. Combining this with the optimistic guesses above, we find that no more than about one suitably advanced civilization would exist at any given time: Ourselves! In principle, it could be located anywhere in our galaxy, just as long as its message arrived between now and the year 3000. If we wished to indicate our presence to them, we could do so, but unless they were no more than 500 light years away, no being would be there capable of receiving our message.

Becoming pen pals requires someone to be next door to us. To exchange five messages, to ask questions and receive answers five times, they must be nearer to us than 100 light-years, assuming, of course, that their 1,000 years fully overlapped ours. Within 100 light years of the Sun, however, there are only about 1,000 stars; so the likelihood that intelligent life is associated with one of them is extremely small. It is clear that the lifetime of a technologically advanced civilization is another critical factor in this discussion, and one about which we can only guess.

Though the absolute values of these numbers are uncertain in the extreme, it is obvious that unless advanced civilizations can last for very long times, the number that we might be able to detect is several orders of magnitude smaller than the number of planets that simply support life of any sort. If the pessimistic values are more nearly correct, there is essentially no chance of finding anything to communicate with. This does not make for enormous optimism when undertaking a search. If we don't try, however, (say the optimists), we have no chance whatsoever of succeeding.

Local Circumstances

How important might local situations be? For example, because our Moon is unusually large with respect to its primary, it raises significant tides on Earth. A few billion years ago these tides brought fresh nutrients twice a day to the pools often invoked as the nurseries of primitive life. In addition, the large Moon has kept the Earth's orbital parameters from varying much, and so prevented the kinds of climatic changes that may have affected Mars. Only one other planet, Pluto, has such a relatively large satellite, but it is far out of the Sun's life-supporting zone. Could our Moon have been critical for life here?

Another interesting local wrinkle has arisen since the 1980s. It has been estimated that perhaps as many as 4 billion species have so far existed on Earth, but only several million are alive today. Thus, the vast majority of organisms that ever lived on Earth are now extinct. Though many extinctions must have occurred independently and gradually over long time spans, evidence is accumulating that there have been several relatively short periods during which large numbers of species died. The largest of these extinctions took place about 250 million years ago, when perhaps 95 percent of all species disappeared in only a few million years. Perhaps the best known extinction happened 65 million years ago when 75 percent of all species, including the dinosaurs, disappeared. What could have caused these is not clear: various geological processes, volcanoes, or a nearby supernova are some of the mechanisms suggested.

In 1980 another possibility was put forward, namely, that 65 million years ago the Earth was struck by a sizable asteroid that produced huge quantities of dust-blocking sunlight for months or even years. The heat of the shock wave from the impact would have caused continent-wide forest fires. In addition, vast quantities of nitrous and nitric oxide were produced that asphyxiated air-breathing animals and caused extremely acid rain. It is suggested that these disasters led to the observed extinction.

What is the evidence for such an impact? Iridium, rare in the Earth's crust but much more common in some meteorites, was found in a thin layer of 65-million-year-old rocks from many parts of the world, with an abundance as much as 100 times greater than normal. In addition, minerals were found in this layer that showed the effects of high shock pressures. A 65-million-year-old crater on the Yucatan coast, 180 kilometers in diameter, has been identified as the likely impact crater. If, as has been suggested, its diameter is really 300 kilometers, then it would be not only the largest impact on Earth, but the largest in the inner solar system since the end of the major bombardment period nearly 4 billion years ago. (Only the crater Mead on Venus is of comparable size.) Unfortunately, the Yucatan crater is buried under 2 kilometers of sediment and its properties have to be ferreted out by mapping subtle changes in the Earth's gravity and magnetic field. The larger diameter is not yet generally accepted. In any event, that a major impact occurred is certain, but what its effects were, are not. Some paleontologists believe that the impact was not the single cause of the extinction, arguing that it simply contributed to the environmental stress caused by the change in sea level and by the 2-million-year-long volcanic eruptions that occurred in India during this period.

What does all this have to do with us? Although small mammals and dinosaurs coexisted for a long time, it has been suggested that the mammals proliferated in number and species as a consequence of the disappearance of the dinosaurs. Thus we might owe our existence to a catastrophe that befell the dinosaurs! The point of all this is that not only biological factors, but also special astronomical events, impossible of generalization, can play a significant role in the development of life in any particular situation.

Searching for Intelligence

Interstellar Travel

Because of the immensity of space, our quest by direct exploration for evidence of any sort of life will be limited to our own solar system for the foreseeable future. A direct search for any kind of life by jumping into a spaceship and going from one star to another is incredibly inefficient. The reason for this is simple: interstellar distances are enormous. At present, the greatest speed we can sustain for a long period of time is that of the planetary probes (Pioneer or Voyager), which travel from planet to planet at about 10 km/sec. At this rate it would take more than 100,000 years to reach the nearest star, α Centauri!

So let's go faster. In fact, let's assume that for our trip to α Centauri, we can achieve 0.99 of the speed of light. This we will do by accelerating at a constant rate of 1 g (980 cm/sec^2). Obviously, that is a comfortable acceleration for humans; this is an important point when we must accelerate for long periods of time, and not just for the few minutes needed to reach the relatively glacial speed for Earth orbital or lunar travel. However, the greater the acceleration, the more rocket fuel is required.

It will take about one year to reach 0.99 light speed, during which time our spaceship will have traveled about 0.5 light-years (see Figure 25.7; make sure you understand why it will have traveled this far). It then simply coasts at 0.99 c until it is within about 0.5 light-years of α Centauri, when it would slow down in the same way that it accelerated, taking another year. Thus the trip there would take about 5.2 years. After a few months spent orbiting the star, our travelers return home in the same way—accelerating, coasting, decelerating—arriving about 11 Earth years after their departure.

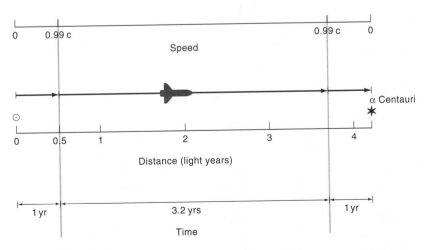

Figure 25.7. A high-speed trip to the nearest star, α Centauri, 4.2 light-years away. If the spaceship accelerates at 1 g (980 cm/sec/sec), then after one year, it will be traveling at essentially the speed of light, c, and will have traveled 0.5 light-years. It then coasts for 3.2 years, after which it must decelerate (again at 1 g) for one year to reach its destination at zero speed. Such a one-way trip would take 5.2 years.

Because they are traveling so rapidly, however, their clock will have measured less time than would an Earth-bound clock. Sounds good.

But what fuel will be used to propel the rocket to such a high speed? As you know, most of the mass of a rocket at lift-off is the fuel; so to achieve enormous velocities, we will need the most efficient fuel imaginable, namely, matter combining with anti-matter. In this process *all* the interacting mass is converted to energy. (Remember that in the nuclear fusion of hydrogen to helium, only 0.007 of the mass is converted to energy.) Even so, it has been calculated that to fly to α Centauri as we described above would require about 40,000 times as much fuel as payload: crew quarters, instruments, food, oxygen, etc. If the latter amounted to 10^5 kg, then an incredible several billion kg of matter and antimatter would be required! And don't forget the problem of manufacturing antimatter, and keeping it from interacting with matter before you want it to. To put it mildly, these are not trivial matters.

So a compromise of sorts has to be reached between realistic quantities of fuel and the time required for the trip. In particular, this means that long trips would be made by self-sustaining space colonies that would in effect become independent entities, with little connection to Earth. If Earth-bound beings are going to search for evidence of life, another approach must be taken.

Searching for Signals

It is far more effective (despite *Star Trek*) to stay here and listen for radio signals, or return the possible favor by sending some out into space ourselves. Such signals travel with the speed of light, and can be directed anywhere the sender desires. How would a distant civilization announce itself to us, or how would we inform them of our existence? Many interesting considerations enter this discussion.

Electromagnetic radiation at radio frequencies is the most efficient way to declare our presence across the galaxy. Radio wavelengths have several advantages over other regions of the spectrum. Radio photons are at the low-energy end of the spectrum, and so are cheaper to produce than the higher-energy visible or x-ray photons. Technological societies such as ours know how to build extremely powerful radio transmitters and exquisitely sensitive receivers. Furthermore, it would be extremely difficult to detect an optical signal from a planet, because its parent star would be so much brighter. With radio signals, it is relatively easy to outpower a solar-type star over a narrow radio-frequency band. Furthermore, radio waves can penetrate the interstellar dust that pervades the plane of the Milky Way and other galaxies.

Which Frequency?

The radio spectrum is broad, extending from wavelengths as long as meters to those as short as millimeters. So the next question is, at what wavelength(s) should the search be made? The low radio frequencies (those of AM radio, for example) are reflected by the Earth's ionosphere, whereas the highest radio frequencies are absorbed by our atmosphere. This suggests that we should use frequencies between about 10^6 and 10^{10} Hz. Various sources within our galaxy radiate low-frequency synchrotron energy, however, which produces a confusing, "noisy" background against which we must attempt to detect weak signals. This consideration leads us to raise the low-frequency limit from 10^6 up to about 10^9 Hz, leaving about a decade range of wavelengths from about 30 to 3 centimeters, or 1 to 10 GHz (Figure 25.8).

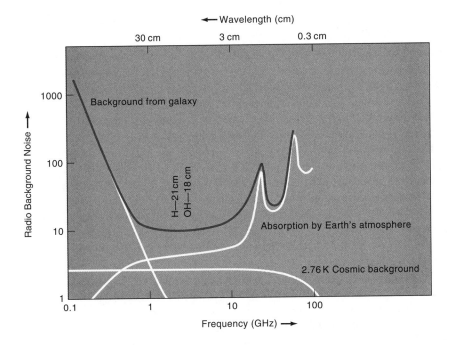

Figure 25.8. The contributions of various sources to the radio brightness of the sky background. Note the minimum from about 1 to 10 GHz (10 to 30 cm). Because lines of H and OH are emitted in this region of the spectrum, the minimum is sometimes called the water hole.

Recall that atomic hydrogen has a transition at 21 cm. Since this is a prominent line of the most abundant element, it seems obvious that any technologically advanced beings would know about it; consequently, the first searches were made at this frequency. Later, when other radio lines were discovered, they too were suggested as reasonable search frequencies, for example, the line at 18 cm due to OH. Suppose, however, we settle on the 21-cm line. Just how do we tune our receiver? To exclude as much extraneous electrical "noise" as possible, and so maximize our sensitivity to 21 cm, we would like to make the bandwidth of frequencies accepted by the receiver as narrow as possible. This has the disadvantage, however, of excluding the 21-cm line whenever it is appreciably Doppler-shifted to longer or shorter wavelengths. Such shifts could be caused by the radial velocity of the parent star with respect to the Sun, as well as by the motion of the planet around its parent star. Ideally, then, we would like to divide the spectrum around 21 cm into as many closely spaced frequency bands as possible (so that a signal with a large Doppler shift will be detected), each of which is narrow (to reduce noise). It would take far too long to scan a broad spectral region one frequency at a time with a single channel receiver. A multichannel receiver capable of measuring many million channels simultaneously must be used.

An example will show what is involved. Let's assume that we want to be able to detect the 21-cm line over a range of velocities of +50 km/sec to −50km/sec. From the Doppler formula we find that this velocity range corresponds to about 5×10^5 Hz. If we take 0.1 Hz to be the width of each channel (the bandwidth used by some radars), then the total number of channels to be measured is about five million. This is a large number, but it is possible to make receivers that scan this many channels simultaneously. The technology to do this has developed remarkably in recent years.

What to Look For?

Finally, what would we look for? After all, 21-cm radiation is copiously produced in our galaxy. How would we know that a signal is artificial—produced and controlled by thinking beings—rather than naturally occurring? (Remember that for a short time, astronomers thought that pulsar signals just might be artificial.) The simplest sort of message might be the first few prime numbers (numbers divisible only by one or by themselves). For example, a series of pulses representing 1, 3, 5, 7, 11, 13, 17, 19, 23, repeated over and over again, would certainly be eye-catching and difficult to interpret as a consequence of some natural process. Another possibility would be the repetition of the value of a basic constant such as 3.1415926 . . . , over and over again.[4]

Messages containing information beyond the mere indication of presence could be sent in such a way as to be decipherable, if that was the intent. For example, this could be done by describing natural phenomena, like the structure of hydrogen, helium, carbon, and other common elements certain to be known by anyone capable of receiving the signals in the first place.

Of course, it might be possible to detect signals that were not intended for us or anybody else, but simply were a consequence of normal activity. Consider the situation here. Powerful television stations and military radar are not uniformly distributed across the Earth but are concentrated in North America, western Europe, and Japan. Because of the location of television viewers, most of the energy is radiated along the Earth's surface rather than being emitted uniformly in all directions. As seen at a distance from the Earth, these transmissions produce a varying amount of radio energy as the Earth rotates and causes the TV stations to rise and set. The pattern is repeated every day, but will be slowly Doppler-shifted as the Earth revolves around the Sun. At the very least, these radio signals would indicate that something unusual was going on around an otherwise completely undistinguished G2V star. (Just think of some distant civilization trying to make sense of *Wheel of Fortune*!) Any advanced civilization capable of hearing us must be nearby, within a sphere of radius about 40 light-years, which expands by one light-year per year (since TV stations were becoming common around 1950). Even the most optimistic estimates do not put advanced life this close to us.

The Search So Far

The year 1960 marks the beginning of modern searches for extraterrestrial intelligence when an astronomer at the then newly formed National Radio Astronomy Observatory at Green Bank, West Virginia, turned an 85-foot radio telescope in the direction of two stars less than 4 parsecs away, ε Eridani and τ Ceti. He used a single channel receiver tuned to 21 cm. All told, the two stars were observed for about 200 hours, with no indication of artificial signals.

Why would these two stars be chosen? Obviously, you begin your list of candidates with the nearest stars to maximize the chance of detecting a weak signal. Appendix D lists the stars within five parsecs of the Sun. Ten (not counting the Sun) are main-sequence stars with spectral types between F5 and K5; three of these are too far south to be observed from Green Bank. Of the remainder, ε Eridani (K2V) is the nearest, about 3.3 pc distant, and τ Ceti (G8V) is in its spectral type most similar to the Sun. Hence these two stars are reasonable choices for a first try.

---

[4] A genuine example of π in the sky!

Such searches gained respectability among astronomers in the 1970s after the U.S. and Soviet Academies of Science sponsored a conference on the topic in the Soviet Union, and after a series of meetings and studies had been conducted by NASA.[5] As of 1985, about 120,000 hours of observing had been carried out with facilities in Australia, Canada, France, Germany, Holland, the Soviet Union, and the U.S. Two long-term efforts have been underway for several years, one at Ohio State University, the other at Harvard. The former telescope operates with 50 frequency channels at 21 cm; the latter has been tuned to an OH line as well as to the hydrogen line using a receiver having 65,000 channels. In 1985 a receiver having 8.5 million channels was put into operation.

After several years of trying, NASA received approval from Congress to undertake a Search for Extraterrestrial Intelligence (SETI). It began on Columbus Day 1992 and was to continue until the turn of the century. Although only a year after operations began Congress cut the funding, it is interesting to see how a modern search could be carried out. Two strategies were planned: an examination of about 800 nearby stars and a less-sensitive survey of the whole sky. Both searches would have had unprecedented spectral resolution, bandwidth, and sensitivity. The targeted program was to observe the spectrum of each star from 1 to 3 GHz with a resolution of less than 1 Hz. The broad bandwidth would take care of the Doppler-shift problem mentioned earlier. The search was sensitive to pulsed signals (somewhat like those from a pulsar) as well as to continuous-wave signals. After the program was canceled, some observing time with this equipment was obtained on two Australian radio telescopes. Efforts are underway to raise private funds to enable this search to continue. The all-sky survey was designed to observe the frequency band from 1 to 10 GHz at about 30 Hz resolution. In just a day or two of operation these programs processed more SETI data than in the 30 preceding years of such work. Had the full program been performed, even a negative result might have been useful in reassessing the optimistic estimate.

It is certain that signals from intelligent beings would have been detected—our own! Some scientists believe that time is rapidly running out for high-sensitivity radio astronomy from Earth.[6] The radio-frequency spectrum is increasingly filled with commercial and military signals, all incomparably stronger than those from extraterrestrial sources. Satellite broadcasting and, more recently, the use of portable telephones will increase the pressure to allocate more of this frequency band to commercial uses. A major challenge in any SETI program is to be able quickly to distinguish our own signals from possibly more interesting ones. Enormous amounts of data must be processed each second, requiring extremely fast computers and data-handling equipment.

The technology being applied to these searches has improved dramatically, significantly improving the chance for success. Nonetheless, most people feel that the likelihood of a positive detection is still extremely small.

A Negative View

In fact, some scientists say flatly that the likelihood is zero, because no intelligent life is out there. This is sometimes called the "Where are they?" argument and is based on technological, rather than biological considerations. In its outline, it is simple: eventually, intelligent life elsewhere would achieve such a high level of technological

---

[5] There are still many skeptics who claim that bioastronomy or exobiology, as it is sometimes called, is a science without a subject matter!

[6] Except for all the obvious difficulties, the far side of the Moon would be an excellent site for a radio observatory, completely shielded from Earth-generated electromagnetic pollution.

sophistication that such beings would be able to explore their surroundings and ultimately the entire galaxy; since they *could* have done so, they *would* have done so, and consequently would have been here, but we see no evidence of their visit; therefore they don't exist.

Some proponents of this argument suggest that the exploration would be carried out not by the intelligent beings themselves, but by nearly as intelligent robots. Such a robot would be sent off in a spaceship and, after a journey of great length, arrive at a possible parent star, explore the surroundings, and perhaps find an asteroid-like body. There the robots would build—entirely from scratch—all the equipment necessary to mine the ore required to build the facilities that would eventually produce all the bits and pieces that would then be assembled into other spaceships and robots who would then fly off in the newly constructed spaceships on the next leg of their exploration, and so on. Because the exploration would be carried out by robots, the lifetime of the advanced society that sent the first one out would be completely irrelevant. (This, of course, raises the question of why the originating civilization would bother.) According to this and similar scenarios, the galaxy would be explored in this way in less than a few hundred million years.

As you might imagine, this argument has been questioned in many ways. None of its premises can be stated with absolute certainty, but only with degrees of probability. Can the nearly countless steps in this process be made to be truly self-replicating? Or would the process eventually founder because of errors that could not be corrected? Hence the conclusion itself is no more than a probability. Even if the premises are conceded, how can it be said that there is no evidence of a visit? We have explored only a minuscule fraction of the solar system in the detail that might be required to discover any artifacts left behind. Even the Earth, mostly covered by water, has not been searched sufficiently carefully to say that no artifacts exist.

This is a particular example of the familiar idea that "absence of evidence is not evidence of absence." More broadly, is it obvious that beings intelligent enough to undertake the exploration of the galaxy, even by robots, would actually do so? Even more, is it necessarily true that if anything is possible, it will in fact be done? The proponents of the "Where are they?" argument say that the simplest assumption to make is that "they" will behave as we do. We are explorers; so they will be too. This may be the simplest assumption, but it is also a highly questionable one.[7] Indeed, at the present stage of our technological development, we are like children in a toy shop: we want everything. Is it impossible to speculate, however, that as we "grow up," we will become mature enough to make choices, to decide to do some things, but not others? Many astronomers are less than enthusiastic about searches for intelligence, citing the biological problems; how many base their skepticism on the "Where are they?" argument is not clear.

Implications of Success or Failure

Despite the "Where are they?" argument, most scientists feel that it would be extremely difficult, in fact, probably impossible, to establish that *no* intelligent life existed somewhere out there. (Perhaps a mark of "their" sophistication is to want nothing to do with us.) But suppose that we do find indisputable evidence that life does indeed exist elsewhere. That single fact would be, it seems to me, by far the most important result of our search. Ever since we have been conscious of ourselves, we have had the tendency to think that we were, in a fundamental sense, special, unique, the purpose and end of creation.

---

[7] Would we undertake a costly program that had a payoff only in the unimaginable future?

Figure 25.9. Comments by Porky, the noted philosopher.

Copernicus removed us from the very core of the solar system, which itself was held to be the center of the universe. Newton and his successors showed that we were not unique, in that we were connected with the rest of the cosmos through the same physical laws. Herschel, Shapley, and Hubble demonstrated how unimaginably huge was the universe, and how in no sense were we located at its center or, in fact, at any point of significance whatsoever. Each of these developments forced us to change our ideas about ourselves and to reconsider what our place in the universe is. The discovery of life elsewhere would strip us of our last vestige of false pride, and make us face the corresponding philosophical consequences. As the price of our lost isolation, we would have to reassess our basic nature and our place in this changed universe. It would require our ultimate, perhaps our most profound, readjustment.

We might even hope that the discovery of life elsewhere would make us more conscious of our common humanity with everyone on this planet. If that idea could penetrate our consciousness to an emotional level, is it too much to hope that we might then think differently about annihilating each other? It probably is, but the other side of the coin gives us the same message. If we are alone, think of the responsibility we have to preserve life in all its forms! At the present time these thoughts receive lip service only. Perhaps we need a profound jolt, such as the discovery of life elsewhere, to change the way we look at ourselves and the world.

Now let's take a further step. Let us suppose that not only did we find evidence of life, but that we actually were to make contact by radio with an advanced culture, one that was willing to communicate with us. What then? First of all, we must remember that even if these beings were nearby, say, 100 light-years away, the "conversation" would not be too sparkling. We would have to wait at least 200 years for an answer (during which time we would have probably forgotten the question). If they were so inclined, however, they could speed things up by transmitting to us the *Encyclopedia Galactica*. Being able to check out books from the galactic library would, of course, be fascinating, but would it be terribly helpful with our immediate problems, as some people have claimed? Just consider, for example, what that implies about "their" social and political institutions. To be of any practical use to us, their institutions would have to be at least recognizably similar to our own. But could we even speak in such terms or would their societal structures be impossibly alien? Would we really learn how to avoid destroying each other in a nuclear war, or discover how to keep our environment from becoming intolerably poisoned? It seems to me that one can far more easily imagine situations in which the sudden acquisition of advanced knowledge would quickly lead to disaster. Would we be able to handle this risk?

Many of the difficulties we face now are not caused by a lack of technology, but by our apparent inability to use technology sensibly. These are political and social problems that (barring imposition of a new political order by bug-eyed monsters) we earthlings have to solve for ourselves and solve soon. In this kind of practical sense, it seems to me, spaceship Earth is alone, and we bear the full responsibility for its continued health and that of all of its inhabitants.

What do you think?

Terms to Know

Panspermia; amino acids, nucleotides, proteins, nucleic acids, DNA; convergent evolution.

Ideas to Understand

Early speculations about life elsewhere; why Mars was such a popular locale for life; the Urey-Miller experiment and what it showed. The biological requirements for life; the astronomical requirements; how to search for other planetary systems. Relevant factors in estimating the probability of intelligent life; possible implications of the dinosaur extinction. How to search for and recognize intelligence; how the search has been conducted so far; the "Where are they?" argument; implications of success or lack of it.

Questions

1. A planet suitable for life should presumably have a reasonably stable temperature for several billion years. Would the following likely be good parent stars? Explain your answers briefly.

(a) A red giant
(b) A blue main-sequence star
(c) A white dwarf

2. Suppose the effective temperature of our Sun were 1.5 times greater than it actually is. What consequences would this have had for the development of life on this planet? Be quantitative when you can.

3. What are the difficulties and implications with any collision theory of the origin of the solar system? Which do you think is the most significant? Do you consider this an important objection?

4. Would life be more or less likely to form on a planet whose parent star was located in a globular cluster? Explain your answer.

5. Suppose, as some scientists claim, that the rapid extinctions of life on Earth were periodic, occurring at intervals of about 25 to 30 million years over the last 250 million years. They suggest that this happened because the Sun has a faint companion star that periodically stirs up the Oort comet cloud, sending some in to collide with the Earth. If the period of the supposed solar companion (given the name Nemesis!) is 30×10^6 years, what is the semimajor axis of its orbit? What fraction of a parsec is this?

6. Show that a planet in orbit around α Centauri (1.3 pc distant) at the same distance as Jupiter is from the Sun (5 AU) would be no more than about 4 arcsecs from it as seen from the Earth.

7. Why—or why not—do you think public funds (say, $5 million per year) should be spent on SETI?

8. Suppose we did make contact with another civilization in our galaxy. Given the numbers developed in this chapter, why would it nearly certainly be far ahead of us technologically?

Suggestions for Further Reading

The following books give a good historical review:

Crowe, M., *The Extraterrestrial Life Debate: 1750–1900*. Cambridge: Cambridge University Press, 1986. Begins where the volume below ends.

Dick, S., *Plurality of Worlds: The Origins of the Extraterrestrial Life Debate from Democritus to Kant*. Cambridge: Cambridge University Press, 1982. A fascinating account of the early ideas.

Lowell, P., *Mars*. Boston: Houghton, Mifflin and Company, 1896. A discussion and interpretation of many of Lowell's drawings in terms of life on the planet.

Searle, G., "Are the Planets Habitable?," *Mercury*, **12**, p. 110, July/August 1983. A reprint of an article written in 1890. Very interesting.

"Life in the Universe," a special issue of *Scientific American*, **271**, October 1994, begins with the origin of the universe, but concentrates on the evolution of life and intelligence on Earth.

Just a few of the many books dealing with modern views of the possibilities of and the search for life elsewhere are listed below:

Bracewell, R., *The Galactic Club: Intelligent Life in Outer Space*. San Francisco: San Francisco Book Company, Inc., 1976.

Drake, F. and Sobel, D., *Is Anyone Out There? The Scientific Search for Extraterrestrial Intelligence*. New York: Delacorte Press, 1992.

Goldsmith, D. (ed.), *The Quest for Extraterrestrial Life*. Mill Valley, CA: University Science Books, 1980. A collection of articles.

Goldsmith, D. and Owen, T., *The Search for Life in the Universe, Second Edition*, Reading, MA: Addison-Wesley Publishing Company, 1992. Gives a lot of background astronomy and biology.

Horowitz, N., *To Utopia and Back: The Search for Life in the Solar System*. New York: W. H. Freeman and Company, 1986.

Rood, R. and Trefil, J., *Are We Alone?* New York: Charles Scribner's Sons, 1981.

The next several articles have to do with the solar system, beginning with the Earth. The impact that killed much of Earth's life 65 million years ago is described in the first two articles:

Beatty, J., "Killer Crater in the Yucatan?," *Sky & Telescope*, **82**, p. 38, 1991. Was this the impact responsible for the extinctions 65 million years ago?

Dietz, R., "Demise of the Dinosaurs: A Mystery Solved?," *Astronomy*, **19**, p. 30, July 1991.

Russell, D., "The Mass Extinctions of the Late Mesozoic," *Scientific American*, **246**, p. 58, January 1982. The extinction that did in the dinosaurs and its possible cause.

Knoll, A., "End of the Proterozoic Eon," *Scientific American*, **265**, p. 64, October 1991. Why animals suddenly appeared about half a billion years ago.

Chyba, C., "The Cosmic Origins of Life on Earth," *Astronomy*, **20**, p. 28, November 1992. The possibility that the essentials of life on Earth may have been extraterrestrial in origin.

Comins, N., "Life on a Metal-Poor Earth," *Astronomy*, **20**, p. 40, October 1992. Consequences for the Earth if it had formed long ago when the metals were less abundant.

————, "Life on an Older Earth," *Astronomy*, **21**, p. 40, March 1993. Suppose intelligent life had taken 9 billion years to evolve on Earth.

Carroll, M., "Digging Deeper for Life on Mars," *Astronomy*, **16**, p. 6, April 1988. A post-Viking view of the possibility of life on Mars and how to look for it.

Mood, S., "Life on Europa?," *Astronomy*, **11**, p. 16, December 1983. Speculation about life on what would seem to be an unlikely object.

Strand, L., "The Search for Life on Mars: Shots in the Dark," *Astronomy*, **11**, p. 66, December 1983. A description of the Viking experiments and results.

The articles below are concerned with the search for extrasolar planets and intelligence:

Black, D., "Worlds Around Other Stars," *Scientific American*, **264**, p. 76, January 1991. How planets form and how to look for them.

Bruning, D., "Desperately Seeking Jupiters," *Astronomy*, **20**, p. 36, July 1992. Good discussion about how to look for planets.

Croswell, K., "Does Barnard's Star Have Planets?," *Astronomy*, **16**, p. 6, March 1988. The evidence (for and against) planets around Barnard's star.

Finley, D., "The Search for Extra Solar Planets," *Astronomy*, **9**, p. 90, December 1981. Some of the ways it is done.

Goodman, A., "The Diplomatic Implications of Discovering Extraterrestrial Intelligence," *Mercury*, **16**, p. 56, March/April 1987. Thinking ahead!

Horgan, J., "In the Beginning . . . ," *Scientific American*, **264**, p. 116, February 1991. New ideas about how life might have begun.

Naeye, R., "SETI at the Crossroads," *Sky & Telescope*, **84**, p. 507, 1992. A good overview of the search for artificial signals from space.

Olson, E., "Intelligent Life in Space," *Astronomy*, **13**, p. 6, July 1985. A good look at the biological aspects. Very interesting.

Papagiannis, M., "The Search for Extra Terrestrial Civilizations—A New Approach," *Mercury*, **11**, p. 12, January/February 1982.

———, "Bioastronomy: The Search for Extra Terrestrial Life," *Sky & Telescope*, **67**, p. 50, 1984.

Parker, B., "Are We the Only Intelligent Life in Our Galaxy?," *Astronomy*, **7**, p. 6, January 1979. A review of the prospects.

Schechter, M., "Planets in Binary Star Systems," *Sky & Telescope*, **68**, p. 394, 1984. Possible orbits when two stars and a planet are at similar distances from each other.

Schorn, R., "Extraterrestrial Beings Don't Exist," *Sky & Telescope*, **62**, p. 207, 1981. A brief account of the negative view.

Shostak, S., "Listening for Life," *Astronomy*, **20**, p. 26, October 1992. Describes the NASA search program (since canceled).

Tarter, J., "Searching for Them: Interstellar Communication," *Astronomy*, **10**, p. 6, October 1982. Includes a list of all the searches undertaken through 1981.

A Little Arithmetic

Scientific Notation

Numbers should be your servants, not your masters. The math in this book is straightforward, and certainly not beyond your abilities. Depending on how much you remember from high school, you may have to give it a little effort, but don't panic: you can understand it! You will also find that the ability to do the kinds of manipulations described in this appendix will serve you well throughout your life in all sorts of ways, from dealing with a fast-talking salesperson to deciding whether a politician is talking nonsense or not.

Let's begin with so-called **scientific notation**. Very large and very small numbers abound in astronomy, for example:

(a) the distance of the Earth from the Sun: 93,000,000 miles;

(b) the speed of light: 186,000 miles per second;

(c) the density of an interstellar cloud: 0.000,000,000,000,000,000,000,0017 times that of water.

Scientific notation is a convenient way of writing numbers so that they can be manipulated (multiplied, divided, etc.) easily. The idea is to express a number as the product of two numbers; the first is usually between 1 and 10, and the second is the appropriate power of ten (called the exponent). By the latter we mean just the number of times 10 must be multiplied by itself to give the desired number. For example, $100 = 10 \times 10 = 10^2$ (the exponent of 10 is 2); or $1,000,000 = 10 \times 10 \times 10 \times 10 \times 10 \times 10 = 10^6$. Note that

$$1 = 10^0$$

$10 = 10^1$	$1/10 = 0.1 = 1 \times 10^{-1}$
$100 = 10^2$	$1/100 = 0.01 = 1 \times 10^{-2}$
$1,000 = 10^3$	$1/1,000 = 0.001 = 1 \times 10^{-3}$
etc.	etc.

So to express any number in scientific notation, we need only determine what small number must be multiplied by what integer power of 10 to obtain the desired number. For example, $150 = 1.5 \times 100$; in scientific notation it would be written as 1.5×10^2; another example, $3,500 = 3.5 \times 1,000 = 3.5 \times 10^3$; similarly, $0.0035 = 3.5 \times 1/1,000 = 3.5 \times 10^{-3}$. The Sun–Earth distance is $93,000,000 = 9.3 \times 10,000,000 = 9.3 \times 10^7$ miles; $0.000,93 = 9.3 \times 1/10,000 = 9.3 \times 10^{-4}$.

Now suppose we want to multiply 3,500 by 20:

$$3,500 \times 20 = (3.5 \times 10^3) \times (2 \times 10^1) = 3.5 \times 2 \times 10^3 \times 10^1 = 7 \times 10^4 \ (= 70,000).$$

See how that's done: you *multiply* the small numbers, and *add* the exponents. To divide the two, you simply *divide* the small numbers and *subtract* the exponents:

$$\frac{3,500}{20} = \frac{3.5 \times 10^3}{2 \times 10^1} = \left(\frac{3.5}{2}\right) \times \left(10^3\right) \times \left(10^{-1}\right) = 1.75 \times 10^2 \ (= 175).$$

A couple of examples: how long does it take for light to travel from the Sun to the Earth?

$$\text{time} = \frac{\text{distance}}{\text{velocity}} = \frac{93,000,000 \text{ miles}}{186,000 \text{ miles per second}}$$

$$= \frac{9.3 \times 10^7}{1.86 \times 10^5} = \frac{9.3}{1.86} \times 10^{7-5} = 5 \times 10^2 \text{ seconds}$$

$$= 8 \text{ minutes, 20 seconds.}$$

Another example: what is the volume of the Earth? The Earth is very nearly a perfect sphere, for which the volume is given by the formula $V = 4/3\pi R^3$, where $\pi = 3.14$ and R, the Earth's radius, is about $4,000 = 4 \times 10^3$ miles. Thus

$$V = \frac{4}{3}\pi R^3 = \frac{4}{3} \times 3.14 \times (4,000)^3 = \frac{4}{3} \times 3.14 \times 4 \times 4 \times 4 \times 10^3 \times 10^3 \times 10^3$$

$$= 267 \times 10^{3+3+3} = 267 \times 10^9 = 2.67 \times 10^{11} \text{ cubic miles.}$$

Checking Your Answer.   *Always look at your answer, think about it, and see if it makes sense!* Admittedly, sometimes it might be hard for you to know if your answer is reasonable, because the quantity involved may be unfamiliar. If, however, the volume of the Earth came out to be 1,000 cubic miles (a cube 10 miles on a side), or the light-travel time from the Sun was a billion years, you should realize that something is drastically wrong. As you get a better "feel" for the magnitudes of various quantities in this course, you will be better able to judge if an answer is sensible or not.

Another way of checking your answer is to pay attention to the units—feet, miles per second, pounds, etc. Just as the values of the numbers on both sides of an equation must be the same, the units on both sides must be the same. (After all, pounds can't equal feet per second!) Suppose you are calculating a speed. The answer should be in appropriate units, say miles/hour. If, to get your answer, you multiplied a length by a time, something is wrong; you must divide a length by a time to get a speed. If you drive your car 3 hours at 55 miles/hour, you will have gone a distance of (55 miles/hour) × 3 hours or 165 miles, because the "hours" cancel

$$\left(55 \ \frac{\text{mi}}{\text{hr}} \times 3 \text{ hr} = 165 \text{ mi}\right).$$

Significant Figures

Suppose you wanted to know what your average speed was during an automobile trip. According to your car's odometer, the length of the journey was 133 miles and it took you 2 hours and 37 minutes or 2.62 hours. Your calculator says your average speed was 133/2.62 = 50.76335878 miles/hour. Does this answer make sense? Consider: you know the distance to the nearest mile and the time to the nearest minute, or both quantities to roughly 1 percent (one mile in 133 and one minute in 157). Therefore, your answer is good to only about 1 percent or a few tenths miles/hour, or 50.8 miles/

hour (0.76 rounds up to 0.8). Not only are all the additional numbers meaningless, but they imply an absurdly high accuracy. After all, your calculator cannot make an answer more accurate than the numbers that went into it.

The digits in a number that are derived directly from a measurement are called "significant figures." In the example above, the measurements were good to three significant figures and so their quotient is valid to no more than that.

A number such as 0.0035 has only two significant figures; the two zeros simply locate the decimal point. This becomes clear when you write that number in scientific notation: 3.5×10^{-3}. Note also that a measured value of 3 contains one significant figure, whereas a value 3.0 contains two (the number is not 2.9 or 3.1, but 3.0 and the zero is a significant figure).

As a general rule, the result of any calculation should have no more significant figures than the least accurate number that went into the calculation. Look at the examples below.

$$\begin{array}{r} 48.73 \\ -\ 1.94 \\ \hline 46.79 \end{array} \text{ becomes } 46.8 \qquad \begin{array}{r} 38.9 \\ +\ 3.6954 \\ \hline 42.5954 \end{array} \text{ becomes } 42.6$$

$$5.62 \times 31.37 = 176.2994 \text{ becomes } 176.$$

Similarly, when you are calculating a quantity with a constant (such as π in $C = 2\pi R$), there is no point in taking for π the value of 3.14159 if R is known to only three significant figures; 3.14 is all you need. In fact, most calculations in this book involve no more than three significant figures.

Keep these precepts in mind when you do calculations.

Order of Magnitude

Given the enormous ranges of sizes, distances, times, and other quantities that we encounter in astronomy, and the often large uncertainty in some of those quantities, a few words concerning the idea of **order of magnitude** are appropriate. When two quantities, expressed in scientific notation, are given to the same power of ten, we say they are of the same order of magnitude. For example, 252 and 397 (or 2.52×10^2 and 3.97×10^2) are of the same order of magnitude; so are 358,937 and 493,799. Note that their actual values can differ by a large amount. All we are saying is that within a factor of no more than ten, they are the same. One hundred and 999 are just within the same order of magnitude; they differ by a factor of 9.99.

It is often very useful to make approximations of this sort. For example, recall the comment made in Chapter 2 that spaceflights to even the nearest star will not be made in your lifetime. One way of emphasizing this is to say that Venus, Mars, and Jupiter are, to within an order of magnitude, all the same distance from the Earth; the nearest star, however, is about five orders of magnitude (five factors of ten) farther away. We don't need to know the exact numbers to make the point.

Another example: sometimes all we want is a rough estimate of a particular quantity to see, for example, if the idea on which the calculation is based is worth pursuing further. The arithmetic will be done without worrying about the exact values of the numbers that enter, since all that we want is an approximate answer. In some examples in this book, you will see that we don't know some of the relevant numbers well enough to make anything but an order-of-magnitude calculation!

Approximate calculations can also be useful in your everyday life, for example, in quickly estimating whether a purchase paid for over many months is really as great a bargain as it may first appear to be. With only a little practice you can become proficient at making such estimates. They are not only useful, but fun to do as well.

Angular Measure

Go around a circle once and you will have gone through an angle of 360 degrees, written as 360°, regardless of how many meters or kilometers you actually traveled. Each degree is divided into 60 minutes of arc (60′), and each minute is divided into 60 seconds of arc (60″). Thus in 360° there are $360 \times 60 \times 60 = (3.6 \times 10^2) \times (6 \times 10^1)^2 = (3.6 \times 10^2) \times (36 \times 10^2) = 129.6 \times 10^4 = 1.296 \times 10^{6\prime\prime}$.

The system of dividing things by 60 as in 60′ = 1° seems a little strange to us, because our *number* system (but not the system of *units* we use in everyday matters) is based on 10. That is, $3.452 = 3 + 4/10 + 5/(10)^2 + 2/(10)^3$. Our system of angular measure, however, comes to us from the Babylonians, and their base number was 60, not 10. Thus $5° 39^1 46″ = 5 + 39/60 + 46/(60)^2$.

In this book you will often see references to the angular size of an object, for example, that the Moon has an angular diameter of about half a degree. What is meant is that two straight lines drawn from opposite sides of the Moon and meeting at your eye (or at the telescope) make an angle of about half a degree (see Figure A-1). The

Figure A-1.   As seen from Earth, the Moon subtends an angle of about 1/2 degree.

shorthand way of saying this is that the Moon **subtends** an angle of half a degree. Note that by itself the angular size of an object tells you nothing about its linear size, that is, its diameter in kilometers. Solar eclipses occur because the Sun also subtends half a degree, but it is much farther away than the Moon; in fact, it is as many times larger in diameter than the Moon as it is farther away from us than the Moon.

Small Angle Formula

Knowing that the Moon's angular diameter is about 0.5°, how do we find its "real" or linear diameter, that is, its diameter in kilometers? If the Moon were twice as far away as it really is, its angular diameter would be only 0.25°. Thus, there is a relation between its angular diameter (θ), its distance (r), and its linear diameter (D). Imagine a circle of radius r (in this case equal to the Moon's distance), centered on your eye, that passes across the Moon. Because the Moon's linear diameter is very nearly equal to the circular arc crossing it, we can write the following ratio:

$$\frac{D}{2\pi r} = \frac{\theta°}{360°} \text{ so } D = \frac{2\pi r\theta°}{360°} = \frac{r\theta°}{57.3°} = \frac{(3.84 \times 10^5 \text{km})(0.5°)}{57.3°}$$

and $D = 3{,}350$ km.

Or, since the angular diameters of astronomical objects are often given in arcseconds,

$$D = \frac{r\theta''}{206{,}265''}$$

since there are 206,265 arcseconds in 57.3° (called a "radian"). For example, when Uranus is 18.4 AU from the Earth, its angular diameter is 3.81 arcseconds. Its linear diameter is

$$D = \frac{(18.4 \text{ AU})3.81''}{206{,}265} = 3.40 \times 10^{-4} \text{ AU} = 51{,}000 \text{ km.}$$

Note that the derivation of the parsec formula given in Figure 12.9 is just a slightly different version of the small angle formula given here.

Units of Measurement

Mass and Length

When you say you were on the highway doing 55, people here will understand you to mean 55 miles/hour, but in Europe, for example, they will take that to mean 55 kilometers/hour. The numerical value of a quantity is meaningless unless the units in which it is measured are given. So what units are we going to use? The metric system is in worldwide use for scientific work, and is in fact in general use nearly everywhere. Liberia, South Yemen, and the U.S. are the holdouts. (The conversion to the metric system in this country is painfully slow. Only a few road signs are given in kilometers; soft-drink bottles are now given in liters rather than quarts.) We will use the metric system in this course.

In the English system the units of length, mass, and time are the foot, the pound, and the second, respectively. The larger and smaller units of the English system are clumsy: the mile is 5,280 feet, the foot is made up of 12 inches, there are 16 ounces in a pound, etc. In the **metric system** the units of length, mass, and time are the centimeter (cm), the gram (gm), and the second (the same in both systems), and all related quantities are multiples of 10 of each other; for example, 1 meter (m) is also 100 cm and also 1,000 millimeters (mm); and 1 kilogram (kg) is 1,000 gm. (Actually, the "official" system of units now takes as the units of length and mass meters and kilograms, respectively, but this is a detail that need not concern us. The important point is to use the much more convenient metric rather than the English system.)

The relationship between the two systems is as follows:

length: 1 cm = 0.01 m = 0.39 inches; so 1 inch = 2.54 cm
1 foot = 30.5 cm.

1 kilometer (km) = 1,000 m = 100,000 cm = 0.62 miles; so 1 mile = 1.61 km.
mass: 1 gm = 0.0022 pounds; so 1 pound = 454 gm
1 kg = 2.2 pounds.

Examples: One meter is 100 cm/2.54 cm per inch = 39.37 inches or just a little over 1 yard. (Note that the "cm" cancel, so that the units are inches.) The speed of light is 186,000 miles/sec, so in the metric system it is 1.86×10^5 miles/sec $\times 1.61$ km/mile = 3×10^5 km/sec = 3×10^{10} cm/sec. The mass of a 10-pound ball is 10 lbs $\times$ 454 gm/lb = 4,540 gm = 4.54 kg.

Temperature

Another inconvenient unit we use in this country is the Fahrenheit temperature scale (after the German who defined it). Its defining cold point is a mixture of salt and ice, which, in the eighteenth century, when the scale was introduced, was the coldest temperature attainable in the laboratory; the warm point was to be normal human body temperature. The cold point was given the value 0, the warm point 100 degrees (as you know, it turned out to be 98.6 degrees!). This results in the freezing and boiling points of water having the values of 32 and 212 degrees, respectively.

The defining points for the Celsius scale (after an eighteenth-century Swedish scientist), a temperature scale often used in scientific work, are the freezing and boiling points of water, which are given the values of 0 and 100 degrees. Because the zero points of the Fahrenheit and Celsius scales differ by 32 degrees, and the difference between freezing and boiling water is 180 degrees and 100 degrees, respectively, the relationship between the two scales is given by

$$°F = \frac{9}{5}°C + 32; \quad °C = \frac{5}{9}(°F - 32).$$

Check that the temperature of boiling water on the Fahrenheit scale is 212° (put C = 100 in the first equation) and that freezing on the Celsius (or Centigrade) scale is 0 degrees. Show that normal body temperature on the Celsius scale is 37°C.

Another temperature scale used in science is named after the nineteenth-century British physicist Lord Kelvin. The size of the degree is the same as on the Centigrade scale (100 divisions between the freezing and boiling points of water), but the zero point of the Kelvin scale is set at "absolute" zero, the lowest temperature possible, where molecules stop moving. This occurs at about −273°C, so that K = °C + 273 (the convention is not to write the degree symbol when using the Kelvin scale). A warm room temperature is about 300 K, or 27°C, which you can check is 80.6°F.

The temperature scales are summarized below:

	Kelvin (K)	Celsius (°C)	Fahrenheit (°F)
Absolute zero	0	−273	−459
Freezing point of water	273	0	32
Body temperature	310	37	98.6
Boiling point of water	373	100	212
Effective temperature of Sun	5,750	5,477	9,891
Conversion formulae	$K = °C + 273$	$°C = \frac{5}{9}(°F - 32)$	$°F = \frac{9}{5}°C + 32$

Note that stellar temperatures are essentially the same on the Kelvin and Celsius scales (273 is small compared with several thousand). Also, high temperatures on the Kelvin or Celsius scales are roughly half (5/9 = 0.56) their value on the Fahrenheit scale.

How to Simplify Computations

Often in this book we will be interested not in the absolute value of a quantity (e.g., that the distance of Jupiter from the Sun is 7.8×10^8 km), but rather in the fact that the quantity in question is a number of times greater than some other quantity (that Jupiter is 5.2 times as far away from the Sun as the Earth is).

Another example: if Circle A has a radius of 2 m, and Circle B a radius of 4 m, then the circumference of B is how many times larger than that of A? Or, more simply, what is the ratio of C_B to C_A? You could plug in the values 2 m and 4 m in the formula for the circumference, $C = 2\pi R$, multiply things out, and then take the ratio of the answers, but that is unnecessarily cumbersome. Instead, notice the following:

$$\frac{C_B}{C_A} = \frac{2\pi R_B}{2\pi R_A} = \frac{R_B}{R_A},$$

since the 2 πs cancel.

So the answer is immediately seen to be

$$\frac{C_B}{C_A} = \frac{4}{2} = 2,$$

or

$$C_B = 2C_A.$$

That is, the circumference of a circle is proportional to its radius, R; double R and you double C. The factor 2π is simply the constant of proportionality between C and R.

Similarly, since the area of a circle is $A = \pi R^2$, the ratio of the *areas* of the two circles is

$$\frac{A_B}{A_A} = \frac{\pi R_B^2}{\pi R_A^2} = \frac{R_B^2}{R_A^2} = \left(\frac{4}{2}\right)^2 = 4,$$

and

$$A_B = 4A_A.$$

So the area of a circle is proportional to R^2, and π is the constant of proportionality between the area and the radius of a circle.

The volume of a sphere is given by $V = 4/3\pi R^3$, so the volume goes as the cube (third power) of the radius, and $4/3\pi$ is the constant of proportionality. We can generalize these results and say that among similarly shaped objects (squares or circles, etc.), linear distances on those objects (e.g., the circumferences of circles) are directly proportional to ($\propto$), the *linear* dimension of the object, i.e., $C \propto R$; the surface areas of similar objects will be directly proportional to the *square* of their linear dimension, $A \propto R^2$; and their volumes will be directly proportional to the cube of their linear dimensions, $V \propto R^3$. Thus, if Jupiter has 10 times the radius of the Earth ($R_{2\!\!\!/} = 10R_\oplus$), we can say immediately that its circumference is also $10C_\oplus$, its surface area = $10^2A_\oplus$ and its volume = $10^3V_\oplus$.

Sometimes quantities depend on more than one factor, e.g., the force of gravity,

$$F \propto \frac{M_1M_2}{R^2},$$

is directly proportional to the product of the two masses involved, and *inversely* proportional to the square of the distance between them. We say *inversely* (and not *directly*) because here if you *increase R*, you *decrease F*; if R is doubled, F is only 1/4 of its original value; if R is tripled, F is 1/9 of its original value, etc. If one of the masses is doubled, however, F is doubled; if both masses are doubled, F is increased by a factor of 4.

If we wanted to calculate the absolute value of the gravitational force in a given situation, then we would have to use

$$F = G\,\frac{M_1M_2}{R^2},$$

where G is the constant of proportionality between F on the one hand, and M and R on the other. As stated at the outset, however, generally we will be more interested in how the force *changes* when M and R are changed, or the ratio of the forces under different circumstances, than in the absolute value of the force.

The Nearest Stars

Name	Parallax (arcsec)	Luminosity ($L_\odot = 1$)	Spectral type	Proper motion (arcsec/yr)	Radial velocity (km/sec)
Sun		1.0	G2 V		
Proxima Cen	0.77	0.000,06	dM5e	3.85	−16
∝ Cen A	0.75	1.6	G2 V	3.68	−22
∝ Cen B		0.45	K0 V		
Barnard's star	0.54	0.000,45	M5 V	10.31	−108
Wolf 359	0.42	0.000,02	dM8e	4.70	+13
BD+36°2147	0.40	0.005,5	M2 V	4.78	−84
L 726−8 = A	0.39	0.000,06	dM6e	3.36	+29
UV Cet = B		0.000,04	dM6e		+32
Sirius A	0.38	23.5	A1 V	1.33	−8
Sirius B		0.003	DA		
Ross 154	0.34	0.000,48	dM5e	0.72	−4
Ross 248	0.31	0.000,11	dM6e	1.60	−81
ε Eri	0.30	0.30	K2 V	0.98	+16
Ross 128	0.30	0.000,36	dM5	1.38	−13
61 Cyg A	0.29	0.082	K5 V	5.22	−64
61 Cyg B		0.039	K7 V		
ε Ind	0.29	0.14	K5 V	4.70	−40
BD+43°44A	0.29	0.006,1	M1 V	2.90	+13
+43°44B		0.000,39	M6 Ve		+20
L 789−6	0.29	0.000,14	dM7e	3.26	−60
Procyon A	0.28	7.65	F5 IV–V	1.25	−3
Procyon B		0.000,55	DF		
BD+59°1915A	0.28	0.003,0	dM4	2.29	0
+59°1915B		0.001,5	dM5	2.27	+10
CD−36°15693	0.28	0.013	M2 V	6.90	+10
G 51−15	0.28	0.000,01		1.27	

Name	Parallax (arcsec)	Luminosity ($L_\odot = 1$)	Spectral type	Proper motion (arcsec/yr)	Radial velocity (km/sec)
τ Cet	0.28	0.45	G8 V	1.92	− 16
BD+5°1668	0.27	0.001,5	dM5	3.77	+ 26
L 725 − 32	0.26	0.000,20	dM5e	1.32	+ 28
CD − 39°14192	0.26	0.028	M0 V	3.46	+ 21
Kapteyn's star	0.26	0.003,9	sdM0 p	8.72	+ 245
Krüger 60A	0.25	0.001,6	dM3	0.86	− 26
Krüger 60B		0.000,4	dM5e		

E

The Twenty Brightest Stars

Star	Name	Distance (pc)	Proper motion (arcsec/yr)	Spectral* type
α CMa	Sirius	2.7	1.33	A1 V + wd
α Car	Canopus	30	0.02	F0 1b–II
α Cen	α Centauri	1.3	3.68	G2 V + KO V
α Boo	Arcturus	11	2.28	K2 IIIp
α Lyr	Vega	8.0	0.34	A0 V
α Aur	Capella	14	0.44	G III + M1 V + M5 V
β Ori	Rigel	250	0.00	B8 Ia + B9
α CMi	Procyon	3.5	1.25	F5 IV–V + wd
α Ori	Betelgeuse	150	0.03	M2 Iab
α Eri	Achernar	20	0.10	B5 V
β Cen	Hadar	90	0.04	B1 III
α Aql	Altair	5.1	0.66	A7 IV–V
α Tau	Aldebaran	16	0.20	K5 III + M2 V
α Cru	Acrux	120	0.04	B1 IV + B3
α Vir	Spica	80	0.05	B1 V
α Sco	Antares	120	0.03	M1 Ib + B4e V
β Gem	Pollux	12	0.62	K0 III
α PsA	Fomalhaut	7.0	0.37	A3 V + K4 V
α Cyg	Deneb	430	0.00	A2 Ia
β Cru	Mimosa	150	0.05	B0.5 IV

* A "p" after a spectral type indicates that the spectrum is peculiar. An "e" after a spectral type indicates that emission lines are present. When the luminosity classification is rather uncertain, a range is given.

Planetary Data

Orbital Elements of the Planets

Planet	Mean distance from Sun (AU)*	Period of revolution		Average orbital velocity (km/sec)	Eccentricity	Inclination to eclipse (degrees)
		Sidereal (yrs)*	Synodic (yrs)			
Mercury	0.387	0.2408	0.3172	47.9	0.21	7.0
Venus	0.723	0.6152	1.599	35.0	0.006,8	3.4
Earth	1.00	1.00	—	29.8	0.017	—
Mars	1.524	1.881	2.135	24.1	0.093	1.8
Jupiter	5.203	11.87	1.092	13.1	0.048	1.3
Saturn	9.539	29.46	1.035	9.64	0.056	2.5
Uranus	19.18	84.01	1.012	6.81	0.047	0.77
Neptune	30.06	164.8	1.006	5.43	0.008,6	1.77
Pluto	39.44	247.7	1.004	4.74	0.25	17.1

* 1 AU = 1.496×10^{13} cm; 1 sidereal year = 365.256 days

Some Properties of the Planets

Planet	Mass* $\oplus = 1$	Equatorial radius (km)	Maximum angular diameter (arcsec)	Escape speed (km/sec)	Sidereal rotation period	Visual albedo
Mercury	0.055,8	2,439.7	11.0	4.3	58.65^d	0.12
Venus	0.815,0	6,051.8	60.2	10.3	244.3^d	0.59
Earth	1.000	6,378.1	—	11.2	$23^h56^m04.1^s$	0.39
(Moon)	0.012,30	1,737.4	1,864.2	2.38	$27^d07^h43.2^m$	0.11
Mars	0.107,4	3,397	17.9	5.0	$24^h37^m22.6^s$	0.15
Jupiter	317.9	71,492	46.9	61	$9^h50^m30^s$	0.44
Saturn	95.15	60,268	19.5	35.6	$10^h13^m59^s$	0.46
Uranus	14.54	25,559	3.9	22	17^h14^m	0.56
Neptune	17.23	24,764	2.3	25	16^h013^m	0.51
Pluto	0.002,2	1,151	0.08	1.2	$6^d9^h21^m$	0.3(?)

* $M_\oplus = 5.974 \times 10^{24}$ gm

Constellations

Name	Possessive form	Abbreviation	Meaning
Andromeda	Andromedae	And	Andromeda* (the chained princess)
Antlia	Antliae	Ant	Air Pump
Apus	Apodis	Aps	Bird of Paradise
** Aquarius	Aquarii	Aqr	Water Bearer
Aquila	Aquilae	Aql	Eagle
Ara	Arae	Ara	Altar
** Aries	Arietis	Ari	Ram
Auriga	Aurigae	Aur	Charioteer
Boötes	Boötis	Boo	Herdsman
Caelum	Caeli	Cae	Chisel
Camelopardalis	Camelopardis	Cam	Giraffe
** Cancer	Cancri	Cnc	Crab
Canes Venatici	Canum Venaticorum	CVn	Hunting Dogs
Canis Major	Canis Majoris	CMa	Big Dog
Canis Minor	Canis Minoris	CMi	Little Dog
** Capricornus	Carpricorni	Cap	Goat
Carina	Carinae	Car	Ship's Keel
Cassiopeia	Cassiopeiae	Cas	Cassiopeia* (seated)
Centaurus	Centauri	Cen	Centaur*
Cepheus	Cephei	Cep	Cepheus* (the king)
Cetus	Ceti	Cet	Whale
Chamaeleon	Chamaeleonis	Cha	Chameleon
Circinus	Circini	Cir	Compass
Columba	Columbae	Col	Noah's Dove
Coma Berenices	Comae Berenices	Com	Berenice's Hair*
Corona Australis	Coronae Australis	CrA	Southern Crown
Corona Borealis	Coronae Borealis	CrB	Northern Crown

* *Proper names*

** *Zodiacal constellations*

Name	Possessive form	Abbreviation	Meaning
Corvus	Corvi	Crv	Crow
Crater	Crateris	Crt	Cup
Crux	Crucis	Cru	Southern Cross
Cygnus	Cygni	Cyg	Swan
Delphinus	Delphini	Del	Dolphin
Dorado	Doradus	Dor	Swordfish
Draco	Draconis	Dra	Dragon
Equuleus	Equulei	Equ	Little Horse
Eridanus	Eridani	Eri	River Eridanus*
Fornax	Fornacis	For	Furnace
**Gemini	Geminorum	Gem	Twins
Grus	Gruis	Gru	Crane
Hercules	Herculis	Her	Hercules*
Horologium	Horologii	Hor	Clock
Hydra	Hydrae	Hya	Hydra* (water snake)
Hydrus	Hydri	Hyi	Sea Serpent
Indus	Indi	Ind	Indian
Lacerta	Lacertae	Lac	Lizard
**Leo	Leonis	Leo	Lion
Leo Minor	Leonis Minoris	LMi	Little Lion
Lepus	Leporis	Lep	Hare
**Libra	Librae	Lib	Scales
Lupus	Lupi	Lup	Wolf
Lynx	Lyncis	Lyn	Lynx
Lyra	Lyrae	Lyr	Harp
Mensa	Mensae	Men	Table (mountain)
Microscopium	Microscopii	Mic	Microscope
Monoceros	Monocerotis	Mon	Unicorn
Musca	Muscae	Mus	Fly
Norma	Normae	Nor	Level (tool)
Octans	Octantis	Oct	Octant
Ophiuchus	Ophiuchi	Oph	Ophiuchus* (serpent bearer)
Orion	Orionis	Ori	Orion* (the hunter)
Pavo	Pavonis	Pav	Peacock
Pegasus	Pegasi	Peg	Pegasus* (winged horse)
Perseus	Persei	Per	Perseus*
Phoenix	Phoenicis	Phe	Phoenix

* *Proper names*

** *Zodiacal constellations*

Name	Possessive form	Abbreviation	Meaning
Pictor	Pictoris	Pic	Easel
**Pisces	Piscium	Psc	Fishes
Piscis Austrinus	Piscis Austrini	PsA	Southern Fish
Puppis	Puppis	Pup	Ship's Stern
Pyxis	Pyxidis	Pyx	Ship's Compass
Reticulum	Reticuli	Ret	Net
Sagitta	Sagittae	Sge	Arrow
**Sagittarius	Sagittarii	Sgr	Archer
**Scorpius	Scorpii	Sco	Scorpion
Sculptor	Sculptoris	Scl	Sculptor
Scutum	Scuti	Sct	Shield
Serpens	Serpentis	Ser	Serpent
Sextans	Sextantis	Sex	Sextant
**Taurus	Tauri	Tau	Bull
Telescopium	Telescopii	Tel	Telescope
Triangulum	Trianguli	Tri	Triangle
Triangulum Australe	Trianguli Australis	TrA	Southern Triangle
Tucana	Tucanae	Tuc	Toucan (bird)
Ursa Major	Ursae Majoris	UMa	Big Bear
Ursa Minor	Ursae Minoris	UMi	Little Bear
Vela	Velorum	Vel	Ship's Sails
**Virgo	Virginis	Vir	Virgin
Volans	Volantis	Vol	Flying Fish
Vulpecula	Vulpeculae	Vul	Little Fox

* Proper names

** Zodiacal constellations

Star Chart 1

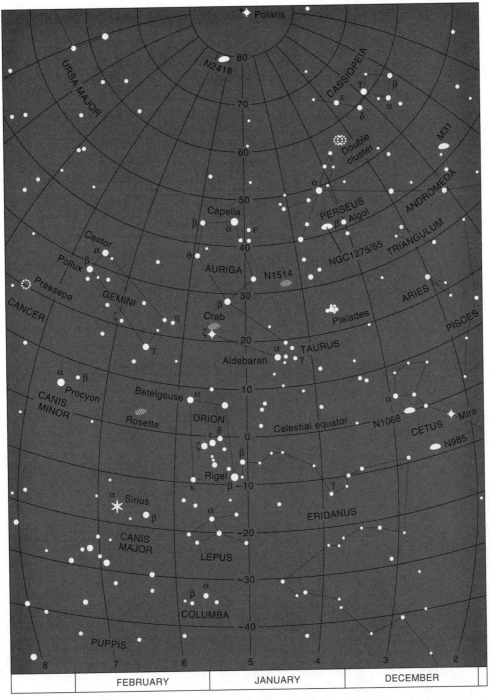

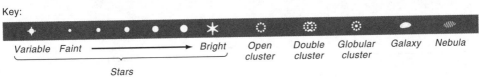

Key:

Variable Faint ——————→ Bright   Open cluster   Double cluster   Globular cluster   Galaxy   Nebula

Stars

A17

Star Chart 2

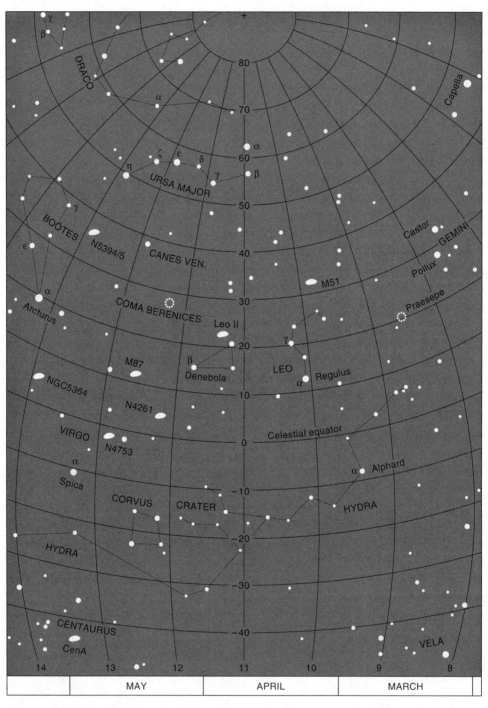

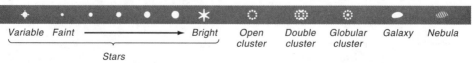

Key:

Variable Faint ⟶ Bright Open cluster Double cluster Globular cluster Galaxy Nebula

Stars

Star Chart 3

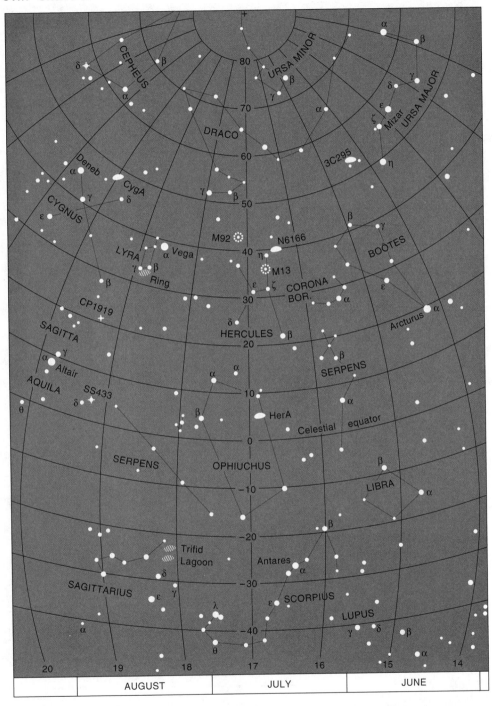

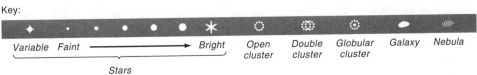

Key:

Variable Faint ———————————➤ Bright Open cluster Double cluster Globular cluster Galaxy Nebula

Stars

Star Chart 4

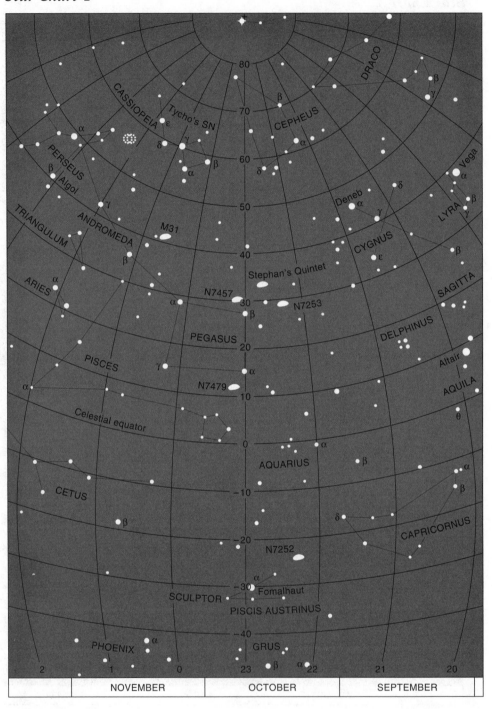

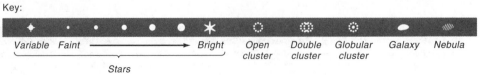

Key:

| Variable | Faint | → | Bright | Open cluster | Double cluster | Globular cluster | Galaxy | Nebula |

Stars

Star Chart 5. North Polar Region

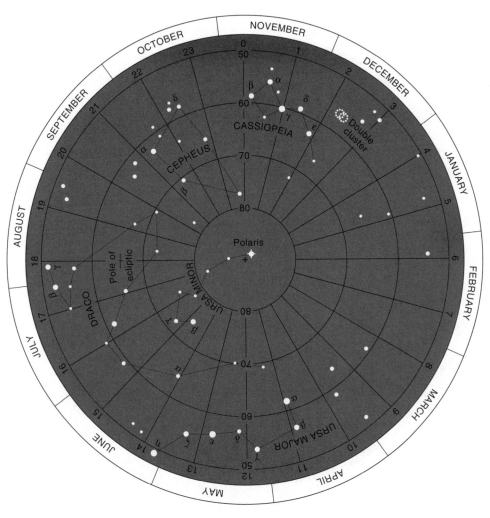

Key:

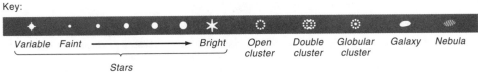

Variable   Faint ⟶ Bright   Open cluster   Double cluster   Globular cluster   Galaxy   Nebula

Stars

Star Chart 6. South Polar Region

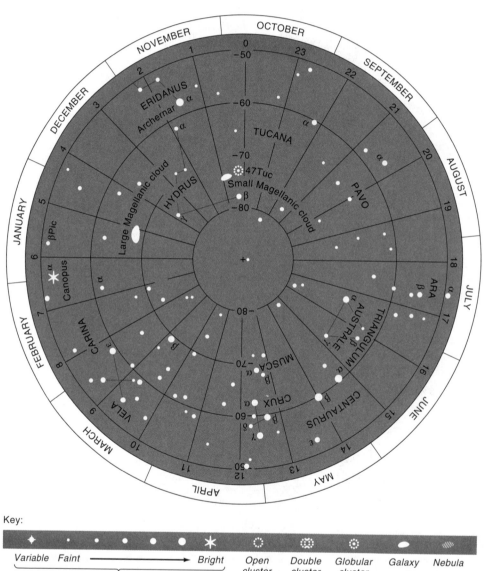

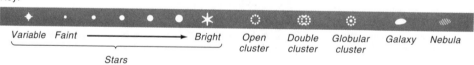

Key:

Variable Faint ⟶ Bright   Open   Double   Globular   Galaxy   Nebula
                            cluster   cluster   cluster

Stars

Illustration Credits

Chapter 1
Opener, NASA.

Part I
Opener, HST/NASA.

Chapter 2
Opener, NOAO photo; Fig. 2.6, NOAO; Fig. 2.16, Lick Observatory Photograph/Image; Figs. 2.18a, 2.18b, Celestial Delights; Figs. 2.19, 2.20, NOAO photo; Figs. 2.21, 2.23, 2.26, Celestial Delights used by permission of Celestial Arts, P.O. Box 7327, Berkeley, CA 94707; pg. 41, Wide Field Camera 2. HST/NASA. R. Beebe.

Chapter 3
Figs. 3.2, 3.3, E. Ersland; Fig. 3.6, adapted from Aveni, A., *Skywatchers of Ancient Mexico*, Austin, University of Texas Press, 1980, p. 279; Fig. 3.7A, J. Eddy; Fig. 3.7B, adapted from J. Eddy, 1974, *Science*, **184**, 1035; Fig. 3.10, J. W. Percival; Fig. 3.11, E. Richards; Fig. 3.13, Yerkes Observatory photograph; Pg. 77, HST/NASA.

Chapter 5
Opener, Yerkes Observatory photograph; Fig. 5.12, O. Gingerich and by permission of the Houghton Library, Harvard University.

Chapter 6
Opener, Yerkes Observatory photograph; Fig. 6.1, The Bayeux Tapestry—11TH Century, by special permission of the City of Bayeux; Fig. 6.2, P. Capretz; Fig. 6.4, Art Resource, NY.

Chapter 7
Opener, O. Gingerich and by permission of the Houghton Library, Harvard University; Figs. 7.1, 7.7, Yerkes Observatory photograph; Fig. 7.8, O. Gingerich and by permission of the Houghton Library, Harvard University; Fig. 7.9, Yerkes Observatory photograph; Figs. 7.10, 7.11, O. Gingerich and by permission of the Houghton Library, Harvard University; Fig. 7.12, Yerkes Observatory photograph; Fig. 7.14, O. Gingerich and by permission of the Houghton Library, Harvard University; Fig. 7.20, 7.21, Yerkes Observatory photograph; Fig. 7.22, O. Gingerich and by permission of the Houghton Library, Harvard University; Fig. 7.23, Yerkes Observatory photograph; Copernicus quotations on pgs. 105, 106 from Toulmin, S. and Goodfield, J., *The Fabric of the Heavens*, NY, Harper Torchbooks, 1961, pgs. 170 and 170–171, respectively; Kepler quotations on p. 122 and Fabricius quotation on p. 123 from Koestler, A., *The Watershed*, Garden City, NY, Anchor Books, 1960, pgs. 133–134, 147, and 164, respectively.

Chapter 8
Figs. 8.1, 8.2, Yerkes Observatory photograph; Fig. 8.3, O. Gingerich and by permission of the Houghton Library, Harvard University; Pg. 169, HST/NASA, J. Hester, P. Scowen.

Part II
Opener, HST/NASA, C. R. O'Dell.

Chapter 10
Opener, Astronomical Society of the Pacific; Fig. 10.10, U.S. Naval Observatory; Figs. 10.12, 10.13, Yerkes Observatory photograph; Figs. 10.14, 10.15, California Institute of Technology; Fig. 10.16, D. J. Schroeder; Fig. 10.19, Lick Observatory Photograph/Image; Fig. 10.21, E. B. Churchwell; Fig. 10.22, NRAO; Fig. 10.23, E. B. Churchwell; Fig. 10.24, National Academy of Sciences; Fig. 10.26, NRAO/AUI; Fig. 10.28, Smithsonian Air & Space Museum; Fig. 10.29, Author; Figs. 10.30, 10.31, Goddard Space Flight Center/NASA; Fig. 10.32, European Space Agency; Fig. 10.33, HST/NASA, European Space Agency; Fig. 10.35, NASA; Fig. 10.36, HST/NASA; Fig. 10.40, NOAO photograph; Fig. 10.42, T. Houck, A. Code; Fig. 10.44, adapted from NOAO diagram; Pg. 239, HST/NASA, P. James, S. Lee.

Chapter 11
Fig. 11.1, Yerkes Observatory photograph; Fig. 11.2, Author; Fig. 11.4, R. Garrison; Fig. 11.9, Yerkes Observatory photograph; Fig. 11.11, NOAO photograph; Fig. 11.12, McMath-Hulbert Observatory; Figs. 11.13, 11.14, NOAO photograph; Fig. 11.16, OSO/NASA; Fig. 11.17, NOAO photograph; Fig. 11.18, NOAO photograph; Fig. 11.20, adapted from J. Eddy; Fig. 11.27, Skylab/NASA; Fig. 11.28, M. J. Taylor.

Chapter 12
Figs. 12.1, 12.2, 12.3, Lick Observatory Photograph/Image; Figs. 12.4, 12.5, NOAO photograph; Fig. 12.16, Yerkes Observatory photograph; Fig. 12.19, Lick Observatory Photograph/Image.

Chapter 14
Opener, NOAO photograph; Fig. 14.1, NOAO photograph; Figs. 14.3, 14.4, 14.6, Lick Observatory Photograph/Image; Fig. 14.7, NOAO photograph; Fig. 14.9, J. Cardelli; Fig. 14.10, Lick Observatory Photograph/Image; Fig. 14.11, M. Meade.

Chapter 15
Fig. 15.2, adapted from Kutner, M., et al., 1977, *Ap. J.* **215**, 521; Fig. 15.3, NOAO photograph; Fig. 15.14, adapted from Hesser, J., et al., 1987, *PASP*, **99**, 739; Fig. 15.5, HST/NASA; Fig. 15.7, NASA; Fig. 15.15, NOAO photograph.

Chapter 16
Fig. 16.3, NOAO photograph; Fig. 16.4, Yerkes Observatory photograph; Fig. 16.10, *L.A. Times*, Jan. 19, 1934; Fig. 16.11, adapted from *Sky & Telescope*; Fig. 16.15, NOAO photograph, Dr. Nigel Sharp; Fig. 16.16, High Speed Photometer, HST/NASA; Fig. 16.18, Wide Field Camera, HST/NASA, J. Hester, P. Scowen, Arizona State University; Fig. 16.19, adapted from McCulloch, P., et al., 1983, *Nature*, **302**, 319; Fig. 16.20, NRAO/AUI, S. P. Reynolds, R. A. Chevalier; Fig. 16.21, ROSAT; Fig. 16.22, NRAO/AUI, P. E. Angelhofer, R. Braun, S. F. Gull, R. A. Perley, R. J. Tuffs; Fig. 16.25, Lick Observatory Photograph/Image; Fig. 16.26, NOAO photograph; Fig. 16.27,

NASA, European Space Agency, Space Telescope Science Institute, C. Burrows; Fig. 16.31, GRO/NASA.

Part III

Opener, (right) WFPC-2 Science Team, HST/NASA, R. Sahai, J. Trauger; (left) WFPC-2, HST/NASA, C. Burrows.

Chapter 17

Figs. 17.2, 17.3, 17.4, NOAO; Fig. 17.5, Lick Observatory Photograph/Image; Fig. 17.6, NOAO photograph; Fig. 17.7, California Institute of Technology; Fig. 17.8A, Lick Observatory Photograph/Image; Fig. 17.8B, NOAO photograph; Fig. 17.9, NOAO photograph; Figs. 17.11–17.14, Yerkes Observatory photograph; Fig 17.16, 1850, *Philosophical Transactions*, p. 499, and NOAO photograph; Figs. 17.17, 17.20, Yerkes Observatory photograph; Fig. 17.22, Hale Observatory, courtesy of AIP Emilio Segrè Visual Archives; Fig. 17.23, Neale Watson Publications.

Chapter 18

Opener, R. C. Kraan-Korteweg, A. J. Loan, W. B. Burton, D. Lahov, H. C. Ferguson, P. A. Henning, D. Lynden-Bell, Isaac Newton Telescope; Fig. 18.1, Yerkes Observatory photograph; Fig. 18.2, NOAO photograph; Fig. 18.3, Yerkes Observatory photograph; Fig. 18.4, Lick Observatory Photograph/Image; Figs. 18.5, 18.6, 18.7, NOAO photograph; Fig. 18.8, COBE/NASA; Fig. 18.9, NOAO photograph; Fig. 18.10, California Institute of Technology; Fig. 18.11, Margaret J. Geller, John P. Huchra, Luis A. N. da Costa, and Emilio E. Falco, Smithsonian Astrophysical Observatory, © 1994; Pg. 465, HST/NASA, R. Williams.

Chapter 19

Opener, HST/NASA, A. Dressler, M. Dickinson, D. Macchetto, M. Giavalisco; Fig. 19.4, V. Rubin; Fig. 19.9, Yerkes Observatory photograph; Figs. 19.12, 19.13, COBE/NASA; Fig. 19.14, Hale Observatory, courtesy of AIP Emilio Segrè Visual Archives, *Physics Today* Collection.

Chapter 20

Fig. 20.1, California Institute of Technology; Fig. 20.2, Lick Observatory Photograph/Image, A. Martel, D. Osterbrock; Fig. 20.4, NOAO photograph; Fig. 20.5, NRAO/AUI, J. O. Burns, S. A. Gregory; Fig. 20.6, NRAO/AUI, J. W. Dreher, E. D. Feigelson; Fig. 20.7, NRAO/AUI, R. A. Perley, J. W. Dreher, J. J. Cowan; Fig. 20.7 Inset, Hale Observatory; Fig. 20.8, NRAO/AUI, C. P. O'Dea, F. N. Owen; Fig. 20.9A, NOAO photograph; Fig. 20.9B, European Space Agency, NASA, F. Duccio Macchetto; Fig. 20.9C, NRAO/AUI; Fig. 20.10, adapted from Burns, Feigelson, and Schreier, 1983, *Ap. J.*, **273**, 128; Fig. 20.11A, Royal Observatory Edinburgh, U.K., Schmidt Telescope Unit; Fig. 20.11B, NOAO photograph; Fig. 20.13, left, California Institute of Technology and NRAO/AUI; right, HST/NASA, Space Telescope Science Institute, W. Jaffe, Leiden Observatory, H. Ford, Johns Hopkins University; Fig. 20.14, Space Telescope Science Institute, HST/NASA, T. Lauer, S. Faber; Fig. 20.16, HST/NASA, T. Lauer.

Chapter 21

Fig. 21.9, HST/NASA, European Space Agency, R. Ellis, Durham University; Fig. 21.11, NOAO photograph; Fig. 21.12, NRAO/AUI, Very Large Array, G. Langston.

Chapter 22

Opener, HST/NASA, K. Borne; Fig. 22.3, left, California Institute of Technology, T. Small; right, HST/NASA, European Space Agency, P. Conti; Fig. 22.5, NOAO photograph, N. A. Sharp; Fig. 22.6, NOAO photograph, W. Schoening, N. Sharp; Figs. 22.7, 22.8, NOAO photograph; Fig, 22.9, HST/NASA, F. Schweizer and B. Whitmore; Fig. 22.10, HST/NASA; Fig. 22.11, HST/NASA, Space Telescope Science Institute, J. Holtzman; Fig. 22.12, NOAO photograph; Fig. 22.13, NOAO photograph, W. Harris, N. A. Sharp.

Part IV

Chapter 23

Opener, Air and Space Museum, Smithsonian Institution; Fig. 23.1, Lick Observatory Photograph/Image; Fig. 23.2, NOAO photograph; Fig. 23.3, JPL/NASA; Fig. 23.4, JPL/NASA; Fig. 23.11, Lick Observatory Photograph/Image; Fig. 23.19, Apollo 16/NASA; Pg. 603, HST/NASA, R. Beebe.

Chapter 24

Opener, HST/NASA, K. Seidelman; Fig. 24.3, adapted from *Earthquakes*, van Rose, S., published by Her Majesty's Stationery Office for the Institute of Geological Sciences, 1983; Fig. 24.6, adapted from R. S. Deitz and J. C. Holden, *Scientific American*, **223**, 30 Oct., 1970; Fig. 24.7, Apollo 17/NASA; Fig. 24.8, NASA; Fig. 24.9, Apollo 17/NASA; Fig. 24.10, Apollo 15/NASA; Figs. 24.11, 24.12, California Institute of Technology; Fig. 24.13, Lunar Orbiter IV/NASA; Fig. 24.14, NASA; Fig. 24.17, Magellan/NASA; Fig. 24.18, D. Morrison; Fig. 24.19, NASA; Fig. 24.20, Mariner 10/NASA; Figs. 24.21, 24.22, NASA; Fig. 24.23, Viking/NASA; Figs. 24.24, 24.25, 24.27, NASA; Figs. 24.28, 24.29, 24.30, 24.31, Magellan/NASA; Fig. 24.32, Venera, Soviet Union; Figs. 24.33, 24.34, 24.35, 24.36, 24.37, Magellan/NASA; Fig. 24.38, Mariner 10/NASA; Fig. 24.39, Voyager 1/NASA; Figs. 24.40, 24.41, NASA; Fig. 24.42, Voyager 2/NASA; Fig. 24.43, NASA; Figs. 24.44, 24.45, Galileo/NASA; Fig. 24.47, California Institute of Technology; Figs. 24.49, 24.51, Fig. 24.52, NASA; Fig. 24.53, adapted from Ingersoll, A. in *The New Solar System*: Beatty, J. and Chaikin, A. eds., CUP and *Sky & Telescope*, 1990; Figs. 24.54, 24.55, NASA; Figs. 24.56, 24.57, 24.58, HST/NASA; Figs. 24.60, 24.61, 24.65, 24.66, 24.67, Voyager 2/NASA; Figs. 24.68–24.81, NASA; Fig. 24.82, Voyager 1/NASA; Fig. 24.83, Lick Observatory Photograph/Image; Figs. 24.84–24.89, Voyager 2/NASA; Fig. 24.90, Faint Object Camera, HST/NASA, European Space Agency.

Chapter 25

Fig. 25.1, NASA; Fig. 25.2, Lick Observatory Photograph/Image, R. Trumpler; Fig. 25.9, © June 20, 1959, 1987 O.G.P.I., used by permission.

Color Plates

Plate C1, O. Gingerich and by permission of the Houghton Library, Harvard University; Plate C3, NOAO photograph; Plate C4 (all), NOAO photographs; Plate C5 (upper and lower), R. J. Dufour; Plate C6 (upper and lower), NASA; Plate C7, NASA; Plate C8, NASA.

Index

Pages on which terms are defined or have their first significant use are given in **boldface type.**